DRILLING

DRILLING
A SOURCE BOOK ON OIL AND GAS WELL DRILLING FROM EXPLORATION TO COMPLETION

J.A. "Jim" Short

PennWell Books
PennWell Publishing Company
Tulsa, Oklahoma

Copyright © 1983 by
PennWell Publishing Company
1421 South Sheridan Road/P.O. Box 1260
Tulsa, Oklahoma 74101

Library of Congress cataloging in publication data

Short, J. A.
    Drilling: a source book on oil and gas well drilling from exploration to completion.

    1. Oil well drilling.   2. Gas well drilling.
I. Title.
TN871.2.S5368   1983   622'.338   83-13314
ISBN 0-87814-242-8

All rights reserved. No part of this book may be reproduced, stored in a retrieval system, or transcribed in any form or by any means, electronic or mechanical, including photocopying and recording, without the prior written permission of the publisher.

Printed in the United States of America

1 2 3 4 5   87 86 85 84 83

DEDICATION

The drilling industry was created by pioneers. Two of them were a rancher's son and a merchant's daughter who joined together and spent a lifetime in the effort. This book is dedicated with love and respect to Mother and Dad.

                    Margaret A. and Arthur E. "Slim" Short

CONTENTS

Dedication	v
Preface	x
Acknowledgments	xi
Introduction	xii

CHAPTER 1: HISTORY OF DRILLING — 1
Types of drilling methods — 1

CHAPTER 2: GEOLOGY AND EXPLORATION — 17
History — 18
Geological cycle — 19
Origin, migration, and accumulation of oil and gas — 36
Oil and gas traps — 44
Exploration operations — 53
Problem formations — 55

CHAPTER 3: RESERVOIRS AND RESERVES — 59
Well logs — 59
Fluid flow — 81
Producing mechanisms — 99
Secondary, tertiary, and enhanced recovery — 104
Volumes — 105
Oil reservoir — 106

CHAPTER 4: DRILLING PROSPECTS, PROGRAMS, AND PROCEDURES — 117
Drilling prospects — 117
Prospect submittal — 121
Drilling programs and procedures — 126
General information on the drilling program — 127
Geological prognosis — 134

Casing and cementing program	134
Bit program	149
Mud program	155
General equipment specifications	162
AFE and contracts	163

CHAPTER 5: DRILLING PERSONNEL AND EQUIPMENT — 169

Personnel and services	169
Drilling equipment	173
Rig classifications	173
Rig parts and functions (component systems)	175
Marine rigs	232

CHAPTER 6: MOVING IN, RIGGING UP, AND DRILLING THE CONDUCTOR HOLE — 241

Activities before move-in	242
Moving the rig	250
Move-in and rig-up procedure	253
Drilling the rathole and mousehole	262
Spud-in and conductor hole section	265
Drilling problems	279
Running and cementing conductor casing	283
Nippling up and drilling out	286
Air-gas mist drilling operations	287
Helicopter rig operations	289
Marine operations	290

CHAPTER 7: SURFACE HOLE SECTION — 295

Drilling procedure	295
Tripping the drillpipe assembly	306
Surface hole drilling problems	321
Landing casing and nippling up	342

CHAPTER 8: INTERMEDIATE HOLE SECTION — 347

Daily operations and crew duties	347

Drilling the section	352
Tripping and related activities	361
Drilling problems	369
Stuck assemblies and fishing	382
Logging, running, and cementing casing	403

CHAPTER 9: PRODUCTION HOLE SECTION 407

Drilling the section	407
Drilling problems	413
Deviated holes	427
Obtaining reservoir and productivity information	433
Plugging and abandoning a dry hole	451
Production casing and liners	454

CHAPTER 10: COMPLETIONS 463

History	463
Factors affecting completions	464
Types of completions	466
Tools and equipment	467
Completion design and procedures	486
Operations	503
Surface equipment	535
Appendix	555
Bibliography and Suggested Readings	569
Index	573

PREFACE

Where does oil come from? How do you find it and get it out of the ground? I have been asked these questions numerous times over the years by many people both in and out of the industry. Those in the industry ask for both general and detailed information to find out how their work fits into the overall picture. The answers to these questions invariably lead to other questions searching for more information.

These questions indicate a widespread interest from a variety of people. They also emphasize the need for one source that (1) covers the entire industry, (2) contains detailed, specific information and answers, (3) allows the reader to select and review areas of special interest, and (4) presents the material in a clear, understandable manner for anyone interested in the industry.

The purpose of drilling is to drill and complete a commercial oil and gas well. The purpose of this book is to explain, describe, and illustrate the process.

ACKNOWLEDGMENTS

The Coastal Corporation's permission to publish this text is gratefully appreciated. Blocker Drilling Company, Four Flags Drilling Company, San Patrico Corporation, and Loffland Brothers Drilling Company gave permission to photograph and publish pictures of their drilling rigs and equipment. Various service and supply companies supplied material and information. These and others have made a major contribution that is sincerely appreciated.

Many people have contributed to this text by teaching, guiding, and helping me throughout my career. I wish to take this opportunity to acknowledge their help and to express my sincerest thanks.

INTRODUCTION

Drilling started as the simple process of digging a hole. This text begins with a historical review. Oil and gas hydrocarbons formed from organisms in the geologic past. Natural geological processes created favorable conditions for migration and accumulation in underground reservoir traps. Areas where these traps occur are located by various methods of exploration. The right to drill wells and produce oil and gas is obtained from the landowner. Then a drilling program containing information about the prospect and instructions on drilling the well is prepared.

A drilling contractor has drilling rigs with operating personnel. The rig has a rotating system to turn the drilling bit to drill the hole, a tower and hoisting system to run tools in the hole and pull them out of the hole, and other equipment needed to drill the well. Then the rig is moved to the drillsite and assembled.

A large-diameter hole is drilled by rotating a joint of drillpipe with a bit fitted on the bottom end. Fluid is circulated down the inside of the pipe and up the outside to remove the pieces of earth drilled by the bit. Additional joints of drillpipe are connected to drill the hole deeper. Large pipe or casing is placed in the hole with cement around the pipe. A slightly smaller hole is drilled deeper through the casing. Slightly smaller casing is placed in this hole and cemented. The sequence is repeated using successively smaller sizes until the last, smallest casing is run through the oil trap thousands of feet below the surface.

Many drilling problems occur such as loose pieces of formation falling in the hole and sticking the drill tools. A particularly dangerous problem occurs when the hole is drilled into high-pressure formations. Fluids from the reservoir can flow up the wellbore and blow hundreds of feet into the air causing a blowout. These types of problems must be handled to drill the well successfully. Tests provide information about the oil and gas in the formations and are taken during and after drilling. Many holes are drilled that do not encounter oil and gas. These dry holes are plugged with cement and abandoned. Wells with good oil and gas potential are completed.

The well is completed by perforating holes through the casing into the formation. Natural forces cause the oil and gas to flow from the reservoir through the perforations into the casing where they flow upward to the surface. Other oil traps in the well can be perforated to make multiple completions. Contaminating agents in the reservoir may cause a low flow rate from the well. The flow rate can be increased by removing these with special treatments. Different kinds of treatments are used to improve low flow rates. Oil production from reservoirs with low pressures can be increased by placing a pump in the well. The top of the casing is fitted with valves that control pressures and flow rates. Impurities are removed, and produced volumes are measured in surface facilities. Gas is piped to market; oil is piped or hauled in tanker trucks.

The text expands this abbreviated summary explaining why the procedures are needed and how the operations are performed. Over 200 photographs and illustrations provide additional clarification and information. The material is presented in a building-block manner following the natural sequence of events. Basics are covered first, followed by more advanced subjects in a stepwise fashion. This serves a dual purpose. The novice can start at the beginning, laying a foundation and building upon it. The more advanced reader can go directly to topics of interest. Many special and sometimes colorful terms are used in the industry. These are defined as they are encountered and then are used as common terminology.

A complete list of contents provides easy, quick access to sections of interest. Specific topics can be located in an expanded index. The appendix also contains listings of additional sources of information.

CHAPTER 1
HISTORY OF DRILLING

Where and when did the petroleum industry begin? One can only conjecture; the beginning is lost in the distant past. A prehistoric man cleaning rubble out of a spring to increase its flow could have founded the industry. The pointed stick he used for cleaning the spring could have been the first drilling tool.

Genesis 21:12 contains one of the earliest records of digging a well: "... you may be a witness that I [Abraham] dug this well." Early water wells were dug with common hand tools. These hand-dug water wells were the predecessors of today's oil wells.

Early drilling activities were conducted in different areas for many reasons, presumably to drill or dig water wells in most cases. The Egyptians drilled core holes, probably in rock quarries, in 1000 B.C. The Chinese drilled brine wells to recover salt in the same era and by the year 1200 A.D. had drilled holes 1,500 ft and deeper. A Greek historian, Herodotus, reported that a well produced asphalt, salt, and oil about 450 B.C. But the first "true" oil wells weren't drilled until the 1700s in France, and hand-powered rotary drilling techniques weren't used until the 1800s in the Larderello, Italy, geothermal fields.

Brine wells were first drilled in the United States in the early 1800s. Some reports of oil and gas wells exist, but it is questionable whether these wells were drilled for oil and gas or were brine wells that encountered hydrocarbons. The Drake well, drilled in Pennsylvania in 1859, is generally credited as the first well in the United States drilled for the purpose of finding oil. Milestones in drilling achievement are listed in Table 1–1.

TYPES OF DRILLING METHODS

The history of drilling can be divided into four general categories according to the equipment used: hand dug, spring pole, cable tool, and rotary. Each method was replaced by a better, newer method because of the demand for deeper, faster drilling.

2
DRILLING

TABLE 1–1 Milestones in Drilling Achievement

Year	
pre 1300 B.C.	". . . you may be a witness that I [Abraham] dug this well."
pre 1600	Chinese drill 1,600-ft wells; kept drilling reports
1745	Oil wells at Pechelbronn, France
1808	First well drilled in America, 58-ft brine well at Charleston, West Virginia
1814	"American well" at Cumberland, Kentucky, drilled to 475-ft and produced up to 1,000 b/d of oil
1841	French drill with dry-hole rotary tools to 1,900 ft
1858	U.S. Army Engineers drill Pecos Valley well to 1,047 ft with steam rig; Pecos water corrodes boiler; rotary geothermal drilling at Larderello, Italy; depth records (2,193 ft, St. Louis, and 2,086 ft, Louisville) set with percussion-type rod tools
1859	Drake well produced 20–30 b/d from 69 ft
1895	First rotary-drilled oil well in U.S. at Corsicana, Texas
1897	Overwater drilling from pier at Santa Barbara
1901	Hamill Brothers cap Spindletop in first high-flow well kill operation
1903	First cementing job—casing "puddled" to shut off water above pay in Lompoc field, California
1910	Two-plug cementing method used near Taft, California
1911	Overwater drilling by Gulf Refining in Caddo Lake, Louisiana; steam rotary rig on cypress platform
1919	Diamond drilling
1925	API standards for chains and for tool joints; first API stamp of approval
1933	Texaco barge Giliasso drills in Lake Pelto, Louisiana; first "balanced" rig—all parts had comparable capacities; deviation clause in drilling contract; method of running pipe in tension
1937	First on-site mud logging; Superior and Pure drill 1 mile off Cameron, Louisiana
1938	Rotary air-drilling in oil field
1939	12,000-ft turnkey
1940	AAODC organized; J.E. Brantly president
1947	Kerr-McGee et al., drill and produce out of sight of land—Ship Shoal Block 32
1953	Congress defines ownership of U.S. offshore areas; Gulf of Mexico leases follow in 1954
1954	Hole direction maintained successfully by varying bottom-hole assembly
1955	Drilling from drillship
1957	Pan American uses 90,000-lb bit weight for extended interval
1957	Offshore Rig 54 drills in 100-ft of water
1960	Weight/speed calculations allow simplified drilling optimization
1963	IADC Sound Business Practice Standards developed Present incidence of drilling blowouts reduced to less than 0.1% through industry training and safety measures.

Source: *Oil & Gas Journal Petroleum/2000,* August 1977, p. 169

3
History of Drilling

Hand-Dug Wells
Hand-dug and *spring-pole* drilling methods are obsolete and of little interest except for their historical significance. Hand-dug wells were literally holes in the ground dug with common hand tools such as a pick and shovel. These holes were either square or circular in shape and large, about 4–6 sq ft in diameter. The size of the hole allowed people to work in the bottom of the hole. Earth was picked loose, scooped up with a shovel, and thrown out of the hole. Baskets suspended from ropes were used to remove earth from deeper holes. Hand-dug wells seldom exceeded 20–30 ft in depth because of the slow method of digging and because of caving, where earth from the sides fell into the hole.

Spring-Pole Drilling
Spring-pole drilling used a long, limber pole, fixed at one end and supported near that end with a forked pole or similar support (Fig. 1–1). Drilling tools were suspended from the longer, limber end of the pole by rope or wooden rods. Iron rods were used on some late spring-pole rigs. The tools were reciprocated by hand or by a foot sling to provide drilling action.

The basic improvement of spring-pole drilling over hand-dug wells was the method of drilling. The procedure permitted much deeper drilling than common or even specialized hand tools. A blunt or wedge-shaped bit was pounded repeatedly against the bottom of the hole as the drill tools were reciprocated. This churning action crushed and pulverized the earth, which was mixed into a thin mud with a small amount of water poured into the hole. This slurry was removed with a bailer or sand pump—a hollow tube sealed on the bottom and fitted with

FIG. 1–1 Spring pole

4
DRILLING

a bail on top. Spring-pole drilling was crude and limited by today's standards, but it did permit drilling holes much deeper than common dug wells.

The spring pole was the forerunner of the cable-tool rig. Spring-pole drilling equipment was improved in various ways, including the use of steam power (power-pole drilling), before the method was replaced by cable-tool drilling. Thus, the tools and drilling techniques used on early cable-tool rigs were developed in spring-pole drilling. By the mid 1800s, most spring poles had been replaced by cable-tool rigs.

Cable-Tool Rigs

The need to drill deeper, faster, and more economically brought about the gradual evolution of the cable-tool rig from the spring-pole rig. By the 1850s, most rigs were cable-tool or modified spring-pole units. Although both types used the same basic drilling procedure, horse or mulepower replaced manpower. The development of steam power was a significant advancement, and this in turn was replaced by the more efficient internal combustion engine. Equipment, tools, and operating techniques were also developed and improved.

The cable-tool rig gradually evolved into a highly efficient machine capable of drilling over 10,000 ft deep in special cases but normally limited to 3,000–5,000 ft. The earliest cable-tool rigs used wooden derricks or tripods (Fig. 1–2). The spring pole was replaced with a wooden walking beam, a mechanical hoisting device to lift or lower the drilling tools. The walking beam was set on a sampson post and was moved up and down (much like a child's seesaw) by a motor-driven concentric attached to a pitman arm at one end of the beam. A temper screw that held the drilling line, the cable that lowered the drilling tools, was connected to the other end of the walking beam.

The cable-tool drilling technique was quite simple. The walking beam reciprocated (lifted and dropped) the tools to drill the hole. The temper screw was turned to let out more line at the proper tension as the hole was drilled. Large powered drums, or reels, held the drilling line and sand line. The sand line was used to raise and lower the bailer, a bucket-like device used to remove mud from the hole. A larger, special casing line was used to run casing, the hollow tubing that lined the hole and prevented the sides from caving in. This casing was an important development that distinguished the cable-tool rig and the late-model spring-pole rigs and allowed deeper drilling.

Early cable-tool rigs were frequently left on successful wells and were used to pump the crude oil. The walking beam supplied the reciprocating action needed to pump the well. However, as cable-tool rigs became more modernized and specialized, they became too costly to leave on the wells as pumping units. The rigs were gradually developed into portable drilling units. To this day,

5
History of Drilling

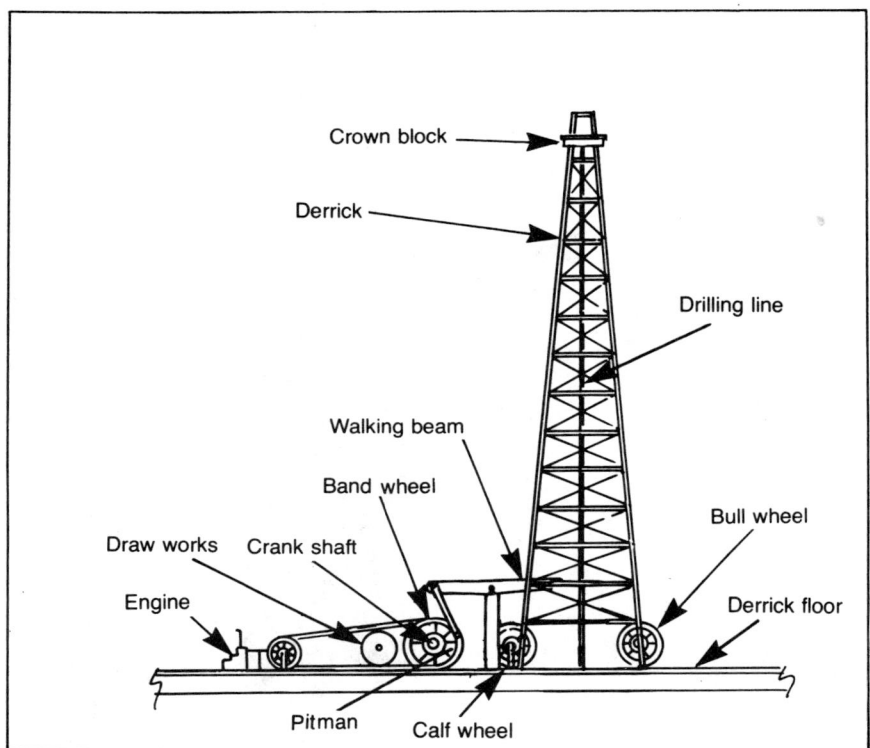

FIG. 1–2 Cable-tool rig

though, pumping units have a walking beam and sampson post—names derived from the early cable-tool rigs.

Although modern cable-tool techniques are similar, equipment has changed (Fig. 1–3). The wooden stationary rig has been replaced with a rig mounted on a skid or a trailer. A collapsible two-pole mast is mounted on the front of the skid and lies back over the top of the skid for moving. The mast is reinforced and supported in place by a framework of small pipe connecting the mast to the skid base at various points. Additional support is obtained with guy-wires.

Cable-tool rigs are rated by their depth capability. The mast capacity can be increased by about 30% using a *third leg*. This girder or pipe extends from a point in front of the mast to the crown or top and effectively forms the third leg of a

6
DRILLING

FIG. 1-3 A Wichita spudder, a form of portable cable-tool rig, shown on location in Wichita County (courtesy Walker-Neer Manufacturing)

tripod. It is used when lowering or running casing or when the drilling depth is extended below the normal capacity of the two-pole mast.

The components of a cable-tool rig are relatively simple. On modern cable-tool rigs, a drum or reel holds the drilling line. This line passes from the bull wheel under the walking-beam pulley and up over the crown. The walking beam is actuated or moved by an arm extending from the beam to a large sprocket with a series of holes that change the length of the stroke. Standard chains and sprockets with dog clutches transmit power from the engine to the bull- and calf-wheel drums, which also have clutches. The calf wheel holds the sand line or bailing line. This line passes from the calf wheel over the crown and is attached to the bailer, sometimes called a sand pump. A traveling block, which helps raise and lower the equipment, is strung up with a casing line. This line may be run on a separate drum but is usually used on the calf wheel.

The cable-tool downhole drilling assembly, from the bottom up, includes a bit, a drillstem, jars, and a rope socket with standard diameters of 3, 4, 5, and 6 in. The hole is drilled with the bit. The drillstem provides additional weight to improve the drilling action. Jars help the drilling action and release stuck tools. The rope socket connects the drilling tools to the drilling line. The assembly is connected by tool joints fitted with a square area (squares). Square-jawed wrenches are fitted on these squares, and the tool joints are tightened by a circle jack fastened to the rig floor. The jack pushes the end of the lead tong around toward the backup tong, which butts against a backup post.

New bits are about 8 ft long and are used until they are about 2½ ft long. Standard drillstems are 20 ft long, and the rope socket is about 2½ ft long. Jars range from 6–8 ft, depending on the stroke length.

Cable-tool bits are usually wedge or chisel shaped. The taper is about 25° from the horizontal; a sharper taper (higher angle) is used to drill in softer formations, and a flatter angle is used in harder formations. A 6-in. long ring of metal, called a bit gauge, is located immediately above the tapered area of the bit (wedge). Its diameter determines the diameter of the hole. The ring has two V-shaped notches, called water courses, on opposite sides. These notches allow fluid to pass by as the bit is reciprocated during drilling.

With 6–12 hr of use, the wedge on the face of the bit will flatten and the ring will swell. At the rig site the dull bit is heated in a forge and is reshaped with sledge hammers. The wedge angle is restored, and the ring is hammered back to size and is checked with a ring gauge. Then the bit is tempered by reheating and quenching it in used engine oil. Since the sharpening process can take so much time, two bits are normally used: one is sharpened while the other is drilling.

A bit will drill a hole 1–1½ in. larger than the diameter of the gauge ring because of the swinging action that occurs while drilling. The amount of oversize depends on how tight or loose the tools are run, i.e., the amount of slack in the drilling line.

In operation, the bit is run—repeatedly dropped and raised—for 1–3 hr and will drill 2–8 ft of hole in a relatively dry well. As with the spring-pole technique, 20–40 ft of water is poured into the hole. After the run the bit is pulled from the hole and is set aside. A bailer is run or lowered on the sand line to clean out the fluid (mud) in the bottom of the hole that contains suspended drill cuttings. Then water is poured into the clean hole, the old bit or a sharpened bit is run, and the process begins again.

If sloughing-type shales—shales that swell when contacted with water—are present, the water is carried into the hole with the bailer to prevent it from wetting the sides. If the formation does not make enough mud to suspend the cuttings, such as when drilling in sand, then bentonite mud balls are dropped into the hole to suspend the cuttings. Likewise, excess water from a producing water sand will flood the hole, restricting the penetration rate as low as 20% of maximum.

Cable-tool rigs were relatively simple to operate and easy to repair. They drilled very slowly compared to rotary rigs and were limited by depth. However, the early oil industry started in the northeastern U.S., where most of the shallow formations are very hard. Cable-tool rigs drill at maximum efficiency in very hard, tough, abrasive formations. They were economically competitive with rotary rigs that drill with mud in hard formations at depths to 3,000 ft prior to air-drilling techniques.

When drilling activities expanded to other areas in the U.S., such as the Midcontinent area and the Texas Gulf Coast, use of cable-tool rigs declined dramatically. Softer formations were usually encountered at shallower depths. The cable-tool rig drilled very slowly in these softer formations because its pounding, crushing type of drilling was much less efficient. Rotary rigs were developed to drill these types of formations effectively.

Cable-tool rigs had another disadvantage in many cases. As deeper wells were drilled, higher formation pressures were encountered. Again, because of the drilling method, the cable-tool rig could not control these pressures efficiently. The wells would literally blow the drilling tools out of the hole (a blowout) and were allowed to flow uncontrolled until the pressure and flow rate declined so the well could be controlled and completed. This obviously created a hazardous condition for personnel and caused great losses of valuable equipment, oil, and gas. It also wasted the energy in the reservoir and reduced the amount of oil ultimately recoverable from the reservoir.

The cable-tool rig evolved from the spring-pole rig and provided a means of drilling holes deeper and faster. It was used to drill most of the wells from

1850–1900. It was gradually replaced by the more efficient rotary rig early in the 20th century.

Rotary Rigs

The need to drill deeper and faster in a safe, economical manner led to the development of the rotary rig. The rotary rig first gained prominence when it drilled the discovery well in the prolific Spindletop field in East Texas in the early 1900s. Although rotary rigs had been used before then, the Spindletop well was one of the first true "gushers"—a highly productive well that literally gushed oil or gas.

The first rotary rigs were known as "rotaries" or "steam rigs." These, like the cable-tool rigs they replaced, were powered by engines that converted steam into mechanical work. Steam was generated in locomotive-type boilers. Wood or

FIG. 1–4 An early rotary rig mounted under a wooden derrick with wood cross bracing (courtesy Loffland Bros. Drilling Company). At one time it was customary for the drilling company to give a picnic or barbeque when a well was spudded. Then the landowners gave one when a successful well was completed.

coal was used as boiler fuel on the earlier rigs, replaced by oil as it became available. Oil contained a much higher energy content per pound, and ash-removal problems were almost eliminated. And it was easier to transport, permitting the rigs to drill exploration holes in more isolated locations and in areas where wood and coal were not available. Later, natural gas was used.

Fuel was burned in a duck's nest, a small pit lined with fire brick inside the firebox of the boilers. It provided a mixing chamber and helped to contain the heat of the burning fuel and to direct it into the fire tubes. It was relatively easy to convert the boilers to use either oil or gas by changing the burners and the shape of the duck's nest. Some of the later rigs used superheated boilers that provided steam with a higher energy content. As larger rigs were built, power requirements increased and more boilers were used.

As power sources steam engines were highly efficient. They had a high torque (power output) at low speed; they had a wide, variable speed range; and they could be built large enough to provide sufficient horsepower. The steam engine used dual-reciprocating—double-action—pistons. When circular motion was needed, the reciprocating motion of the engine was converted to circular motion similar to the crankshaft-and-piston arrangement in the modern internal combustion engine. Energy was transmitted from the flywheel by belts or chains to power the drawworks. Larger rigs used a jack shaft and dog clutches to change the speed and power ratio on the drawworks drum. Electrical generators and centrifugal water pumps also were powered by steam-driven turbines.

The drawworks, or cable hoist, had a main shaft fitted with a spool or drum. Iron or steel cables, known as wire lines, were wound or spooled on the drawworks drum. The drawworks shafts were fitted with smaller, smooth spools or catheads for light auxiliary lifting and pulling jobs. The main drum was fitted with a brake to control movement. Later rigs were also fitted with a hydromatic on the end of the drum shaft. This used a system of paddles, fixed to the shaft and rotating in a water bath, to restrict the unspooling drum and to conserve the brakes. A rotary, driven by chains and sprockets from the drawworks, rotated the drillpipe to drill the hole.

The rotary rig was an innovation in drilling equipment and techniques. A drill head or bit was connected or screwed onto the end of a section of iron pipe known as drillpipe. The hole was drilled by rotating the pipe and bit. The bit drilled the formation with a scraping, tearing, crushing action like a carpenter's bit or auger. This was more efficient, especially in softer formations, than the reciprocating cable-tool method of pounding and crushing. Drillpipe was heavier and sturdier than rope, so more force—bit horsepower—could be applied to the formation. Thus, the hole could be drilled more quickly, and drilling operations could be continued for a longer time, increasing the overall penetration rate.

After the hole was drilled as deep as the length of a joint of drillpipe, another joint would be connected to the drillstring by means of threaded couplings or tool joints on the ends of the pipe. A much deeper hole could be drilled using lengths of strong pipe than using cable or rope as on the cable-tool method.

A drilling fluid called mud was circulated down the inside of the drillpipe, through the bit, and up the annular space between the outside of the drillpipe and the wall of the hole. The mud carried the pieces of formation, or bit cuttings, out of the hole in a continuous, efficient manner. This eliminated the need of shutting down drilling operations to remove the cuttings with a bailer—a major improvement.

The mud served many other purposes. Formation fluids (liquids or gases present in the rock) under high pressure could be controlled by increasing the density or mud weight. Many formations that caused problems by breaking off and falling into the hole could be controlled with chemical additives. The mud also decreased friction and cooled the bit, increasing overall efficiency of the drilling operation.

Drilling equipment and techniques have been improved continually, particularly in the development of drilling bits. The first bits were one solid piece and had massive, sharp blades. These drag bits drilled with a scraping, tearing action as they rotated in the bottom of the hole. These bits drilled well in soft formations but had a limited life and were inefficient in hard formations. Roller-cone bits overcame many of the disadvantages of the drag bits. These bits had 1–4 movable rollers on the bottom that rotated individually as the entire bit rotated. This allowed a larger cutting area, which increased the bit life—the time the bit could drill efficiently. The rotating cutting faces provided a cutting, tearing, crushing action that was very efficient, especially in softer formations. Later bits were fitted with jets. The jets directed a high-velocity stream of mud onto the roller cutters and the bottom of the hole. The hydraulic and erosional action of the jet stream helped erode and drill the softer formations and cleaned the bottom of the hole and the roller teeth, increasing penetration rates.

It is interesting to note that the original method of drilling with cable-tool rigs was to strike the formation with a pointed object, drilling with a crushing action. This is one of the most efficient methods of drilling in very hard, tough, abrasive formations. To a great extent, this crushing principle is used in modern roller bits. However, they have more points to wear, and they strike the formation harder and faster.

The bit is replaced when it is dulled to the point that it will not drill efficiently. The process of removing the drillpipe and replacing the bit is called *tripping*. The drillpipe is pulled from the hole, disconnected, and set aside. The dull bit is replaced with a new bit, and the drillpipe is run back into the hole by reconnecting the joints. On earlier rigs each joint, called a single, was laid down when it was disconnected and later was picked up to run the bit in the hole, a time-consuming process. On modern rigs two or three joints—called doubles or thribbles, respectively—are pulled at a time and are placed back in the mast or derrick. This method of pulling stands of drillpipe is a faster way to trip out of the hole.

From an operational viewpoint steam rigs were highly efficient. Nevertheless, they had their problems. Steam rigs used large amounts of water and had boiler scaling problems. The rigs also used large amounts of fuel in a relatively inefficient manner. This was not a problem with cheap fuel, except in the case of isolated wildcat wells with higher fuel transportation costs. These factors limited the use of steam rigs to areas where water and fuel were readily obtainable.

The demand for deeper, faster, more economical drilling and the increasing disadvantages of the steam rig led to the development of the mechanical rig. The significant difference between the late-model steam rigs and the early mechanical rigs was the power source and method of power transmission. Mechanical rigs are powered by an internal combustion engine, either gasoline or diesel. Power was transmitted through clutches, sprockets, and chains or through pulleys and belts.

The mechanical rig was an efficient drilling machine; however, it had several undesirable features. The engines and compound had to be lifted to the rig floor and precisely fitted together during moving. The rig floors were built at a higher level to accommodate the larger stacks of blowout preventers necessary to handle increasingly higher pressure at deeper depths. Installing, maintaining, and dismantling equipment under these conditions created a problem. The engines, near the rig floor, were a fire hazard in case of a blowout. They created a continuous loud noise in the immediate work area. The power coupling and transmission system were the best available but needed improvement.

These disadvantages led to the development of the electric or diesel-electric rig. Although many mechanical rigs are in use and are still being built, there is a tendency to convert mechanical rigs—especially large ones—to electric rigs. An increasing number of new rigs are electric, again, especially the larger ones.

The basic difference between the electric rig and the mechanical rig is the method of coupling and transmitting the power from the driver to the driven units. A direct-current (dc) generator is coupled directly to a diesel engine (unitized) and is mounted on a skid. Direct-current motors are installed on the drawworks and pumps. Electrical power is transmitted directly from the generators to the motors by large electric cables.

The dc rig has a number of advantages over the mechanical rig. The skid-mounted, unitized engine-generator units can be moved easily and set up at ground level at any convenient location on the drillsite. This reduces rig-down, moving, and rig-up time. The high-torque, variable-speed electric motor reduces the amount of mechanical equipment needed in the drawworks and permits additional flexibility and smoothness in operation.

The dc rig retains several disadvantages of the mechanical rig. The prime movers cannot be split to provide power to more than one dc motor. Separate generating units are needed to supply alternating-current (ac) power to the lighting system and to small motors such as the shale shaker and mud mixer. Considerable electrical maintenance is required, especially on dc generators.

Silicon-controlled rectifier (SCR) rigs are the latest generation of electric rigs. An ac generator is coupled directly to a diesel engine and mounted on a skid. Direct current motors are installed on the drawworks and pumps, similar to the dc rig. Electric power is transmitted from the ac generators to a busbar by flexible electrical conduits. The busbar acts as a converter-distribution center. The electrical power is converted by transformers and rectifiers and is distributed to each electric motor (ac or dc) and to other electrical requirements, such as the rig lights, at the required voltage and frequency.

The SCR system permits efficient distribution of power in the form needed up to the maximum capacity of the prime movers. It has all of the advantages of the dc system but eliminates most of the disadvantages.

The basic drilling rig was improved simultaneously with the improvement of other drilling equipment and operating techniques. These changes increased the rate of penetration, improved efficiency, and permitted drilling and completion in the hostile environment of higher pressures and higher temperatures at deeper depths. Significant advancement has been made in bit design and drilling techniques, so that in some softer formations, 2,000+ ft have been drilled in one day. By comparison, an average of 25 ft/day or less was achieved with the cable-tool rig. New types of mud have been developed, greatly facilitating the drilling operation. Mud-treatment and solids-separation equipment has been greatly improved. Modern well-logging equipment provides many details about the formations that permit improved completions. Special steel and coupling designs permit the drilling and casing of ultradeep holes.

Special modifications permit drilling large holes (more than 10 ft in diameter) for mine shafts and atom-bomb test holes. Hole-deviation techniques have advanced so that a number of holes can be drilled from one surface location; each hole enters the producing formation at a precise, predetermined point. Several different formations in the same borehole can be completed and produced indi-

14
DRILLING

FIG. 1–5 A 30,000-ft class rig capable of drilling over 6 mi deep (courtesy Loffland Bros. Drilling Company). Note the large amount of pipe located in front of the rig that is needed to drill deep holes.

vidually. Blowout and completion equipment can control formation pressures in excess of 20,000 pounds per square inch (psi). By comparison, the hydraulic lift used in service stations needs less than 500 psi.

Rotary rigs and drilling techniques have been continuously improved, resulting in the development of new modern rigs capable of drilling to depths of over 30,000 ft. As exploration indicates offshore production potential, the rigs have been modified for man-made islands, barges, platforms, jackups, semi-submersibles, and drillships that can drill in deep ocean waters.

CHAPTER 2
GEOLOGY AND EXPLORATION

Geology and the drilling industry have developed simultaneously and are interdependent. The continual improvements in drilling techniques and equipment have been paralleled by an increase in exploration techniques and equipment and in general geological knowledge. Geological studies are of prime importance because they are the main method of finding and evaluating drilling prospects, an area or location where the underlying formations are expected to contain commercially productive hydrocarbons.

A good knowledge of geology is important in designing the drilling program and conducting drilling operations. Knowledge of formation lithology helps in drilling-bit selection and in determining casing setting depths. Many formations and their contents can adversely affect the drilling operation. Some of these problem areas are heaving and caving formations, steeply dipping formations, lost-circulation zones, saltwater flows, high-pressure zones, and contaminants such as salt, gypsum, and anhydrite. It is important to have a basic understanding of these formations when drilling and completing the well.

Many formations contain hydrocarbons that are not commercially productive. The oil shales in western Colorado and the tar sands (or heavy oil deposits) in Utah are two examples. Many similar deposits exist throughout the world. The volumes of hydrocarbons in these known deposits are many times greater than the total amount of oil (and gas equivalent) that has been found and produced to date. However, at current prices and with today's technology, these hydrocarbons cannot be produced commercially. The cost of drilling, producing, and upgrading is more than the price obtainable for the product. Therefore, the geologist must use his expertise not only to find hydrocarbons but to locate them in conditions where they can be produced economically.

Most common geological exploration methods are indirect. The geologist looks for oil traps—reservoirs where hydrocarbons can accumulate. These traps must be drilled to determine if they contain commercially productive hydrocarbons. All traps do not contain hydrocarbons, and some that do are not necessarily

18
DRILLING

commercially productive. There are direct methods of searching for hydrocarbons, such as geochemical surveys. Such surveys are in the early stages of development but they show considerable promise. However, the ultimate test to locate commercially productive hydrocarbons is drilling a well.

The terms "well" and "hole" are often used interchangeably. "Well" usually signifies a commercial producer, although this may not always be the case; "hole" refers to the wellbore.

HISTORY

Early drilling prospects were near oil and gas seeps. The operator was frequently the owner, geologist, driller, engineer, leaseman, and whoever else was required to locate, drill, and complete the well. He may have noticed that if he drilled on the downdip side of the seep, his chance of success was better. This could have led to the occupation of petroleum geologist, commonly called geologist.

Surface geology is the study of surface geological conditions. Locating drillsites near oil seeps is a form of surface geology. Most, if not all, of these more favorable locations were drilled early in the industry.

Hydrocarbons are found below the surface, which led to the development of subsurface geology. As wells were drilled farther from oil seeps and in other areas, additional subsurface information was needed. Some of this was obtained from drill cuttings, pieces of formation (rock or sand) released by the drilling action of the bit. The combination of surface and subsurface geological studies directed at finding hydrocarbons led to the field of petroleum geology.

The early geologist was severely hampered by the lack of data. Good formation lithology, a complete description of the formation, was difficult to determine because bit cuttings were broken into very small pieces. The depth in the hole where the cuttings originated was often difficult to determine because they were contaminated by caving material from shallower formations. These problems were alleviated with the development of well logging. The well log identified the different types of formations in the hole, recorded their depths and thicknesses, and gave some lithological information. As well-logging technology advanced, additional information was obtained. These data helped the engineer plan a completion program and calculate reserves. The technology of obtaining information from the wellbore was further advanced by using mud logging, coring, and drillstem testing, techniques that are discussed in following sections.

Although well logging was a major contribution to geology and engineering, it did not solve the problem of selecting drilling sites (the exact surface location where the well is to be drilled). Logging was of little value in selecting wildcat drilling prospects—wells drilled in outlying, virgin, undrilled areas away from known hydrocarbon production. The geologist could use regional surface geology to define large areas with hydrocarbon potential. However, these areas were often flat with few surface features such as formation outcrops and hills to help predict the subsurface geology, like the productive but featureless Kansas and Oklahoma plains and the Gulf Coast area, including the Continental Shelf, out to water depths of 200–500 ft. It was impractical (uneconomical) to drill a well every few miles over these broad areas to obtain subsurface information. But with the advent of exploration tools, extremely complex subsurface geology was found in many of these areas.

Geological subsurface interpretation was advanced significantly with the development of seismic exploration techniques. The geologist could now map the subsurface formations below large featureless areas in an economical manner. This technique alone is responsible for many major discoveries in the Gulf Coast and inland areas, including the recent discoveries in the overthrust areas. Advances in techniques combined with computerization have made seismic a valuable tool. Magnetic and gravimetric techniques are also used. Original mapping procedures are supplemented by computer-derived maps using different parameters. Today these geological exploration techniques first used on land have been modified and are used in offshore waters.

GEOLOGICAL CYCLE

The crust of the earth is a thin layer a few tens of miles thick. This crustal section has undergone many changes in the geologic past and is still in the process of change. Continental land masses have moved over the face of the globe. The surface has been wrinkled by folding. The crustal section has been punctured by volcanoes and wrenched by faulting—shifts in the earth's crust. Inland areas have subsided, forming inland seas that were filled with evaporates and sediments. Land masses have alternately uplifted and subsided, allowing the seas to transgress and regress. Mountains have been thrust up and eroded away. Throughout geological time the weathering-erosional action of water and wind has eroded the highlands, grinding, breaking up, and pulverizing the rocks as they are transported and deposited in lower areas.

This ever-changing crust is composed of a variety of naturally occurring compounds and elements, all commonly referred to as rocks. Rocks are further classified as igneous, metamorphic, and sedimentary.

Igneous rocks originate from molten rock material, or magma, deep in the earth. The magma in the underground magma chamber moves upward as an intrusive or vent to the surface of the earth by volcanic action where it cools and forms igneous rock. Through time, the intrusives and other crustal material can be uplifted by crustal movements. This mountain-building process creates highland areas. All types of rocks occur in the uplifted areas, but igneous and metamorphic rocks are more common.

A fundamental process of the geologic cycle is erosion–transportation–deposition. Wind, moving water and ice, and limited chemical action break up the rocks on the landmass and transport the broken material to the oceans or other low areas, which are usually large bodies of water. As the material is transported, it is broken up into smaller particles. The final particle size depends on the hardness of the rock material and the distance it is transported. For example, quartz and feldspars can be transported from the Rocky Mountains (source material) through tributaries into the Mississippi River and down the Mississippi into the Gulf of Mexico. The harder quartz may be broken up into sand-size particles. At the same time the softer feldspars will be broken and worn to smaller silt or clay-sized particles.

When the rock particles are deposited in the ocean waters, they tend to settle down to the ocean floor—hence, the name sediments or sedimentary rocks. They can be deposited in any low-lying area, but in the usual case the final depository is the oceans or large bodies of water.

The transportation-deposition process tends to sort the sediments by size and deposit them in layers. The thickness of these layers or formations ranges from a fraction of an inch to hundreds of feet. The areal extent of the formations ranges from small channel deposits to blanket deposits covering hundreds of square miles.

As the deposition process continues in one area, the accumulation of sediments becomes thicker and heavier. This causes subsidence, and the sediments are buried deeper in the earth. As the process continues, the sediments are moved downward and are exposed to increasingly higher pressures and temperatures. The process of exposing the sediments to higher pressures and temperatures changes or metamorphoses the sediments to metamorphic rocks. Clay is compacted to shale and slate, sedimentary rocks, and then under continued heat and pressure is metamorphosed to a schist and gneiss, metamorphic rocks. The soft, unconsolidated clay is changed into a dense, hard, compact metamorphic rock with a similar but different chemical composition—again caused by the heat and pressure.

As the burial process continues, the pressure and temperature can increase until the metamorphic rock is melted. The different rocks melt at slightly different temperatures and pressures. This provides a means of separating the rocks into their various components. For example, in a molten mixture of quartz and gold, the quartz with its higher melting point would solidify first and the gold with its lower melting point would solidify later, giving a mixture of gold veinlets in quartz, or possibly disseminated gold. The rate of cooling also affects the texture of some rocks. For example, if quartz is cooled very slowly, it will form large crystals; smaller crystals form with faster cooling.

During the burial or subsidence process, the metamorphic rocks may be moved laterally for long distances. They can also be moved upward, such as in thrust faulting. Molten metamorphic rocks can mix with deep lava and can be extruded toward the surface, beginning the cycle again.

Different parts of the earth are in different stages of the geologic cycle. On a miniature scale the Hawaiian island chain is a good example. The big island of Hawaii is still growing and expanding as a result of volcanic action. The northwestern-most island, Kauai, is an older island that shows evidence of considerable erosion and is developing a sedimentary coastal plain and an offshore shelf area.

The geologic cycle is not always completed in a smooth cycle but may be interrupted or suspended for a long period. For example, during the burial process sediments that are partially metamorphosed can be uplifted as a result of crustal movements to alter or change the geologic cycle. A shale or slate can be uplifted before it has completely metamorphosed into a gneiss caused by crustal movements. Sands can be deposited and buried, the sand grains cemented together with calcareous (containing calcium) or occasionally siliceous (containing silicate) material to become a hard, dense sandstone. Clays and sands can be deposited together and later become shaly sands.

There are various geological subprocesses. Reefs and other sedimentary deposits can be formed in the seas from the shells and skeletons of marine life. Deposits of evaporates such as salt, anhydrite, and gypsum are formed in shallow inland or isolated seas when the water evaporates. Salts become partially plastic under higher heat and pressure and flow upward, pushing and deforming upper layers to form salt domes, or before the sediments have been completely ground and pulverized in the erosional-transportation cycle, they may be deposited in areas with high subsidence giving rise to conglomerate formations containing boulders and rock particles of various sizes.

The history of the earth is measured in geological time units covering millions of years. Major units are eras. Eras are subdivided into periods, which are further subdivided into epochs. Further subdivision into ages and stages may be used. A geological section for an area will show all of the formations (rocks) underlying the area. The formations are usually listed by a local name. In the geological section, the formations are further divided into the major geological time units (Fig. 2–1).

The geological time units help correlate the formations within an area of interest and from area to area. For example, in a major geological province

ERA	SYSTEM	LIVING THINGS	YEARS AGO
PRE-CAMBRIAN		ALGAE	2,000,000,000
PALEOZOIC (Ancient Life)	CAMBRIAN	SPONGES	520,000,000
	ORDOVICIAN	SNAILS	430,000,000
	SILURIAN	CORALS	360,000,000
	DEVONIAN	SHELLFISH	330,000,000
	MISSISSIPPIAN	SHARKS	280,000,000
	PENNSYLVANIAN	FORESTS	250,000,000
	PERMIAN	INSECTS	230,000,000

FIG. 2–1 Geologic ages

(basin, large area of sedimentation, etc.) the upper Cretaceous formation may be a sand-shale sequence called the Mesa Verde formation. This type of classification will help the geologist follow or trace the formation throughout the geological province. In an adjacent basin the same type of sand-shale sequence may occur at the top of the Cretaceous and have a different name. When deposition conditions are similar, equivalent formations in different basins may have similar characteristics. Therefore, the geologist can recognize the formation in the next basin and some indications of the formation lithology.

In the geologic cycle the sediments are normally deposited over an area that can range in size from very small to very large and with various irregular shapes. The thickness of the sediments and, to some degree, their aerial extent depend upon the rate and length of time the sediments are deposited. The type of sediment will change over geologic time due to a change in the source of the sediments, a change of the rocks at the source, or a change in depositional conditions. The overall depositional process gives rise to a layering effect where each layer or formation is different from the adjacent formations.

The formation or individual layer is sometimes referred to as a rock unit, or facies. The surface (usually roughly horizontal) where two different formations

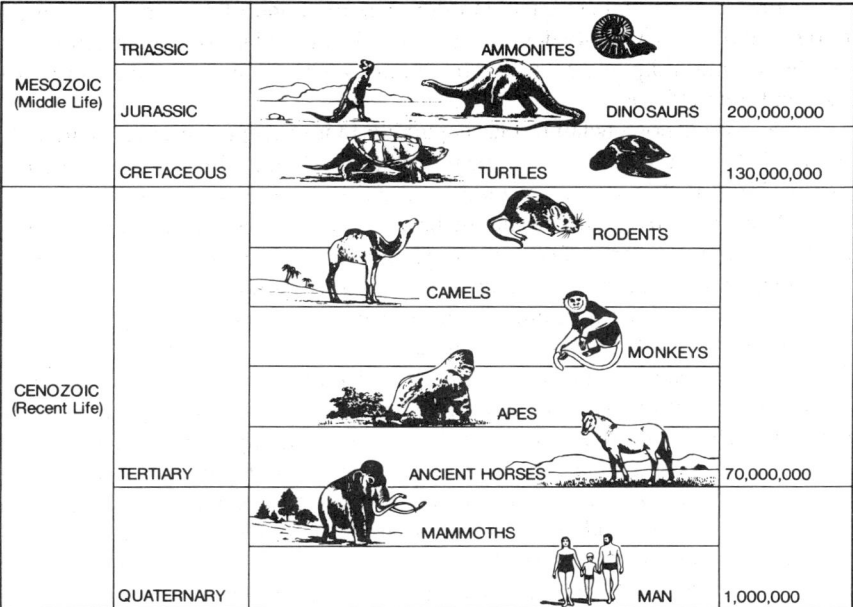

FIG. 2–1 *continued*

are in contact is known as the formation contact or the bedding plane. Some formations, such as shales, tend to be deposited in laminations, and these are also sometimes referred to as bedding planes.

The character or lithology of an individual formation may change in either the vertical or horizontal direction. Changes in the vertical direction are normally called formation changes. A gradual change in the horizontal direction is usually known as a facies change, for example when a sand may gradually change to a shale.

Maps and Cross Sections

Most formations are relatively flat and roughly horizontal. However, many formations have curvature and changes of curvature. This is described by means of strike and dip information. In a small area the formation is considered to be a plane. The strike is a line made by the intersection of the plane of the formation with a horizontal plane. The strike-line direction is oriented to the compass points. The dip of the formation is the angular measurement in degrees of the angle between two lines perpendicular from and intersecting the strike line, one line in the plane of the formation and one in the horizontal plane.

Fig. 2–2 illustrates strike and dip. The direction of the strike line in the horizontal plane is north 45 degrees east (N45°E). The plane of the formation dips southeast with a dip angle of 20°. In summary, the formation strikes north 45° east and dips 20° east.

A shorter nomenclature that is gaining more acceptance is to identify the dip angle and the horizontal component of the dip angle. For example, the formation illustrated in Fig. 2–2 could be described as dipping 20°S45°E.

Maps and cross sections are the basic geologic tools used to show regional geologic conditions and illustrate geological features. The two most commonly used maps are structure or contour maps and isopach maps.

Nearly everyone is familiar with surface contour maps. They contain lines of constant elevation and show mountains, valleys, plains, and other surface features. The maps provide a scale so that horizontal distances can be measured. They are conventionally contoured in feet above sea level. Geological structure maps are very similar to conventional surface contour maps. The main difference is that instead of mapping the surface of the earth, the geological structure map shows the surface of a specific formation of interest. It is often called a contour map but more commonly is called a structure map because it is an effective method of illustrating structures.[1]

The depths on the structure map are a reference to sea level as on common surface contour maps. The geologist works with formations that may be either above or below sea level, depending on formation depth. The contours above sea level are positive values, and contours below sea level are negative values. Various contour intervals and horizontal scales are used, depending on the coverage and purpose of the map.

Isopach maps show formation or interval thicknesses and help determine reservoir size, calculate reservoir volumes, and illustrate other formation and structural features (Fig. 2–3). Isopach maps are similar to surface contour or structure maps, except that the contour lines are drawn over points of equal thickness. For example, the formation is 30 ft thick at every point the 30-ft contour is drawn. Horizontal scales and contour intervals are selected for similar reasons, as are those for structure maps.

Frequently, both structure and isopach maps will be constructed to illustrate the same feature. Conventionally, the same horizontal scales are usually used.

Cross sections are a side view of a cutaway section of the formations. For example, if a cake made of several layers were sliced and one-half were removed, the cut-off section would be a cross section showing the alternating layers of cake

---

[1]Fig. 2–4a shows an example structure map.

Geology and Exploration

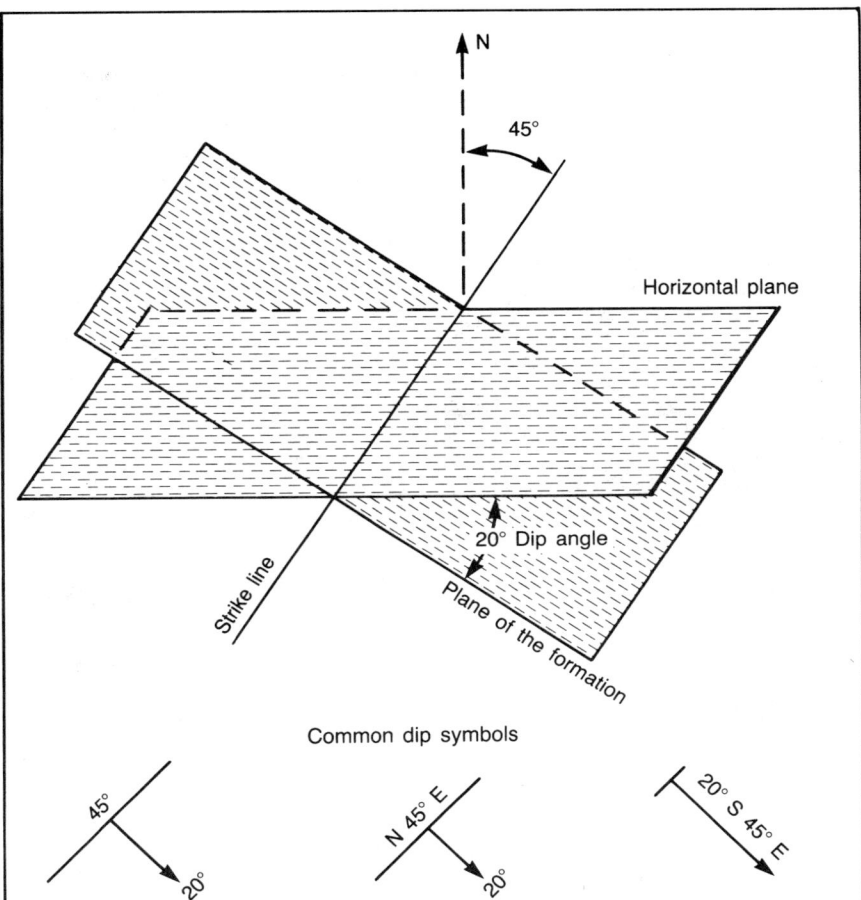

FIG. 2–2 Formation strike and dip

and icing. Cross sections have horizontal and vertical scales. Usually, different scales are used in the horizontal and vertical directions for clarity. Cross sections correlate formations and zones, and they illustrate and clarify structures and special features such as gas caps, oil-water contacts, and facies changes.

Fig. 2–4 shows a structure map and cross sections of an anticline contoured on top of a sand that is expected to be a reservoir sand. The data for Fig. 2–4a would be obtained from well data, seismic, and other available sources. On larger-scale maps all of the data points, such as the two dry holes with the depth to

26 DRILLING

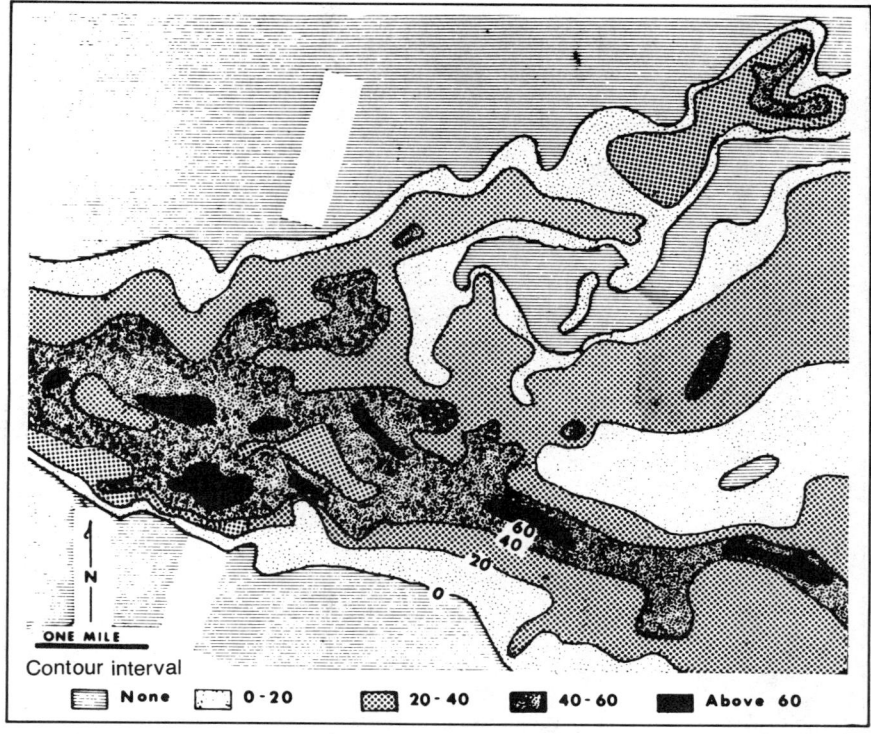

FIG. 2–3 Isopach map

the top of the sand noted below the well symbol, would be put on the map. Line *mn* represents a seismic line; the small circles indicate data points (shot points). Most of the contour lines are solid, but some on the northwest side are dashed. This indicates a lack of control (insufficient data) in that area. The dashed contours represent the geologist's estimate of the structure.

Fig. 2–4b is a cross section of the anticline along the line AA' marked on the structure map. The crest is the line along the highest part of the anticline. Fig. 2–4c is an axial cross section through the axis of the anticline.

Many illustrations of formations are not to scale. If they were drawn to true scale, the vertical dimensions would be too small for illustrative purposes. Also, maps are conventionally oriented so the north direction is toward the top of the page.

Geology and Exploration

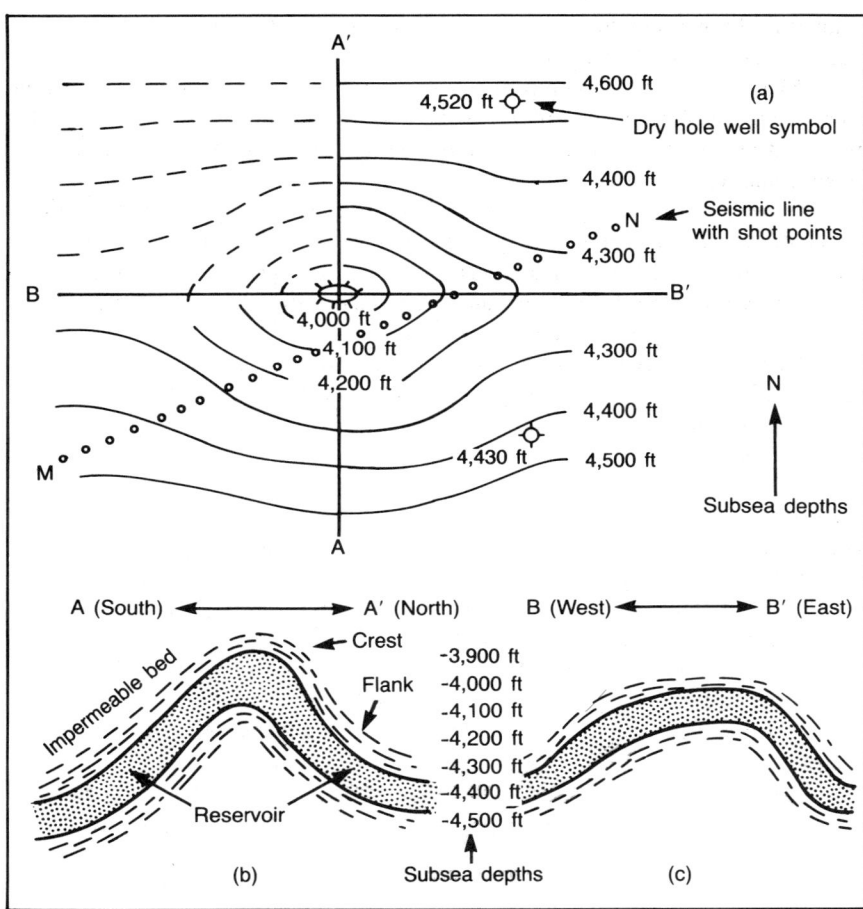

FIG. 2–4 Anticlines (a) structure map, (b) south-north (AA') cross section, and (c) west-east (BB') cross section

Surface and Well Data

The best sources of data for subsurface geological mapping are data from wells that have been drilled. Formation depths and thicknesses can be taken from well logs, plotted on maps, and contoured to construct structure and isopach maps. The data are plotted on the maps by using the map scale and referencing the data point from known boundaries, such as lease lines, section lines, and reference markers. This type of mapping is widely used.

Many formations outcrop at the surface, often at a number of different points on the surface. The thickness, strike, and dip of the formation can often be measured where the formation outcrops. These data can be plotted and used to make a map of the formation.

However, there may be a limited number of outcrops. The formations frequently change direction underground, and this change of direction cannot be accurately projected from surface measurements. The petroleum geologist usually works with deeper formations, and in most cases these do not outcrop, which further complicates the problem.

There is usually very little well data available in wildcat areas. The exploration geologist working (studying) these areas must look for information from other sources. Surface data are used when available, but they are generally of limited value.

Seismic

The development of seismic tools and interpretation techniques was a major contribution to oil-field exploration geology. The geologist now had a tool that could be used on the surface to determine the depth to the formation(s). The tool was more economical than drilling wells, and seismic quickly became the single most important exploration tool as geologists gained experience with the method and improved technology increased use and reliability.

Seismic methods were first used to study earthquakes. When an earthquake occurs, energy is released. This energy travels through the earth in the form of seismic waves. These waves are compressional, primary (P) waves, and sheer, secondary (S) waves. The movement of these waves through the earth; their different abilities to travel through solids, liquids, and gases; the velocity changes through the different materials; and the reflection and refraction of the waves as they encounter different formation boundaries are highly complex. Nevertheless, significant information about the interior of the earth can be obtained from a study of these waves and their differences in arrival time at various points on the surface. Although this natural phenomenon could not be used directly to aid the exploration geologist, the principles and tools were modified and utilized to develop seismic exploration tools and techniques.

The first seismic tools used explosives detonated in shallow holes to generate the energy for seismic waves. Today, thumpers—trucks with units that strike the surface—are often used as well.

As Fig. 2–5 shows, the wave travels downward until it hits a prominent formation (reflecting bed), which reflects part of the wave back to the surface.

Geophones (seismic wave detectors), or jugs, are spaced out in linear patterns (geophone array) leading away from the hole where the explosive is detonated. The time of arrival of the reflected wave is detected at the geophone and recorded. The travel time, measured in milliseconds (thousandths of a second), is the time period from the start of the wave at the explosion until its arrival at the geophone. The distance that the wave travels from the surface down to the formation and back again is calculated from the travel time and the known seismic wave velocity through the formation(s). The depth to the formation, or reflecting bed, at a specific point then can be calculated. The geologist uses these data to construct maps and cross sections of the formation. The computer printout of the seismic data also provides a type of cross section that can be used.

This description is very simple; many complexities are involved. When the seismic wave encounters a reflecting bed, part of the wave is reflected back to the surface and part of the wave is refracted into the formation. When the refracted wave encounters another formation boundary, part of the wave is reflected to the surface and part of the wave is refracted deeper into the formation. The wave that is reflected to the surface may be affected by bed boundaries in its travel. The seismic wave will encounter an increasing number of formations as it travels deeper, thus increasing the complexity of interpretation.

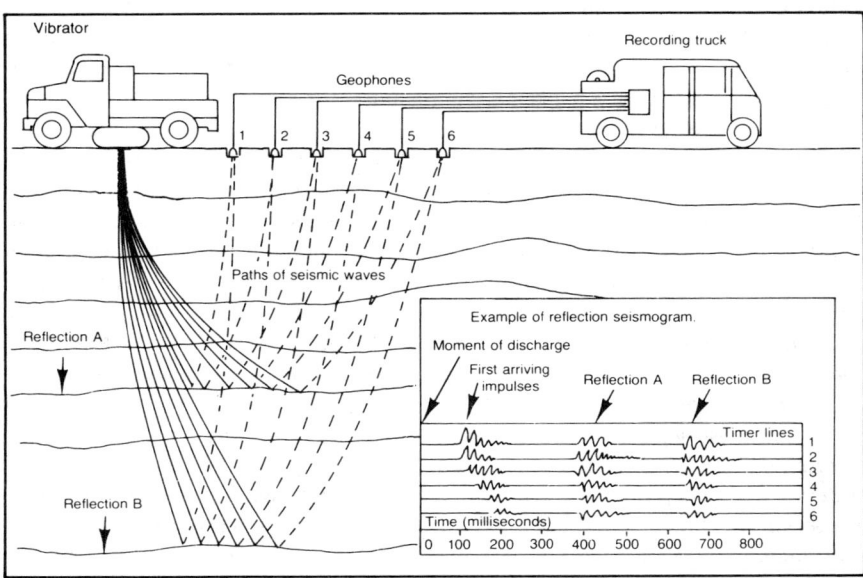

FIG. 2–5 Seismic lines

Determining the seismic wave velocity through the rocks can be a problem, especially in wildcat operations. The earth's crust is layered with many formations of varying thicknesses and is made of different types of rocks ranging from loose, unconsolidated formations (loose, uncemented particles), where seismic velocities are low to hard dense rocks with high velocities. The seismic wave velocity is not constant through these formations but changes from one to another. Also, some of the formations reflect the seismic wave better than others. However, as the seismologists (people who work with seismic) gain experience, they are able to use estimated composite travel times through the different formations.

One method of illustrating seismic data is to construct maps using seismic travel times for data points. One common map is a time structure map, which uses travel time in lieu of feet (Fig. 2–6). Where there is a prominent reflecting bed, the travel time to the bed is plotted and contoured in a conventional manner using appropriate contour intervals. Shorter travel times indicate higher areas, and longer travel times indicate lower areas on the map. Travel times through the overlying beds will usually be approximately the same over a relatively small area. Therefore, the differences in travel times portray structures on top of the reflecting bed, and formations above and below the reflecting bed normally have similar structures. The approximate depth to the reflecting bed can be calculated by using known travel times from other areas that are believed to be similar to the area under consideration. Sometimes several good reflecting beds will be encountered, and each can be mapped in this manner.

Many problems are encountered; however, the seismologist is innovative. Results can be improved by using different seismic arrays or varying the explosive charge. For example, an explosive charge can be detonated in a deep, dry hole and seismic wave arrival times recorded at the surface to determine the average travel time through a thick group of formations. This value can then be used in other areas with similar rock types. Experienced seismologists can recognize formation parameters from the strength and other characteristics of the returning seismic wave. In some cases the difference between the absorption of the seismic waves in the rocks has led to the detection of bright spots, which are believed to be direct indications of hydrocarbon deposits.

The first seismic data were hand processed, a tedious, time-consuming operation. This was revolutionized with the development of computers. Computers decreased processing time and allowed the seismologist to use different processing techniques and test a variety of variables to obtain the best results. Continual improvements in data processing have been made possible by the development of faster, higher-capacity computers.

The development of microcircuitry has also led to significant improvements in seismic equipment. Geophones are more sensitive, so weaker signals

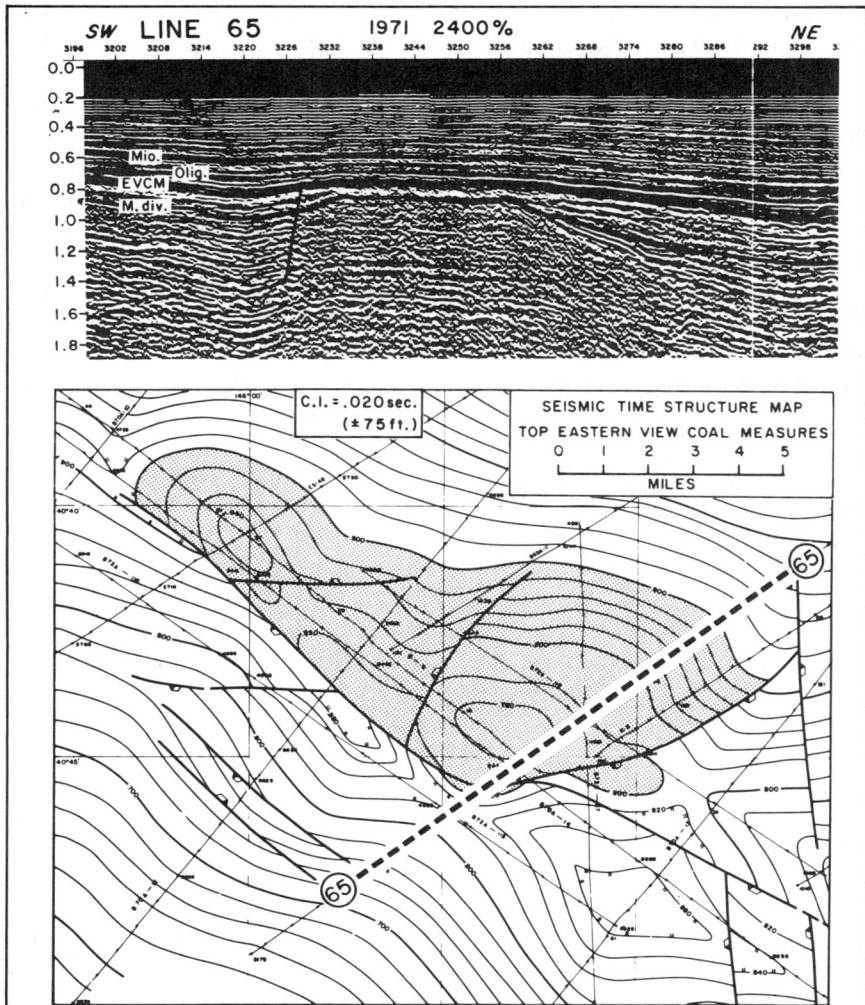

FIG. 2–6 Seismic section and time structure map (after Weaver et al., "Bass Basin Set for New Exploration," *OGJ,* January 4, 1982, p. 156)

can be recorded and analyzed. This allows reading to deeper depths with more accuracy. In some cases less energy is needed, and the conventional explosive technique is replaced with a thumper, a truck that generates seismic wave energy by dropping a large weight onto the surface. Where applicable, this tool greatly increases the productivity of seismic field crews. Data can be transmitted from the geophones by short-wave radio, eliminating large reels of connecting cables and increasing equipment portability.

Seismic techniques originally used on land have been modified and improved so they can be used in offshore waters. The equipment is sealed (waterproofed). Geophone arrays are distributed and positioned with boats and buoys. Explosives are detonated near the ocean floor or bay bottom, eliminating the need to drill holes. Satellite navigation devices are used to locate areas and seismic lines precisely.

Gravity

One of the first tools used to unravel the mysteries of the earth's crust was the gravity method. Gravity is the attractive force between objects or masses. This force is proportional to the product of the masses and inversely proportional to the square of the distance between the masses.

To illustrate, the earth's mass can be considered to be concentrated at the center of the earth. The earth is not exactly round but has an equatorial bulge due to rotation. Therefore, a person at the north pole would be closer to the center of the earth and would weigh more than he would at the equator. The difference is very small but measurable—about 1 lb for a 200-lb person.

The crust of the earth contains a mixture of materials of different densities. If the materials are more dense in one area, one would expect slightly higher gravity readings, and this is the case. If the materials are less dense, the gravity readings are lower.

Fig. 2–7 is a cross section illustrating three buried intrusives associated with each other. The first (*a*) is a massive plug, possibly some type of buried volcanic intrusive composed of denser material than the surrounding formations. The second (*b*) is a shale plug or feature with a density similar to the surrounding formations. The third (*c*) is a salt dome, which is less dense than the surrounding formations. As the figure illustrates, the gravity increases above normal over the dense material, is not affected by the average-density material, and decreases over the less dense salt dome. The detection of these intrusives is important because the formations located around the intrusive can trap hydrocarbons.

Early gravity methods required taking precise measurements at points (stations) over an area on a grid basis and then correcting and plotting the data. Gravity measurements can now be made from airborne equipment.[2] The data are corrected by computer and can even be illustrated by computer plotters. Gravity

---

[2]Sigmund Hammer, "Airborne Gravity Is Here." *Oil & Gas Journal* (January 11, 1982), pp. 113–125.

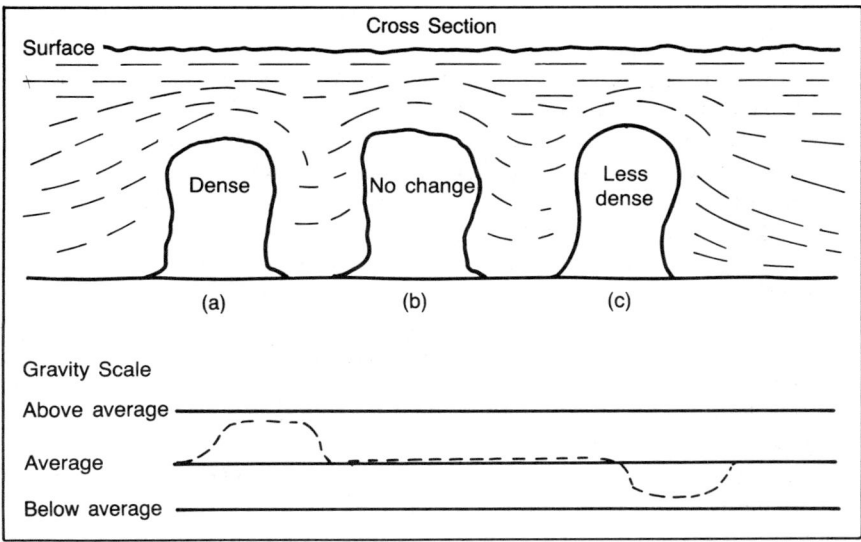

FIG. 2–7 Cross section of three buried intrusives and their relation to a gravity scale

methods are usually used for regional surveys over wide areas. An anomalous condition, such as a suspected formation or a structure that may contain hydrocarbons, is investigated by other tools or methods.

Magnetic

The earth acts as a huge bar magnet with north and south poles. These magnetic poles are near but do not coincide with the earth's geologic poles, which are referred to as true north or true south. The magnetic poles are referred to as magnetic north or magnetic south. The magnetic poles are not constant; they have varied in position over geological time and are still moving slowly.

The magnetic declination is the horizontal angle between the magnetic and geologic north poles. The magnetic poles appear to originate deep in the earth. The magnetic inclination is the angle made by two intersecting lines, one pointing northward and downward at the magnetic pole (direct line of force) and one as the horizontal component of the direct line of force similar to the line of the compass needle. These measurements differ at different places on the surface of the earth.

Early magnetic measurements were taken at stations, similar to gravity measurements. They are more commonly recorded by airborne instruments, which permit widespread, economic coverage. The data are corrected, processed, and handled by computer techniques.

Magnetic measurements are affected by magnetic objects. If you move a metal object such as a knife blade near a compass, the compass needle will be deflected. Airborne magnetic measurements are made with the measuring instrument suspended from a cable to keep it away from the metallic mass of the airplane for this reason.

The formations of the earth contain iron, usually in the form of oxides such as magnetite. The amount of iron in the formations ranges from a trace to a high percentage in iron ores. The deep basement rocks are rich in iron.

Magnetic methods are used primarily for wide, regional reconnaissance. It is a good tool for mapping basement structures. Intrusives and other formations that contain a larger than normal amount of iron can be detected by a magnetic survey. The intrusives and formations that change direction are anomalous geological conditions where hydrocarbons can accumulate. When these anomalies are encountered, they are investigated by other exploration techniques. Magnetic exploration is also widely used in mineral exploration.

Other Indirect Exploration Techniques

Other exploration techniques are in various stages of development and use. Early surface contour maps were used to locate surface features, which in turn helped detect subsurface structures. These have been supplemented and in some techniques replaced by aerial photographs that provide wide coverage quickly and economically.

Airborne radar, including Side Looking Radar (SLR), is used. One application of SLR is to remove the masking effects of heavy ground cover such as high trees, brush, and vines in jungle areas so the geologist can look at the ground surface. Infrared photography helps distinguish various formations due to their different capacities to absorb heat. Satellite imagery is gaining wider acceptance as it becomes more available.

Another interesting, indirect method of searching for hydrocarbons has been proposed and used to a limited extent. Uranium is present in most shallow groundwater in minute concentrations of one to two parts per billion (1–2 ppb). The uranium is in a soluble form and moves with the groundwater. When the mobile uranium contacts reductants such as humus or hydrogen sulfide (H_2S), it is reduced to the insoluble oxide form and precipitates, causing an accumulation of uranium oxide. Hydrogen sulfide is commonly associated with oil and gas deposits. The theory is that hydrogen sulfide in the gaseous form migrates upward, possibly partially along fault lines into the shallow groundwater where it contacts the mobile uranium and reduces it to form a deposit. Uranium and its decomposition, or daughter products, emit gamma rays. These are easily absorbed in deeper deposits but can be detected by airborne instruments in exposed or very shallow formations. Therefore, searching for uranium may provide an indirect method of searching for hydrocarbons.

Faults frequently provide a trapping mechanism for hydrocarbons. Faults are also important in the occurrence of geothermal energy, especially in more active (tectonic) areas. Telluric surveys, measuring waves in the ground originated by thunderstorms and other atmospheric disturbances, have been used. Microseismicity, tiltmeter, and resistivity (with induced electric currents) surveys have also been conducted. All of these exploration techniques either have or may have application in the exploration for petroleum hydrocarbons.

Geological concepts and theories are constantly being reviewed and revised as more information is obtained. Mountain-building concepts are a good example. It was once thought that the granitic basement found near some mountains was the depth limit for finding hydrocarbons. Some mountains are associated with an overthrust of many miles. Reviews of these theories and advanced seismic have led to drilling prospects below these formations, which were once considered basement rock.

Direct Exploration Methods

The exploration methods discussed to this point have been indirect methods. The geologist first explores for structures or other geological conditions favorable for the accumulation of hydrocarbons. Then a hole must be drilled to determine if the hydrocarbons are present.

Direct exploration techniques concentrate on exploring directly for hydrocarbons. The bright spot concept of locating hydrocarbons by seismic exploration is one form of direct exploration for hydrocarbons. However, like other techniques it has not been completely proven or verified to date.

The geochemical survey is a direct exploration method that has had some success and shows considerable promise. Samples of the earth are taken at depths ranging from a few inches (microlayer) to a few feet. The samples are analyzed for various hydrocarbons, associated compounds, and some isotopes. The data are plotted and contoured for analysis.

A similar method uses a sniffer. The sniffer has a vacuum arrangement to take air samples through a perforated tube driven into the ground to various depths. The air samples are analyzed in a manner similar to the surface samples, and the data are plotted and contoured for analysis.

These methods have a reasonable theoretical basis.[3] Studies have indicated that gaseous hydrocarbons migrate upward through the formations much faster

---

[3] Leo L. Hörvitz, "Near-Surface Evidence of Hydrocarbon Movement from Depth," *AAPG Studies in Geology*, No. 10 (1980), pp. 241–270.

than originally believed. Gas samples from oil have a higher concentration of heavier hydrocarbons, and gas samples from natural gas have a higher concentration of lighter hydrocarbons. The methods have a number of obvious problems. For example, methane is associated with hydrocarbons but can also be given off by decomposing animal and vegetable matter. Analysis of the isotopes may indicate the nature of the methane source. Hydrogen sulfide is common in hydrocarbons but is also generated from other natural sources. The accuracy of these methods may increase with better sampling techniques and more detailed laboratory analysis.

One very interesting advantage of geochemical surveys is that surveys can be run over known productive areas. This provides a model or pattern for exploration in untested areas. The sampling procedure has been modified so it can be used offshore and in deep ocean waters.

ORIGIN, MIGRATION, AND ACCUMULATION OF OIL AND GAS

The petroleum geologist is primarily interested in sedimentary rocks. They are the source beds for petroleum hydrocarbons and contain most of the oil and gas deposits. Most of the surface of the earth's land masses and continental shelf areas are covered by sedimentary rocks. The thickness or depths of these range from 0 ft near areas where the igneous rocks outcrop to depths of over 35,000 ft with some projections to over 50,000 ft in the deep sedimentary basins and shelf areas. The sedimentary rocks are normally underlain by igneous or highly metamorphosed rocks sometimes called basement rocks. Some basement rocks are of Precambrian age and are referred to as Precambrian rocks. Very few if any hydrocarbons are found below the top of the basement.

Several lithologic qualities of the formation are fundamental to a discussion of the origin, migration, and accumulation of oil and gas.

Porosity and Reservoir Rocks

Porosity is the pore or void space in the rock.[4] This space is usually filled with oil, gas, or water—most frequently salt water since the originating sediments are usually deposited in sea water. Porosity is measured as a percentage of the unit volume. The concept of porosity can be understood by considering a gallon container filled with dry sand. If water is poured slowly into the sand-filled container, about a quart will be required to fill the void spaces between the sand grains. In this case the sand has a porosity of 25%.

[4]Walter H. Fertl, "Knowing Basic Reservoir Parameter First Step in Log Analysis," *Oil & Gas Journal* (May 20, 1978), pp. 98–118.

$$(1 \text{ quart})(100)/(4 \text{ quarts/gal}) = 25\%$$

If 1½ quarts of water were required to fill the container of dry sand, the porosity would be 37.5%. Thus, unconsolidated sands would have porosities in these ranges (Fig. 2–8).

The size of the sand grains in the gallon container was not specified. Porosity is not directly related to sand-grain size. The porosity of the sand in the gallon container would have been about the same with either large or small sand grains if they had approximately the same shape. Sand grain sizes are normally classified as large, medium large, medium, medium fine, fine, and very fine for descriptive purposes.

The shape of the sand grains does affect the porosity. Sand-grain shapes are classified as well rounded (almost spherical), rounded, irregular, subangular, and angular. The highest porosity occurs with well-rounded sand grains and reduces to the minimum with angular shapes. The shape is related to the packing factor; angular sand grains can be packed into a more dense mass with less porosity than well-rounded sand grains can.

Consolidated, or cemented, sands (sandstones) have sand grains cemented together, usually with calcareous cementing material and less frequently with siliceous cementing material during the sedimentation-burial part of the geologic cycle. This solid cementing material occupies part of the void space between the sand grains; so, as would be expected, sandstones have less porosity than unconsolidated sands. Sandstone porosities range from a very few percent for angular, well-cemented sandstones up to about 25% for rounded, poorly cemented sandstones.

There are two types of porosity. One is effective or communicated porosity (commonly called "porosity") as described above for sand and sandstone. Most of the water and fluids in the pore space are free to move under the proper conditions of pressure differential. A second type of porosity is sealed, ineffective, or uncommunicated porosity. The fluid cannot be moved appreciably through the pore spaces under normal pressure differentials. During subsidence-burial the fine clay particles that were deposited in water are compacted and sealed to form shale. Shale normally contains 25–35% porosity with the pore spaces filled with water. However, the shale must be pulverized, heated, or subjected to other actions to remove the fluid from the pore spaces.

Porosity measurements are normally obtained from well logs and measurements on cores. Rocks that have porosity (effective) are frequently referred to as reservoir rocks. This indicates that the rocks have void space to hold reservoir fluids. The fluids in the reservoir are called reservoir fluids, whether oil, gas, or water.

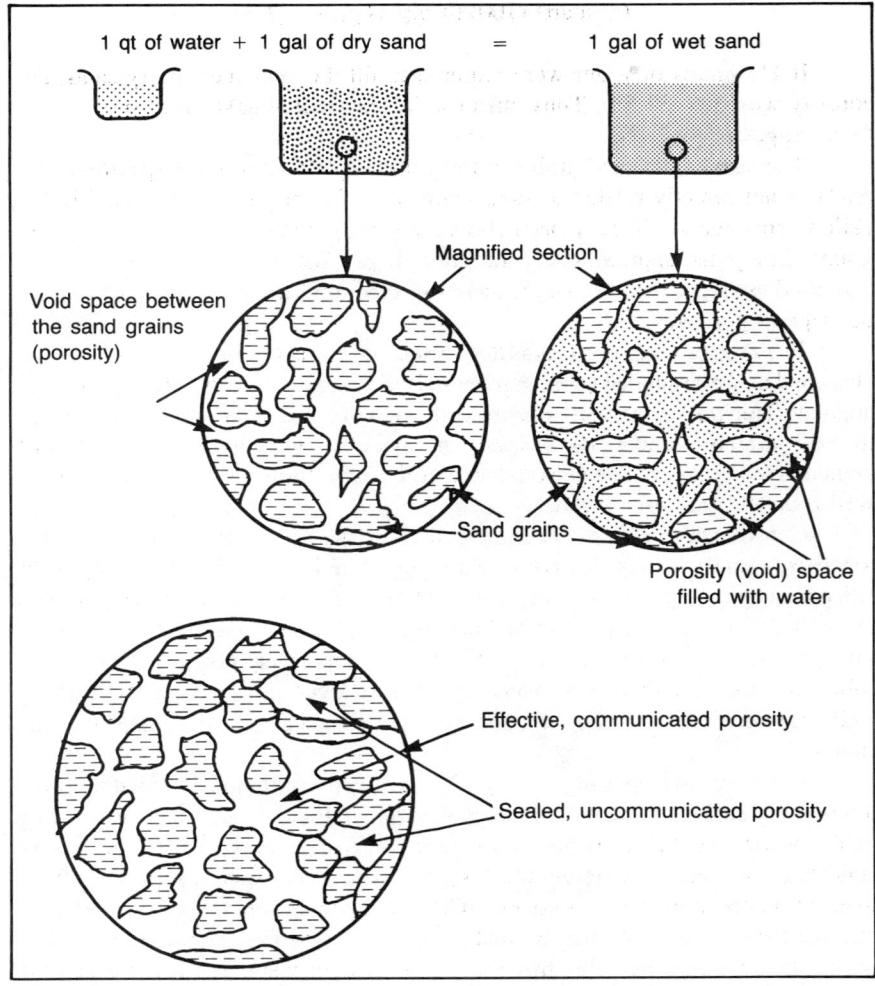

FIG. 2–8 Concept of porosity

Sands and sandstones (also commonly called sands) are almost always considered reservoir rocks or formations. Carbonates such as limestone and dolomite may or may not be reservoir rocks. Pure, dense limestones and dolomites do not have porosity. In some cases they may have vugular porosity—pore spaces or channels that may or may not be communicated.

Limestone reefs are made of marine shells and frequently have a very high porosity. Limestones are often mixed with sands (sandy limestone) and may be a reservoir depending on the grain size and the amount of sand. Oolitic and intergranular or crystalline type limestones may make good reservoirs.

Dolomites are limestones that have been subjected to heat and pressure and chemical changes during the geologic cycle. Dolomites range from dolomitic limestone (limestones containing dolomite) to pure dolomite depending on their progress in the geologic cycle. Therefore, many of the lithologic features that cause limestones to be reservoir rocks may apply to dolomites but usually to a lesser degree since dolomite is often more dense.

The porosity described above is sometimes referred to as primary porosity. Primary porosity is developed during deposition. Secondary porosity is developed after deposition. A common form of secondary porosity is fracture porosity, created by fracturing. During crustal movements, the rocks are subjected to stresses. When these stresses exceed the rock strength, the brittle, hard rocks break and fracture. In a few cases extensive fracture systems are created and can act as reservoirs. Softer rocks tend to be more plastic and either flow with the stresses or fill the fracture systems. Limestones can have secondary porosity as a result of solution channels. Flowing groundwaters and chemicals can leach or dissolve void spaces in the rock.

Origin of Oil and Gas

Many theories have been advanced for the origin of oil and gas (petroleum hydrocarbons).[5,6,7] These range from chemical reaction theories to the existence of hydrocarbons when the earth was formed. The most accepted theory is that most if not all of the petroleum hydrocarbons originated from the residue of marine organisms.

In the erosion-transportation process, various minerals and salts were leached and transported in liquid or gaseous (soluble) form to the oceans. These supplied nutrients that enhanced the growth of marine organisms. At the end of their life cycle, the marine organisms were deposited on the ocean floor along with clay-sized sediments. The sediments were converted to marine shales in the geologic cycle. The heat and pressure during subsidence, as the formation moved downward and compacted, gradually converted the organic remains of the marine

---

[5]Cecil G. Lalicker, *Principals of Petroleum Geology* (New York: Appleton-Century-Croft Inc.), pp. 57–68.
[6]"Geological Aspects of Origin of Petroleum," *AAPG Bulletin*, No. 11 (November 1964), pp. 1,755–1,803.
[7]Lester Charles Uren, *Petroleum Production Practices*, third edition (New York: McGraw-Hill Book Co., 1946), p. 11.

organisms to petroleum hydrocarbons. This theory has not been proven; however, it does appear to fit most of the observable physical facts.

Migration and Accumulation

The migration of oil and gas is subject to question as was the origin. However, the generally accepted opinion is that most oil and gas migrated from the source beds, or the marine shales, and accumulated in their present position in traps. This migration theory has not been proven, but it provides the best fit of all of the facts.

One of the most important factors affecting the movement of fluids in the subsurface is formation permeability. Permeability is broadly defined as the resistance of the formation to the flow of fluid. For the same type of fluid, more can pass through a high-permeability formation and less can pass through a low-permeability formation. Permeability is a characteristic of the formation and is discussed in more detail in the section on reservoirs. It is not directly related to porosity, but rocks with higher porosity generally have higher permeability.

Most sediments were ultimately deposited in the oceans or inland seas during the geologic cycle. The pore spaces would be expected to be filled with salt water. This almost invariably occurs except for the very shallow formations, which are filled with fresh water. Even when the pore spaces contain oil and gas, they usually have at least 20% connate water—water present when the sediments were originally laid down.

The average hydrocarbon fluid is considerably lighter than salt water. A 40°API oil will weigh about 50 lb/ft^3 compared to about 65 lb/ft^3 for salt water.[8] Therefore, oil and gas would be expected to migrate upward through the water.

Consider again the gallon container (Fig. 2–9). If a pint of oil is poured into the container and the container is then filled with sand, the oil will stay on the bottom. If the container is then filled with water and allowed to stand for a period of time, most of the oil will move upward through the water-sand mixture and will collect on top of the water. The rate at which the oil moves upward and displaces the water depends on a number of factors. If the container is filled with large, well-rounded sand grains (high permeability), the oil will move faster than if the grains are angular and fine (low permeability). If the fluids are warm, the oil will move upward faster than if colder fluids are used. This is due to viscosity, or thickness, effects.

---

[8]Oil is conventionally measured in degrees API (American Petroleum Institute). The oil gravity is measured with a hydrometer. If the measuring temperature is other than 60°F, the oil gravity is corrected to 60° and is known as API gravity. The API gravity is related to the specific gravity by the following formula:

$$\text{specific gravity} = (141.5 - °\text{API})/(131.5)$$

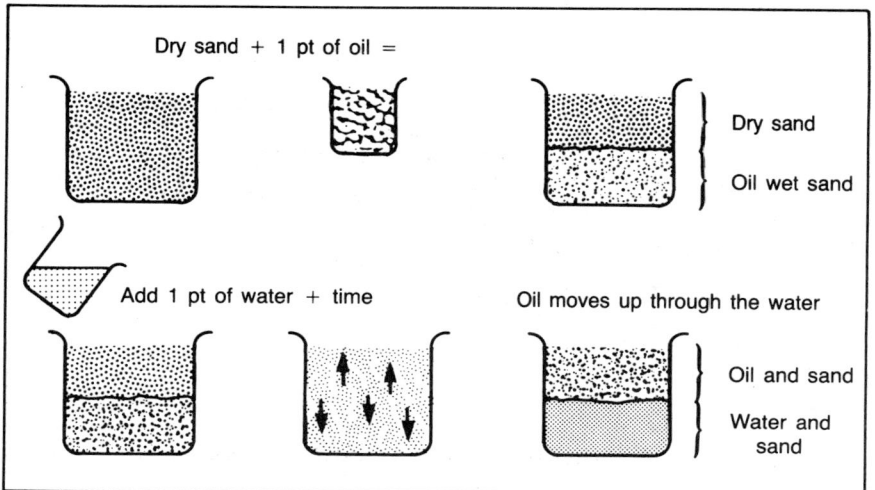

FIG. 2–9 Oil-water interchange

Most oil and gas originated in marine shales. These shales have a lower permeability than sands. For practical purposes they are considered impermeable. However, the shales lose some water, which is squeezed out as a result of pressure during subsidence. Also, although shales are relatively impermeable, over millions of years and with high pressures, it is reasonable to expect migration over appreciable distances. The hydrocarbons probably migrated updip (upward angle) in the shales until they encountered a porous rock such as a sandstone. The fluids would migrate much faster through sands because of the higher permeability. The permeability in most formations is higher in the direction parallel to the bedding planes (horizontal permeability) than in the direction perpendicular to the bedding planes (vertical permeability). The formations also tend to dip downward toward the ocean (or basin, in the case of inland seas). Therefore, oil and gas would tend to migrate both upward and inland (toward the land mass) (Fig. 2–10). The oil and gas would continue to migrate until they encountered an impermeable barrier or trap where they could accumulate. Although oil and gas are found in many areas, there is a definite tendency for accumulation on the edges of basins and updip on broad structural features, as indicated by producing fields.

A valid question at this point is, if the oil and gas were not trapped in the impermeable shale source beds, why do they accumulate in traps that are sealed with similar impermeable barriers? There are many theoretical-conjectural reasons for this, although none have been proven. One fact is that oil and gas are

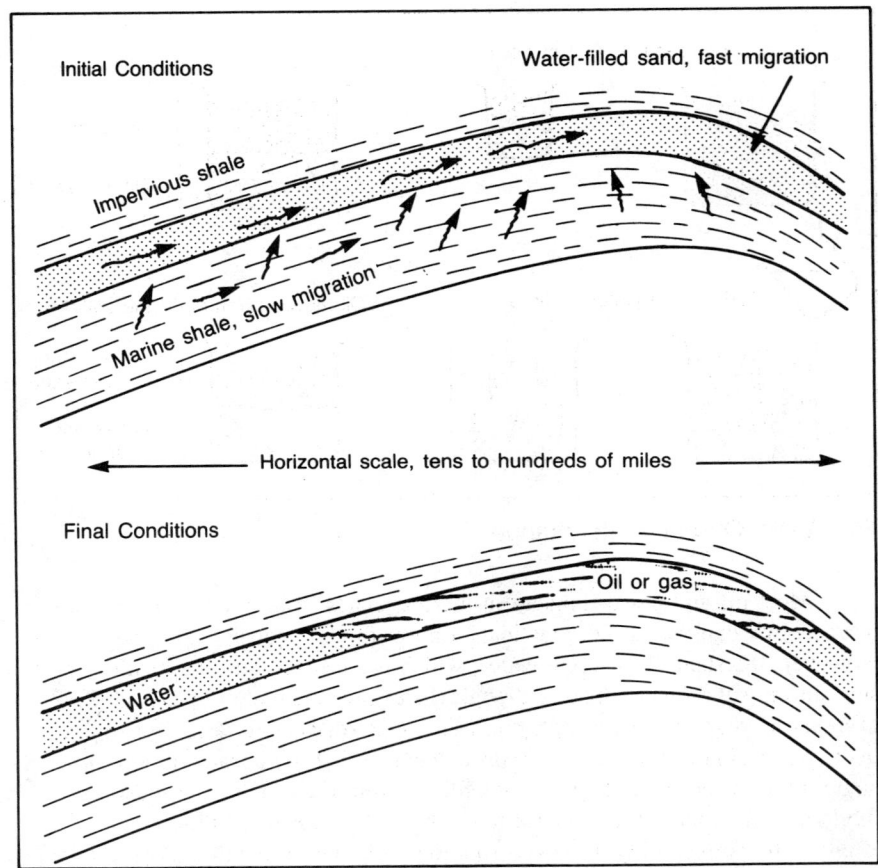

FIG. 2–10 Oil or gas migration and accumulation

found in traps. Some of these traps are clean (pure quartzitic) sands where there is no reasonable explanation for the oil to have been formed in place. The logical conclusion is that the oil and gas migrated into the trap. It is also possible that the oil is still in the process of migration. This is partially verified because some oil traps contain only residual oil. Residual oil usually is thicker, more viscous oil occurring in limited quantities and filling about 20–30% of the pore space.

The oil could have migrated through the deeper, impermeable formations as a result of the higher hydrodynamic forces and associated pressures that occur deeper. Temperatures are higher deeper, and the fluids would have a correspondingly reduced viscosity. Therefore, there would be a greater tendency for the fluids to flow at lower depths. The traps could be more effective at shallower depths because of reduced pressure and temperature. Most of this reasoning is valid, but the salient fact is that oil and gas are generally found in traps.

The Occurrence of Oil and Gas

Petroleum hydrocarbons occur as oil and gas, including gas condensate—a high-gravity hydrocarbon liquid commonly occurring with gas. Free salt water is also often found in the reservoir rock. When oil, gas, and water occur in the reservoir, they are usually separated with gas on top, oil in the middle, and water on the bottom (Fig. 2–11). This separation is caused by the difference in the densities of the fluids, with the lighter-weight gas on top and the heavier water on the bottom. Where only two fluids are found, they occur in the same relative order. It is also common to refer to the position of the fluids as phases or columns. For example, the oil column or oil phase would refer to that section of the reservoir filled with oil. The same terminology applies to the gas and water phases or columns.

In the normal average sequence of events, heavy oils (20°API or less) and tar sands (usually 10–15°API or less) occur at shallow depths down to about 4,000 ft with a relatively high concentration above 2,000 ft.

Oil gravities (degrees API) tend to increase with increasing depth. There is no definite reason for this except that the heavier hydrocarbons (lower-gravity oils) may be separated into lighter hydrocarbons as a result of the heat and pressure encountered at greater depths. The occurrence of oil and gas is about the same at depths of 10,000–12,000 ft. More oil is found above these depths; more gas is found below. The depths are not a definitive dividing line, since gas is found at shallower depths and oil is found at deeper depths, but it is a point of change.[9]

The maximum depth at which hydrocarbons can be found has not been determined. Commercial gas fields have been found at depths of about 25,000 ft. However, there has generally been very little exploration drilling below 20,000 ft, and many areas have not been explored below 15,000 ft. It is possible that hydrocarbons may occur throughout the sedimentary section (35,000+ ft). Very little oil or gas has been found in igneous or highly metamorphosed rocks, and most of that probably migrated.

There is some belief that the maximum depth for mobile-producible hydrocarbons is 25,000–35,000 ft. There are three reasons for this. (1) The heat and pressure at deeper depths would force all of the mobile hydrocarbons to migrate upward above depths of 25,000–35,000 ft. (2) At the greater depths the rocks

[9]H. D. Klemme, "One-Fifth of Reserves Lie Offshore," *Oil & Gas Journal* (August 1977), pp. 102–128.

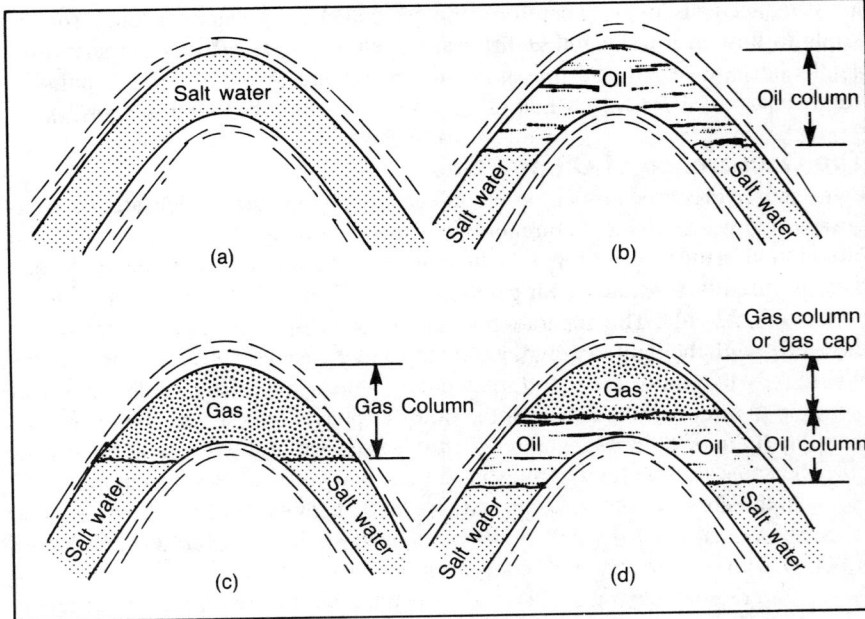

FIG. 2–11 Occurrence of oil, gas, and water in an anticline (a) salt water only, (b) oil and salt water, (c) gas and salt water, and (d) gas, oil, and salt water

may be semiplastic due to a combination of heat and pressure. This could literally squeeze the sand grains or other rocks together into a dense mass so there would be no porosity or void space available for hydrocarbon accumulation. (3) Deeper wells have not been as successful in finding hydrocarbons as would be expected (or possibly desired). However, there has only been a limited number of test penetrations.

OIL AND GAS TRAPS

Most hydrocarbons originate in marine shales. They migrate from the source beds through the path of least resistance until an impermeable barrier is encountered. The combination of hydrocarbons in a porous, permeable formation trapped and isolated by a less-permeable formation is called an oil and gas trap. The reservoir is the porous, more-permeable formation, and the less-permeable formation is the trapping formation. The reservoir area may be as small as 40 acres to over 100 sq miles. Thicknesses range from a few feet to over 100 ft, with an average of 15 ft.

Traps are usually named for the trapping structure or mechanism, but the naming procedure is not standardized. Common classifications include structural, fault, and stratigraphic traps.

Structural Traps

Structures can be defined broadly as any horizontal or vertical change in the continuity of the formation(s). This can range from a minor change in the dip angle to a highly folded, possibly inverted condition. Structures may be completely or partially formed. They can be exposed at the surface and partially eroded or buried with no surface indication of the structure.

Structures can be created by tectonic movements of the earth's crust, such as warping and folding. They may be created by actions of the geologic cycle, such as subsidence, uplift, and deposition. Many structures are created by faulting. Or the nature of the structure may be subtle, especially buried structures and where data are limited.

The geologist is interested in all structures since most of them have potential as an oil trap or provide information that can lead to a trap.

Structural closure is the maximum surface area of the reservoir that may contain trapped hydrocarbons. It is conventionally expressed in acres or, less frequently, in square miles for very large structures. Fig. 2–4 is a structure map of a reservoir. The areal closure is the area inside a 0-ft contour line, which is about the same as the area inside the −4,200-ft contour on Fig. 2–4. If the area were planimetered, it would contain about 1,000 acres.[10] The vertical closure (height of closure) is the thickness of the reservoir from the lowest point of closure (−4,200 ft) to the highest point inside the area of the closure (about −3,800 ft) or about 320 ft. In practice, the word *closure* is frequently applied to either area or height. The more common practice is to use the word closure to mean areal closure and to identify vertical closure.

Anticlines Anticlines are one of the more common, important structural traps. They are caused by uplifting, folding, and other tectonic actions. Anticlines are also formed by deposition over an uplifted area.

Many different anticline shapes are encountered. If one end of the anticline is higher than the other, it is frequently referred to as the nose and the formation is called a plunging anticline. A symmetrical anticline dips the same on both sides. The flanks often have different dip angles or slopes.

Excess oil or gas in an anticline spills over or moves out of the anticlinal area. If both gas and oil migrate into an anticline, excess gas tends to fill the

---

[10] A planimeter is a drafting tool used to measure areas on maps, schematics, etc.

anticline replacing oil, which would spill over and migrate as noted. It is interesting to note that this process of natural segregation of oil and gas may to some extent account for gas being found at deeper depths and oil at shallower depths.

Anticlines and other structures may contain various combinations of salt water, oil, or gas. Normally, the reservoir would be filled with one or a combination of these fluids. Gas and oil can occur singularly or in combination, but gas always rises above oil. The reservoirs are almost never found to be void or air filled. A few reservoirs are filled with carbon dioxide although normally it is generally present only as a contaminant. Other common contaminants are nitrogen, helium (very rare), and hydrogen sulfide.[11]

When gas occurs over oil, it is referred to as a gas cap. When is occurs over salt water, it is referred to as a gas reservoir or gas-filled reservoir. The vertical height or thickness in feet of the oil or gas zones are frequently referred to as the oil column or gas column, respectively. When the oil column is very thin relative to the gas column, the oil zone is sometimes called an oil ring or oil rim.

The terms closure, oil column, gas column, and oil ring, apply to the reservoir. An anticlinal reservoir trap has been used to illustrate and describe these terms. However, the general terms apply to any reservoir, not just anticlinal reservoirs.

Salt Domes Salt domes are a common structural feature that frequently create hydrocarbon traps. Other types of domes such as igneous-metamorphic rock and clay or shale domes (rare and sometimes called plugs) are either less common or do not usually occur in areas favorable for the accumulation of hydrocarbons.

In the geologic past seas periodically moved over the land mass, their outlet was blocked, and the water evaporated, depositing salt. This process was repeated until in some areas a thick layer of salt accumulated. Some of the salt flats in desert areas in the western United States are examples of these. After the salt deposit accumulated, sediments were deposited over the salt and the area subsided. The salt formation was subjected to increasing heat and pressure with increasing subsidence.

Salt acts as a slow-moving plastic mass under conditions of high pressure and temperature. It has a lower density than other common sediments. With time, the salt will flow slowly into a weaker section of the overlying strata and force its way upward, creating a salt dome (Fig. 2–12).

---

[11]Hydrogen sulfide is a deadly gas, even in very low concentrations and minimum exposure times. It is barely detectable by odor because it tends to deaden or desensitize the olfactory nerves. It should be avoided. Only experienced personnel with adequate training and safety equipment should work in areas where there is potential exposure to hydrogen sulfide.

As the salt dome thrusts upward, it first uplifts and then ruptures the overlying formations. The formations adjacent to the salt dome tend to slope upward in the direction of the salt dome as a result of the upward thrusting action of the dome and the dragging force of the upward moving salt. Since the salt dome is impermeable, hydrocarbons from porous formations may become trapped. The formations overlying the salt dome may be thrust upward and have closure to become dome traps.

Salt domes occur in various sizes and are sometimes several miles in diameter. The top of the dome may be relatively shallow or deeply buried. The reservoirs may occur on one side of the dome or extend around it, depending on local geologic conditions. Salt water, oil, and gas can occur in the reservoirs singularly or in combination, similar to that in anticlinal traps. Salt domes also can be leached out with fresh water to create large caverns for hydrocarbon storage.

Other Structural Traps There are various other structural traps, all created by crustal movements. Folds are formed as a large wrinkle in the crust. They are similar to an anticline except usually smaller and more elongated. Synclines are structures similar to troughs, the reverse of an anticline. The upward sloping sides may contain a trap if there is a closing feature such as a discontinuity (formation change) or fault.

Formations in basin areas usually have a regional dip toward the center of the basin. Regional dip signifies a widespread formation that may have local dip variations but over a wide area generally dips in one direction. Traps often occur

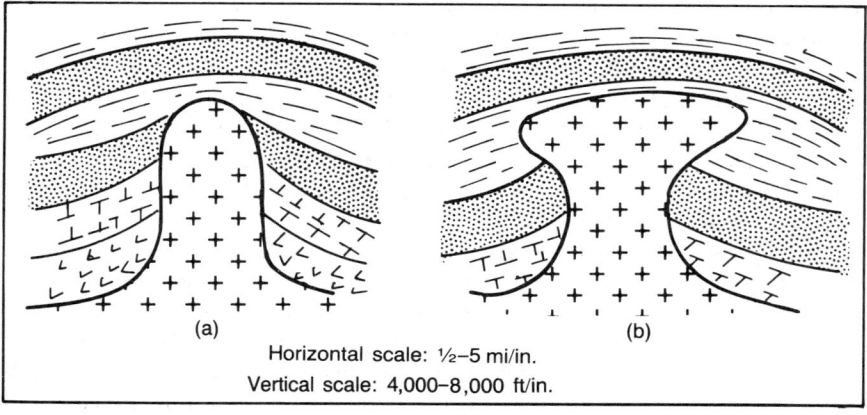

FIG. 2–12 Salt domes

in the dipping formations on the edge of the basin as a result of rollover or a change of dip. Sometimes, trapping is caused by benching or terracing.

Fault Traps

Faults are planar sections where the crust has broken and ruptured and the formations on opposite sides of the fault plane have moved relative to each other. They are important in the overall study of the earth's crust and are of special interest to the petroleum geologist because a large number of oil and gas traps are associated with and/or are caused by faults.

Fault planes are measured and identified in terms of strike and dip in the same way as formations. When the fault plane is exposed at the surface, the surface trace is known as the outcrop or fault line. Most faults do not outcrop to the surface. These buried faults occur frequently in sedimentary rocks.

Movement on the fault can be vertical, horizontal, diagonal, or rotational in the plane of the fault, or the walls or faces can move away from each other. The movement can be caused by many actions, such as thrusting, deposition, or continental drift. Active or live faults are moving, and dead faults are still. In general, active faults move very slowly, usually less than a few inches a year, although they may move long distances over long periods of time.

There are many types of faults (Fig. 2–13). The most common faults and those with which most of the oil and gas traps are associated are normal (gravity), reverse (thrust), and to a lesser extent overthrust. In a normal fault the hanging wall moves down relative to the standing or foot wall. In a reverse fault the hanging wall moves up relative to the standing wall. In coastal areas the normal fault is often called a down-to-the-coast fault, and the reverse fault is an up-to-the-coast fault. High-angle faults are normal or thrust faults where the dip angle of the fault plane is over 45°. Low-angle faults have a dip angle less than 45°. An overthrust fault is a thrust fault with a low dip angle, usually 10°–15° or less. Thrust faults frequently move a long distance (measured in miles).

Faults often occur in trends or zones. These can be from a few miles to over 100 miles long and from a few hundred feet to several miles wide. The faults tend to occur parallel to the trend, but some cross faulting may occur. Fault zones tend to parallel the older coastline where deposition is still in progress, as in the Texas-Louisiana Gulf Coast area. These faults are caused by sedimentation and subsidence—burial of the sediments. Or fault zones may occur parallel to the edge of sedimentary basins, such as the Powder River Basin in Wyoming. They also occur on the edges of uplifted areas such as parallel to mountain fronts and to a lesser extent around small uplifts such as domes. Some fault structures occur as combinations of faults.

Geology and Exploration

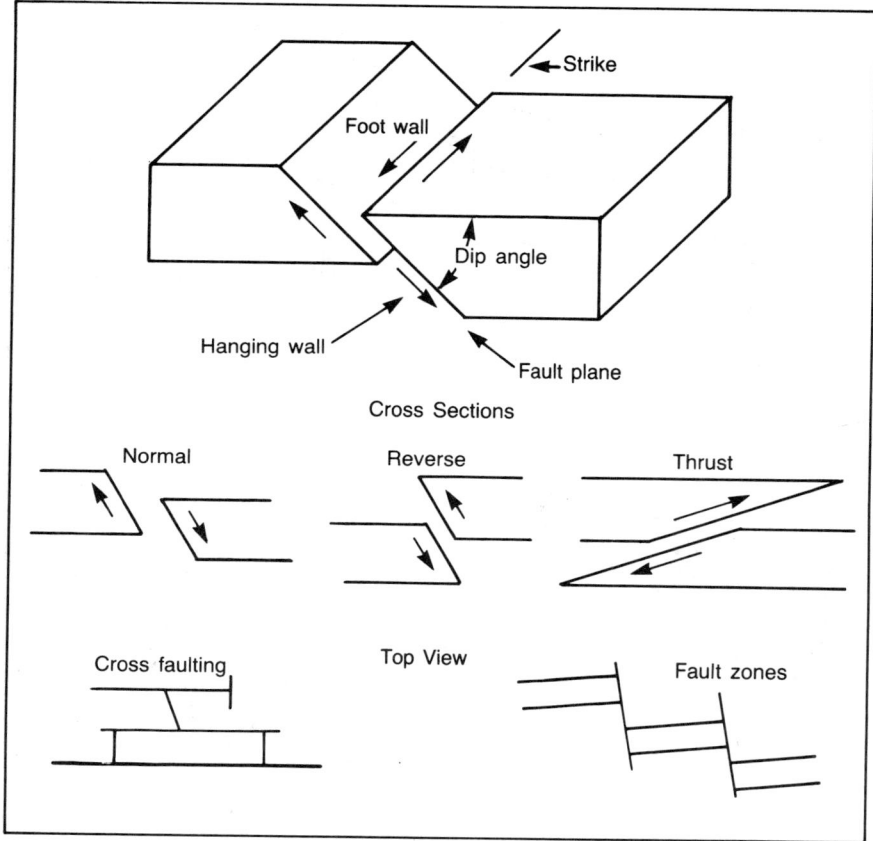

FIG. 2–13 Types of faults. Note that the strike, dip, and fault plane are identified in the same manner as the formations

The formation of a common, normal (gravity) fault trap is illustrated in Fig. 2–14. This is a west-east cross section of a sand formation striking north and dipping to the west. The sand is bounded above and below by impermeable shale beds but is not considered a trap since it is open on the eastern edge to allow spillover. Any hydrocarbons that enter from the downdip western edge would continue to migrate through the sand and out the east (updip) boundary.

After faulting, the east side has moved down relative to the west side. This movement has literally severed the sand and moved an impermeable shale across the sand face. This effectively blocks fluid movement in the updip (eastern)

50
DRILLING

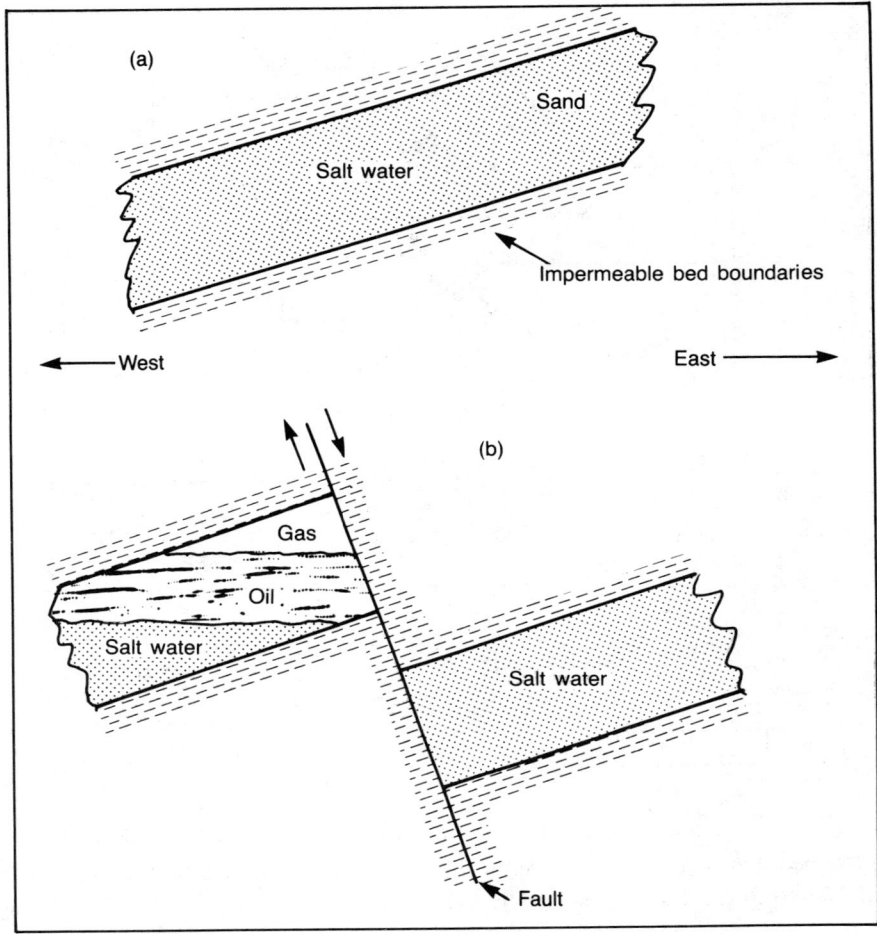

FIG. 2–14 Normal fault trap (a) original and (b) after faulting

direction past the fault plane. The sand on the west half of the illustration has now become a trap. If oil or gas migrates from the west, it will be stopped at the fault plane and will accumulate in the reservoir sand.

The direction of the dip of the fault plane is not important in this case. The illustration does not show end closure. But to be a trap, both ends or the updip end must be closed with a cross fault, pinchout, or other type of closing mechanism. The average fault trap is smaller than the average structural trap, but many fault traps contain millions of barrels of oil.

When faulting occurs, the formations may drag against the face of the opposite formation across the fault, creating a folding effect. In Fig. 2–15, the regional dip is west. A regular trap would occur on the west side of the reverse fault because of the regional dip. However, without the drag-fold effect, there would not be a trap on the east side of the fault. The hydrocarbon would migrate updip. The drag effect creates a closure for a trap on the east side of the fault. Fault-drag folds are relatively common, especially with larger faults, and account for many oil traps.

Stratigraphic Traps

Stratigraphic traps are created by deposition and the stratigraphy, or types and contents, of the formations. They are generally similar to other types of traps in that a porous formation serves as the reservoir and is bounded by impermeable formations (Fig. 2–16). Hydrocarbons accumulate in stratigraphic traps in the same general manner as in other traps. The average stratigraphic trap is smaller than structural traps, although many contain large volumes of hydrocarbons and they account for a significant part of the oil and gas production.

Stratigraphic traps are somewhat more difficult to locate than other traps. Part of this is due to their average small size and the subtle nature of the trapping mechanism. The trapping mechanism may only be a slight formation change, such as a pinchout or facies change. Sometimes these small changes are very

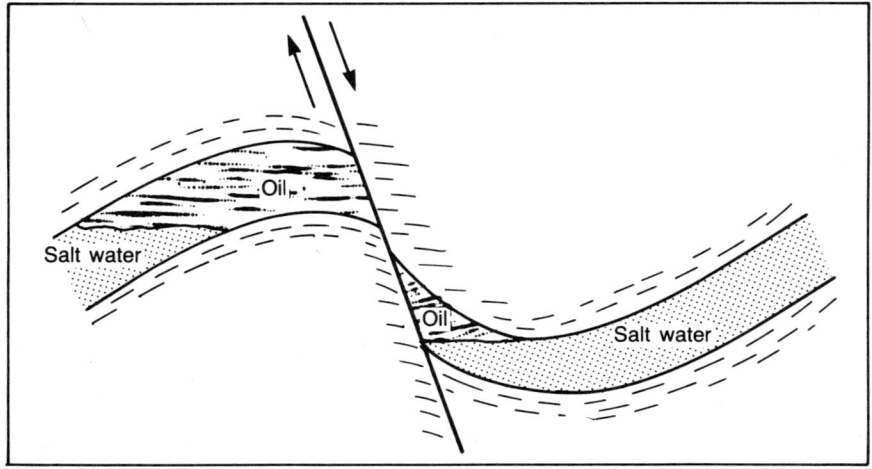

FIG. 2–15 Fault drag

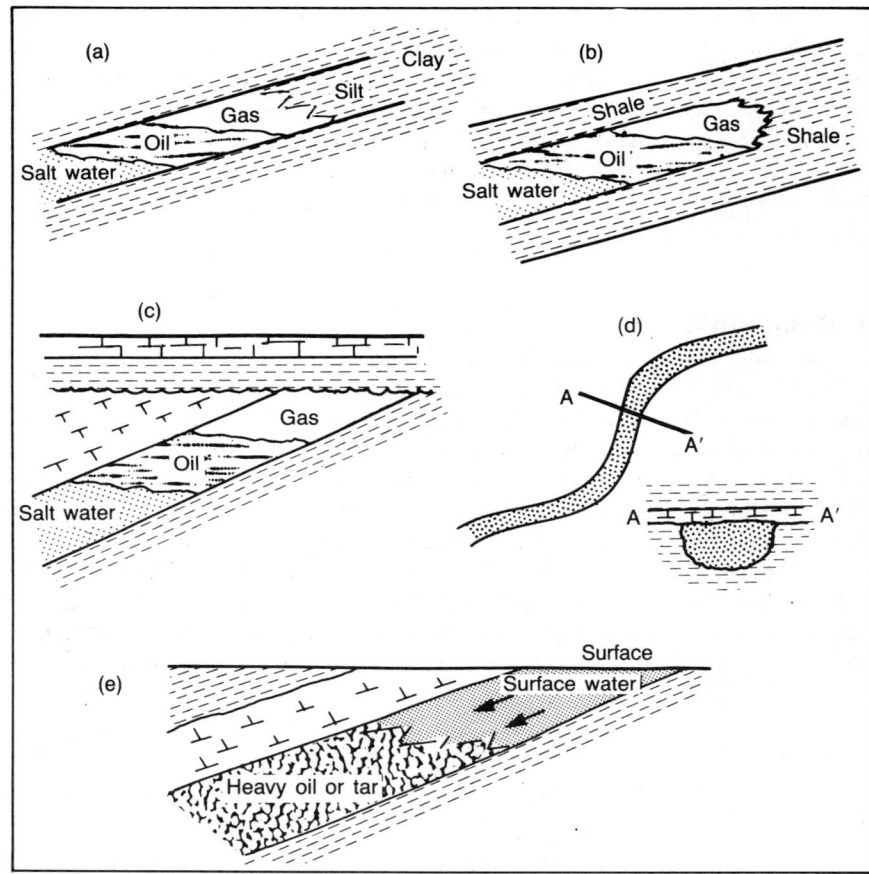

FIG. 2–16 Stratigraphic traps: (a) facies changes, (b) pinchout, (c) discontinuity, (d) top view of channel sands, and (e) chemical or heavy oil trap

difficult—at times almost impossible—to detect, even with the best exploration tools. Frequently, the only way they can be found is by drilling a well.

There are many types and kinds of stratigraphic traps, and they lead to single, dual, and triple zones that can be encountered in a borehole in stacked formations such as sand lenses. In Fig. 2–17, if a well is drilled at position X, it will encounter one productive sand. At position Y it will encounter two productive sands and three at position Z. Multiple zones can be encountered any time two or more oil traps, regardless of the type, are located in vertical alignment.

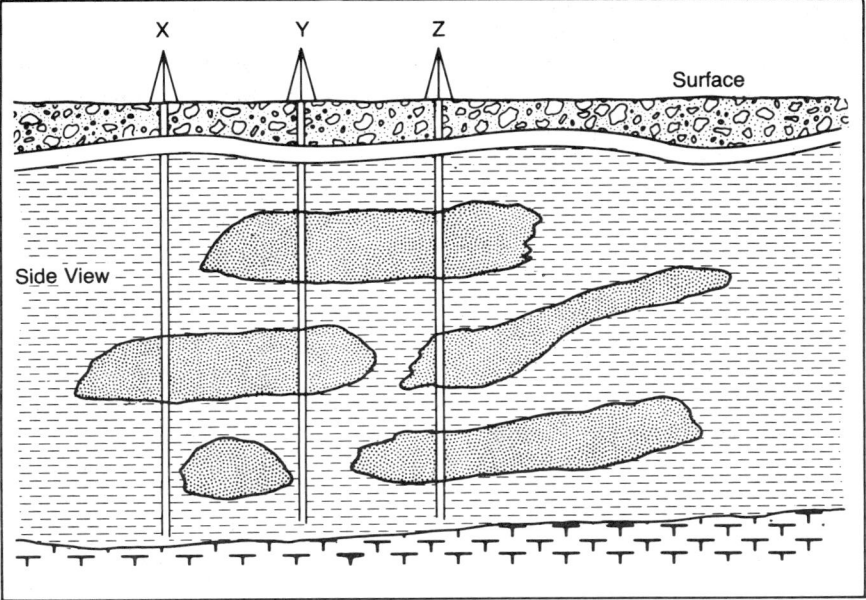

FIG. 2–17 Sand lenses—multiple completions

EXPLORATION OPERATIONS

The primary function of the exploration geologist is to locate prospects that can be drilled to produce salable hydrocarbons in commercial volumes. In order to do this, he must know the geologic cycle, how hydrocarbons originate and migrate, the nature and character of oil and gas traps, and exploration methods and techniques. Exploration is broadly subdivided into wildcat and development exploration. By definition, wildcat exploration is conducted in virgin, unexplored areas. Development exploration is conducted in and near producing areas. Both are conducted in a generally similar manner and with similar tools. The areal coverage of the two exploration approaches sometimes overlap and are not clearly defined in the industry.

Wildcat Exploration

The wildcat explorationist looks at broad areas. He first reviews surface geology, aerial photos, regional magnetics, and other exploration tools that give a wide areal coverage. The geologist looks for large areas of thick sediments. Areas where igneous or highly metamorphosed rocks outcrop and are exposed on the surface, such as mountains and granitic uplifts, are eliminated. However, the edges of these areas are checked for possible trapping.

Prime areas of interest are large inland sedimentary basins such as the Uinta, Piceance, and Wind River basins in the Rocky Mountains, the Appalachian and Michigan basins in the eastern United States, and the large basins of West Texas, which have thousands of feet of sediments spread over large areas. Many basin areas occur under marine waters, like the Gulf of Alaska. Other areas of interest are the large plain areas of Kansas and Oklahoma and the coastal plains of Louisiana and Texas, including the continental shelf areas that can extend several hundred miles offshore.

The geologist studies these areas, looking for structures and faults that can lead to oil and gas traps. When an area of special interest is located, exploration efforts are concentrated on it. Seismic is often one of the more common methods used, but all will usually supply the information needed to determine if a drillable prospect—an economically justifiable prospect—has been found.

Regional exploration is costly and time consuming. One method of regional exploration that is becoming more common is seismic group shoots. A group of companies define a widespread area and contract a geophysical company to run seismic over the area. The costs are shared equally, and each company has access to the data. Another exploration method is for two or more companies to form a mutual area of interest. One company or a third-party company, another company or individual who is not a member or party to the joint agreement, explores the area. The data and cost are shared equally among the participants.

Early in the industry, strat tests or strat holes were often used as an exploration tool to learn the stratigraphy of the area. Strat or slim-hole drilling declined as seismic improved and became more accurate and usable. However, strat drilling is still used in areas where seismic is less effective or the oil and gas traps cannot be defined by seismic techniques. It is also interesting to note that the number of seismic crews operating, which is reported in industry trade journals, is a method of determining the level of exploration activity. This is also used to help predict the future level of drilling activity and the number of operating rigs.

Development Exploration

The objective of the development exploration geologist is the same as the wildcat explorationist—to find drilling prospects. He confines his activities to areas in and near developed or developing fields. Frequently, the same geologist will do both wildcat and development exploration, but there is an increasing tendency to specialize.

After the wildcat well has been drilled and depending on the size and type of trap, it is often necessary to drill one or more holes, even if the first test well is dry. For example, in most oil and gas traps hydrocarbons accumulate on top of salt water. If the first hole or the test well, is drilled into water and there is closure updip, the logical action is to drill another test hole further updip. Since oil rises above water, only oil present should be updip. If the first hole is productive, then confirming offset test wells must be drilled, followed by development drilling.

Where the wildcat geologists work with broad areas trying to locate structures, etc., the development geologist is concerned with the structure and immediate surrounding area. He is interested in more precise details, such as oil-water and gas-water contacts, formation lithology, reservoir characteristics, and exact structure and isopach maps.

The development geologist builds and adds to the basic data collected by the wildcat explorationist. In addition, he has well data from cores—samples of rock from test well boreholes—and well logs. He may elect to improve the quality of the regional seismic data by using a well to measure seismic wave travel. The regional seismic can then be added to or reprocessed with the new data to give improved information. All of this is necessary to develop the prospect in the most economical manner.

Frequently, development drilling will lead to stepouts or on-structure wildcats that extend the productive area of the field and may find nearby new fields. Stepouts are wells drilled into the same formation produced in the development or discovery wells, but the stepout is usually two or more offset locations away. On-structure wildcats are drilled on the same structure as the producing wells, but it is designed to explore for another, usually deeper formation. All of this type of work requires expert, experienced geological guidance.

PROBLEM FORMATIONS

Problem formations are difficult to drill or create problems during or after drilling. These problems are caused by the physical and chemical properties of the formations. Problem formations include fluid-sensitive shales, geopressured sections, fractured strata, lost-circulation zones, contaminants such as gypsum and anhydrite, salt sections, and steeply dipping, layered formations that cause crooked or deviated hole. These formations can cause problems that increase the time and cost required to drill the hole. In severe cases they may cause a fishing job or lost hole, requiring sidetracking or redrilling with still higher costs. Therefore, it is important to know how to identify the formations, the problems associated with them, and ways to handle the problems.

Fractured Formations

Fractures can occur in nearly all formations but are more common in the harder, more consolidated, competent rocks. There is also a higher incidence of fracturing near faults, uplifted or downwarped areas, and where rocks have been subjected to tectonic forces and stresses.

Fractures tend to part the formation but, unlike faults, there is no relative movement of the formations. Fractures can be closely spaced and oriented in different directions. One conception of a fractured formation is a large number of different-sized blocks stacked together to form a formation. When a hole is drilled through a fractured formation, the loose blocks may fall into the wellbore. This sloughing action can block the hole so that it must be redrilled. In severe cases the sloughing material can fall into the hole and stick the drilling assembly, requiring a fishing job and possibly a lost hole. Fractured formations cause enlarged holes, which in turn cause other problems such as difficulty in cleaning the hole, sections that cause the drillpipe to stick, points where the hole may sidetrack, and completion problems.

Fractured formations can usually be safely drilled by using good drilling procedures. In more severe cases it may be necessary to increase the mud weight or to cement and redrill the formation. In very severe cases it may be necessary to run an extra string of casing.

Fractured formations are unconsolidated formations. However, the term "unconsolidated formations" is normally applied to formations where the materials have not been consolidated by cementing or other natural processes, such as loose gravel and boulders. These formations are often difficult to drill and create problems similar to those that occur when drilling fractured formations.

Fluid-sensitive Shales Shales are often a major drilling problem for several reasons. Normally, over 50% of the formations drilled are shales. Shales are often associated with other formations, which further increases the exposure to shale problems. The most common, economical drilling mud is a water-base fluid, but many shales cause problems when they are exposed to it. A number of different drilling problems, some very severe, can occur while drilling shales.

Shale chemistry is complex. When shales are exposed to fluids, they undergo various chemical and physical changes that occur at various rates. Most of these are detrimental to and hinder the drilling operation.

Shales can be classified as fractured, geopressured, and fluid sensitive. Fractured shales act like fractured formations, cause similar problems, and are handled similarly. Geopressured shales are similar to geopressured formations. Fluid-sensitive shales can be subdivided into swelling shales and disintegrating shales.

Swelling shales contain a high percentage of montmorillonite or bentonitic clays and are frequently known as mud-making shales. These shales usually occur in younger formations at shallower depths. They are composed of microscopic-sized, thin-layer or stacked platelets. These platelets have unbalanced ionic charges on the sides and ends. Exposure to certain fluids such as fresh water causes them to separate. When the shales start to separate, they exhibit swelling. This action can close the drillhole, requiring redrilling, or can stick the drilling assembly. Bentonite, commonly called gel, is a similar solid. It is commonly used in controlled amounts as a base for some water-base muds, hence the name mud-making shales. When mud-making shales are drilled, excess volumes of gel can enter the mud system and can change the physical and chemical properties of the mud so that it cannot perform satisfactorily as a drilling fluid, thus hindering the drilling operation.

Disintegrating shales are sometimes known as nonswelling shales. They tend to occur at deeper depths in the older, more consolidated formations. Shales are deposited in bedding planes and frequently are highly laminated and often fractured. During the drilling operation, fluid tends to move into these fractures and laminations. The fluid dissolves and/or reacts between shale platelets and laminations, releasing pieces of shale that fall into the hole. The net result is drilling problems similar to those that occur when drilling fractured formations.

Fluid-sensitive shales occur in varying degrees of severity and are handled accordingly. Good drilling techniques must always be used. Drilling muds with low fluid loss will limit the amount of fluid available to contact and react with the shales. Inhibited muds have mud filtrates with chemical properties that eliminate or minimize the reactions of the filtrate with the shale. Inert mud systems such as oil muds do not react appreciably with the shales.

Lost Circulation Lost circulation occurs when the mud in the drilled hole flows into the formation. Natural lost circulation is attributed primarily to formation conditions. Drilling fluid can be lost in low-pressure, unconsolidated formations, fractured formations, vugs, caverns, or fault planes. Induced lost circulation is usually caused by some operational action and may be augmented by low-pressure formations. Operational actions that induce lost circulation include excess mud weight, high-mud gels, surge pressures caused by running the pipe in the hole too fast, and overpressuring.

Lost circulation is often a major problem in drilling operations. The severity of the problem ranges from lost time with associated increased costs to stuck pipe, a fishing job, a blowout, or a lost hole—all of which are extremely costly.

Lost circulation frequently occurs with other problems and may in turn cause other problems. One of the most severe problems is lost circulation with high-pressure zones exposed in the borehole. This can lead to a blowout with a high risk of personnel and equipment loss, and a lost hole. Lost circulation occurs at all depths but is slightly more common at shallower depths. Depending on the severity of the lost circulation problem, treatment includes allowing the hole to heal with time, using lost circulation materials or cement to plug the zone, or as a last resort, placing large pipe (casing) in the hole to prevent losing mud.

Miscellaneous Formation Problems Geopressured formations have a combination of pore pressure and physical characteristics that cause pieces of the formation to spall or pop off and fall into the wellbore. This happens when the combination of the formation pressure and strength of the formation is greater than the equivalent of the mud pressure. These formations cause problems like those caused by fractured formations. It is often difficult to diagnose the problem, but the condition frequently improves with time. One method of treatment is to increase the mud weight.

Salt is highly soluble in fresh water. If a freshwater mud is used to drill a salt bed, the salt will dissolve, creating a large cavern or out-of-gauge (enlarged) hole. This creates problems in hole cleaning and cementing casing properly. The high concentration of salt in the mud will affect the physical and chemical properties of the drilling mud. Heavy, high-strength casing must be set through salt beds to prevent collapse. The salt tends to become plastic with increased heat and pressure and can exert very high forces on the casing, causing it to collapse if it does not have sufficient strength.

Gypsum and anhydrite usually occur as stringers (very thin formations) but can occur in thicker sections or as contaminants in other formations. Gypsum and anhydrite both affect certain types of water-base fluids, requiring special, more expensive mud treatment. They also affect logging tool responses so that it is more difficult to analyze the logs correctly.

The crustal zone, which has undergone numerous changes in the geologic past and is still changing, is the province of the petroleum geologist. He must unravel its mysteries and select areas (drilling prospects) that are favorable for the accumulation of hydrocarbons. If the geologist is an oil finder, his discovery ratio—the ratio of discovery wells to dry holes—will be favorable. If he discovers a successful wildcat, his discovery may lead to a commercial field with many wells. But the odds of finding a successful wildcat are low; less than 5% of the wildcats lead to a commercial field discovery. Nevertheless, geologists and geophysicists continue to improve their methods to increase the chances of locating oil and gas.

CHAPTER 3
RESERVOIRS AND RESERVES

Development geologists and drilling, production, and reservoir engineers study reservoirs in order to drill and produce the maximum amount of oil and gas in the most efficient, economical manner. One of the main tools are well logs, which show lithology, porosity, resistivity, and other information. Additional information on the reservoir can be obtained by sidewall and conventional cores and wire-line and open-hole drillstem tests. Analyses of the drill cuttings and other data obtained during drilling are also used.

The movement of the fluids in the reservoir (fluid flow) is calculated using permeability, pressure, saturations, and viscosity. Radial flow normally occurs in either single or multiple fluid phases. The natural reservoir energy drives the fluid to the wellbore so it can be produced. This is the reservoir drive mechanism. Gas is produced by fluid expansion. Oil is produced by fluid and gas-cap expansion, solution-gas drive, or gravity drainage. A natural water drive will increase the productivity greatly. The original amount of hydrocarbons in place in the reservoir and other factors are used to project recovery. Gas recoveries range from 50–80%, and oil from 5–40% of hydrocarbons originally in place. Additional oil may be recovered after the reservoir energy has been depleted by applying secondary, tertiary, and enhanced recovery techniques. In many cases the total recovery from the reservoir can be increased if these techniques are applied before the reservoir energy is completely spent (depleted).

WELL LOGS

Early geologists and engineers were severely hampered by the lack of precise details of the formations exposed in the wellbore. If the well were a gusher and blew in, then obviously the formation was productive. However, many formations are capable of producing oil and gas but do not blow out, especially with rotary drilling methods where the wellbore is filled with mud. The development of well logs resolved this problem.

Well logs are the printed output or display of measurements made by logging tools run into the borehole. Early logging methods were limited to electrical measurements. Modern methods measure gamma rays and use neutron emission and other methods to obtain information about the formations. However, in all cases electrical methods are used to transmit the information from the logging tool to the surface and record it. Therefore, the term "electric logs" is commonly accepted.

Most logging is done by service companies. They have logging trucks that are completely equipped with all of the equipment needed to run the electric logs. The logging truck has a reel with a brake driven from the truck motor. The reel contains 10,000–30,000 ft of electric cable with a number of insulated electric wires or conduits. The number of conduits ranges from 1–10, depending on the type of logs being run and other factors. The cable is run through a system of pulleys, and the logging tool is connected to one end of the cable. The logging truck has depth measuring equipment that measures the exact amount of cable unreeled as the logging tool is lowered into the hole, which has been drilled to logging depth. The electrical conduits on the drum end of the cable in the logging truck are connected to instruments in the logging truck that record the depth and electrical responses as the tool is moved in the hole. The logging data are recorded on film, along with the depth at which the data were taken. The film is later developed and used to make copies of the well logs. If the data determine the well will be productive, casing is run and cemented, and other logs may be run.

Different types of logging tools are used to obtain different well logs. Frequently, two or more tools are used in each log run—one trip in and out of the hole with the logging tool. Modern logging equipment can run three or more logs in one run.

Early well logs were interpreted by hand calculations. Many service trucks now record the logging data on tape that is then processed by computer to give the desired output. Newer trucks have computers so the finished log can be obtained at the wellsite. The taped data can be transmitted by telephone so that it is rapidly available in central offices, often located a long distance from the wellsite.

The data from the well logs are usually the determining factor in whether a well will be completed or plugged and abandoned. Often a drilling rig is on site and costs thousands of dollars per day. It is important to obtain the electrical logs and evaluate them as quickly as possible in order to minimize the amount of rig waiting time and associated standby costs.

Some logs are designed to be run only in the open, or uncased, hole (open-hole logs), and others can be run only in the cased hole (cased-hole logs). Some very important formation information such as resistivity can only be obtained from open-hole logs.

Electric logs are displayed on long, narrow strips of paper known as prints or logs. The log heading is located at the top of the strip and is subdivided into two sections. The first (upper) section gives all of the information about the well, including owner and location. The second half of the heading gives specific information about the log run, including depth, type of tool, tool spacing, and corrections. The body of the log is divided into three vertical columns, sometimes referred to as traces, and their units are noted at the top. A narrow depth column is located between the first and second trace (reading from the left-hand margin). This column contains the depth of the logging data and is noted by horizontal lines as the depth scale. Conventionally, lithology and caliper logs are listed in the left-hand trace, and other logs are listed in the right-hand traces.

Logs are conventionally scaled in inches of paper per 100 ft of hole. For example, a 1-in. log would contain the data from 100 ft of hole in one vertical inch on the log display. One-inch and two-inch logs are called small-scale logs and are generally used for correlation purposes. Five-inch logs (5 in. of paper/100 ft of hole) are called detail logs and are used for detailed calculations, selecting perforations, and close-scaled lithology studies.

In interpreting electric logs it is very important to recognize that the interpretations differ from area to area. For example, in some places in South Texas, producing oil wells can be made from sands with resistivities as low as 3 ohm-meters (ohm-m). In other areas formations with these low resistivities will produce salt water, and higher resistivities on the order of 10 ohms or more are needed to make a productive well. Therefore, it is very important that electric logs be interpreted by a person familiar with the geographic area.

Three important formation parameters obtained from electric logs are lithology (type of formation), porosity (to show if the formation can be a reservoir), and resistivity (to show the type of fluid in the porosity). Other data can be obtained from logs depending on the type of formation and equipment available.

Lithology Logs

Lithology logs show the depth and thickness of the various formations in the wellbore and help identify the type of formation. They are used to locate prospective zones and for stratigraphic studies and correlation. If the geologist identifies a sequence of formations from lithology logs on a number of different wells, he can pick a prominent marker bed and use it for mapping purposes over the area.

Lithology is very important in locating prospective productive formations in the wellbore. For example, limestone changes into dolomite in the metamorphic process. When the magnesium ion replaces the calcium ion, the matrix volume is reduced, which can increase the porosity (possibly secondary porosity). Therefore, it is important to recognize the difference in the two formations and investigate them further by porosity logs.

This example illustrates another important fact about electric logs and their interpretation. One log will give some information about a formation; other logs will give different information. All of the logs must be analyzed to obtain the maximum amount of data about the formation.

One of the more important uses of the lithology logs is to identify shales. Shales can occur as a clean (100%) shale formation, but shale also occurs in varying percentages in other formations such as sandy shales and shaly limes (limestones). Producible hydrocarbons are not found in shales except in a few cases of fractured shales. Therefore, identifying shale formations helps to locate barren areas in the wellbore. Some porosity logs are affected by shales, generally causing the logs to read optimistically high. To obtain correct porosities, it is important to recognize the shale content of the formation and to correct the porosity accordingly.

Spontaneous Potential (SP) Logs The spontaneous potential (SP) was one of the first electric logs developed to log oil and gas wells. Where applicable, it is a highly effective tool and is still one of the basic logging tools used today.

Salt (sodium chloride, NaCl) is commonly found in many formations, very predominantly in shale formations. Sodium chloride will ionize into positive sodium ions and negative chloride ions to a limited extent, depending on the concentration of sodium chloride and other factors. When two solutions with different concentrations of sodium chloride are separated by a membrane that is permeable to ionic movement, sodium ions will move from the solution with a low concentration to the solution with a high concentration. This is equivalent to a small flow of electrical current. When saline formation fluids intermingle with mud filtrate, this may also create a localized electric current.

The net effect is that downhole conditions in the wellbore are frequently favorable for the generation of these weak currents between the formation fluids and the mud filtrate, and the currents tend to change at bed boundaries. These currents can be detected by grounding one electrode at the surface and moving another electrode through the borehole. This movable electrode is the SP logging tool.

Fig. 3–1 illustrates an idealized SP curve drawn opposite a log of the formations. Note that the curve returns to the shale line whenever it is opposite a shale formation. Over long sections of hole, the shale line will drift, i.e., change directions slightly with depth as a result of changing formation conditions. Over the short interval under consideration, the shale line is assumed to be constant. One of the more important characteristics of the SP log is its variation from the shale line.

Reservoirs and Reserves

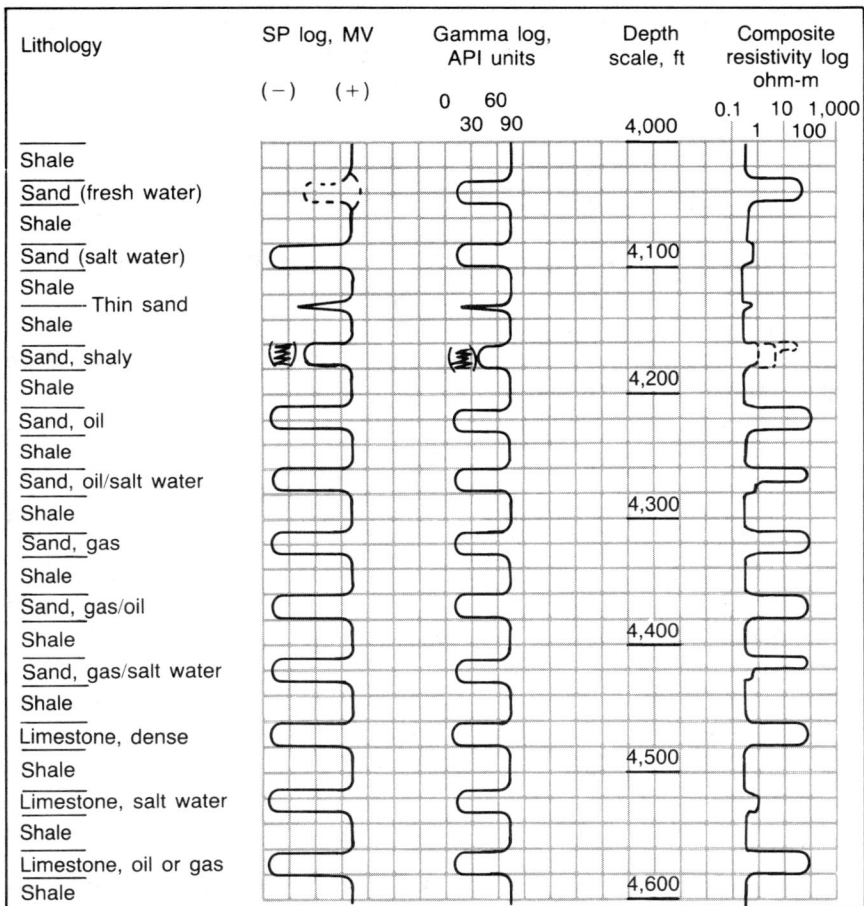

FIG. 3–1 SP-gamma-resistivity log

The normal sand on the SP log will be similar to the sand at 4,080–4,100 ft. The SP trace has moved from the shale line to the clean sand line. The curving nature of the SP trace from the shale line to the sand line is caused by the tool starting to read the current change before it completely enters the sand. The same thing occurs in the reverse at 4,100 ft when the tool passes from the sand back into the shale. The midpoint of the change is the point of inflection and would be picked as the bed boundary.

The sand from 4,030–4,050 ft is not well defined by the SP log. If the log deflects to the left, this would be the case where the salinity of the formation water is greater than the salinity of the mud filtrate. If the log deflects to the right, known as a reversal, this would indicate that the formation salinity is less than the salinity of the mud. Since this is a freshwater sand, the tool does not define the bed boundaries of the sand as well as the saltwater sand from 4,080–4,100 ft.

The thin sand at 4,130–4,135 ft illustrates another characteristic of the SP log, which is sometimes unfavorable or difficult to evaluate. The sand can be a good, clean sand; however, since it is thin, the SP tool may not have time to completely respond. In this case it might be misinterpreted as a shaly sand.

A true shaly sand occurs in the interval from 4,160–4,180 ft. The SP trace moves about halfway up to the clean sand line while logging the section and then returns to the shale line. The relative difference between the shale line and sand line can give a rough indication of the amount of shale in the formation. If the logging trace in the shaly sand is relatively smooth, the shale is finely disseminated in the sand. If the trace through the shaly sand interval is jagged or erratic, the shale could be in layers within the sand section or it could be layers of sand within the shale section.

The SP does not differentiate between a saltwater sand and an oil sand or oil-water sand.

The SP log is run in the open hole filled with a conductive fluid such as water or water-base mud. It cannot be run in a cased hole or in a hole filled with nonconductive fluid such as air or oil-base mud.

Gamma Ray (GR) Log The gamma ray log is a basic lithology tool. It is generally used for the same purpose as the SP log. The tool operates on a different principle from the SP log and to a great extent has overcome many of the disadvantages of the SP log.

The gamma log measures natural gamma radiation. The tool has a gamma ray detector that measures the gamma rays in counts per seconds (cps) and transmits these to the surface by electrical methods. The data are recorded on the logging trace as API units (APIU).

Clean quartz sands have a low level of radioactivity, on the order of 15 APIU. Granites and feldspars have a higher level of radioactivity. In the erosion-transportation-deposition cycle, these and other materials are broken up, pulverized, and later deposited in low-lying areas. Pure quartz is slightly harder than the other minerals and for the same distance of transportation remains a larger-sized particle (sand size) than the granites, feldspars, and similar rocks that are ground and pulverized to a finer particle (silt size). Radioactive minerals tend to accumulate with the silts and clays, which are later changed to shales in the burial process. These shales have an average radioactivity on the order of 75 APIU. When a gamma ray log is run through these formations, the clean sand line is about 15 APIU and the shale line is about 75 APIU.

The gamma ray and SP logs are very similar (Fig. 3–1). However, the gamma log has several subtle but important advantages. The freshwater sand at 4,030–4,050 ft that was not defined by the SP log is clearly defined by the gamma ray log. The thin sand bed at 4,130–4,135 ft that was not clearly defined by the SP log is more clearly defined by the gamma ray log, which is better at identifying thin zones. One of the main advantages of the gamma ray log is that it can be run in either open or cased holes filled with conductive or nonconductive fluids.

Resistivity Logs

Electrical resistivity (and SP) logs were the earliest methods of logging oil and gas wells. The basic principles are still used although the equipment and techniques have been modified and improved over the years.

Electrical resistance, commonly called resistivity, is a measure of resistance of a substance to the flow of an electric current. The ability to impede or resist the flow of electricity is a natural characteristic of the substance. Substances that are good conductors of electricity, such as copper wire and salt water, have a low resistivity. Substances that are poor conductors, such as fresh water, dry rocks, oil, gas, and air, have high resistivities. For logging purposes resistivities are measured in ohms per cubic meter, usually referred to as ohm-meters. The more common terminology is to use ohms, for example, "The formation has 5 ohms resistivity."

Most massive rocks (formation matrix material) such as quartz, granite, chert, feldspars, limestone, and dolomite are highly resistive. If the rocks are broken and pulverized (erosion and transportation), the resistivity of the rocks will not be changed. If the pulverized rocks are deposited in a fluid environment, the resistivity of the composite formation will be a measure of the combined resistivity of the rocks and fluid. Since the rocks have a very high resistivity, the resistivity of the combination of rocks and fluid is controlled primarily by the resistivity of the fluid as well as by the configuration of the porosity and the type of formation.

Formation waters contain salts, principally sodium, calcium, and potassium. The salts ionize and conduct the electric current through the fluid like the electrolyte (battery acid) in a car battery. The resistivity of formation waters is affected by the type and concentration of salts in the water and the temperature. Water with a high concentration of salt (salt water) has a low resistivity; water with a low concentration of salts (brackish to fresh water) has a higher resistivity. Hydrocarbons, both oil and gas, are highly resistive. Therefore, if the pore space in the formation is filled with salt water, it should have a low resistivity; if it is filled with hydrocarbons, it should have a high resistivity.

Unfortunately, the problem is more complicated. Nearly all formations have some salt water, an irreducible minimum saltwater saturation. This is normally 25–35% of the porosity. Also, the salinity of the water varies from highly saline to brackish or fresh.

In practice, formation resistivities—including fluid contents—range from a few tenths of an ohm to over 1,000 ohms. The amount of salt water in the pore space can be calculated if the type of formation and the porosity and resistivity of the water in the formation are known. The water content of the formation is conventionally calculated as a percentage of the pore space. Any remaining pore space (100 minus the percent water saturation) is assumed to be filled with hydrocarbons. Oil and gas have similar resistivities, so the log does not identify them separately.

Shale resistivity is somewhat different. Shale contains up to about 30% water. The water is not in pore spaces between sand grains but is tied up or bonded in the internal shale structure. Electric current moves as an ion exchange process on the surface of the shale or clay particles. Thus, shales normally exhibit low resistivities.

Resistivity logs help determine lithology by mainly identifying shales and hard, tight, highly resistive formations. These formations can sometimes be used as marker beds to correlate formations between different wells. Some formations have a definite character in the resistivity log trace, and this is also used to help correlate formations between wells. Resistivity logs are also used to help identify and calculate mud filtrate invasion—the amount of drilling fluids absorbed into the formation.

Electrical Resistivity Logs Electrical resistivity logs are commonly called resistivity logs. One electrode is grounded (makes contact with the earth) at the surface, and the other electrode is on the logging tool that is moved down through the borehole. The logging tool also contains other measuring electrodes. A current is passed between the two electrodes, and the resistivity log is made by measuring the voltage drop (potential difference) in different sections of the current flow path through the formation.

The electric log must be corrected for many affecting conditions such as mud filtrate resistivity, depth of invasion, and formation thickness. This is one disadvantage of the resistivity log since some of the corrections are either unknown or are difficult to make. The resistivity log, like all resistivity tools, must be run in an open hole. Additionally, the electrical resistivity log must be run in an open hole filled with conductive fluid.

Various types of current-emitting and voltage-measuring electrode spacings and configurations are used. These range from tools designed for shallow investigation (close to the wellbore) to deep investigation (measuring resistivity deep in the formation) focusing tools.

Some common resistivity logs include normal, lateral, dual lateral, and spherically focused logs. Resistivity logs are also measured in microspacing for special shallow investigation.

Induction Logs The induction log measures formation resistivity. It was developed to overcome some of the disadvantages of the electrical resistivity log and to measure formation resistivities in holes where the electrical resistivity log could not be used. The induction log can be run in open holes filled with nonconductive fluids. It can also be used in open holes that contain a conductive fluid, but it gives better results with decreasing conductivity.

The induction tool uses a high-frequency, alternating current transmitter coil that induces secondary currents in the formation. These currents in turn induce a current in a receiver coil in the tool. The measurement of this current gives the induction log trace. Various tool configurations are used for different logging objectives, and all base data must be corrected for downhole conditions.

A resistivity curve is illustrated in Fig. 3–1 and can be analyzed with earlier lithology log results. The tracing shows a low-resistivity shale line. The freshwater sand at 4,030–4,050 ft has a higher resistivity due to the lack of conductive ions in the water. The saltwater sand at 4,080–4,100 ft has a low resistivity, slightly higher than the surrounding shale. This sand can be compared with the oil sand at 4,210–4,230 ft, which has a higher resistivity that indicates hydrocarbons. This oil sand and the fresh (brackish) water sand at 4,030–4,050 ft could be completely similar except for the nature of the fluid in the pore space. In this case the resistivity of the formation fluid must be known before a final evaluation can be made.

Usually, the waters become more saline with depth. After saltwater sands have been encountered, normally the sands below this point will all contain salt water or hydrocarbons. Therefore, the high resistivity in the saltwater sand at 4,030 ft would be questioned as possible fresh or brackish water. However, the resistivity of the sand at 4,210 ft is probably caused by hydrocarbons because it is below the saltwater sand at 4,080 ft.[1]

The thin sand at 4,130 ft has a low resistivity and probably contains salt water. The shaly sand at 4,160–4,180 ft has some resistivity that could indicate hydrocarbons. The higher resistivity that would be expected in a hydrocarbon-bearing sand is repressed because of the shale content of the sand. The small

---

[1]Sands are referred to by the depth at the top of the sand.

high-resistivity peak at the top of the sand could be a hard streak. These are frequently encountered in or near the top of the sand and, as the name implies, are hard, dense sections with very low porosity and higher resistivity.

The sand at 4,260–4,280 ft has oil in the upper 10 ft above a 10-ft water column. This is clearly indicated on the resistivity log, which has a higher resistivity in the upper half of the sand and a low resistivity similar to a saltwater sand in the bottom 10 ft. The oil-water contact at 4,270 ft is clearly shown on the resistivity log. The sand at 4,410–4,430 ft has a 10-ft gas column on top of a 10-ft water column. The gas-water contact is shown at 4,420 ft on the resistivity log. The overall sand is very similar to the sand at 4,260 ft, which has an oil column on top of a water column. The gas sand from 4,310–4,330 ft looks the same as the oil sand at 4,210 ft. This illustrates that the resistivity log does not differentiate between oil and gas.

Porosity Logs

Porosity logs were developed later in the history of the industry. Since porosities also can help calculate fluid saturations, the experienced geologist or engineer can use these fluid saturations and the porosity to make a very good estimate of the producibility of the formation. The formation porosity can often be read directly from the porosity log by making the proper corrections. The only method of determining porosity prior to the development of this tool was by coring the formation (drilling with a hollow, doughnut-shaped bit) and analyzing the core— a time-consuming and expensive operation filled with problems in wildcat areas. The development of the sonic logging tool alleviated a number of these problems.

Three basic porosity logs are used. Although each has advantages and disadvantages, they generally tend to complement each other. Therefore, it is often justified to run at least two and possibly three of the porosity tools.

Sonic Logs The sonic log measures the speed of a sound wave through the formation. This is recorded as transient or travel time in microseconds through a fixed distance.

The speed of sound in a massive (no porosity) dense material is a constant. If the material is porous, the speed of sound through the material will be reduced. The amount of reduction depends on the extent and nature of the porosity and other factors. Empirical relationships have been developed to relate velocity and porosity for various materials. These relationships are obtained by measuring the speed of sound through the same type of rock at different porosities. The process is then repeated for the various types of normal reservoir rocks. If the formation material is known from the lithology log, the porosity can be calculated from the data obtained by the sonic log. If the type of formation is not known, the data can be entered into standardized charts with other types of porosity data (crossplotting) to determine the porosity and type of formation.

The sonic log is a pad-type tool. A spring-loaded pad that contains the sound transmitters and receivers is held against the wall of the borehole. The readings must be corrected for the type of fluid in the hole and other conditions. Modern tools make these corrections and are known as compensated or borehole compensated sonic logs. The tool has a very shallow depth of investigation; it only reads the formation lithology near the borehole and cannot be used in a cased hole.

Formation lithology generally must be known and is usually obtained from other logs as well as crossplots for sonic log interpretation. Empirical curves relating travel time to porosity are used, and the data are recorded on the printed log in microseconds or porosity. The log is strongly affected by shale or shaly formations and may be used with other logs to help determine secondary porosity such as vugs and fractures. The log is subject to cycle skipping—erratic readings, usually higher than normal, due to aerated or gaseated muds and loss of padwall contact that can occur on rough (rugose) wellbore walls or in fractured areas.

Density Log The density log is sometimes called a gamma density or gamma-gamma density log. The pad-type tool can also be used to obtain a caliper log just like the sonic log.[2] The pad has a gamma-emitting source and measures the returning scattered gamma rays at one or more detectors. Short-spaced detectors are used for shallow investigation, and long-spaced detectors are used for deeper investigation. The measurement of the returning scattered gammas gives a relative measure of the formation bulk density the actual density of the formation.

For example, consider a formation of pure quartz. The bulk density would be 2.65. Now consider a formation of pure quartz sand that has 25% porosity. The bulk density would be 2.274 calculated as follows:

$$\begin{aligned}
\text{Porosity} &= 0.2865 \\
0.75 \times 2.65 &= 1.9875 \\
0.25 \times 1.146 &= 0.2865 \\
\text{Bulk density} &= 2.2740
\end{aligned}$$

As can be seen, the formation has 25% porosity that is filled with salt water, and 75% of the formation is pure quartz, giving the total bulk density of 2.274. If the porosity were 10%, the bulk density would be 2.4996:

---

[2]The caliper log records the diameter of the hole, usually in inches, on a regular logging depth scale.

$$
\begin{aligned}
\text{Porosity} &= 10\% \\
0.9 \times 2.65 &= 2.385 \\
0.1 \times 1.146 &= 0.1146 \\
\text{Bulk density} &= 2.4996
\end{aligned}
$$

If the pores were filled with oil, the bulk density would be slightly different in the above cases. If they were filled with a combination of oil and water, the bulk densities would again be slightly different. Thus, bulk densities can become highly complex when one considers the number of different types of formations, fluids, and porosities.

The density log readings must be corrected for various downhole conditions. Modern tools make these corrections and are known as compensated density logs. Density logs record total porosity, including primary or effective porosity, sealed or unconnected porosity, and secondary porosity. Other logs must be used to help distinguish these. If the pore space is filled with salt water or oil, the porosity readings are about the same, but gas gives an optimistically high porosity reading. Heavier materials such as anhydrite and less frequently magnetite can also cause erroneous readings.

Neutron Logs Neutron logs are basic porosity logging tools. They are used in conjunction with other logs to determine porosity and lithology, including sand shaliness. If secondary porosity is significant, other logs must be used to help distinguish between primary and secondary porosity. Neutron logs can be used to help pick oil-gas and water-gas contacts.

The basic neutron log has a neutron-emitting source and a receiver that measures (counts or captures) gamma or neutron rays, depending on the type of tool. Various types of neutron logging tools are available for use in both open and cased holes. The neutron readings must be corrected for various downhole parameters. The newer tools, called compensated neutron logs, make these corrections and print out the corrected data on the log display.

Any logging tool can be lost in the hole while being run. When a tool is lost in the hole, it is usually recovered without difficulty. The logging tools are expensive, so fishing or recovery, operations for lost tools are justified. However, fishing is expensive, and sometimes the most economical method is to leave the tool in the hole. The neutron log presents a special problem in this case. The neutron source has a long half-life and under certain conditions can be dangerous. Therefore, when it is lost in the hole, regulatory agencies require that all reasonable efforts be made to recover the tool. If it cannot be recovered, then it must be cemented (entombed) according to prescribed rules and regulations.

Other Well Logs

Many other well logs are used in drilling and completion operations. Some of these are used to supplement the basic well logs used for evaluation. Others are used for special situations that may not occur on every well.

Caliper log tools measure the diameter of the wellbore and record this on a log display opposite the depth of the logging tool. Some of the pad-type tools such as the gamma-gamma density porosity tool often record a caliper log with the porosity log. The data from the caliper log is used to correct the porosity data. Pad tools are often known as two-arm calipers. Other caliper log tools use three spring-loaded arms and are sometimes known as three-arm tools. These tools are more accurate than two-arm caliper tools.

One of the main uses of caliper logs is to calculate the hole volume in order to determine the volume of cement slurry needed for cementing casing in place in the hole. The log also shows the general condition of the hole, including out-of-gauge hole sections where the hole diameter is larger than the bit diameter. This information may help in selecting the type of perforating to be used during completion. Undergauge sections where the hole diameter is the same as or less than the drill-bit diameter may need to be reamed to prevent sticking the drilling assembly and causing other problems. The caliper log may show the condition of the wall of the wellbore and help determine whether cycle skipping is caused by lack of pad contact or other reasons. Overall, the caliper log is often a highly valuable open-hole logging tool.

Formation dip logs measure the formation dip at various points in the wellbore. This open-hole log is used mainly in geological studies for mapping the formations and helping to locate faults.

Hole inclination logs measure the inclination of the wellbore from the vertical, measured in degrees, and the compass direction in the horizontal plane at various depths in the wellbore. This log gives the course and direction of the wellbore with increasing depth. It is used when the well must be deviated or drilled around certain areas, such as drilling around other wellbores. It is also used when the well must be drilled into a specific target at a specific depth. They are used when drilling kill wells—wells drilled into or near the bore of a well that is blowing out in order to kill the blowout well. Inclination surveys are used where a number of deviated holes are drilled from one surface location.

Temperature logs measure the temperature in the wellbore at various depths. These logs are used in the open hole to locate lost circulation zones and water flows. In cased holes the logs are used to locate cement tops, casing and tubing leaks (similar to lost circulation), and points of fluid and gas entry into the wellbore. Maximum reading thermometers are often run on the basic logging tools to give a maximum temperature reading at the bottom of the hole. Temperature readings are used to help determine the correct resistivity of the formation fluids.

Free-point log tools are run inside pipe (usually drillpipe or tubing) to locate the point where the pipe is stuck. The pipe can be backed off or cut above the stuck point and the free pipe pulled out of the hole. The tool is very important in fishing operations.

Cement logs are run in the casing after it has been cemented. They locate the top of the cement to show that the cement is properly placed and not lost as a result of lost circulation or miscalculated volumes. Some cement logs use sonic techniques to measure the cement bonding to the pipe and to the formation. This shows areas where the zones are properly isolated for perforating and completion. It also shows areas where the cement is not properly bonded or may be absent because of channeling or for other reasons. If these occur near prospective productive zones, they must be filled with cement to provide good zone isolation.

Perforation logs are run in the cased hole to locate precise points to be perforated. The most common tool uses a gamma ray log and a collar locator to locate pipe collars. The hole is logged with the gamma ray log, and the collar locator locates the fixed casing collars and marks them on the gamma ray log. The zones to be completed are located from the basic log analysis and are referenced to the open-hole gamma ray log. These zones are then marked on the cased-hole gamma ray log and are referenced to the casing collar depths. The perforating gun can be run in the hole and moved the correct distance from the casing collar to perforate the preselected formation interval. Perforation logs are frequently run at the same time the cement logs are run.

Casing inspection logs are run in the cased hole to show the condition of the casing. They can be used to detect split or parted casing, worn casing due to drillpipe wear, and pitted casing due to corrosion. These logs are frequently used in deeper wells, older wells before workovers, and in wells exposed to corrosive fluids or conditions. Similar logs are sometimes run in tubing, smaller diameter pipe used to produce the completed well.

Production logs measure the volume and type of fluid entering the cased hole from different points. Various tools are used, but the most common types measure the fluid velocity and density at various points in the wellbore as the fluid is produced. The amount of production from different points in the wellbore (different perforated intervals) is determined by the difference in fluid velocities. The type of production (gas, oil, or water) is calculated from the density of the fluid. These types of production logs are run in flowing wells.

Flood-level logs are run on pumping wells to determine the depth to the top of the fluid in the annular space. Both static and dynamic (pumping) fluid levels are recorded and used to calculate bottom-hole pressures and production efficiency. The tool uses a sound source that projects a sound wave down the annular space between the tubing and casing. The sound waves are reflected off of each tubing collar with a large reflection at the fluid level. The reflected sound waves are picked up by a detector (microphone) and are recorded on a tape. The number of tubing collar reflections is counted and multiplied by the average tubing joint length to give the depth to the top of the fluid. Foaming fluid can create a false high fluid level. This can sometimes be eliminated by dumping a small volume of liquid hydrocarbons downhole before running the fluid-level survey. These surveys are very important for the correct, efficient operation of pumping wells.

Other types of logs and surveys are in various stages of development and use. Fracture logs are used to locate fractures in the open drill hole. Single and multiple deviation surveys (some directional) are run while drilling. Dynamometer surveys are run on pumping wells to check the efficiency of the pumping units.

Log Evaluation

Log evaluation is the process of reviewing the well logs and making necessary calculations to determine what types of formations are exposed in the borehole and their individual characteristics. The primary purpose of log evaluation is to locate productive oil and gas zones. The basic tools are the well log prints and an understanding of how to use them. The logs must be analyzed by a person familiar with the area since formation characteristics and the way they are shown on the well logs change geographically. All other available data are also used in well log analysis, including other types of well logs, coring and drilllstem-test data, and information from mud logging.

Fig. 3–2 shows a simplified log printout with three composite logs (lithology, resistivity, and porosity), log traces, which are evaluated to show the type of formation and fluid content. Logs are read from top to bottom and from left to right.

The lithology log, generally the gamma log, trace is followed and, as long as it is in the shale line, this indicates there are no zones of interest. When the trace deviates, usually to the left, it indicates the start of a zone of interest.

In the figure, Zone A from 4,030–4,050 ft is the first zone of interest. It is a clean sand on the gamma log and has resistivity and porosity, which makes it a possible oil and gas reservoir. However, the SP log would probably show that the sand contained fresh or brackish water.

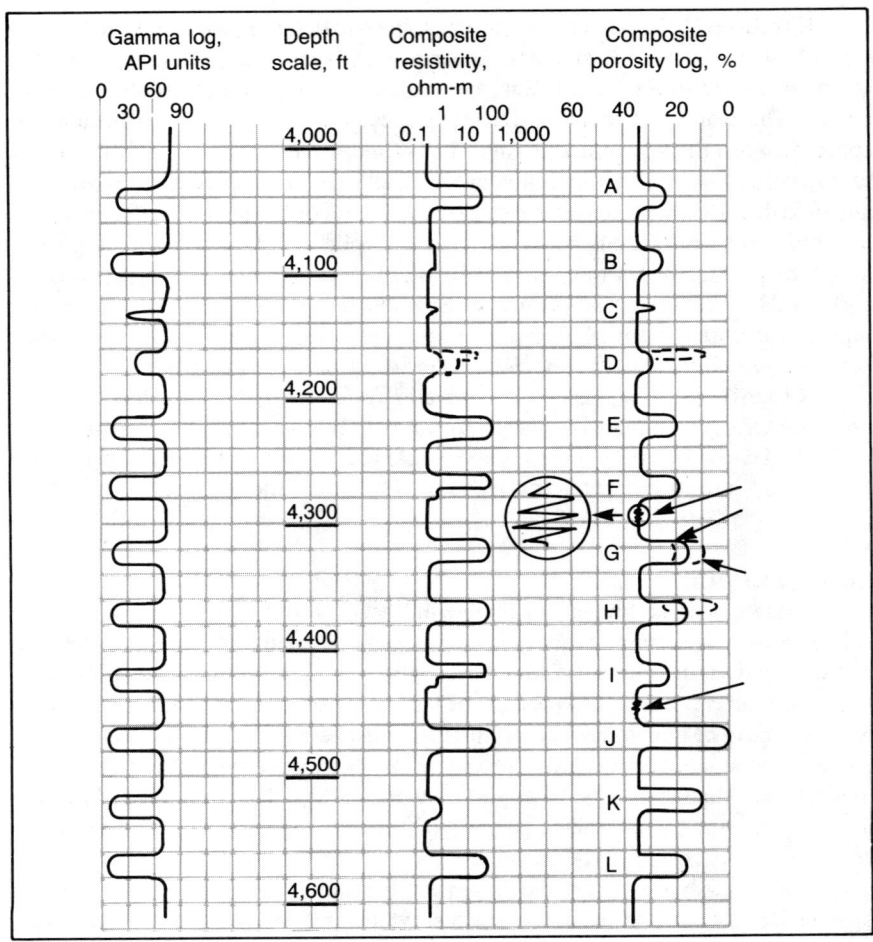

FIG. 3–2 Simplified well log printout

Continuing down the log column, all log traces return to the shale line, indicating shale, until the next section of interest at Zone B. Zone B is a clean sand on the gamma and has good porosity but a very low resistivity. This indicates a saltwater-bearing sand.

The thin sand at C is of questionable interest because it is very thin and does not have a significant resistivity. It could be either a hard streak or possibly a shaly sand.

Section D at 4,160–4,180 ft is of interest because it has some resistivity and good porosity. However, the relatively low resistivity and the gamma trace indicate a shaly sand. The hatched trace on the resistivity could indicate hydrocarbons at the top of the sand, but the low porosity indicates this is a hard streak.

Sand E is a clean oil or gas sand.[3] It has good porosity, high resistivity, and clean gamma.

Sand F is very similar to Sand E except the lower half of the sand has a low resistivity. The upper half of the sand is a clean oil or gas sand, and the lower half of the sand is water saturated, indicated by the low resistivity. The oil- or gas-water contact is clearly indicated at the inflection point of the resistivity log at 4,270 ft.

The gamma density trace shows cycle skipping as a result of gaseated or aerated mud, lack of pad contact due to a rough wall on the borehole, or fracturing. Generally this phenomenon is unimportant except when fracturing occurs with possible secondary porosity to make a reservoir. Cycle skipping frequently obscures the other log traces and makes the logs more difficult to analyze. The cycle skipping effect can occur anywhere in the wellbore where conditions are favorable.

Sand G is a clean gas sand, similar to Sand E. The hashed neutron and gamma density traces illustrate crossover. Crossover sometimes occurs in gas sands and is caused by the tendency of the neutron log to read a lower porosity in the presence of gas.

Sand H shows an oil sand with a gas cap. The logs cannot clearly distinguish between gas and oil, but a gas-oil contact can be inferred at 4,370 ft at the bottom of the gamma density-neutron crossover. Sand I shows a gas column on top of a water column. The gas-water contact is shown on the resistivity log at 4,420 ft. This sand is very similar to Sand F and illustrates the difficulty of distinguishing gas and oil on the logs.

A fracture is shown at 4,445 ft. Fractures are very difficult to locate on the logs and generally must be inferred.

Sections J, K, and L are limestones (carbonates). Section J is a dense limestone and does not have any porosity. If this were a sand, it would generally have at least a small amount of porosity, probably filled with salt water, and would show a much lower resistivity. Section K is a porous limestone with some porosity filled with salt water, as indicated by the low resistivity. Section L is a limestone with oil or gas in the pore space.

---

[3]Generally, logs cannot define clearly between oil and gas.

This analysis has used clean formations with definite clean fluid contents, the best way to illustrate log analysis. However, formations are often mixtures of sands, shales, and, less frequently, carbonates of varying porosities and fluid contents. Although most logging companies have computerized log displays that generally simplify log analysis, it is important to understand basic log analysis since many assumptions must be used in the computerized log displays.

Mud Logging

Mud logging is the process of studying, analyzing, testing, and recording the mud, drill cuttings, and related drilling parameters while the well is being drilled. The primary purpose is to obtain information that will help find a commercially productive formation. The secondary purpose is to obtain information that will help the operator drill the well in the fastest, safest, most economical manner.

The earliest mud logging units analyzed the returning mud and drill cuttings, the small pieces of drilled rock from downhole. Later, mud logging units improved, and they now perform a number of other functions. The data obtained by the mud logging unit is printed on a mud log, which is slightly similar to the well log.

Mud logging units are not used on all wells. However, they are used on most wildcats and deeper wells. The operator may elect to use part or all of the mud logging unit services.

The rate of penetration, how fast the hole is being drilled, is recorded in minutes per foot or feet per hour. Changes in the penetration rate can indicate formation changes, since all formations do not drill exactly the same. The penetration rate normally increases more in porous formations than in nonporous formations. Porous formations can be hydrocarbon reservoirs. Therefore, a drilling break or increase in the penetration rate can indicate a possible productive formation. Depending on the type of formation, the penetration rate during a drilling break will be from 1½ to 4 times the drilling rate before entering the drilling break. The amount of footage drilled during the drilling break will indicate the thickness of the prospective productive zone.

Drilling breaks are often circulated out. Drilling operations are stopped, and the drill cuttings that were drilled out during the drilling break are circulated via the drilling fluid to the surface where they are caught and analyzed. Lag time is the time required for the circulating mud to move from the bit to the surface of the hole. This is recorded as time in minutes at a certain pumping rate (strokes per minutes, spm) or the total number of pump strokes.

The mud logger, who operates the logging unit, catches samples of the drill cuttings. These are washed, dried, examined under a microscope, analyzed with a fluoroscope (ultraviolet light), and tested for density, resistivity, and volume of

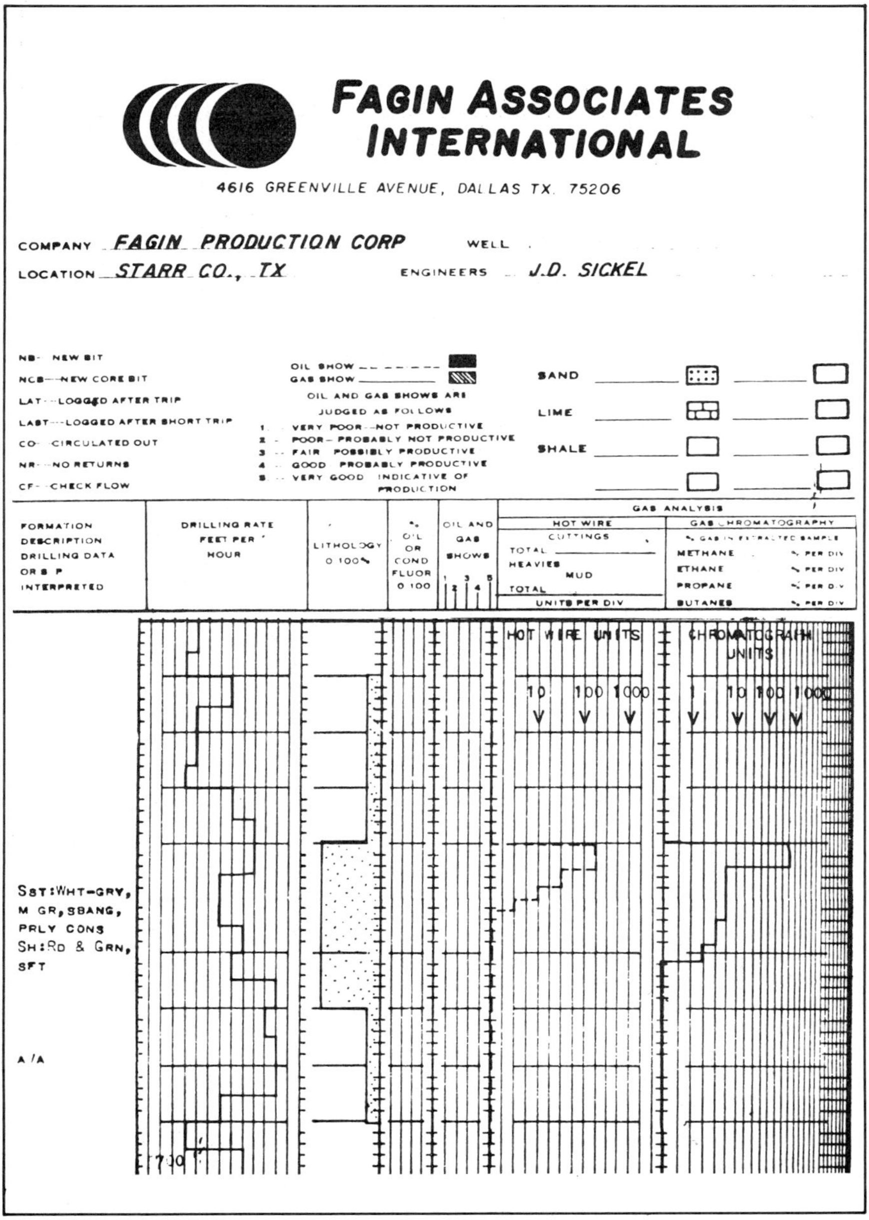

FIG. 3–3 Mud log (courtesy Fagin Associates International)

FIG. 3–3 *continued*

Reservoirs and Reserves

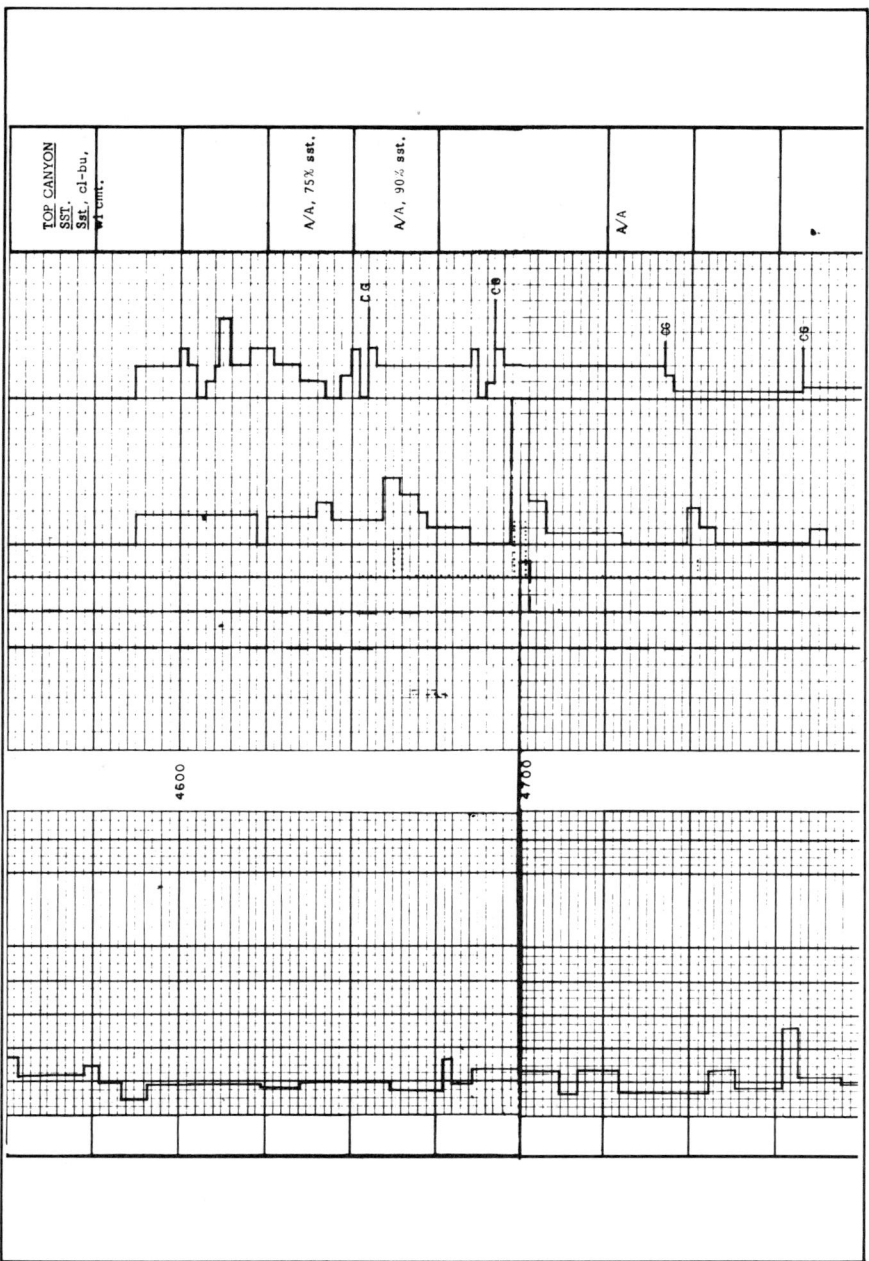

FIG. 3-3 continued

cuttings gas. These tests help describe the complete lithology of the formation being drilled. These data are recorded on the mud log (Fig. 3–3).

The data obtained from the cuttings and other drilling data may be used to construct curves of the d exponent, corrected d exponent—a function of lithology and drillability—shale density, and conductivity. These curves, frequently called pore pressure plots, help predict the pore pressure (formation pressure) in the formations below the point where the hole is being drilled. These data help select casing points, determine the correct mud weight to be used, and give advance warning of possible high-pressure formations where a blowout may occur.

Mud returns, mud circulated out of the borehole, can be analyzed with a gas chromatograph, or hot wire, to detect methane, ethane, propane, butane, isobutane, isopentane, hexanes plus, helium, hydrogen, carbon dioxide, hydrogen sulfide, and sodium chloride. If the hydrocarbon (methane through hexanes plus) content of the mud is high or increases, this can indicate that a hydrocarbon-bearing formation has been drilled. A high methane content in the mud indicates the formation is gas bearing. A mud with a high percentage of heptanes plus indicates an oil-bearing formation. Large volumes of helium, hydrogen, or carbon dioxide in the mud could indicate a high content of these inert (less valuable) gases in the reservoir. High percentages of carbon dioxide and hydrogen sulfide could indicate the need to add inhibitors to the mud system to prevent corrosion of the casing and drilling assembly.

If the gaseous hydrocarbon content of the mud increases so the mud is saturated, this usually indicates that the mud weight is too low and should be increased to control the formation pressure. If the percentage of hydrogen sulfide increases, safety precautions must be reviewed and taken as necessary because hydrogen sulfide is a deadly gas. An increase in sodium chloride content could indicate a saltwater flow.

The mud samples are checked carefully after drilling breaks for hydrocarbons. If these gases are detected, the formation drilled during the drilling break has a show—it contains hydrocarbons. A special set of well logs may be run or a formation test may be justified if the show is good or very good.

A common practice is to drill 5 ft into a drilling break, circulate out samples, and analyze these samples and the mud. If there is a show, then the formation is cored to recover large samples of the formation that can be tested for porosity, permeability, and fluid saturations (volume and types of fluid).

Other information that may be recorded on the mud log includes bit data and feet drilled per bit, mud properties (weight, viscosity, gels, fluid loss), record of the drilling assembly, pump rate and pressure, flow rate, mud surface volume (and changes in volume), flow line temperature, net weight, rotary speed, and trip fill-up volumes. Some mud logging units have computers programmed to provide special information to aid in the drilling operations.

FLUID FLOW

Fluid flow, properly termed reservoir fluid mechanics, is a technical, complex subject covering the movement of liquids and gases. This is very important in the completion process and helps explain some of the activities in drilling operations.

"Fluid" is used for both gas and liquid (oil or water). This usage has developed because the substance in the reservoir may be in either a liquid, gaseous, or intermediate state or it may be in another form at the surface. For example, some hydrocarbons are in the gaseous form at the surface, but they also can occur as liquids (condensates) or can be dissolved in crude oil in the reservoir. Therefore, the term "fluid" is used to describe all of the states of hydrocarbons in the reservoir.

Geothermal Gradients

In a normal geologic section, the temperature of the formations increases with depth. This is attributed to (1) residual heat retained in the earth when the earth formed from hot, molten, gaseous material, (2) atomic disintegration of granitic material, and (3) heat due to compaction (pressure) from overlying sediments. This increase in temperature with increasing depth is measured as a geothermal gradient in degrees Fahrenheit per hundred feet of depth. The geothermal gradient generally is linear with depth within the depth range of current drilling operations.

The average geothermal gradient is about 1.5°F/100 ft and varies from as low as 1.1 to 1.8°F. To calculate the temperature at depth, first determine the ambient temperature, the average annual temperature in an area taken over a long period. A measurement of the temperature in the ground at shallow depths but substantially below the depth of the frostline will closely approximate the ambient temperature. The temperature at depth is then equal to the ambient temperature plus the depth in hundreds of feet multiplied by the geothermal gradient. If the gradient in an area is known, it should be used.

For example, assume an ambient temperature of 80°F and a geothermal gradient of 1.5°/100 ft. The temperature at 9,000 ft would be 215°F (1.5 × 90 + 80).

(80°F, ambient temperature) + (1.5°F/100 ft)(9,000 ft/100 ft)

The geothermal gradient can also be calculated from temperature measurements taken in the wellbore during drilling:

$$\frac{(\text{temperature, °F, at specific depth}) - (\text{ambient temperature, °F})}{(\text{depth, ft})/(100 \text{ ft})}$$

The wellbore temperature is usually obtained during logging operations.

Formation temperatures are used in various ways in the drilling and completion process. Increased temperatures cause deteriorating mud properties. Other factors such as contaminating formations also adversely affect the mud. The operator must know what causes the mud to go bad so it can be treated properly to restore it to a good working condition. Special high-temperature muds may be needed in areas where very high temperatures are encountered because the condition of the drilling mud affects the success of the drilling operation.

Temperature affects the length of time required for cement to harden. Therefore, the temperature of the formations must be known in order to design the proper cement blend used in cementing casing and squeezing. Temperature affects corrosion rates, which can cause problems during drilling and later producing life. Specialized test and production equipment such as packer rubbers are needed for high-temperature conditions. Special logging tools are required for wells with very high temperatures. Thus, temperatures are a fundamental parameter used in all reservoir calculations.

Permeability

One reason that fluid flow and general reservoir engineering appear overly complicated is the terminology. This is due in part to the need to be exact and for convenience. For example, it is much easier to say "permeability" than it is to say "a measure of the resistance of a unit volume of reservoir rock to the flow of fluid through the rock, measured in darcies and conventionally used in millidarcies (md) or one one-thousandth of a darcy." Generally, the reason for using millidarcies is that average good oil reservoirs have permeabilities in the range of 20 to over several hundred millidarcies. This is easier to say, understand, and write than "two-hundredths (0.02) to over two-tenths (0.2) darcies."

Permeability may be better understood by considering two containers; one is filled with very fine sand, and the other is filled with marbles or gravel. Water is poured into both containers to fill the pore spaces. If a screen is placed on the top of the containers and they are both turned over, the water will flow out of the containers (Fig. 3–4). The container filled with large rocks will empty rapidly,

Reservoirs and Reserves

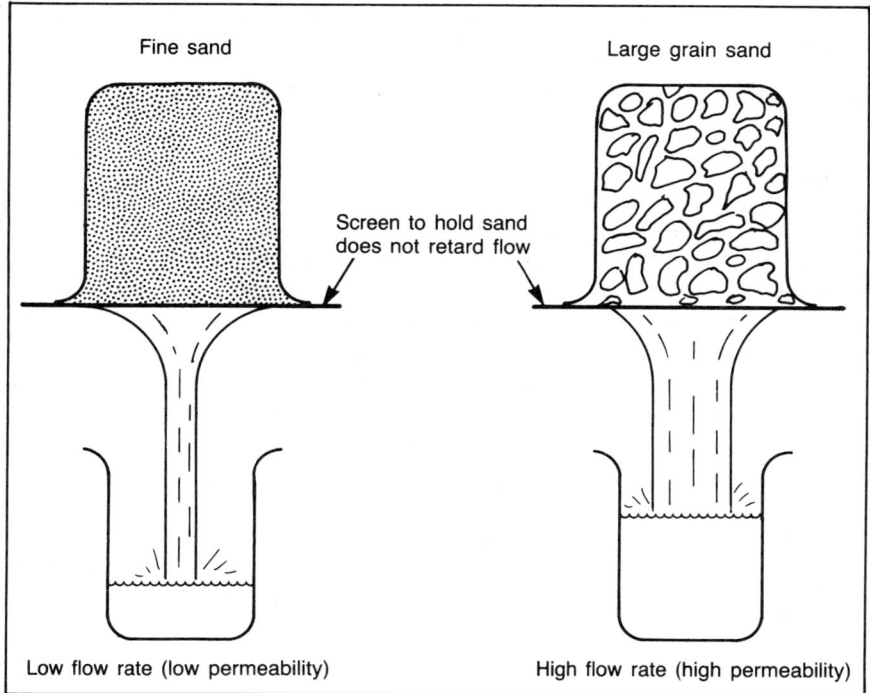

FIG. 3–4 Experiment for observing permeability

showing that it has a high permeability. The water will flow much more slowly out of the container filled with fine sand, showing that the sand has a much lower permeability. Hydrocarbons must flow through the reservoir into the wellbore to be produced. All other factors being equal, oil and gas can be produced faster from a high-permeability reservoir than from a low-permeability or tight reservoir.

Permeabilities can also be measured from reservoir cores. Sometimes these measurements are misleading because the core may change as a result of being removed from the reservoir and may not exactly represent the reservoir. However, this is often the best information available. Permeabilities can also be measured by production tests after the well has been completed.

Flow capacity, also referred to as formation or well conductivity, is similar to permeability except that it is applied to the thickness of the reservoir exposed in the hole. For example, if a 20 ft thick reservoir had a permeability of 7 md, the flow capacity of the reservoir would be:

(reservoir thickness, 20 ft) × (permeability, 7 md) = 140 md ft

Viscosity

Viscosity is a measure of the liquidity or fluidity of a fluid. Viscosities are measured under standard conditions (60°F and 1 atmosphere of pressure; average pressure at sea level is about 14.65 psia). Hydrocarbon viscosities are measured in centipoises (cp).

Hydrocarbon viscosities are affected by temperature and pressure, hence the need of measuring viscosity at standard conditions. The viscosity of a fluid tends to decrease with increasing temperature. This is easily demonstrated with a fluid-like thick syrup. When the syrup is cold, it is thick and will hardly pour. When it is heated, it thins and will pour rapidly. In this case the cold syrup is said to have a high viscosity, and the thin, hot syrup is said to have a low viscosity.

Viscosities are affected moderately by pressure. When a viscosity is given without indicating the pressure or temperature, the viscosity is assumed to be at standard conditions. For example, the viscosity of salt water is about 1 cp, but at average reservoir conditions it is about 0.7 cp. In this case the viscosity of 1 cp is at standard conditions, and the lower viscosity, 0.7 cp, is at reservoir conditions with a higher pressure and temperature than standard conditions.

Viscosities are important because they are a direct measure of the ability of the fluid to flow through the formation. This can be illustrated with the example of the two containers. Assume in this case the containers are filled with the same kind of sand. The pore space in one container is filled with water (viscosity equals 1 cp), and the pore space in the other container is filled with a fluid that has a viscosity of 2 cp, perhaps water mixed with a little syrup. If the two containers are inverted, the container filled with plain water will drain empty exactly twice as fast as the container filled with a 2 cp viscosity fluid. If the second container had been filled with a fluid with a viscosity of 3 cp, it would take about three times as long to empty as the container filled with plain water.

Average reservoir oils have viscosities in the range of 1–3 cp at reservoir conditions. Very heavy oils can have viscosities of several thousand centipoise or more. Therefore, a light oil with a lower viscosity is much easier to produce than a heavier oil with a higher viscosity. These concepts are important in determining if the reservoir can produce oil in commercial amounts and in planning the completion.

Gases in the reservoir have viscosities on the order of 0.015 cp. This is much lower than oil and accounts for the reason that gas is generally easier to produce than oil.

Pressure

Pressure affects all phases of the industry. The exploration geologist is indirectly concerned with pressures since they are fundamental to the accumulation and production of oil and gas. Drilling personnel constantly monitor formation and hydrostatic mud column pressures to obtain the maximum penetration rate and prevent lost circulation and blowouts. The completion engineer uses pressures to help design the most efficient completion. Production and reservoir engineers use pressures to determine the most efficient production rate, prevent the loss of hydrocarbons (and reservoir energy), and calculate reservoir volumes. Pressures are fundamental to secondary, tertiary, and enhanced recovery processes. Therefore, it is important to have a good understanding of pressures and their cause and effect.

Pressures as used in the industry are generally hydraulic pressures. Hydraulic simply means that the pressure is applied by a fluid (liquid or gas), sometimes called fluid pressure. Pressure is commonly defined as force per unit area, or pounds per square inch (psi). If a 150-lb person has 30 sq in. of area under each foot, this person will always exert a force of 150 lb, no matter where he is standing. The floor supports this force. If the person distributes his weight on both feet, he is exerting a pressure of 2.5 psi:

$$\text{Pressure} = (150 \text{ lb})/(30 \text{ sq in.} + 30 \text{ sq in.})$$
$$= 2.5 \text{ psi}$$

If the person begins walking, the weight is alternately shifted from foot to foot. The pressure on the bottom of each foot would alternate from 5 psi to 0 psi as the foot alternately supports the complete weight.

Fig. 3–5 shows a simple hydraulic lift used in garages and service stations to lift cars so the underside of the car can be repaired or inspected. A cylinder is mounted in the floor and has a moving piston with a lifting ramp fixed to the top of the piston. The car is driven onto the lifting ramp, and air is pumped into the pressure chamber. The car is lifted when the force on the bottom of the piston is equal to the weight of the piston and car. In the illustration, the piston has an area of 50 sq in. and the car, ramp, and piston together, weighs 3,000 lb. The car will move upward when the pressure in the chamber reaches 60 psi, and the force on the bottom of the piston is equal to the weight of the car and piston. If a larger car is to be lifted, more pressure is required.

The pressure is exerted equally against all surfaces inside the pressure chamber. However, the cylinder is immobile, and the piston is free to move. Therefore, the piston moves upward when sufficient pressure is applied.

The cylinder is analogous to casing in the hole; the plunger is equivalent to the wellhead controls. In this case the casing and wellhead controls must be strong enough to support the pressures that may be encountered. Pressures of

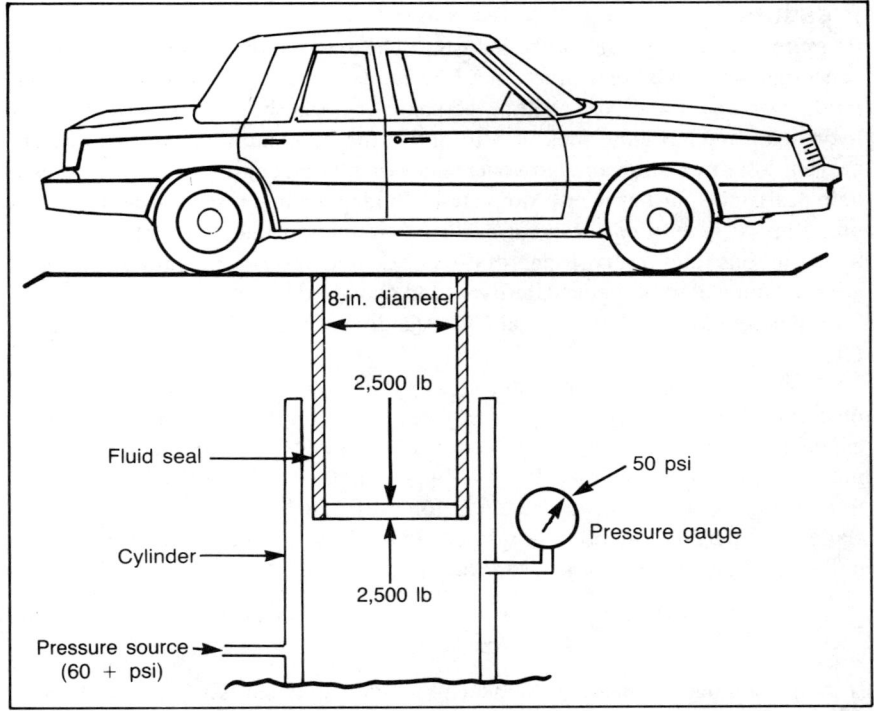

FIG. 3–5 Pressure lift—60 psi acting on the 8-in. diameter piston gives a lifting force of 2,500 lb

5,000–10,000 psi are relatively common, and pressures in excess of 15,000 psi have been encountered. Ten thousand pounds of pressure could lift two hundred cars (2,500 lb/car) or about 500,000 lb. It would take very special equipment to hold this force. Therefore, high formation pressures can create dangerous situations and require special equipment and techniques to handle them safely.

Pressures are conventionally measured in psi. The abbreviations *psig* and *psia* are sometimes used. The term psig is the abbreviated form for pounds per square inch gauge and is the same as the abbreviated form psi. The term psia is the abbreviation for pounds per square inch absolute. The absolute pressure (psia) is the gauge pressure (psi) plus the atmospheric pressure at sea level (14.65 psi, frequently rounded off to 15 psi). Therefore, if the pressure on a system is 600 psi, the absolute pressure would be 615 psia (600 + 15 = 615). Normally, the difference between gauge and absolute pressures is smaller than the margin of error in the measurements. Usually gauge pressure is used, but it is customary to use absolute pressures when performing detailed reservoir calculations.

Oil and gas operations deal with long (high) columns of fluid. The fluid exerts hydrostatic pressure. The amount of hydrostatic pressure depends on the density or weight per unit volume of the fluid and the vertical height of the fluid column. The fluid exerts a pressure or fluid gradient, measured in psi/ft. This is merely the pressure exerted by the fluid column per foot of depth. Therefore, the pressure gradient times the depth (or height) of the fluid column will give the pressure at that point.

Fig. 3–6 shows a fluid column with fresh water that has a fluid gradient of 0.433 psi/ft. At 300 ft the pressure is 130 psi; at 1,000 ft, it is 433 psi. In this case the fluid column is measured from the surface to the depth where the pressure is required. The size of the fluid column does not have any effect on the pressure. For example, if a town has a water tower 100 ft high, the water pressure in the houses would be about 43 psi.

The figure is also similar to a wellbore filled with fluid. The weight of the fluid is 62.4 lb/cu ft or 8.4 lb/gal (fresh water). These two numbers are used because they are common methods of measuring the weight of fluid used in oil-field operations. Either number can be easily converted to a pressure gradient:

$$\begin{aligned}\text{Pressure gradient} &= (\text{fluid weight, lb/cu ft})/(144 \text{ sq in./sq ft}) \\ &= (62.4 \text{ lb/cu ft})/144 \text{ sq in./sq ft}) \\ &= 0.433 \text{ psi/ft}\end{aligned}$$

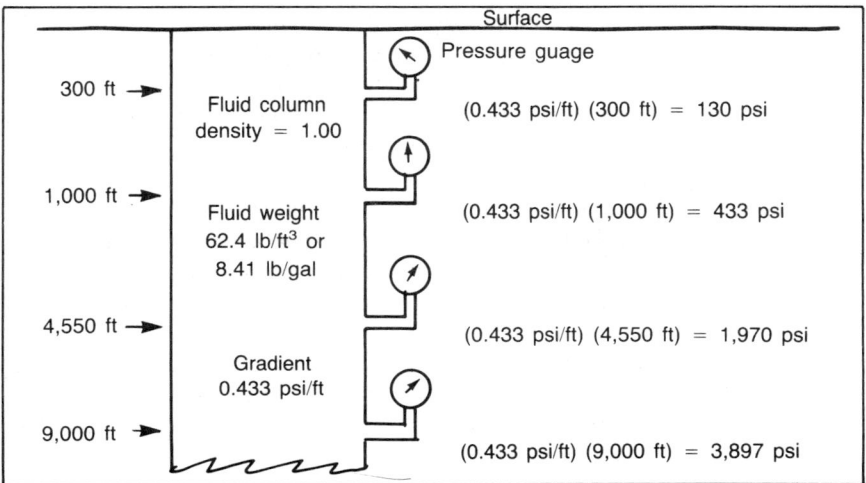

FIG. 3–6 Fluid column pressure

$$\text{Pressure gradient} = (\text{fluid weight, lb/gal})(0.0518)$$
$$= (8.4 \text{ lb/gal})(0.0518)$$
$$= 0.433 \text{ psi/ft}$$

The numbers 144 sq in./sq ft and 0.0518 are conversion constants used to convert fluid weight to a pressure gradient. This procedure can be used to calculate the pressure at any depth for any type of fluid. For example, assume that a hole is filled with 12-lb/gal drilling fluid. The pressure gradient would be:

$$(12 \text{ lb/gal})(0.0518) = 0.624 \text{ psi/ft}$$

With the pressure gradient, the pressure can be calculated at any depth. For example, at 1,500 ft the pressure would be 936 psi, $(0.624) \times (1,500 \text{ ft})$. By the same method the pressure at 7,850 ft would be 4,898 psi, $(0.624 \text{ psi/ft}) \times (7,850 \text{ ft})$.

In the geologic cycle most of the sediments are deposited in salt water that fills the pore spaces. In the average case, over geologic time the pore spaces are pressure connected. The term pressure connected indicates that a very limited flow may occur between the pore spaces, but there is sufficient movement so a pressure gradient occurs.

Therefore, the entire sequence of formations in one area acts like a gigantic fluid column filled with the formation material and salt water in the pore spaces. The normal pressure gradient would then be the pressure gradient of the sea water in which the formations were deposited. The normal seawater gradient is about 0.44 psi/ft. The saltwater gradient occurring in the formations ranges from the seawater gradient to an average slightly higher number of 0.465 psi/ft.

A formation fluid column and a regular fluid column are similar (Fig. 3–7). Pressures are calculated by multiplying the pressure gradient times the depth to the point where the pressure is being calculated. Therefore, the pressure can be calculated in any normal pressured area by using the depth and pressure gradient. If a pressure gradient is not precisely known, then an average pressure gradient of 0.465 psi/ft can be used for normal pressured areas.

Pressures are not normal in some areas. Subnormal (lower-than-normal) pressures range downward from the normal pressure gradient to minimum pressure gradients of about 0.25 psi/ft. Abnormal or geopressures occur in areas where the pressure gradient is greater than normal. Abnormal pressure gradients in excess of 1.00 psi/ft have been encountered.

It is not uncommon for the pressure gradient to change with depth. This change can be either to a higher or lower pressure gradient. Sometimes, the gradient will change again deeper in the hole, causing a pressure reversal. Subnormal and geopressured formations are caused by geologic conditions, with geopressure formations slightly more common than subnormally pressured formations.

Reservoirs and Reserves

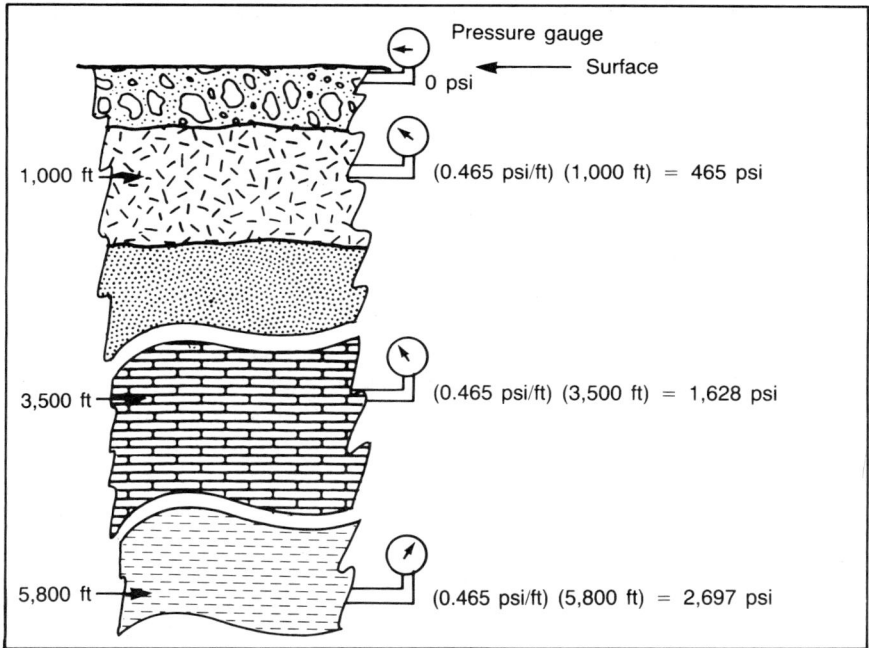

FIG. 3–7 Formation fluid column

Pressure gradients help determine the correct mud weight to be used for drilling the well and are a guide in designing the casing program. Both occur, but geopressured formations are potentially more severe because of the possibility of a blowout.

Geopressured formations are related to the overburden pressure, usually expressed as a pressure gradient and sometimes as specific gravity. The normal overburden pressure at a point in the earth is the total weight of the overlying formations, including any fluids that may be contained in the formations. Sometimes, an abnormal overburden pressure occurs when formations are subjected to additional special stresses.

Normally, formations tend to support themselves by a bridging effect of the matrix rock where the weight of the overlying rocks is supported by the underlying rocks. Then the fluid pressure in the pore space is not affected by the overburden pressure, rather by the weight of the overlying fluid column. Geopressured conditions occur when the higher overburden pressure is applied to the pore space. The average overburden pressure gradient is about 1.0 psi/ft in the depths encountered in normal drilling practices. There is some evidence that the overburden pressure gradient increases with depth, again within the range of depths normally drilled.

Crustal movements (tectonic activities) are a major cause of geopressured reservoirs. Fig. 3–8 illustrates two ways that crustal movements cause geopressured formations. In Fig.3–8a, the center section at B has been uplifted. The formation is normally pressured at positions A and C. If the uplifted portion is properly sealed, it can contain the same pressure as A and C. Since it is now closer to the surface, it represents a geopressured situation.

Figure 3–8b illustrates how a thrust fault can cause a geopressured formation. A and B are the same formation, which has been separated by a thrust fault. Thrust faults have low (small) angles but often have a long horizontal movement so that in some cases the formations are shifted a substantial distance. The formation at B is normal pressured. The formation at A has been uplifted and, assuming the pressure has not escaped, could represent a geopressured reservoir since it has been uplifted closer to the surface. If the formations are less competent, the added weight of the overthrust could compress the lower formations and cause B to be geopressured.

Other geologic conditions lead to different types of geopressure. The average overburden pressure gradient is about 1.0 psi/ft. This could lead to the erroneous belief that the maximum hydraulic pressure gradient is also 1.0. This is incorrect; hydraulic pressure gradients exceed 1.00 psi/ft under certain geological conditions. Fig. 3–8c illustrates a dome or anticline structure that can occur from a high rate of sedimentation on the east and west (right and left) sides, downwarping caused by thrusting, or an uplift in the center of the structure as a result of an intrusive, salt-dome formation or other geologic action. If a thick competent bed occurred as illustrated in the figure, this could cause compressive forces below the bed that would exceed the normal overburden pressure gradient—in some cases by a considerable amount.

Shallow formations can be pressurized to a geopressured condition by communication through a fault with a deeper, higher-pressure zone. Steeply dipping, widespread, permeable formations may be overpressured at shallower depths as a result of excess pressures at lower depths and communication throughout the zone.

Erosion and subsidence may also cause geopressured formations. Fig. 3–8d illustrates a widespread (blanket) sand that outcrops in the west at A. The original surface is changed to the present surface by differential erosion, where the east side erodes more rapidly, by differential subsidence, where the east side subsides more rapidly, or by a combination of these. Assume a well is drilled at point B and intersects the top of the formation at point D. Point D has a hydrostatic head

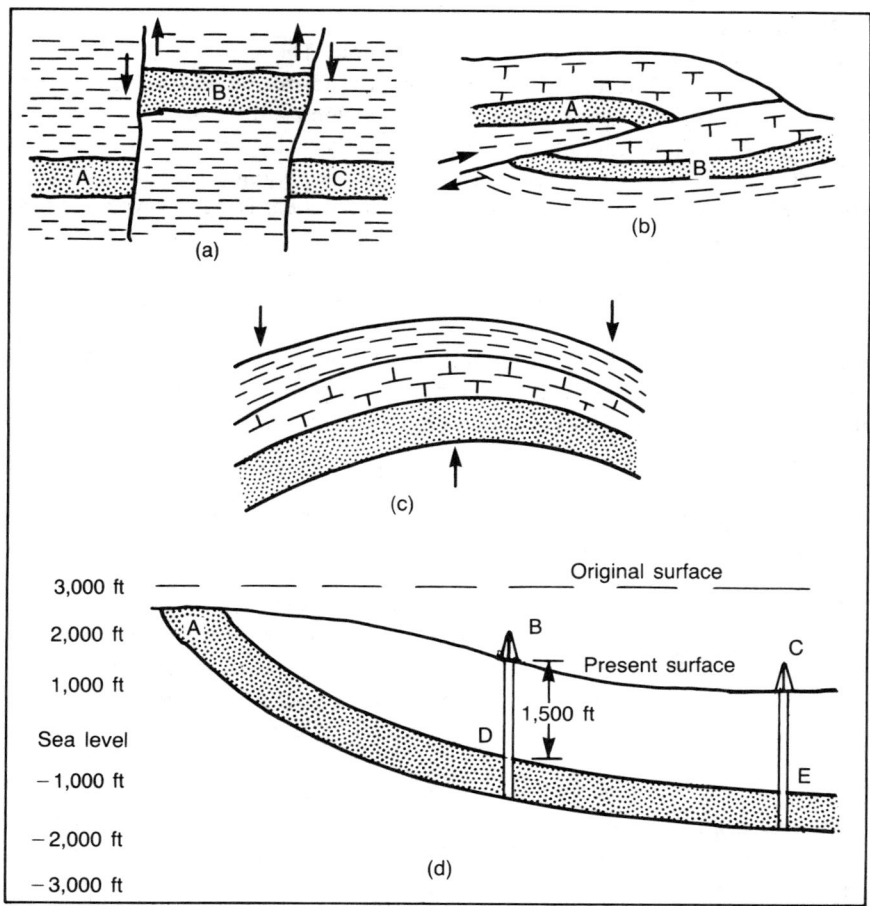

FIG. 3–8 Abnormal geopressures (a) horst fault block, (b) thrust fault, (c) dome or anticline, and (d) widespread (blanket) sand

equivalent to 3,000 ft, or the vertical distance between point D and point A. Assuming a fluid with a pressure gradient of 0.45 psi/ft, the pressure at point D would be 1,250 psi. At a depth of 1,500 ft, this would represent a pressure gradient of 0.9 psi/ft, a very high geopressured condition. If a well is drilled at point C and intersects the top of the formation at point E, the pressure gradient would be 1.05 psi/ft.

Saturations

Porosity is the void or pore space in the reservoir. The term "void" is used to show that the space is not filled with rock or other formation material. However, the space is usually filled with salt water. In reservoir traps the pore space may be filled with salt water, gas, or oil, either singularly or in combination.

The term "saturation" identifies the type and amount of fluid in the pore space (Fig. 3–9). It is always preceded by a word identifying the type of fluid, such as water saturation, oil saturation, or gas saturation. The amount of fluid in the pore space is listed as a percentage of the total pore space.

If the pore space is assumed to be filled with salt water, the reservoir has 100% water (salt) saturation, commonly written $S_w = 100\%$. Likewise, if we say another reservoir has a high oil saturation in the order of 70%, then 70% of the pore space is filled with oil. If the remainder of the pore space is filled with water, the formation has a 30% water saturation (100% minus 70%). The total of the saturation must always equal 100%.

Most reservoirs are water wet. A thin, microscopic layer of water surrounds the sand grains, so the oil does not contact the formation. Although this water occupies pore space, it is immobile and cannot be produced by normal means, sometimes referred to as the irreducible water saturation (S_{wi}). Most reservoirs are water wet, but there are some oil-wet reservoirs.

The word "phase" is sometimes used to define the different types of fluids in the reservoir. For example, a reservoir containing oil and water would have a water phase and an oil phase. If the oil phase were continuous, it would continue or move through the formation. On a single-phase reservoir, the single phase is also the continuous phase.

The term "mobile" indicates that the fluid can be moved and produced. Immobile fluids do not move. For a reservoir condition where the water-wet reservoir has both mobile oil and mobile water phases, the water layered around the sand grains would still represent irreducible water saturation. Free (mobile) water suspended in the continuous oil phase as small droplets would tend to move with the oil, so both oil and water would be produced.

This free, movable water is the mobile water saturation, and it ranges from less than 1% up to about 35%. At about 35% mobile water saturation, the oil saturation would also be about 35%. If the mobile water saturation exceeds 35%, then it will probably become the continuous phase and droplets of oil will be suspended in the water. As the mobile water saturation continues to increase, the mobile oil saturation decreases and a reducing amount of oil (and correspondingly a larger volume of water) is produced. In some cases the water and oil will separate because they have different densities. However, many cases of mobile water saturation are encountered where the water is distributed in the oil.

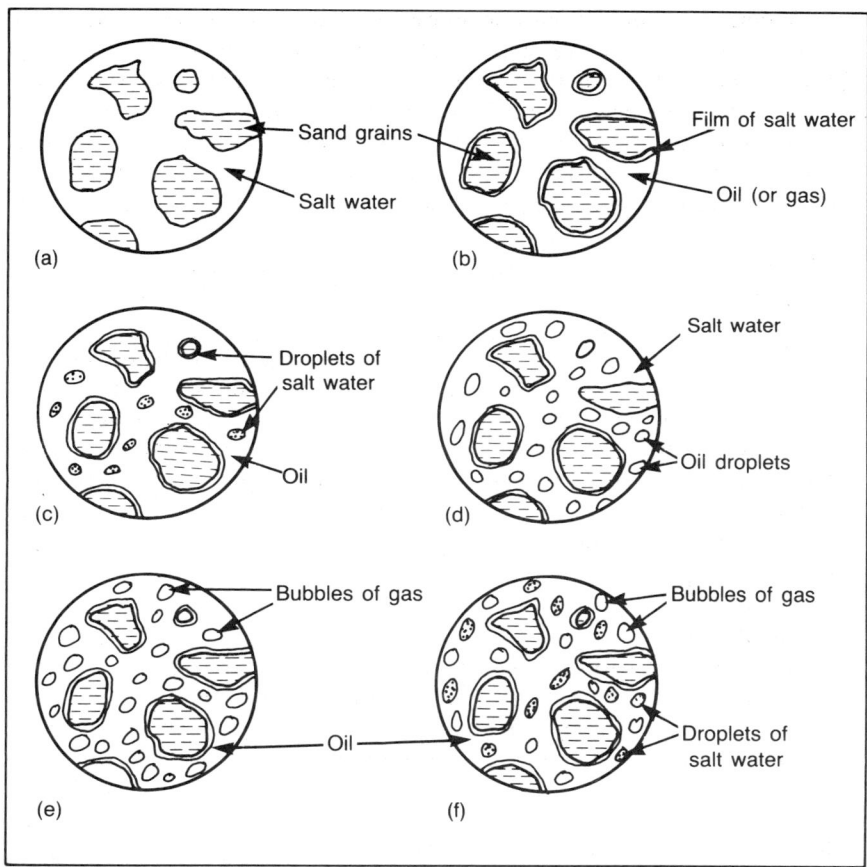

FIG. 3–9 Saturation, phases, and mobility (a) 100% water saturation, (b) mobile oil or gas, (c) continuous oil phase with mobile water, (d) continuous water phase with mobile oil, (e) continuous oil phase with mobile gas, and (f) continuous oil phase with mobile gas and water

Fig. 3–9e illustrates a common combination of fluid saturations that frequently occurs in a reservoir. The basic reservoir has 30% irreducible water saturation and a continuous oil phase. Droplets of free gas are suspended in the oil. The reservoir has three fluid phases: water, oil, and gas. Various combinations of oil and gas saturations are encountered, but the total fluid saturation in the reservoir still must equal 100%. The reservoir will produce both oil and gas, and the amount of relative production depends on the fluid saturations.

Another common combination of reservoir fluid saturations is illustrated in Fig. 3–9f. There are three fluid phases and a free, mobile water phase. Assuming a 30% immobile water saturation, the reservoir could have mobile water, oil, and gas phases. If the continuous phase is oil with 40% saturation, the reservoir could then have 20% free mobile water saturation and 10% free mobile gas saturation. In this case the reservoir would produce water, oil, and gas. If the water saturation were equal to or higher than the oil saturation, water might be the continuous phase.

Within reason, there can be any combination of fluid saturations, as long as the total fluid saturation is 100%.

When the reservoir contains oil, there can be an irreducible oil saturation (S_{oi}) just as there is an irreducible water saturation (S_{wi}). The irreducible oil saturation is effectively immobile. Small, isolated globules of oil are entrapped in the reservoir and are left behind as the oil moves through the pore channels of the reservoir toward the producing well. These may be all or part of the irreducible oil saturation. If the formation is oil wet, some oil will be retained as a thin film around the sand grains, similar to the irreducible water saturation. This oil is also part of the immobile, irreducible oil saturation, which normally range from 20–35%.

Flow Mechanics

Rock, salt water, and hydrocarbons combine to form an oil and gas reservoir. The oil and gas are a complex mixture of hydrocarbons. These are contained in the pore space, often in a maze of tiny, almost microscopic pore channels. These pore channels occur in the formation rocks and may be so small that the rock appears solid. The surface area of these pore channels can be as much as 100,000 sq ft/cu ft of rock.[4] Fluids must flow through these pore channels and into the wellbore to be produced. The flow mechanism is properly termed flow of viscous fluids in porous media. Some of the forces that affect the flow of these fluids through the pore channels include pressure differentials, interfacial and surface tension, and capillarity and gravitational forces.

Fluids tend to flow from a high-pressure area to a low-pressure area. The difference between the pressures is called the pressure differential, ΔP. This is illustrated in Fig. 3–10 where two containers are connected by a tube or flow channel. Container B has a higher fluid level than Container A and, consequently, a higher pressure at the base of the container. Fluid will flow from B to A

---

[4]Interstate Oil Compact Commission, "A Study of Conservation of Oil and Gas in the United States" (1964), p. 48.

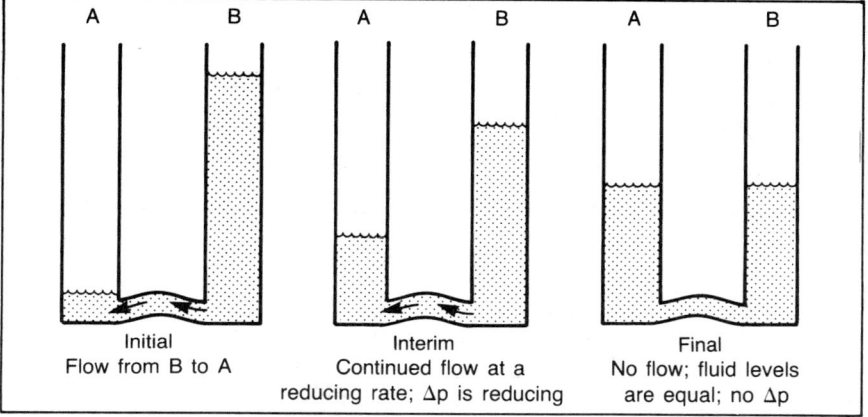

FIG. 3–10 Pressure differential and fluid flow

until the fluid has the same level in both containers. This is rudimentary but is the basis of fluid flow in the reservoir.

Imagine a single-phase fluid flowing through a pore channel under a pressure differential (Fig. 3–11). A globule of a different fluid phase, such as water or oil, enters the pore channel. It starts to distort as it reaches a restriction between two sand grains and is further distorted as it is forced through the restriction. The movement of the globule through the pore channel restriction requires a higher pressure gradient because of surface and interfacial tension effects. Surface tension is the attraction between the molecules in the surface of a liquid; interfacial tension is the molecular forces at the interface between two liquids or a liquid and a solid. The pressure required to move one globule through the restriction would be infinitesimal. However, there are literally millions of pore channel restrictions, so in some cases very high pressure differentials are required to move two fluid phases through the pore channel at the same time.

Other problems can be encountered in multiphase fluid flow through pore channels (Fig. 3–12). The pore channel can be further restricted by a film of water on the sand grain. This reduces the size of the opening and increases the pressure differential required to move the fluid through the restriction. If the fluid globule is water, in some cases it can combine with the fluid around the sand grain and block the pore channel at the restriction. If several mobile fluid phases are flowing through the pore channel, higher pressure differentials are required and/or the pore channel can be blocked.

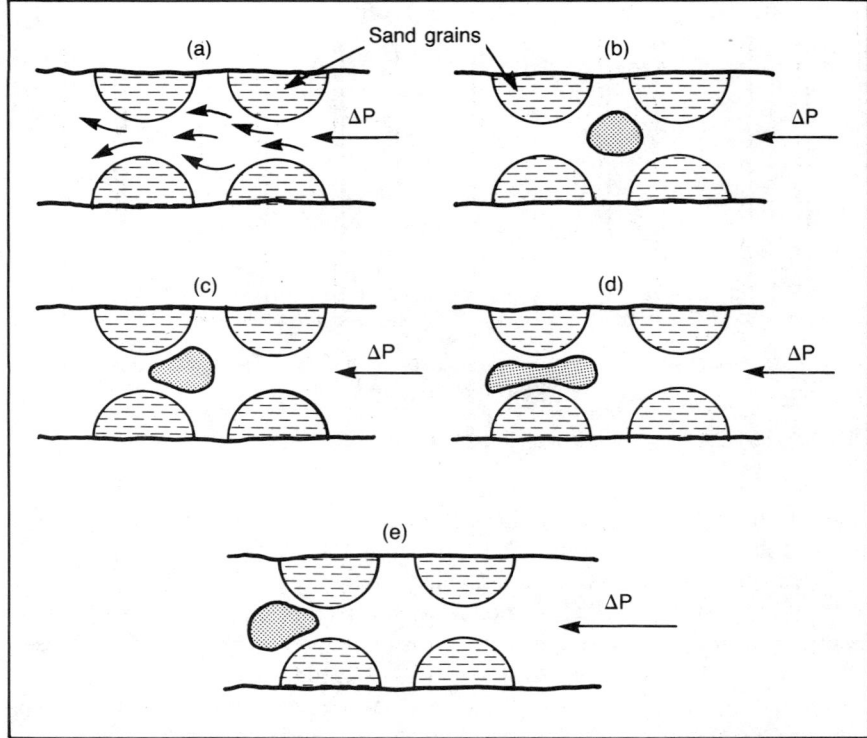

FIG. 3–11 Flow through pore channels: (a) single phase flowing through a pore channel, (b) droplet of different fluid; (c) droplet entering restriction, (d) droplet distorts, and (e) droplet exiting restriction

The flow of fluid through the reservoir is further complicated by the pattern or type of flow. In the normal case fluids flow radially from all points in the reservoir inward toward the producing well, like water in a sink flowing into a drain. This is called radial flow. The fluids at some distance from the wellbore have a relatively large area to flow through and are less restricted. As the fluid moves closer to the wellbore, the area through which the fluid flows reduces. Therefore, the fluid flows faster. This increases the frictional forces and requires a higher pressure differential. Thus, the pressure differentials adjacent to the wellbore can be very high.

Although fluid movement in the reservoir is complicated, fluids can move through the reservoir under the proper conditions and with an adequate pressure

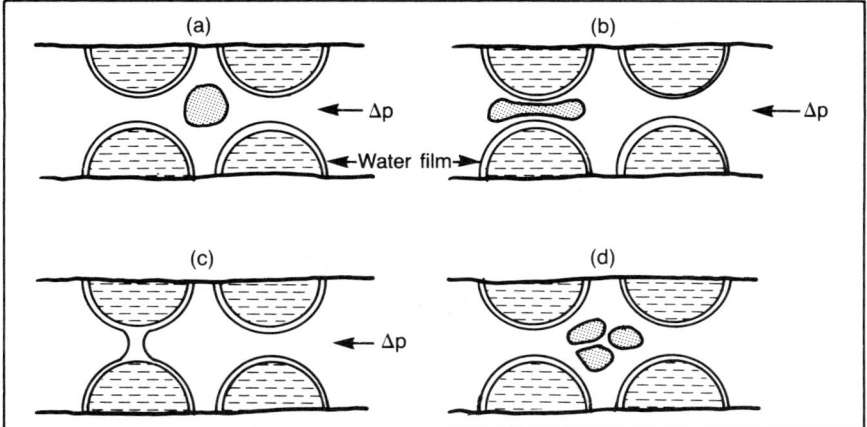

FIG. 3–12 Pore channel distribution: (a) water droplet in oil, (b) droplet distorts, (c) sand grain and fluid block pore channel, and (d) several mobile fluid phases

differential. The next step is to see how the hydrocarbon fluids move from the reservoir to the surface. The basic process again involves pressure differentials and the tendency for the fluid to flow from a high-pressure region to a region of lower pressure.

To help understand this, first consider a U-tube. When the two legs of the tube are filled with the same type of fluid, the fluid will move until both fluid levels are at the same height. If one leg has a heavier, more dense fluid than the other leg, the fluid level will move downward in the leg with the heavier fluid until the pressure at the bottom of the leg is the same in both legs.

The U-tube principles and pressure differentials are analogous to a reservoir producing by a water drive. Fig 3–13 illustrates an oil trap in a dipping sand reservoir. Well A is completed in the gas cap, Well B is completed in the oil column, and Well C is a dry hole drilled into the water column. The saltwater gradient is 0.465 psi/ft, and the reservoir pressure is 2,790 psi, which is equivalent to a 6,000-ft water column. This is one leg of the U-tube.

The oil well (B) is the other leg of the U-tube. Assuming the reservoir has a 40° API gravity crude oil and Well B is filled with this crude, the gradient of the oil fluid column is 0.356 psi/ft. The pressure at the bottom of the column due to the weight of the oil in the column is 2,136 psi. The actual pressure at the bottom of the column must be the same as the reservoir pressure of 2,790 psi. Therefore, the closed-in wellhead at the top of the well has a pressure of 654 psi (2,790 psi −

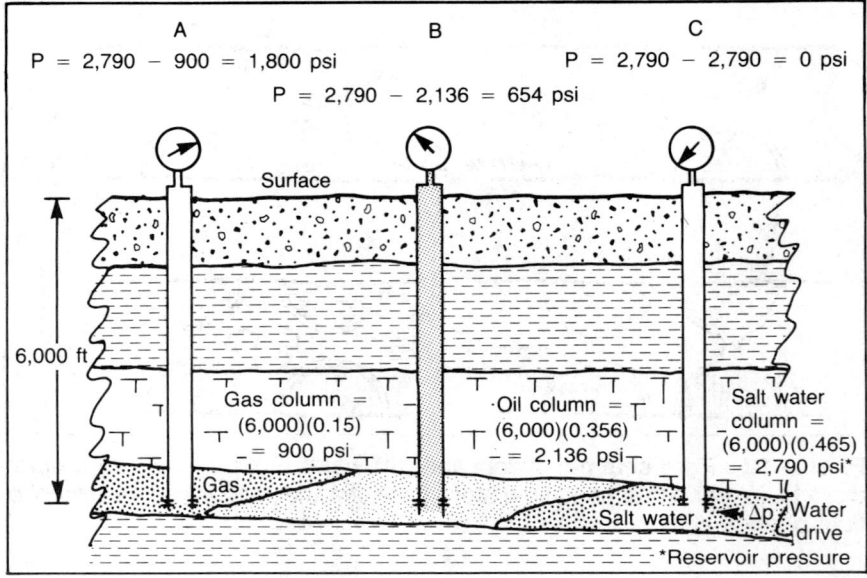

FIG. 3–13 Wellhead pressures

2,136 psi = 654 psi). Oil will flow out of the wellhead at B when the control valve is opened. Water will move updip from the water column to maintain the pressure, and the oil will continue to flow upward.

The gas well at A can be analyzed in a similar manner. The gas has a low density and under pressure is estimated to have a gradient of 0.15 psi/ft. This is equivalent to a pressure of 900 psi at the reservoir caused by the gas column. When the wellhead valve is closed, the wellhead pressure is 1,890 psi (2,790 − 900 = 1,890) and the pressure in the reservoir at the bottom of the well (the bottom of the U-tube leg) is 2,790 psi. When the wellhead valve is open, the pressure reduces. Then fluid (gas) will flow from the reservoir into the wellbore and to the surface in the same manner as the oil well (B) when the surface control valve was opened.

The wells will continue to flow as long as there is adequate reservoir pressure. Well productivity reduces as the reservoir pressure depletes. Declining production is caused by the reduced pressure differential. Assuming there is no gas dissolved in the oil, when the reservoir pressure declines to 2,136 psi, Well B will stop flowing. At that time the well could be placed on pump to increase its productivity. The pump reduces the pressure in the wellbore and increases the pressure differential, causing more fluid to flow.

PRODUCING MECHANISMS

The producing mechanism is the method by which the reservoir energy is utilized to expel the hydrocarbons from the formation into the wellbore and thus to the surface. Reservoir energy is a combination of pressure, fluid solubility and compressibility, and/or latent displacing fluids. The efficiency with which the hydrocarbons are produced depends on the properties of the hydrocarbons and displacing fluids, if applicable; the type and kind of reservoir energy, formation porosity, permeability, wettability, and well spacing; and the rate and method at which the reservoir fluids are produced.

Although the details of reservoir drives (producing mechanisms) are complex, the basic principles are relatively simple. The four basic producing mechanisms are fluid expansion, solution-gas drive, gravity drainage, and natural water drive. Most fields (reservoirs) are produced by a single mechanism, although it is not uncommon for two drive mechanisms to be active. When this occurs, usually one drive mechanism predominates.

Fluid Expansion

Some fluids, such as gas and to a lesser extent highly gas-saturated crude oil, are very compressible. When a highly compressible fluid is confined under high pressure, its volume is greatly reduced. If a gas column in a reservoir has a pressure of 2,790 psi, under normal conditions if the pressure was released to atmospheric pressure, the gas volume would expand by a factor of over 200. One cubic foot of gas in the reservoir would occupy over 200 cu ft of space at the surface. Thus, the compressed gas contains energy that is directly related to the confining pressure. When a well is completed in the gas reservoir, the natural pressure of the gas causes it to expand and flow into the wellbore and to the surface.

This is slightly analogous to an inflated automobile tire with the verticle jacked up to hold the tire off the ground. If the valve stem is opened, air will flow out of the tire at a continual reducing rate until the pressure inside the tire is equal to the pressure outside the tire.

In a general sense the reservoir produces gas through the wellbore in a manner similar to air flowing through the valve stem of an automobile tire. The flow rate decreases with time due to the decreasing reservoir energy (reservoir pressure). Some pressure is expended, pushing the gas to the wellbore through the reservoir restrictions—the pore channels.

Theoretically, the reservoir is not depleted until it is at atmospheric pressure. In practice, it is above this pressure. Some shallow reservoirs are produced on a vacuum until the reservoir pressure is very low at depletion.

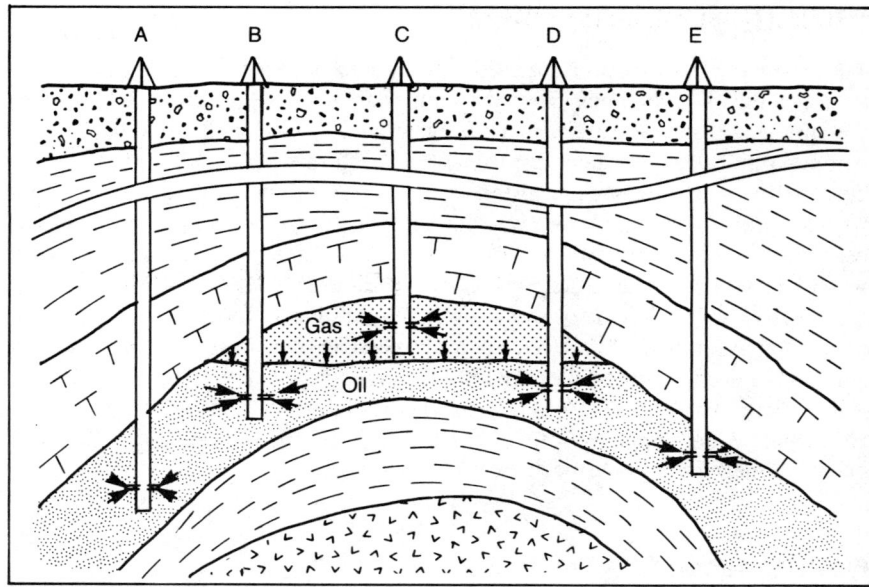

FIG. 3–14 Gas-cap expansion

Generally, gas reservoirs are highly efficient. Recoveries will often approach 85% of the original gas in place. This is much more efficient than some of the other recovery mechanisms.

One special case of fluid expansion occurs when the reservoir has a gas cap, free gas above an oil column, and various wells producing the reservoir (Fig. 3–14). The gas cap acts as a pressure source and pushes the oil downward into the producing wellbores. If the gas in the gas cap is produced through Well C, a major part of the reservoir energy will be wasted and total oil recovery will be reduced. The correct way to produce the oil in the reservoir is to produce Wells A, B, D, and E. As these wells are produced, the gas-oil contact will move downward. When the contact reaches the perforations in Wells B and D, they should be shut in while Wells A and E continue to produce. After all of the oil is produced, the gas is then produced from all of the wells. This is known as blowing down the gas cap.

Solution-Gas Drive

Solution-gas drive, sometimes known as dissolved-gas drive, is the most common oil-producing mechanism. It is a moderately efficient producing mechanism, but the actual production efficiency depends to a great extent on the amount of gas dissolved in the oil, the reservoir pressure, the configuration of the reservoir trap, the producing well spacing, and the manner in which the oil and gas are produced from the reservoir.

Before discussing this producing mechanism, it is necessary to understand how gas dissolves in crude oil. When crude oil and gas are placed in contact with each other under low-pressure conditions, they separate due to gravitational action and their different densities. The gas will exist as a gas on top of the liquid crude oil. As the pressure is increased, part of the gas will dissolve in the crude oil. The amount of gas that will dissolve is a function of pressure, temperature, and the physical and chemical characteristics of the gas and crude oil.

Gas will continue to dissolve in the crude oil under increasing pressure until the pressure reaches the bubble-point pressure of the crude oil, the minimum pressure at which the gas and oil become a homogeneous mixture—neither gas nor liquid. This mixture is conventionally called a fluid.

The reverse action also occurs. Assume that a mixture of oil and gas is at a high pressure (above the bubble point). The pressure is reduced by expanding the fluid. This is analogous to the fluid-expansion producing mechanism. When the pressure declines to the bubble point, some gas will start to come out of solution. At that time there will be two fluid phases: gas and liquid. The gas first appears in the liquid as tiny, microscopic bubbles or globules. As the pressure continues to decline, these gas bubbles enlarge and more gas comes out of solution. This process continues until the pressure is very low and most of the gas that was originally in solution is liberated.

Gas solubility in crude oil is similar to carbon dioxide solubility in common soda water or carbonated beverages, with carbon dioxide dissolved in the flavored fluid. Carbonated beverages are under a slight pressure, which is indicated by the hissing escape of carbon dioxide gas when the container is open. When the open container sits for a period of time, the carbon dioxide will escape and the beverage becomes flat.

The energy in the carbonated beverage is evident when the gas escapes as the container is opened. If the container is resealed and agitated, the fluid will spurt out of the container. If the same action is tried with a container that has been opened for a long time, the fluid will not spurt out. This is because the carbon dioxide—the source of energy—has escaped.

The action of carbon dioxide dissolved in a beverage is analogous to gas dissolved in crude oil. The reservoir pressure in a solution-gas-drive reservoir decreases as oil and dissolved gas are produced from the reservoir. As the pressure declines, dissolved gas is liberated from the oil in small bubbles. These gas bubbles migrate with the fluid into the wellbore and are produced. However, since these bubbles occupy space in the pore channels, there is less space for oil. Oil production reduces correspondingly. As the reservoir pressure continues to decline, more gas is liberated from solution and the amount of oil production is reduced correspondingly. This process can continue until the well is only producing solution gas that has bypassed the oil. Usually when this occurs, the produced volumes are relatively small, so continued production is usually uneconomical.

When the solution-gas-drive reservoir is a thin, flat structure, generally very little can be done to improve maximum recovery from the reservoir except secondary recovery. Completing the wells near the bottom of the reservoir and restricting the production rate may help obtain a maximum efficient recovery.

Improved recoveries can be obtained when the reservoir is relatively thick and if it has room for a gas cap, such as an anticlinal and dipping pinched-out structure. Here, there is room for the gas to migrate vertically toward the top of the reservoir due to its lesser density. The higher-density oil will flow into the wellbore. The rising gas will accumulate in the top of the reservoir and form a secondary gas cap. In this manner the gas is retained in the reservoir so its expanding characteristics can be utilized to help drive the crude oil toward the producing wells. If a well is located in the forming gas cap, it should not be produced until all of the crude oil has been produced. At that time the gas cap is produced in a blowdown operation near the end of the producing life of the reservoir.

The recovery efficiency of solution-gas-drive reservoirs varies widely—from less than 5% to about 30% of the original oil in place. The wide variance depends on the reservoir configuration, reservoir pressure, and other factors that affect the productivity of solution-gas-drive reservoirs.

Gravity Drainage

Gravity drainage, as the name implies, depends on gravity to help drain the oil to the lowest part of the reservoir. Gravity acts on the fluids in the reservoir at all times. However, the force of gravity is normally smaller than the other forces available to produce the hydrocarbons in the reservoir. Therefore, it is usually relatively ineffective.

Gravity drainage is more effective in steeply dipping reservoirs or reservoirs where there is a thick productive section. The effectiveness of gravity drainage depends on the height of the oil column, and these types of reservoirs often have sufficient room for a higher fluid column. Gravity drainage is the only type of producing mechanism available after all of the other types of reservoir energy have been expended, excluding secondary recovery.

Water Drive

A good natural water drive is without question one of the most efficient producing mechanisms to recover oil from a reservoir. Oil frequently accumulates on top of salt water in an oil trap. In some cases the salt water may have considerable volume from an exterior source. This source could be a very large body of salt water that can expand slightly as pressure is released and may be aided in expansion by a small amount of dissolved gas. Another exterior source of the salt water would be a connection with a body of salt water located at a higher elevation to give the U-tube effect described earlier. In some cases the saltwater sand may continue for a long horizontal distance and outcrop at the surface, where it can be recharged with surface water. Shale compaction may be another source of salt water.

When there is an active water drive, the salt water will tend to push the oil updip toward the producing wells. As the oil is withdrawn from the reservoir, the salt water moves to succeeding higher levels in the reservoir, displacing the oil as it moves.

The natural water drive is an efficient producing mechanism because the salt water acts as a driver fluid to move the oil to the wellbore and literally sweeps the oil out of the formation. The oil in the reservoir is maintained at a higher pressure so it loses less solution gas. This action helps the oil to retain a low viscosity so it can move through the reservoir with less difficulty, thus increasing overall productivity and recovery. As the moving oil-water contact approaches the producing wells, the wells will begin to produce larger volumes of salt water and decreasing volumes of oil until it is uneconomical to produce the well.

Sometimes, a reservoir will have a limited water drive. The water drive is limited, so it does not sustain the reservoir pressure. Several actions may be taken in this case. The production can be restricted so that the withdrawal of oil and associated gas from the reservoir equals the volume of water entering the reservoir from the limited water drive. If the water drive is very limited, this procedure may restrict the production rate to an uneconomical level. Another procedure is to inject water into the downdip wells to supplement the natural water drive.

Water drives are sometimes referred to as edge-water drives and bottom-water drives. These usually occur in thicker reservoirs that have an original oil-water contact. As oil is withdrawn from the reservoir, water enters from the sides or bottom. The oil-water contact tends to remain relatively level and moves vertically upward. The oil-displacing mechanism as described earlier is the same in both types of drives.

When a reservoir is being produced under a very strong water drive, sometimes the encroaching water can bypass the oil and channel into the producing wells. When this occurs, it is often necessary to restrict production and take other action to prevent the water from bypassing the oil and reducing the overall recovery from the reservoir.

Sometimes, water will enter a high-productivity well at a point near the oil-water contact. This is known as coning or water coning. Some actions that may help in this case include recompleting the well higher in the reservoir and limiting the production rate from the well.

Sometimes, a tilted oil-water contact occurs. This is slightly similar to water coning but generally occurs over a wider area of the reservoir and is caused by either a stronger water drive in one section of the reservoir and/or several higher-productivity wells in one area. Sometimes, the tilted oil-water contact is referred to as a tilted water table.

Oil recovery from a water-drive producing mechanism depends on the strength of the water drive, the reservoir configuration, the pressure and viscosity of the oil in the reservoir, and the producing well spacing. The average recovery from a good water drive is in the range of 50% of the oil originally in place in the reservoir. These high recovery factors for water drive reservoirs are important. Many solution-gas-drive reservoirs have over 20 million bbl of oil originally in place. With a recovery factor of 5–30%, these fields will produce from 1–6 million bbl of oil and will probably average about 3 million bbl of oil. If these reservoirs had a good natural water drive, they would produce an average of about 10 million bbl per reservoir, or about twice the amount produced by solution-gas drive. This difference is significant.

SECONDARY, TERTIARY, AND ENHANCED RECOVERY

The terms secondary, tertiary, and enhanced oil recovery are sometimes used incorrectly or interchangeably. The industry has not standardized the definitions of these terms, which is why they are sometimes used indiscriminantly. The term "primary recovery" is used to indicate that the reservoir is being produced by the naturally occurring energy originally in the reservoir. Solution-gas drive, natural water drive, and the other methods discussed earlier are primary producing mechanisms.

Secondary recovery techniques were developed when it was recognized that all of the oil in the reservoir was not produced by primary means. Secondary recovery was developed to recover the mobile oil remaining in the reservoir. Recall that solution-gas-drive recovery was about 25% of the oil originally in place (OOIP). If the irreducible oil saturation was 30% and the reservoir originally had 70% oil saturation, then mobile oil saturation would be:

$$(0.70) - (0.25)(0.70) - (0.30) = 22.5\%$$

This represents: $(0.225)(100)/(0.70) = 32\%$ of the oil originally in place, or more than the primary recovery. An old rule of thumb is that successful secondary recovery will equal 75–100% of primary recovery.

The basic secondary recovery methods include gas cycling, waterflooding, repressurization, and sweeping with gas, air, or inert gas. Waterflooding is by far the most common, efficient, and economical method and is used whenever applicable.

All waterfloods are not successful and this, in combination with the large amount of oil in the irreducible oil saturation, led to tertiary recovery techniques. Tertiary recovery was designed to recover any oil left in the reservoir after secondary recovery, including part of the oil in the irreducible oil saturation. The earlier tertiary recovery methods included solvents, steamflood, fireflood (in situ combustion), and miscible flooding.

Many of the early tertiary projects met with only limited success. At the same time efforts were made to complete and produce reservoirs that had very little if any primary production. A number of other recovery techniques were developed or proposed, and some were field tested and later used with varying degrees of success. The terms "enhanced recovery," "improved recovery," and "unconventional recovery" were used to cover part of these processes and to some extent are synonymous with tertiary recovery. The term "enhanced recovery" is gaining acceptance for general usage to mean any type of recovery mechanism after primary recovery.

Enhanced recovery operations should normally be started as early in the producing life of the reservoir as conveniently possible so the reservoir energy can be utilized efficiently. Most of these methods involve injecting fluids into the reservoir to displace the oil and push it to the producing wells. The oil viscosity is normally at a minimum value early in the producing life of the reservoir and increases as gas is evolved from the oil. A lower-viscosity oil can be displaced more efficiently than a higher-viscosity oil. The cost of initiating the enhanced recovery project may be reduced if the requirements are considered when the reservoir is developed. And the reservoir can be depleted in a much shorter period if the enhanced recovery project is initiated early in the reservoir producing life.

VOLUMES

The volumes of oil and gas in the reservoir must be known for several reasons. First, there must be a sufficient amount to justify completing the well. If the reservoir has a large volume and the well has a low productivity rate, then steps must be taken to increase the producing rate if possible. The volume of oil and gas in the reservoir helps determine development well spacing, i.e., the distance between wells and the part or fraction of the total reservoir that can be produced economically by one well. The type of secondary-enhanced recovery will depend in part on the volume of oil and gas in the reservoir at the end of the primary producing life. The owner-operator needs to know how much oil and gas is in the reservoir and how much can be recovered over a period of time so future income and expenditure can be budgeted.

OIL RESERVOIR

Oil in place and recoverable reserves should be calculated for each reservoir to determine if an area can be produced economically. For example, an oil field has 960 acres and is developed with 24 wells on 40-acre spacing (40 acres/well). Drilling, well logs, coring, formation tests, and reservoir fluid analysis have given the following data:

Reservoir depth, ft	7,000
Thickness, ft	20
Pressure, P, psi	3,220
Porosity, ϕ, %	25
Water saturation, S_w, %	30
Oil gravity, °API	35
Gas gravity, air = 1.00	0.9
Gas-oil ratio, GOR, cu ft/bbl	800
Reservoir temperature, °F	190
Formation volume factor, B_g, formation bbl/stock-tank bbl (stb)	1.45

The formation volume factor is obtained from industry tables using the listed data.[5] It relates the combined volume of oil and dissolved gas in the reservoir to a barrel of gas-free oil at the surface—often referred to as a stock-tank barrel. The apparent reduction in oil volume is termed shrinkage.

The basic reservoir volume unit is acre-feet and is used for convenience and to reduce the overall size of the numbers. An acre-foot is the volume of one acre 1 ft thick (43,560 cu ft). This same unit is commonly used to measure large volumes of water, such as a lake behind a dam.

[5]M.B. Standing, *Volumetric and Phase Behavior of Oil Field Hydrocarbon Systems* (New York: Reinhold Publishing, 1952).

The original oil in place (OOIP) calculated in stock tank barrels per acre-foot (stb/acre-ft) is calculated from the above data as follows:

$$\text{stb/acre-ft} = (43{,}560 \text{ cu ft})(\phi)(1 - S_w)/(B_g)(5.62 \text{ cu ft/bbl})$$
$$= (43{,}560)(0.20)(1 - 0.30)/(1.45)(5.62)$$
$$= 748 \text{ bbl}$$

Remember that these are barrels in place, not necessarily recoverable barrels. Note that the term $(1 - S_w)$ is equal to the oil saturation (S_o), and that could have been used. It is assumed that the void space (pore volume) in the reservoir that is not filled with water contains oil or gas, in this case oil. The gas dissolved in the oil is considered part of the oil in the reservoir. The number 5.62 is used to convert cubic feet to barrels (42 gal/bbl), the common unit of measurement.

The next step is to calculate the volume of oil in place per well. The wells are on a 40-acre spacing and the reservoir is 20 ft thick, so each well has 800 acre-ft (40 acre × 20 ft). The oil in place per well is then stb/well = (748 bbl/acre-ft)(800 acre-ft/well) = 598,400 bbl.

The original oil in place (stb) in the entire reservoir is calculated in a similar manner: stb/reservoir = (598,400 bbl/well)(24 wells) = 14,361,600 bbl. It can also be calculated by acre-feet: (960 acre/reservoir)(20 ft thick)(748 bbl/acre-ft) = 14,361,600 bbl.

The amount of oil that can be recovered cannot be calculated at this time because of the unknown irreducible oil saturation and the efficiency of the oil recovery mechanism. It can only be estimated, although experienced geologists and engineers can usually make accurate estimates. Table 3–1 lists approximate primary recoveries.

Table 3–1 Oil Recovery

Type of Reservoir Drive	Recovery Efficiency, %	Oil Recovery, bbl × 1,000	
		Per Well	Total Reservoir
Solution-gas drive			
Low	10	60	1,436
High	30	180	4,309
Gas-cap expansion			
Low	20	120	2,872
High	40	239	5,745
Water drive			
Low (limited)	35	209	5,027
High (full)	60	359	8,617

Actual recoveries depend on various reservoir and fluid characteristics. If secondary or enhanced recovery procedures are applicable, they should be started as soon as possible. Secondary-enhanced recovery will be approximately equal to primary recovery up to a total of about 75% of the original oil in place. The recoveries listed in Table 3–1 are low in terms of recovery efficiency but are probably realistic. As another point, the recovery efficiency is referenced to the total volume of oil in the reservoir. Before enhanced recovery the irreducible oil saturation is on the order of 30%. Therefore, this reduces the amount of mobile oil available to be produced by primary means so the recovery efficiency in terms of mobile oil would be considerably higher.

The well producing rate must be determined by actual production tests. Again, experienced geologists and engineers may make good estimates, but actual rates must be determined by production testing. As the well is produced, records of produced volumes (oil and gas) and pressures are maintained and plotted on graphs (Fig. 3–15). After sufficient production history the production curves can be extrapolated—projected or extended for the same trend—to project how much oil will be recovered in the future. The accuracy of these curves increases with increasing production history.

A reservoir material balance is another method used to calculate the volume of oil and gas in the reservoir. The reservoir fluid properties and pressure are analyzed early in the producing life of the reservoir and again after 15–25% of the primary recovery is produced. These analyses and the volumes produced are used to calculate the oil and dissolved gas in the reservoir.

Computer modeling is also used to calculate oil in place. All of the parameters that affect production can be entered into a computer in the form of complex equations. Well production can then be simulated on the computer, and the resulting volumes, pressures, and other parameters can be compared with the actual well and reservoir performance. The computer data are modified until the results fit or match the well performance—sometimes called history matching or reservoir simulation.

One or a combination of the above methods will give a good value of the oil in place early in the producing life of the field and will permit the operator to plan for secondary-enhanced recovery methods.

Substantial amounts of gas are often produced with the oil. The amount of gas produced can also be calculated with the material balance and computer modeling techniques. Before this, the best estimate is the gas-oil ratio (cu ft/bbl) multiplied by the projected oil production.

The amount of gas that will be produced can be calculated by estimating the formation volume factor at depletion. Assume solution-gas drive, 30% recovery, and a formation volume factor of about 1.05 at depletion. The dissolved gas in the

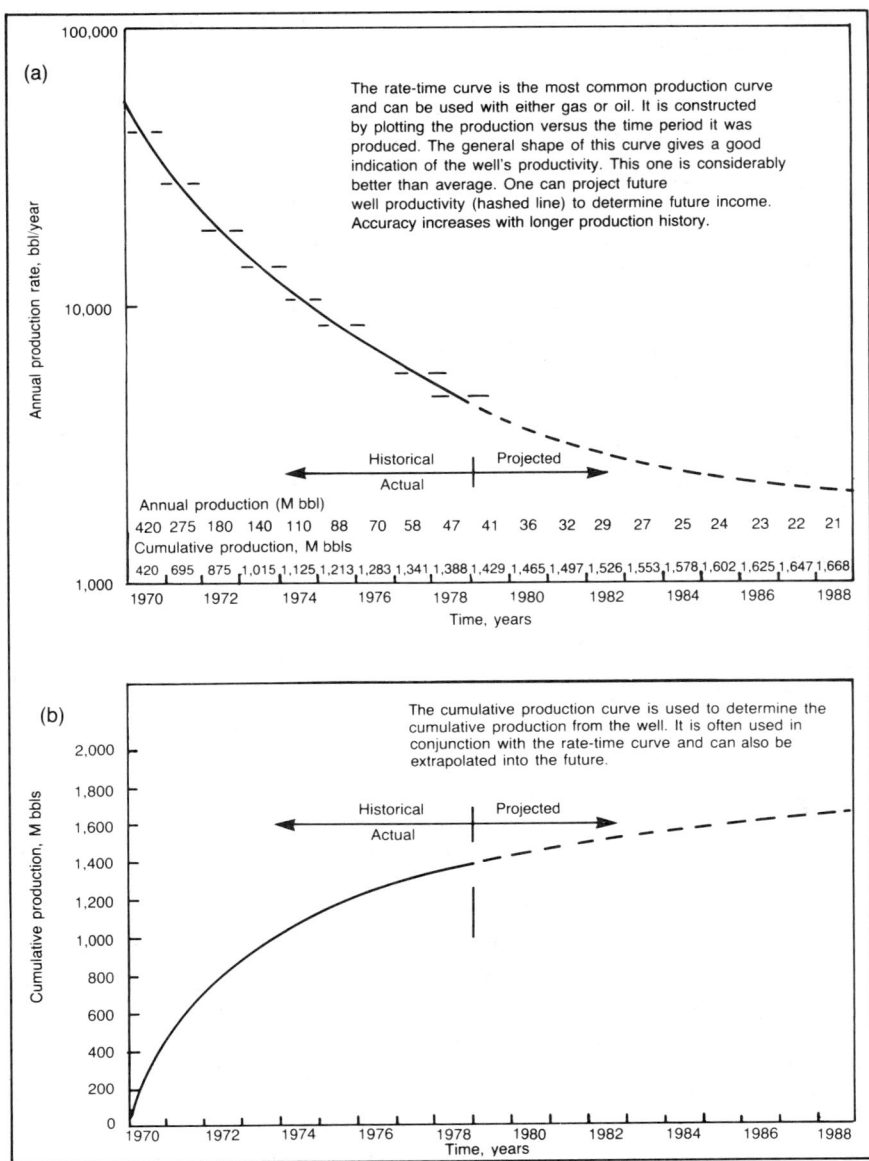

FIG. 3–15 (a) Rate-time curve for oil production and (b) cumulative production curve

reservoir will be about 250 cu ft/bbl. The amount of gas produced per well is calculated as follows:

Gas originally in place:

$$(598,400 \text{ bbl})(800 \text{ cu ft/bbl}) = 478,720,000 \text{ cu ft}$$

Oil in place at depletion:

$$(598,400 \text{ bbl}) - (180,000 \text{ bbl}) = 418,400 \text{ bbl}$$

Gas in place at depletion:

$$(418,400 \text{ bbl})(250 \text{ cu ft/bbl}) = 104,600,000 \text{ cu ft}$$

Produced gas is equal to original gas in place less gas in place at depletion, or 374,120,000 cu ft. The lifetime producing gas-oil ratio is (374,120,000 cu ft)/(180,000), or 2,078 cu ft/bbl.

The reservoir originally contained 800 cu ft/bbl. The high producing gas-oil ratio is caused by gas coming out of the solution from all of the oil in the reservoir.

Gas Reservoir

Various combinations of gas and oil can be produced from the reservoir. If the principal production is oil, the reservoir is considered an oil-producing reservoir and the oil and gas volumes are calculated as described in the previous section. When the principal production is gas, the reservoir is called a gas reservoir. If relatively high volumes of liquids are produced with the gas, it may be called a gas condensate (gas liquids) or rich gas reservoir.

Gas gravity is the density or weight of the gas. It is measured and reported as a decimal fraction of the weight of an equal volume of air where both are measured at standard conditions. As mentioned, standard conditions are defined as an absolute pressure of 14.65 psi, 1 atm at sea level, and a temperature of 60°F. Average produced dry-gas gravities are in the range of 0.60–0.70, i.e., 60–70% of the weight of an equal volume of air.

Gas is highly compressible, and the gas volume is affected by pressure and temperature. Therefore, gas volumes are always reported at standard conditions. To illustrate the compressibility of gas and assuming standard pressure and temperature gradients, the gas contained in 1 cu ft of reservoir pore volume at 5,000 ft would expand to 159 cu ft at the surface (standard conditions). At 10,000 ft the equivalent would be 233 cu ft, and at 20,000 ft it would be 301 cu ft.

Gas volumes are normally measured in cubic feet. Since very large numbers are often encountered in calculating these volumes, it is common practice to use abbreviations as follows:

Thousand cubic feet, Mcf
Million cubic feet, MMcf
Billion cubic feet, Bcf
Trillion cubic feet, Tcf

Gas and oil are sometimes combined on an equivalent energy basis for reporting purposes. One barrel of oil contains about 5.6 million British thermal units (btu). These are standard units of heat measurement. A cubic foot of gas contains about 1,000 btu's, and 5.6 Mcf of gas contains 5.6 million btu's or about the same amount of btu's as in a barrel of oil. For example, a reservoir may contain 1 million bbl of oil and 500,000 Mcf (500 MMcf) gas. The 500,000 Mcf is equivalent to 89,286 bbl of oil, based on 5.6 Mcf/bbl of oil equivalent. Therefore, the reservoir would contain (1,000,000 + 89,286) or 1,089,286 bbl of oil and oil equivalents.

As an interesting note and to bring these numbers into perspective, the average home in the U.S. that uses gas for space heating and water heating uses an average of 500 cu ft of gas per day or 183 Mcf annually. This is approximately equal to 33 bbl or 1,372 gal of oil equivalents annually.

The formula used to convert reservoir gas volumes to volumes at standard conditions is:

$$V_s = (V_r)(P_r)(T_s)/(P_s)(T_r)(Z)$$

where:

V_s = volume at standard conditions
V_r = volume at reservoir conditions
P_s = pressure at standard conditions (14.65 psi)
P_r = reservoir pressure, psia
T_s = temperature, °R at standard conditions (460 + 60 = 520°R)
T_r = reservoir temperature, °K (460 + °F)
Z = compressibility factor (obtained from standard charts)

Take the example reservoir used earlier to calculate oil volumes and replace the oil with gas. The 70% oil saturation $(1 - S_w)$ will be called gas saturation (S_g). The gas compressibility factor from standard tables is 0.92 for 0.6 gravity gas. The gas in place per acre-foot at standard conditions is calculated as follows:

V_r (per acre-foot) = (43,560)(0.20)(0.70)/(1,000)
           = 6.098 Mcf
V_s = (6.098)(3,220 + 14.65)(460 + 160)/(14.65)(460 + 190)(0.92)
   = 1,170 Mcf/acre-ft

The reservoir pore volume per acre-foot (V_r) was first calculated in Mcf (1,000 in the divisor). The volume at standard conditions was calculated by substituting actual pressures and temperatures into the formula.

Gas wells in this case would be drilled on at least 160-acre spacing or possibly 320-acre spacing. The gas in place would be calculated for the entire reservoir (960 acres, 20 ft thick or 19,200 acre-feet):

$$\begin{aligned} \text{Gas in place} &= (\text{Mcf/acre-ft})(\text{acre-ft}) \\ &= (1{,}170 \text{ Mcf})(19{,}200) \\ &= 22{,}464{,}000 \text{ Mcf (or } 22{,}464 \text{ MMcf or } 22.5 \text{ Bcf)} \end{aligned}$$

The next step is to calculate the recoverable gas. In normal gas-field operations, compressor suctions operate down to about 40 psi. This is equivalent to a reservoir pressure of about 50 psi at the wellbore, allowing for the hydrostatic head (gas gradient) of a 7,000-ft column of gas. The average reservoir pressure at depletion would be about 700 psi after allowing for pressure differentials from the wellbore to the edge of the reservoir. The reservoir volume per acre-foot at depletion is:

$$\begin{aligned} V_s &= (6.098)(700 + 14.65)(460 + 60)/(14.65)(460 + 190)(0.95) \\ &= 251 \text{ Mcf/acre-ft} \end{aligned}$$

Total reservoir volume at depletion is $V_s = (251)(19{,}200) = 4{,}819{,}000$ Mcf (rounded off to thousands). Gas recovery, the gas originally in place less the gas in place at depletion, is (22.5 − 4.8), or 17.7 Bcf. Recovery efficiency is (22.5 − 4.8)(100)/(22.5), or 78.7%.

This is an average recovery for a good gas reservoir. It may be possible to produce the reservoir to a lower pressure. However, this would probably require extended time and possibly would cause some problems as a result of water production—often a problem later in the life of a producing gas well.

Gas condensate, light hydrocarbon fluid, may be produced with the gas if it occurs in the reservoir. This is conventionally measured in gallons per thousand cubic feet of gas produced (gal/Mcf) or in barrels per million cubic feet of gas (bbl/MMcf).

It is not uncommon for a good gas-rich reservoir to produce 2 gal of condensate per thousand (2 gal/Mcf), which is equivalent to 48 bbl per million (48 bbl/MMcf). Assuming that the reservoir produces this amount, total condensate production would be (48 bbl/MMcf)(4,819 MMcf), or 231,312 bbl.

Condensate is light hydrocarbons in the range of the hexanes plus (C_6H_{14} plus). Liquid petroleum gas (LPG) is the ethane through pentane series, although there may be some pentane in the condensate. LPG should not be confused with condensate. Condensate separates naturally from the gas in the surface producing facilities. LPG is normally separated by refrigeration processes in specially designed LPG stripping plants located on the main gas lines. The LPG content of natural gas is measured in gallons per thousand cubic feet of gas (gal/Mcf) and is normally in the range of 0.5–3 gal/Mcf.

Reserve calculations for gas reservoirs, like those for oil reservoirs, are more accurate as more data are obtained. Gas reserves can be calculated by a material balance. Rate-time and cumulative production graphs are used similar to oil production. Gas in place can also be projected accurately after about 20% of the reservoir volume is produced by using P/Z curves (Fig. 3–16). The reservoir pressure, P, is measured periodically during production. P, divided by the compressibility factor, Z, at the pressure being used is plotted as the cumulative gas volume produced at the time the pressure is taken. These points plot as a straight line on regular coordinate graph paper. If the straight line is extrapolated (extended) to a pressure of 0 psi, this will give the total volume of gas originally in place. The abandonment pressure is determined, and the intersection of the two lines gives the cumulative recoverable reserves as long as conditions are constant. About 20% of reserves should be produced to give an accurate estimate.

An interesting comparison can be made between the oil and gas reservoirs discussed above—both of which had the same reservoir volume. The method is to compare the oil equivalents produced by each. Assuming a 30% recovery, the oil reservoir produced 4,309,000 bbl of oil and 374,120 Mcf of gas.

$$\text{Oil equivalents} = 4{,}309{,}000 + (374{,}120)/(5.6)$$
$$= 4{,}376{,}000 \text{ bbl (rounded to thousands)}$$

The same size reservoir as a gas field produced 17,645,000 Mcf gas and 231,312 bbl of condensate.

$$\text{Oil equivalents} = (17{,}645{,}000)/(5.6) + 231{,}312$$
$$= 3{,}382{,}000 \text{ bbl (rounded to thousands)}$$

Therefore, the oil field in this case produced more energy (btu's). If secondary-enhanced oil recovery is obtained, the oil field will produce substantially more energy.

Reservoir analysis and a good understanding of the formations is fundamental to the successful drilling-completion-production process. Information on the formations and hydrocarbon reservoirs is obtained from drilling information, core and formation, test data, and mud and well logs. The geologist and engineer must know which types of data to collect, how to analyze the data, and how to use the results. The data can be used to construct pore pressure plots that predict high-pressure and low-pressure zones deeper in the hole. An advance warning of these allows the operator to take the necessary precautions to prevent lost circu-

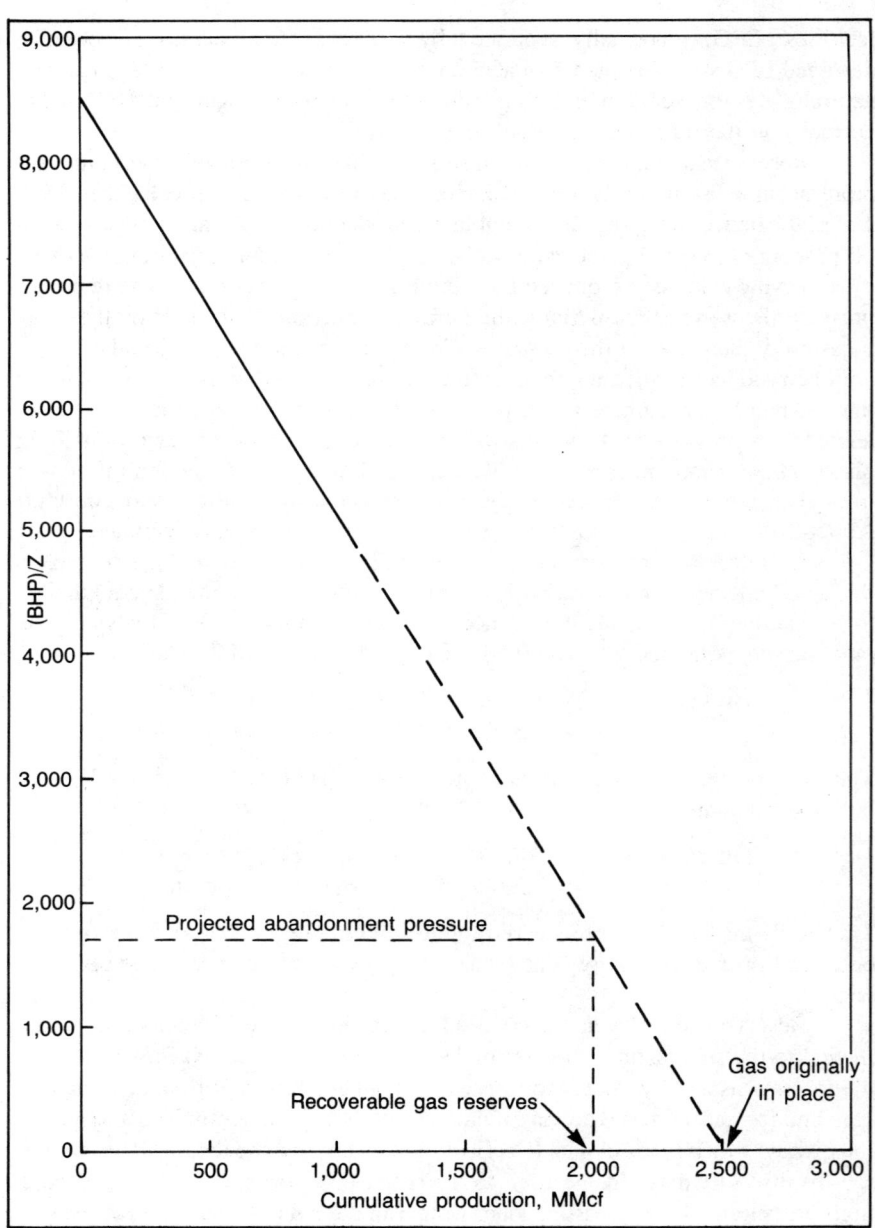

FIG. 3–16 Bottom-hole pressure/Z curve

lation or a blowout, which normally causes appreciable losses. The well logs are important because they provide data on the formations that frequently cannot be obtained by any other means. The physical properties of the formations, porosity and permeability, and the pressure and physical properties of the fluids in the pore space must be known in order to drill the well in the most efficient manner.

Fluid flow in the reservoir is controlled by the permeability of the rock and the pressure and viscosity of the fluids. These must be considered during drilling. They are very important for a successful completion and also help determine well spacing for future development of the reservoir. Producing mechanisms such as fluid expansion, gas-cap drive, solution-gas drive, water drive, and gravity drainage depend on the pressure, type, and nature of fluids in the pore space. The reservoir energy must be conserved in order to obtain the maximum oil and gas production from the reservoir in the most efficient manner. Well spacing and completion and production practices are important factors in conserving reservoir energy.

The volumes of oil and gas in the reservoir are calculated by volumetric calculations, material balance, and reservoir modeling and by projecting various types of production graphs. The recovery efficiency is the percentage of the oil and gas originally in place that is recovered. The actual recovery efficiency obtained depends on the natural producing mechanism and how well the reservoir energy is conserved. The natural reservoir energy is expended during the primary recovery period. This energy may be supplemented to obtain additional oil recovery by secondary-enhanced recovery techniques.

In summary, the primary purpose for drilling the well is to complete a commercial producing well. In order to accomplish this successfully, the operator must have a good understanding of the formations and reservoir rocks, fluid flow in the reservoir, producing mechanisms, reservoir volumes, recovery efficiencies, and secondary-enhanced recovery techniques.

CHAPTER 4
DRILLING PROSPECTS, PROGRAMS & PROCEDURES

Natural geological processes have formed hydrocarbons and have trapped them in the reservoir rock. A well must be drilled to determine if the trap is productive, i.e., if the oil and gas in the pore spaces will move through the reservoir to the well. Before drilling can commence, however, the prospect must be developed and rights must be acquired for drilling.

DRILLING PROSPECTS

The term *drilling prospect* refers to a point or tract of land whose underlying formations are expected to contain commercially productive hydrocarbons. Drilling prospects are divided into two general categories: *development* and *exploration* (wildcat) prospects. The true development prospect is a low-risk venture for an infield or onstructure well that is close to producing wells. The wildcat prospect is located in an unexplored area—usually on a geologic structure that has not been tested or drilled. It is generally a high-risk venture.

In practice, drilling prospects may not be this well defined. Depending on the operator's definition, the prospect can be a true infield development well located between two adjacent producing wells, an onstructure or drill-deeper prospect on the same structure, a step-out or field extension well located several miles from existing production, or a rank wildcat on an untested structure.

Evaluating the risk of drilling prospects also can be difficult. For example, an infield development well is normally considered a low-risk prospect. However, fields with pinchouts, faults, or other geological features that make the prospect questionable may be very high risk even in development drilling. Therefore, the practice of locating and evaluating prospects has become a highly competitive science.

Source of Prospects

Most companies that conduct exploration and drilling activities have three main departments involved in generating prospects: geology, land, and engineering. Depending upon the size of the company, the departments may be subdivided into different areas and activities. For example, the geology department may be subdivided into exploration and development (exploitation) geologists. The geology department may also have a geophysics group that specializes in seismic exploration. Land and engineering departments may have similar subdivisions. The engineering department also may be subdivided into drilling, production, and reservoir engineering.

The geology, land, and engineering departments work closely together, coordinating their activities to generate most drilling prospects. The special training and experience of these people are required to generate prospects, evaluate them correctly, present them adequately, and perform the subsequent drilling and completion operations. Sometimes, smaller companies do not have clearly defined departments. However, they have personnel with similar training and experience.

Geology Department   Geologists are frequently assigned to geographical areas. They monitor drilling operations and exploration activities of other companies in the area. They also work the area, studying geological conditions such as sand trends, investigating new and old wells, and compiling all other available data. The geophysical group conducts seismic exploration activities and works with other companies in running regional seismic lines in areas of mutual interest (group shoots). Geochemical surveys and other exploration techniques may also be used.

These studies locate structures, possible stratigraphic traps, anomalies, anticlines, fault traps, and other geological features that may indicate areas of potential hydrocarbon accumulation. Geologists concentrate on these areas of interest. They study all available information, conduct additional surveys such as running detailed seismic lines on close spacing, and build maps as needed to prove or disprove the existence of an oil trap.

Computerized data banks and data processing programs often enhance geological studies. Well data recorded on computer tapes includes information such as the dates the well is begun and completed, bit and casing data, formation depths and thicknesses, and production test data. Any of this information can be retrieved in tabular form or illustrated on various kinds of maps, including structure, isopach, and well productivity maps. Most programs also can draw maps with various degrees of smoothing, construct residuals, or make isometric projections. the information can be presented in many forms and on a variety of scales.

Most prospects are generated by the geology department, and support is often requested from other departments to explore and develop the area further to determine if it should be recommended for drilling.

Land Department The land department also observes general exploration and drilling activities and maintains contact with other land departments. It also may generate prospects through trades or joint ventures with other companies. When general exploration activities increase significantly in one area (a hot area) that appears to have considerable potential, the land department will attempt to obtain land there. It usually does this by taking farmins (acquiring land position from another company), forming joint drilling units or ventures (jointly drilling the well in participation with one or more companies), and leasing acreage directly from the landowner.

The land department is trained and experienced in the legal complexities of exploration, drilling, and production activities. It searches titles to ensure there is a clear title to the property. Landowners are contacted to lease acreage, grant road and pipeline rights-of-way, and negotiate settlement of damage and other claims. It negotiates trades and joint venture operating agreements covering trades. Federal, state, and county regulatory agencies are contacted to obtain leases, establish federal units, and negotiate and write proposed development plans, unit operating agreements, and other required documents.

Engineering Department The engineering department prepares drilling programs, costs, and reserve estimates and usually supervises or aids the operating department in supervising the drilling and completion of the well. It also aids in the economic evaluation of drilling prospects.

When preparing drilling programs for a drilling prospect, an engineer will study wells in the area. His research may turn up formations that had productive potential but were not tested. He may decide that new well stimulation techniques justify completing zones in older wells that were not commercially productive with older completion techniques. His production and reservoir studies may indicate that additional wells are needed to obtain the maximum economic production from the reservoir. Pressure-buildup tests and reservoir limit studies may imply field extensions and justify drilling additional wells. These observations may provide additional data to verify the prospect.

Other Sources Most drilling prospects are worked up or generated as noted. If the prospect is found and drilled within one company, it is internally generated. If the prospect is submitted to another company, the company receiving the submittal considers it externally generated.

Small companies, independents, and consultants may generate a prospect but may not have adequate funds to drill the well. Nevertheless, the groups follow regional exploration activity. When activity increases in an area, they will lease acreage and offer it to larger companies with sufficient funds for exploration. In these cases the offering company usually receives a cash consideration and retains an interest in the prospect.

Many wells are promoted, and the company offering the well receives promotion, and becomes a promoter. Promotion means paying a disproportionate share of the costs to participate in these activities. For example, a prospect offered on a promotional basis might be accepted by a company that would pay ¾ of the cost to earn a ½ interest. That is, the company pays a disproportionate share of the cost. Since the offering company has invested money to work up the prospect and acquire a land position, it is reimbursed by this method.

Many types of promotion are used. A company may offer one or more prospects to a group of investors on a promotional basis, frequently called a *drilling package* or *drilling fund*. The offering company invests time and money in a staff with the expertise to develop the prospect, acquire prospective acreage, and drill the wells. It is justified in receiving compensation through promotion.

Prospects are acquired from other sources outside the company. Areas of interest may be established with other companies for exploration. Leasing may be done directly or through a third party, such as a lease broker, with both companies bearing the leasing costs according to a predetermined agreement. Joint ventures may be formed when two or more companies combine their acreage and select an operator to explore an area under the terms of a joint venture agreement.

Land departments conduct lease checks or lease take-offs, of an area to determine who owns the acreage. This provides the preliminary information needed to determine if sufficient acreage can be acquired to cover a prospect if it is developed.

Drilling prospects—especially wildcat prospects—are normally drilled on substantial blocks of acreage where the owner has leases on all or most of the acreage. This is done for two reasons. Sometimes the oil trap will cover a large area because the geological information is not precise enough to determine which part is productive. A wildcat well drilled in this area is expensive, normally costing two to three times more than a comparable development well. Second, the operator needs to have all or a substantial part of the prospective area leased so there will be sufficient room to drill development wells if the wildcat is successful. This is the best way for the operator to earn a fair return on investment.

An operator—who supervises and performs the drilling operation—will drill many dry holes. Therefore, the successful prospect must pay for what it costs as well as for the costs of acquiring and drilling unsuccessful prospects.

PROSPECT SUBMITTAL

The drilling prospect submittal is the conventional manner to initiate drilling a well. The submittal may range from a short explanatory letter to a formal presentation complete with area and well location maps, land ownership, geological justification for drilling the well including maps and cross sections, objective (prospective productive) horizons and depths, drilling program and drilling costs (AFE–Authority for Expenditure), projected recoveries of gas and oil (reserves), and all other items relating to the prospect.

The completeness and detail of the prospect submittal depends in part on the type of prospect. A development prospect is usually relatively simple and straightforward. For example, a submittal for an infield well located between two producing wells could be relatively brief. The prospect would be justified easily, based on the production history of the offsetting wells. The costs would be easily determined, again based on the offsetting wells.

Wildcat prospects are more complex, especially the deeper and more expensive exploration wells. Additional documentation is needed to justify these prospects because of the higher degree of risk and the larger capital investment. A wildcat well will cost two or three times more than a comparable development well drilled to the same depth in similar formations. Therefore, the geological data should provide specific information; the drilling program and cost must be detailed and complete. Operating costs of the completed well and methods of marketing the hydrocarbons should also be considered. Any applicable rules and regulations of the federal, state, and county regulatory agencies are included.

The drilling prospect submittal is of primary importance since it is the document that can initiate expenditures from thousands to millions of dollars. Therefore, it should be given detailed, careful consideration both in preparation and presentation. The data must be compiled, studied, and presented clearly and concisely with complete and supporting details that can be investigated and verified by all interested parties.

Economics and Evaluation

The basic reason for evaluating a prospect is to determine if it will make a profit for the owner. It must pay out and generate a rate of return consistent with the financial objectives of the owner. Often, the owner has a number of prospects that require expenditures in excess of available capital. Economic evaluation provides one method for the owner-operator to determine which prospects should be selected for participation, the amount of that participation, and those prospects that should be deferred or rejected.

Several economic criteria evaluate a drilling prospect. Capital investment is the total amount of investment required to drill and complete a well usually in cash dollars. Future net revenue is the total future gross revenue less operating expenses, usually less state and local taxes (ad valorem, severance, business, and occupation). The discounted future net revenue is the total future net revenue discounted by some annual percentage rate, usually taken as the cost of money. The cost of money is normally considered as the interest rate paid on borrowed funds and is often $1/2-1 1/2\%$ over the prime rate. Payout is the time in years from the start of the project until the discounted future net revenue equals the capital investment. Sometimes, the future net revenue is used, but the more correct procedure is to use the discounted future net revenue. Present worth, sometimes called the present value, is the discounted future net revenue less the discounted capital investment. The rate of return generally is the average annual interest rate that the project returns on the initial capital investment; however, other methods are available for calculating this rate.

Each company gives different weight or emphasis on factors used to evaluate prospects. Companies with a large amount of capital may look at larger prospects and those with a higher present value or rate of return. Other companies may be short of funds and look for smaller-sized prospects with a faster payout. Growth-oriented companies may place a strong emphasis on current income. They may select prospects that have a high early income. Usually, the companies will look for some balance between the various factors used for evaluation.

Economics and evaluation are closely related and are interdependent. Economics usually refers only to the profitability of the single well or prospect. Evaluation uses economics as a basis and considers risk and other intangible items in determining whether a prospect is profitable and if funds to be expended are within the owner's objectives to earn the greatest return.

Economic analyses are normally run on computers. This eliminates tedious, time-consuming hand calculations. The possibility of error is reduced, and various alternative cases can be run for different possible situations to provide a more detailed analysis.

Risk evaluation is one of the more important steps in an economic analysis. It too frequently is overlooked or underestimated or is a vague procedure that leaves too much latitude for interpretation and argument. There are many methods of evaluating risks, and generally they have one common denominator: all available information about the prospect must be obtained. This is important since it could uncover facts that would prove or disprove the prospect. To some degree this may account for the success of select methods of evaluation.

Risk includes both the risk of successfully drilling the well to total depth and completing it for the projected cost and the risk of encountering a reservoir that can be produced in accordance with the projected schedule of production in the economic analysis.

One simple and effective method to evaluate risk is to apply it as a percentage chance of success (fraction) to the investment cost. The obvious difficulty or disadvantage in this procedure is determining the chance of success. The originator of the prospect is the first logical choice to evaluate the risk since he has studied the prospect and should be knowledgeable. However, he has probably evaluated the project by recommending it, so he may be too close to it. He should have a vote but not the final decision. Others knowledgeable in the area should make the final decision on risk. Upper-level management usually makes the final choice; but unless it is intimately familiar with the area, its decision on risk should be guided by experts.

A number of other considerations, many intangible in nature, may affect the evaluation of the prospect. For example, a gas prospect may be located near a company gas transmission line. This prospect might be selected over other prospects even though it offers a lower rate of return because the company may be able to generate additional revenues by transporting the gas in its transmission system. Similarly, prospects located near a company transmission system may be rejected because the company plans to discontinue the use of the line or sell the system. A prospect with a low deliverability (productivity) and less-favorable economic parameters but with high reserves may be selected because it would increase the company's reserve position.

Management may elect to divide capital expenditures between development and exploration drilling. Expenditures may be further divided into geographical areas. The investment capital also may be divided according to risk with a certain percentage to low-risk projects and the remainder to high-risk projects. All of these factors enter into the evaluation of the prospect.

The company's federal income tax position may be a major consideration in the economic evaluation. Prospect economics are frequently run based on results before federal income taxes (BFIT) and after allowance for federal income taxes (AFIT). For evaluation purposes these taxes should be and usually are included. The tax rate used in the evaluation should be the actual or estimated rate expected to be paid in the years that the investment is made and income is generated. This may be difficult, but the best estimate should be used.

In some cases part of the initial capital expenditure may be entered as a tax credit, depending on the tax position of the company, and thus enhance the economics. Tax carry-forwards may affect the economics and subsequent evaluation. The overall tax position depends on a number of factors that require a special evaluation. Therefore, tax effects must be evaluated by personnel who are knowledgeable in taxes, project evaluation, and analysis.

Leases and Trades

In the United States property can be owned by individuals, companies, states, countries, municipalities, and the federal government. The original surface landowner—whether an individual or a government—is considered to be the original owner of the oil and gas minerals (the mineral owner). In some cases the ownership of the oil and gas minerals may have been separated (severed) from the surface ownership in a prior transaction. In this case the mineral owner normally retains the right of surface entry to explore for and develop the minerals, subject to paying the surface owners a fair compensation for damages. In a few cases the surface owner may sell or lease the oil and gas minerals with a nondrilling clause that expressively forbids the mineral interest owner from having access to the surface. This type of lease or mineral interest ownership is often used where the surface cannot be changed, such as a highway or industrial complex. In these cases the mineral interest owner or lessor must drill a directional or angled hole under the property from an adjacent tract to obtain the hydrocarbons, or he must drain the reservoir from a well near the property.

The standard procedure when oil and gas leases are taken is to make a cursory check of the title to be reasonably sure that the lessor (the party owning the mineral rights) owns the property and has the right to lease it to the lessee (the party obtaining the lease). A detailed title check would be prohibitively expensive, especially for companies that have very large lease holdings. However, before the well is drilled the common practice is to obtain a title opinion on the wellsite. This includes a detailed investigation of all records of ownership. A title abstract is either prepared or an older abstract is updated. The title opinion is obtained to verify ownership of the land before the well is drilled. If the well were drilled on another owner's land, the other owner could take possession of the well. Therefore, the title is researched in detail to ensure that the operating company has legal title to the property.

Drilling Prospects, Programs, and Procedures

When a property is leased for oil and gas, the party obtaining the lease (the lessee) purchases or rents the right to explore for, develop, and market the oil and gas from the owner (the lessor), subject to the terms and conditions of the lease agreement. The lessee normally pays a per-acre bonus for the lease and an annual rental. Bonus payments can be very high in special, highly prospective areas but normally are $25–$250/acre. Rentals are normally $1–$5/acre/year.

Most leases have a primary term of 3–5 years and can be extended for a secondary term of 3–5 years by paying an additional bonus. During the primary and secondary terms of the average lease, the lessee can conduct operations at his election. However, at the end of the secondary term (or primary term, if there is not a provision for a secondary term), the lessee often can hold the lease only by either continuous operations or production. Continuous operations are usually defined in the lease agreement and normally allow the lessee 60 days from the time one well is completed to spud or begin the next well.

In older leases if production was encountered, the lessee could often hold the entire lease as long as a well produced. In newer leases the trend is to let the owner reserve only a normal well spacing around the producing well. The rest of the acreage reverts to the mineral interest owner (lessor) unless it is held by continuous operations.

Conventionally, the mineral owner or lessor receives a $1/8–3/16$ royalty interest. The royalty interest is negotiated between the lessor and the lessee; in some cases higher royalties are reserved. The lessor (royalty interest owner) receives his royalty share (in this case $1/8$ or $3/16$) of the market value of the oil and gas produced from the property. He does not pay any exploration, development, or operating costs but may pay his share of the product transportation costs (e.g., trucking). He is also responsible for all taxes on the sale of his share of the oil and gas. The royalty interest owner normally does not have any control over operations unless otherwise stated in the lease agreement.

The lessee's or owner's retained ownership in the oil and gas (e.g., $7/8$ or $13/16$) is sometimes referred to as net revenue interest or mineral ownership interest. Since the lessee must pay all costs and all taxes on installations and equipment, this is effectively 100% of the cost; so the lessee has a 100% ($8/8$) working interest.

Sometimes, an oil and gas lease will have an overriding royalty interest (ORRI), also called an override which is similar to a royalty. Overrides range from as low as $1/64$ to $1/8$ or more. The overriding royalty interest usually carries the same terms and conditions as a royalty interest. When a lease has an overriding royalty interest, it usually indicates that the lease has changed hands or has been sold at least once. It is common practice for a lessee to sell a lease to another party and retain an override.

Various types of lease trades are made. One of the most common is a farmout. In a farmout the lessee sells the lease rights to another party to drill the well or otherwise exploit the lease. Various types of farmout terms are negotiated, and often the party selling the lease—making the farmout—retains an override. The prospects could be farmed out to another party to drill the well, and the original owner would reserve an override—possibly with some type of back-in or conversion. For example, the seller might reserve a $1/16$ override with a back-in of a $1/16$ override when the purchaser had recovered 200% of his costs (i.e., at two times payout). In this case the seller would receive $1/16$ of the oil and gas until the purchaser had recovered 200% of his costs, at which time the seller's override would convert, back-in, or escalate to a $1/8$ ORRI.

Conversions can be triggered on income, production, or both. Sometimes a second conversion provision may be included. The seller may reserve an override that later converts to a working interest position, or he may have the option to convert an original override into a larger override and a working interest position where he pays part of the costs and receives a larger percentage of the income.

In another type of trade, the seller may reserve a carried working interest. The purchasing company will supply or carry the seller's share of the working interest until the working interest had paid out. Then the seller can begin receiving income. The seller is not liable for the working interest expenditures except to the extent that they are covered by his net revenue interest. Sometimes, interests are changed when the well reaches total depth. This is known as casing point election.

DRILLING PROGRAMS AND PROCEDURES

The drilling program contains all of the basic data that an experienced operator needs to drill and complete the well. It describes how to drill the well only in special cases because the experienced operator is familiar with general drilling practices and procedures. However, the program does contain the necessary data and instructions to guide the operator in drilling the well. Much of this data is presented as subprograms for casing, bits, mud, and hydraulics.

Drilling programs are generally prepared by drilling engineers experienced in drilling, casing, and completion operations and familiar with the area where the well will be drilled. The process of preparing a drilling program is sometimes referred to as "drilling a well on paper" or "drilling a well with a pencil." In a broad sense this is exactly what the engineer does. For example, he knows that certain formations must be cased, or sealed, off before the hole can be drilled deeper. This helps the engineer select casing points—the depth to which a string of casing is to be run, set, and cemented. He knows how fast a bit will drill in a given formation, which helps determine the number of bits and how much time will be needed to drill the well. Thus, many factors must be considered in preparing the drilling program.

Many services are available to help plan the drilling program. Most companies submit proposals covering their area of expertise. These can be used as a guide, or they can be incorporated into the general drilling program. Drilling programs on other wells in the area can also be used as a guide. Well records, drilling reports, and all other types of local drilling information are used as well. Since very little information is available on wildcat areas, it may be necessary to use information from another locale where conditions are believed to be similar. Therefore, it is very important to have a well-thought-out, complete, detailed drilling program.

The detail of a drilling program depends on the depth and type of well (development or wildcat) and general drilling problems in the area. The program for a shallow, infield development well may take 2–5 pages. On the other hand a program for a deep wildcat well with many drilling problems may need a complete, detailed booklet. As a rule, it is better to have too much than too little information. The driller may not have access to any other information than that included in the drilling program.

GENERAL INFORMATION ON THE DRILLING PROGRAM

The name and address of each owner (if more than one) is listed on every drilling program. One or possibly two people within the company are designated as the company contact. Their names, addresses, and office and home telephone numbers are also included. They will receive reports, copies of logs, and all other data obtained during drilling.

The name and address of the operator—who supervises the overall drilling operation and is usually one of the owners—are also listed. There are usually one or two people in the operator's organization who are directly responsible for the drilling operation, and their names, addresses, and telephone numbers (office and home) are included. The operator is responsible for obtaining all reports from the rig and distributing copies of them to other parties.

A summary of the lease may be included, especially if the terms relate directly to the drilling operation. A wide-scale area map shows the location of the well with respect to nearby airports, towns, and roads and includes detailed instructions on how to reach the well. The lease boundaries may also be included on this map.

A summarized schematic of the hole is also included. This shows the hole sizes and depths, casing sizes, and other general summarized information. The estimated drilling times are shown on rate-of-penetration (ROP) curves (Fig. 4–1). Drilling time and total time on the well are included.

The drilling time curve shows only time spent drilling. The total time curve shows both the actual drilling time and time required for all other operations, such as tripping the drilling assembly, running and cementing casing, and logging. The difference between these two times is a measure of the efficiency of the drilling operation.

In a normal drilling operation about 70% of the time is actually spent drilling; the remainder of the time is used for tripping (lifting the pipe out of and lowering it into the hole), coring and testing, logging, running casing, and performing the miscellaneous work required to drill the well. If the total drilling time is less than about 70% of the total time, this usually indicates that drilling problems such as fishing, lost circulation, heaving and caving formations, and other problems were encountered and reduced the overall efficiency of the drilling operation.

A more efficient drilling operation is indicated by the higher percentage of total time drilling. The 70% figure is an average number. A higher percentage of time will be spent drilling in the shallower holes, and a lower percentage will be spent in the deeper holes because of the increased amount of time spent tripping and performing other operations and the increased exposure to hole problems as a result of both the depth of the hole and the time spent drilling.

Permits and Reports

Many different permits may be required before drilling the well. Various federal, state, county or parish, and municipal regulatory agencies may have jurisdiction over some or all parts of the drilling operation, depending on where the well is drilled and who owns the surface. All states have a regulatory agency that supervises drilling-completion-production within the state (see appendix). Various states have different requirements, but all generally require a drilling permit before the well is drilled, a completion report or plugging and abandonment report after the well is drilled, and periodic production reports during the producing life of the well. The size and complexity of these reports vary. Monthly or quarterly reports may also be required.

Permitting and reporting procedures to regulatory agencies are frequently complex and require a substantial amount of time and increased costs to the operator. The foregoing is only an abbreviated summary. These permits and reports may or may not be included in the drilling program. However, all of those

Drilling Prospects, Programs, and Procedures

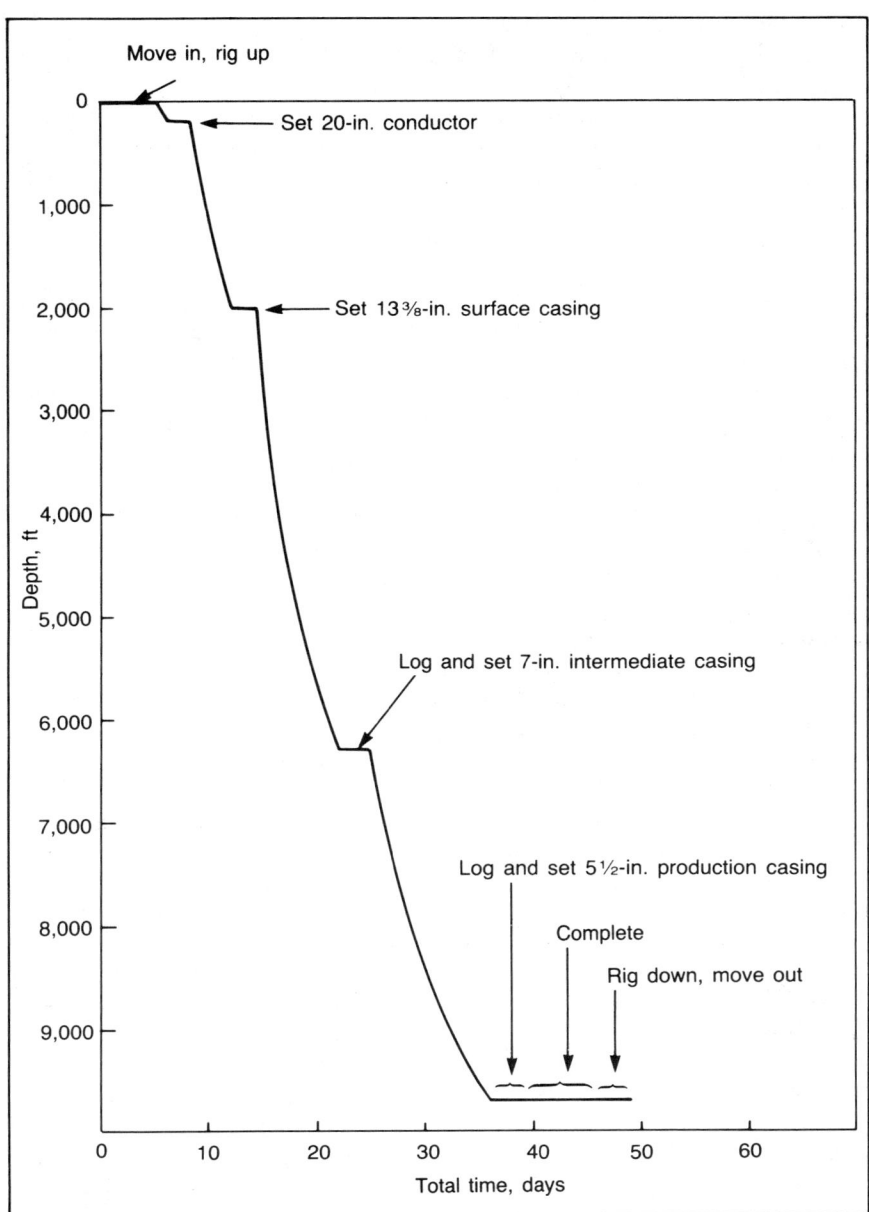

FIG. 4–1 Rate of penetration curve

that pertain directly to the drilling operation are at least listed or noted in the drilling program, often with comments on necessary actions that may be required by the drilling or office supervisory personnel.

Wells drilled on federal and Indian lands require similar permitting and reporting procedures, in many cases to two or more federal agencies. Environmental impact studies and archeological surveys are required in many areas before the well is drilled, including construction of both the wellsite and the access road. Wildcat wells often have long roads in isolated areas, which further complicates the problem.

It is frequently necessary to obtain permits for the relatively large volumes of water used in drilling operations. Special programs such as casing and cementing procedures may be required to protect the shallow aquifers (potable water sands). Rules, regulations, and reporting procedures also are often required for the installation and testing of blowout preventers. Special safety procedures and reporting requirements may be required in areas where hydrogen sulfide (H_2S) and other dangerous gases are encountered. Regulatory agencies must frequently be advised prior to testing and cementing in order to have an inspector on site to monitor the operations. Highway permits are often required to move the large, heavy loads containing the drilling equipment.

Deviation and Dogleg

The maximum allowable deviation and dogleg angles are specified in the drilling program. Buildup angle, hole course and direction, and targets are included for deviated holes. These are very important in the drilling operation because uncontrolled angles cause fishing-jobs and lost holes, and target directions must be supplied for deviated holes.

Deviation and dogleg are measurements relating to the direction of the drilled hole (Fig. 4–2). Deviation is the minimum angle measured in degrees made by the intersection of a straight line through the center of the wellbore and a true vertical line. The coordinates of the direction of the deviation may also be included. For example, if the hole was 2° from the vertical in a northeasterly direction (30° east of north), then the deviation would be recorded as 2°N30°E. Similarly, a well with a 3° deviation in the south-westerly direction would have a deviation of 3°S45°W.

Dogleg is the change in deviation, conventionally measured over 100-ft intervals. For example, if the hole had a deviation of 2° at one point and 100 ft deeper it had a deviation of 4°, then the dogleg would be 2° (4° − 2°) over the depth interval where the deviations were measured. Absolute dogleg is the dogleg angle with allowance for the change in the horizontal component of the deviation. For example, if at a point in the hole the deviation is 3° east and 100 ft deeper the

Drilling Prospects, Programs, and Procedures

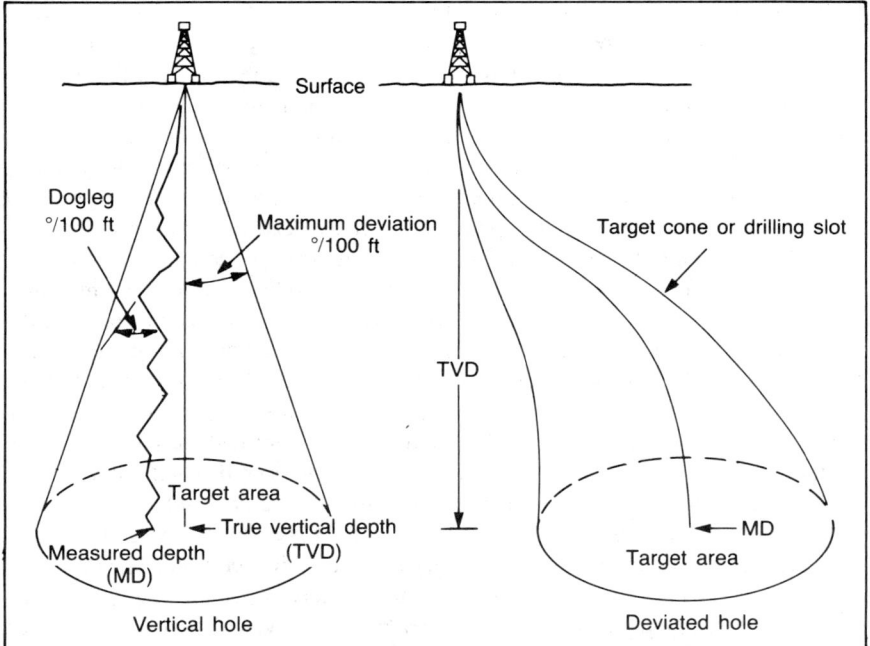

FIG. 4–2 Drilling target, deviation, and dogleg

deviation is 2° west, then the absolute change in the hole angle is 5°/100 ft. Standard tables give the absolute dogleg for various combinations of vertical and horizontal angles.

Limiting deviation and dogleg angles are normally specified in the drilling program. For example, in a normal shallow hole of 8,000–10,000 ft the deviation angle might be limited to 3° and the dogleg angle to 2°/100 ft. In a deeper hole, say 15,000 ft, the deviation and dogleg angles might be limited to 2° and 2°/100 ft, respectively, for the hole section down to 8,000 ft and then increased to 4° and 3°/100 ft for the hole section fom 8,000 ft to total depth (15,000 ft). If the deviation and dogleg limits are restricted excessively, additional expenses can be incurred during drilling to maintain the low, conservative angle. On the other hand, if very high angles are allowed, then excessive hole problems with attending increased costs may be encountered. Experience, knowledge, and good judgment are required in order to select the correct deviation and dogleg angles.

Most holes are drilled as straight holes. A straight hole is drilled straight and vertical with a deviation of 5–10° or less; the average is probably 3°. A hole drilled within these limits will normally enter the objective horizon (reservoir) within the areal limits required by the operator.

Crooked-hole formations tend to cause the hole to deviate. Deviation can cause problems during drilling, especially in high-angle holes (10° or more). The severity of the drilling problems usually increases as the angle of deviation increases. Therefore, hole deviation is monitored during drilling. If it begins to increase, it is controlled by equipment and special drilling techniques to prevent the angle from increasing to the point where it will cause drilling problems.

High-angle holes can cause problems during the producing life of the well. For example, one of the most economical ways to produce shallow and medium-depth oil wells is by rod pumping. The rods will travel a distance in excess of the equivalent of one hundred thousand miles over the average producing life of more than 15 years. If the hole is highly deviated, the rods will lie on one side of the hole and cause rod and casing wear. A similar wearing action occurs during drilling and while drilling below high-angle, deep intermediate casing strings. Methods of minimizing these types of problems have been devised but it is more economical, when possible, to operate in a straight, vertical hole.

Normally, the most economical hole to drill is a straight hole. However, there are many cases where it is necessary to drill a deviated hole. The hole may be deviated because the surface location immediately above the reservoir is not accessible for practical purposes. For example, a deviated hole may be used to produce a reservoir located under an industrial area or to control a blow-out well.

The equipment and techniques for drilling deviated holes are highly developed, and in many areas deviated holes are drilled as a routine operation. However, deviated drilling has higher risks than straight-hole drilling. Consequently, deviated holes cost about twice as much as regular, straight holes to drill.

Doglegs create serious problems during drilling, primarily due to the formation of keyseats—sections in the wellbore that deviate abruptly from the vertical. If the dogleg angle is controlled so that it does not create a problem during drilling, then it normally will be too low to affect production operations adversely.

Keyseats are formed on the inside bend of a dogleg while the hole is being drilled deeper (Fig. 4–3). A recessed hole section is cut or worn into the wall of the wellbore by the rotating and reciprocating action of the drilling assembly. When the larger drill collars and bit are pulled through this section, they can wedge and become stuck, thus sticking the drilling assembly. The drilling assembly can become further stuck as a result of wall sticking—when the pressure differential between the mud column and the formation cause the pipe to stick to the side of the borehole. The net result is that an extensive, costly fishing job may be required to release and recover the drilling assembly. Sometimes these fishing

133
Drilling Prospects, Programs,
and Procedures

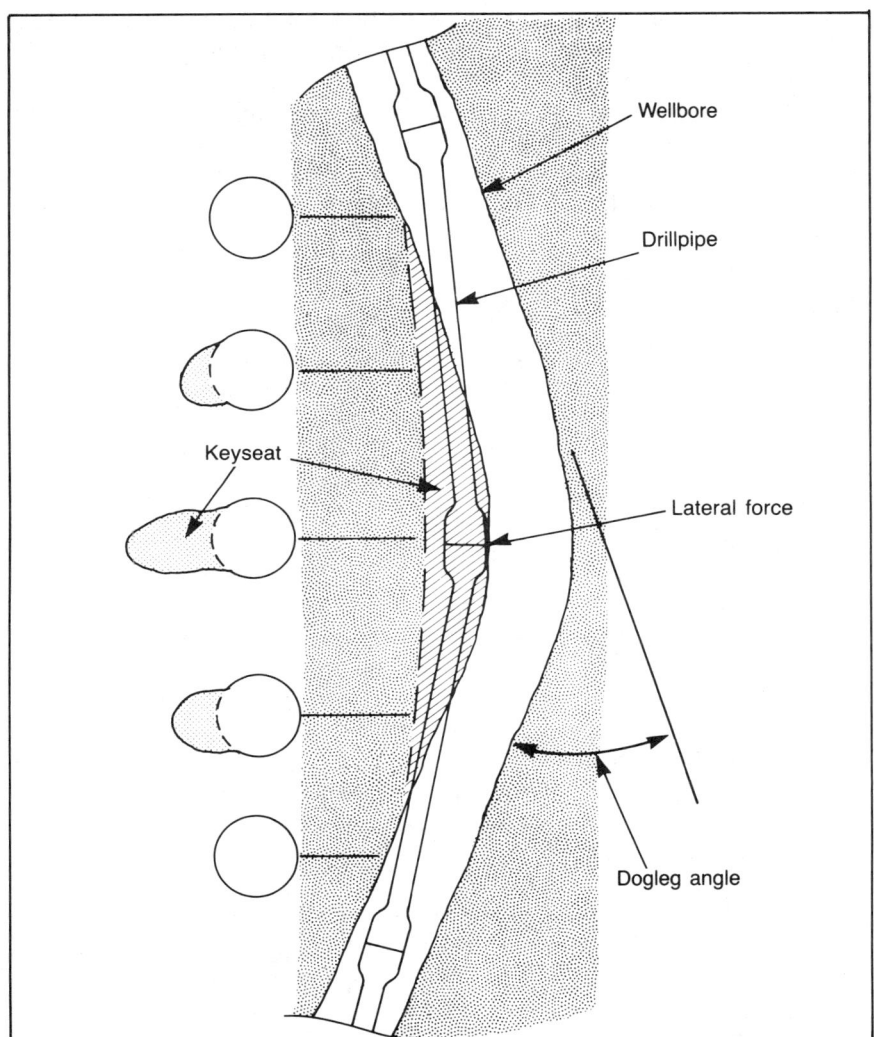

FIG. 4–3 Keyseat formation

operations are unsuccessful. Then the hole must be sidetracked or abandoned and a new hole drilled with resulting additional drilling costs.

Dogleg angles of less than 2°/100 ft seldom cause problems, of 2°–4°/100 ft may cause drilling problems, and exceeding 4° frequently cause severe drilling problems. The severity of the problem generally increases as the depth of the hole and the dogleg angle increase. Doglegs also can cause drilling problems in both straight and deviated holes.

When the hole is to be deviated, the course of the projected hole is included in the drilling program. This includes both the direction of deviation (coordinates) and the angle of deviation at various depths. The amount of dogleg allowed during drilling is also defined for deviated holes in a similar manner to straight holes.

GEOLOGICAL PROGNOSIS

A complete geological prognosis is included in the drilling program. This will range from a one-page summary to a detailed report, depending on the depth of the well, the number of expected productive horizons, and the complexity of the formations and associated formation problems. This prognosis differs from that in the drilling prospect, which is primarily devoted to exploration, and existence of an oil trap. The geological prognosis contained in the drilling program is written to help the geologist, drilling engineer, and operational personnel successfully drill the well and analyze and test the prospective productive horizon.

One preferred method is to list all of the formations or show them in a cross section with the formation depths and thicknesses noted. The formation characteristics that could affect the drilling operation are listed along with suggested remedial action if available. This would include lost circulation zones, water flows, fractured sections, fluid-sensitive and geopressured shales, high-pressure zones, salt sections, mud contaminating zones such a gypsum and anhydrite, and zones that might contain hydrogen sulfide (H_2S) and carbon dioxide (CO_2). The prognosis will also indicate data reliability. Pore pressure data and plots may be included if available.

The geological prognosis also includes instructions on sampling, collecting data, circulating out shows, coring, testing, mud logging, and well logging. The amount of sampling and testing depends to a great extent on whether the well is a development well or a wildcat well. Less testing is required on development wells since most of the information about the formations is known from prior wells. More work of this nature is required on wildcat wells—often an extensive amount on deep wildcats.

CASING AND CEMENTING PROGRAM

The basic purpose of casing and cementing is to (1) seal the upper formations so the well can be drilled deeper, (2) provide a means of controlling formation fluids

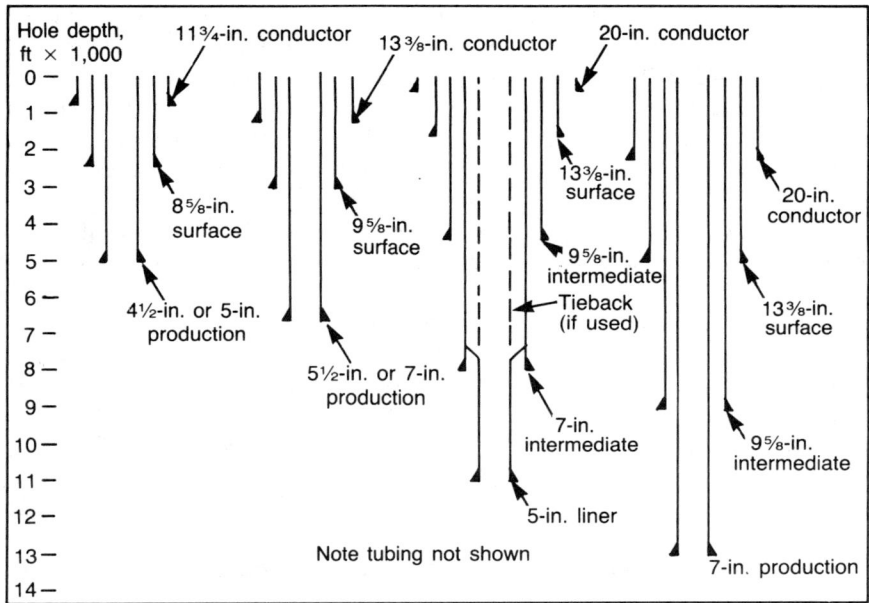

FIG. 4–4 Common casing designs

(control blowouts) while drilling, (3) act as a conduit or pipeline for oil and gas to flow from the reservoir to the surface during production and (4) control the pressure and flow rate. Casing and cementing are an integral, important part of the drilling-completion-production process. Casing sizes, weights, grades, and strengths are very important in casing design.

Oil-field casing looks like steel pipe. A single piece of casing is referred to as a joint of casing, or more commonly a joint. The joint has a tube or body with screw-type connectors on each end, commonly a pin, or male, connector on the bottom and a box, or female, connector on the top of the tube. (Top and bottom here refer to the position of the casing after it has been run into the hole.) Various types of connectors are used on the ends of the casing, but the most common is threaded and coupled (T&C) casing. When a number of joints of casing have been screwed together and run in the hole, they are known as a string of casing or a casing string. The process of screwing the joints of casing together (making up the casing), running it into the drilled borehole, and cementing the casing in place is called casing the hole, sometimes abbreviated to casing (used as a verb).

The dimensions and strengths of casing have been standardized by the API in its "Bulletin on Performance Properties of Casing and Tubing." Casing is initially identified by its outside diameter (OD), measured in inches. Casing sizes range from 4½ to over 30 in. OD. Casing lengths are generally supplied in three ranges. Range 1 casing lengths range from 19–26 ft; Range 2, 26–32 ft; and Range 3, 32–42 ft. The most common lengths are 31 ft and 42 ft. Longer casing joints are generally preferred because, since there are fewer connections per foot of casing, there are fewer points of possible failure. Running longer casing is also more economical because making the connection adds to the cost of the casing.

Some of the earlier types of casing were spiral or lap-welded, but all casing currently used in oil-field operations is seamless and has an even, consistent body thickness.

Casing is further classified by weight and grade. Casing weights are given as pounds per foot (lb/ft) with the extra weight of the coupling averaged over the joint of casing. For example, 5½-in. casing is a common size that can be obtained in casing weights ranging from 13–26 lb/ft. Since the OD of the casing is always held constant for one size, the inside diameter (ID) varies according to the casing weight. For example, 5½-in. 13-lb/ft casing has an ID of 5.044 in., compared to 4.548 in. for a 5½-in. 26-lb/ft casing. These small differences in inside diameter are very important in selecting bit sizes and other tools to be run inside the casing. For the same size casing, heavier-weight casing is stronger with greater collapse and tensile strengths. Casing collapse occurs when the external forces and pressures cause the sides of the casing to flatten and push together. Tensile strength is a measure of the actual steel strength of the pipe, which shows how much weight or loading the casing can support. Heavier casing weighs more and also costs more, since there is a larger amount of steel in the casing on a per-foot basis.

Casing is also supplied in various grades. The grade refers to the quality of the steel. Low-grade casing has a lower quality of steel and consequently less tensile strength. Higher-grade steels have greater strengths. Common steel casing grades are H-40, J&K-55, C-75, N-80, C-95, and P-110. The numbers refer to the tensile strength of the steel in thousands of pounds per square inch. For example J-55 steel has a tensile strength of 55,000 psi. Put another way, a rod with 1 sq in. of cross-sectional area made with J-55 steel would support 55,000 lb; a similar rod made with P-110 grade would support 110,000 lb. Higher grades up to V-150 (150,000 psi) are available upon special order.

Tensile strengths are normally based on the minimum yield of the steel. If the loading is increased above the tensile strength, the steel will begin to yield (elongate) and will continue to do so with added loading until it fails at maximum tensile strength.

Basic Casing Strings

Four general casing strings are used in drilling: conductor, surface, intermediate, and production. Depending on the hole depth and formation conditions, these may be run to different depths in different areas (Fig. 4–5). One or more of the casing strings may be omitted, such as running surface and production casing only—a common practice in shallow holes. Two or more of the same type of strings may be used, such as running two strings of intermediate casing. They may be run as liners, in combination with liners, or as liners and tieback liners. Liners are short strings of casing run from the bottom of the hole to an intermediate point uphole. Casing is run from the bottom of the hole back to the surface.

The section of hole in which the casing is to be run normally is referred to by the same name as the casing string. For example, conductor casing is run in the conductor hole section, and intermediate casing is run in the intermediate hole section.

Conductor The conductor hole is the first section of hole drilled. The conductor casing, or conductor, is run and cemented in this hole. This pipe is the first and largest string of casing run into the hole. Sizes range from 8⅝ in. OD to over 30 in. OD, and the casing is run to depths of 10 ft to over 1,000 ft.

One main purpose of this casing is to hold back the unconsolidated surface formations and prevent them from falling (caving or sloughing) into the hole. The conductor also provides a flow path for the drilling mud, which is used to drill the hole below the conductor casing. The mud flows up the annular space between the drillpipe and casing to a flow line and then is returned to the mud pits, which are above ground level.

Shallow water or gas flows are sometimes encountered while drilling below the conductor. Water flows rarely cause a problem, but shallow gas flows can blowout and destroy a rig in less than 10 min.

The conductor is often fitted with a rotating head or similar diverting device above the flow-line outlet to control or reduce the severity of the problem. These tools divert the flow of mud and other fluids away from the rig. Regular blowout preventers used to seal the annulus cannot be used on the conductor because the conductor is not normally set deep enough to provide sufficient holding force. If a flow of gas or water occurred with the conductor completely shut in, the force of the gas or water would either lift the conductor casing out of the hole and flow wildly around the rig or the fluids would initially channel or fracture through the unconsolidated surface formation around the conductor and flow up around the rig. Experience has indicated that the best approach is to divert these flows.

Deeper conductor may also be used as a base or to provide partial support for the suspended weight of other casing strings run later.

Surface The surface hole section is drilled below the cased conductor section and is then cased with surface casing. The surface casing is the second string of casing run in the hole. The outside diameter of the surface casing is less than the inside diameter of the conductor casing through which it is run.

The surface section is normally drilled through the remainder of the unconsolidated shallow formations and into the more consolidated formations. It holds back the unconsolidated shallower formations and prevents them from falling into the hole and causing problems similar to those in the conductor hole. It isolates or cases off shallow water zones and prevents them from flowing into the wellbore and contaminating the drilling fluid. This casing also prevents heavier drilling mud (when it is used) from invading into and contaminating shallower aquifers that may contain potable water.

If an extra-long hole section is to be drilled before the final casing string is run, surface casing will reduce the amount of uncased, or open, hole. The surface casing is designed to have a burst pressure in excess of any pressures expected to be encountered while drilling below the surface casing and to the next casing seat. The surface casing serves as a base or anchor to hold the blowout preventers. The combination of the surface casing and BOPs allows the operator to control any blowouts that may occur while drilling deeper to the next casing point.

The surface casing normally acts as a base to support the suspended weight of all subsequent casing strings that are run. In later operations the surface casing may provide a pressure backup or relief outlet for the next casing string in case it ruptures, usually as a result of excess pressure.

Surface casing sizes range from 7-in. OD to over 18 in. diameter set at depths of 300–5,000 ft. It is normally cemented from the bottom back to the surface by filling the annular space with cement. This seals the annular space so all mud and formation fluids will flow through the casing.

Intermediate After the intermediate hole section has been drilled, intermediate casing is run through the surface casing to the bottom of the intermediate hole section and is cemented. The outside diameter of the intermediate casing (after allowance for couplings or upset connectors) is slightly less than the diameter of the bit used to drill the intermediate hole section and is less than the inside diameter of the surface casing. The intermediate casing is sometimes called protective casing or protection pipe because it is often set to protect the upper hole and to permit deeper drilling. This casing is set at intermediate depths between the surface and production casing depths.

One of the main reasons for setting intermediate casing is to case or shut off a hole condition that will prevent the well from being drilled safely to a deeper depth. It is often used to hold back caving, sloughing shales and to isolate water and gas flows that can hinder or prevent drilling the hole deeper. One of the more common reasons is to shut off lost circulation zones, especially when the zones at greater depths are expected to have higher pressures. Intermediate casing is also used to reduce the amount of uncased hole, especially if a long section must be drilled before the next casing string is run and where formation problems are expected.

Intermediate casing may be used as part of the completion casing when a production liner is used. It also can be set as a liner (drilling liner), especially when two strings of intermediate casing are required (as often occurs in deeper holes). Or it may be used as a backup for the production casing. The design criteria for intermediate casing is similar to that for surface casing after allowing for deeper depths, higher pressures, and heavier mud weights.

Production After the production hole section has been drilled, the production casing or, long string, is set through the prospective productive formations. In open-hole or barefoot completions, the pipe usually is set immediately above or a few feet into the top of formation. This casing pipe is designed with sufficient strength in burst to hold the maximum shutin pressure (the formation fluid pressure) of the producing formation. It may be designed to contain higher stimulation pressures used during completion.

In cases where very long, heavy production strings are run, the lower half of the string may be run as a liner and the upper half run as a tieback string—a short string of casing connected to the top of the liner and extended back to the surface. When a deep intermediate casing string has been run, usually as a drilling liner, a production liner may be run and hung in the bottom of the drilling liner. The drilling liner is extended back to the surface with tieback completion (production) casing. The drilling liner then becomes part of the completion casing.

Casing Design

The basic principles of casing design are relatively simple and straightforward. A wide variety of wells have been cased in the past, so a large number of casing programs are available to fit most formation and drilling conditions that may be encountered. Computer programs, a significant help in casing design, are also used. In addition, there is a wide range of casing sizes, weights, grades, and connectors available that further aid in casing design.

Designing casing requires determining how many casing strings will be needed and the size and setting depth of each. Many factors are involved, but the depth to the producing horizon is probably one of the more important factors.

The length of open-hole section that can be safely drilled and cased depends on the type and kind of formations, hole size, and other parameters. In the normal case the length of the open-hole section should be limited to about

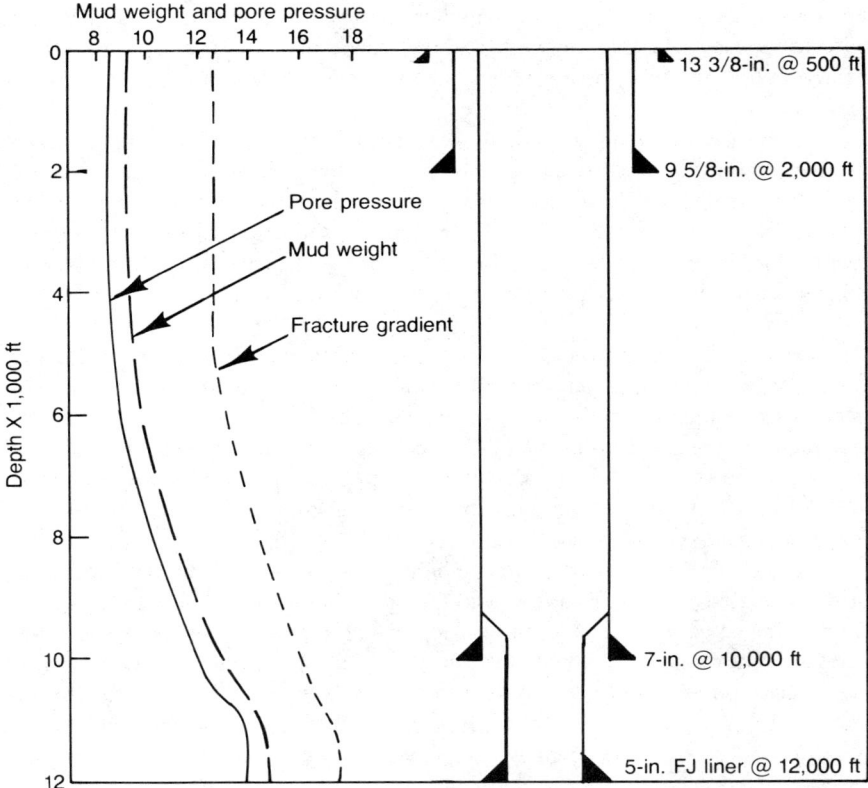

FIG. 4–5 In this example the pore pressures increase with depth to about 11,000 ft. The mud weight is increased as the hole is drilled deeper to equalize pore pressures and to prevent a blowout. The mud weight required at about 10,000 ft exceeds the fracture gradient of the shallow formations immediately below the surface casing at 2,000 ft. If this happens, circulation will be lost in the shallow formations; the top of the mud column will drop, reducing the bottom-hole pressure; and there would be a risk of a blowout. Therefore, the program is designed to set 7-in. casing at 10,000 ft to prevent lost circulation at 2,000–3,000 ft.

12,000 ft for 7⅞-in. and larger holes, 6,000 ft for holes from 6¼–7⅞ in., 3,000 ft for holes from 4½–6½ in., and about 2,000 ft or less for smaller holes. The length of open hole may be further reduced due to lost circulation, caving formations, high-pressure zones, and other formation problems. For example, if a lost circulation zone occurs at a shallower depth and a high-pressure zone is expected at a lower depth, then it is usually necessary to run casing through the lost circulation zone before drilling the high-pressure zone. If this is not done, there is a risk of a blowout.

Hole sizes must be large enough to accommodate an efficient drilling assembly. When possible, the hole size should be large enough so the drill-collar assembly can be washed over (a special fishing operation) if necessary. Sometimes when very large holes are drilled in hard, abrasive formations, it is necessary to drill a pilot hole, a smaller-diameter hole that is later opened up to the required hole size with a special drilling bit called a hole opener. The pilot hole can be drilled and opened at an optimum rate. Such a drilling rate may not be possible when drilling a larger hole because of the lack of weight on the drill bit. In other cases it may be necessary to drill a pilot hole in order to run an open-hole drillstem test. This is a special operation conducted during drilling to test a prospective formation to see if it is capable of producing oil and gas.

The casing program must be designed to allow for changes in the pore pressures and fracture gradient (Fig. 4–5). Pore pressure gradients average about 0.46 psi/ft and range from 0.25–1.00 psi/ft. Fracture gradients are measured similarly, and the normal fracture gradient exceeds the pore pressure by a few pounds per gallon. The formation fracture pressure, indicated by the fracture gradient, is the pressure at which the formation will fracture. When the formation fractures, it literally parts (splits) and allows fluid to escape from the wellbore into the formation. When this happens, there is a high risk of possibly losing the hole due to sticking the drilling assembly caused by caving or a blowout. The casing program must be designed to eliminate as much of this risk as possible.

Pressure gradients are usually relatively constant over long intervals, and constant mud weight can be used. In some cases the gradient increases with depth, and this normally requires increasing the mud weight. If the pore pressure and the mud weight required to contain the pore pressure continue to increase, the mud weight can reach a point where the weight of the mud column (hydrostatic pressure) approaches or exceeds the fracture gradient of the formation in the upper hole section. This can cause fracturing and associated problems. Before reaching this mud weight, it is necessary to case off the upper hole section so that the higher mud weight needed to control the pressures in the lower hole can be safely contained.

Pore pressure gradients and associated fracture gradients can also increase rapidly as the drill bit approaches a high-pressure zone. This is normally monitored during the drilling operation by using pore pressure plots.

A reverse condition can occur where the pore pressure decreases in the lower hole, known as a pore pressure reversal. Usually a certain mud weight is required for the upper formations and cannot be decreased because of possible

142
DRILLING

SIZE	DEPTH
13³⁄₈, 13½	8,000 ft

MUD WEIGHT
9.6 lb/gal × 0.0519 = 0.4982 psi/ft

CASING DESCRIPTION TOP TO BOTTOM

ITEM NO.	FEET/ SECTION	LB/ FOOT	GRADE & END FINISH	SETTING DEPTH	COLLAPSE[1] LOAD	STRENGTH	DESIGN FACTOR	BURST[2] LOAD	STRENGTH	DESIGN FACTOR	TENSION[3] ACCUM WEIGHT	STRENGTH 1,000 LB	DESIGN FACTOR
1	3,700	68.00	N80 BTC	3,700	1,843	2,089	1.133	3,986	5,020	1.259	580,000	1,585	2.733
2	2,300	72.00	S95 BTC	6,000	2,989	3,385	1.132	3,986	6,390	1.603	328,400	1,935	5.892
3	2,000	81.40	S95 BTC*	8,000	3,986	4,860	1.219	3,986	7,140	1.791	162,800	1,885	11.579

*Item 3 has a 13½-in. OD.
Drift in design is 12.250 in.

SIZE	DEPTH
9⅝ in.	11,000 ft

MUD WEIGHT
12.0 lb/gal × 0.0519 = 0.6228 psi/ft

DESIGN CONDITIONS
1. Tension effect on collapse considered
2. Burst load = wellhead pressure plus depth × (Inside minus outside pressure gradient)
   Outside pressure gradient 0 (Zero for production strings, 0.500 for intermediate surface strings)*
   Inside pressure gradient 0 (Zero)*
   Wellhead shutin pressure 3,986 (Bottom collapse pressure)*
3. Buoyancy not considered*

CASING DESCRIPTION TOP TO BOTTOM

ITEM NO.	FEET/ SECTION	LB/ FOOT	GRADE & END FINISH	SETTING DEPTH	COLLAPSE[1] LOAD	STRENGTH	DESIGN FACTOR	BURST[2] LOAD	STRENGTH	DESIGN FACTOR	TENSION[3] ACCUM WEIGHT	STRENGTH 1,000 LB	DESIGN FACTOR
1	7,600	43.50	S95 LTC	7,600	4,733	5,346	1.129	6,851	7,510	1.096	496,900	960 1,193	1.932
2	2,400	47.00	S95 LTC	10,000	6,228	7,018	1.127	6,851	8,150	1.190	166,300	1,053 1,289	6.332
3	1,000	53.50	S95 LTC	11,000	6,851	8,850	1.292	6,851	9,410	1.374	53,500	1,235 1,477	23.08

Drift in design is 8.500 in.

SIZE	DEPTH
7 in.	16,500 ft

MUD WEIGHT
14.7 lb/gal × 0.0519 = 0.7629 psi/ft

DESIGN CONDITIONS
1. Tension effect on collapse considered
2. Burst load = wellhead pressure plus depth × (Inside minus outside pressure gradient)
   Outside pressure gradient 0 (Zero for production strings, 0.500 for intermediate surface strings)*
   Inside pressure gradient 0 (Zero)*
   Wellhead shutin pressure 6,851 (Bottom collapse pressure)*
3. Buoyancy not considered*

CASING DESCRIPTION TOP TO BOTTOM

ITEM NO.	FEET/ SECTION	LB/ FOOT	GRADE & END FINISH	SETTING DEPTH	COLLAPSE[1] LOAD	STRENGTH	DESIGN FACTOR	BURST[2] LOAD	STRENGTH	DESIGN FACTOR	TENSION[3] ACCUM WEIGHT	STRENGTH 1,000 LB	DESIGN FACTOR
1	3,300	32.00	LS125 LTC	3,300	2,518	9,811	3.987	9,660	14,160	1.466	537,900	996	1.852
2	4,900	32.00	S95 LTC	8,200	6,256	9,077	1.451	10,747	10,760	1.001	432,300	779	1.802
3	5,000	32.00	LS125 LTC	13,200	10,071	11,364	1.128	11,856	14,160	1.194	275,500	996	3.615
4	3,300	35.00	LS125 LTC	16,500	12,588	14,330	1.138	12,588	14,440	1.147	115,500	1,106	9.576

Drift in design is 5.879 in.

DESIGN CONDITIONS
1. Tension effect on collapse considered
2. Burst load = wellhead pressure plus depth × (Inside minus outside pressure gradient)
   Outside pressure gradient 0 (Zero for production strings, 0.500 for intermediate surface strings)*
   Inside pressure gradient 0.222 (Zero)*
   Wellhead shutin pressure 8,928 (Bottom collapse pressure)* Based on 0.6 sp gr gas column and BHP of 12,588 psi
3. Buoyancy not considered*

In collapse:
- inside of casing is empty
- pipe is perfectly vertical
- buoyancy has not been considered for tension effect on collapse

In burst: **Production casing**
- inside of casing contains a 0.6 specific gravity gas column
- burst load (production string) = internal gas gradient × depth plus the wellhead pressure
- no external gradient

Surface and intermediate casing
- bottom-hole hydrostatic pressure = burst load

In tension:
- buoyancy is not considered

*Unless otherwise specified

FIG. 4–6 Casing design program (courtesy Lone Star Steel)

pressured zones, formation problems, or water flows from the upper zones. If the pore pressure decreases in the lower hole section, it can drop to the point where the formation can fracture and take fluid, causing lost circulation. This in turn will cause problems in the upper hole. When this happens, it is often necessary to case off the upper zones so the mud weight can be reduced to drill the lower formations.

The type and number of completions also affects the casing size. Gas wells can normally use smaller-size completion casing than oil wells can. Oil wells require comparatively larger casing to accommodate tubing, pump rods, and other types of lifting equipment used to pump the well in its later producing life after the reservoir energy has been depleted to the point where the well will not flow naturally at an economic rate.

If the well is completed in two or more zones, larger-size production casing may be needed to accommodate the additional downhole completion equipment required for dual, triple, and multizone completions. Deeper wells must have larger production casing to accommodate the larger work strings used in completion and remedial operations, a rework, workover, or other type of downhole job done on the well after it is completed.

The casing program must also provide for drillpipe wear (usually intermediate casing) on deep holes. Many formations contain corrosive agents such as hydrogen sulfide and carbon dioxide. Electrolytic corrosion may also be a problem. These conditions must be provided for in the casing program.

Another and sometimes critical factor in casing design is the flexibility of the casing program and allowance for an extra string of casing. An extra casing string means extra costs. However, if it is needed it must be included; otherwise, there is a high risk of losing the hole and wasting the entire expenditure of drilling the well. On the other hand if it is not needed, obviously it should not be used. The real problem is determining when it is needed and making allowances accordingly. Sometimes this cannot be determined until the well has been partially drilled. At that time it is too late to plan for an additional string of casing unless it was included in the original casing program because hole and casing sizes have already been determined. On wildcat wells, deep holes and other wells where the formations are less predictable, the usual procedure is to allow for an extra string of casing. If drilling results indicate that the extra string of casing is not needed, then it is omitted.

For example, a well could be designed for conductor casing at 200 ft, surface casing at 2,000 ft, a questionable intermediate casing string at 8,000 ft, and completion casing at 12,000 ft. If the hole is drilled to the intermediate casing point and the drilling indications are that the intermediate casing is not needed, then the hole size is reduced and the well is drilled to total depth where completion casing is run. In this case some additional expenditure was required for larger-than-normal surface and intermediate casing, but this was good insurance since it gave the operator the option of running the intermediate casing string if it was needed. If the extra string of casing had been needed and was not provided for in the program—that is, if the upper casing sizes had been too small—then the hole would probably be lost. This illustrates some of the factors that must be considered in selecting the number of casing strings and setting depths.

Casing and hole sizes are selected in the opposite order in which they are drilled and run. The production casing is first selected, and this in turn determines the size of the production hole section. The next-larger uphole casing—usually intermediate—is selected and sized so operations can be conducted through it to drill and case the lower production hole section. The next-larger uphole casing—usually surface—is selected and sized to allow drilling and casing the intermediate hole section. This is repeated until all casing strings are designed. All of this information is included as the casing design in the drilling program (Fig. 4–7).

After the number of casing strings and setting depths have been determined, the next step is to select the size of the completion casing. This depends on depth, type of completion, amount of testing, tubing size, and other factors.

The weight and grades for each string of casing are then selected. The weight and grade are selected so the casing string will have enough tensile strength to support its own weight. Burst and collapse pressure rating must be higher than those expected to be encountered in the well. Tension is at a maximum in the top part of the casing, while maximum burst and collapse pressures usually occur in the lower part of the casing. Safety factors have been established for these three criteria (tension, burst, and collapse). Thus, the weights and grades of casing are selected based on the required strength and after allowing for the proper safety factor, although lighter weights and lower grades are used whenever possible to reduce costs.

The next step is to determine the size of hole that the completion casing will be run in. This will be the smallest size hole that can be efficiently and safely drilled considering the hole depth and that is large enough to accommodate the completion casing. In the normal case the diameter of this hole will be 1–2 in. larger than the outside diameter of the casing collar or semiflush-joint connector or the diameter of the casing if flush-joint casing is used.* This 1–2-in. difference is known as the casing clearance. The amount of casing clearance required in order to safely run the casing depends on the formations. Normally larger clearances are used in softer formations or formations that tend to cave. Less clearance is needed in harder, more compact formations. These are averages; more or less

---

*Flush joint casing has threads cut in the casing body and does not use collars or couplings. Semiflush-joint connectors are similar except the casing diameter is increased slightly near the end of the joint.

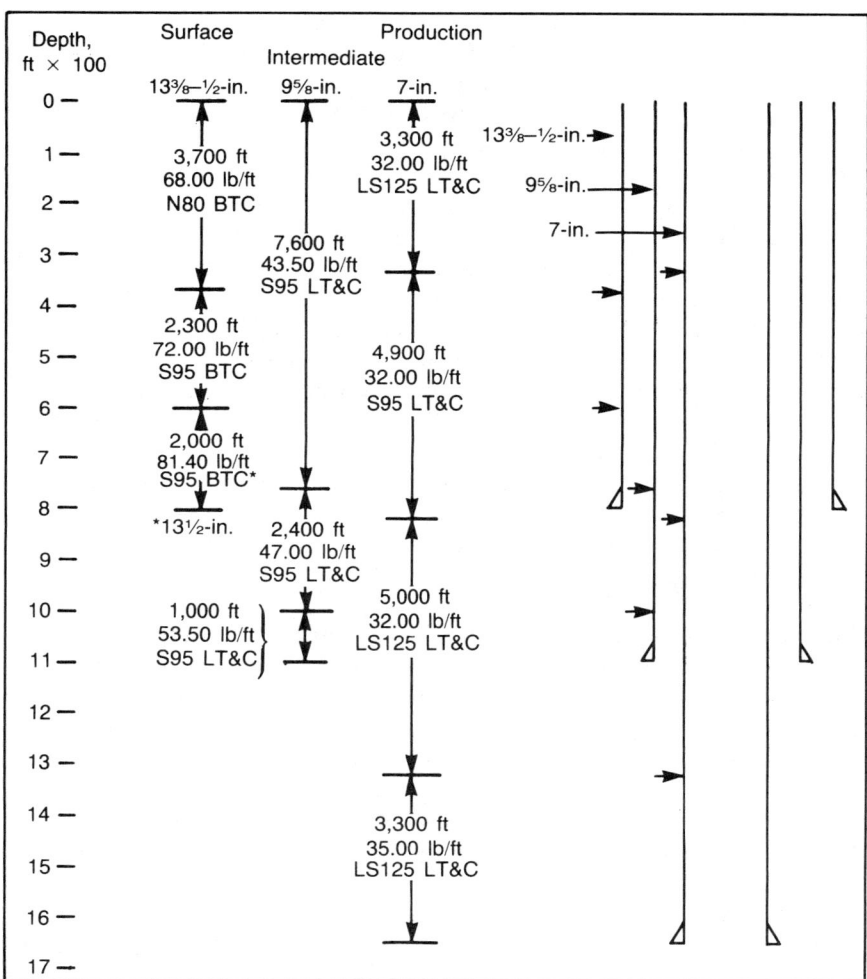

FIG. 4–7 Casing design (courtesy Lone Star Steel)

clearance may be needed in some areas depending on the formations. Selecting the correct casing clearance requires considerable judgment and experience in the area.

The next upper string of casing will have the smallest inside diameter that can be safely run and still have a large enough ID for the bit that will drill the lower (production) hole section.

Casing tables list two diameters for casing. One is the actual inside diameter and the other is the drift diameter (sometimes called drift), which is slightly less than the inside diameter. The drift is one of the construction specifications of the casing, and it indicates that a long tool with the specified drift diameter can pass freely through the casing. Normally the largest standard bit size (OD) that is still smaller than the drift diameter of the casing is selected. For example, 7-in., J-55, 20-lb/ft casing has a 6.456-in. ID and a drift diameter of 6.331 in. The largest standard size bit with a diameter smaller than the drift diameter is a 6¼-in. bit. In this case the bit clearance would be 0.206 in., or the actual inside diameter less the bit diameter.

In special casing designs it may be necessary to use a bit diameter slightly larger than the drift diameter. This has been done many times without problems, but the recommended procedure is to select the bit size as noted except when there are special reasons to use the larger bit size.

This procedure determines the casing size for the next upper string of casing. The hole size is then determined in the same manner as that for the production casing, as are the casing and hole sizes for the succeeding upper casing strings. The process is repeated until all casing and hole sizes have been determined.

Centralizers and Scratchers The casing program will also have the recommended placement of centralizers and scratchers. All holes have some inclination, and the casing will lie against one side of the hole. When it is cemented, there will be very little if any cement between the casing and the wall of the hole on the side where the casing touches the wellbore (Fig. 4–8). So centralizers are placed on the casing when it is run. They are designed to center the casing in the hole so that, when it is cemented, there will be a cement sheath completely around it. This also helps prevent communication behind the pipe in the completion zones.

The wellbore normally has a layer of mud cake deposited from the drilling mud. When the casing is cemented, this mud cake may limit bonding between the cement and the wellbore. Scratchers are run on the casing to help remove this layer so a better cement bond to the formation can be obtained. The scratchers literally scratch the walls to remove the mud cake. Both rotating and reciprocating scratchers are available. As the names imply, one type scratches while the casing is rotated and the other scratches while the casing is reciprocated.

Centralizers are generally run on the bottom two or three joints of all casing strings. Centralizers, scratchers, or both may be run on the casing string opposite the prospective producing horizons.

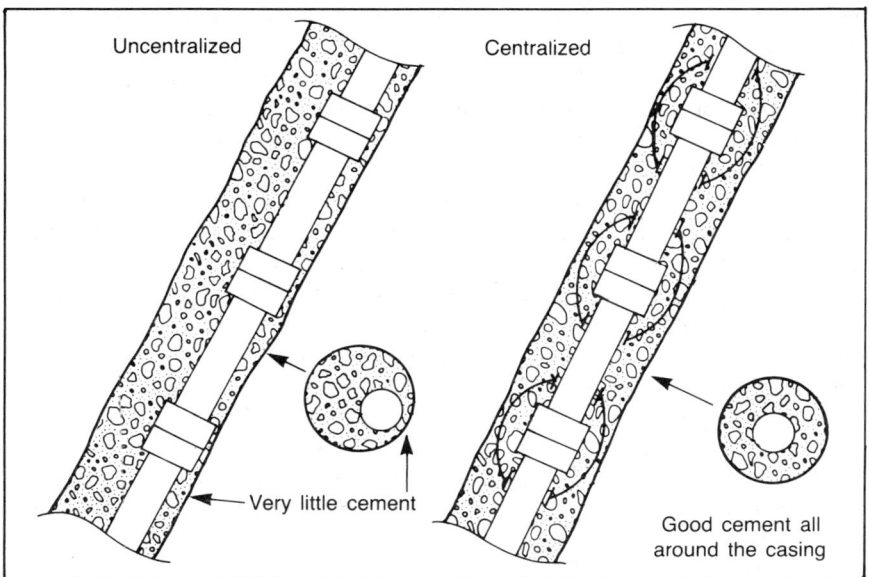

FIG. 4–8 Centralized casing

Casing is normally inspected at the steel plant or rolling mill where it is made. Sometimes it is inspected again at the pipe storage yard. There is an increasing tendency to make a final casing inspection at the wellsite prior to running the casing. Different kinds of inspections may be made, ranging from an end-area inspection through a complete body inspection and possibly a burst test. The decision to test the casing and the type of inspection to be made depend to a great extent on the type of casing string and the depth to which it will be run. For example, shallow casing strings such as conductor and surface casing are seldom inspected. Deeper casing strings, which may be run at or near their rated strength and with minimal safety factors, are inspected often.

Cementing

After drilling the hole and running casing, the next step is to cement the casing in place. The casing is cemented to hold it in place and to prevent migration of formation fluids in the annular space between the outside wall of the casing and the wellbore. When a section of hole has been properly cased and cemented, it is considered safe. The risk of losing the hole as a result of caving formations, lost circulation, and high-pressure zones has either been eliminated or reduced to the point where it is insignificant. There is usually a sense of relaxation and relief when the hole has been cased and cemented.

Sometimes the conductor casing will be cemented with concrete. Concrete is a mixture of cement and sand, and it sometimes contains gravel. All other casing strings are cemented with cement slurry, commonly called cement. This is a mixture of pure cement and water.

Various types of cement are available for different uses. Lightweight cement is used in areas where a low-weight slurry is needed to help prevent lost circulation. Some cements are resistant to chemical action. Other cements are designed for very high-temperature applications. Oil-field cements have a very carefully controlled chemical composition and particle size (grind). This is important so the pumping time and ultimate hardness of the cement can be accurately controlled and predicted.

Pumping time is the amount of time the cement slurry remains liquid and pumpable prior to hardening or setting-up, sometimes referred to as thickening time. This time is important because the slurry must remain liquid and pumpable long enough to mix the slurry, pump it downhole, and displace it into the annular space between the casing and the wellbore.

Various cement additives are available. Retarders slow the cement from setting-up and increase the pumping time. Accelerators help the cement set up faster. Fluid loss additives control the fluid loss. Silica flour increases the cement's strength. Gel is used to increase the volume of the cement slurry and reduce its density. Other additives are also available to reduce the cement density. Lost circulation materials are used to help prevent lost circulation. Many other additives are also available for specific purposes.

A cementing program is calculated for each string of casing. The amount of fill-up is first determined. Fill-up is the distance in feet measured from the bottom of the hole where the annular space is to be filled with cement. The volume in the annular space is calculated from the estimated hole diameter, the known casing outside diameter, and the length of the section to be filled with cement. Excess cement volumes are normally used. The average excess ranges from about 25–50% for production casing to 200–300% for surface casing. The type of cement and additives is selected, and the volume required is calculated. This information for each casing string is included in the cementing program.

Float equipment is run on the casing. This usually consists of a guide shoe, or float shoe, on the bottom of the casing and a float collar one or two joints above the bottom. These are normally constructed of drillable material so that they can be drilled without difficulty. The guide shoe has a rounded bottom and is run on the bottom of the casing to help guide it into the hole. If it is a float shoe, then it has a float or check valve so that fluids can be pumped down the casing but cannot flow back up through the casing. This is important to prevent cavings and cuttings suspended in the mud from flowing up into the casing when it is run. These materials could plug the bottom of the casing so that it could not be cemented properly. The float collar serves as a second or double check valve in this case.

When regular float equipment is used, the casing must be filled with mud as it is run in the hole. Automatic float equipment is often used. It allows a controlled amount of mud to flow into the casing as it is being run, thus eliminating the problem of filling the casing with mud.

Sometimes the casing may be cemented in two different sections. The lower section cemented is called the first stage; the upper section is called the second stage. A cement diverting tool is run on the casing and positioned so that when the casing is in the hole the tool will be positioned near the bottom of the section where the bottom of the second stage of cement is to be placed.

The cementing volumes listed in the cementing program are used as a guide. The actual cement volumes are calculated from caliper logs after the hole section has been drilled. Actual hole diameters can be read from the caliper logs to give a more accurate cement volume. When possible, the data from the three-arm caliper should be used since it is more accurate than the two-arm caliper.

BIT PROGRAM

The rotary drill bit, commonly called the bit, is the tool on the bottom of the drilling assembly (drillstring) that actually drills the formations. The bit converts the rotating action of the drillstring into a pounding-scraping-tearing action. This in combination with the weight applied on the bit (and subsequently on the formation) by the drillstring and the jetting, erosional action of the drilling fluid drills the hole through the formations. The tool must be extremely tough and ruggedly built since bit weights in excess of 25,000 lb are common and bit weights over 50,000 lb have been used. The bit weight is the amount of weight applied to the bit by the drilling assembly and that in turn is transmitted through the bit and applied to the face of the formation to help drill the hole. Therefore, the bit cutting structure must be extremely durable and wear resistant in order to support the bit weight and drill the formation, especially hard, tough abrasive formations.

The average rig spends about 70% of its time drilling and about 30% of its time in other work. About ⅔, or 20% of the 30%, is spent tripping to change bits. When the bit becomes worn to the point where it will not drill efficiently, then all of the drilling assembly must be pulled out of the hole, the worn bit replaced with a new bit, and the assembly run back into the hole. This procedure is known as tripping to change bits. The operation is necessary but in a sense is not an actual drilling operation. Therefore, it is important to use a bit with a long bit life that will make a long drilling run in order to reduce the amount of nondrilling time. A bit that has an efficient cutting structure designed for the particular formation being drilled will drill more hole in the same period of time than a bit with a less-efficient cutting structure.

Roller Bits

The roller bit has a heavy metal body (Fig. 4–9). The top of this body has a pin connector so the bit can be connected to the drilling assembly. The lower part of the bit has three metal extensions called bit shanks. An axle or bearing shaft is connected to each bit shank. There is one-half of a bearing race on the axle and a mating half of the bearing race on the inside of a bit cone. The bit cone fits over the bearing shaft and rolls on bearings in the bearing race(s). The outer face of the cone, or cutting surface, has different types of teeth.

The outer cutting surface is large, greatly increasing the life of the roller bit. Tooth design and offset and pitch design also help increase the drilling rate. Offset refers to the distance a line through the axis of the bit bearing is from the true center of the bit. If the bit does not have any offset, the cone will have a true rolling action. The dragging action of the cone increases as the offset increases.

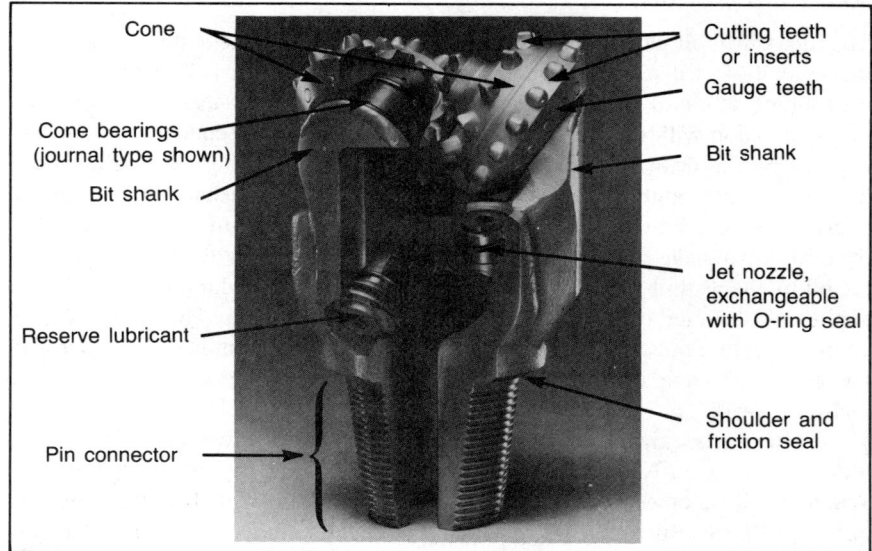

FIG. 4–9 Three-cone bit cutaway (courtesy Hughes Tool Company)

Usually a bit with more offset or dragging action will drill faster, but the teeth will also wear faster, reducing the bit life. Pitch refers to the spacing of the teeth on the circumference of the cone and includes both the spacing between the teeth on a row of teeth and the spacing between adjacent rows of teeth. Pitch helps determine the point of impact of the bit tooth on the formation with each successive revolution of the bit cone and bit.

The first roller bits were called conventional roller bits and had a mud course (conduit or channel for fluid passage) in the center of the bit. Jet bits were later developed. These bits have jet nozzles that increase the velocity of the drilling fluid by restricting the flow area. A high velocity (jet) of fluid is directed onto the face of the formation. Most roller bits used in drilling now are jet-type bits although a few conventional bits are used, usually in the larger bit sizes. Bit construction is illustrated in Fig. 4–9.

Roller bits are subdivided into two broad classifications based on the type of teeth. One type has teeth that are milled as an integral part of the bit cone and are commonly referred to as a mill-tooth bit. Different tooth configurations are used, generally ranging from long, sharp, pointed teeth for softer formations to shorter, stubbier (stronger) teeth used for harder formations. The first roller bits developed were mill-tooth bits (Fig. 4–10).

The second type of classification based on teeth is the insert tooth bit, commonly called insert bit. The teeth are separately formed from a very hard, tough metal such as tungsten carbide. These teeth or "inserts" are pressed, "inserted," into precisely sized holes drilled in the cone. Various shapes and sizes of teeth are used and are placed on the cone in various configurations. Generally a long, medium or short chisel shape is used for harder formations. Conical or hemispherical button-shaped inserts are used for very hard, abrasive formations and are often called button bits. Insert bits are also illustrated in Fig. 4–10.

As can be seen from the foregoing description of roller bits and depending on the bit design, these bits drill with a combination of crushing, scraping, and gouging and fluid erosion. Some actions are more efficient for drilling one type of formation and other actions are more efficient drilling other types of formation. For example, the combined actions of gouging with long bit teeth and fluid erosion caused by the jetting of the mud can drill very soft formations at penetration rates in excess of 1,000 ft/day. Extremely hard, tough abrasive formations are drilled primarily by a crushing action with button bits with a minimum amount of offset.

The earlier rotary bits used only ball bearings. Since ball bearings have single contact points, they are less efficient and more subject to wear with the heavy loading due to high bit weights. In later bits part of the ball bearings were

FIG. 4–10 Rotary drill bits (courtesy Hughes Tool Company)

replaced with roller bearings. This increased the contact area and subsequent bit life. These roller-bearing bits still contained ball bearings. Bit bearings were later prelubricated and fitted with a seal—sealed-bearing bits—which further extended bit life. The roller bearings were later replaced with sleeve bearings or journals. These journal-bearing or premium bits have a much longer bit life than the earlier bit designs. The roller bit has developed into a carefully constructed, precisely machined, highly efficient drilling tool.

Diamond and Other Bits

Diamond bits were developed to drill very hard formations that could not be drilled efficiently with roller bits. The diamond bit has a metal body with a pin-type connection on top to connect it to the drilling assembly. The bottom of the bit contains a tough durable matrix material with commercial diamonds partially imbedded in the matrix material. The diamonds are placed in rows, spirals, or other configurations on the face of the bit. Different body shapes are used ranging from a full-gauge round body to a lobe-type configuration with two to four lobes. Different sizes and types of water courses also are used. Diamond bits are illustrated in Fig. 4–10.

Diamond bits drill primarily by a scraping action and secondarily by a crushing force. The point of the diamond is imbedded in the formation a few thousandths of an inch by the weight of the drilling assembly on the bit. As the bit is rotated, each diamond scrapes or cuts a groove in the face of the formation. The imbedment depth is determined by the size and number of the diamonds, the force exerted on the formation (bit weight), and the formation hardness. The size of the groove cut by the diamond is determined by the size of the diamond and the depth of imbedment. The drilling rate is determined by the rate of rotation and the size and depth of the groove cut by each diamond. Normally the diamond bit has a long life because the diamonds are much harder than the formation.

A very hard, polycrystalline material has been developed recently to make a drag-type drilling bit. A disk of the material is bonded to a tungsten carbide type material, which is then inserted into the bit body leaving the polycrystalline material exposed as a cutting surface. The bits drill by a combination crushing and scraping action. The bits have only been recently developed but preliminary reports indicate that they drill soft and medium-soft formations efficiently.

Percussion bits are a special bit used with percussion tools (drilling hammer). These bits are similar to regular roller bits except the bit shank has a heavier construction to withstand the shock of the hammer tool.

Coring bits, more commonly called core heads, are used to cut formation cores, which are retrieved for inspection and analysis. The first core heads used drag-type teeth. Roller cutters were developed for the harder formations. Diamond core heads were developed for very hard formations and are now used also for softer formations because they have an efficient cutting action. Diamond core heads have been considered for standard drilling operations. The concept is that drilling with a core head will not require cutting as much formation and the core head will drill faster and stay in the hole longer than a regular diamond bit. This concept has not been proven, and the practice is seldom used.

Bit Selection

In theory, bit selection is relatively simple. The ideal roller bit will wear out the cones and bearings at the same time. The diamonds will flatten on a diamond bit as it wears out. The ideal bit will drill more hole at a faster, more efficient, and cost-effective rate than any other bit available. Therefore the main factors in selecting an ideal bit are (1) the type of formations to be drilled, (2) the drilling fluid being used, (3) the type and strength of the drilling assembly, and (4) the capability of the equipment (pumps, rotary, drawworks, etc.) to supply the power needed to drill.

For drillability purposes formations are classified as soft, medium, hard, very hard, and abrasive. Each classification is further subdivided in varying degrees of drillability. The bit manufacturers classify their bits according to formation drillability classifications (Fig. 4–11). This simplifies bit selection to some extent; however, an obvious problem is determining exactly what formations are to be drilled. When there has been considerable drilling in the area, bit records and other formation data will be available as a guide. Computerized bit-selection programs, some with a built-in data base, are gaining acceptance.

The problem of bit selection is much more difficult in areas where no drilling data are available. Geological information is the main source of help in this case. Sometimes the area will be similar to another area where data are available and can be used as a guide. The performance of the bits being used may also help select bits for deeper drilling.

General information and experience with drilling and bits will also help in the bit selection process. For example if soft or medium formations are being drilled, then only roller bits should be considered. Always use the largest bit possible since it will have larger bearings that usually give a longer bit life than smaller bearings. Good bit stabilization will improve bit life. Use premium-grade bits in deeper holes to reduce the risk of bit failure and a resulting fishing job to recover bit junk—a costly operation at that depth.

Both roller and diamond bits should be considered for the harder formations. The roller bit will generally drill a wider variety of formations more efficiently than a diamond bit. Diamond bits have one advantage over roller bits: they do not have moving parts. If the bit becomes extremely worn, there is no risk of leaving bit debris in the hole such as with a roller bit. However, diamond bits

STEEL TOOTH BITS		
SOFT	**MEDIUM**	**HARD**
Formations normally drilled— soft shales, clays, unconsolidated sands, red beds, salt. Bit types to use: Journal Bearings J1, J2, J3, JD3 Std. Bearings OSC-3AJ, OSC-3J, OSC-1GJ Sealed Bearings X3A, X3, X1G, XDG	Formations normally drilled — hard shales, sandstones, limestones, Bit types to use: Journal Bearings J4, JD4 Std. Bearings OWV-J, OW4-J, WO Sealed Bearings XV, XDV	Formations normally drilled — hard sands, cherty limestones, granite, chert. Bit types to use: Journal Bearings J7, J8, JD8 Std. Bearings W7J, W7R-2J

CARBIDE BITS		
SOFT	**MEDIUM**	**HARD**
Formations normally drilled — shales, clays, red beds, salt, sands, soft limestone. Bit types to use: J11, J22, J33, HH33	Formations normally drilled — limestones, sands, dolomites, hard shales Bit types to use: J44, J55R, J55, HH44, HH55	Formations normally drilled — limestone, dolomite, chert, sandy shales. Bit types to use: J77, J99, HH77, HH99

FIG. 4–11 Bit selector chart (courtesy Hughes Tool Company)

must be run in a clean hole, and they cost more. A small piece of metallic junk such as a nut or bolt will ruin the diamond bit. Both diamond and roller bits are subject to failure or a shorter-than-normal bit life if they do not have an adequate mud circulation rate.

The drilling program contains a recommended bit program that includes a list of recommended bits usually grouped by the depth interval the bits are to drill, estimated bit life, penetration rate, number of bits, operating conditions (bit weight and rotary speed), and recommended mud pressures and volumes. Drilling or formation conditions that may affect bit life or penetration rate should also be included.

MUD PROGRAM

The mud program is one of the more important parts of the drilling program since the drilling fluid, commonly called mud, directly affects the success of the drilling operation. Mud is almost a misnomer since it is a carefully prepared fluid maintained and treated with chemicals and additives in order to control precisely the chemical and physical properties. The hole is normally spudded with a low-quality mud, and each succeeding hole section (conductor hole, surface hole, etc.) is drilled with a higher-quality and sometimes different type of mud.

The mud program will contain the type of mud to be used for drilling each section of hole and the physical and sometimes chemical characteristics of the mud. Some of these are density (mud weight in pounds per gallon or pounds per cubic foot), viscosity (funnel-seconds or sometimes in centipoises), initial and ten-minute gel strengths, fluid loss (cubic centimeters per half hour), sand content (percent), total solids (percent), oil content (percent), chlorides (parts per million), and other properties depending on the type of mud being used.

The program will also have the name of the mud supplier and the mud engineer(s) servicing the well. A total list of the estimated amount of mud additives may also be included. If any special mud problems are expected to be encountered, such as high-temperature gellation, gypsum and anhydrite contaminants, corrosive agents, or fluid-sensitive shales, they are noted with recommended mud treatment procedures if the problems occur.

The mud program will frequently contain recommendations on mud treatment and solids removal equipment. This includes items such as special high-speed shale shakers, desanders, desilters, mud cleaners, degassers, mud mixing unit, mud centrifuge, and mud storage capacity.

Purpose of Mud

The basic purpose of the drilling fluid is to help drill the hole in the fastest, safest, most efficient manner. The mud accomplishes this by performing a number of different functions. Some of these functions are dependent upon the formations and will vary from formation to formation and area to area, and different types of mud perform different functions. Therefore, the mud must be selected and designed so that it can fulfill all of the requirements necessary to drill the formations.

One of the basic purposes common to all muds is to carry the drill cuttings from the bottom of the hole to the surface. The cuttings must be removed from the bottom of the hole so the bit teeth can contact and drill new formation rather than running on cuttings that have already been drilled and released. Bit life and penetration rate are greatly increased with efficient bottom-hole cleaning. The cuttings must be carried to the surface in order to prevent them from "stacking up" and sticking the drilling assembly or possibly causing a fishing job. If the hole is not properly cleaned with mud, the cuttings can settle around the drilling assembly and stick it in the hole, much like tamping a fence post in a hole in the ground. When this occurs, an expensive fishing job is often needed to release the drilling assembly and sometimes the hole is lost (junked), requiring sidetracking or drilling a new hole.

The mud also transmits hydraulic horsepower to the bit to do useful work. This is accomplished in several different ways. One method is removing cuttings as described. Another method is to transform the mud flow into high-velocity jets (by using jet bits) to help erode and drill the formation. This action can be very efficient in soft formations; efficiency decreases with harder formations. There is an increasing use of downhole motors and turbines to rotate the bit without rotating the entire drilling assembly. This method of rotating the bit has many advantages over conventional methods. Many problems have been encountered, but the method has high potential. The energy (or work) to operate these devices is transmitted downhole with the drilling fluid.

A considerable amount of work and energy is expended at the bit drilling the formation. This creates a large amount of heat. The natural heat of the formations increases with depth, and downhole temperatures of over 200°F are relatively common. The bit and other downhole equipment must be cooled to prevent damage. During drilling there also may be a tendency for the bit cutting face to become packed and coated with cuttings, thereby decreasing its efficiency. The drilling mud cools the equipment and cleans (in some cases lubricates) the bit.

High-pressure formations are often encountered in drilling operations. These pressures must be contained in order to prevent personnel injury and equipment damage and to allow the drilling-completion operation to be conducted in a normal manner. The column of mud exerts a hydrostatic head, and the mud density (mud weight) can be increased or decreased as necessary to control high or abnormally pressured formations.

Some formations are unstable as a result of fracturing or other conditions. These formations will frequently cave or slough into the hole. This caving material can stick the drilling assembly in a manner similar to the "stack-up" of cuttings with similar results. In many cases the hydrostatic head exerted by the mud column will hold these formations back and stabilize the wellbore.

Fluid-sensitive formations tend to swell and cave or slough when they come into contact with certain fluids. For example, bentonitic (or montmorillonitic) shales swell and then cave and slough in the presence of fresh water. The action can be minimized or eliminated by selecting the type of mud and controlling its physical and chemical properties. Therefore, another purpose of the mud is to reduce or minimize any adverse effects of the formations being drilled.

The mud must have sufficient gel strength to carry the cuttings out of the hole and a low enough gel strength to allow the cuttings to be easily separated from the mud on the surface. These are functions of the initial gel strength of the mud. The mud must also have sufficient final gel strength so that, when circulation is stopped such as to make a connection, the cuttings will remain suspended in the mud system and will not settle out and stick the assembly. At the same time the final gel strengths must be low enough so that circulation can be started without excessive pressures.

Some highly treated muds are corrosive. Corrosion must be minimized in order to prevent damage to the drilling equipment. Corrosive agents (O_2, CO_2, and H_2S) are encountered in drilling operations. The mud must be compatible with inhibitors, mud additives that reduce or eliminate corrosive action.

The mud must also have sufficient lubricity (act as a lubricant) to reduce drag so the drilling tools can slide easily as they are run in and pulled out of the hole.

The basic purpose of drilling is to complete a producing well. Therefore, the mud must have the proper physical and chemical properties so the operator can obtain information about the formation and its contents. The mud must complement cuttings recovery, coring, and open-hole drillstem testing operations. The composition of the mud should be such that when formation fluids migrate into the mud column, they can be detected without difficulty.

The mud is carefully prepared and maintained. It must be tested periodically to ensure that it is in good condition or to determine what additives are needed to restore and maintain it.

Types of Mud

There are three basic types of drilling fluids that are classified according to the fluid in the continuous phase: air (or gas), water-based muds, and oil-based muds. Some of these are further subdivided into different types within the classification to serve different purposes. The type of mud selected for a drilling operation will be the most economical fluid to accomplish the drilling objective. Economics in this case is based both on the cost of the mud and additives and a cost comparison of the effect of the mud on drillability and control of hole problems.

Air or Gas Air or gas is without question the best drilling fluid if evaluated based strictly on obtaining a maximum drilling rate. It has the lowest viscosity, density, gel strength, and solids content. A maximum drilling rate is obtained when these parameters have a minimum value. It is not uncommon to drill a hole with air 2–5 times faster than with liquid mud, especially in very hard formations.

Air usually does not affect the formations, and lost circulation is not a problem. The method is highly effective for detecting gas shows when drilling low-pressured formations.

Air or gas has several disadvantages that limit its use. It does not have sufficient density to control permeable, high-pressure formations. A downhole explosion may occur if flowing gas zones are encountered while drilling with air. More caving may occur in the air-drilled hole due to the lack of hydrostatic head normally supplied by the drilling mud column. Air has a low density and a resulting low buoyancy so that hook loads are higher. The hook load is the total weight of the downhole drilling assembly suspended from the hook attached to the bottom of the traveling block. Higher hook loads cause additional wear on the equipment. Air has very little lubricating effect, so there is increased wear on the downhole tools. It can be highly corrosive, causing damage to downhole tools.

Many formations will flow small amounts of liquid into the air-drilled hole. This causes the cuttings that are finely pulverized (almost dust) to stick to the drillpipe and the walls of the hole, forming mud rings that can stick the assembly. Air mist or foam may be used for drilling when small amounts of liquids are encountered. The mist or foam is created by injecting a small volume of liquid soap into the injected air or gas stream. The soapy foam lifts the liquids and cuttings in the air stream.

Air or gas drilling techniques usually are limited to about 10,000–12,000 ft, less when formation conditions are encountered that require a liquid mud system. At deeper depths air pressure requirements become increasingly higher with associated higher costs.

Water-Based Muds   Water-based muds use liquid water as the continuous phase; the mud filtrate is also water. Very fine clay or gel particles are suspended in water to make a colloidal suspension. The clay particles are known as active solids because they have valence bonds that may or may not be broken and unsatisfied, requiring chemical treatment. The clay particles are obtained from the formations being drilled (mud-making formations) and additions of commercial bentonite.

Water-based muds are subject to both physical and chemical contamination during drilling. Finely ground barite is added as a weighting material. Fine drill cuttings such as sand can remain suspended in the mud. These types of materials are known as inert solids because they do not react chemically with the mud. An excess of inert solids will cause unfavorable mud properties, thus requiring the removal of the solids by dilution or physical separation at the surface. The mud properties are also affected by an excess of active solids, which are removed by dilution or separation at the surface or treated chemically.

Some formation materials such as salt, gypsum, and anhydrite will contaminate the mud, upset its chemical balance, and cause deteriorating mud properties so that the mud cannot accomplish its primary purposes. If small amounts of these materials are encountered during drilling, they are usually treated by a combination of chemical treatment and removal by dilution. When large amounts of these materials are encountered during drilling, usually the most economical method is to change the mud system to a type of mud that will tolerate the contaminating material.

For example, freshwater muds cannot tolerate large volumes of salt. This is one of the reasons that a saltwater flow is dangerous because it causes deteriorating mud properties that lead to other problems. In order to drill salt formations, a saltwater mud is used. The basic mud is somewhat similar to freshwater mud except salt water is used with a different type of clay (attapulgite) that is effective in salt water. This mud is slightly more difficult to control than regular freshwater mud, but it can be controlled to drill salt sections satisfactorily.

Sometimes massive salt sections are drilled with saturated salt muds in order to prevent the salt from dissolving in the mud system and creating a huge cavern. Other muds that are used for a similar reason but for different contaminating formations or similar conditions are gyp (gypsum) muds, lime-based mud, and high pH muds.

During the drilling process the mud deposits a wall cake on the wall of the borehole. The wall cake is made of very fine, small particles in the mud that are filtered out on the wall and in pore spaces immediately adjacent to the wall. These particles move to the wall and initially become stuck or plastered because of the pressure differential between the weight of the mud column and the pore pressure of the formation. The wall cake may be further strengthened by the constant rotating, whipping movement of the drilling assembly, which strikes the wall cake. Mud filtrate moves into (invades) the pore spaces while the wall cake is being formed. After the wall cake has been formed, there is very little fluid movement through the wall cake into the formation pore spaces. Standard API fluid loss tests are run to determine the quality of the wall cake, commonly called filter cake or mud cake, and the amount of filtrate loss, commonly called fluid loss or water loss.

Fluid-sensitive formations may cause drilling problems (caving and sloughing) when they contact contaminating fluids such as those in the mud filtrate. The wall cake acts as a barrier to prevent the escape of mud filtrate into the formation. When fluid-sensitive formations are drilled, the first approach is to improve the quality of the wall cake and reduce the volume of the fluid loss. A common method is to use fluid-loss additives such as carboxyl-methyl-cellullose (CMC). Another approach is to increase the amount of good wall-building, small-sized particles in the mud system, such as dispersing a nondispersed system. Dispersing reduces the size of the solid particles in the mud; nondispersed mud systems have larger particles. Thinners such as tannins and quebroxins may also be used. These types of muds are collectively known as low-fluid-loss muds.

Sometimes the fluid loss cannot be reduced low enough, or the shales are extremely sensitive so that other methods must be employed. One approach is to use special polymers added to the mud system or polymer muds. These long-chain molecules—many molecules bonded together in long chains—coat the fluid-sensitive shale surfaces so that the formation filtrate does not contact the shale, thus reducing or eliminating the problem. These types of mud also have other uses.

Any of these methods can be unsuccessful for a variety of reasons, such as extremely fluid-sensitive shales, fractured shales, or the expense in reducing the water loss. Another approach is to use an inhibited mud. An inhibited mud is a system where the mud filtrate is either chemically unreactive to the formations causing problems or one that reacts in such a manner as to prevent the problem. Certain complex phosphates can make an inhibited mud system. In some cases lime-based mud and saltwater muds may be inhibited systems.

Oil is frequently mixed with (emulsified in) water-based muds. These are known as oil-in-water emulsion muds or, more commonly, oil emulsion muds. The oil content averages about 5–10%, but higher oil contents have been used. Oil increases the lubricity of the mud and improves the fluid loss.

Clear water is the best water-based drilling fluid where drillability is the main criteria. Drillability increases with lower mud weights and viscosities and with higher fluid losses. Salt water is also a very good drilling fluid where it can be used safely.

Oil-Based Muds Oil-based muds have oil as the continuous phase and in the mud filtrate. They are normally not considered as a regular, conventional mud used in drilling for several reasons. The muds are very expensive, often costing over 3–5 times more than a comparable water-based system. General field experience has indicated that drilling rates are lower in oil muds than in water-based muds. However, new oil-based muds have been developed that may not reduce the penetration rate. These muds are susceptible to aeration, foaming, and water contamination and may create a fire hazard. The muds may be difficult to maintain and definitely present a cleanup problem with both personnel and rig equipment. In many areas the crews are paid special wages when using oil mud.

Invert muds or water-in-oil emulsion muds contain a relatively high percentage of water, which tends to reduce the overall mud cost. Since oil is the continuous phase, invert emulsion muds may be used in most cases where oil-based muds are used, providing that the emulsion remains stable and does not break down releasing the water. For example, oil-based muds will usually withstand considerably higher temperatures than invert muds.

Oil muds generally are used in special drilling situations such as drilling in extremely high-temperature areas that create problems with water-based fluids. Oil-based muds are relatively inert and can be used to drill fluid-sensitive formations where water-based fluids cannot be used or cause problems. They are frequently used to drill productive zones that may be damaged by water-based fluids. And they are often used in extremely difficult fishing jobs, especially when washover operations are being conducted and when formation problems may occur with water-based fluids.

In summary, the selection and treatment of the drilling mud is an important contributing factor to efficient drilling and completion operations. The mud program must be well planned and all details included in the drilling program.

GENERAL EQUIPMENT SPECIFICATIONS

A large amount of specialized equipment is required to drill and complete a well. The personnel who have studied the well and prepared the drilling program usually are the most qualified to list the specifications on equipment that will be needed. This may be summarized and abbreviated on shallow and infield development wells where the drilling program is relatively simple and straightforward. More detail should be included on deeper wells that may require a wider variety of equipment.

Most rigs are rated by depth and will have the capacity and associated equipment needed to drill and case a hole to that depth. Therefore, in most cases all that is necessary is to state the depth capacity of the rig. However, different rigs may have different equipment, even though they have the same depth capacity rating. Therefore, the drilling program should specify such items as the drawworks and pump horsepower, hook-load rating with and without setback (pipe racked back in the mast), pump capacity (pressure rating for various liner sizes), mud solids separation equipment, and any special items that might be required but are not normally included with a standard rig with the depth capacity required. This information will allow the operator to check the rig inventory against the actual drilling requirements to ensure that any rig being considered has sufficient horsepower and capacity.

Bit sizes for the different hole sections are always included. The drilling assemblies required to drill each hole section usually are included in deep well drilling programs but may be omitted in programs for standard, shallow wells. In the shallower holes the drilling assembly usually is supplied with the rig. However, in the deeper, more complex holes a number of different drilling assemblies may be required. These are often rented, especially when they are specialized items that the drilling contractor does not normally keep in stock. The drilling program should also contain recommendations for the inspections of the drilling assemblies and the time interval between inspections. These are usually included as recommendations only, since the final inspections must be determined by the operating conditions.

Blowout preventer and choke manifold specifications and the number and position of the control stations should always be listed. The procedures for testing this equipment and the time interval between tests should also be noted. Blowout drill procedures should also be established, and standard and special blowout detection equipment should be listed.

AFE AND CONTRACTS

The authority for expenditure (AFE) contains the estimated total cost to drill and complete the well (Fig 4–12). It is normally subdivided into a dry hole and a completed well cost. The costs are subdivided into tangible and intangible costs and are further subdivided and grouped under standardized accounting codes. The data from the drilling program are used to help prepare the AFE, and a copy normally is included in the drilling program.

The contracts for the various services normally are obtained after the drilling program has been prepared. The main contract is the drilling contract, covering the services and obligations of the contractor who supplies the rig and the company or owner for whom the well is being drilled. This binding legal document is negotiated by both parties. Other equipment and services, such as mud supplies, logging, cementing, equipment rentals, and testing, are supplied under a contract, price list, and/or rental agreement.

The current practice is to keep the original documents in the company office and copies at the well site. This allows the wellsite supervisor, who is more closely associated with the operations, to ensure that the terms of the agreements are complied with by all parties.

Any other documents or information that affects the drilling of the well is also given to the wellsite supervisor. This procedure helps the wellsite supervisor control costs and keep an accurate, current accounting of expenditures. In the normal sequence of operations, the wellsite supervisor reports the daily and cumulative costs on a daily basis with the drilling report.

In summary, the drilling program contains all of the information, data, and planning for drilling and completing the well. The drilling program is designed to be flexible and can be changed as indicated by operations. Experienced operating personnel can take the drilling program and successfully drill and complete the well.

Summary

Open: _Little Pecan Lake, Ltd._   Date: _14-Jun-82_

Lease: _Denex # 1_   Field: _Go Around Bayou Field_

Sec. _29_   Twp. _14S_   Rng. _4W_   County: _Cameron Parish_   State: _LA_

Expenditure	Dry hole (24.5 days)	Completed (32.5 days)
Intangible costs		
100 Location preparation	30,000.00	65,000.00
200 Drilling rig and tools	298,185.75	366,612.94
300 Drilling fluids	113,543.19	116,976.37
400 Rental equipment	77,896.37	133,784.75
500 Cementing	49,534.68	54,368.73
600 Support services	152,285.44	275,647.50
700 Transporation	70,200.00	83,400.00
800 Supervision and administration	23,282.50	30,790.50
Subtotal	814,927.94	1,126,581.00
Tangible costs		
900 Tubular equipment	406,100.87	846,529.44
1000 Wellhead equipment	16,864.00	156,201.00
1100 Completion equipment	.00	15,717.00
Subtotal	422,964.87	1,018,447.44
Subtotal	1,237,893.00	2,145,028.00
Contingency (15.0%)	185,683.94	321,754.25
Total	1,423,577.00	2,466,782.00

Detailed summary

Expenditure	Dry hole (24.5 days)	Completed (32.5 days)
100 Location Preparation		
110 Permit	500.00	2,500.00
120 Survey	2,500.00	7,500.00
130 Right of way, special permit, etc.	2,000.00	2,000.00
140 Physical location preparation	20,000.00	48,000.00
150 Cleanup	5,000.00	5,000.00
Total	30,000.00	65,000.00

FIG. 4–12 Authority for expenditure (after Adams and Frederick, "How to Estimate Well Costs," courtesy *OGJ,* December 13, 1982)

Detailed summary

Expenditure	Dry hole (24.5 days)	Completed (32.5 days)
200 Drilling rig and tools		
210 Move in and out	57,135.37	57,135.37
220 Footage bid	.00	.00
230 Straight day work bid	182,327.06	241,862.44
240 Fuel	32,915.79	41,018.13
250 Water	5,000.00	5,000.00
260 Bits	20,807.50	21,597.00
270 Completion rig	.00	.00
Total	**298,185.75**	**366,612.94**
300 Drilling fluids		
310 Drilling fluids	113,543.19	113,543.19
320 Packer fluids	.00	3,433.16
330 Completion fluids	.00	.00
Total	**113,543.19**	**116,976.37**
400 Rental equipment		
410 Well control equipment	29,852.00	43,262.00
420 Rotary tools and accessories	6,794.22	22,425.67
430 Mud related equipment	19,475.00	23,856.87
440 Casing tools	21,775.16	44,240.16
450 Miscellaneous	.00	.00
Total	**77,896.37**	**133,784.75**
500 Cementing		
510 Conductor casing	.00	.00
520 Surface casing	20,121.85	20,121.85
530 Intermediate	15,619.91	15,619.91
540 First liner	.00	.00
550 Second liner	.00	.00
560 Production casing	.00	18,626.97
570 Squeezes	.00	.00
580 Plugs	13,792.92	.00
Total	**49,534.68**	**54,368.73**
600 Support services		
610 Casing crews	11,759.15	23,536.71
620 Logging		
621 Mud logging	18,000.00	18,000.00
623 Wire line		
624 Logging	77,656.56	109,083.94
625 Perforating	.00	11,447.00
626 Testing	14,480.00	14,480.00
627 Completion services	.00	33,597.00

FIG. 4–12 *continued*

Detailed summary

Expenditure	Dry hole (24.5 days)	Completed (32.5 days)
630 Tubular inspection		
631 Surface casing	4,896.45	4,896.45
632 Intermediate casing	14,643.30	14,634.30
633 First liner	.00	.00
634 Second liner	.00	.00
635 Production casing	.00	18,213.00
636 Tieback string	.00	.00
637 Tubing	.00	13,960.00
638 Miscellaneous	.00	.00
640 Galley	.00	.00
650 Welding, labor, rental equipment	10,850.00	13,790.00
660 Formation testing	.00	.00
670 Fishing and directional consultants	.00	.00
680 Acidizing, fracturing, and gravel pack	.00	.00
690 Miscellaneous	.00	.00
Total	**152,285.44**	**275,647.50**
700 Transporation		
710 Trucking	70,200.00	83,400.00
720 Marine	.00	.00
730 Air	.00	.00
Total	**70,200.00**	**83,400.00**
800 Supervision and administration		
810 Field supervision	16,250.00	20,250.00
820 Office supervision	7,032.50	10,540.50
830 Insurances, bonds	.00	.00
Total	**23,282.50**	**30,790.50**
900 Tubular equipment		
905 Drive pipe	7,498.00	7,498.00
910 Conductor casing	.00	.00
915 Surface casing	71,006.56	71,006.56
920 Intermediate casing	321,156.31	321,156.31
925 First liner	.00	.00
930 Second liner	.00	.00
935 Production casing	.00	325,291.06
940 Tieback string	.00	.00
950 Tubing	.00	113.048.50
960 Casing equipment		
961 Drive pipe	230.00	230.00

FIG. 4–12 *continued*

Detailed summary

Expenditure	Dry hole (24.5 days)	Completed (32.5 days)
962 Conductor casing	.00	.00
963 Surface casing	3,500.00	3,500.00
964 Intermediate casing	2,710.00	2,710.00
965 First liner	.00	.00
966 Second liner	.00	.00
967 Production casing	.00	2,089.00
Total	406,100.87	846,529.44
1000 Wellhead equipment		
1010 Casing head	3,220.00	3,220.00
1020 Intermediate spool	13,644.00	13,644.00
1030 Tubing spool	.00	55,465.00
1040 Tree	.00	83,872.00
1050 Miscellaneous	.00	.00
Total	16,864.00	156.201.00
1100 Completion equipment		
1105 Packers	.00	2,059.00
1110 Blast joint and landing nipples	.00	3,955.00
1115 Special liners	.00	.00
1120 Safety joints	.00	796.00
1125 Subsurface safety devices	.00	4,388.00
1130 Seal assembly	.00	4,519.00
1135 Gaslift equipment	.00	.00
1140 Gravel packing equipment	.00	.00
1145 Miscellaneous	.00	.00
Total	.00	15,717.00

FIG. 4–12 *continued*

CHAPTER 5
DRILLING PERSONNEL & EQUIPMENT

Two of the most important factors in the industry are personnel and equipment. They are interrelated and interdependent. The best personnel available cannot perform efficiently without the proper equipment. Conversely, the best equipment available is useless if it is not developed, selected, and utilized efficiently. Personnel have learned from past experience and have developed new, innovative techniques, practices, and procedures. Equipment has been developed in a parallel fashion. The combination has drilled holes over six miles deep and is technically capable of drilling to depths over ten miles. Personnel and equipment have grown and improved with time and are directly responsible for the current, successful oil and gas industry that supplies the world's energy requirements.

PERSONNEL AND SERVICES

Only a very few people performed all of the exploration, drilling, and completion work during the early history of the oil industry. The industry expanded rapidly geographically, and new equipment mirrored the growth. This created a greater demand for people with widely varied talents and expertise. With technological advances, more specialization was required. The net result was an industry that used a large number of people with varying background, training, and experience.

Company/Owner

The company, frequently referred to as the owner or operator, usually has geological, land, and engineering personnel who locate the drilling prospects. The company then drills the well with a drilling rig owned by a drilling contractor.

Numerous other details must also be handled in the drilling operation. These include materials movement and control, invoicing and payment procedures, clerical work and recordkeeping, and obtaining services of third parties, usually on a contract basis (see appendix).

Most companies have an operations department, sometimes called a drilling department, that handles all of the operational details associated with drilling and completing the well. The department may have two or three people or several dozen, depending on the area of operations and the average number of wells drilled. The department is headed by a manager, drilling manager, or drilling superintendent who is thoroughly experienced in drilling operations and frequently has a B.S. degree in petroleum engineering. There will usually be an assistant manager (assistant drilling superintendent) and several drilling engineers and/or drilling supervisors.

A materials supervisor arranges for the purchase and delivery of casing, bits, mud, and all other materials and equipment needed at the wellsite to drill the well. Other office personnel may include administrative assistants, invoice analysts, secretaries, clerk-typists, and file clerks.

In the normal drilling operation the field personnel at the rig communicate daily with the office personnel, giving reports on the drilling operation, the status of the materials and supplies, operating problems, and generally all details relating to drilling the well. The field and office personnel coordinate activities and discuss problems in order to drill the well most efficiently.

A drilling report is usually made out at the rig and telephoned to the office each morning covering the drilling activities for the past 24 hr. This drilling report is circulated in the office to higher-level management and to partners and other interested parties. Most companies maintain daily operating costs, so the report contains the operating costs for the prior 24-hr period and cumulative costs on the well to date.

Many companies also have field personnel who supervise drilling operations at the wellsite in conjunction with the contractor's personnel. The drilling foreman, or *company man,* has direct responsibility for all drilling operations at the wellsite. He also maintains contact with office personnel, helps coordinate the movement of material and supplies, and supervises third-party services.

In summary, the operations department or owner takes a drilling prospect, prepares a drilling program, negotiates a drilling contract, obtains a drilling rig and third-party services as needed, orders materials, pays invoices, and generally performs all of the duties required in order to drill and complete the well.

Drilling Contractor

Earlier in the industry, companies or owners had their own rigs and drilled their own wells. When the rigs were not busy drilling on the company's properties, they were used to drill wells for other companies under a drilling contract. As the industry evolved, some companies began to specialize in drilling operations and were known as drilling contractors, or simply contractors. The contractors became more efficient and were able to drill wells more economically than the companies could with their own rigs. As a result the companies began to sell their company-owned rigs and use the drilling contractors for drilling. A few companies still have rigs, and some contractors drill their own wells. However, the majority either specialize in owning the wells or drilling the wells.

The drilling contractor supplies to the owner the basic drilling rig and accessory equipment to drill the well. He also supplies the supervisory personnel and drilling crews to operate the rig and furnishes all of the equipment, materials, and supplies (expendables) used on the rig. The owner or operator, who is usually a part owner of the well and represents the other owners, in turn furnishes equipment, material, and supplies such as drilling bits, casing, and mud that are used in drilling the well.

The contractor drills the well for the owner under the terms of a drilling contract. This is a legal instrument negotiated between the owner and contractor. It contains all of the terms and provisions that govern the use of the rig in drilling the well and defines the materials, supplies, and services to be supplied by each party and their respective responsibilities.

Many wells, especially with land rigs, are drilled on a *footage basis*. The contractor is paid a footage rate, i.e., so many dollars per foot for the number of feet drilled. He is normally responsible for all operations, including downhole, while drilling on a footage basis. A move-in, rig-up, rig-down, and move-out charge is either included in the footage rate or stated in the contract as a fixed amount.

During drilling, many operations are normally done on a *day-work basis*. These include logging, running and cementing casing, and testing. The cost of drillpipe and drill collars is one of the contractor's major expenses. Therefore, the average contract will normally have two day rates: one with drillpipe and one without drillpipe. The contractor bills the owner for (1) the amount of hole drilled at the footage rate, (2) for day work with drillpipe, and (3) for day work without drillpipe. Day work with drillpipe includes drilling, casing, and similar operations. Day work without drillpipe includes logging, running casing, cementing, and other operations performed without drillpipe.

Drilling rigs are sometimes contracted on a straight day-work basis. One of two day-work rates may be used, depending on whether the contractor supplies the drillpipe and drill collars or whether these are supplied by the owner. When the rig is operating on a day-work basis, the contractor has little risk. If a fishing job occurs, the owner is responsible for recovering the pipe lost in the hole. On a footage rate contract, the contractor would be responsible for the fishing job. The owner is normally responsible for any other third-party tools—such as reamers, stabilizers, and core barrels—that are run in the hole.

In a few cases contractors will drill wells on a *turnkey basis*. In the normal turnkey contract the contractor drills the hole to a predetermined turnkey or contract depth that can be logged and cased for a fixed price. The contractor assumes responsibility for all downhole problems and usually receives a higher price as compensation for assuming these additional risks.

The drilling contractor usually has a central office with a drilling superintendent and an assistant superintendent. The office will also have regular clerical help with personnel to handle materials and bookkeeping services. Very large drilling contractors may have a central office with district or regional offices in the areas where the rigs are operating.

Most drilling contractors have yards with storage and repair facilities. Extra equipment is stored in the yard. The rigs or component parts are brought in for repairs when needed. Some contractors maintain a complete fleet of oil-field trucks to move the rigs, but most contractors use trucking contractors who have trucks that are specially designed for moving rigs and oil-field equipment.

A *toolpusher,* sometimes called the rig manager or rig superintendent, is in charge of each drilling rig. He supervises all of the work done by the rig in accordance with the drilling contract. He is also in charge of maintaining the rig, ordering materials and supplies as needed, and supervising the drilling crews.

The drilling crew consists of a driller, a derrickman, a motorman, and two rotary helpers, sometimes called floor men or roughnecks. The driller is directly in charge of the crew and the operation of the rig. The driller makes out reports as required, operates the drawworks, and supervises all other activities around the rig.

The derrickman is normally the second most-experienced man on the rig. He works in the derrick when the pipe is being pulled or run in the hole. He also repairs and maintains the mud pumps, checks and treats the drilling mud, and helps the driller as needed. The motorman is responsible for the maintenance, upkeep, and repair of all motors and fills in on other jobs as required. The rotary helpers work around the rotary, making connections during drilling and tripping the drillpipe. They also do cleanup work and help the derrickman and motorman as needed. Most rigs have two rotary helpers, but some of the larger rigs have either three rotary helpers or an extra cleanup man. Some drilling contractors have electricians, mechanics, and welders who work on the individual rigs as they are needed.

The drilling crew—driller, derrickman, motorman, and two rotary helpers—is listed in the descending order of authority, experience, and rate of pay. Normally each man in the crew is directly responsible to the driller.

The traditional line of promotion is from floorman to motorman to derrickman and then driller, although there are exceptions. Frequently the motorman is a specialist in maintaining and repairing the motors; he remains on this job and does not move in the normal line of promotion. The derrickman's job requires the special ability to be able to work up in the derrick, normally about 85 ft above the rig floor and 100–120 ft above ground level. In some cases capable personnel who fear heights are promoted directly from rotary helper to driller, but the normal procedure is rotary helper to derrickman to driller.

Promotion above the driller's level in the contractor's organization is rig superintendent, assistant drilling superintendent, and drilling superintendent.

DRILLING EQUIPMENT

All types of rigs drill a hole similarly. A bit is simultaneously rotated and forced against the formation, drilling a hole. Mud is circulated through the bit to wash out the bit cuttings and perform other functions. Therefore, all rigs must have a means of rotating the bit while pushing it into the formation and simultaneously circulating mud.

The rig must also be able to perform other functions such as replacing the worn bit with a new bit, running casing, and controlling formation pressures. Therefore, the rig will have a number of parts, each of which performs one of the required functions—either singularly or in combination with other parts of the rig.

Overall, though, the rig must be typed or classified based on the depth of the hole, the type of power source, and where and how the rig will be used.

RIG CLASSIFICATIONS

Rigs are basically classified by their depth rating. This is the average maximum depth a hole can be drilled with a rig using standard equipment and tools. A large, heavy-duty rig can drill a deeper hole than a lightweight truck-mounted unit. Shallow-drilling and trailer-mounted rigs are usually limited to drilling depths of 5,000–8,000 ft. A 10,000-ft rig normally is used to drill holes from 7,500–10,000 ft. A rig with a 25,000-ft depth rating is used to drill wells from 18,000–26,000 ft deep.

The size, weight, and amount of equipment and tools increase as the depth rating increases. For example, an 8,000-ft rig might carry 8,000 ft of 3½-in., 13.30-lb/ft drillpipe with a total weight of 106,400 lb. A 15,000-ft rig could carry 15,000 ft of 4½-in., 16.60-lb/ft drillpipe, or a total weight of 249,000 lb—70 tons more pipe than the smaller rig. Other equipment would be correspondingly heavier and larger.

Therefore, rig portability decreases with increasing depth capacity. Heavy, deep rigs will be large and difficult to move. But for shallower holes a rig will be relatively small, lightweight, and mobile. Mobility is important because it makes the rig more economical. Since the rig is not actually working while being moved, the move must be made as fast and economically as possible. Thus, small rigs are often mounted on trucks or trailers to increase portability and decrease the cost of rig-down, moving, and rig-up.

Rig rental rates (either footage or day-work) also increase with increased depth rating because of the corresponding larger investment and increased operating maintenance costs.

The operator should select a rig with the proper depth rating for the well to be drilled. A 10,000-ft rig should be chosen for a hole dug to 7,500–10,000 ft. A smaller rig should not be used because of insufficient capacity and possible failure as a result of overstressing. This could result in a fishing job (retrieving broken equipment downhole) or a lost hole with resulting higher expenditures. Likewise, a larger rig should not be used because it would cost more. If the correct sized rig cannot be found, normally the next larger size rig is used.

The depth rating is the main method of classifying drilling rigs. There is also a tendency to classify rigs by the amount of horsepower available to run the drawworks (Fig. 5–1). There can be a wide variance in rig horsepower for a given depth. This is caused by the type of rig, the area where the rig is operating, and the operator's preference.

The first rotary rigs were powered by steam. Later, power or mechanical rigs were developed that used internal combustion engines as a power source. In the meantime, the diesel-electric, or dc electrical, rig was developed, followed by the SCR rig. Presently, both mechanical and electric rigs are used, although there is an increasing tendency to build the new, larger rigs with the SCR system. Both mechanical and SCR systems are used on medium-sized rigs; most smaller rigs are mechanical rigs.

Conventional land rigs were modified to make a new classification of marine rigs as drilling operations were expanded to the submerged coastal areas and offshore waters. Marine rigs were further subdivided and classified as barge or platform, jackup, semisubmersible, and drillships. They are also classified by their depth rating. These rigs are basically similar to large land rigs and use similar equipment. The main differences are (1) they are permanently mounted on the barge or platform and (2) they often have extra (backup) equipment, such as three mud pumps instead of the two pumps used on a similar land rig. They also have special equipment, such as motion compensators, that are required for marine operations.

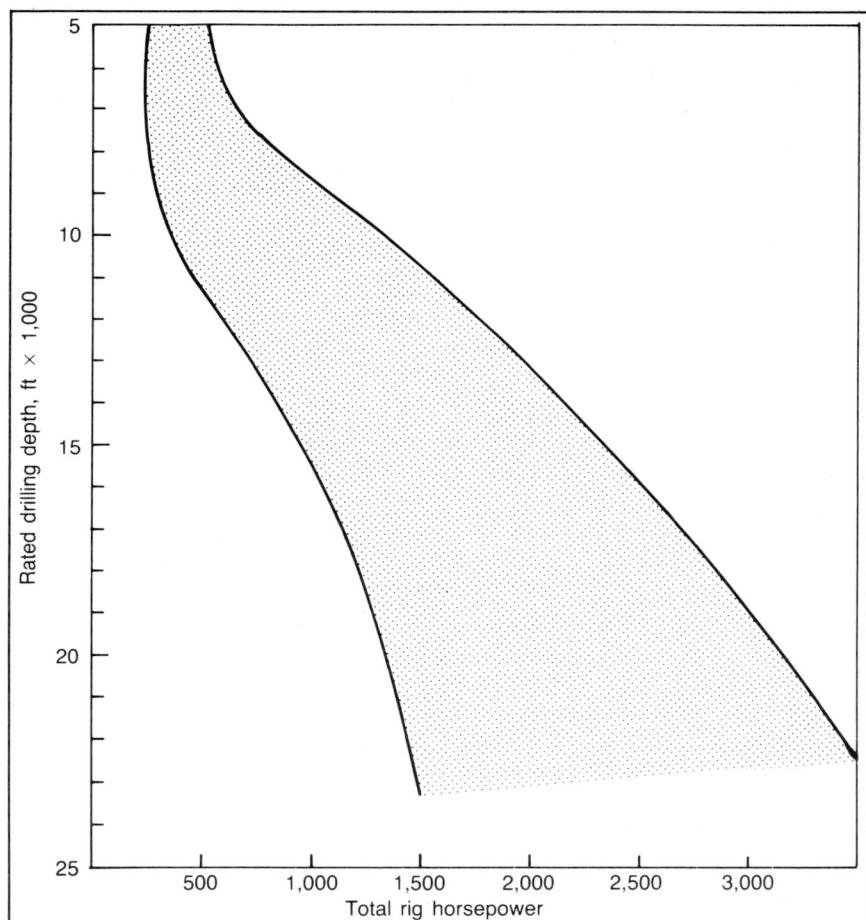

FIG. 5–1 Drilling rig horsepower

RIG PARTS AND FUNCTIONS (COMPONENT SYSTEMS)

The common land drilling rig is made up of various component systems such as the structure (substructure and mast or derrick), hoisting system, prime movers and power train, rotating equipment, pumping and circulating system, blowout control equipment, metering-monitoring equipment, and accessory-miscellaneous equipment (Fig. 5–2). Each system has a number of different parts and pieces of equipment that operate together to complete a function. The

176
DRILLING

1. Accumulator
2. A-frame
3. Air compressor
4. Annular (bag) preventer
5. Annulus
6. Base
7. Bell nipple
8. BOP control
9. Bit (drill)
10. Bradenhead
11. Burning pit
12. Casing-hanger spool
13. Cathead
14. Cat line
15. Catwalk
16. Cellar
17. Centrifuge
18. Chemical barrel
19. Choke line
20. Choke manifold
21. Choke manifold control
22. Compound
23. Conductor casing
24. Crown block
25. Cyclone desander desilter
26. Dead line
27. Degasser
28. Discharge line
29. Doghouse
30. Drawworks
31. Drill collars
32. Driller's console
33. Drilling line
34. Drillpipe
35. Drill tool storage (junk box)
36. Dynamatic or hydromatic
37. Elevators
38. Engines
39. Fast line
40. Fill-up line
41. Flow line
42. Fuel line
43. Fuel tank
44. Generating unit (light plant)
45. Gin pole
46. Hoisting line
47. Hook
48. Intermediate casing
49. Kelly
50. Kelly bushing
51. Kelly (rotary) hose
52. Kill line
53. Ladder
54. Line guide
55. Mast
56. Mast lifting line
57. Mixing (mud) pit
58. Monkey board
59. Mousehole
60. Mud
61. Mud-gas separator (gas buster)
62. Mud gun (submerged)
63. Mud gun (surface)
64. Mud hopper
65. Mud line
66. Mud logging unit
67. Mud (paddle) mixer
68. Mud-mixing plant
69. Oil and grease storage
70. Pipe rack (floor)
71. Pipe racks
72. Pressure (mud) gauge
73. Preventer control lines
74. Preventer (BOP) ram type
75. Production casing
76. Pump drive
77. Pump, mud mixing
78. Pumps, mud
79. Ram wheel
80. Ramp
81. Rathole
82. Reserve drilling line
83. Reserve (mud) pit

Detail for 113

FIG. 5–2 Drilling rig schematic

177
Drilling Personnel and Equipment

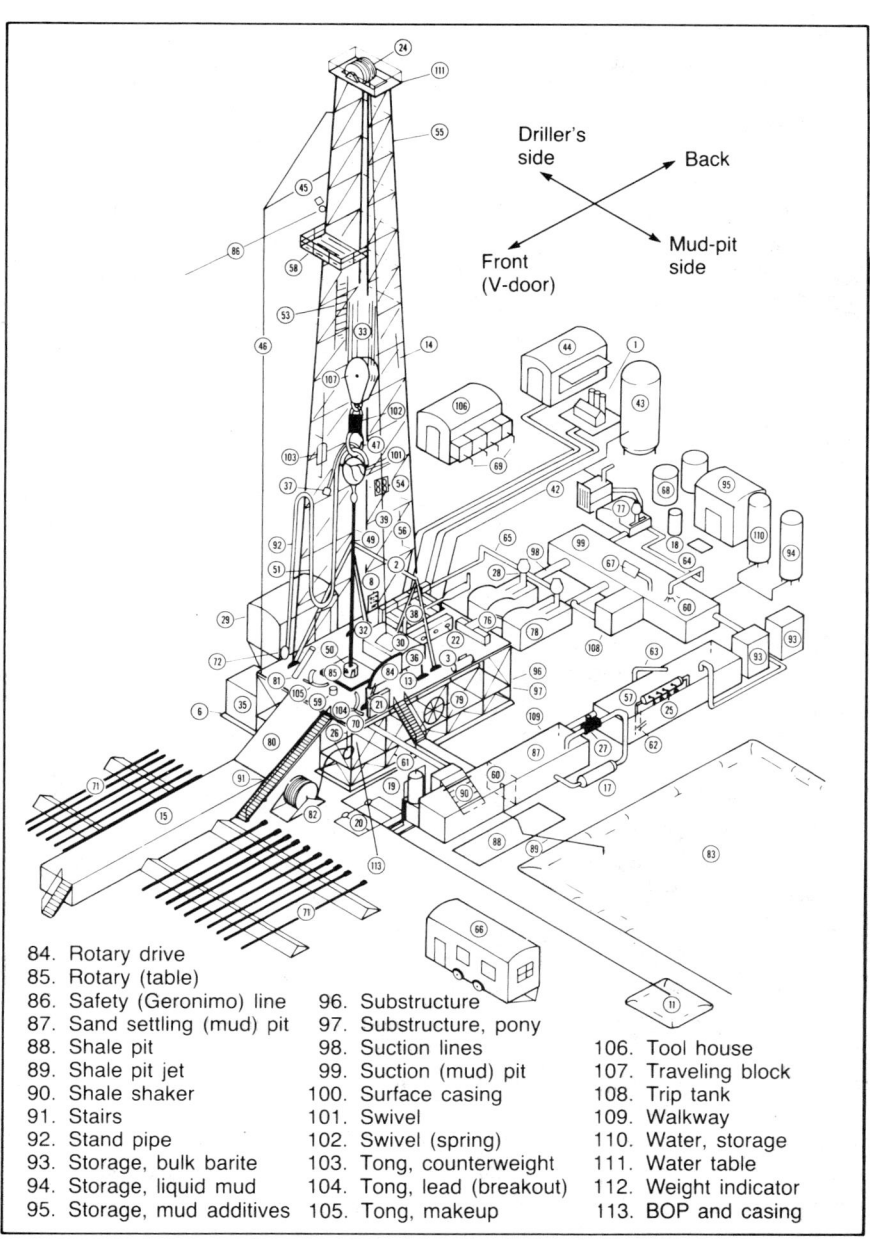

84. Rotary drive
85. Rotary (table)
86. Safety (Geronimo) line
87. Sand settling (mud) pit
88. Shale pit
89. Shale pit jet
90. Shale shaker
91. Stairs
92. Stand pipe
93. Storage, bulk barite
94. Storage, liquid mud
95. Storage, mud additives
96. Substructure
97. Substructure, pony
98. Suction lines
99. Suction (mud) pit
100. Surface casing
101. Swivel
102. Swivel (spring)
103. Tong, counterweight
104. Tong, lead (breakout)
105. Tong, makeup
106. Tool house
107. Traveling block
108. Trip tank
109. Walkway
110. Water, storage
111. Water table
112. Weight indicator
113. BOP and casing

FIG. 5–2 *continued*

various component systems combine to provide all of the functions needed in the drilling-completion operation.

Structure

The structure of a drilling rig is like the human skeleton. It is the centralized, basic frame that holds, supports, and provides a working surface for all of the equipment and component systems that comprise the drilling rig.

Early in the drilling industry, a wooden floor was laid on a level section of ground and a wooden derrick was constructed on top of it. The equipment was moved in, the well was drilled, the equipment was moved out, and the wooden derrick was left in place to be used as a production derrick to help service the well during its producing life. These wooden derricks were replaced by steel derricks, which had a greater structural strength required by deeper drilling depths and larger derrick loads. Early steel derricks were also left over the well as a production derrick, but today they are moved.

Higher formation pressures were encountered as the wells were drilled deeper. Blowout preventers (BOPs) were developed to control these pressures. The blowout preventers had to be placed below the derrick floor in order to be effective. The first BOPs were relatively small and were placed in a cellar. The cellar was a large, square hole usually about 8 ft square and 6 ft deep, centralized over the wellbore. The cellar was boarded-in with heavy 2-in. by 12-in. planks and the floor was cemented. Drilling mud, washdown water, and other fluids collected in the cellar and were removed with a jet.

As the wells were drilled deeper, increasingly higher formation pressures were encountered, which in turn required more blowout preventers with heavier construction. This equipment could not be conveniently placed in a cellar because of the large size of the equipment and the difficulty of installing and maintaining it in that confined space. So, it was necessary to build the working rig floor surface at a higher level, thus necessitating a substructure. The substructure was built with heavy-duty construction in order to support the weight of the equipment located on top of it at the working level, or derrick floor. This heavier-duty substructure was incorporated into the base of the drilling derrick. A separate substructure was attached to the back of the rig to hold the engines and part of the drawworks.

The standard drilling structure became larger and heavier and required extra men and equipment to move. Modifications of this type of equipment are still used on some marine rigs where the entire drilling rig with its component parts is moved as a unit. However, it has generally been replaced by unitized substructures, A-frames, and masts on most land rigs. This equipment can be disassembled into truck-size loads, moved, and reassembled with the regular rig crews. This increases portability and decreases overall costs.

In the normal configuration two substructures are used. Each substructure is about 10 ft wide, 10 ft high, and 30 ft long. Both are built of a combination of H-beams, angle iron, and metal plate. Box beams and plated I-beams are used where heavier, stronger construction is needed (Fig. 5–3). The substructures normally are more heavily reinforced on the front end, which supports the heaviest loads—the mast, rotary, pipe, and casing.

The top of the substructure is covered with a steel plate that serves as a working surface, the rig floor. Sections inside the substructure may be enclosed and fitted with doors to be used for equipment and tool storage (tool house) and a dressing area for the crew (doghouse). Other sections in the substructures may be enclosed, fitted with pipe inlets and outlets, and used for fuel and water storage. They also may be fitted with storage bins, racks, and mounts for storing large tools, subs and pickup nipples, mud and water lines, flow lines, a shale shaker rack, blowout preventer racks, and extra drilling line on a storage spool.

In the normal drilling rig configuration, two substructures (subs) are used. The subs are set parallel at a predetermined fixed distance 10–20 ft apart. They are connected by steel crossbeams and diagonal brace beams to make a rigid structure. The beams are attached to each other and to the substructure by mating flanges and large connecting steel pins with retainer keys to prevent the pins from

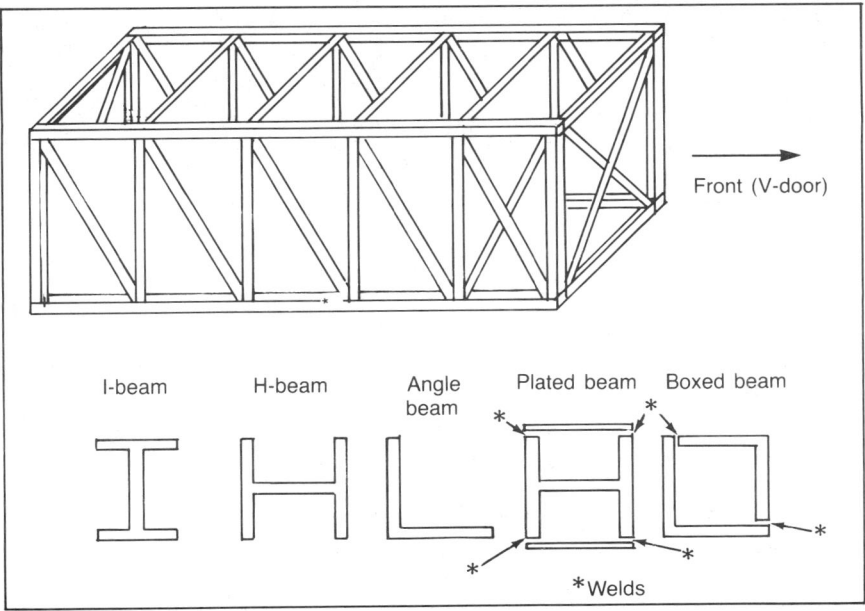

FIG. 5–3 Substructure construction

FIG. 5–4 Substructure and blowout preventer (courtesy *OGJ*)

slipping out. Mating flanges are similar to tongue and groove lumber. The tongue on the beam fits in a groove on the sub, and the retainer pin goes into a drilled hole to lock the two pieces securely together.

Extra heavy-duty beams, called rotary beams, usually with double-pinned flanges, are used near the front of the substructure. The rotary sits on these beams, so they must be strongly constructed and reinforced to support the heavy loads. Heavier-duty beams are also used to connect the front end of the substructure where heavier loading occurs.

Sometimes it is necessary to have a higher rig floor, either to allow additional room for more and larger preventers, to comply with states that require all preventers be above ground, or to provide more clearance when using rotating heads. Pony substructures can be used to provide more space between the base of the rotary and ground level. The pony sub has the same width and length as the regular sub but usually is only 3–6 ft high. In the conventional substructure assembly the pony sub goes on the bottom, and the regular sub sits on top of it. The two subs are connected with large bolts or locking devices. Two of the regular sub-pony combinations are then set down parallel and apart from each other and are connected with pinned cross members as described earlier.

When additional space is needed below the rotary—usually for large rigs with a deeper depth rating—double subs can be used. Two similar-sized subs are set one above the other; two of these combinations are then set and pinned together with cross members as described earlier.

The mast is constructed in a factory from H-beam and angle iron as well as tubulars. All of the welding can be done under favorable conditions, properly heat treated, and X-rayed to ensure that there are no slag inclusions (residue from welding). The mast is constructed in sections, which are connected with large bolts and pins to make one long, unitized, rigid member.

The A-frame used as a base for the mast is mounted with two legs on one substructure and two legs on the other substructure. The legs are connected to the substructures with large pins held in place with retainer keys. The bottom or foot of the mast is connected to the top front of the substructure near the front A-frame legs with large pins positioned so the mast can swivel from the horizontal to the vertical as it is raised or lowered when rigging up or down.

When the substructures, A-frame, and mast are raised and locked, they become one rigid structure. Then steel floor plates are laid over the connecting beams between the substructures to form a large solid floor area.[1]

The structure (substructure, A-frame, and mast) is serviceable for 10–15 years under normal operating conditions with good maintenance. Subs and masts are subject to failure and require inspection at periodic intervals. The main failures are bent bracing, warped or bent floor beams, cracked welds, bent or worn pins and flanges, and corrosion. Most failures are caused by improper handling during moving. The braces and cross members in the mast can be bent when hit by a swinging traveling block or other heavy load. Pinholes become worn and develop excessive clearances. Pins become scarred and the pinheads enlarged (smashed) and cracked due to heavy usage, especially when being inserted when the holes are slightly out of alignment.

[1]Recently, there has been a trend to change substructure construction. The floor is assembled at ground level and is attached to the substructure by a system of connecting legs that can swivel. During rig-up, all of the equipment is assembled on the floor while it is at ground level. This is easier, faster, and more economical than hoisting the equipment up to the floor on top of the substructures, which is 20 ft or higher above ground level. The floor with the assembled equipment is then pivoted upward and pinned in place to form a rigid structure. The procedure is reversed when disassembling the rig.

FIG. 5–5 Structural elements of the rotary rig

183
Drilling Personnel and Equipment

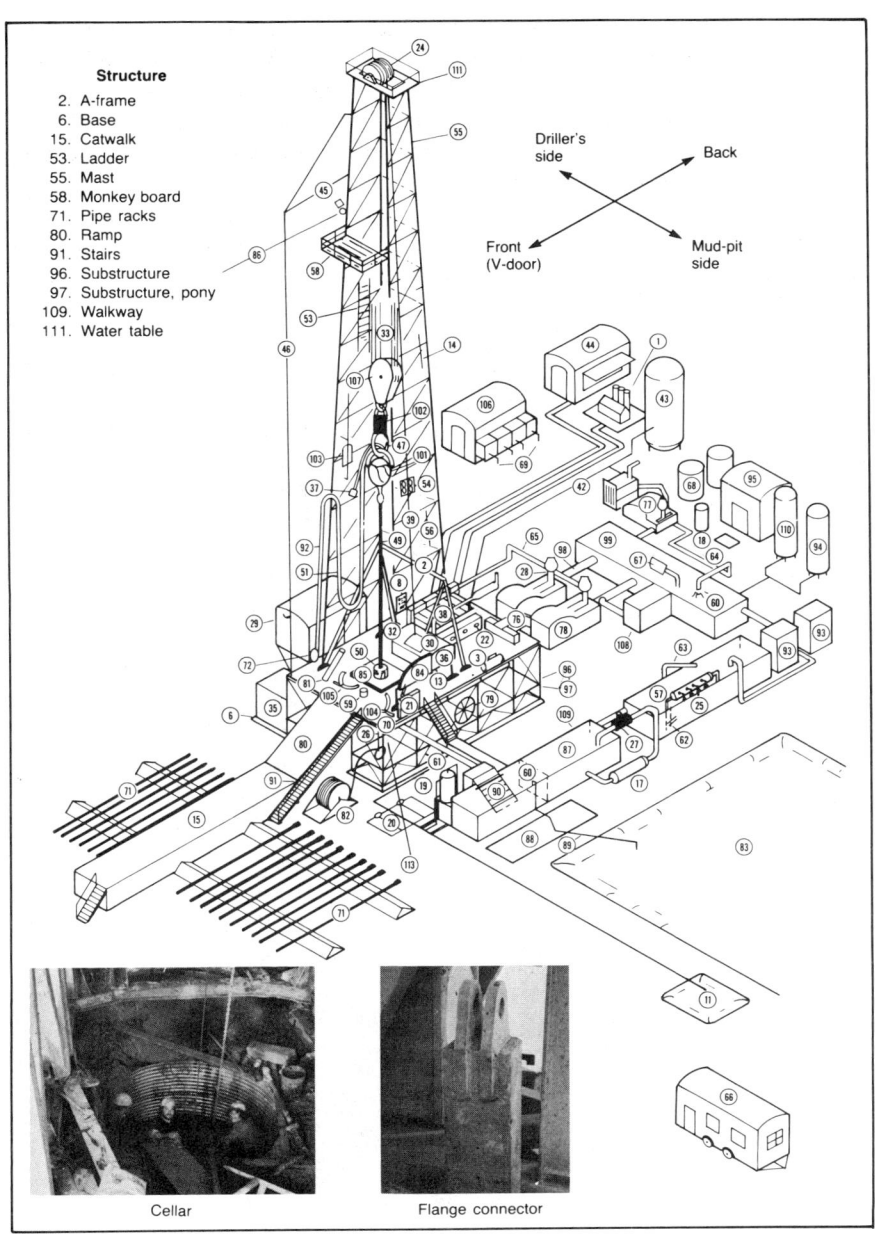

Structure
- 2. A-frame
- 6. Base
- 15. Catwalk
- 53. Ladder
- 55. Mast
- 58. Monkey board
- 71. Pipe racks
- 80. Ramp
- 91. Stairs
- 96. Substructure
- 97. Substructure, pony
- 109. Walkway
- 111. Water table

Cellar

Flange connector

FIG. 5–5 *continued*

The substructures are repaired by either straightening slightly bent members or replacing severely bent or damaged members. The usual method is to cut out the member at a welded connection or joint and replace the entire member. These welding repairs can be made in the field by a competent, experienced welder. Mast repairs are usually made by employees of the mast manufacturer, who are experienced in this type of work.

Masts are seldom dropped while raising or lowering. This is usually caused by some severe mechanical malfunction, such as a broken lifting line, drawworks failure, or split roller guide. When the mast is dropped, it must usually be replaced.

The entire mast and substructure are subject to corrosion. The chemicals contained in the mud, water-treating chemicals, treating fluids, produced salt water, and other fluids commonly used around the rig are very corrosive. The substructure has the highest exposure since it is below the rig floor where most of these fluids are used. Closed areas where fluids can accumulate should have drain holes whenever possible, and the substructure should be washed frequently and painted periodically.

The substructure must be set on an adequate base or foundation. The base under the front end of the substructure must be specially prepared because the area around the wellbore supports the weight of the substructures and equipment plus the weight of pipe racked in the mast and hook loads. These hook loads alone average 200,000 lb and have exceeded 1,000,000 lb.

Assuming average soil conditions, small rig substructures are set on leveled earth. Sand, matting boards, or a combination of these may be used on medium-sized rigs. One of two types of matting boards may be used. Wooden planks (3 in. thick by 12 in. wide) that are about 4 ft longer than the width of a single substructure are laid under the substructure about 3–4 in. apart. Two layers of matting boards may be laid crosswise to each other and bolted together to form a large matting pad. These pads are laid solidly under both substructures. Matting boards are used for better soil conditions and lighter rigs. Matting pads are used for heavier rigs or where the surface soils are less stable.

Very large, heavy rigs are set on concrete bases. Depending on soil conditions and rig weight (including anticipated hook loads), these bases can range from small concrete slabs under each substructure to a large, thick slab containing heavy reinforcement (rebarb) material.

A ramp with a smooth, flat surface extends from the edge of the rig floor on the V-door side, or front, of the rig and tapers down or slopes to the front end of the catwalk. A catwalk is positioned in front of the rig and connected to the ramp. The catwalk is a flat surface 4–6 ft wide and 40–50 ft long, extending from the ramp away from the rig. It is elevated about 3 ft above ground level and is frequently built on top of a framework or metal rack used to store equipment and tools during a rig move. The catwalk and ramp provide a convenient area for pulling or sliding equipment up to the rig floor or lowering equipment from the rig floor with a cat line or other lifting line.

The pipe racks are heavy-duty racks about 20 ft long located in pairs perpendicular to and on each side of the catwalk. They are spaced about 20 ft apart to hold pipe, which is usually 30–40 ft long. Normally, a single set of pipe racks is used on the driller's side of the catwalk and double racks are set on the mud-pit side. Pipe is stored on the pipe racks. When pipe is needed in the rig, it is rolled onto the catwalk and pulled up to the rig floor with a cat line. When pipe is laid down out of the derrick, the singles are slid down the ramp and catwalk and are rolled onto the pipe racks.

NOTE: There is a pipe rack on the rig floor that is used to hold the stands of pipe when they are set back in the mast. This pipe rack is normally referred to as the pipe rack or floor pipe rack. It should not be confused with the pipe racks, which are always referred to in the plural and are located at ground level to be used for pipe storage.

Hoisting System

The hoisting system includes the drawworks, the crown block, the traveling block, the drilling line, and provisions for extra drilling line storage (Fig. 5–6). The basic installation is similar to a conventional block-and-tackle lifting system. The crown block, or crown, sits on top of the mast. The drilling line is reaved over the crown and the traveling block sheaves so four or more lines are between the crown and traveling block. One end of the drilling line, the fast line, is connected to the drum of the rig hoist, more commonly called the drawworks. The other end of the drilling line, the dead line, is securely clamped to the substructure. The remaining part of the drilling line is coiled on a storage spool. Drilling loads such as drillpipe and casing are connected to a becket on the bottom of the traveling block. As the drawworks drum is turned, the drilling line reels or unreels, raising or lowering the traveling block, which in turn raises or lowers the hook load—the equipment or material suspended from the traveling block.

The number of lines strung between the crown and traveling block generally depends on the size of the rig, the single line pull capacity of the drawworks, and the operation (Fig. 5–7). For example, fewer lines are used during shallower drilling operations while more lines are used in deeper drilling operations. Additional lines are sometimes strung up to run heavy casing and liner loads. In the conventional dead-line/fast-line configuration, neglecting the weight of the drilling lines, the single-line load and the load on the mast decrease with an increasing number of lines. So the main reason for increasing the number of lines is to

FIG. 5-6 Hoisting system of rotary rig

187
Drilling Personnel and Equipment

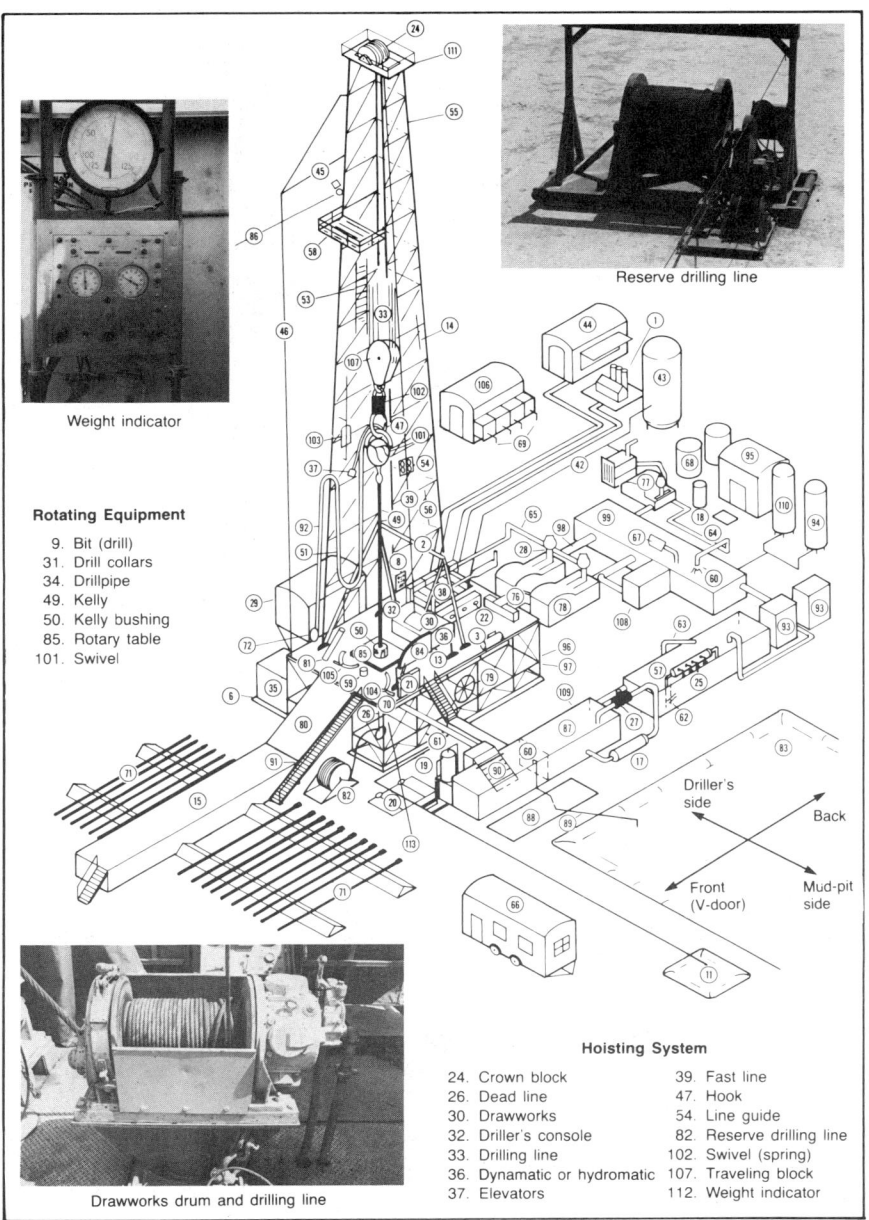

Weight indicator

Rotating Equipment
- 9. Bit (drill)
- 31. Drill collars
- 34. Drillpipe
- 49. Kelly
- 50. Kelly bushing
- 85. Rotary table
- 101. Swivel

Reserve drilling line

Drawworks drum and drilling line

Hoisting System
- 24. Crown block
- 26. Dead line
- 30. Drawworks
- 32. Driller's console
- 33. Drilling line
- 36. Dynamatic or hydromatic
- 37. Elevators
- 39. Fast line
- 47. Hook
- 54. Line guide
- 82. Reserve drilling line
- 102. Swivel (spring)
- 107. Traveling block
- 112. Weight indicator

FIG. 5–6 *continued*

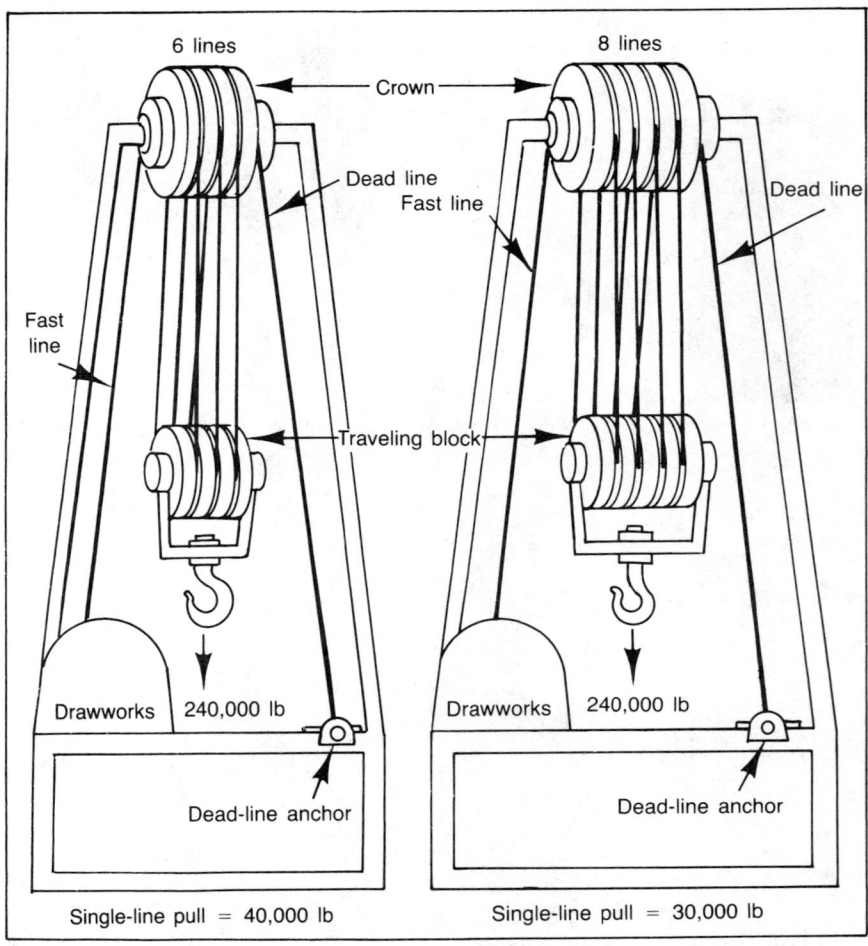

FIG. 5–7 Hook loads

decrease the single-line load. This in turn increases the safety factor on the drilling line and reduces the load on the drawworks so the hook load can be handled more easily and smoothly.

The rig hoist or drawworks is a basic part of the hoisting system. To a great extent, the entire rig is built around it. In the past most rigs were rated by the depth capacity of the drawworks. The current trend is to rate the rig by the drawworks' input horsepower, single-line pull (pounds), and speed. The remainder of the component parts of the drawworks, such as the shafts, sprockets, chains, clutch sizes, and brake drum, are designed and sized to be compatible with the drawworks capacity.

The drawworks takes power from an external source and transmits it to the drum shaft through clutches, chains, and sprockets and extra shafts. Then the shaft can be rotated at different speeds, depending on the hook load and the speed that the traveling block can be safely moved (Fig. 5–8). The drum shaft is fitted with a wire-line reel or spool that holds the drilling line. The drum is sized for the particular drilling line being used. If the drum diameter is too small, the wire line—more commonly known as the drilling line—will wear excessively and break. A grooved liner is fitted over the drum, and the grooves are sized according to the size of the drilling line. Both ends of the wire-line drum are fitted with large, smooth brake drums. Brake bands with brake blocks are fitted over the brake drums to hold the drum and in turn control the movement of the hook load.

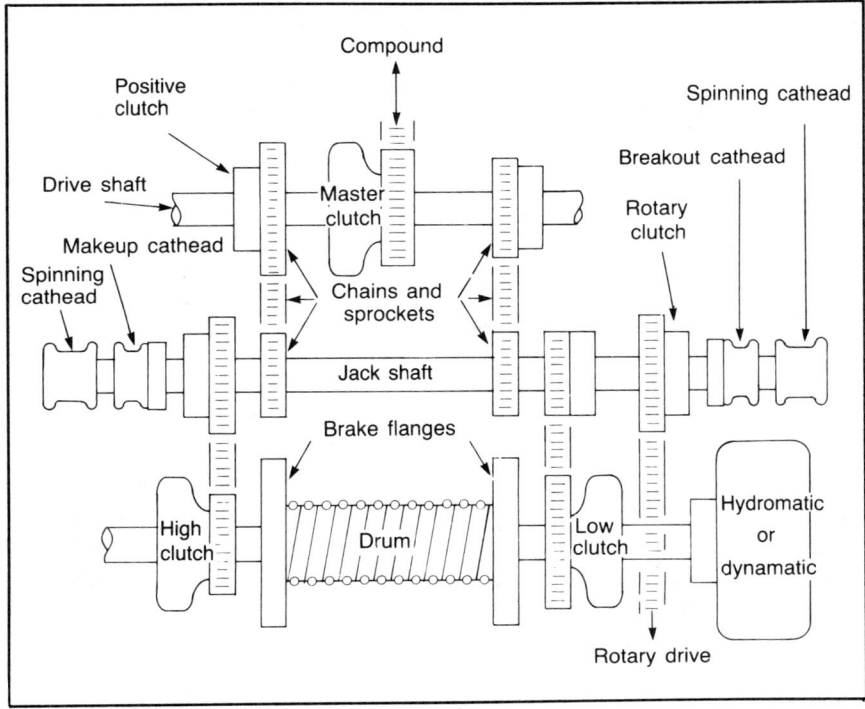

FIG. 5–8 Drawworks

The brake bands and drums are subject to excess wear when large heavy loads are lowered. They are water cooled to dissipate heat. To reduce wear further, a hydromatic or dynamatic is connected to the end of the drum shaft with a clutch. Hydromatics are hydraulically actuated and are generally used on smaller and older rigs. They have paddles connected to the drawworks shaft that rotate in a water bath to retard the downward movement of the traveling block. Dynamatics are electrically actuated and are used on newer and larger rigs. On dynamatics electromagnetic forces similar to those in a generator provide the retarding force. When these tools are connected (clutch engaged), they retard the rotation of the drum shaft so the hook load is lowered at a predetermined slower speed. This reduces brake wear and allows the heavy weights to be lowered at safe, controlled speeds.

One of the extra shafts in the drawworks is the jack shaft. The ends of the jack shaft are fitted with spinning catheads. Manilla lines can be wrapped around the catheads with one or more wraps to increase their pulling force for lifting heavy loads. The jack shaft is also fitted with clutch-operated automatic catheads to work jerk lines, which in turn pull on the end of the pipe tongs to connect and disconnect tool joints in the drilling assembly.

Some drawworks are fitted with a second drum shaft and wire-line reel used for swabbing and core recovery lines. These double-drum drawworks are more common on smaller-sized drilling rigs and workover rigs.

The drawworks normally has four shaft speeds. When tripping, or pulling all of the drillpipe out of the hole, to change the bit, the trip is started with the drawworks in a lower gear. The hook load decreases as each successive stand of drillpipe is pulled out and set back in the mast. When the hook load has been reduced a sufficient amount, a higher gear and faster speed are used. This is repeated as the hook load is decreased until a maximum safe pulling speed is reached.

When a well is drilled deeper, drillpipe is added to the drilling assembly and the hook load increases. There is a corresponding increase in the single-line load and resulting load on the drawworks. When the single-line load approaches the efficient drawworks single-line load capacity, additional lines are strung between the crown and traveling block. This reduces the single-line load and the drawworks load. The traveling block speed is also reduced, but the load can be handled more safely. Additional lines are strung for the same purpose before running heavy strings of casings and liners.

The traveling block is nominally sized by the maximum safe working load that can be suspended from it, rated in tons. For example, a 100-ton traveling block in good condition can safely support a 200,000-lb hook load.

The crown block is matched to the traveling block. The sheaves or pulleys on the crown and the traveling block are sized for the drilling line that is used. They are strung between the crown and the block. This includes both the sheave diameter and the size of the line groove in the sheave. Standards have been established for the minimum size diameter sheave for each size of drilling line. The line is subject to excess wear, flexure, and possible breakage if smaller-diameter sheaves are used. The grooves in the sheaves should also be properly sized. If the grooves are too small, the line can become pinched and/or wear excessively. If the grooves are too large, the drilling line will not be supported properly and will flatten, causing wear and further increasing the sheave groove size.

Drilling lines wear with use. Most of the wear occurs where the line passes over the sheaves and the drawworks drum. The drilling line always wears faster on the fast-line end because more line travels through the sheaves than through the dead-line end. When a new drilling line is installed, the excess line is normally stored near the dead-line clamp. Line wear is normally recorded in the driller's reports in units of ton-miles. Ton-miles are based on the amount of drilling and tripping and the hook-load weight.

Cutting and/or slipping practices are used to remove worn drilling line. For those who use cutting practices only, after a predetermined number of ton-miles of use a length of line (100–300 ft) is cut off at the drawworks drum. At the same time an equal amount of line is slipped through the dead-line clamp from the new line reserve. This cutting procedure is repeated after additional periodic ton-miles of use until the entire reserve (new) drilling line is used. Some operators use the alternate slip-and-cut method. After a predetermined amount of ton-miles of use, about 50 ft of new line is allowed to slip from the reserve line through the dead-line clamp. The extra line accumulates on the drawworks drum. This procedure may be repeated several times at periodic ton-mile intervals. At the next interval of wear, the line is cut and the excess is removed.

The driller's work station is normally at the right front corner of the drawworks near the drawworks brake handle. This position provides maximum visibility of all operations. The various controls, engine throttles, clutch controls, dials, and gauges showing pump pressure, engine rpm, engine temperatures, and hook-load weight are conveniently located on panels near the driller, called the driller's console.

The hoisting system is strongly built and handles heavy loads under relatively constant use. It must be properly maintained including greasing and oiling. Worn parts are replaced periodically, such as chains, sprockets, clutch plates, and brake blocks. The drawworks, crown, and block are overhauled, replacing sheaves, bearings, and other major items after 3–5 years of use. Properly maintained equipment is normally serviceable for 10–15 years.

Prime Movers and Power Trains

All modern rigs use internal combustion engines as prime movers or power sources. The efficient, reliable diesel engine is by far the most common. Some earlier rigs used gas-gasoline engines, which could be fueled by gasoline or gas (natural gas, butane, or propane). A few of these are still used on smaller rigs, but in most cases diesels are used. Engine sizes range from about 100 hp to over 500 hp. A few small rigs have one engine, but most rigs have at least two engines and large rigs may have five or more.

The prime mover or engine power is transmitted to the load (drawworks, pumps, rotary) by the power train (Fig. 5–9). This item of drilling equipment is important because it must transmit the power smoothly in the amounts needed and with minimum loss.

The development and use of mechanical rigs was delayed because of problems in obtaining an efficient coupling, or connection, between the engine and the load. Friction clutches were first used and were later replaced with more efficient fluid couplings. Diesel-electric rigs were developed later, followed by the more efficient SCR (silicon controlled rectifier) rigs. Mechanical rigs and both types of electric rigs are currently in wide use.

Mechanical Rigs Mechanical rigs or power rigs use internal combustion engines for power. Internal combustion engines were developed in the early 1900s but were not widely used on rigs until the mid-1940s. The main reason for the delay was the problem of coupling the engine to the load. This problem and the concept of power coupling and transmission deserve some discussion because they explain (1) why the steam engine was efficiently fitted for use on drilling rigs, (2) the problem of coupling the internal combustion engine to a heavy load, and (3) why electric motors (DC) are also an efficient power source for drilling rigs.

The steam engine used a large piston. Steam pressure applied to the piston created a force that was transmitted through the engine crank as torque. This torque could be very high, depending on the steam pressure and piston area. Therefore, the steam engine developed a high torque at a stalled (zero) or very low speed. The engine could be coupled to a heavy load, such as a drawworks picking up pipe, steam could be applied to the engine, and the load could be moved easily at increasing speeds as the amount of steam flowing to the engine was increased. This process was easily controlled with a common throttle valve that increased or decreased the flow of steam.

The internal combustion engine presented a different problem. When the engine was still (not rotating), it did not develop power. As the engine began to run, it developed power. The power increased as the engine speed increased. When the drawworks was connected to a load ready to be picked up, the drawworks was still or at a zero speed. The power to pick up this load could require engines running at about 1,000 rpm. If the engine was slowed down to several hundred rpm, it did not have sufficient power to pick up the load. Basically, the drawworks required power without rotation, and the engine could not supply power until it was rotating. Any instantaneous coupling of power would break chains, sprockets, and shafts, and would cause other damage.

Early automobiles had a similar problem. When the car was still, the wheels did not move. The engine had to turn to develop horsepower in order to start moving the car. The engine could not be started when directly connected to the wheels except by pushing. The problem was solved by placing a friction clutch between the engine and the rear wheels. The engine was started with the clutch disengaged. The clutch was allowed to engage slowly and would slip until the rear wheels and drive shaft were turning fast enough so that the clutch could be completely released to engage the engine directly to the rear wheels. As the car speed was increased, the clutch was used while changing gears. There was less clutch slippage since both the engine and rear wheels (drive shaft) were rotating.

Those of us who learned to drive with cars equipped with clutches can remember how difficult it was to release the clutch smoothly so the car would begin moving forward slowly without jerking. The mechanical rig had the same problem as a car with a friction clutch, except it was much greater. Where a car might be using a few horsepower, hundreds of horsepower were transmitted from the internal combustion engine to the drawworks. The early power rigs had severe clutch problems. Bigger clutches with larger contact areas and other improvements were devised that helped to alleviate the problem, but it was never completely solved while using friction clutches.

The clutch problem on rigs was finally solved in a similar manner to that in the automotive industry. Cars were equipped with automatic transmissions that effectively eliminated the friction clutch and associated problems. Power could be applied smoothly to the rear wheels. The rig manufacturers solved their power transmission problem similarly by developing a fluid coupling or torque converter.

The automatic transmission on a car is a fluid coupling. The fluid coupling or torque converter used on the rig is similar in principle to an automatic transmission except that it has only one speed and uses a somewhat different and more heavy-duty design and construction to transmit heavy loads (high horsepower). The torque converter allows the rig engine to run at a low speed while being coupled directly to a stalled load. As the engine speed increases, additional torque

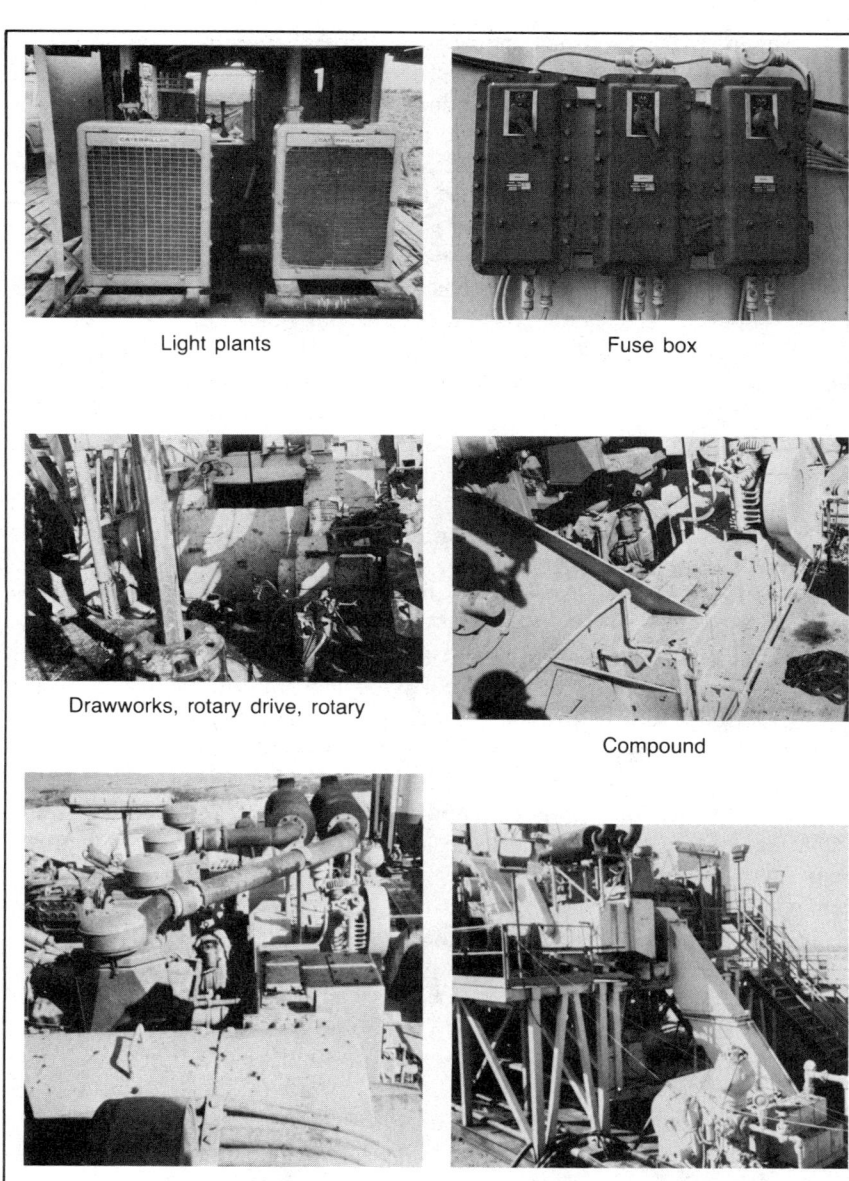

FIG. 5-9 Prime mover and power train

195
Drilling Personnel and Equipment

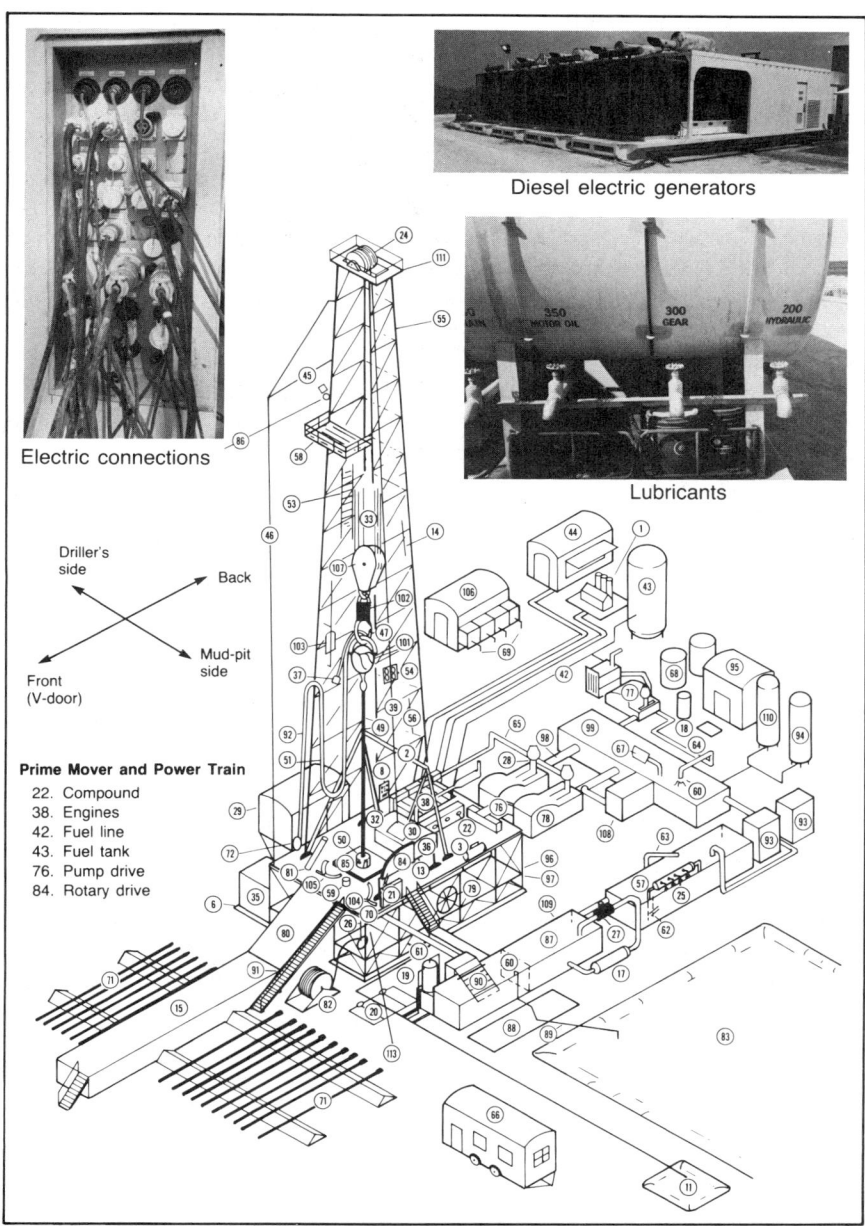

Electric connections

Diesel electric generators

Lubricants

Driller's side ↔ Back

Front (V-door) ↔ Mud-pit side

Prime Mover and Power Train
- 22. Compound
- 38. Engines
- 42. Fuel line
- 43. Fuel tank
- 76. Pump drive
- 84. Rotary drive

FIG. 5–9 *continued*

is applied to the load until it moves as required. Different drum speeds are obtained by changing gears in the drawworks.

The compound is the basic power train on a mechanical rig. It is made up of chains, sprockets, shafts, bearings, positive-action dog-type clutches, friction clutches, and fluid couplings. The compound is enclosed in heavy-gauge metal, and the chains run in an oil bath. Most of the horsepower requirements for an average, medium-depth power rig are for the drawworks and two mud pumps. On a standard prime mover and power train assembly for a medium-depth rig, three engines are connected to a three-engine compound. The front end of the compound is connected to a four-speed drawworks. The rear end of the compound has a pump shaft that drives two tail-driven mud pumps, which supply the pumping requirements to circulate the drilling fluid.

The positions of the automatic and spinning catheads on the jack shaft are illustrated in Fig. 5–10. The forward-drive chain and sprockets going into the drawworks and the rear-drive chain and sprockets going into the pump shaft may be larger than the other chains and sprockets in the compound because they may be carrying the power from two or more engines, depending on the operation. Normally, engine 1 is always connected to the drawworks and engine 3 is connected to the pumps. Engine 2 is a swing engine. When the rig is drilling, more horsepower is needed on the mud pumps. The swing engine is connected to the pumps by disengaging the inboard clutch and engaging the outboard clutch. This couples engines 2 and 3 to the pump drive. One of the pump clutches is then engaged to drive the pump. Both pumps can be run in parallel by engaging both pump clutches, but normally only one pump is run.

Engine 1 is coupled to the drawworks during drilling. By engaging the proper clutches, the power from this engine is transferred from the drive shaft to the jack shaft and through the rotary clutch on the jack shaft to drive the rotary. If pipe must be raised to make a connection, the rotary clutch is disengaged. Usually one engine will have enough power to pick up the pipe in low gear, so the low-gear drum clutch can be engaged to pick up the pipe and make the connection as necessary. After the connection is made, the low-gear drum clutch is disengaged and the rotary clutch is re-engaged to resume drilling operations.

When making connections, the mud pump is shut down. If for some reason one engine does not have sufficient power to lift the pipe for making a connection, the swing engine is disconnected from the pump shaft and is connected to the drawworks with engine 1 by disengaging the outboard clutch and engaging the inboard clutch. The swing engine is reconnected to the pump shaft to resume drilling operations after the connection.

If one of the engines breaks down, it can be disconnected using the master engine clutch. The other engines can be connected through the compound as needed in order to drive the pump or drawworks.

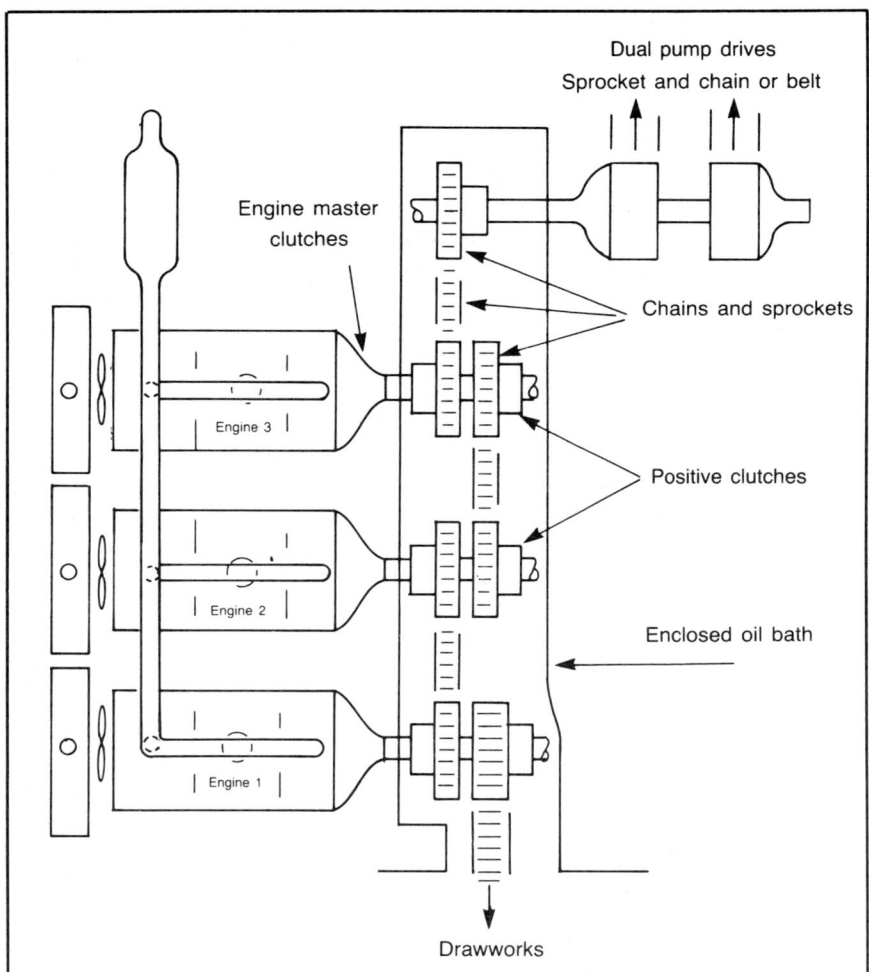

FIG. 5-10 Schematic of the compound

Various drum speeds are obtained by engaging the different clutches in the drawworks. For example, if the low-gear clutch on the drive shaft is engaged, then engaging the low gear on the drum shaft would be first gear. Engaging the high-gear clutch on the drum shaft would be second gear. If the low-gear drive shaft clutch is disengaged and the high-gear drive-shaft clutch is engaged, then engaging the low-gear clutch on the drum shaft would be third gear and engaging the high-gear drum clutch would be fourth gear. When the drum shaft clutches are disengaged, the jack shaft rolls freely. The spinning or automatic catheads can then be used. When the pipe is being run in the hole, the hydromatic or dynamatic can be connected to the drum shaft to control the lowering speed and conserve the brakes.

The sprockets, chains, and clutches in the compound and drawworks and the pulleys on the pump shaft and pumps must be properly sized to have enough strength to carry the horsepower load and give the proper speed reduction from the engines to the various driven components. For example, the rotary is normally turned at speeds ranging from a few rpm to as much as 150 rpm. Conventional oil-field motors have a speed range of 600–1,500 rpm. Therefore, the sprockets in the compound and drawworks must be properly sized so the rotary can run at its given rpm range while the engines are running at their normal rpm range. The same requirements apply to the drum shaft and the pumps.

The prime mover and power train illustrated in Fig. 5–9 are relatively common for mechanical rigs, but other configurations are also used. For example, smaller rigs may only have two engines or, in some cases, one engine. The size of the compound would be reduced accordingly. Sometimes the pumps are driven from individual engines mounted on the pump skid, or unitized engine-pump installation. In this case the pump shaft would be eliminated.

The mechanical rig also has a light plant unit with smaller engines coupled directly to ac generators. These supply electrical current for the rig lighting system and the various small electric motors, such as the shale shaker motor and mud mixing motors.

The power rig is an efficient drilling rig, but it has several disadvantages. Generally the equipment is big, heavy, and somewhat difficult to rig-down, move, and rig-up. The prime movers and power train must be set on the rig floor, which in some cases requires lifting heavy loads to heights of 20 ft or more.

The drawworks were first positioned on the rig floor during rigging up. The compound was precisely positioned to fit the drawworks, and the engines were precisely positioned to fit the compound to align the connecting shafts and clutches. The pumps were then positioned to align precisely with the pump shaft pulleys. All of this required extra rig-up time. In some cases the engines and compound were unitized on one skid, but this was a very large, heavy load, especially to lift to the rig floor.

The engines were relatively near the hole and represented a potential fire hazard in case of a blowout or escape of gas or oil. Spark suppressors and dampened exhaust helped alleviate this problem. The engines created a loud noise so that at times it was difficult for the crew members to communicate, creating a possible hazardous condition and noise pollution. Considerable maintenance and upkeep were required to keep all of the equipment in good operating condition.

Despite these disadvantages, the power rig is a milestone in the evolution of drilling rigs. An estimated 90% of the small rigs and 60% of the larger rigs in operation today are conventional mechanical rigs.

Electric Rigs Electric rigs are a continuing step in the evolution of drilling rigs. First called diesel-electric rigs, today's dc (direct current) electric motors can supply a very high torque (force applied in a circular direction) at low operating speeds. The most common example of a dc motor is the starter in an automobile. This relatively small motor can deliver a high torque to turn the car engine until it starts. In some respects the torque and power characteristics of a dc electric motor are similar to the steam engine.

Direct current electric power is generated by a dc generator driven by an internal combustion engine, usually a diesel engine. The engine and generator are mounted on a skid as a unitized component called an engine-generator set. The average rig has 3–4 engine-generator sets connected to a switch panel by flexible electric conduits (large insulated copper wires).

A direct current motor is mounted on the pump skid at the back of the drawworks skid and is coupled to the drawworks drive shaft with sprockets and chains as a single skid-mounted unit. Sometimes two direct current motors are mounted on the back of the drawworks, depending on the size of the rig. A direct current motor also is mounted on the pump skid and coupled to the pump drive shaft with sprockets and chains as a single skid-mounted unit. All of the motors are connected to the switch panel by electrical conduits. Solenoid-type switches, which are similar to those in cars but are larger, are used on the switch panel and are controlled by a bundle of small wires connecting the switch panel to the control panel on the driller's console.

Since the power is transmitted by flexible electrical conduits, the engine-generator sets can be positioned at ground level a convenient safe distance from the rig. The amount of equipment that must be positioned on the rig floor is reduced to the drawworks with the attached dc motors. The mud pumps can be positioned relative to the mud pits.

In operation, the motor-generator sets are used like those in the power rig except that instead of changing clutches in a compound, the motor-generator sets are connected to the various power units by electrical switches controlled by the driller at the console. The diesel-electric rig uses light plant units similar to the mechanical rig.

A late modification of the diesel-electric rig is to use separate direct current motors on the drawworks attached to the low- and high-gear drive trains. The motors and drawworks are geared to match, thus increasing the flexibility of the drawworks.

The diesel-electric rig is an improvement over the mechanical rig, especially the larger rigs with a greater depth rating. A dc rig retains one disadvantage of the mechanical rig: A complete motor-generator set or sets must be connected or addressed to an individual dc motor. For example, consider a rig with three motor-generator sets. At times it is desirable to connect 1½ motor-generator sets to one dc motor and the other 1½ motor-generator sets to another motor. However, this cannot be done with the conventional diesel-electric rig.

The latest-model rigs are SCR electric rigs. SCR rigs are similar to diesel-electric rigs with a few significant changes. The motor-generator sets use ac (alternating current) generators. The ac current from these generators is fed into a bus bar, or switch panel. The alternating current is rectified to a direct current, which is then fed into dc motors like those on the diesel-electric rig.

There are two main advantages to this system. First, the total power from the motor-generator sets can be infinitely distributed to the motors where it is needed. In other words, 1½ motor-generator sets can be addressed to one motor and the remainder addressed to another motor or several motors. Second, ac generators are much more efficient and trouble free than dc generators.

The analogy with an automotive generating system can be used here. Early cars used direct current generators, which often caused problems and seldom ran over 10,000–20,000 miles without repairs or replacement. Later model cars use alternators (ac generators) and rectifier units. These generating systems were relatively trouble free. Similar results were obtained with ac generators on rigs.

As an additional improvement, the conventional light plant unit used on the earlier rigs is eliminated on the SCR rig. Alternating current is taken directly from the input side of the bus bar and is transformed into voltages as required for lighting and powering auxiliary ac motors.

In summary, the prime movers and power trains of drilling rigs have advanced from the steam-powered rig to the SCR rig. Most of the new, larger rigs presently being built are fitted with an SCR system, and many of the older, larger rigs are being converted to SCR systems. Most of the new, small rigs use a conventional mechanical power system, while the new medium-depth rigs are built with both mechanical and SCR power systems.

Rotating Equipment

The rotating equipment provides a method of turning the bit to drill the hole. In order to accomplish this, the rotating equipment must be connected to the bit in such a manner that the bit can be lowered into the hole, rotated to drill new hole, and pulled from the hole for replacement when it becomes worn. The equipment must also provide for a channel or flow path so drilling fluid can be pumped from the surface to and through the bit. Additionally, it must have a configuration so that all fluid movement into and out of the wellbore can be sealed off or flowed at controlled rates in the event of a blowout.

The rotating equipment ranges in length from a few hundred feet to over five miles. Therefore, in some cases the equipment must have high strength in order to support weights in excess of 300,000 lb if it becomes stuck in the hole. Mud circulating pressures can exceed 3,000 psi, and these must be contained.

The rotating equipment is divided into surface equipment, which includes everything above the bottom of the kelly, and downhole equipment, which includes the equipment below the kelly. Downhole equipment is referred to as the downhole assembly.

Surface Equipment The parts of the moving, rotating surface equipment listed in order from the top include the spring swivel, the swivel, the kelly spinner (if used), the safety shutoff valve, the kelly, the kelly drive bushing, the inside blowout preventer (if used), and a kelly thread-saver sub fitted with a protector rubber. The rotary table, sometimes abbreviated to rotary or table, is also part of the surface rotating equipment.

The spring swivel has a heavy cylindrical body with a becket on top that attaches the spring swivel to the bottom of the traveling block. The body contains a plunger spring attached to a large hook. The end section of the hook is fitted with a swivel pin and a locking device so the hook can be opened or closed and locked. The hook is connected to the elevators for moving and tripping the downhole equipment. During drilling operations, the hook is connected to the swivel bail.

The body of the spring swivel is prevented from rotating by the fixed connection to the traveling block. The hook can either rotate freely or can be locked in the desired direction by a locking device on the base of the spring-swivel body. The hook is normally allowed to swivel freely when pulling equipment out of the hole in order to prevent possibly backing off (disconnecting or unscrewing) the downhole equipment. It is usually locked when running tools into the hole or during drilling operations to prevent it from turning.

The spring is fully compressed when connected to normal drilling loads. However, it supplies a cushioning effect along with the small amount of elasticity in the drilling lines to minimize the shock loading of the metal-to-metal contact when heavy loads are picked up and set down. The spring swivel is used for a minimal rotation and is not designed for continuous, heavy-duty rotating action.

The swivel serves a dual purpose: it supports the rotating equipment and provides a fluid channel from the nonrotating circulating system to the rotating equipment. The swivel has a metal housing fitted with a bail to connect the entire

FIG. 5–11 Closeup of the swivel assembly (courtesy National Supply Company, an Amoco Group)

swivel assembly to the hook in the bottom of the spring swivel. This is a non-rotating member, although it may be rotated to a fixed position and locked by the spring-swivel or moved vertically. The rotating member in the swivel has a left-hand threaded connection in the bottom and a polished-bore receptacle in the top. This rotating member called the swivel sub, is fitted with a bearing race to rotate freely on roller bearings in an oil bath.

The gooseneck on top of the swivel provides a channel for drilling fluid to flow from an external source into the swivel assembly. A wash pipe with a polished outer surface is connected to the base of the gooseneck by a threaded connection and is inserted into the polished-bore receiver in the top of the rotating member. The space between the polished surfaces of the rotating member and the wash pipe is filled with packing elements to contain pressure and prevent leakage of drilling fluid between the polished surfaces. A properly designed sealing assembly can easily hold over 3,000 psi.

A kelly spinner may be attached to the bottom of the swivel. This is a light-duty rotating device usually operated by a small hydraulic motor. It rotates the swivel sub and kelly where light-duty rotation is required, such as when making a connection.

A safety shutoff valve, sometimes called a nut because of its circular shape, is connected to the bottom of the swivel sub and has a left-hand connection on the bottom. This full-opening, ball-type valve shuts off flow inside the rotating equipment. The only time this valve is normally used is when the well is blowing out and there is fluid flow in the upward direction or excess pressure inside the rotating equipment caused by high formation pressures.

The kelly is a long tubular with a left-hand, screw-type female or box tool joint connection on top and a male or pin right-hand connection on bottom. The average kelly is 42 ft long (overall) and has a 38 ft-long center section with a square configuration on the outside surface and a round hole (fluid channel) completely through the center (end to end). This center section is known as the square. Octagonal or hexagonal outer configurations are used, but the square shape is more common.

A kelly drive bushing, more commonly called a kelly bushing, fits over the kelly square and rotates with the kelly. It is free to move from one end of the kelly square to the other. The bushing has a square-shaped bottom that fits in the top of the rotary table. It transmits torque from the rotary to the kelly and in turn to the entire rotating equipment. It also provides an easy, quick method of disconnecting the rotary from the rotating equipment. The operator need only to pick up the rotating equipment until the bottom of the kelly square contacts the drive bushing. This action lifts the drive bushing out of the rotary and disconnects the rotary from the rotating system.

The rotary bushing is dismantled to install it on the kelly. The locking ring and roller drive bushings are removed from the body section. The kelly is inserted through the body and through a kelly wiper rubber, if it is used. The roller drive bushings are placed inside the body around the kelly square and are locked in place with the locking ring. These bushings transmit the torque to the square and allow the kelly to pass or roll through the drive bushing.

All connections in the rotating equipment located above the kelly drive bushing are left-hand connections; all connections below the kelly drive bushing are right-hand connections. There is a special reason for this. All of the torque is applied to the rotating system on the kelly square by the drive bushing. This torque is normally applied smoothly. In some cases though, it is relatively high and may fluctuate, causing a jerking action on the rotating equipment. Torque is normally applied in the right-hand or clockwise direction (looking downward). If the connections in the rotating equipment above the drive bushings are right-hand connections, they could tend to loosen and leak or unscrew, causing the assembly to drop. By using left-hand connections above the drive bushings and right-hand connections below the drive bushings, all of the normal torque applied to the rotating equipment acts to tighten the joints in the rotating equipment so they will not disconnect under normal operations.

An inside BOP is often installed on the bottom of the kelly. This tool is similar to the safety shutoff valve located above the kelly except that the outside diameter (OD) of the tool is the same size or nearly the same outside diameter as the tool joints. The inside BOP serves the same purpose as the safety shutoff valve and actually is used as a backup or safety valve in case the safety shutoff valve fails.

The tool joint on the bottom of the kelly (or the bottom of the inside BOP, if it is used) is connected and disconnected many times during the drilling operation. This causes an accelerated rate of wear, and the entire kelly must be taken to the machine shop for repair when the connection becomes worn. A thread-saver sub is screwed onto the bottom of the kelly (or the inside BOP) so that it is worn as the joint is connected and disconnected instead of the kelly. It is much easier and more economical to repair the connection on the bottom of the thread saver sub than the connection on the bottom of the kelly.

The rotary table supplies torque to the kelly drive bushing to turn the rotating equipment and provides a working surface (table) for the slips, which hold the downhole rotating equipment while making connections and performing other work when the downhole assembly is not being rotated. In special cases the downhole assembly can be rotated by setting it in the slips, which are seated in the rotary, and then turning the rotary. The vertical movement of the rotating equipment is limited when it turns in this manner.

The rotary table, as the name implies, is a large metal table fitted with a ring gear and seated on heavy-duty roller bearings. The ring gear is driven by a pinion gear connected to a rotary drive sprocket. The sprocket is connected to a sprocket in the drawworks by a rotary chain. Some rotaries are driven by a separate engine or motor. (Fig. 5–12).

There is a recessed, square area in the top of the rotary and a hole from the base of this square area through the rotary. The hole may be cylindrical or tapered (smaller at the bottom), depending on the construction of the rotary. Split rotary bushings, or split bushings are fitted into this hole. The split bushings can be pulled so that there is a larger hole through the rotary for running larger tools such as bits and large-diameter casing. The split bushings also have a tapered surface so that the slips can be inserted to act as wedges and hold the downhole equipment.

FIG. 5–12 Driller's console and rotary table (courtesy *OGJ*)

The rotary table is seated on heavily reinforced rotary beams (*see* "Substructures"). All downhole equipment used in drilling and completion operations is run through the rotary and is often seated or suspended in the rotary with the slips. Therefore, the rotary must be strongly constructed to support these weights, which can exceed one million pounds in deep casing strings.

Downhole Equipment Downhole rotating equipment includes, from the top down, drillpipe, drill collars, and the bit. The general term "downhole equipment" also includes any tools used in the hole, such as fishing tools.

The drillpipe is connected to the bottom of the kelly thread saver sub. Drillpipe is available in various lengths, but the average is about 31 ft. Each joint, or single, is made up of a body or tube fitted with connectors on each end. The connectors, known as tool joints, make a threaded, butt-shouldered-type connection. The joint has a box-type or female tool joint on top and a pin-type or male tool joint on bottom. Older-style drillpipe used box-shouldered tool joints connected to the drillpipe tube with a threaded connection. Modern drillpipe uses an upset-type tool joint connected to the tube by special welding techniques where the diameter of the tool joint is larger than the pipe body.

Various types of tool joints are used, depending on special conditions where the drillpipe is to be used. The different types of tool joints have different names, such as internal flush, regular or open hole, full hole, slim hole, and extra hole. Generally the same tool joint designations are used for different sizes of pipe, although all types may not be available except on the more commonly used pipe sizes.

Drillpipe, like casing, is normally referred to by the outside diameter of the tube. It is also specified by the weight per foot, which in turn fixes the inside diameter (ID). The tool joints are larger and heavier than the body or tube on a per-foot basis, so the extra tool joint weight is prorated over the length of the joint to give an adjusted weight used when calculating the total weight of the drillpipe. For example, 4½-in., 16.60-lb/ft nominal weight drillpipe can have an adjusted weight ranging from 16.66–18.40 lb/ft, depending on the inside diameter.

Various grades of steel are used in making drillpipe, ranging from a tensile strength of 55,000 pounds per square inch (psi) to 135,000 psi (high-strength steel). The size, weight, and grade of drillpipe that is selected for a drilling operation depends on the specific conditions of the drilling operation.

Conventionally, most drilling strings—the downhole rotating equipment—are designed so that the drillpipe will have a 100,000-lb overpull. This specification requires that the drillpipe have a strength rating equal to the sum of the total weight of the downhole assembly plus 100,000 lb. The minimum overpull may be increased in deeper holes.

Tapered drillstrings are sometimes used when designing drillpipe assemblies for deeper holes that have correspondingly larger downhole equipment weights. When one size of drillpipe is used in the drillstring, the topmost joint of drillpipe is under the greatest stress since it supports the total weight of all drillpipe and drill collars suspended below. If larger and heavier drillpipe is used for the entire drillstring, the load is increased correspondingly. The tapered drillstring design overcomes this problem by using smaller, lighter pipe in the lower part of the drillstring where the suspended pipe weights are less and using larger drillpipe in the upper part of the hole where suspended weights are heavier. Sometimes the same size of drillpipe will be used and a higher-grade (stronger) steel will be used in the upper part of the drillstring.

Drill collars serve the same purpose as the drillpipe and provide additional weight for the bit and rigidity in the lower part of the drilling assembly so the hole can be drilled as straight as possible. The drill collars connected to the bottom of the drillpipe collectively are called the drill-collar assembly. Drill collars are generally similar to drillpipe except that they are larger and correspondingly heavier. The ends of the collars are fitted with tool joints that are similar but larger than drillpipe tool joints. Drill-collar tool joints are sometimes referred to as connectors. Most drill collars do not have upsets and have the same OD over the entire length of the collar. Drill-collar connectors must be larger than the drillpipe tool joints in a matched drill-collar-drillpipe assembly because the drill collars are closer to the bit, are subject to higher torque and shock loadings, and are run partially in compression.

Various types of drill-collar assemblies are used, depending on the type of formation, hole size, and deviation. Special drill-collar tools such as keyseat wipers, reamers, stabilizers, and short drill collars may be used.

The common method of drilling is to rotate the entire downhole assembly, which in turn rotates the bit to drill the formation. Special tools such as mud motors and turbines have been developed that are run in the drill-collar assembly immediately above the bit. These tools use the hydraulic force of the drilling mud to rotate the lower part of the tool and the bit. The rest of the drillstring remains stationary or is rotated slowly. These tools are generally used in special drilling situations because they are very expensive.

Various standards for the selection, operation, and maintenance of downhole assemblies, including allowance for wear, are included in the American Petroleum Institute's "API Recommended Practice for Drill Stem Design and Operating Limits."[2]

---

[2]Drillstem is drilling pipe or the drilling assembly.

Circulating System

The physical and chemical characteristics of the drilling fluid, or mud, and the pressure and circulation rate strongly affect the overall efficiency of the drilling operation. Having good physical and chemical characteristics means selecting the best mud for the well being drilled and having a good treating and mud solids control program. This in turn requires that the rig have good equipment to separate low-gravity solids and recover barite as needed, sufficient pit volume to ensure that the mud is in the surface system long enough to be mixed properly, and treating equipment and facilities to treat the mud as required.

In order to have the correct pressure and circulating rate, several conditions must be set. The mud must have good physical characteristics. The pump must be fitted with the correct size plunger and liner or piston and have sufficient horsepower driving the pump so it can develop adequate hydraulic horsepower. All of the flow channels, including the surface piping and the downhole assembly, must have a large enough area so that friction pressures, which raise pumping pressure to dangerous levels and restrict flow rates, are not excessive. Most of this equipment is incorporated in the hydraulics program, which is used to calculate the pressure drop of the drilling mud throughout the circulating system. The hydraulics program also includes the sizes of the nozzles in jet bits, jet velocities, annular velocities, drill-cuttings velocity, and circulating and lag times (Fig. 5–13).

Mud Pumps   It would be difficult to select one piece of equipment as the most important item on the rig. However, there is no question that undersized, poorly maintained mud pumps will reduce overall efficiency. Properly sized and well-maintained pumps are important.

One type of mud pump is a duplex mud pump. The power end of the pump has a crankshaft driven by a large bull gear. The crank drives a crank arm-slide-pony rod connected to a crosshead. The pump rod and piston are connected to the crosshead and reciprocate as the crank shaft turns. Mud from the suction tank passes through the suction hose into the pump suction chamber. As the pump piston moves back toward the gearbox-power end, the back suction valve is closed and the moving piston forces mud through the open discharge valve into the discharge chamber. At the same time the front discharge valve is closed and the front suction valve opens to allow mud to flow from the suction chamber into the liner chamber. When the piston moves forward, the procedure is reversed and mud is pumped through the front discharge valve into the discharge chamber. Simultaneously, a second mud piston performs the same sequence of operations.

As the piston moves back and forth, it creates surge pressures that can cause excess stresses on all of the circulating system. A surge chamber or pulsation dampener is connected to the pump discharge. This chamber usually has a gas-filled rubber bag containing a predetermined gas pressure. The pulsating mud acts against this flexible bladder, which in turn dampens or reduces the severity of the pulsations.

The pump suction is connected to the mud suction tank by a large-diameter, heavy-duty, flexible suction line. A butterfly-type shutoff valve closes the opening into the suction tank if it is necessary to change the suction hose or repair the mud pump. Valve caps are located over each valve and can be removed for repair. A larger liner cap is fitted to the end of the pump and can be removed to repair or replace the piston or liner.

On butterfly valves the gate or closing device is shaped like butterfly wings. These valves are opened or closed with a quarter turn of the handle; they are used in low-pressure applications. Regular high-pressure valves use a tongue (closing device) that is raised (opened) or lowered (closed) with a screw-type handle somewhat like a common garden faucet.

The gear end of the pump and the connecting rods are designed for a specified amount of force on the piston. This is conventionally designated as a maximum pressure (in psi) for a certain size piston. If more pressure is required, a smaller-diameter piston and liner must be used. For example, a pump could have a maximum pressure rating of 1,500 psi with 7-in. liners and pistons. The same pump could put up a pressure of about 2,000 psi with 6-in. liners and pistons and 2,900 psi with 5-in. liners and pistons.

For a given pump, the amount of fluid pumped is decreased as the pressure is increased. The pump has a fixed stroke length and a maximum rpm or strokes per minute (spm). When a rig is selected to drill a well, one of the main factors to be considered is the pump capacity, usually rated in gallons per minute (gpm), in order to ensure that the pump has sufficient capacity to supply the fluid requirements (volume and pressure) to drill the well. An estimated 75% of the drilling rigs currently operating use one or more duplex mud pumps.

Plunger pumps are often called triplex pumps (three working plungers) because the first models were built as triplexes. The triplex is still the most popular model, although quadriplex pumps with four plungers are also used. The pump is basically similar to the duplex pump except it is single action. The discharge valve closes as the plunger moves toward the crank end and fluid enters the chamber through the suction valve. As the plunger moves away from the crank end, the suction valve closes and fluid is pushed through the discharge valve into the discharge chamber. The plunger is driven in a reciprocating motion by a gear box similar to that on the duplex piston pump.

The pump has valve and plunger caps to facilitate repair and replacement. Many of these pumps, including the duplex pumps, use a screw-type cap with O-ring seals. O-ring seals are special rubber seals with circular cross sections similar to round rubber bands. They are highly efficient at high pressures. The

FIG. 5–13 The circulating system

Drilling Personnel and Equipment

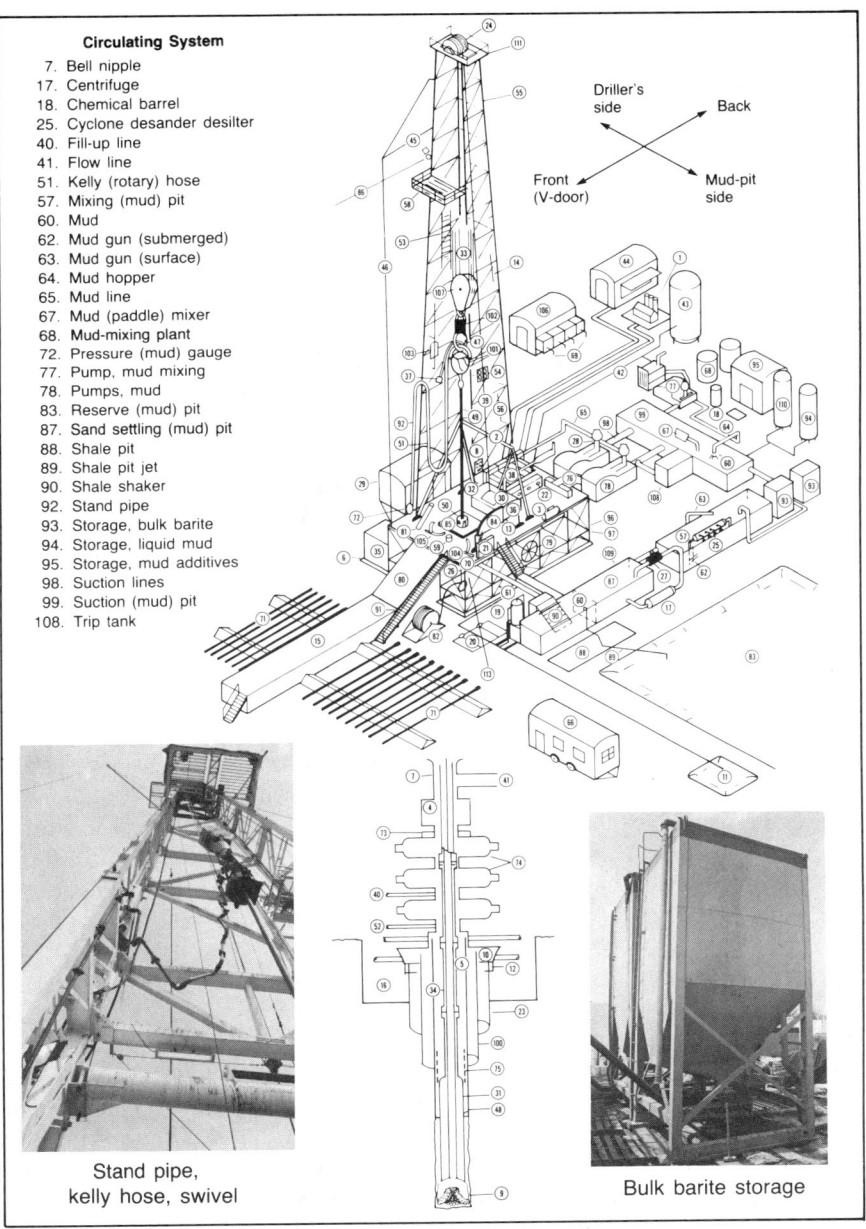

Circulating System
7. Bell nipple
17. Centrifuge
18. Chemical barrel
25. Cyclone desander desilter
40. Fill-up line
41. Flow line
51. Kelly (rotary) hose
57. Mixing (mud) pit
60. Mud
62. Mud gun (submerged)
63. Mud gun (surface)
64. Mud hopper
65. Mud line
67. Mud (paddle) mixer
68. Mud-mixing plant
72. Pressure (mud) gauge
77. Pump, mud mixing
78. Pumps, mud
83. Reserve (mud) pit
87. Sand settling (mud) pit
88. Shale pit
89. Shale pit jet
90. Shale shaker
92. Stand pipe
93. Storage, bulk barite
94. Storage, liquid mud
95. Storage, mud additives
98. Suction lines
99. Suction (mud) pit
108. Trip tank

Stand pipe, kelly hose, swivel

Bulk barite storage

FIG. 5–13 *continued*

pump is mounted on skids to simplify moving, similar to the duplex pump. It can be driven by a diesel engine or an electric motor or from the rig compound.

Triplex or plunger pumps generally run at higher speeds than duplex-piston pumps. Therefore, they can develop more hydraulic horsepower with less overall size and weight.

Mud pump fluid efficiency, a percentage, is calculated as the product of the actual volume of fluid pumped times 100 divided by the theoretical volume of fluid pumped. Efficiencies depend upon the pump piping, the condition of the valves and plunger, and the physical characteristics of the fluid being pumped. Most triplex-plunger pumps are a centrifugal priming pump. The priming pump maintains a slight pressure in the pump suction chamber and helps ensure that the main chamber is completely filled with fluid as the plunger is withdrawn. Most new, larger rigs use plunger pumps. As the piston pumps become worn, many are replaced with more efficient plunger pumps.

Flow Path The average drilling rig circulating system has 700–1,200 bbl of mud in the surface circulating system and 600–1,500+ bbl in the hole, depending on hole size and depth. Under certain operating conditions there may be an additional 500–1,000 bbl of liquid mud storage. The mud is circulated continuously in a closed cycle during drilling, cleanout, reaming, and circulating. At other times the mud is often circulated through the surface system to keep it in good condition. In various parts of the cycle, the mud is alternately transported, used, cleaned to remove impurities, and then treated to restore the chemical balance and maintain good physical properties.

Various circulating times are referred to in operations. A complete circulation is the total time required to circulate the mud from the pumps through the entire system and back through the pumps. The hole circulating time, sometimes referred to as a circulation, is the time required for the mud to move from the pumps through the surface facilities (mud line and standpipe), downhole, and back to the surface. The lag time is the time required for the mud to move from the bit through the annulus to the surface.

The mud moves continuously through the circulating system, and a logical point to begin tracing the flow path is at the mud pumps. Clean, treated mud moves from the mud suction pit into the mud pump where it is pressurized and passed through a manifold into the mud line. A manifold is a combination of valves and piping that diverts fluids from one or more sources to one or more outlets. In this case the line pump delivers mud into the manifold; there, the valves are adjusted so the mud moves from the manifold into the mud line. The pumps can be switched by changing the manifold valves. Mud from a second pump can be delivered into the mud line, and the first pump can be isolated for

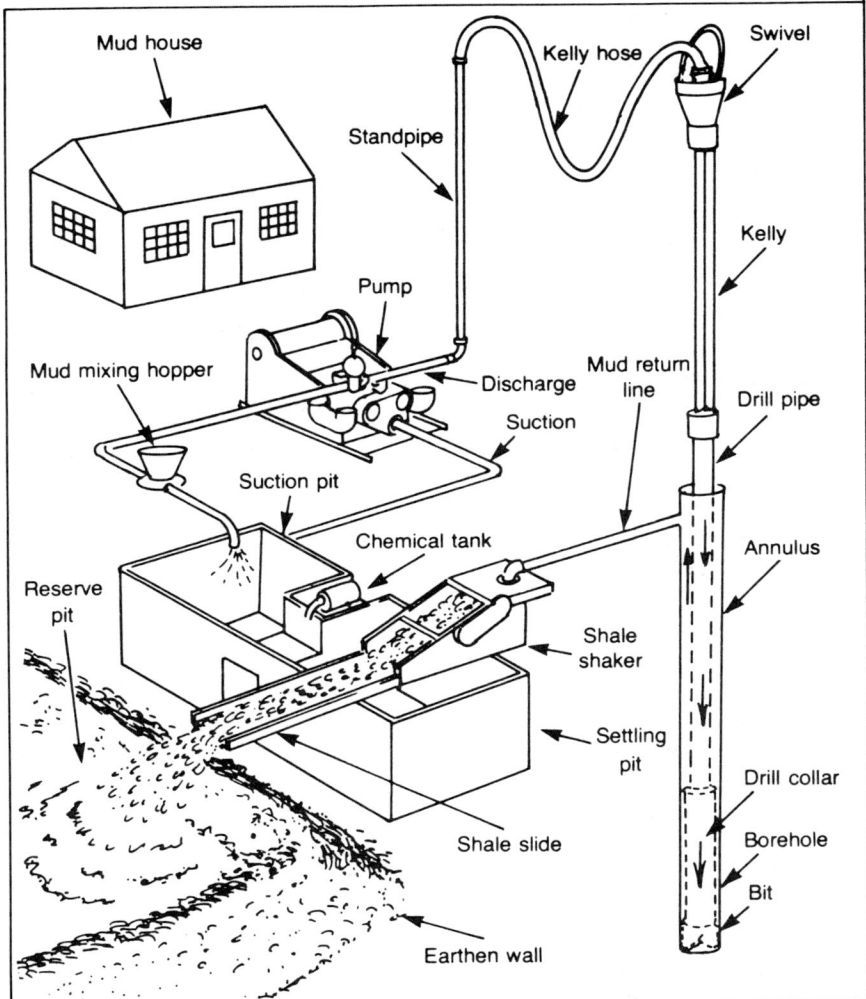

FIG. 5–14 Fluid flow path (after Berger and Anderson, PennWell)

repairs. Or both pumps can continue pumping, a procedure known as running the pumps in parallel.

In the entire mud circulating system, the pumps pressurize the mud and deliver it to the mud line. Mud pressures range from 500–3,500 psi, depending on the size of the pumps and hole depth. A surge chamber is installed on the pump discharge manifold or the mud line near the pumps. This chamber reduces the pulsations in the mud caused by the action of the mud pumps. If the pulsations are not dampened, they can damage the lines that carry the mud. A vibration hose connects the pumps to the mud line. This flexible, high-pressure hose helps reduce pulsations and decreases wear on the mud line and connections.

The mud line is connected to the rig floor circulating system. Mud flows from the mud line into the standpipe manifold. In normal drilling operations, the standpipe manifold valves are arranged so that the mud flows up through the standpipe, kelly hose, gooseneck, and swivel and down through the kelly into the drillpipe. The standpipe manifold has a pressure gauge so that the mud pressure can be observed. The driller may have another pressure gauge on the driller's console, but there will always be a mud gauge on the standpipe manifold. The standpipe manifold is connected to a fill-up line, which is connected to the drillpipe-casing annulus in the lower section of the preventer stack. During tripping, as each stand of drillpipe is pulled out of the hole, mud is pumped into the hole through the fill-up line to replace the drillpipe volume and maintain the hole full of mud. This is a very important function, as discussed later in the text.

During drilling, reaming, circulating, and similar operations, the mud flows from the standpipe into the drillpipe. The mud continues flowing downhole through the drillpipe, drill collars, and bit. It then flows up through the annular space between the drill collars-drillpipe and the walls of the hole to the surface and out the flow line.

The mud performs useful work at the bit and while returning to the surface. There is usually a relatively large pressure drop at the bit, especially with jet and diamond bits. Jet nozzles on the bit direct a high-velocity jet stream onto the bottom of the formation. This helps erode softer formations and improves penetration rates. The mud cleans the bit cutters and cutting face on all types of bits and sweeps across the formation face, removing bit cuttings. This helps ensure that the bit drills on the formation and not on cuttings, which would cause unnecessary wear and increased drilling time. Mud carries the cuttings out of the hole so that new hole can be drilled, i.e., so the well can be deepened. The returning mud also helps build a mud cake on the walls of the open hole, exerts a hydrostatic head (or column of liquid) against the formation, lubricates the hole, and dilutes and carries reservoir fluids to the surface.

The returning mud flows by gravity from the bell nipple through the flow line to the shale shaker. The front end of the shale shaker has a settling pit where bit cuttings, commonly called shale or formation cuttings, settle to the bottom and are discarded. Mud overflows from this pit onto an inclined vibrating screen. Larger mud solids (another name for bit cuttings) are separated from the mud. They roll and slide down the inclined surface of the screen and are discarded into a shale pit. The mud with finer-sized mud solids (the size depends on the screen size) passes through the screen into the mud pan at the bottom of the shale shaker.

A mud-gas separator, or gas buster, may be installed on the flow line when drilling high-pressure zones that flow gas into the circulating system. The gas buster is connected to the flow line and butterfly valves divert the mud flow either through the flow line or into the upper section of the gas buster. Another line from the bottom of the gas buster carries mud back to the flow line. The top of the gas buster has a gas flow line laid to a flare pit a safe distance away from the rig. The gas-mud mixture enters the gas buster, and the mud flows downward through a series of baffles. The gas separates from the slow-moving mud and passes out the top of the gas buster into the gas flare line, where it is flared, or burned. Degaseated mud flows from the bottom of the gas buster back into the flow line.

A rotating head may be used on top of the bell nipple to divert the total flow of gas and mud into the flow line. The rotating head is a low-pressure device (500–1,000 psi) that seals the annnular space between the kelly and the top of the bell nipple and allows the kelly to rotate. The bell nipple has an inverted bell-type opening—like a funnel—so tools can be run in the hole without catching on the edge of the nipple.

A magnet may be installed in the flow line to help catch small particles of metal circulating in the mud system. These steel particles, often called iron filings, can cause excessive wear on the pump and circulating system. They also indicate casing wear and other potential failures.

A sample catcher, or sample trap, may be installed on the flow line to catch samples of the drill cuttings.[3] The trap is emptied periodically and the samples are washed, cleaned, and stored in sample bags. The drilling depths over the period when the samples were collected are recorded, along with the date, well name, and other information on a tag affixed to the sample log. Lag time may also be recorded. It may take an hour or more for the cuttings in deeper wells to be carried from the bit to the surface. When the samples are actually taken, the bit will be drilling deeper than the point where the samples originated. Lag time determines the cross section of the hole where the cuttings originated. These are sometimes known as lag-corrected cuttings.

Mud flows from the shale shaker through a mud ditch into the mud pits. Small rigs may only use one pit; large rigs may use as many as four. The standard, medium-depth rig uses three mud pits. The first mud pit (from the shale shaker) is the sand settling pit, the second is a mixing pit, and the third is the

---

[3]Mud logging units are often used, especially on deeper wells. Usually a small hose will be connected at the sample catcher to carry mud to the mud logging unit.

suction pit. The pit and shale shaker are connected by a mud ditch fitted with gates so that mud can flow from the shale shaker into any pit and from that pit to other (downstream) pits as required.

The volume of mud in the pits is known as the surface mud volume or, more commonly, surface volume. The capacity surface volume of the mud tanks (pits) should be approximately equal to ½ or ⅔ of the volume of mud in the hole. Experience has indicated that this ratio will allow sufficient mud retention time at the surface so the very fine mud solids can be separated and removed from the mud system. The mud can be treated with chemicals, weighting material, or bentonite (gel) and then mixed so that it will have the required chemical and physical properties before it is recirculated downhole.

Mud chemistry is the study of drilling fluids. Mud treating costs can be very high, and it is not uncommon for the drilling mud to cost over $50/bbl with some mud costs exceeding $100/bbl. Drilling mud represents an appreciable investment, especially with larger systems containing over 2,000 bbl.

Solids control is a very important factor in maintaining a good mud system. These solids are fine, very small drill cuttings sometimes referred to as low-gravity solids to distinguish them from the heavier barite (high-gravity solid), which is used as a weighting material. These low-gravity solids will be carried in the mud system and recirculated if they are not removed. Particle size continues to decrease with additional circulation as a result of the mechanical action of the pumps and the downhole rotating equipment. The difficulty of removing these small particles increases as the particle size decreases. Larger particles are removed by the shale shaker and the settling pit. However, smaller particles will continue to circulate through the system unless special solids removal equipment is used.

The hydrocyclone is a basic tool for removing solids from the mud (Fig. 5–15). Hydrocyclones are often mounted in banks containing six or more cyclones. A centrifugal pump takes mud from the mud pit and pumps it through the hydrocyclones. The mud enters the side of the cyclone and spins very rapidly as it passes down to the inner part of the cyclone and out the discharge tube. This rapid rotation creates a centrifugal force on the particles, forcing them out against the inner walls of the cyclone body. The solid particles slide down the tapered inner wall of the cyclone. They are removed at the solids discharge and discarded. Mud from the fluid discharge is returned to the mud system.

The dimensions of the hydrocyclone and the fluid volumes must be relatively precise. Each size of hydrocyclone will remove a relatively narrow range of particle sizes. Desanders are larger hydrocyclones, 8–10 in. in diameter, and designed to remove sand-sized material. Desilters are smaller hydrocyclones, 6–8 in. in diameter, used to remove silt-size material. Smaller hydrocyclones,

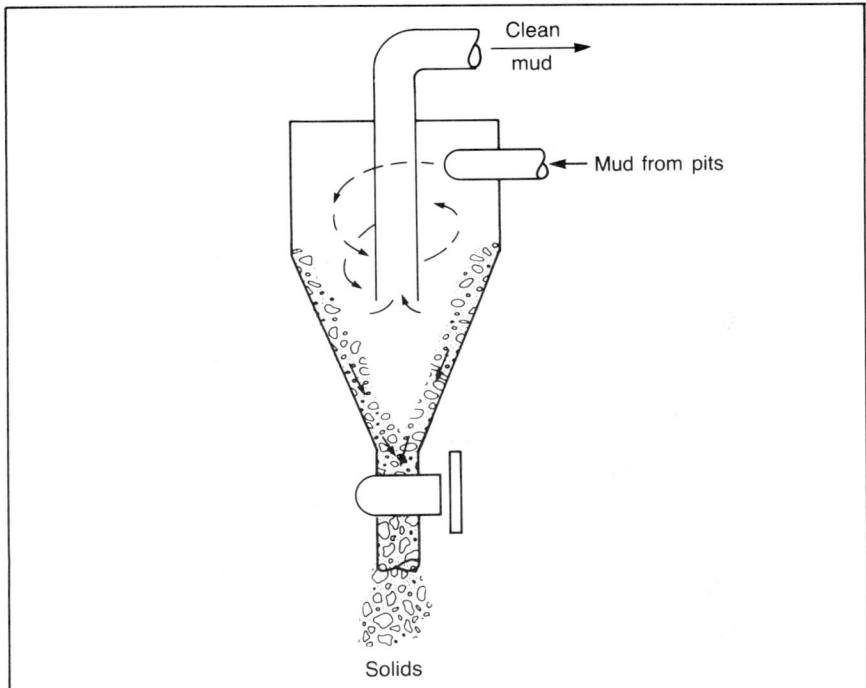

FIG. 5–15 Hydrocyclone

4–6 in. in diameter, are used to recover weighting material (barite) from mud that is to be discarded.

Mud entering the cyclone is usually diluted with a small amount of water. Mud from the cyclone fluid discharge is discarded to the waste pit, and the high-gravity barite solids from the solids discharge is returned to the mud system. The mud system is opposite the desander and desilter where mud is returned to the pit and low-gravity solids are discarded. There is some disadvantage in using the hydrocyclone in this manner since it does cause a loss of mud. However, excess mud is frequently available and the hydrocyclone will allow an appreciable part of the expensive weighting material to be recovered from the mud before it is discarded.

Centrifuges separate solids from the mud and operate somewhat similar to the hydrocyclone, except the bowl of the centrifuge is rotated at high speed. This speed imparts a centrifugal force to the solid particles so they can be separated from the mud.

The gas buster, as mentioned earlier, separates large volumes of gas from the mud. However, in some cases gas will be trapped in the mud in the form of very fine, small bubbles that cannot be efficiently removed with the gas buster. A degasser removes this entrapped gas. Various types of degassers are used. One of the most efficient types circulates the mud continuously through a vacuum chamber. When the gas bubbles entrapped in the mud enter the vacuum chamber, they enlarge as a result of reduced pressure, separate from the mud by gravity action, and are discharged to the flare line. The degassed mud reenters the system.

Some mud-treating chemicals are mixed with water and are added to the mud system. A chemical barrel is used for this purpose. The chemical barrel is a small tank that holds 1–5 bbl of fluid. It has a drain connection into the mud system controlled by a valve. The tank has a water inlet to fill it with water and may have a small mixing system. In operation, the required amount of water is placed in the tank and treating chemicals are poured into and mixed with the water. The discharge valve is then opened to allow the chemical to enter the mud system at the desired rate. Chemicals are normally added to the mud in even circulations. For example, a barrel of chemical would be mixed and then allowed to run into the mud slowly so that all of the chemical would be run into the mud in one, two, or three complete circulations. This allows the chemical to be evenly distributed throughout the mud system.

Dry solids are often added to the mud. Barite, weighting material, increases the density or weight of the mud. Bentonite makes new mud and treats the mud in the system. Lost circulation material reduces lost circulation, i.e., mud loss into the open, highly permeable or fractured formation. Fluid loss additives build a thicker filter cake and reduce the mud seepage through the cake into the formation. All of these additives are mixed with the mud in a mud-mixing hopper (Fig. 5–16). A large funnel is connected to a mixing chamber that has a mud jet. Mud is pumped through the jet, and dry solid material is poured into the funnel. The jetting mud mixes the dry additives with the mud and pumps it to the mud pit.

Some mud systems have a trip tank with a fluid capacity of 15–30 bbl. A connection from the mud-mixing pump is used to fill the tank. The tank discharge is connected to the mud pump suction manifold. When the drillpipe is pulled, a fixed amount of mud must be pumped into the hole to replace the volume of the pipe pulled out of the hole. This amount of mud is calculated, pumped into the trip tank, and pumped into the hole as the pipe is pulled. This procedure allows the operator to know exactly how much mud is pumped into the hole. It is important for the hole to take the exact amount of mud equivalent to the volume of pipe pulled out of the hole. The trip tank allows the operator to determine the exact volume of mud used. If the hole takes more or less, this can indicate pending problems that must be attended to.

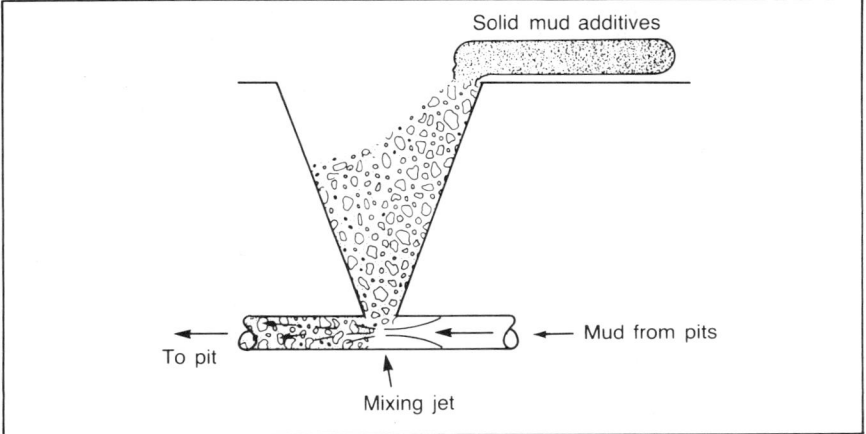

FIG. 5–16 Mud hopper

The mixing and suction pits are fitted with a system to mix and blend mud additives and chemicals and to ensure that the mud has an even weight distribution and consistency throughout the system. The three common methods of mixing the mud are with a surface mud gun, with a submerged mud gun, or with a paddle mixer. Some rigs use two of these mixing systems. Most of the mixing is done in the mixing pit with some action in the suction pit.

In normal operation the mud flows from the shale shaker to the mud settling pit. This pit is not mixed and may be fitted with baffles so that the mud has an extended travel path. This allows more time for the low-gravity particles to settle to the bottom. The pit is cleaned periodically by bypassing mud around the pit and opening a lower drain line to wash the solid particles out of the bottom. The mud is then rerouted through the pit. Sometimes this pit is divided into two parts by a partition and only the front (upstream) end is cleaned.

In summary, the mud is circulated continuously through the circulating system. In parts of the system, the mud performs work; in other parts, the mud is cleaned and treated to keep it in a good condition. The entire mud circulating system is an integral part of the overall drilling operation.

Blowout Control Equipment

A blowout is unquestionably the worst action that can occur in a drilling operation. Blowouts create a very high risk of injury or death to the personnel. All of the rig equipment can be destroyed. Additional damages and expenses are incurred in killing, or ending, the blowout, such as repairing and paying for surface damages and redrilling the well.

A blowout occurs when formation fluids (water, oil, or gas) flow from the formation into the wellbore. This can occur when the hydrostatic pressure exerted by the mud column is less than the formation pore pressure (reservoir pressure) and the formation has enough permeability to allow the formation fluids to flow into the wellbore. The formation fluids normally have a lower density than the mud in the wellbore. As these fluids enter the wellbore, they displace mud in the borehole. This in turn decreases the hydrostatic pressure of the mud column, which serves as a back pressure to hold the formation fluids in the formation. Therefore, the fluids flow into the wellbore at an increasing rate as the back pressure is reduced.

If the flow rate is not controlled immediately, the fluids will continue to flow into the wellbore and up the hole to the surface at an increasing rate. When the mud has flowed or been blown out of the hole, these fluids spurt into the air and can reach heights of several hundred feet. Rocks and other formation debris are carried out of the hole with the uncontrolled flow. These rocks can strike each other and the metal in the derrick, causing a spark that ignites the oil and gas. The result can be a fire with a flame hundreds of feet into the air. Any equipment that is not damaged by the fluid flow will be damaged by the intense heat from the fire. By this time, the mast will normally have collapsed. If the well continues to flow uncontrollably, it will frequently create a large crater. The remainder of the drilling equipment can fall into this crater and become buried. There have been cases where a well blew all of the mud and drillpipe out of the hole and caught fire in a period of less than 10 min. This example emphasizes the importance of detection and immediate preventive action.

Blowouts can be prevented by maintaining sufficient mud weight so the hydrostatic pressure of the mud column is higher than the formation pore pressure. However, in some cases this is easier said than done. Drilling rates increase with decreasing mud weight, and increased mud weights increase mud costs. Increased mud weights can cause lost circulation, which can in turn create a potential blowout condition. A low fluid level may occur while pulling the pipe out of the hole, creating a potential blowout condition. This is the reason for stressing the importance of the trip tank and maintaining a hole full of mud, as noted earlier in the text.

Swabbing may occur while pulling out of the hole with a balled bit, creating a potential blowout condition. The balled bit, a bit enlarged by a coating of cuttings, acts like a plunger and literally lifts or swabs the mud out of the hole. Formation fluids replacing the mud can cause a blowout. Sometimes the formation pressures will increase over a short distance so the hole is inadvertently drilled into a higher-pressured formation with insufficient mud weight. Pore pressure plots help predict pressures in the undrilled formations below the bit. In summary, many conditions can occur that create a possible blowout situation, and considerable experience and training are required for the operator to select the best course of action.

The blowout control equipment is designed to control the blowout condition to the wellbore by sealing off the top of the hole. Formation fluids are contained in the wellbore and are prevented from escaping to the surface (Fig. 5-17). The equipment also provides a method of killing the blowout so normal operations can be resumed. The equipment must be designed and constructed to hold high pressures, in some cases over 15,000 psi. It also must be securely connected to the casing to withstand lifting forces that may exceed 500,000 lb.

One main problem in designing blowout preventers is providing a method of making a positive annular pressure seal for the various sizes and shapes in the downhole rotating equipment. This equipment usually has a circular shape but with different sizes for the drillpipe, tool joints, and drill collars. Some equipment, such as the kelly, has a square or hexagonal shape that further complicates the problem of obtaining an effective pressure seal. These problems are resolved by using two types of preventers. A ram-type preventer seals around specific pipe sizes or the open hole when the pipe is out of the hole (Fig. 5-18). A bag-type preventer, sometimes called an annular preventer, seals around other, different-sized circular and noncircular shapes, such as the kelly (Fig. 5-19).

Various types of construction are used on a ram-type BOP. Each side of the preventer has two pistons connected to a ram arm, which in turn is connected to a ram in the inner preventer cylinder. In operation, pressure is applied through the closing lines and forces the opposite pairs of pistons to move together toward each other. This causes the rams—either pipe rams or blind rams—to move toward the center of the hole and seal around the pipe. When the rams are closed, the space is sealed. Pipe rams close around circular parts of the downhole assembly (pipe and collars); blind rams seal the open hole, much like a large massive valve. The preventers are opened by releasing the pressure on the closing line and applying pressure to the opening line. This fluid pressure acts against the pistons to push them outward and move the reams away from the center of the hole.

The rams can be changed by removing the bonnet—the massive cover over the end of the ram chamber—opening the ram piston to move the ram outward, removing the ram, replacing it, reversing the procedure to close the reams, and reinstalling the bonnet. It normally takes about 1 hr to change rams.

The rams can be closed manually in the event of a failure in the closing-opening hydraulic system. Hydraulic indicates that the system is operated by fluid pressure. In manual operation the rams are closed by turning a large closing wheel that in turn rotates a ram screw and closes the ram. A second closing

222
DRILLING

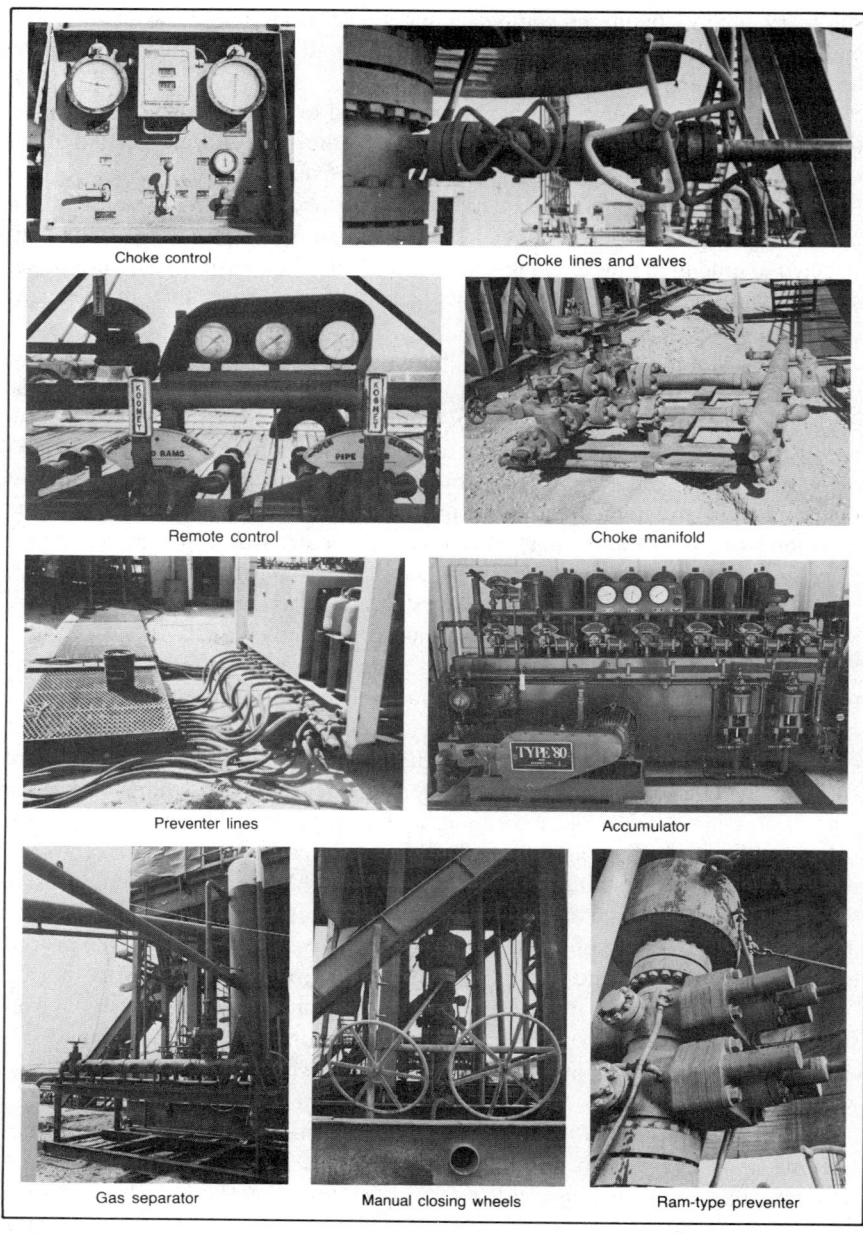

FIG. 5–17 Blowout control equipment

223
Drilling Personnel and Equipment

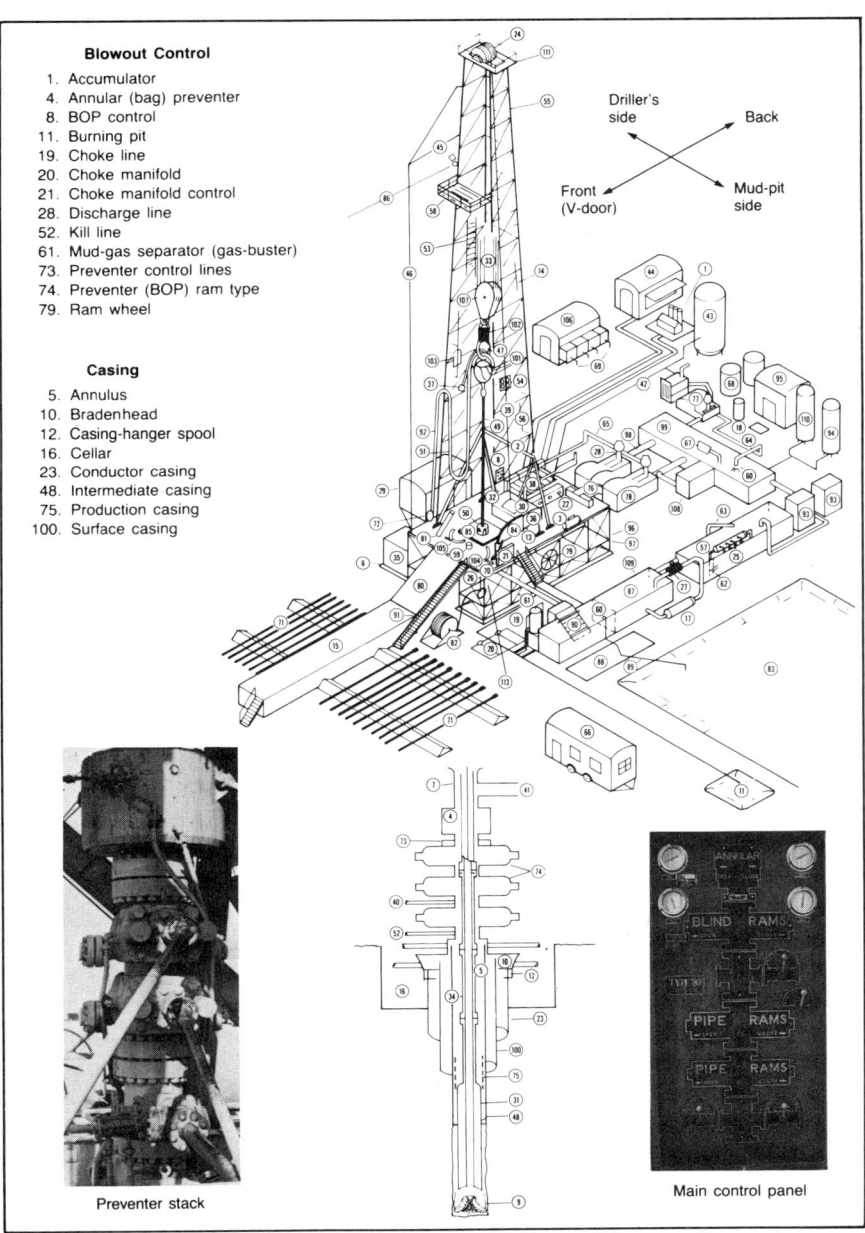

Blowout Control
1. Accumulator
4. Annular (bag) preventer
8. BOP control
11. Burning pit
19. Choke line
20. Choke manifold
21. Choke manifold control
28. Discharge line
52. Kill line
61. Mud-gas separator (gas-buster)
73. Preventer control lines
74. Preventer (BOP) ram type
79. Ram wheel

Casing
5. Annulus
10. Bradenhead
12. Casing-hanger spool
16. Cellar
23. Conductor casing
48. Intermediate casing
75. Production casing
100. Surface casing

Preventer stack

Main control panel

FIG. 5–17 *continued*

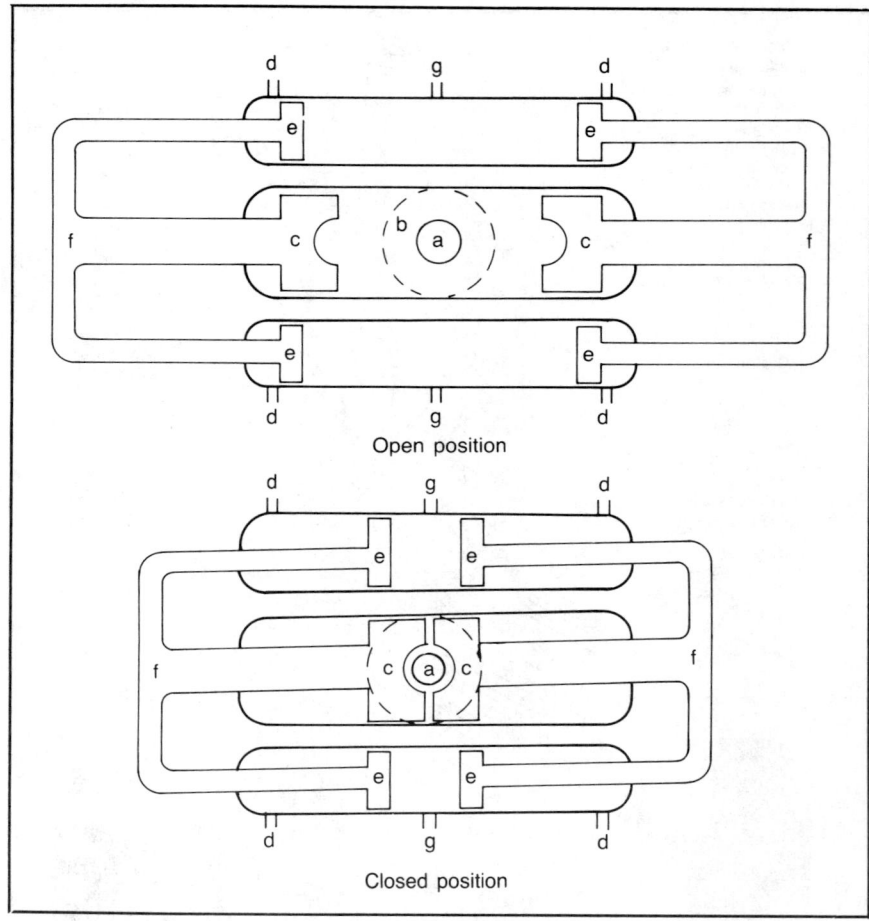

FIG. 5–18 Ram-type blowout preventer. The drillpipe (a) is inside the annulus (b) with the preventer rams (c) in the open position. The accumulator (not shown) applies fluid pressure through the preventer closing lines (d). The pressure acts on the back side of the pistons (e) causing them to move inward, toward each other. This causes the ram arms (f) to move inward, which forces the rams inward around the drillpipe. This seals the annular space, thus preventing fluids from flowing upward around the drillpipe. The preventers are opened by applying pressure to the preventer opening lines (g).

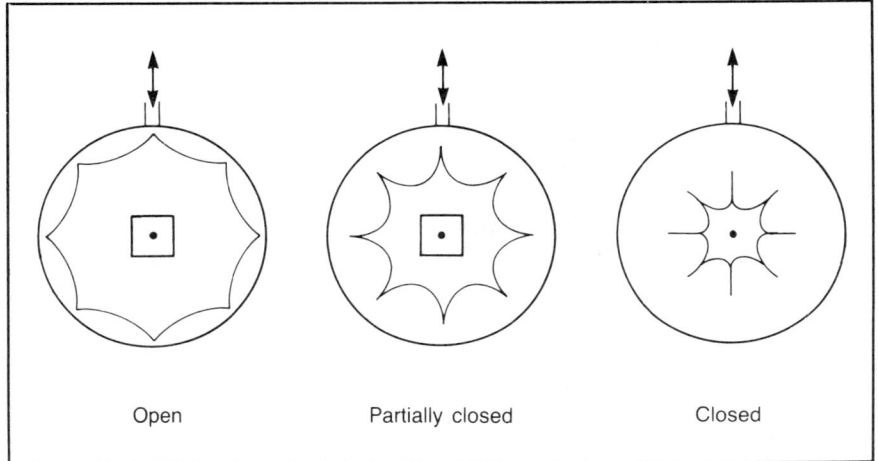

FIG. 5–19 Annular (bag-type) preventer

wheel-ram screw is used to close the ram on the opposite side of the preventers. In normal practice after the preventers have been closed hydraulically, they are locked with the manual closing wheels.

The bag-type preventer has a metal body that contains a rubber bag or sleeve sealed at both ends of the preventer. Fluid pressure is applied to the closing line that in turn causes the rubber sealing element to expand inward and close around the kelly, pipe, or tubular in the hole. The preventer is opened by releasing the pressure from the closing line; the rubber element moves outward to its original position. The rubber sealing element cannot be built as strongly as the rams, so the bag-type preventer generally has a lower pressure rating.

On a conventional blowout preventer-choke manifold installation, the preventers are seated one above the other and are connected by heavy-duty flanges. This configuration is called a preventer stack or stack (Fig. 5–20). The bag-type preventer is located above the ram-type preventers. The ram-type preventers can be fitted with pipe rams and blind rams, sometimes called blank rams. Lower pipe rams are used as a safety or reserve set of rams, and upper pipe rams are working rams. The lower set of pipe rams may be omitted on smaller rigs.

In operation, if the well begins to kick, or blow, when the pipe is out of the hole, the blind rams are closed. If the pipe is in the hole, the bag-type preventer or upper drillpipe rams are closed, depending on the position of the pipe. Normally, the pipe will be picked up to pull the kelly out of the hole so the upper drillpipe rams can be closed. If the well is flowing strongly while picking the pipe up, the

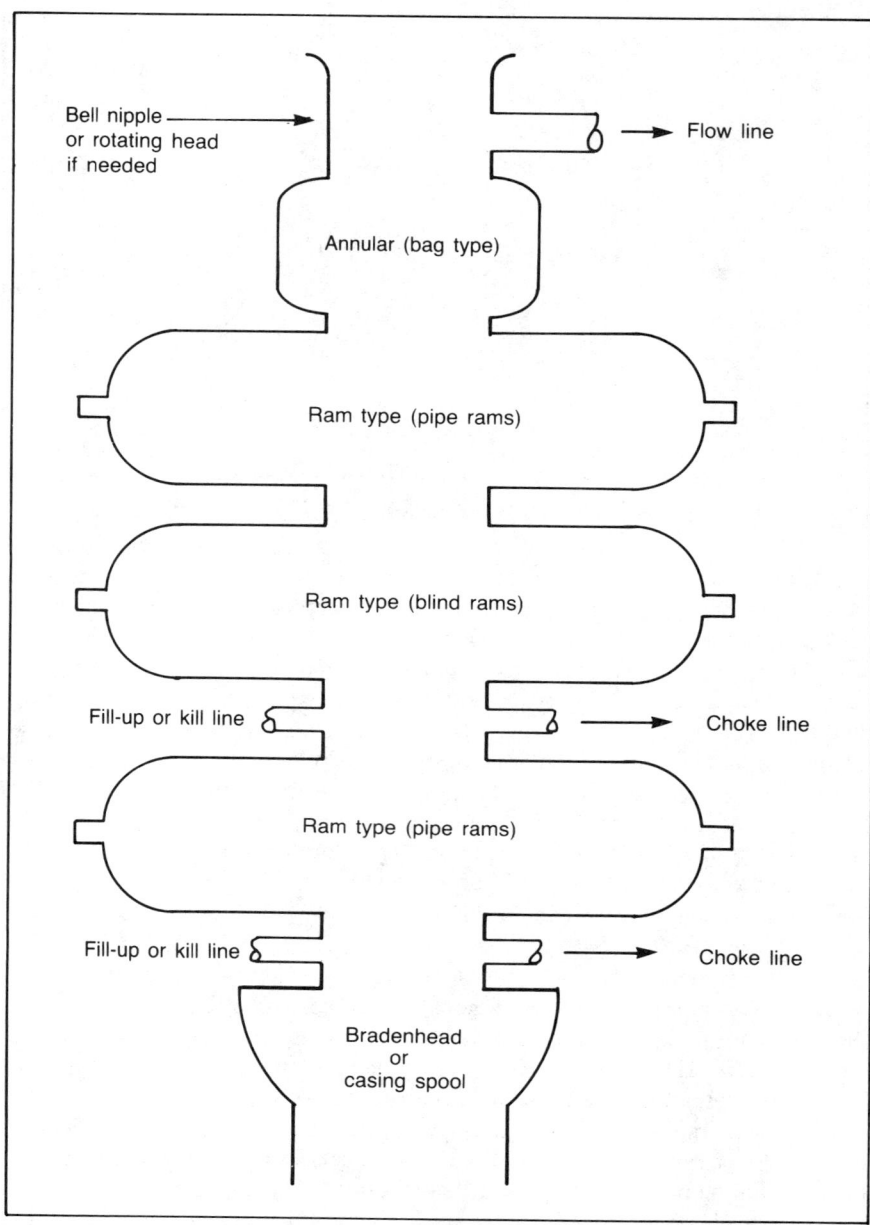

FIG. 5–20 Blowout preventer (BOP) stack

bag-type preventer can be closed and the pipe allowed to slide through it until a joint of drillpipe is opposite the pipe rams where they can be closed. It is always preferable to use the pipe rams because they are much stronger (have a higher pressure rating) than the bag-type preventer. When the rams are closed, the well is safely shut in and the high-pressure fluids are contained in the wellbore below the preventers.

If the upper pipe rams begin to leak, the lower pipe rams can be closed to contain the pressure in the wellbore while the upper pipe rams are changed for a new set. When only two ram-type preventers are being used, the pipe rams are always placed below the blind rams to provide a method of changing rams in case the working rams begin to leak. For example, if the pipe rams began to leak, the blind rams could be changed for pipe rams and used as the working rams with the lower pipe rams held as a safety reserve.

The choke manifold contains and controls the flow of high-pressure fluids from the wellbore annulus while the blowout is being brought under control, or killed. The normal procedure for killing a blowout is to pump heavy, weighted mud down the drillpipe and up the annulus. This heavier mud displaces the lighter mud and reservoir fluids in the annulus, which are allowed to flow out through the choke manifold at a controlled rate. If the flow rate of these fluids were not controlled, additional reservoir fluids would flow into the wellbore and effectively prevent killing the well.

The choke manifold is fitted with a full-opening connection, an adjustable hydraulic choke, and a manual adjustable choke that are used singularly or in combination to control the flow of fluid from the wellbore. The manual adjustable choke is normally not used except as a safety control in case the hydraulic system fails on the hydraulically actuated adjustable choke. The fluid from the wellbore can be flowed to a flare or waste pit or back to the mud system, depending on the type of fluid.

The actual procedure for killing a blowout is more complicated than the simplified description used here to illustrate the equipment. The weight of the kill mud, or heavy mud, must be precisely calculated and mixed to the exact weight needed. The mud must be circulated under carefully controlled pressure conditions by coordinating both the pump pressure and the size of the openings in the choke manifold.

An accumulator located at a convenient point 50–150 ft from the wellbore supplies the fluid and pressure required to activate (close and open) the blowout preventers through small-diameter, high-pressure pipelines. The accumulator has a fluid reservoir and high-pressure pump that maintains a constant pressure to the control valves, which divert the fluid flow as necessary to close and open the preventers. The control valves are on a control panel. One control panel is always located on the rig floor near the driller's console; another is located at the accumulator. One or more additional control panels may be located at remote positions away from the rig so the preventers can be closed should blowout fluids around the rig floor prevent the driller's control panel from being used.

The choke manifold is operated from a control panel located on the rig floor near the driller's console. This panel contains pressure gauges, pump stroke counters, regulators, and control valves needed to operate the choke manifold while circulating out a kick or killing a blowout. The choke manifold can also be operated manually in case of a hydraulic or electrical failure.

Blowout control equipment is designed to control the well safely under all foreseeable conditions. Reserve or backup equipment is available in case of failures. Other monitoring equipment such as pit volume totalizers, flow-line flow-rate recorders, detectors to measure the gas content in the mud, and drilling rate recorders (for drilling breaks) provide advance notice of a pending blowout situation. Early detection is important, so blowout equipment is operated and tested frequently and periodic blowout drills are conducted.

Accessory and Miscellaneous Equipment

Rigs are equipped with a wide variety of accessory and miscellaneous equipment. The amount and type of equipment depends on the type of rig, its size and depth classification, and to some extent the area where the rig is working. For example, smaller rigs may have only one mud pump used for drilling and a smaller mud-mixing pump that also serves as a standby pump for drilling when the main pump fails. Very large rigs have at least two main pumps, sometimes three, and one or more mud-mixing pumps. In some areas less pump capacity is required, so rigs will often have reduced pump capacity.

It is relatively common to use at least two different sets of preventers and two different-size drilling assemblies when drilling a well. If the rig is located a long distance from the source of supply, the extra equipment must be carried with the rig. However, if the rig is located near the contractor's storage yard, the extra equipment will be moved to the rig before it is needed.

Rigs have many different kinds of equipment. Most repair parts, supplies, and expendable items must be kept at the rig to prevent shutdowns. These materials are sometimes called rope, soap, and dope. A partial listing of these includes fuel, oil, grease, valves, connections, seals, packing, gauges, screens, filters and air cleaners, drive chains, belts, tongs and slip dies (segments), and pump repair parts.

Air-Gas Drilling One modified procedure is drilling with air, commonly called air drilling. The basic drilling procedure is similar to regular drilling except air is pumped down the hole in place of fluid mud. The air is supplied by a portable, skid-mounted unit containing an air compressor usually driven by a diesel engine. The compressor unit is designed to deliver 300–500 cu ft of air/min at a pressure of about 300–600 psi. Three to five compressor units are used, depending on depth and hole size. Sometimes the air compressor units are mounted on large floats (fifth-wheel truck trailers) for increased portability. Larger volumes of air and/or higher pressures are required for drilling deeper or larger holes, in which case a booster compressor may be used. The booster compressor is a special air compressor designed to compress large volumes of air from atmospheric pressure to a low pressure. The discharge from the booster compressor is connected to the suction of the regular compressors. The booster literally boosts the air volume and pressure into the regular compressors, increasing the capacity of the regular compressors.

Most rigs rent air compressors from service companies. The air compressors are installed at a convenient place on the location and are connected to the mud line of the rig mud circulating system. The flow line is disconnected from the bell nipple, and a blooie line is connected at the flow-line connection on the bell nipple. The blooie line extends away from the rig to a blooie-line pit. In operation, the air compressors deliver air to the mud line and the air circulates through the mud line, following the same flow path as the liquid mud—down the drillpipe and up the annulus to the blooie line. At this point the air with entrapped, finely ground shale cuttings passes through the blooie line to a blooie-line pit located on the edge of the location.

The top of the bell nipple is fitted with an air drilling head to prevent air from flowing out the top of the bell nipple. Air drilling is sometimes referred to as "dusting" when there is no water entry into the borehole. The name is derived from the large plume of dust emitted from the end of the blooie line and caused by finely ground bit cuttings.

Air drilling operations are basically similar to normal fluid mud drilling operations. A percussion tool may be used to drill very hard formations. The percussion tool is connected to the bottom of the drill collars, and a bit is connected to the bottom of the percussion tool. The tool has a heavy plunger that moves a short distance up and down rapidly by the air flow. The action is similar to a common jackhammer and imparts rapid blows to the bit and in turn to the formation to increase the drilling rate. Special percussion bits that have heavily reinforced shanks are used for drilling with a percussion tool.

Gas drilling is sometimes used. The procedure is similar to air drilling except that natural gas (principally methane) is circulated in the system instead of air. The gas may be obtained from high-pressure gas lines or from low-pressure gas lines and boosted to the required pressure for drilling with compressors. The amount of gas drilling has decreased appreciably in recent years as a result of the increased cost of gas and has been replaced by air drilling.

Small amounts of water may enter the borehole while drilling with air or gas. If the amount of water is too much to dry up with the air, then the combination of water and finely ground bit cuttings creates mud that sticks to the downhole drilling assembly, and the bore of the hole. These mud rings create a danger of sticking the drilling assembly in the hole. Foam drilling techniques may be used in this case. A soap tank with 10–20 bbl of capacity and an injection pump are installed near the mud line. A dilute soap-water mixture is injected slowly into the moving air or gas stream. As this mixture moves downhole and returns to the surface, it creates a foam that helps lift the drill cuttings and water out of the hole. Environmentally acceptable biodegradable soap solutions are normally used.

Winterized Rigs   Rigs in colder climates must be winterized. Moderate winterization is needed in latitudes equivalent to the central United States with increasing winterization north of this latitude. Maximum winterization is needed in the northern latitudes and in Arctic waters where the rigs must operate at subzero temperatures for long periods of time. Winterization is the process of modifying the rig and its component systems for operation in cold, inclement weather.

Various degrees of winterization are used, depending on the average temperature of the area where the rig will be working. Mild winterization may include temporary windbreaks around the rig floor and the monkey board where the derrickman works and provisions for heating these areas. Heating is usually provided by automatic boilers. Steam is circulated from the boiler through radiators where it condenses to water and is returned to the boilers. The hot air from the radiators is drafted across the work areas and other places where needed with large fans driven by electric motors.

Permanent shelters or windbreaks are used and cover larger areas of the rig in colder climates. All lines that carry mud or water are located either in a sheltered, warm area or are fitted with trace lines and insulated. Trace lines are small pipelines placed adjacent to the mud and water lines, sometimes called a pipe bundle, and covered with insulation. Steam is circulated through the trace line to keep the other lines warm or above freezing. If the lines freeze, fluid cannot be pumped through the lines, and the rig operations must be shut down. The mud tanks are often enclosed in a housing, both as a protection from cold and to prevent rainwater from diluting the mud. Blowout control equipment, especially the preventers and choke manifold, must be given special attention to prevent freezing because these are primary safety devices.

Automatic Rigs   There has been considerable work done by a few companies to develop an automatic rig. Several shallow automatic rigs are presently in operation, and these may be the first prototype of the next major development and improvement in drilling equipment. These rigs are not completely automatic in the true sense of the word and require at least a two-man crew. Many normal operations have been automated by the use of hydraulics and electronics. A major part of the hard physical labor required on today's rigs has been eliminated. These rigs are still undergoing field testing but have a high potential.

Fly-in and Helicopter Rigs Exploration has developed drilling prospects in remote areas that may be tens to hundreds of miles from the nearest all-weather road, seaport, railroad, or landing strip. Furthermore, the drilling site is frequently in a marsh or swamp where normal earth and load-moving equipment cannot work without becoming mired down and stuck. The tropical, often swampy jungles of South America and Southeast Asia are typical examples of these remote, inaccessible locations. The need to drill in these areas led to the development of the earlier fly-in rig followed later by the helicopter rig.

The average-depth land rig weighs about 1–2 million lb (500–1,000 tons) completely equipped with drillpipe and drill collars. Furthermore, 2–4 million lb (1,000–2,000 tons) of fuel, equipment, materials, and supplies may be needed while drilling the well. The average rig is designed to be dismantled into normal highway loads of 10–20 tons with a few loads such as drawworks and pumps that may exceed this. Load sizes are in the range of 8 ft wide, 8–10 ft high, and 30–40 ft long with some loads such as substructures and the dismantled mast exceeding these. The scheduling and movement of this equipment is a tremendous logistics problem, especially in remote areas and over long distances.

The first attempts to drill these remote locations used fly-in rigs. These rigs required that an airfield be available at or near the location. Normal, smaller land rigs were used. Standard derricks were sometimes used in lieu of masts because they could be dismantled into smaller loads. Steps such as increasing the number of drilling lines, using several smaller pumps in parallel, and using slim-hole drilling techniques were used to maximize the depth capacity of the rig. This procedure allowed many locations to be drilled but left many others undrilled.

The highly mobile, portable helicopter rig was developed to permit drilling the more inaccessible locations by utilizing the special load-carrying and hovering ability of the helicopter. The helicopter can ascend and descend vertically, but its load-carrying ability is increased if it has an 8 to 1 distance versus height take-off and approach pattern. The load-carrying ability increases with forward speed. The heavier loads are picked up with the helicopter partially supported by the ground cushion, the slightly compressed air between the rotors and the ground. Then it is lifted higher as the aircraft moves forward.

Helicopter rigs are basic land rigs where the component parts have been redesigned and modified to reduce them to the minimum size so they can be joined at the wellsite to make a complete rig. More smaller equipment is sometimes used. For example, the two large mud pumps and three mud tanks on a conventional land rig are replaced with about six small mud pumps and eight mud tanks. Other innovations are used. Helicopter rigs have been developed with a depth capacity of 15,000 ft, extendable with smaller drillpipe. The rig, including the camp for personnel, is packaged into about 300 loads with a maximum weight of 2 tons/load.

MARINE RIGS

There are many types of drilling rigs, such as water well, blast hole, core drill, and big hole. These rigs generally have the basic component systems of a standard rig, although some of the systems may be highly modified and changed for the special drilling conditions for which the rigs are designed. This text deals primarily with drilling rigs used in the oil and gas industry, and the rigs that have been discussed to this point are basic land rigs, which comprise over 80% of all rigs currently operating.

From a geological viewpoint many offshore areas, such as the U.S. Gulf Coast, the Gulf of Alaska, the Bay of Campeche, the Arabian Gulf, and the North Sea, are underlain by potential petroleum-bearing sediments. The fact that these areas are covered with water ranging from inland marshes to areas where the water is over 1,000 ft deep does not mask their potential. This is a relatively shallow depth when one considers that hydrocarbons are found at depths exceeding 20,000 ft. However, this water cover does increase the difficulty of exploration, drilling, and production. This has led to the development of equipment and techniques for operating in marine waters.

Marine rigs are basically similar to land rigs with rig parts (component systems) modified to drill in water. The basic difference is that marine rigs (excluding platform rigs) are permanently mounted on a mobile base, and the rig and base are moved as a unit. They are broadly classified as barge, jackup, semi-submersible, and drillship. The latter two types are sometimes referred to as floaters.

Although there are are fewer marine rigs than land rigs, an average marine rig costs about 10 times more than an average land rig. Marine rigs also cost much more to move and operate. Because of these much higher drilling costs, drilling prospects must be carefully evaluated.

Marine rigs are designed to drill in protected or unprotected waters and are broadly classified accordingly. Generally, a rig designed to drill in one type of water cannot drill in another. Unprotected waters are the open oceans and seas, including those areas near the coastlines. These areas are subject to high waves and the water is often deep.

Protected waters are inland waters such as tidal flats, salt marshes, shallow bays, and estuaries. They are known as protected waters because of the limited wave action and shallow depth. High waves do not develop in protected waters because of barrier islands and adjacent land masses and because the water depth is too shallow. These areas are covered with salt and brackish water ranging from a few inches periodically during the day (due to tidal action) to 10–20 ft of water in the deeper bays. Wave heights average 1–2 ft and seldom exceed 5 ft except during major storms.

The nature of the bottom or sea floor in both protected and unprotected waters ranges from sandy to very soft mud, especially in the marshes and delta areas near river mouths. The land surface below the water is commonly referred to as the "bottom" and sometimes as the bay bottom, sea bottom, or ocean floor in deeper marine waters. When drilling in marine areas, the load-bearing strength or soil strength must be taken into account because of the high weight of the drilling equipment.

The first marine operations were conducted in shallow, protected waters. These areas, or the fringes of these areas, were drilled by building dikes or levies with a dragline—a large, track-mounted piece of equipment that moves earth. The dragline was often run on matting boards, large wooden planks laid out in layers to form a "mat" to distribute and support heavy weights. The dragline would dredge up material from both sides and stack it ahead to create a dike or levy. The machine would then advance over this to extend the road.

The location for the drilling rig was constructed in a similar manner. The road and location were then covered with boards. This provided a surface that the trucks could drive over to move the rig in and out and carry supplies during the drilling operation.

Platforms were built on top of piles driven into the bay bottom in deeper protected waters. The rig was moved to the platform by barges and supplied in a similar manner during the drilling operation. If the well was a dry hole, the high cost of building the large drilling platform was wasted. If the well was a producer, though, the platform could sometimes be used for the wellhead facilities. Generally, though, another platform had to be constructed for production facilities. Since the majority of the wildcat wells are dry holes, the high expenditure for drilling platforms emphasized the need for and helped justify developing and building barge rigs, the first marine rigs.

Barge Rigs

The barge rig was the earliest marine-type rig. It consists of a drilling rig mounted on a large barge (Fig. 5–21). The barge is floated to the location and then ballasted down by flooding the ballast tanks. This allows the 8–12-ft tall barge to set

234
DRILLING

FIG. 5–21   Barge rig drilling in a canal dredged across tidal flats

on the bottom and provides a fixed platform to drill the well. After drilling the well, the water is pumped out of the ballast tanks and the barge is moved to the next drilling location.

If water depths are too shallow to float the barge, dredges are used to open a shallow canal to the location.

In deeper protected waters double barges are used; one barge sits on top of another barge to provide additional height. In some cases a substructure arrangement is mounted on top of a larger barge so the barge can stand in deeper water and keep the rig above the water level.

Production facilities are constructed after the well is drilled and determined to be a producer. The production facilities are constructed on top of a platform installed near the wellsite or are placed on a nearby land surface that can be reached by roads. In the latter case the wellhead is connected to the production facilities by pipelines laid on the bottom or in a trench in the bottom and is covered to protect the pipeline. If the production facility is installed on a platform near the wellsite, pipelines are constructed from the production facilities to an accessible point onshore.

Barge rigs are always operated in protected waters. They are not designed for high waves. Small waves 5–10 ft, which are very common in unprotected waters, create tremendous forces against a flat surface such as the side of a barge. These larger waves also make it difficult to ensure that the barge is ballasted down and remains securely seated on the bottom. The combination of these waves, partially floating and thrusting against the side of a barge, can move it, in turn breaking all of the casing and drilling assembly. Therefore, the combination of waves and water depth prohibits the use of barge-type rigs in unprotected waters.

Jackups

The first offshore wells were drilled with land-type rigs set on drilling platforms. The same costs and problems occurred as in drilling from platforms in the deeper protected waters. This led to the development of the jackup rig (Fig. 5–22). The earlier jackup rigs, some of which are still in use, were similar to large, modified barge rigs. The barge was reinforced with steel construction. Each corner of the barge was fitted with a long, strong leg that could be moved up or down with large jacks. The bottom of the leg was fitted with a broad plate, or foot, to distribute the weight of the barge on the ocean floor.

The modern jackup, with the legs in the highest position, is moved to the drilling site, usually with seagoing tugs. The legs are jacked downward until the feet touch the ocean floor. Continued jacking lifts the barge or platform up and away from the water at least 30–50 ft, well above the top of high waves. Waves pass under the barge so that it is unaffected. The waves strike against the legs of the barge, but these legs are strongly constructed and designed to offer a minimum surface to the wave action.

Jackups can drill in water depths up to about 250 ft. Different hull shapes and leg configurations are used. Some jackups are self-propelled, eliminating the need for a tug. The working floor is high above the water level. Permanently mounted cranes lift equipment and supplies from the supply vessels to the platform floor.

On some jackups the mast is mounted on a cantilevered floor, so the hole is drilled from the side of the jackup body. Cantilevered jackups can also be used to drill wells on small platforms. The small platform can be constructed economically. The jackup is floated in and raised so the rig is over the platform. The well is then drilled through the platform, and the platform is used to support the casing and wellhead.

FIG. 5–22 Jackup rig (courtesy Keydrill Company)

Jackups are frequently used to drill offshore wildcat wells. If the well is a dry hole, it is plugged below the sea bottom and the rig is moved off. If the well is a producer, it may be temporarily plugged at the sea bottom and the jackup moved off. A drilling platform can then be constructed over the existing well. After the platform has been constructed, the discovery well is connected to the platform. A rig is moved onto the platform to drill additional wells into the reservoir by deviating, or angling, the wellbores.

Deviated drilling is a procedure used to change and control the direction of the hole so that it can be drilled into a reservoir at a predetermined direction and distance away from the first well, which is usually drilled vertically from the surface. It is relatively common to drill eight or ten deviated wells from one platform because drilling multiple wells from one platform reduces overall costs. Many more wells may be drilled, depending on the depth to the reservoir and other factors. Sometimes a slant-hole rig is used, which increases the number of wells that can be drilled from one platform, especially in shallow reservoirs. Normal deviated holes are started by drilling vertically, and the amount of curvature is limited. Slant-hole rigs begin drilling the hole at an angle, which allows the well to be deviated further horizontally at a shallower depth.

Oil, gas, and condensate produced at offshore production facilities are transferred to accessible points onshore by pipelines laid on the ocean bottom.

Floaters

Continued exploration in deeper waters led to the development of floaters. Waters below the depth capacity of the jackup rig down to 1,000 ft or more could be explored by building drilling platforms. However, the same problems as with exploring in shallower waters occurred, especially the very high cost of building and installing a platform for an exploration well that might be a dry hole. The floater provided a method of drilling the hole to test the structure without building a platform, much like the jackup rig in shallower waters.

The floater is a marine rig where the base floats while drilling operations are conducted. Floaters are broadly subdivided into two classes, based on the manner in which they are maintained in position while drilling. One type remains in position by six or more massive anchors connected to the floater by chains and wire rope. Wire rope is another term for steel cable or wire line but is usually reserved for large wire lines several inches in diameter. These rigs can drill in water depths to about 2,000 ft.

Semisubmersibles are generally used as an anchored-type marine rig. These vessels have two or more large chambers or bodies positioned one above the other and connected by heavy-duty legs, usually tubular in construction. The lower section, which is normally submerged during drilling, contains tanks and provides buoyancy. The drilling rig and other peripheral equipment are mounted on top of the upper deck.

The main advantage of this type of construction is that the large surfaces buffeted and struck by wave action are either submerged where the wave action is negligible (buoyancy section) or are located above the average wave height (upper body and working deck). The connecting members or leg sections are exposed to the wave action but have a relatively small area, so movement caused by wave action is minimized. This configuration also gives a maximum stability and resistance to movement caused by changing buoyancy created by wave action.

The second type of floater, often called a drillship is dynamically positioned by thrusters, or propeller units. Several drillships have been built by converting regular oceangoing ships. A rig is mounted on the ship with peripheral drilling equipment, and the ship is modified so that it can be dynamically positioned (Fig. 5–23). These ships can drill in water over 15,000 ft deep. Some drillships are equipped with anchors and drill in shallower waters.

To date, most marine drilling operations are conducted in water depths to about 1,200 ft. A floater is used to drill the wildcat well and frequently two or more confirmation-development wells to ensure that the reservoir has sufficient productive potential to justify building the expensive drilling-production platform in these deeper waters. The platform is built in a yard that specializes in constructing offshore platforms. The platform is towed to the site and installed. Drilling equipment is placed on the platform and is used to drill the remaining wells required to develop the structure. Many of these wells are deviated in order to penetrate the structure with the correct well spacing so that it can be produced efficiently and economically. Production facilities are then installed and connected to the nearest onshore point by pipeline.

One of the more severe problems encountered in marine operations is inclement weather. The rigs are designed to withstand very high waves and wind speeds. Generally, the waves are more damaging and dangerous than the wind, but both usually occur together. It is impractical to design the equipment for the very highest waves and winds that may occur. Therefore, depending on the size of the rig and the predicted severity of the storm, when the rig is in the projected path of a large storm, it is battened down and personnel are evacuated to a safe point on land.

Icebergs are also a potential hazard in northern waters. Iceberg movement is charted to determine if they may be on a collision course with the rig. Larger icebergs may scrape the bottom, creating scour channels. Mapping these helps to determine iceberg movement. Small icebergs on a collision course with the rig may be broken up with explosives and/or diverted by towing. Very large icebergs cannot be diverted, except by tremendous natural forces such as water currents.

239
Drilling Personnel and Equipment

FIG. 5–23 Drillship (courtesy Petrodrill Private Limited)

When these icebergs are on a collision course with the rig and approach within a predetermined distance, the rig must be disconnected from the hole and moved until the iceberg passes.

An especially hazardous condition can occur with a gas blowout where gas is escaping from the bottom (ocean floor) of a well drilled by a floater. The gas will gasseate the water, causing an apparent reduced density that in turn can cause the floater to lose buoyancy and in extreme conditions to sink.

Marine rigs use normal drilling equipment with modifications for drilling in the marine environment. For example, floaters use a motion compensator connected between the traveling block and the drilling assembly. This tool compensates for the vertical movement caused by wave action so it is not imparted to the downhole assembly. A riser must be connected from the subsea wellhead to the drilling rig to act as a wellbore or string of casing so drilling fluid can be circulated back to the surface. Tools also have been designed for setting liners and performing other specialized operations. Overall, though, the techniques for drilling, completing, and producing wells drilled in very deep marine waters are still in the development-testing stage.

CHAPTER 6
MOVING IN, RIGGING UP, & DRILLING THE CONDUCTOR HOLE

Modern drilling rigs range from small, portable truck-mounted units that drill to shallow depths to very large rigs designed to drill over 25,000 ft deep and which may weight over 1 million lb with all of the peripheral equipment. After the drilling operation is completed, the mast is laid down and dismantled. All of the rig equipment must be dismantled, disconnected, unassembled, stacked, stored, and otherwise broken down into truck-size loads. Trucks normally are used to move the equipment to the next location where it is assembled (rigged-up) to make an operational drilling rig. The old location must be cleaned and restored, which may include hauling off mud, contaminated water, and shale cuttings and disposing these at an approved site. Sometimes the mud and water are left to evaporate and dry out in the pits. The pits are then covered, and the location is leveled or restored to the original ground contours.

In the meantime the next drilling site is prepared, and the drilling rig is assembled again. Then drilling can begin at the new location. The rig-down, move, and rig-up process is repeated for each hole drilled and many times during the life of the rig. Experienced crewmembers have learned to perform these operations safely and efficiently.

ACTIVITIES BEFORE MOVE-IN

In normal operations the rig is moved as fast as possible. Trucks and other moving equipment are expensive and must be used efficiently. The rig makes money only when it is drilling or conducting associated operations. Therefore, it must be moved efficiently, including rigging-down and rigging-up, to minimize downtime (when the rig is not working) and allow the rig to earn a maximum income.

Planning and preparatory work are required in order to move the rig and drill the next well efficiently. The amount of preparatory work and planning depends on the length of the move, the location and depth of the well, and related factors. Much of this work is done weeks or months before the rig is moved, and other work is done immediately before or during the rig move. Personnel at all levels from management to field personnel are involved in this.

Drilling Program

The drilling program normally is prepared well in advance of drilling. The program should be reviewed and checked by all personnel involved in the drilling operation, including both the owner or operator and the contractor. The hydraulics program is of special importance and should always be checked and reviewed, especially if the rig used to drill the well is not the same rig that was considered when the hydraulics program was prepared.

It is always a good practice to make another check of the area for offset well data before drilling the well. Sometimes the drilling program will be prepared months in advance. It will normally have information on other wells drilled in the area at the time the program was prepared. However, new wells may have been drilled since the program was prepared, and data on these wells may help improve operations on the well to be drilled. Special items to look for include problem formations, casing seats, type of mud used, and bit selection.

Permits and Regulatory Agency Reports

Various permits and reports may be required, depending on the location of the well, the number of regulatory agencies that have jurisdiction over the wellsite, and special conditions pertaining to the well. Each state has a regulatory agency that supervises oil and gas operations. The drilling permit must be obtained from this agency before drilling operations are started. Subsequent reports are submitted during and after drilling the well. Most states have rules and regulations governing oil and gas operations. Some states have inspectors who must be notified in time to witness certain operations, such as running and cementing casing.

If the well is located on state lands, the state land commissioner or similir regulatory agency may also have jurisdiction over the drilling operation.

The BLM (Bureau of Land Management) has jurisdiction over wells drilled on public lands. When the well is drilled on Indian land, either the BLM, the Bureau of Indian Affairs, the Indian Agency, or all three may have jurisdiction. In some areas counties and cities may have jurisdiction over the wellsite. The Coast Guard has jurisdiction over some parts of the drilling operation when the wells are drilled in open water or in or near navigable waterways.

An environmental impact statement (environmental study) may be required. Some regulatory agencies require an abbreviated drilling plan. Other regulatory agencies call for an abbreviated report before the location is surveyed, followed by a detailed land-use plan before a permit will be given to drill the well. An archeological study also may be needed before the wellsite can be cleared.

Wells may be drilled in fields that have field rules. Field rules are established by a regulatory agency, usually the state regulatory agency, to govern drilling-completion-production operations.

Sometimes it is necessary to obtain an exception to these field rules. For example, field rules may establish field spacing of 80 acres/well. The operator may have an odd-sized tract containing less acreage, perhaps 70 acres. In this case the operator would request an exception to the field rules. The exception may be granted administratively by the regulatory authorities, or a hearing may be necessary. Then the operator presents his case at the hearing, including the reasons for requesting the exception to the rules, similar to a court trial. Two or three regulatory agency officials usually preside. The agency makes a decision and either grants or denies the exception.

These hearings are lengthy, time-consuming processes that require testimony, usually by expert witnesses, exhibits, and other material and evidence as needed to support the operator's request. Similar procedures are used before all regulatory agencies when the operator's plans do not conform with established rules and regulations.

Immediately before the rig move and depending on the size of the rig, it may be necessary to obtain special permits such as overweight or oversize permits. The route over which the rig is to be moved must be checked to ensure that the bridges will support the loads and there are no limited clearance underpasses or other overhead obstructions such as telephone lines or electric power lines. In some cases these overhead lines must be removed or repositioned. Depending on local traffic conditions, truck movement may be restricted to certain times.

Equipment Repairs and Inspections

Most drilling contractors have scheduled preventive maintenance programs so that the rig is always in good operating condition. Many routine repairs are made during normal interruptions in the drilling operation. However, some inspections and repairs cannot be made until the rig is completely shut down.

For example, minor repairs can be made to a standby mud pump while the other pump is online (pumping). However, the standby pump normally would not be pulled completely out of service for major repairs unless the repairs were absolutely necessary because the standby pump would be needed if the online pump should fail. The normal procedure would be to repair or replace the pump during the rig move.

Many contractors keep spare repaired units such as pumps and engines to replace operating equipment. This equipment is then repaired and replaced on the rig or on another rig. Rotating the equipment in this manner is an optimum method of maintainance.

Some of the rig equipment, such as the mast, crown block, parts of the substructure, and mud pits cannot be repaired except when the rig is shut down. One good practice is to clean and inspect as much of the rig as possible before shutting down for moving. Any repairs or replacements needed can then be scheduled during the moving operation. For example, the crown sheave bearings might be scheduled for replacement during a move, or welding repairs could made on a faulty mud pit. Other welding repairs may also be needed, such as mud-mixer bearing housing in the bottom of the mud pits, bent or broken rotary beams, cracked equipment bases, and repairing or replacing the mud line and flow line. The drawworks and power train are inspected periodically, and repairs are scheduled during the rig move. These include replacing chains, belts, bearings, brake blocks, clutch plates, seals, and other worn equipment. The brake bands, equalizer bar, concentrics, and levers may be removed and inspected for cracks, sometimes using sonic-type inspection tools. Bent chain and belt guards and other equipment housings are repaired.

The rotating equipment is normally repaired and maintained during the drilling operation. However, spare equipment is often used and damaged equipment accumulates, including drillpipe that is bent or has damaged boxes and pins and drill collars with damaged connectors. This damaged equipment is often shopped or taken to a machine shop for repairs during the rig move. Bent drillpipe is straightened. Drillpipe tool joints and drill collar connectors are either rerun through the dies, to correct any damaged part or may be recut—the old threads removed and new threads and shoulders cut. (Drillpipe tool joints cannot be recut more than once or twice due to the short length of the tool joint, but drill collars may be recut several times.)

At periodic intervals all of the rotating equipment is inspected by inspection service companies with portable equipment. Most drillpipe has hard banding on the tool joint to reduce tool-joint wear during drilling. As the hard banding material wears, it must be replaced. This is frequently done during the rig move or at the wellsite by service companies with portable equipment.

245
Moving In, Rigging Up, and Drilling the Conductor Hole

Location (Drilling Site) Preparation

The drilling site, commonly called the location, must be found and prepared for the drilling rig. The first step is to locate the site on a map relative to the nearest accessible public road. The location is then surveyed from known reference points to ensure that the site selected on the ground is exactly the same as the drilling site selected on the geological maps. An access road route from the public road to the drilling site is also selected and surveyed or otherwise marked (flagged). The location normally requires an acre or more, depending on the size of the rig and peripheral equipment. Since this area must be leveled, in extremely rugged terrain an alternate location may be selected after consultation with the geologist. The access road must be an all-weather road capable of supporting heavily loaded trucks. The drilling foreman, drilling supervisor, or other personnel familiar with drilling equipment coordinate with the surveyor in selecting the location and access road.

Locations in remote areas, in shallow marine waters (swamps, tidal flats), and in the far northern regions present special problems. These may involve dredging canals, building dike and board roads, or constructing landfill or ice islands.

FIG. 6–1 In swampy areas working locations are stabilized with board platforms

In normal land locations earth-moving contractors are selected (sometimes by bid) to construct the access road and clear and level the location. After the location has been leveled, the position of the various major items of equipment such as the substructure, mud pits, pumps, and pipe racks are marked and staked with small markers. The position of the equipment is determined by measurements from the borehole according to a rig layout.

The position of the rig on the location is determined by several factors. It is preferable to have the V-door facing the north or northeast so the driller will not be blinded by the rising or setting sun and so he has a full view of the rig floor and pipe rack areas. The V-door may be set downwind, pointing away from the direction where high winds and gale-type weather originate. This position allows any oil or gas escaping from the wellbore to be blown in the direction of the pipe racks and away from the engines and other rig machinery that could start a fire. It also helps the derrickman, since the prevailing winds will push the pipe toward him on trips in and out of the hole. In hot, arid climates the engines (and rig if necessary) are positioned so the engine radiators face the prevailing winds to obtain optimum cooling.

The location is leveled, but normally the hole will be at a slightly higher point on the location with a gradual slope of about 1 ft/100 ft in all directions away from the rig. This prevents an accumulation of liquids on the ground around the rig and provides for gravity drainage to move these liquids away from the rig where they are collected by trenches and flow into a large waste pit, sometimes called the reserve pit. The position of the shale and reserve pits is staked, and the pits are dug out or opened with earth-moving equipment (Fig. 6–2).

The base for the rig substructures is prepared. This may be leveled ground, leveled ground with a layer of sand several inches deep, or leveled ground with a layer of sand and matting boards or a reinforced concrete slab up to 12 in. thick, depending on the weight of the rig and the load-bearing capacity of the soil. A cellar may be dug and walled-in with wooden planks or a large conduit (Fig. 6–3).

The rathole, mousehole, and shallow conductor (or mud riser) may be installed at this time or after the rig is moved in. The rathole is a slanted hole drilled near the well's borehole to hold the kelly when not in use. The mousehole is a hole drilled to the side of the wellbore to hold the next joint of drillpipe to be used.[1] A rathole digger drills the holes for the rathole, mousehole, and shallow conductor. It is a small, portable, truck-mounted drilling unit that drills large-diameter, shallow holes with an auger, similar to the equipment used to drill holes for telephone poles and foundations. The shallow conductor or mud riser pipe is placed in the hole, aligned vertically, and cemented in place, usually with grout-type concrete. The other two holes are left open until the rig is moved in.

[1]R. D. Langenkamp, *Handbook of Oil Industry Terms and Phrases,* third edition (Tulsa: PennWell, 1981).

FIG. 6–2 Jets transfer waste mud from the shale pit (a) to the reserve pits (b)

FIG. 6-3 Building and cementing the cellar

A large volume of water is required in drilling operations—5–10 bbl/ft of hole drilled. A reliable supply of water must be obtained. This may be nearby commercial sources, water wells in the area, lakes, rivers, and ponds. A water source well may be drilled on the edge of the location. Water is pumped to the location through a pipeline (water line) or is carried in tank trucks. The most economical, convenient method normally is selected.

Miscellaneous

Equipment must be ordered for delivery to the wellsite by the time it is needed. The normal shelf items are maintained in stock at supply points so they can be delivered to the location in a relatively short period of time. Nonstock items must be special ordered from the steel mill. A considerable delay may be encountered if the pipe must be rolled and threads cut. Delivery times on these tubulars can be as long as 6–9 months, so the equipment must be ordered well in advance of the time when it is needed. Certain types of wellhead equipment and other specialty items must be handled in a similar manner. Remote and foreign locations must also allow for an extended transportation time. Drilling bits are sometimes in short supply, especially the less-common sizes. This must also be considered when ordering equipment for the well.

When the rig is moved into a new area, sources of services and supplies must be located. Service and supply companies are accustomed to providing their services over long distances. However, this is costly and time consuming, and the nearest satisfactory source of service and supply should be used.

It is often advisable to have a spud conference before a new well is started, especially a deep, complicated operation. The spud conference familiarizes everyone with the drilling, mud, hydraulics, and casing programs and any other specialty items relating to the well. The conference is normally called by the owner-operator, who also presides. Personnel attending the conference include representatives of the drilling contractor, the various service and supply companies, regulatory agencies, and any other parties associated with the well.

The conference is usually conducted informally but is controlled so the information can be presented. The operator normally prepares a small handout covering the summarized plans for drilling the well. This is discussed by the operator's representative, and comments and suggestions are requested from the attendees. The conference should give everyone a clear understanding of the overall operations, problems expected, and how they are to be handled.

In summary the entire rig move and the drilling of the next well must be carefully planned before the rig move is started. The purpose of this planning is to move the rig and drill the next well in the fastest, most economical manner. These plans include preparing the drilling program, obtaining permits as needed, making all of the necessary inspections and equipment repairs on the rig before or during the move, preparing the next drilling site so that it will be ready before the rig arrives on location, and ordering equipment and making other plans as necessary to drill the next well.

Coordinating Move-Out with Move-In

Drilling rigs are made up of a number of component parts. When the rig is dismantled and loaded on trucks, this is known as rigging-down and moving-out. The dismantled rig is transported to the new location. Unloading, positioning, and connecting the various parts of the rig together is known as moving-in and rigging-up. The rigging-up procedure also includes drilling the rathole and mousehole if these were not drilled prior to the move-in. The rig-up procedure is completed when the rig is ready to spud-in and begin drilling.

The rig is assembled at the new drillsite in the reverse order in which it is disassembled and rigged-down at the old wellsite. The substructures are the first pieces of equipment needed at the new wellsite, and they are the last piece of equipment rigged-down at the old wellsite. Therefore, the rig-down and move-out operation must be coordinated with move-in and rig-up.

MOVING THE RIG

The process of rigging-down, moving, and rigging-up is collectively referred to as moving, or moving the rig. The rig is designed to be dismantled into loads that can be carried by a normal heavy-duty oil-field truck. Sometimes the loads may exceed the length, width, height, and/or weight limits that can be legally moved on the public roads, so special permits are required. A number of personnel and various kinds of moving equipment are required to move the rig. The entire move must be coordinated to use personnel and equipment efficiently.

Three rig crews normally work to move the average rig. If other crews are assigned to the rig, then they will usually be on their days off or rest days. All three crews begin dismantling the rig. As the amount of rigging-down work decreases, one crew goes to the new location to begin rigging-up. A second crew goes to the new location a short time later. The third crew remains at the old location until the remainder of the rig has been dismantled and loaded.

Various types of moving equipment are needed. The type of equipment used depends on the size of the rig, the length of the move, the terrain, and other related factors. Oil-field trucks are the main moving equipment used to transport drilling rigs (Fig. 6–4). Various-sized trucks are used, ranging from small dual-wheel trucks to large tandem-axle, heavy-duty trucks. Most of the trucks have flatbeds, heavy-duty winches and cables, and a rolling tailgate. These trucks can pick up one end of a heavy load with the winch, pull it over the rolling tailgate and slide it onto the truck bed. Some trucks are fitted with steel poles at the rear

FIG. 6–4 Special trucks (a) used to move heavy oil-field equipment have heavy-duty beds, headache posts, and winches. They are equipped with chains and boomers to tie down large loads (b) and have fully instrumented cabs (c).

end of the bed (gin-pole trucks) (Fig. 6–5). The combination of the winchline and poles is used to lift and move moderately heavy loads.

Crawler tractors, dozers, or bulldozers are used to drag or push skid-mounted equipment and heavily loaded trucks. They are especially useful on muddy locations where the trucks may become stuck. The dozers are usually equipped with heavy-duty winches and cables to pull very heavy loads. Some dozers are fitted with gin poles for lifting equipment. The dozer blade is used to smooth the location when it becomes rutted by the heavily loaded trucks.

Most of the trucks have trailer hitches or fifth wheels to haul trailers. Different kinds of trailers are used. Pipe trailers are used for pipe and long, rigid loads such as pipe racks. Flatbed trailers, or floats are used to carry large pieces of

FIG. 6–5 Gin-pole trucks can move equipment near the rig

equipment or pipe (Fig. 6–6). Trailers with low bed heights, lowboys, are used to carry tall, heavy loads so they can be loaded more easily without a high lift and will pass under low overhead obstructions such as overpasses and power lines while being transported to the new location.

More trucks may be needed in special cases. Larger rigs use larger equipment and need more trucks. More trucks are usually needed for longer moves because of the distance and because smaller loads must be carried due to legal limits on height, width, length, or weight. The time available for moving also affects the amount of moving equipment needed. Normally there is an optimum number of trucks that can be used for each rig move. When the move must be hurried, more trucks are required.

In the past when derricks were used, it was common practice to skid them whenever possible. Skidboards—10 to 15-ft long 3×12s—were used for tracks, and the derrick was moved on rollers made of short lengths of heavy drillpipe. The derrick was pulled by one or more bulldozers. As the derrick passed over the skidboards, the skidboards behind the derrick were picked up, carried to the front of the derrick, and laid back down to provide a track for the rollers. As the derrick skids passed over the rollers, the rollers were picked up from behind the skids and placed on the skidboards in front of the skids. On level terrain under good weather conditions, a rig could be skidded as far as two miles in one day.

Desert rigs are designed for skidding. They use a specially designed and reinforced substructure, heavy-duty I-beams (reinforced by steel plating), and crawler trailers, athey wagons. The I-beams are inserted through the lower part of the substructure and are then jacked up with heavy hydraulic jacks. Athey wagons, special trailers with oversized rubber tires, are set under the ends of the I-beams. Crawler tractors are then used to pull or skid the rig to the next location. The rig is set down by reversing the lifting procedure.

Modern land rigs are seldom skidded. They are designed to be dismantled (broken down) into loads that can be moved by trucks. Sometimes the mast will be moved in one piece by using several trucks where the move is over a road or area that is not limited by low overhead obstructions or side clearance.

MOVE-IN AND RIG-UP PROCEDURE

Each rig has a slightly different move-in and rig-up procedure, depending on the size of the rig and the way the various parts have been unitized. A general procedure for moving-in and rigging-up a conventional mechanical land rig with tail-driven pumps is listed in Table 6–1.

In the general procedure, the substructures are positioned and connected. The remainder of the rig equipment is placed on top or to the sides of the sub-

FIG. 6–6 At the old location pipe racks are loaded on top of the drillpipe on a float trailer (a). The pipe racks are unloaded and placed in position with a tandem axle, dual-wheel gin-pole truck (b), and the pipe is unloaded by rolling it off of the float trailer onto the pipe racks (c).

FIG. 6–7 Modern land rig drilling in the desert (courtesy Loffland Bros. Drilling)

structures; therefore, the substructures are the first item of equipment needed at the new wellsite. But when the rig is dismantled at the old location preparatory to moving, the substructures are the last pieces of equipment to be disconnected. When the rig is being dismantled (torn down), the equipment is disconnected and moved out of the way as fast as possible in order to release the substructures so they can be moved to the new location.

As equipment is moved away from the substructures, some double loading occurs—loading the equipment, moving it a short distance, unloading it, and reloading it to move to the new location. One example of double loading is to move the drawworks off of the substructure, then move the subs to the new location followed by the drawworks. If the drawworks were left on a truck during this time, the truck would not be utilized efficiently. Double loading should be prevented whenever possible; it is a waste of time and equipment, even when it is necessary. The use of trailers helps reduce double loading. The equipment is loaded directly on the trailer so it can be moved to the new location, unloaded, and positioned.

The main objectives of rigging-down are to disconnect the equipment and load it for moving to the next wellsite and at the same time to disconnect the

256 DRILLING

TABLE 6-1 Move-In and Rig-Up Procedure*

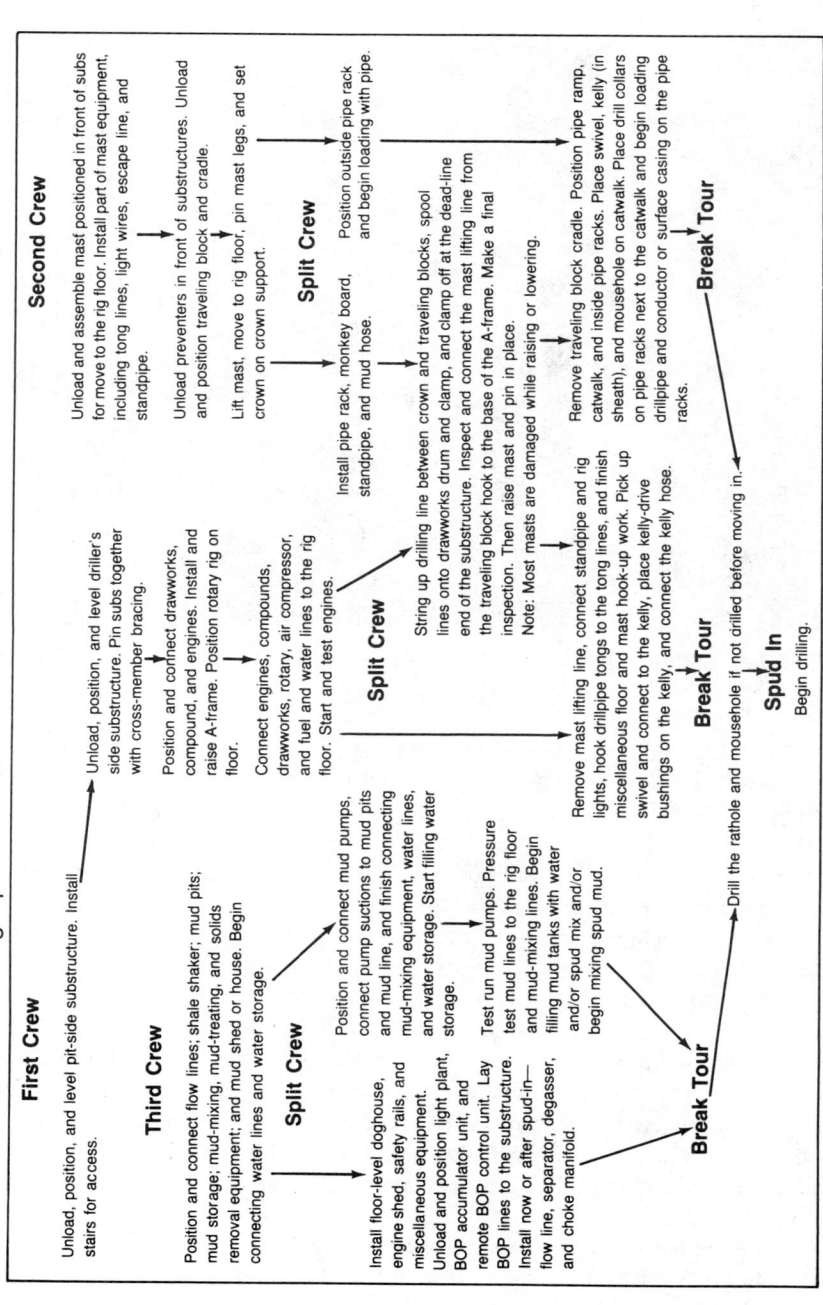

*Mechanical rig with tail-driven pumps

substructures as soon as possible so they can be moved to the new location in order to begin rigging-up.

Work Scheduling

As soon as equipment begins to arrive at the new location, two crews can be used. One crew begins unloading and positioning the pit-side substructure—the side of the substructure nearest the mud and reserve pits. Another crew begins to unload and assemble the mast (Fig. 6–8). The third crew is usually at the old location, loading equipment.

After the pit-side substructure has been positioned, a third crew can begin to unload, position, and connect the flow line, shale shaker, and mud pits. At the same time the second substructure (or sub) is positioned and the first crew pins the substructures together and places the drawworks, engines, and other equipment on the rig floor, the third crew can start to unload, position, and connect the mud pumps and complete rigging-up the mud and water supply systems.

In the meantime the second crew will complete assembling the mast and placing equipment in it. The mast is then moved to the rig floor where the mast legs are pinned to the substructure. Part of the second crew completes adding equipment to the mast, including installing the drilling line (stringing-up), aided by part of the first crew (Fig. 6–9). Part of the second crew will position the outside pipe rack and begin loading it with pipe.

By this time the engines and other equipment on the rig floor have been connected and test run, and the next step is to raise the mast. The first crew continues to install and connect equipment on the rig floor and in the mast. The second crew installs the equipment in front of the rig, including the ramp and pipe racks. At the same time the third crew continues rigging-up the back of the rig, including the pumps, blowout preventer, accumulator, and generators.

Breaking Tour

It normally takes at least two days to rig-down, move, and rig-up. Most of this work is done during daylight hours with three crews working. During the latter stage of rigging-up, after the light plant and lights have been installed, the rig operations are converted to a 24-hr operation with one crew working on each 8-hr shift or tour (pronounced "tower"). This is known as breaking tour.

The day crew continues to work the regular tour or shift from 8 a.m. to 4 p.m. On the day the rig is to go on 24-hr operations, the evening crew will begin work at 4 p.m. and will continue until 12 midnight, their regular tour when not moving the rig. The morning crew will work the regular 8 a.m. to 4 p.m. tour on the last day of rigging up and then return at 12 midnight to work the morning or graveyard tour from 12 midnight to 8 a.m., their regular tour when not moving

258
DRILLING

FIG. 6–8 The mast is pinned together in sections (a) and is set with the legs in hinged floor plates (b). The crown is positioned on A-frame support at the water table (c) so it can be rigged up and raised (d).

the rig. The evening crew has a long break and the morning crew has a short break. These breaks are reversed when the rig breaks tour to go from a 24-hr operation to day work for the next rig move.

Alternate Move-In and Rig-Up Procedures

With some modification, the move-in and rig-up procedure will generally apply to most land rigs. The specific move-in and rig-up operation depends on the type of rig and the equipment.

259
Moving In, Rigging Up, and Drilling the Conductor Hole

FIG. 6–9 The mast is rigged up in the horizontal position. The stand pipe and drilling hose (a), the monkey board (b), the drilling line (c), and the mast lifting line and fast line (d) are installed.

For example, there is an increasing tendency to use double substructure rigs to provide more clearance under the rig floor. This clearance is normally measured from the base of the rotary to ground level. When double substructures are used, the lower substructures are unloaded, positioned, and connected. An additional substructure is set on top of each of the first substructures and is connected to form a rigid base. The remainder of the move-in and rig-up procedure is similar to that described earlier, except that the equipment to be placed on the rig floor must be lifted higher.

260
DRILLING

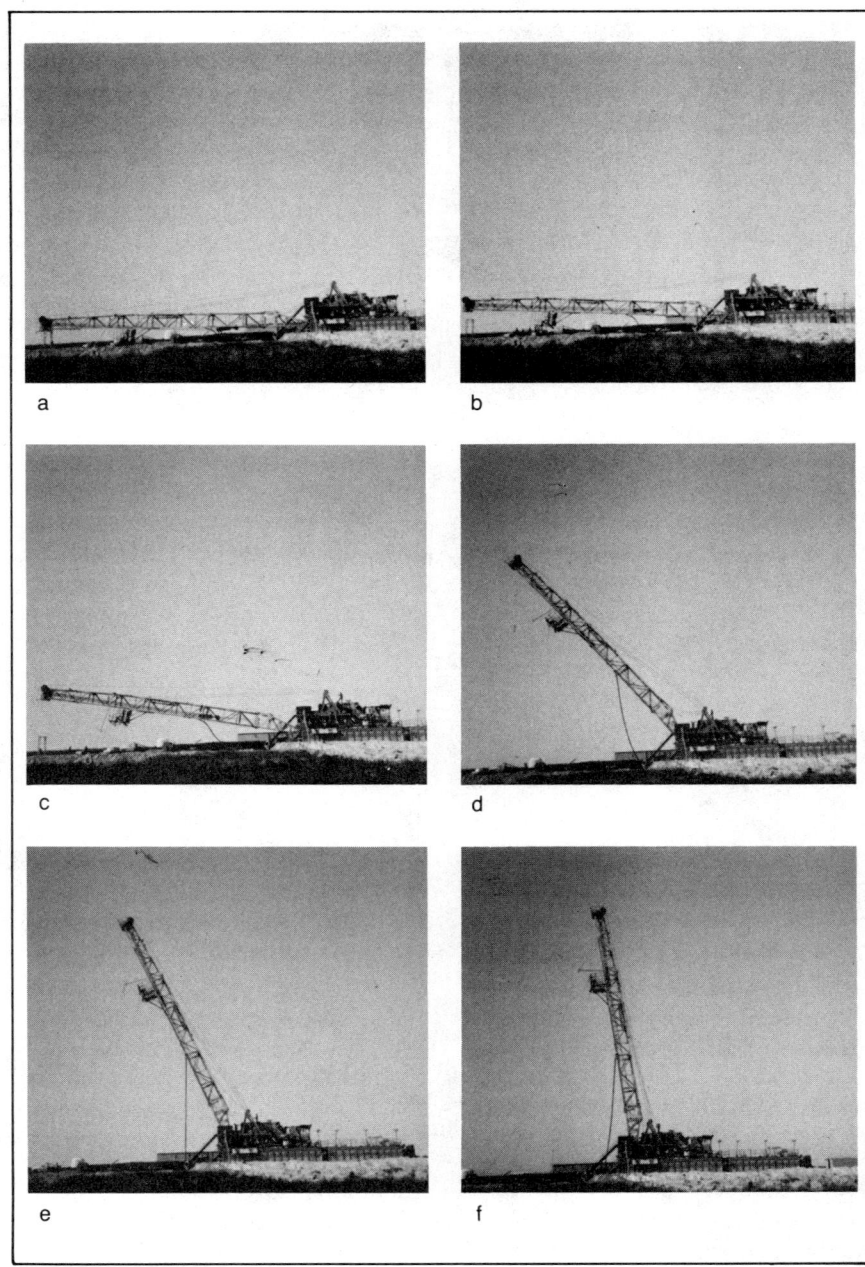

261
Moving In, Rigging Up, and
Drilling the Conductor Hole

FIG. 6–10 (a) Drilling and mast raising lines tighten, and the traveling block is lifted off the catwalk. All equipment is checked one final time (time—1 min). (b) Crown lifts off of support (time—4 min). (c) Crews check to ensure (1) all extra lines suspended from the mast are free, (2) structural members are rigid, and (3) the mast is moving upward in a plane (time—5 min). (d) Raising rate is increased slightly. Note the traveling block is moving upward relative to the mast as the mast is raised (time—7 min). (e) The drilling (fast) line starts to clear the A-frame pulley, the point of high tension on the lines (time—9 min). (f) Raising rate is reduced to the minimum as the mast approaches the balance point (time—11 min). (g) Mast is moved past the center of balance and falls (gently) against the A-frame. Driller continues to tighten the drilling line carefully to hold the mast securely against the A-frame until the mast is in position. The derrickman starts climbing up the A-frame leg to lock the mast into position against the A-frame (time—14 min).

Pony substructures are also frequently used. They are about one-half as tall as regular substructures and are installed similar to double substructures. The pony substructures are positioned, leveled, and connected together. The regular substructures are then set on top of the pony substructures and are connected together. The remainder of the equipment is installed in the normal manner.

Electrically powered rigs have a special advantage over conventional mechanical rigs during the moving phase of the operations, one reason for their increasing popularity and use. The electrical motors that supply power for the drawworks are mounted on the drawworks skid (unitized with the drawworks). There is only a single unit for moving and rigging-up. During rig-up, only one unit must be placed on the rig floor rather than placing the drawworks, engines, and compound on the rig floor of a conventional mechanical rig. The skid-mounted engines with unitized generators are unloaded at a convenient point, usually behind the rig at ground level. This is easier and faster than lifting the equipment to the rig floor and precisely matching the engines, compound, and drawworks. There is always a risk of damaging equipment when it is being moved and a greater risk when the heavy equipment is lifted. Therefore, there is less risk of damaging the equipment on electrical rigs than on mechanical rigs. The reduced amount of work and risk is more pronounced on a double substructure rig. The generators and drawworks motors are connected by flexible electric cables as part of the rig-up procedure.

A similar savings applies to moving and rigging-up the mud pumps on electrical rigs. The electrical pump motors are mounted on the same skid as the pump. Each pump with its unitized motor is moved as a unit. The pump is unloaded and positioned so that the pump suction can be connected to the mud tanks. In the rig-up procedure the pump discharge is connected to the flow line with a high-pressure, flexible hose (vibration hose). The generators are connected to the pump motors by a flexible electric cable. This is a much easier and faster power connection than the connecting drive belts and/or drive shafts and universals on a conventional mechanical rig.

DRILLING THE RATHOLE AND MOUSEHOLE

In conventional oil-field operations the rig is not ready to begin drillings—spud-in—until the rathole and mousehole have been drilled and set. Sometimes the rig is not considered ready to spud-in officially until a shallow conductor or mud riser has been set and cemented, but more often this is considered part of the drilling operation.

The rathole is a large piece of pipe 8–12 in. in diameter and 3–5 ft longer than the kelly with a closed bottom usually connected to a drill bit. The rathole is used as a sheath to protect the kelly when it is moved. The kelly is usually about 42 ft long, is relatively limber, and is subject to bending.

During drilling operations the rathole is set through a hole in the rig floor (rathole port) so that the top of the rathole is about 4 ft above the level of the rig floor. The rathole port is located near the V-door on the driller's side of the floor. When the kelly is not being used in the drilling operations, such as when tripping, it is placed in the rathole—set back for storage.

The mousehole is similar to the rathole except it is smaller, usually 7–10 in. in diameter and about 28 ft long. The mousehole is set through a mousehole port located immediately in front of the rotary (toward the V-door) with the top of the mousehole level with the rig floor. The mousehole is used to hold a joint of drillpipe for a drillpipe connection. The length of the mousehole is adjusted so that the top of the drillpipe tool joint will extend about 2½ ft above the rig floor. The bottom of the mousehole is sealed. Frequently, a heavy-duty, coiled spring is dropped in the mousehole. The spring serves two purposes: it prevents damage to the bottom tool joint of the drillpipe when it is set in the mousehole and makes it easier for the driller to adjust the height of the kelly tool joint when making a drillpipe connection in the mousehole. Both the rathole and the mousehole have small drain holes cut through the pipe walls at about ground level to drain fluid. They are also frequently washed with water to clean out accumulated mud.

The holes for the rathole and mousehole are frequently dug with a rathole digger before the rig equipment is moved on location. When holes have been predrilled, after rigging-up the rathole and mousehole are set in their respective positions and operations continue (Fig. 6–11).

When the rig-up operations have proceeded to this point, the engines and pumps have been tested, the mast has been raised, and the swivel, mousehole, and rathole have been unloaded onto the catwalk (assuming the mousehole and rathole have not been drilled). Two types of lines are used for general lifting operations during normal drilling operations. The pick-up line is a heavy-duty cable usually fitted with chains on each end. It is attached to the traveling block in a single or double-line configuration for lifting heavy loads such as blowout preventers, the rotary, and the swivel. Lifting lines are smaller lines, such as the catline, run from the drawworks cathead or other means and used to lift lighter loads such as joints of drillpipe, pipe tongs, or heavy subs.

The swivel is picked up with a pickup line and set on the rig floor, and the swivel bail is latched into the hook at the bottom of the traveling block. The kelly inside the rathole sheath is picked up with the pickup line and pulled into the V-door. A lifting line is tied to the bottom of the swivel, and the swivel is lifted by the block and lifting line to align the left-hand threaded pin in the bottom of the swivel with the left-hand threaded box in the top of the kelly. The connection is

FIG. 6–11 Rathole (a) and mousehole (b) on the catwalk ready to be picked up.

screwed together and tightened temporarily with a wrench and sledgehammer. The kelly mud hose is connected to the swivel. The swivel and kelly with the attached rathole sheath are picked up and lowered into the rathole port until the bit on the bottom of the rathole is set on the ground.

Depending on the type of kelly-rathole connection, the drain hole may be closed. The hole for the rathole is drilled by lowering the kelly and swivel so that the bit exerts weight on the ground. The rathole is rotated while pumping fluid slowly through the bit. In soft formations the rathole is rotated with drillpipe tongs or casing tongs. The hole is usually drilled in 30 min to several hours. In hard formations the rathole is rotated with casing tongs or a chain from a clamp-on sprocket on the rathole. The hole must be drilled from a few feet to as much as 35 ft below ground level, depending on the height of the substructure. After the hole has been drilled, the kelly is disconnected from the rathole and the rathóle is left in place.

When drilling operations have been completed and the rig is to be moved again, the rathole is pulled out of the hole. The kelly and swivel are disconnected and laid on the catwalk in the reverse order of rigging-up.

After the rathole has been drilled, the kelly is pulled out of the rathole. A bit is connected to the bottom of the kelly with a crossover sub. This is lowered into the mousehole port, and the hole for the mousehole is drilled in a manner similar to drilling the hole for the rathole. After the hole has been drilled, the kelly is set back (replaced in the rathole). The mousehole is picked up with a lifting line and lowered through the mousehole port into the hole. The top of the mousehole is fitted with a steel ring or plate so that it stops flush with the floor.

SPUD-IN AND CONDUCTOR HOLE SECTION

After the equipment is rigged up and the rathole and mousehole have been set, the next step is to begin drilling (spudding in) to drill the conductor hole section. Most wells are spudded with clear water as the drilling fluid. The water mixes with the formation cuttings as the hole is drilled. As the mixture is circulated and recirculated, the formation cuttings are ground into smaller particles. Soon the water has the appearance and consistency of common thin mud, hence the name mud. Water is added to dilute the mud so it will be thin (low viscosity) and pumpable. Mud treating chemicals are seldom used at shallow depths because dilution with water is a good, economical treatment and usually all that is required. The mud is normally discarded after the conductor hole section has been drilled.

In some areas formation conditions require that a better quality mud be used to drill the conductor hole section. When this happens, the hole is usually spudded with *spud-mud*. Spud-mud has a higher viscosity and better capacity to carry cuttings as compared to water. Spud-mud is surplus mud from other drilling operations, mud carried over from the last hole drilled by the rig, or mud mixed prior to spudding-in. Spud-mud is prepared by mixing bentonite, commonly called gel, with water. Water is circulated in the mud mixing tank while adding gel to a concentration of 5–15 lb of gel/bbl of water 42 U.S. gal/bbl. The pit is mixed and circulated until the gel hydrates. A small amount of chemical may be added to improve hydration.

Most shallow formations contain sufficient mud-making clays to form *native mud* with satisfactory characteristics for drilling the conductor hole. Spud-mud is used when the shallow formations do not contain enough clays or gels to make a good native mud and where the mud must have an additional capacity to carry drill, cuttings, and caving material out of the hole.

Drilling the Section
Shallow conductor is frequently preset before the rig is moved in and rigged up. It is often necessary to set another shallow string of conductor casing to case off caving formations, lost circulation zones, and water flows. These zones will frequently occur deeper than the drilling capacity or depth rating of the rathole digger. When this occurs, the conductor hole section must be drilled and cased with the drilling rig.

Another reason for presetting the shallow conductor or mud riser is to provide a return flow path for mud from the hole back to the mud tanks. The flow line is welded to the side of the shallow conductor at a level slightly higher than the height of the mud tanks. The drilling fluid is then pumped down through the drilling assembly and returns up through the annular space to the mud line. The mud flows by gravity action through the mud line, shale shakers, and into the mud pits. In this case the shallow conductor is acting primarily as a mud riser or mud return line.

The hole can be spudded-in (begin drilling) without using a mud riser if the mud is returned to the mud tanks. This can be done by digging a ditch from the point where the hole is drilled to the reserve mud pit. The returning mud flows through this ditch and into the reserve mud pit. A centrifugal pump, sometimes called a Donald Duck or Yellow Boy, pumps the mud from the reserve mud pit back into the mud tanks. The centrifugal pump has a high-speed rotor (impeller) that pumps by centrifugal action like the pump is a common household washing machine. If a cellar is used, an alternate method is to use the centrifugal pump to transfer the mud directly from the cellar to the mud tanks.

Drill cuttings can cause a problem when a mud riser is not used. They tend to stack up in the mud ditch flowing to the reserve pit, causing the mud to overflow. The ditch must be kept clean using water hoses and shovels. When the mud is pumped out of the cellar, the drill cuttings may clog the pump impeller or may accumulate in the mud tank and must be cleaned out later. Because of these problems, a mud riser is usually installed if over one or two days are needed to drill the conductor hole section.

In the earlier history of the industry, open, earthen mud pits were used with a shale pit at the end of the shale shaker to catch the drill cuttings, similar to that used today. The returning mud, carrying drill cuttings, flowed through a ditch to the shale pit. The cuttings settled out of the mud in the shale pit, and the mud flowed through a short ditch into the mud pit. Two mud pits were conventionally used: the first was called the sand-settling pit and the second was the suction pit. The steel mud tanks used on current rigs are named after these.

For descriptive purposes, assume in this chapter that a 15-in. conductor hole will be drilled to about 350 ft and $13\frac{3}{8}$–in. OD conductor casing will be run and cemented. This is a common combination, and the same general procedure applies to different hole and casing sizes and different depths. This includes mud risers, deeper conductor casing, and shallow surface casing.

Pick Up Drill Collar and Bit

The last step before spudding in is to pick up a drill collar and bit and connect these to the kelly. This is the spud-in drilling assembly. It may be necessary to use only a short collar or to pickup several drill collars, depending on the substructure

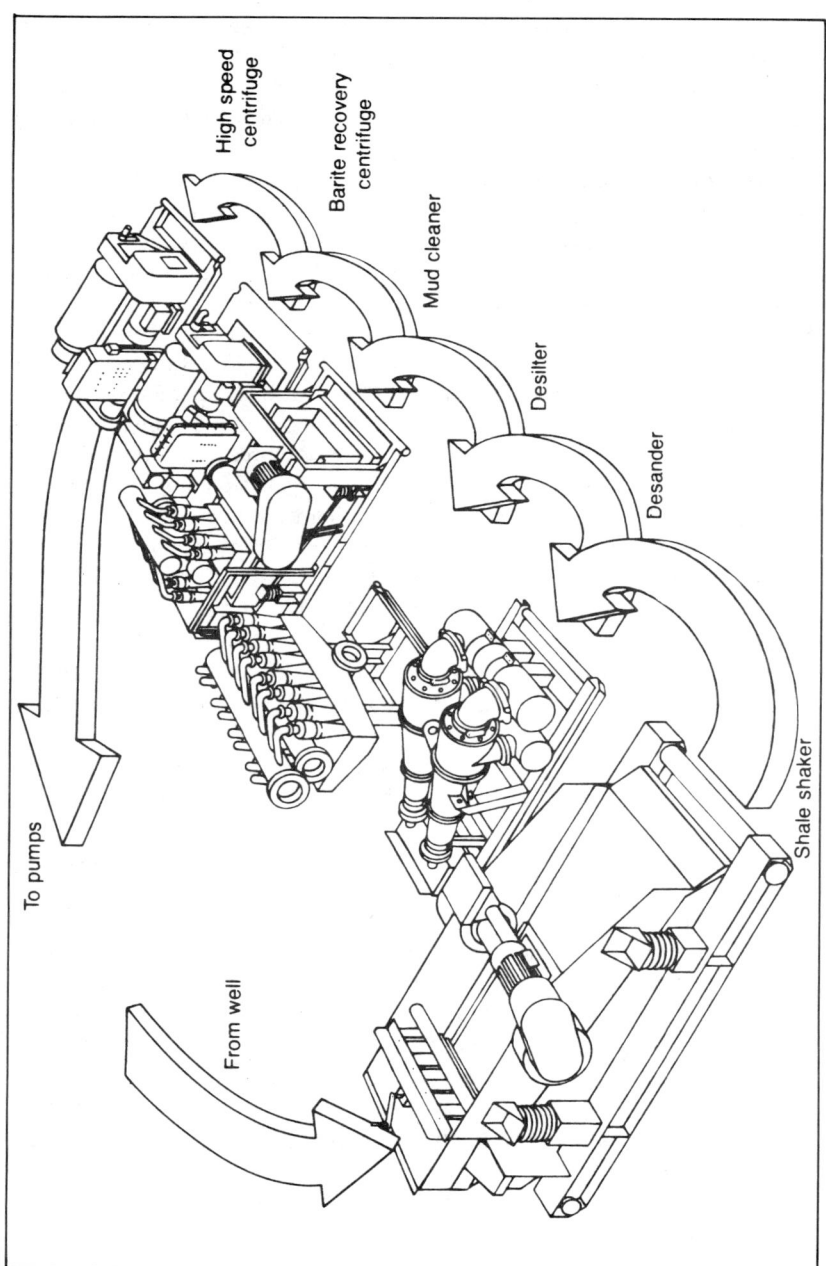

FIG. 6-12 Mud return flow path

clearance, the distance from the bottom of the rotary to ground level, and the depth at which the shallow conductor or mud riser was set, if used. The drilling assembly must be long enough to reach the point where drilling will be started and short enough so the kelly drive bushings can be seated in the rotary. Normally one drill collar will provide the correct length.

Except for special equipment and bits, all of the drilling assembly, including the drill collars, subs, drillpipe, and kelly, are constructed and used with a pin-down, box-up configuration. All threaded tool-joint connections are constructed with a pin end that is connected to a box end. When the drillpipe and drill collar are picked up to run in the hole, they are conventionally picked up by the box end so the pin end points downward. Part of the reason for this may be because there is better visibility and it is faster and easier to put the pin into the box when handling heavy equipment under normal drilling operations.

One exception to the conventional pin-down box-up tool-joint configuration is roller-cone drilling bits, which are constructed with a tool-joint pin on the top of the bit. This construction is commonly referred to as pin up or pin looking up. The bit can be connected to a regular drill collar by using a sub or crossover sub. The sub is similar to a short drill collar or piece of drillpipe with one size and type of connection on one end and another size and type of connection on the other end. It connects parts of the downhole assembly that have different-type tool joints or connectors.

Drill collars have a conventional box-up pin-down tool-joint configuration and usually have a different type of connection than the pin connection on the bit (Fig. 6–13). A common practice is to cut a box connection on the bottom of one drill collar to fit the bit pin. The bit can then be connected directly to the drill collar, which is known as the bottom collar, bit collar, or double-box collar, thus eliminating using a sub.

The bit collar is pulled into the V-door with a lifting line. After cleaning and greasing the threads, a lift nipple, or pickup sub, is connected to the box on the top of the drill collar (Fig. 6–14). The lift nipple is similar to a crossover sub, except it has a recessed body that can be caught by the elevators. Another method of picking up the drill collar is with drill-collar buttons. The drill-collar button has a pin tool-joint thread that fits the top of the drill collar. The button has a wide shoulder that can be used to catch the top of the elevators.

Drillpipe elevators are usually used with lift nipples. Larger-sized drill-collar elevators are used with drill-collar buttons. The elevators are fitted around the lift nipple or drill collars, depending on the lifting system, and are latched. Then the drill collar is lifted into the derrick. When the bottom of the drill collar reaches the level of the derrick floor, a tailback line or snubbing line prevents the bottom of the drill collar from swinging in too rapidly and possibly injuring

269
Moving In, Rigging Up, and
Drilling the Conductor Hole

FIG. 6–13 Drill collars (a) pin end and (b) box end

FIG. 6–14 Lift nipples

personnel or damaging equipment. The thread protector—a light metal half-coupling designed to protect threads from damage—is removed from the bottom of the drill collar, and the threads are cleaned and greased.

A bit breaker block is set in the rotary (Fig. 6–15). This block fits the rotary and has a cutout center section that fits the bit. The block is designed to hold the bit and prevent it from turning so the drill collar can be screwed onto the bit pin and tightened. A bit is set in the bit breaker block, and the bit threads are cleaned and greased. The bit is connected to the drill collar by lowering the drill-collar box slowly over the bit pin and rotating the drill collar to the right. After the connection has been screwed together or made up hand tight, it is tightened with the drillpipe tongs using jaws sized for the drill collar (Fig. 6–16).

The end of the tongs is connected to a makeup cathead on the drawworks by a steel cable. The driller actuates a friction-clutch-drive cathead to pull the steel cable and tighten the tool joint. The steel cable is attached to the end of the tongs through a hydraulic coupling (Fig. 6–17), which in turn is connected to a pressure gauge that is graduated in foot-pounds of torque. With this gauge and the friction-drive cathead, the driller can tighten the joint to the correct torque recommended by the tool-joint manufacturer.

FIG. 6–15 Bit breaker block for $9^{5}/_{8}$–$9^{7}/_{8}$-in. roller bit

FIG. 6–16 Hydraulic (a) and regular or manual (b) drillpipe tongs. Each set of tong jaws and links are designed for one size of pipe.

This is a very important part of the overall drilling procedure. The entire drilling assembly is connected with tool joints. Correct tool joints are strong and serviceable. However, the two joints must be clean, properly greased, and correctly tightened to the recommended torque. If the tool joint is not tightened enough (undertorqued) it can wobble off and cause a fishing job. If it is tightened too much (overtorqued), it may be overstressed and subject to failure, which can also cause a fishing job. So handling tool joints correctly is an important part of successful drilling operations.

After the bit has been tightened, the drill collar and the bit are picked up a short distance. The bit block is removed from the rotary, and the bit and drill collar are lowered through the rotary until the top of the drill collar is about 3 ft above the top of the rotary. The drill-collar slips are set; a drill-collar safety clamp is latched around the drill collar near the top and is securely tightened. The drill-collar clamp prevents the flush-joint top of the drill collar from falling through the slips into the hole, causing a fishing job if the slip hold on the collar is loosened. The clamps are not needed on drillpipe with an upset that can catch in the slips. The drill-collar lift nipple is unscrewed, removed from the elevators, and set aside to be used to pick up the next drill collar.

The kelly, which is set back in the rathole, is then picked up. The elevators are closed and turned so the hook is pointed toward the bail on the kelly. The hook swivel is locked and the hook is opened. Two crewmembers grasp the

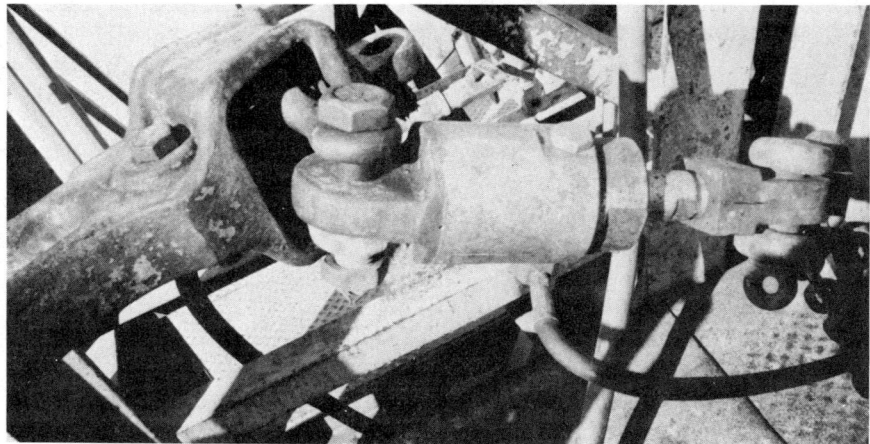

FIG. 6–17 A hydraulic cylinder fits between the tong arm and the pulling line

elevators and swing the block over until the hook is below the swivel bail. As the traveling block is lifted upward, the hook catches the swivel bail. As the block is lifted higher, the swivel bail falls inside the hook, closing and locking the hook. The kelly is then pulled out of the rathole. The pin on the bottom of the kelly has a different thread from the tool-joint box on top of the drill collar so a crossover sub is screwed into the top of the drill collar and the kelly is screwed into the top of the crossover sub. Both connections are then tightened with tongs.

When the bit was being connected to the first drill collar, the rotary was locked to hold the bit breaker block and bit stationary while tightening the drill collar. With the drill collar held by the rotary slips, there is not enough weight to prevent the slips from turning inside the rotary as the kelly is tightened. Therefore, the second set of tongs, back-up tongs, must be used to keep the drill collar and the rotary from turning as the crossover sub and kelly connection are tightened. The assembly is then picked up a short distance, the drill collar safety clamp is loosened and removed, the drill-collar slips (Fig. 6–18) are pulled out of the rotary, and the assembly is lowered slowly through the rotary until the square base of the kelly drive bushing seats in the rotary (Fig. 6–19).

With the bit near bottom, the rotary clutch is engaged to begin rotating the drilling assembly and the pump is turned on to start circulating mud through it. At this point the rig is ready to spud-in and begin drilling.

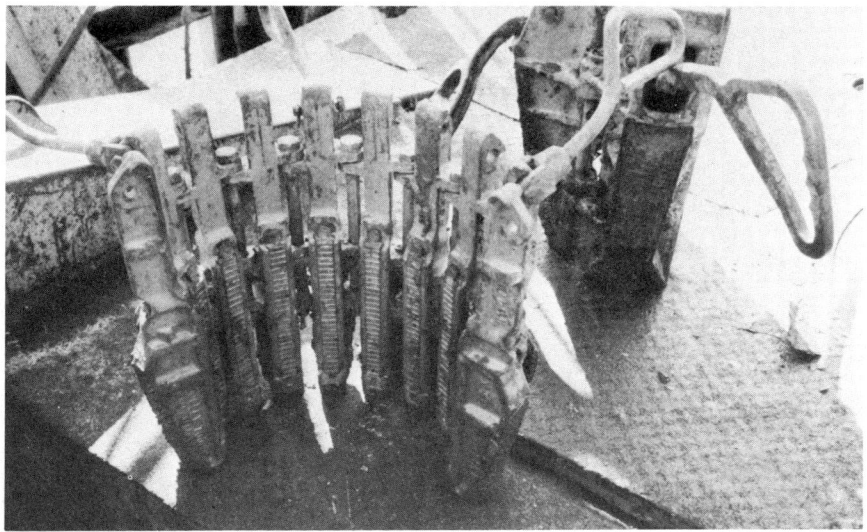

FIG. 6-18 Drill-collar slips

Spud-In (Begin Drilling) and Drilling Procedure

Spud-in is the point at which drilling operations are first started. The drilling process includes applying a controlled amount of the drilling assembly weight through the bit onto the formation while rotating the assembly and circulating mud through the bit. The weight exerted by the bit on the formation and the rate at which the drilling assembly is lowered or slacked off are controlled by the drawworks brake. The amount of weight that the bit exerts on the formation is determined by subtracting the indicated weight of the drilling assembly while drilling from the weight of the drilling assembly when it is suspended with the bit above the bottom of the hole. The weight that the bit exerts on the formation is called the bit weight or drilling weight. These are determined by reading the weight indicator located near or within the sight of the driller.

The weight indicator is calibrated to read the total suspended weight, including the weight of the traveling block. As an example, at spud-in the drilling assembly weight is relatively low, on the order of 25,000 lb with the bit suspended off bottom. As the drilling assembly is lowered and part of the assembly weight is supported by the formation, the weight indicator might show a weight of 20,000 lb. In this case the drilling weight would be 5,000 lb, or the difference

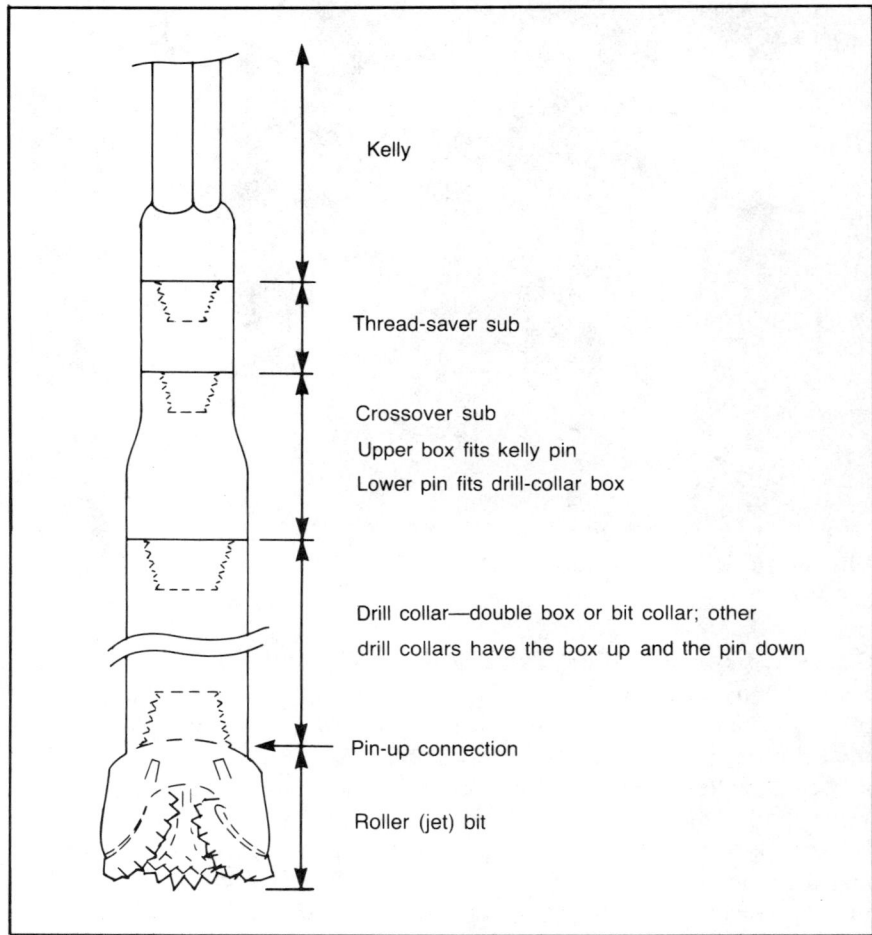

FIG. 6–19 Spud-in assembly

between the weight of the drilling assembly suspended and hanging free with the bit above bottom and the apparent weight of the assembly with the bit setting on the bottom.

The overall drilling process is very simple to understand. As the bit rotates, the teeth on the rollers crush, gouge, and tear the formation, releasing small pieces of formation called drill cuttings. The drill cuttings are circulated out of the hole with the mud. On jet bits the mud is forced through the jets at a high velocity (jet velocity). The jet of mud is directed on the bottom of the hole so that it washes the drill cuttings off the bottom and cleans the bit cutters. This also cleans the bottom so, as the bit rotates, the bit teeth are always penetrating new formation and are not wasting energy and wearing the bit by redrilling bit cuttings. In softer formations the high-velocity mud jetting action will scour and erode the bottom of the hole further, contributing to the *drilling rate* or *penetration rate*. As the bit drills, the assembly is lowered slowly to maintain a constant bit weight on the formation.

When the hole is first spudded-in, very light bit weights are used (1,000–5,000 lb). The assembly is turned at a relatively low speed, compared to deeper drilling, to ensure the hole is started vertical and straight. As the hole is drilled deeper, the walls of the hole provide lateral support for the drilling assembly. This confines the bit and drilling assembly and helps reduce wobbling. The bit weight and rotary speed are gradually increased with depth to the operating conditions recommended by the manufacturer. Experience in an area may show that certain combinations of bit weight and rotary speed give the best penetration rate; such weights and speeds are normally within recommended operating limits.

Drilling a Straight, Vertical Hole

It is important to drill a straight, vertical hole. If the hole is very crooked, running the conductor casing may be difficult. Future problems with crooked and deviated conductor hole may include excessive wear on the casing and drilling assembly, difficulty in placing the kelly drive bushing in the rotary, and various other problems. It is more difficult to drill a straight hole in steeply dipping, harder formations and alternating hard-soft, layered, dipping formations than in flat, soft formations. Formations that tend to cause a crooked or deviated hole are called crooked-hole formations.

Various drilling procedures can be used to drill a straight hole through crooked-hole formations. Some of these, such as drilling with a pendulum, do not apply at shallow depths. The most common method of drilling a straight vertical hole at shallow depths is to reduce the bit weight and increase the rotary speed. This allows the entire drilling assembly to act as a vertically hanging pendulum and increases the drilling action on the low side of the hole—the side nearest the vertical. The side cutting action can be increased by using larger, heavier drill collars. These collars are more rigid. If they are started in a straight hole, they will tend to maintain the direction (straightness) of the hole.

Although it is important to spud the hole and drill vertically, one special condition may cause the shallow hole to deviate, which is not a problem in the deeper hole. The kelly hose is relatively heavy. One end of the hose is connected to the gooseneck on the swivel. The weight of this hose can cause the kelly to bend slightly or lean in one direction. This can cause the hole to be spudded-in at a slight angle from the vertical. To remedy this, a lifting line is tied to the kelly hose to help support the weight of the hose while drilling the first 50–100 ft. Another method is to make one wrap around the upper swivel with the lifting line and to tie the end of the lifting line to the base of the mast. The lifting line is then tensioned so it will counteract the weight of the kelly hose and hold the kelly straight and vertical in the rotary. After one or two drill collars have been added to the drilling assembly, the assembly is usually heavy enough so the offcenter weight of the kelly hose will not affect the hole direction.

Reaming

The kelly can be lowered during drilling until the top of the square or octagonal section is near the top of the drive bushing. This is known as kelly down. Conventionally, the recently drilled hole is reamed. In the reaming procedure the bit is rotated through the recently drilled section to ensure that it is open and drilled out to full gauge (to the size of the bit).

After drilling the kelly down, the driller shuts off the rotary and picks up the kelly as high as possible without pulling the drive bushings out of the rotary. Then he starts the rotary and turns it at a relatively high speed. He begins lowering the kelly at a rate of about 3–5 times faster than when the section was drilled. The driller continues lowering the kelly until it is down. This can be repeated several times if necessary.

An indication of whether the hole needs to be reamed can be determined by watching the weight indicator as the kelly is picked up. If the hole is in good condition, the weight indicator will show a steady, constant load as the kelly is picked up. If the hole has ledges, restrictions, or is otherwise tight, the weight indicator will show an extra weight above the normal weight of the drilling assembly or a high fluctuation in weight. This extra weight when picking up the kelly is known as drag or pipe drag. When there are indications of drag while picking the kelly up, the hole must be reamed more carefully. This includes using less rotary speed and lowering the assembly at a reduced rate.

In some cases it is unnecessary to ream after drilling the kelly down. This will depend on the operator's experience in the area and the dragging action as the kelly is picked up.

The hole is reamed for other reasons, such as wiping out keyseats, cleaning out caving material or excess wall cake, and opening the hole behind an undergauge, worn-out bit. Different downhole assemblies are used for different types of reaming operations. Reaming is normally a low-risk operation. However, reaming can lead to problems such as twisting off the drilling assembly, sticking the drilling assembly, and sidetracking and drilling a new hole. Therefore, reaming operations must be conducted carefully.

Drill-Collar Connection

After the kelly is drilled down and the section is reamed, another drill collar must be added to the drilling assembly in order to drill deeper. The average kelly has 35–40 ft of travel through the kelly drive bushing. Since the average drill collar or joint of drillpipe is about 31 ft long, when the joint is connected and run in the hole, the bit will be 8–12 ft off bottom. This allows room to connect the kelly to the drilling assembly and lower the kelly until the kelly drive bushing enters the rotary to begin drilling.

To make a drill-collar connection after the last reaming pass, the rotary and mud pump are shut down and the drilling assembly is picked up until the tool joint on the bottom of the kelly is about 3 ft above the rotary. The drill-collar slips are set, and the assembly is lowered until all of the assembly weight is supported by the drill-collar slips except the kelly, swivel, and traveling block (about 15,000–25,000 lb). The connection between the kelly and the crossover sub is loosened by latching the makeup tongs below the joint as backup tongs and latching the lead tongs or breakout tongs on the kelly above the tool joint and pulling on the handle of the breakout tongs with a cable attached to the breakout cathead (Fig. 6–20).

The assembly is then picked up until the tool joint on the bottom of the crossover sub is about 3 ft above the rotary. The slips are usually allowed to drag (stay in the rotary) for this short movement of the assembly. The assembly is then slacked off slowly to distribute the weight on the drill-collar slips. A drill-collar clamp is fastened around the drill collar above the casing slips and is tightened securely. The connection between the crossover sub and the drill collar is loosened with the tongs in the same manner as the connection between the crossover sub and the kelly. The lead tongs are removed after the connection has been loosened. Then the rotary is turned slowly to unscrew the drill collar from the crossover sub, which is held by the breakout tongs. After the connection has been unscrewed, the breakout tongs are removed, the kelly is moved to one side of the rotary, and the crossover sub is backed off with a small pair of chain tongs (hand tools).

A pull-back line run over the cathead pulls the bottom of the kelly over to the rathole, and the kelly is lowered into the rathole. A long-handled hook release tool is used to unlock and open the hook to release the swivel bail. The swivel on the block is unlatched, and the elevators are opened and turned to pick up another drill collar.

FIG. 6–20 The cathead or spinning cathead is the smooth spool on the end of the jack shaft. Lifting lines are wrapped around the turning (spinning) cathead, and a slight pull by the operator is multiplied by the friction between the spinning cathead and the lifting line. An experienced operator can exert a few pounds of pull on one end of the lifting line wrapped around the cathead and lift hundreds of pounds with the other end. The automatic cathead—it is not automatic but operated by the driller—has a cable connected to the end of the tong arms to make up (connect) and break out (disconnect) tool joints and other pipe.

In the meantime a drill collar is pulled into the V-door, the threads are cleaned and greased, and a lift nipple is screwed into the top of the drill collar. The elevators are latched around the lift nipple, and the drill collar is picked up. The pin end is cleaned and greased and is then connected to the drill collar sitting in the rotary.

The drill-collar clamp is removed, the assembly is picked up a short distance so the drill-collar slips can be removed, and then the assembly is lowered until the top of the drill collar is about 3 ft above the rotary. The slips are reset, the drill-collar clamp is fastened around the drill collar above the slips, and the lift nipple is removed and set to one side. The kelly is picked up and connected to the top of the crossover sub, the bottom of the crossover sub is connected to the drill collar in the rotary, the drill-collar clamp and slips are removed, and the drilling assembly is lowered to resume drilling operations.

As one can see, picking up the drill collars is a slow, tedious process, especially heavy, large-diameter collars. However, the drill collars are needed in order to have enough weight to drill the hole efficiently. If reamers or stabilizers are used in the drilling assembly, they can be installed as the drill collars are picked up. They are not normally used while drilling a shallow, large conductor hole except when special drilling problems, such as a crooked hole, occur.

Wire-Line Deviation Survey

Deviation must be measured to see if the hole is straight and vertical, so deviation surveys are run at 30-ft intervals in the conductor hole and in the upper part of the surface hole section.

Various types of deviation survey instruments are used. One of the most common has a small plumb bob with a pointed end suspended above a disc that is calibrated in degrees (Fig. 6–21). A timing mechanism releases the plumb bob to perforate the disc after a preset time interval. The point where the disc is perforated is then read directly as the hole inclination, in degrees. The entire mechanism is contained in a slender sealed tube. The instrument is run inside another sealed steel tube called the carrier. A single-strand wire line, the slick line, of high-grade steel is connected to one end of the carrier tube. It is coiled on a small powered reel. The reel is fitted with a device to measure the length of the wire line as it is run in the hole.

In operation, the deviation survey tool is loaded, set, and placed in the carrier tube (sometimes called a bomb). Meanwhile, the drilling assembly is picked up, the slips are set, the drill-collar clamp is fastened around the drill collar, and the kelly is disconnected and set back. A small pulley is attached to the elevators, and the slick line is passed through the pulley. The deviation tool is lowered in the hole to a predetermined depth, near or on top of the bit. The instrument should be run in the hole and be in place about 2–3 min before it activates and punctures the disc (called taking a picture). After waiting another 2–3 min, the instrument is pulled out of the hole and dismantled to recover the disc. The hole deviation is read from the disc and is recorded on the drilling report. The kelly is reconnected and drilling operations are resumed. Normally the deviation or straight-hole survey is run while making a connection.

DRILLING PROBLEMS

The severity and frequency of drilling problems generally increase with depth. The average conductor hole is often drilled without any severe problems, espe-

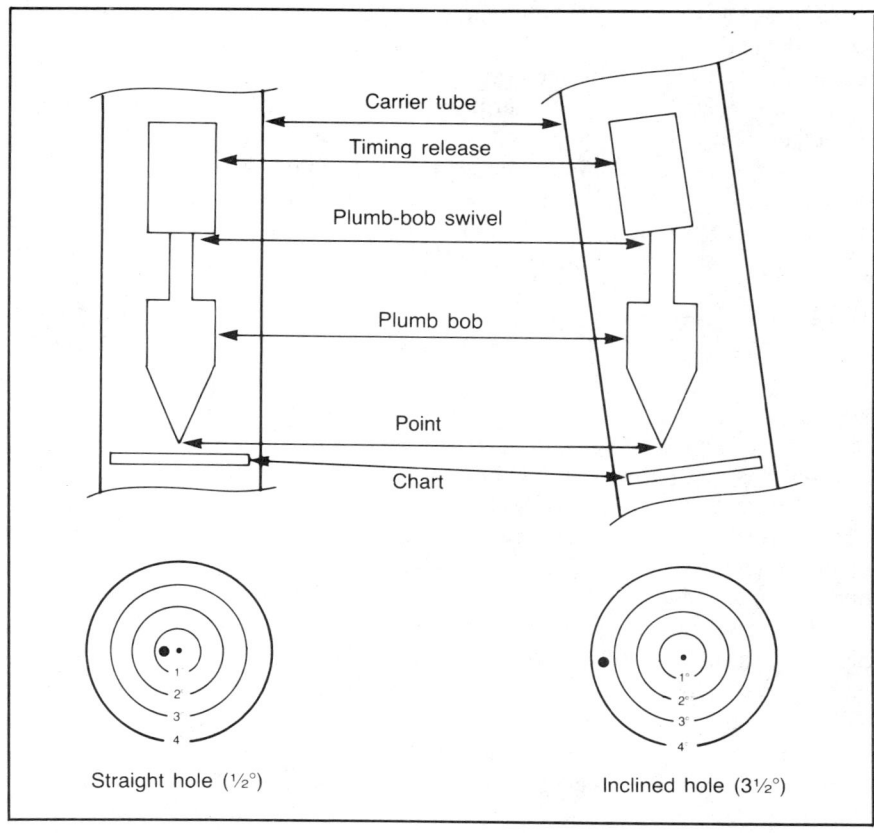

FIG. 6–21 Deviation survey

cially in soft formations. These formations drill rapidly, considering the limited amount of bit weight available for the large hole. However, there are problems that occur more frequently while drilling the conductor hole.

Bit Balling
There is some tendency for the bit to ball up and become coated with formation drill cuttings. The rollers can become clogged and tend to slide rather than roll, thus greatly decreasing their drilling action. As the cuttings accumulate on the

sides of the bit, they restrict the upward flow of the mud. This can increase the mud pressure and cause the formations to break down and lose circulation. The formations break down when the mud pressure exceeds the formation strength.

Bit balling can normally be controlled by using lightweight, low-viscosity mud and reaming and spudding the bit to remove the accumulated cuttings. Spudding the bit is similar to reaming except the action is more forceful. The kelly is rotated and lowered faster.

Slow Drilling in Hard Formations

One main problem with hard formations is a slow drilling rate. At shallow depths before all of the drill collars are picked up, the drilling assembly is relatively lightweight. Therefore, there is not enough weight available to apply to the bit and efficiently drill the formation.

In severe cases one alternative is to drill a pilot hole and open the pilot hole to the required size. The pilot hole is a small hole that can be drilled faster and more efficiently than a larger hole because the available bit weight is concentrated. A hole opener is run to open the pilot hole to the normal hole size. This procedure is seldom used to increase the drilling rate, except in very severe conditions. Its main disadvantage is that the hole must effectively be drilled twice.

Hole Cleaning

Drill cuttings are more difficult to remove from a large hole because of the large cross-sectional area and reduced upward flow rate of the mud. Under normal conditions there will be an adequate mud flow rate to clean the hole. If the mud flow rate is too low to clean the hole, the mud viscosity can be raised to increase its capacity to carry cuttings.

Unconsolidated Formations

Unconsolidated gravels, conglomerates, and boulders frequently occur at shallow depths penetrated by the conductor hole. In some cases the existence of these formations is the reason for running conductor casing. Gravel and small conglomerates are relatively easy to drill but sometimes are very hard to circulate or wash out of the hole, especially if water is being used for the drilling fluid. A thick, viscous spud-mud may be needed to carry the material out of the hole. If gravel is allowed to accumulate on top of the bit, it can stick the drilling assembly.

The normal procedure for drilling gravel and smaller conglomerates is to drill a short distance into the formation and pick up the drilling assembly until the bit is above the formation. The hole is then reamed back to bottom cleaning out caving, and another short section is drilled. The assembly is then picked up, and the operation repeated. It may be necessary to increase the viscosity of the mud to circulate the material out of the hole. The reaming, drilling a short distance, picking up, and circulating procedure is repeated until the gravel stops caving and the hole remains open or clean. The gravel will stop caving into the hole when the gravel in the formation has reached a stabilized angle of repose. It may take from a few minutes to more than 8 hr of cleaning out before the gravel stops caving.

In severe cases of gravel caving, it may be necessary to consolidate the formation by cementing. Drive pipe may be used when the gravel beds occur near the surface. If the problem is known before the rig is moved in, the drive pipe, or large conductor casing, is driven through the formation with a pile driver.

Large conglomerates and boulders can cause a problem similar to gravel. There may be a higher risk of this material staying in place until the bit has drilled past the material. Boulders can then fall into the hole and stick the drilling assembly.

Boulders

Large boulders also cause another drilling problem. They are hard to drill, especially with the soft formation bits used for drilling the conductor hole. Also, the bit tends to drill off the side of the boulder causing crooked or deviated hole problems. Remedial procedures in this case include using harder formation bits to drill the boulders, extensive reaming, and possibly cementing to consolidate the formation.

Lost Circulation

Lost circulation is usually a minor problem, especially if there is a good supply of water. Circulation is seldom completely lost. When partial lost circulation occurs, the normal procedure is to drill ahead and make up the mud loss by adding water to the mud system.

When there is a complete loss of circulation—commonly called *lost returns*—one procedure is to dry drill ahead. In the dry-drilling procedure mud is pumped slowly down the drillpipe without mud returns at the surface, while continuing to drill at a reduced rate. This procedure may require large volumes of makeup water. When drilling in this manner, it is important to pick up the drilling assembly after drilling a short distance to ensure that the drill cuttings are not stacking up or accumulating in the annular space above the bit. If this occurs there is a high risk of sticking the drilling assembly.

In the dry- or blind-drilling procedure, drill cuttings are normally carried with mud into the lost circulation zone. It is not uncommon for these to plug the lost circulation zone so the drilling fluid is circulated back to the surface. In severe cases where there is a shortage of water, the lost circulation zone may be plugged by cementing. Lost circulation material is seldom needed when drilling the conductor hole section.

Sometimes it is necessary to cement the conductor hole to consolidate caving formations or to plug lost circulation zones. The procedure is similar in both cases. The drill collars are pulled out of the hole. Drillpipe is picked up and run open ended to the bottom of the hole and circulated. In the meantime a cement truck is rigged up, and the discharge line from the cement truck is connected to the top of the drillpipe. A volume of cement slurry, commonly called cement, calculated to fill the hole from the total depth to a distance of 50–150 ft above the caving or lost circulation zone is mixed and pumped. The drillpipe is pulled out of the hole, and operations are shut down until the cement hardens. The cement will flow into the lost circulation zone, and the hydrostatic head of the cement will cause it to penetrate a short distance into the unconsolidated formation.

The open-ended drill collars without the drill bit on the end could be used to spot the cement, but it is normally easier to use drillpipe. Also, there is less risk of sticking the assembly. After the cement has hardened, the bit and drill collars are run back in the hole to drill out the cement. Sometimes it may be necessary to recement a second or possibly third time in order to correct the formation problem.

Water Flows

Water flows may occur while drilling the conductor hole. When a water flow is encountered, the standard procedure is to continue drilling. The water flow usually depletes or plugs off with time. High-volume water flows can be restricted or shut off by adding barite weight material to the mud system, if necessary.

Most conductor holes are drilled without appreciable formation problems. When problems do occur, they are usually not too severe and are handled with good operating procedures. The most common problems are caving, lost circulation, and water flows. In some localized areas these problems can be severe.

RUNNING AND CEMENTING CONDUCTOR CASING

Conductor casing must be run and cemented after the hole is drilled. When long strings of casing are run, various procedures and precautionary tests are performed. Some of these include inspecting the casing, conditioning the hole by short tripping, rigging up special equipment to run the casing, and running and cementing the casing. (Running casing is covered in detail in Chapter 7.) From a practical viewpoint the conductor casing is a short string of casing, about 10 joints, that should be relatively simple to run and cement. This small amount of casing does not justify using special casing crews. The drillpipe tongs can be used to make up and tighten the casing, providing the proper-sized jaws are available. Slips and elevators to fit the casing are also needed.

The hole must be conditioned; but if the hole was reamed after each connection, it is probably in good condition. The mud will normally be in satisfactory condition, too. The hole is circulated while working the drilling assembly and preparing to run casing. Working the drilling assembly in this case is similar to reaming, except the assembly is rotated and moved slowly. A very low amount of drill cuttings in the returning mud indicates the hole is clean.

The first step is to pull the drill collars out of the hole. If the drill collars are to be used to drill below the conductor casing, then they are racked back, or stood back, in the mast. Sometimes the drill collars are too large to drill through the conductor casing and are laid down. For the purposes here, it will be assumed that the drill collars are laid down.

When everything is ready for running casing, the next step is to start out of the hole lying down. The crew shuts the pump off and picks up the drilling assembly. It disconnects the kelly and the crossover sub and sets the kelly back in a manner similar to making a drill-collar connection. The crew must be sure that the drill-collar clamp has been installed and tightened on the drill collar in the rotary.

A lift nipple is connected on top of the drill collar in the rotary and is tightened slightly. The elevators are latched onto the lift nipple, the drill-collar clamp is removed, and the drilling assembly is picked up while pulling the slips. The assembly is stopped with the first drill-collar tool joint about 3 ft above the rotary, slips are set, and the drill-collar safety clamp is installed. The drill-collar connection is loosened with the drillpipe tongs, and the drill collar is rotated free with a slow-moving rotary. Some operators feel that using the rotary to free the drill collar can cause thread damage. They require that the drill collar be rotated free with either hand tools (chain tong) or a hydraulic tong. Then a thread protector is placed on the drill-collar pin.

The end of the drill collar is either pushed or pulled out the V-door while being lowered slowly. Depending on the weight of the drill collar, the bottom end may be pushed out the V-door with the rotary helpers or it may be pulled out with a special pulling line. In either case the end of the drill collar is set on the ramp and then lowered slowly until it is lying in the V-door. The elevators are unlatched and moved out of the way, the lift nipple is removed, and a thread protector is screwed into the top of the drill collar.

A pipe dolly is used to lift the lower end of the drill collar so the collar can be slid down the ramp and laid on the catwalk. The drill collar is snubbed back with a lifting line tied to the upper end of the collar so it can be lowered slowly down the ramp to the catwalk. The drill collar is then rolled off the catwalk onto the pipe racks and out of the way. Later, the drillpipe protectors should be removed, the drill collar threads cleaned, inspected, and greased as necessary, and the thread protectors replaced.

While the first drill collar is being slid out the V-door onto the catwalk, a lift nipple is connected to the drill collar in the rotary. The procedure of picking up the drill collar, disconnecting it, and laying it out is repeated until all of the drill collars have been pulled out of the hole. The bit is disconnected from the bit collar before the collar is laid down.

The drill-collar slips and elevators are changed for slips and elevators that will fit the conductor casing. Either the tong jaws are changed for the correct-sized jaws to run the conductor casing, or hydraulic casing tongs are used. For this short string of conductor casing, each joint is picked up and run in the hole in almost the same manner as picking up, connecting, and running drill collars.

A guide shoe is usually run on the bottom of the first joint of conductor casing. Each joint of conductor casing is pulled into the V-door, picked up with the elevators, connected to the joint in the rotary, and run in the hole. Collared conductor casing is normally used so that it is unnecessary to use the drill-collar clamp.

If float equipment is used instead of a guide shoe, a fill-up line is connected to a valve-controlled outlet at the base of the standpipe and fills the casing with mud as it is run. Otherwise, the large-diameter casing may tend to float and must be filled so it can be run in the hole. On long strings of casing, the casing must be filled to prevent possible collapse, but this normally is not a problem with shallow conductor casing.

After the last joint of casing has been run, a cementing head is connected to the top of the casing and the mud fill-up line is connected to the cementing head. The casing is moved up and down slowly while circulating mud to clean the hole.

In the meantime a cement pump and bulk trucks have arrived on location and have been positioned and connected with one flexible line reaching the rig floor. After the hole is circulated clean, the fill-up line is disconnected at the standpipe and is reconnected to the line from the cement pump truck. Cement is mixed and pumped into the conductor casing. After the required volume of cement has been mixed, it is displaced with either mud or water until the top of the cement is about 100 ft above the bottom of the casing. The casing is worked slowly (reciprocated up and down) while the cement is being mixed, pumped, and displaced. Near the end of the cement displacement, the casing is lowered until it is within about 6 in. of the bottom of the hole and is allowed to remain stationary at this point.

After the cement has been displaced, the valve on the cementing head is closed and the cementing equipment is disconnected.

An excess amount of cement is used, normally twice the calculated volume or more, and cement is usually circulated to the surface. If cement is not circulated to the surface, several joints of small pipe are connected together, run down the side of the casing, and used to cement the upper part of the hole. The conductor casing is left stationary until the cement hardens or sets up.

In some cases conductor casing is cemented by using plugs or float equipment. The procedure for using this type of equipment is covered in Chapter 7.

NIPPLING UP AND DRILLING OUT

The time waiting for the cement to harden is called waiting on cement, or WOC. A pressure gauge and pressure relief valve are installed on the cementing head to monitor the pressure inside the casing and to release part of it if it increases an appreciable amount. The inside of the casing will have a small amount of pressure because of the difference in the hydrostatic head of the column of cement outside the casing and the combined columns of water and cement inside the casing. The pressure inside the casing can increase, caused by heating the water as the cement heats during the setting and curing process.

Cementing time is used to complete any rig-up operations that need to be done. Mud pits are usually drained, cleaned to remove all drill cuttings, and filled with water. If mud is needed to drill below the conductor casing, it is mixed at this time. The drillpipe and drill collars to be used to drill the surface hole section are moved as needed so they can be picked up; threads are cleaned and greased. If a mud riser was used, the flow line to the riser is disconnected. The mud riser is cut off with a cutting torch.

After the cement has hardened, the elevators are lowered a very short distance to let the conductor casing set down. The conductor casing is cut off below the rotary with a cutting torch. A thread may be welded on the top of the conductor so that a rotating head or other type of diverter may be connected. Sometimes the upper part of the casing is heated and tapered outward slightly to form a bell nipple. The slightly larger end of the bell nipple permits full-gauge tools such as bits to be run into the top of the conductor casing easier and without hanging up by setting down on the top edge of the conductor casing. A hole that is slightly less in diameter than the diameter of the flow line is cut in the side of the conductor casing about 1–2 ft below the top of the bell nipple and is aligned with the flow line. A short piece of pipe, called a nipple, with the same diameter as the flow line is welded over the hole in the conductor casing. The flow line is connected to this nipple with a flexible coupling. A small connection about 2 in. in diameter is welded to the conductor at a convenient place. This is connected to the mud line as a fill-up line and completes the nippling-up operations.

The next step is to pick up a bit and drill collars and to run these in the hole to drill out the cement. When the bit reaches the top of the cement, drilling operations are started and the cement is drilled out of the conductor casing. The cement can usually be drilled out at about the same rate that the formation is drilled.

The cement is usually drilled out while circulating with water. The water returns that are contaminated by cement are discarded. After the cement has been drilled out, the bit penetrates new formation and drilling operations are resumed.

AIR-GAS MIST DRILLING OPERATIONS

If air or gas is used to drill a hole, it normally is used to drill the conductor hole and the deeper hole section down to the air-drilling limits (the depth of water flows and high-pressure zones). A mud riser or shallow conductor is normally used with a rotating head on top of the riser. In some cases the hole for the mud riser is also air drilled.

The drilling procedure with air is similar to drilling with liquid mud. Air is circulated down the kelly, through the bit, and up the annular space while the drilling assembly is rotated. The formations are drilled by the rotating bit and weight applied by the drill collars. The hole is reamed in the same manner as with fluid mud drilling.

Connections are made in a conventional manner with a few variations. After the kelly has been drilled down and the hole reamed once or twice, the kelly is picked up in the position to make a connection. The standpipe valve is closed to shut the air off to the circulating system. The compressors have pop-off valves so, when the standpipe valve is closed, air from the compressors passes through the pop-off valves to the atmosphere. The pop-off valves are a spring-loaded, pressure-relief valve adjusted so, when the compressor discharge air pressure increases above a certain level (caused by closing the standpipe valve), the pop-off valve opens to relieve the air pressure.

A check valve is installed inside the drilling assembly above the bit or higher in the drilling assembly when drilling deeper. This prevents a large blow-back of air when the kelly is disconnected.

The standpipe valve is opened after the single (drill collar or drillpipe) has been connected. Drilling operations are resumed after the system pressurizes, and there is a steady flow of air out the blooie line.

The hole cannot be drilled efficiently with air when large volumes of formation water flow into the borehole. Large volumes of water require higher air pressure and larger volumes of air to flow or blow the water out of the hole. This requires more compressors with higher pressure ratings. Generally, the additional cost of this equipment does not justify the additional benefits gained by air drilling. Air at higher pressures is also more dangerous and difficult to control. The upward flow in the annular space of two fluid phases (air and water) may decrease the stability of the walls of the borehole and cause increased caving and sloughing, which in turn can stick the drilling assembly and cause a fishing job.

When a small amount of water is encountered, it can often be removed by reducing the drilling rate and allowing more time for circulating and drying the hole. The hole must be dry for normal air drilling. Small amounts of water will dampen the outer surfaces of the drilling assembly and the inner walls of the borehole. Drill cuttings stick to these damp surfaces and continue to build up and increase in thickness. This accumulation on the drilling assembly, sometimes called mud rings, effectively acts to increase the outside diameter of the drilling assembly. At the same time the cuttings collect on the walls of the borehole and tend to reduce its diameter. The combination can cause the drilling assembly to become stuck and prevent additional drilling.

Mist or foam drilling are used as alternative drilling procedures when small-volume water flows encountered prevent conventional air drilling. The same air-drilling equipment is used for mist or foam drilling with the addition of a soap tank and injection pump.

A concentrated soap solution containing about 5 gal of liquid soap/10–20 bbl of water is mixed in the soap tank. This solution is injected slowly into the circulating air stream. The soap solution mixes with the downhole water flow, creating a foam or mist (hence the name foam or mist drilling). This low-density foam can be flowed or blown out of the hole easier than liquid water. This mixture keeps all of the downhole surfaces wet and slippery, which helps reduce the tendency for the drill cuttings to adhere to the surfaces. This reduces the risk of sticking the drilling assembly, and the drill cuttings are removed more efficiently. In some cases foam drilling can handle extremely large volumes of water. In one extreme case a hole was drilled 200 ft below a 4-ft thick mined-out, flooded coal seam (mine).

Downhole explosions have occurred while drilling with air. A gas-bearing zone can be encountered that flows natural gas into the wellbore. A downhole explosion can occur if the gas and air are mixed within their explosive limits. The downhole explosion apparently does not create a hazardous condition at the surface. However, there have been cases where it caused the drilling assembly to be cut, broken, or burned so that it parted, causing a fishing job. The downhole equipment tends to wear out faster in air drilling because air does not have the lubricating effect of liquid mud. For the same depth, drilling assemblies weigh more in air than in mud because buoyancy is lost. This causes higher hook loads.

Air- and mist-drilled holes are logged, cased, and cemented in a conventional manner.

HELICOPTER RIG OPERATIONS

A special logistics problem occurs in moving and drilling operations and supplying a rig for remote, isolated locations especially when the drillsite is in swampy, marshy terrain. Fly-in rigs were first used in some of these and were later replaced by helicopter rigs.

Access to the location is the first problem. The rig is moved by ocean freighter, large cargo airplanes, trucks, or train to the point nearest the location. If an air strip can be built at or near the location, the packaged rig (2 tons or 4,000 lb/package) is flown to the air strip in small cargo planes that can carry 2–3 tons and land on a 3,500-ft runway. Many remote locations are in swampy, jungle areas, and large rivers may lead to or near the drillsite. More economical tug-barge transportation is used whenever possible.

If the landing strip cannot be built at the rig site, the final rig move from the landing field (river bank, road end, etc.) to the wellsite is by helicopter. The reasonable, one-way moving distance by helicopter is about 40–50 miles with an average load, allowing for the fuel load for the return trip. Longer distances may require a staging area to refuel the load-bearing helicopter. In this case helicopter fuel is transported to the staging area by helicopter to extend the range of the load-carrying helicopter.

In most of these types of locations, the drillsite must be prepared after arrangements have been made to fly in the equipment but before it is moved in. If the drillsite is on good soil, it can be cleared and leveled in the normal manner with small earth-moving equipment. The tractors are dismantled, helicoptered to the location, and assembled. The same tractors can be fitted with a gin pole or crane and winch to help move and assemble the rig equipment when it arrives.

In many cases these locations are covered with trees and located in swampy marshy areas. The location is then made by using a swamp platform on piles or a board on riff-raffed trees.[2] A crew carried to the site by helicopter uses chain

---

[2] "Improved rigs, swamp platforms aid jungle-terrain drilling in Peru" by Alvaro Franco, *Oil & Gas Journal* (July 14, 1975).

saws to cut down the trees. These are stacked in layers with the trees laying cross ways in alternate layers. This stack is covered with several layers of matting boards. The matting boards (3 in. thick by 12 in. wide by 10–12 ft long) are either cut on site with a portable saw mill or are flown in. The rig is then helicoptered in and assembled on top of this platform. The equipment is normally carried in a rope sling, cargo net, or hooks and cables suspended from the bottom of the helicopter.

The riff-raff platform is not permanent and will usually deteriorate (rot) quickly. Another method is to build the location on piles. After clearing the location, a small pile driver is helicoptered in and used to drive piles—usually wood or steel. The platform deck is constructed on top of the piles, and the rig is helicoptered in to drill the well. The rig is supplied in the same manner while drilling and is moved out by helicopter after the well is drilled.

MARINE OPERATIONS

Most of the equipment on marine rigs is similar to land rigs. Marine rigs use diesel-generated electrical power. All or most of the drilling equipment is permanently mounted and ready for use, so very little rigging-up is required. Drilling ratholes and mouseholes is not necessary except on very shallow barge rigs.

Most marine rigs have complete living quarters including kitchen facilities because the crews live on the rig. In these cases two crews usually work 12-hr shifts while on the rig. Schedules vary but range from seven days on the rig and seven days off to ten days on and five days off. Barge rig crews are usually transported to and from the drilling barge in small crew boats. Some barge rigs are equipped with helipads, and the crews are transported by helicopter.

Although moving procedures are different on marine rigs, most of the drilling operations are generally similar to drilling operations conducted by land rigs.

Most barge rigs use masts that are laid down when the rig and barge are moved (Fig. 6–22). The mast is raised after the barge is submerged, or set down, at the new location. The drilling equipment is permanently mounted on one or several decks. One end of the barge is fitted with a drilling slot that is opened out through the end of the barge. The well is drilled through this slot. After the well has been drilled and the barge is ready to move, the mast is lowered and water is pumped out of the ballast tank. As the barge gains buoyancy and begins to float, tugs move the barge directly away from the wellhead so the wellhead effectively passes through the slot until the barge is clear.

Depending on the water depth, it may be necessary to dredge a canal to the next location to be drilled. This is done before moving. The sea floor is normally

FIG. 6–22 The mast is laid down on most barge rig moves. Note the cargo barges in the foreground.

level on marine locations used by barge rigs. If the bottom is not level, then shell or other fill material may be used. Tugs tow and/or push the drilling barge to the next location where they temporarily hold the barge in place as it is positioned. Valves are opened to flood the ballast tanks and the barge settles to the bottom.

Cargo barges moved by tugs carry materials and supplies to the drilling barge. The materials and supplies are off-loaded from the cargo barge onto the drilling barge with cranes mounted on the drilling barge.

Barges are also used for additional storage. When third-party services such as cementing trucks and logging units are needed, conventional land-type equipment is used. The equipment is loaded on barges, transported to the drilling barge, used to perform the required services, and then returned to land on the same barge.

Equipment and operations on barge rigs are generally similar to land rigs. One exception is the shallow conductor. After the barge is positioned and ready to drill, the first operation is to drive shallow conductor to a depth of 50–250 ft below the mud line. The well is drilled through the conductor casing in a manner similar to land rig operations.

Wells are drilled in open water from various sizes of platforms constructed by different methods and of different materials. They essentially consist of a platform supported above the surface of the water on legs that extend into the sea floor. The platform is equipped with large cranes. A conventional rig is moved to the platform on barges and large work boats. Cranes lift the disassembled rig from the barge and place it on the platform where it is assembled.

The rig is used to drill a number of directional holes from the platform. One common procedure is to run a number of conductor casings through the legs of the platform. The wells are then drilled through these conductor casings. After the wells have been drilled, the rig is moved back to shore and the platform is used as a production facility.

Platform rigs are serviced by workboats and barges that move personnel, material, and supplies. Larger platforms in deeper waters further from the shore are equipped with helipads. Helicopters are used to move personnel and light loads of materials and supplies.

Jackups are essentially a drilling rig mounted on a floating body equipped with retractable legs. When the jackup is being moved, the legs are in the upward position. Some jackups are towed with tugs; others are self-propelled.

When the jackup reaches the new location, the legs are jacked downward. When the leg footings contact the sea floor, the barge or body is then lifted or jacked upward to a height above the highest wave expected. After the well has been drilled, the legs are jacked upward until the jackup floats. It is then moved to the next location, and the operation is repeated.

Jackups are supplied by workboats and barges. They are usually fitted with helipads, and the crews are transported by workboats or helicopter. Small jackups operating near the shore may use third-party land services, similar to those used on drilling barges. Larger jackups operating in deeper waters frequently have all third-party services on the jackup.

Floaters, as the name implies, are floating drilling platforms that can drill in thousands of feet of water depth. Most are self-propelled. When the floater reaches the drilling site, it is usually held in position by one of two methods. One method is to anchor the floater with massive anchors and chains that are usually tensioned by large winches.

The second method of holding the floater in position is to maintain its position dynamically. Various methods of drives are used. Two or more large, swiveling outdrives may be used. These look like the outboard drive of an inboard-outboard motor used on pleasure boats but are much larger and are propelled by an electric motor. The direction and amount of thrust on the outdrives are controlled by sophisticated techniques using radar, sonar, and other methods.

Another type of drive consists of fixed outdrive units mounted so they can move the floater in any direction. They are actuated and controlled in a manner similar to that used with the swiveling outdrive. These devices can hold the floater in position and point it in one direction under very severe wind and wave conditions.

In summary, many prespud activities must be completed. The drilling program must be prepared or updated. The access road and drilling site must be

Moving In, Rigging Up, and Drilling the Conductor Hole

surveyed and prepared. Regulatory agencies are contacted, reports are written up, and permits are obtained. The availability of equipment, materials, supplies, and third-party services are determined. Inspections, preventive maintenance, and rig repairs also are scheduled during the rig-down, move, and rig-up operations.

The rig is dismantled at the old location, moved with oil-field trucks, and assembled at the new location. The rathole and mousehole are drilled, and shallow conduit may be installed. The conductor hole is spudded and drilled, and conductor casing is run and cemented. The deeper hole is drilled through this casing. Special drilling techniques such as air or mist drilling also may be used. All of these operations are conducted to begin drilling a new well.

CHAPTER 7
SURFACE HOLE SECTION

The surface hole section for an average to moderately deep well usually is from the bottom of the cased conductor hole to about 2,500 ft. with 9⅝-in. casing in a 12¼-in. hole. The surface hole is drilled through and below the conductor casing—350 ft of 13⅜-in. as described in Chapter 6. For deeper holes the surface hole section may be called the deep conductor hole with larger hole and casing sizes. For shallower holes it may be considered as the intermediate hole section, and would probably be a smaller size hole and casing combination. In very shallow holes it could be the production hole section. Softer formations are usually encountered in the surface hole section so it is drilled faster than the deeper sections.

DRILLING PROCEDURE

The surface hole section is drilled in the same manner as the conductor hole with minor changes to allow for the increasing depth. Then the conductor casing shoe and 10–20 ft of formation are drilled. The bit is reamed through the bottom 30–50 ft of conductor casing several times to ensure that all of the cement has been removed and the tools are free to pass back into the casing. Drilling operations are resumed to drill the surface hole section. At this point only drill collars are being used in the drilling assembly.

Drilling operations are continued by drilling the kelly down, adding a drill collar, and drilling the kelly down again. This procedure is repeated with reaming as needed before each drill collar connection until all of the drill collars have been picked up. As the drill collars are picked up, additional weight is applied to the bit to increase the drilling weight.

Bit manufacturers supply a recommended range of bit weights and corresponding rotary speeds for each type and size of bit. This usually is in the range of 2,000–5,000 lb/in. of bit diameter. Higher rotary speeds are recommended at lower bit weight (bit loading).

Experience in the area is an important factor in drilling. For example, some formations drill faster with one combination of bit weight and rotary speed than other formations at the same depth. The bit wears out faster if it is operated at higher bit weights and rotary speeds, so the operator must select the best combination of bit weight and rotary speed to obtain the most economical cost per foot drilled. This is done by evaluating the length of time the bit drills (the bit life), the amount of hole drilled by the bit, the time required to trip the drilling assembly to change the bit, and the average daily rig operating cost.

As a general rule the operator tends to load the bit heavier at shallow depths because the trip time to change the bit is relatively short. At greater depths, the operator may use lighter bit loading to extend the bit life because longer trip time is needed to change the bit.

The operator will hold up on the bit—use less bit weight—while drilling immediately below the conductor casing and other casing strings. To hold up indicates that the hole is being drilled at a lighter bit weight than either the total amount of bit weight available in the drill collars or the manufacturer's recommended bit weight. Bit weights are usually reduced until the drill collars are buried, the top of the drill collars are past or below the bottom of the casing. When the drill collars are opposite the bottom of the casing, the bit is held up and the rotational speed is adjusted so the drilling assembly is running as smoothly as possible. This helps prevent breaking the cement-casing bond and reduces the risk of backing off the bottom joint(s) of casing.

The drilling assembly has a right-hand or clockwise rotation. As the drilling assembly rotates, it can hit and drag on the casing. This is a right-hand or clockwise rotating force on the casing as seen from the top looking down. However, on the bottom joint(s), when looking up, this is a left-hand or counterclockwise force that can cause one or more joints of casing to unscrew and back off. When this occurs, a severe fishing job and a lost hole can result. A lost hole must be sidetracked or redrilled from a different surface location that is usually nearby. Either operation is expensive and takes additional time. To avoid the risk of backing off, the bottom two or three joints of casing are cemented with epoxy cement when the joints are screwed together or the joints are tackwelded to the collars when the casing is run.

An important concept in drilling is the location of the *free point*. The free point is the point in the drilling assembly that is free of vertical stresses. It is sometimes described as the point where the sum of the vertical stresses is zero. As weight is applied to the formation through the bit, the lower part of the drilling assembly is in compression. The upper part of the drilling assembly, which is suspended, is in tension. The point where the drilling assembly is neither in tension nor compression is the free point.

As the drilling assembly hangs suspended with the bit off bottom, the entire assembly is in tension and the free point is effectively below the bit. As the drilling assembly is lowered, the bit is set on the bottom of the hole. If the assembly is slacked off a small distance, part of the assembly weight becomes supported by the formation through the bit. That part of the assembly then is in compression, and the free point is at the top of the section in compression. If the assembly is slacked off another small distance, more of the assembly weight is supported by the formation through the bit and a longer length of the assembly is in compression. Again, the free point is at the top of the section of the assembly that is in compression. The remainder of the upper pipe is in tension.

The distance between the bit and the free point can be calculated by taking the amount of weight set on the formation (bit weight) and dividing this by the weight of the assembly per foot. As an example, assume that a drill-collar assembly is 860 ft long and is composed of drill collars that weigh 70 lb/ft in the drilling fluid. If the bit weight is 49,000 lb, the free point would be 700 ft above the bit—49,000 lb divided by 70 lb/ft.

The concept of the free point is important in drilling and other operations. The drill collars have large, heavy-duty tool joints that can withstand the impact and torsional forces caused by drilling. However, the drillpipe tool joints are smaller and cannot withstand the heavy forces that the drill-collar tool joints are subjected to. This is one reason for running drill collars immediately above the bit.

In order to drill, weight must be placed on the formation, so some of the drill collars are run in compression. However, as the bit drills, there is considerable vibration, movement, bit bounce, and wobble. Therefore, the bit weight is not exactly constant but fluctuates appreciably and rapidly. The net effect is that the free point also moves rapidly up and down the drill-collar assembly within limits, depending on the weight fluctuations caused by the bit as it drills. Over the distance that the free point moves, the drill collars alternate from tension to compression, often very rapidly.

The large drill-collar tool joints are designed to withstand these forces. However, the smaller drillpipe tool joints cannot withstand the movement for an appreciable period without a high risk of failure as a result of metal fatigue and breaking. If this occurs, a fishing job and a possible lost hole can result. Therefore, the free point during drilling must be maintained in the drill-collar assembly. This helps determine the amount of drill collars needed.

Experience has indicated that the free point should be maintained at a distance of about 20% of the drill collar length from the top of the drill collars. The first step to determine the number of drill collars needed, is to estimate the drilling weight necessary. Allow 20% excess, and select the number of drill collars from the known drill-collar weight.

For example, assume that a drilling weight of 50,000 lb is used. The drill collars that fit the particular drilling conditions weigh 70 lb/ft in the drilling mud. Dividing 50,000 lb by 70 lb/ft would give 714 ft of drill collars needed below the free point. Since the distance below the free point represents 80% of the drill-collar length, the entire drill-collar assembly would be 892 ft long, or 714 ft divided by 0.80. Drill collars average about 31 ft in length, so 892 divided by 31 is 28.8, or 29 drill collars. This would be nine stands of *thribbles* (three joints of drill collars/stand), one stand with two drill collars, and a drillpipe single on top. This drilling assembly would allow a drilling weight of 50,000 lb under normal conditions. Under severe operating conditions the drilling weight would be reduced, and under very smooth operating conditions the drilling rate could be increased slightly to possibly 60,000 lb, subject to recommended bit weights.

Drillpipe Connection

After the required number of drill collars have been picked up, drilling operations are continued by adding joints of drillpipe to the drilling assembly. The procedure for adding a joint of drillpipe to the assembly is known as *making a connection* or drillpipe connection.

The drillpipe is stored on the pipe racks. Three joints of drillpipe are rolled onto the catwalk and are pulled into the V-door with a lifting line. The length of the joints is measured and recorded in the pipe tally book.

First, the joints are checked to ensure that they are open and that there is no debris such as sticks or rocks inside the joint that could be pumped down the hole and plug the bit. This process is usually done by using the rig washdown hose to run water through the joint. Sometimes a *rabbet* is used. The rabbet is a short piece of pipe that has an outside diameter slightly smaller than the inside diamter of the drillpipe. The rabbet is dropped through the joint of the drillpipe to ensure that it is open and clean; this is called *rabbeting the pipe*.

The tool joints are cleaned and greased, or doped. They are visually inspected at this time to ensure that the shoulders are smooth and free of nicks and gouges, the threads are not stretched, and there are no visible cracks. Then one joint is picked up with a lifting line and is placed in the mousehole.

In fast drilling sometimes six or nine joints of drillpipes are pulled into the V-door and are measured and cleaned (Fig. 7–1). The joints are normally measured in feet using tenths and hundredths of a foot. Sometimes a number of joints are measured on the pipe rack, and the joint measurement is written on the tool-joint box with chalk and recorded in the pipe tally book when the joint is run. Some operators also measure the drillpipe and stencil the measurement on the tool-joint box. The individual joints of drillpipe in most strings of good pipe are almost the same length within a few inches.

In very fast drilling the pipe is sometimes picked up, cleaned, and run in the hole without measuring and using an average length for each joint run. On the first trip out of the hole with the pipe, usually to change bits, the drillpipe is measured and the hole depth is corrected accordingly. This depth usually is within a few feet of the measured depth.

FIG. 7–1 Drillpipe in the V-door. Note elevator on left.

After the kelly has been drilled down, the drilling assembly is picked up and reamed if necessary. The drilling assembly weight and apparent change in weight as shown on the weight indicator while pulling the kelly up to make the connection can be used to determine if reaming is required (Fig. 7–2). The weight indicator, as the name implies, shows an indicated weight, which is the approximate weight supported by the mast. For drilling operations, two weights must be considered: the true weight or *weight in air* of the drilling assembly and the submerged weight or *effective weight*. The weight indicator shows the effective weight and allows for the buoyancy caused by mud in the hole. The effective weight is always equal to or less than the air weight.

The kelly, swivel, traveling block, and drilling lines are always in air. The kelly may be partially submerged in the drilling mud, but the small variation in weight caused by this is insignificant in comparison to the total weight of the assembly. These items are collectively referred to as *surface weight,* and this weight does not change unless the equipment is changed.

The weight of the drilling assembly is affected by the mud weight, conventionally measured in pounds per gallon. As the mud weight increases, the buoyancy factor decreases and the indicator weight or effective weight decreases correspondingly (Table 7–1).

Table 7–1 Variations in Drilling Assembly Weight Resulting from Buoyancy

	Assembly Weight = 1,000 lb					
	Air-Drilled Hole		10-lb/gal Mud		16.5-lb/gal Mud	
	Air Weight	Indicator Weight	Air Weight	Indicator Weight	Air Weight	Indicator Weight
Surface weight	20	20	20	20	30	30
Drilling assembly	140	140	140	119	250	188
Total	160	160	160	139	280	218

The first comparison in Table 7–1 is an air-drilled hole where air is used as the drilling fluid. The air weight in this case is the same as the effective weight or indicator weight. The same drilling assembly is used in the second comparison, except the hole is filled with 10-lb/gal mud. The air weight is the same as the first case, but the effective weight is 139,000 lb, which is 21,000 lb (13%) less than the first case. As the drilling assembly weight increases and higher mud weights

FIG. 7–2 Reaming before a connection. Note kelly spinner on top of kelly and below swivel.

are used, the difference between the effective weight and the air weight increases. This is demonstrated in the third comparison in Table 7–1. This difference represents a deeper hole using heavier mud. The difference between the air weight and the effective weight is 62,000 lb or 22%.

Sometimes indicator weights—especially weight differences—are referred to as *points*. For example, 21,000 lb difference in weight would be the same as 21 points difference in weight.

The weight indicator normally shows the effective weight plus *drag* as the drilling assembly is moved. Drag, or pipe drag, is caused by the drilling assembly sliding and rubbing on the wall of the hole. The amount of drag depends upon the depth of the hole. The amount of hole that is cased, the deviation, the change in deviation, the mud lubricity, and to some extent the speed or rate at which the drilling assembly is being pulled upward. Normal drag on a good, vertical hole is about 1,000 lb/M ft of hole. For example, a hole being drilled at 5,000 ft with an assembly weight (effective) of 125,000 lb would have a drag of about 5,000 lb. Therefore, as the drilling assembly is picked up, the weight indicator will show about 130,000 lb.

The driller is familiar with normal drag measurements. If the drag is excessive when the drilling assembly is pulled upward, the hole may be tight, or partially restricted, or the bit may be balled up. In this case the hole should be reamed—possibly several times, depending on the drag when the assembly is pulled up. Sometimes the weight indicator shows a fluctuating drag. This can indicate alternating tight and loose places in the hole, ledges, and similar obstructions. Again, the hole should be reamed.

When the drilling assembly is picked up to make a connection, the pump is shut off shortly after starting to pick up the assembly or when the kelly is about halfway out of the rotary. This helps reduce (1) the pressure of the mud inside the drilling assembly and (2) the amount of mud spilled when the kelly is disconnected. The standpipe drain valve can be opened to drain the mud out of the kelly and standpipe.

Then the assembly is picked up until the tool joint on top of the crossover sub is about 3 ft above the rotary. The top of the drillpipe tool joint should not extend over about 2 ft above the top of the mousehole. If it is much higher, the force of the drillpipe tongs pulling sideways on the single can bend it slightly or kink it in the area where the joint is supported by the mousehole. Drill-collar slips are set, the drill-collar safety clamp is installed, the tool joint is loosened with the drillpipe tongs, and the kelly is unscrewed with either the rotary or kelly spinner. Sometimes a mud box is wrapped around the loosened tool joint to prevent drilling mud from spraying on personnel and equipment when the disconnected kelly is picked up. The cylindrical, hinged mud box is fitted over the loose connection.

When the pipe is picked up, the mud is contained by the mud box and flows downward onto the rotary.

The bottom of the kelly is pushed over, and the kelly tool-joint pin is set into the tool-joint box of the joint of drillpipe in the mousehole (Fig. 7–3). While the kelly is being rotated out and swung over by one of the rotary helpers, other personnel place the drillpipe tongs on the drillpipe single in the mousehole. The backup tongs are placed on bottom and are latched and set. The makeup tongs are placed on top but are latched and left loose.

After the kelly pin is set in the tool-joint box, the kelly is screwed into the tool joint by rotating with the kelly spinner. The backup tongs are lifted upward onto the kelley tool-joint pin. The backup tongs are repositioned on the box of the drillpipe single. The tool joint is then tightened to the correct torque.

Some rigs are equipped with hydraulic drillpipe tongs that spin up the kelly to screw the tool joints together. Some hydraulic tongs can also tighten the tool-joint connection, and others have a pulling line from the tong handle to the makeup cathead to tighten the tool joint.

One of the earliest methods of making up the kelly pin (or drillpipe) into the tool-joint box was by using a spinning chain. One end of the spinning chain was

FIG. 7–3 Joint (single) in the mousehole

fastened to the pulling line connected to the end of the tong handle. When used, the spinning chain was wrapped around the tool-joint box in a counterclockwise manner. After the kelly pin was set into the box, the chain was pulled by the cathead. At the same time the loose end of the chain was flipped upward with a rolling-flipping hand motion. This caused the chain to wrap around the tool joint above the pin. As the chain was pulled, the kelly would rotate to screw the kelly pin into the drillpipe box. The connection was tightened with the drillpipe tongs in a conventional manner. Many rigs still use spinning chains.

Whether the tongs or the spinning-chain method is used, after the kelly and single are connected, the kelly with the attached single is pulled up into the mast. The pin on the bottom of the drillpipe single is connected to the box of the tool joint in the rotary in the same manner as the mousehole connection was made, by spinning up and tightening with the tongs (Fig. 7–4).

The mud pump is then restarted, and the drill-collar safety clamp is removed. The kelly and drilling assembly are picked up a short distance so the slips can be pulled out of the rotary. Then the drilling assembly is lowered downward into the hole. The kelly drive bushing is reseated in the rotary, the rotary is started, and drilling operations are resumed.

As the hole is drilled, additional drillpipe singles are added to the assembly by making the mousehole connection. The subsequent mousehole connections can be made somewhat faster than the first one since the drill-collar safety clamp is not used. The drillpipe tong jaws are also used for the complete connection, so it is not necessary to change tong jaws in the middle of the connection to fit the drill collars. Many connections are made as the well is drilled, and the drilling crews should perform these in a fast, efficient manner.

Deviation Surveys

The precautions taken to ensure drilling a straight, vertical hole through the conductor hole section are continued while drilling the surface hole. Under normal conditions it is easier to drill and maintain a straight vertical hole as more drill collars are added to the drilling assembly because the increased weight acts as a pendulum to hold the drillpipe straight and vertical.

In the shallow conductor hole, surveys may be taken at 30-ft intervals. Under normal conditions the survey intervals are increased to 60 ft or more at the bottom of the conductor hole section. At the start of the surface hole section, the surveys are taken at intervals of 60–90 ft. The interval is increased to 200–400 ft while drilling the lower part of the surface-hole section. When the surveys must be taken at closely spaced intervals, they are run on a single-strand wire-line unit, similar to the procedure used in the conductor hole. When the survey interval is about equal to the amount of hole drilled by a bit, the surveys are run at the start of a trip to change bits.

FIG. 7–4 Connecting the kelly

TRIPPING THE DRILLPIPE ASSEMBLY

The downhole rotating assembly, including the drillpipe, drill collars, and other downhole tools, must frequently be pulled from and run back in the hole. This process is called *tripping the pipe* or *making a trip*. The most common reason for tripping is to change bits, to replace a worn drill bit with a new one. Other reasons for tripping include running different downhole drilling assemblies, repairing leaks, running test tools, and conducting fishing operations.

The drilling assembly should always be tripped rapidly and safely. Although tripping is an important operation and contributes to the drilling, it is not a drilling procedure. For example, it is first important to replace a worn bit with a new bit, but this is not drilling. The only time the rig is making money is when it is making a hole. Second, a blowout may occur at any time conditions are favorable. Blowouts can be detected, prevented, or controlled in a more positive manner and with less risk if the drilling assembly is in the hole. There is a higher occurrence of problems while tripping than while drilling. Therefore, tripping operations should be conducted in a fast, efficient manner.

Preparing for the Trip

The rig floor should always be kept clean with all equipment stored in its proper place, and it should be checked before tripping. The equipment to be used during the trip also should be inspected and checked. Worn tong dies and slip segments should be replaced. The tongs, elevator, and slip hinges should be greased. The elevator latch should be checked to ensure that it has a strong closing spring. The threads on the drill-collar lift nipples or buttons should be cleaned and greased. Extra equipment such as the drill collar handling tools should be checked to ensure that it is available and in good condition. Any equipment to be run back in the hole should be brought to the rig floor. For example, on a trip to change bits, the new bit should be brought to the rig floor if the type of bit to be run is known.

The condition of the mud and the hole is an important consideration when preparing for a trip. The mud should always be maintained in the best condition possible. Sometimes, the mud is temporarily in poor condition, for example, after drilling contaminating formations. In this case, the mud should be circulated and conditioned before tripping.

If there is a high concentration of cuttings or cavings in the mud, the hole should be circulated clean before starting the trip. It is normally not necessary to circulate out a low concentration of cuttings. If the mud is gas cut, the hole is caving, or there is lost circulation, these problems normally must be corrected before making the trip. For example, highly gas-cut mud indicates underbalanced mud or a possible high-pressure zone and pending kick or blowout condition. Then it would be necessary to circulate and condition the mud, possibly increasing the mud weight before pulling the drilling assembly from the hole.

It may be necessary at times to make a trip when the hole condition is unstable. For example, if the hole is caving severely, requiring reaming and cleaning out, and the bit is almost worn out, a trip may be justified to change bits rather than continue to drill until bit cones are left in the hole. In this case the caving formation problem could be complicated by a fishing job to recover the cones. Items such as these must be considered while preparing for the trip and before actually starting to pull the drilling assembly out of the hole.

Deviation Surveys

Deviation surveys are often run just before making a trip to change bits, and the survey instrument is recovered when the drilling assembly is pulled out of the hole. When the deviation survey is run in this manner, most bits must be fitted with a *crow's foot*. This centering device is placed on top of the bit before the bit is run in the hole. The bottom of the deviation survey case sits in the recessed center of the crow's foot, which aligns the case with the drilling assembly.

The deviation instrument is loaded, set, and placed in a sealed tube. The pump is shut down, and the drilling assembly is picked up until the kelly tool joint is about 2 ft above the rotary. Slips are set and the kelly is disconnected. The deviation survey tool is dropped in the drillpipe, and the kelly is reconnected. The tool falls through the inside of the drilling assembly and lands in the recessed part of the crow's foot on top of the bit. The drilling assembly is not moved until about 3 min after the deviation survey tool has recorded the hole deviation. The survey tool is left in the drilling assembly and is recovered after the assembly has been pulled from the hole.

Slugging the Pipe

The last step before pulling the assembly from the hole is to slug the drillpipe. If the drilling assembly were pulled out of the hole without slugging, each joint or stand of pipe would be full of mud. The mud would drain out of the stand if the assembly were allowed to set for a long period, but this would greatly delay the tripping operation. If the drillpipe were disconnected, this mud would spray out on the rig floor and would create a work hazard and possible injury to the rig crew, especially when muds are highly treated with chemicals. To prevent this risk, the pipe is slugged with the weighted mud so it can be pulled dry, i.e., the top of the mud inside the pipe is below the tool joint to be disconnected.

The pipe can be slugged in various ways. One convenient method is to mix about 5 bbl of heavily weighted mud in the trip tank and pump it into the drilling assembly. When the top of the weighted mud is a few hundred feet below the kelly, the pump is stopped. The heavy mud inside the drillpipe has a higher hydrostatic head than the lighter mud outside the drillpipe. The heavier mud inside the pipe flows down through the pipe by a simple U-tube effect until the two mud columns are balanced. This leaves a section of dry pipe above the slug of weighted mud. As the drilling assembly is pulled from the hole, the weighted slug tends to stay at the same depth. Then the drilling assembly can be pulled dry.

When a good slug is obtained, there is a small, reducing flow in the flow line for a short period after the mud pump has been shut down. A good slug is also evidenced by a sharp sucking sound when the kelly is disconnected. This is caused by air rushing into the drilling assembly to fill the partial vacuum created when the slug falls downhole. It is important to obtain a good slug. Otherwise, the mud box must be used, which increases the overall trip time. Using the mud box increases trip time because the box must be placed on each tool joint before it is disconnected and removed. The mud on the equipment makes it more difficult for the crew to handle the equipment and perform the work.

Filling the Hole

Two important functions of mud are (1) to exert a sufficient hydrostatic head to prevent the flow of formation fluids (e.g., gas, oil, or salt water) into the wellbore and (2) to maintain a hydrostatic head against the exposed formations in the open hole to improve wellbore stability. These objectives are accomplished by maintaining the proper mud density and keeping the hole full of mud.

Maintaining the proper density is generally not a problem. During drilling operations, excluding lost-circulation problems, the hole always remains full because the mud is circulated. When the drilling assembly or other downhole equipment is pulled from the hole, the fluid level in the hole is lowered. If the fluid level is lowered a sufficient amount, formations can begin caving. If high-pressure, permeable zones are exposed in the open hole, they can begin to flow and possibly may cause a blowout or saltwater flow.

To prevent such an occurrence, a volume of mud equal to the volume of the pulled equipment must be pumped down to fill the hole. This procedure is always important, and in some cases it is critical. It is strongly suspected that most of the hole problems that occur during or immediately after tripping are contributed to or caused by improper hole-filling procedures. The importance of properly filling the hole cannot be overemphasized.

The hole is usually filled by using a trip tank or by measuring the amount of mud by counting pump strokes (Fig. 7–5). The tank is filled with mud just before making the trip. The mud pump is used to pump the mud from the trip tank into the hole to keep it full. The pumps circulate a fixed volume per stroke, depending

FIG. 7–5 Trip tank

on the pump size, liner size, etc. The mud volume displaced as each stand is pulled from the hole can be determined from standard tables. When the hole is filled by counting pump strokes, 5–10 stands of pipe are pulled. Then the required amount of mud is pumped into the hole as measured by counting the pump strokes. Either method is satisfactory, providing the mud is measured accurately.

Sometimes, when the slug is not mixed or displaced properly, the mud level in the annular space is allowed to drop a limited distance to prevent pulling a wet string. If the mud level is not allowed to drop too low, this procedure is probably satisfactory. However, the problem is determining how low to let the mud level drop and if the mud level is at the predetermined depth. The best and safest procedure is to mix and pump a good slug and keep the hole full.

Pulling the Drillpipe

When the slug is mixed and displaced, the pump is shut down. The drillpipe assembly is picked up until the kelly tool joint is about 2½ ft above the rotary.

One method of determining if the slug is effective is to strike the body of the drillpipe lightly with a medium-sized metal object, such as a small hammer. If the pipe is full of mud—indicating the slug was not effective—it will have a heavy, solid sound when struck. If the pipe is empty, it will have a ringing sound.

If the pipe is full, the driller may race the pipe to help drop the slug down hole. The drillpipe is picked up to the full length of the kelly hose. Then it is lowered rapidly for a distance of about 15 ft and is stopped sharply with the brake. This procedure is repeated until the kelly is down, and the entire process may be repeated several times. In general, racing the pipe is more effective at shallow depths because the assembly can be lowered rapidly and stopped faster than at deeper depths with a heavier assembly.

A single in the mousehole is picked up with the lifting line and is laid out of the V-door. The kelly is disconnected in the same manner as when making a drillpipe connection (Fig. 7–6). It is then set back in the rathole, and the hook is opened to release the traveling block from the kelly. Then the hook swivel is unlatched.

The drilling assembly is normally pulled out of the hole in stands (thribbles). If there is an extra single in the drillpipe assembly, it is laid down at this time. The elevators are turned to face the drawworks and are latched around the drillpipe below the tool joint. The slips are pulled out of the rotary as the assembly is picked up. The pipe is picked up until the tool joint on the bottom of the single is about 2½ ft above the rotary where the slips are reset. Then the tool joint is loosened and unscrewed in the same manner as when making a drillpipe connection. The bottom of the single is pushed out the V-door and is allowed to slide down the ramp as the single is lowered. Finally, the elevators are unlatched to release the single and are turned to face the working position of the derrickman.

The platform that the derrickman stands on, the monkey board, is about 85 ft above the rig floor and is usually on the V-door side of the mast. Small steel pipes or *fingers* are located on each side of the monkey board and are used to rack the top of the stands of drillpipe. On shallow to medium-depth holes, all of the drillpipe is normally racked on the pit side of the mast, or the right-hand side of the derrickman. On deeper wells the drillpipe is racked on both sides of the monkey board.

While the rotary helpers set back the kelly, lay down extra singles, and perform other work prior to starting the trip, the derrickman fills the trip tank and lines the pump to fill the hole. This includes opening and closing the proper valves so the mud-pump suction is connected to the trip tank, the main standpipe valve is closed, and the fillup line valve is opened. The shale shaker is bypassed, the shakers are shut off, and the screens are cleaned with a washdown hose. If the screens are not washed, the mud may dry on them during the shutdown period

311
Surface Hole Section

FIG. 7–6 Disconnecting and setting back the kelly. The drillpipe slips (a) are set in the rotary (b). The kelly is disconnected (c) and is set back in the rathole (d). Note the dope bucket and paddle behind the slips (a) used to grease the connections.

while tripping. It is often hard to clean off, this dry mud when circulation is resumed.

The derrickman climbs the ladder located on the side of the mast to the monkey board.*

Two to five bbl of water are often run into the drillpipe with the rig washdown hose as the drilling assembly is pulled from the hole. The water cleans the inside walls of the drillpipe and prevents mud from running on the threads of the tool-joint box, ensuring a cleaner, better connection. Also, if the drilling assembly starts to become wet while being pulled from the hole, water instead of mud will run out of the assembly as a stand is disconnected. The water is less dangerous than mud and can alert the crew of a problem. Then the mud box can be used when subsequent stands are disconnected, or the assembly can be slugged again.

Stands of drillpipe are pulled by latching the elevators below the tool joint of the pipe in the rotary. The slips are pulled as the assembly is picked up, and a stand of drillpipe is pulled from the hole.

As the last single of the stand passes through the rotary, the rotary helpers place the drillpipe tongs on the single with the breakout tong on top. The tongs are latched but are not gripped. As the tool joint is pulled above the rotary, the breakout tongs are gripped on the pin end of tool joint. The makeup tongs, used here as backup tongs, are gripped on the box end of the tool joint. The slips are set around the drillpipe with the tool joint about 2½ ft above the rotary. Then the drilling assembly is slacked off so the weight of the assembly is supported by the slips.

The breakout tong is pulled by the breakout cathead to loosen the friction grip of the tool joint. In some cases hydraulic drillpipe tongs are used to break or spin out the tool joint. In most cases, though, the backup tongs are unlatched, and the tool joint is disconnected by rotating the assembly with the rotary. A light upward tension is maintained on the stand. When it disconnects, the pin jumps out of the tool-joint box.

At this time the top of the stand is about 2–3 ft above the derrickman's head when he is standing on the monkey board in the mast or derrick. The derrickman uses a pullback line with one end tied to the mast to help handle the top of the stand. He throws the free end of this piece of ¾–1-in. rope around the drillpipe and pulls on the free end. In the meantime the rotary helpers swing the lower end of the stand of drillpipe over to the pipe rack while the driller lowers the stand. The bottom end of the stand is set on the pipe rack, and the derrickman pulls the top end of the stand to one side with the pullback rope.

---

*Most rigs are equipped with a safety, counterweighted climbing line. After the derrickman reaches the monkey board, he puts on a safety belt.

FIG. 7–7 Pulling a stand of drillpipe. The elevators are latched onto a stand of drillpipe (a). The stand is pulled and disconnected. The bottom of the stand is set on the floor pipe rack (b) by the rotary helpers. The derrickman on the monkey board unlatches the elevators from the top of the stand (c) and lays the stand back in the pipe fingers.

After the stand sits on the pipe rack on the rig floor, the elevators begin to slide down the drillpipe. When the elevators are about level with the derrickman's shoulders, he opens or unlatches the moving elevators with one hand while holding the pullback line with the other hand. As soon as the elevators are unlatched, the top of the stand is pulled out of and away from the descending elevators and traveling block. The top of the stand is moved into the pipe-racking finger in the finger board where it is locked in place until needed.

The elevators are lowered to the rig floor, and the rotary helpers latch the elevators on the drillpipe below the tool joint. Another stand is pulled from the hole, disconnected, and set back on the pipe rack. This process is repeated until all of the drillpipe has been pulled out of the hole.

Pulling the Drill Collars

The drill collars are pulled from the hole in a similar manner to pulling the drillpipe. Regular drill-collar tool joints do not have a shoulder that can be caught by the elevators. However, the drill-collar button effectively provides a shoulder for the elevators to catch. Since the drill collars are larger in diameter than the drillpipe, larger elevators must be used with a drill-collar button.

Lift nipples that usually fit the drillpipe elevators can be used instead of buttons. Bottleneck drill collars have a recessed area below the tool joint. The reduced outside diameter is usually set in the rotary to be caught by the slips. The area does not normally have sufficient indentation so the elevators can be latched around the recessed area to lift the drill collar. Instead, bottleneck drill collars are usually handled with buttons or lift nipples. Drill-collar clamps should always be used when handling drill collars, even the bottleneck type. The clamp should be latched around the drill collar as near the slips as possible and should be securely fastened. It should be tapped lightly with a hammer as it is tightened to ensure that it is snugly fitted to the drill collar.

Since the drill collars are larger than the drillpipe, different sized slips must be used. Depending on the diameter of the drill collars and the drillpipe tool joints, the drillpipe tongs may fit the drill collars or it may be necessary to use larger jaws or additional links. Except for the differences noted, the drill collars are handled in a similar manner to the drillpipe.

The first drill-collar stand normally has one and possibly two drillpipe singles. The stand is pulled up into the mast in the usual manner. Drill-collar slips are set, and the drill-collar clamp is fastened on the collar. The drill-collar connection is loosened and then is rotated free. The drill collars are rotated very slowly, and the drill-collar stand is held with a minimum amount of upward tension. After the tool joint is rotated free, the drill collar is picked up a short distance for the pin to clear the collar. The stand is then set back similar to a stand of drillpipe.

Since the drill collars are considerably heavier than the drillpipe, it may be necessary to use a lifting line or other means to help move the drill collar to the racking position. One method is to set down the bottom of the drill collar near the front of the drawworks. The top part of the drill collar is allowed to lean over toward the derrickman where it is snubbed with the tieback line. The collar is then picked up a short distance, and it swings over to the racking position because the top of the collar is snubbed over. After the bottom of the collar is set down in the correct racking position, the elevators are unlatched and the top of the collar is moved into position. A lift nipple or drill-collar button is made up on the drill collar in the rotary. The elevators are latched on it and the procedure is repeated until all of the drill collars have been pulled from the hole.

Sometimes, the bit or other tools on the bottom of the drill collars will not pass through the split rotary bushings. Then one or both of the bushings are pulled. After the tool has been pulled through the rotary, the bushings are replaced and the hole is covered with either a bit-breaker block or the hole cover plate. The hole cover plate is a piece of heavy steel usually about ½ in. thick and fitted with handles. The plate is cut so it fits in the square of the rotary. It is important to keep the hole covered to prevent dropping tools or other metal objects in the hole that may cause a fishing job. Blowout preventers are not normally used while drilling the shallow surface hole. However, when preventers are installed and the pipe is out of the hole, the blind or blank preventor rams should be closed.

Most trips are made to change bits. The bit-breaker block is set in the rotary square, and the drill collar is lowered slowly to set the bit in the bit breaker. The rotary is locked, and the bit is loosened by turning the drill collar with the break-out tongs. The drill collar with the loosened bit is then picked up and held over to one side of the rotary while the bit is rotated out by hand. A new bit is set in the bit-breaker block, and the drill collar is set down slowly over the bit pin. The drill collar is rotated with a pair of chain tongs until the bit is hand tight. The makeup tongs are then used to tighten the bit connection to the correct torque. The used bit is graded for out of gauge, bearing wear, and tooth wear according to standard specifications, and grading results are recorded on the drilling report (Fig. 7–8).

Sometimes stabilizers, reamers, keyseat wipers, and other tools are run in the drill-collar assembly. The tools are checked for wear either while the drill collars are being pulled from the hole or when the drill collars are run back in the hole. If the tools are worn to the point where they are undergauge, below the manufacturer's recommendations, i.e., where the outside diameter of the tool is less than the original gauge or new diameter of the tool, they are replaced.

Running the Assembly in the Hole

After the assembly is pulled, the hole is completely filled with mud. The mud

FIG. 7–8 Used bits are graded and stored

level is checked periodically to ensure that the hole is neither losing fluid nor beginning to flow, which could indicate a possible approaching blowout condition. Before running the assembly back in the hole, extra equipment is returned to its correct storage position and the floor is cleaned with the washdown hose. The pins of the racked drillpipe and drill collars are washed thoroughly to remove mud and other debris.

After all of the equipment changes have been made, the assembly is run back in the hole in the reverse order in which it was pulled. Any changes in the drilling assembly are made as the assembly is pulled or run in the hole, whichever is most convenient. If blowout preventers are being used, they are opened and the hole cover plate is removed.

The first stand of drill collars is lowered into the hole until the top of the stand is about 2½ ft above the rotary. Slips are set, and the drill-collar safety clamp is installed. The lift nipple is loosened, disconnected, and lifted out of the drill collar. Then it is replaced in its usual storage location. The elevators are pulled upward and are stopped opposite the derrickman so they can be latched onto the lift nipple of the next stand of drill collars run in the hole. The derrickman passes the pullback line around the next drill collar to be run. The drill collar is untied, and the top is allowed to swing slowly or to move toward the elevators by gradually slacking off on the free end of the pullback line. When it is in position, the elevators are latched.

Large, heavy drill collars may be handled by passing the free end of the pullback line around the drill collar and over a brace behind the derrickman to improve the leverage in slacking the line. This allows the drill collar to move toward the elevators. When the drill collar is in position, the end of the pullback

FIG. 7–9 The drillpipe (right in a) and the drill collars (left in a) are racked (stood) back in the mast on the trip out of the hole. On the trip in the hole, the derrickman latches the elevators around the top of a stand of pipe (b), which is connected and lowered into the hole (c). After the pipe is run into the hole, the kelly is picked up and connected (d) and drilling or other downhole operations are resumed.

line is tied to the brace. The elevators are then latched, and the end of the line is untied with the derrickman still holding the drill collar in place.

The drill collar is raised slowly while several floormen hold the bottom of the collar to prevent it from swinging over to the center of the hole too rapidly. A *snub line* is used to help hold the collar and prevent it from swinging if necessary. If the derrickman does not hold the top of the drill collar while it is being picked up, the top of the heavy collar will swing the elevators over to the opposite side of the mast. The drill-collar pin can hit a metal object on the floor and can damage the threads on the pin. Or the swinging drill collar can hit the collar sitting in the rotary. The vibration may damage the slips and allow the drill collar in the rotary to fall down the hole, resulting in a fishing job.

The drill-collar pin is slowly lowered into the tool-joint box of the drill collar sitting in the rotary. The stand of collars is connected by rotating them in the elevators with a hand chain tong or hydraulic tong. Sometimes the rotary is reversed and is turned slowly to connect the stand. The tool joint is then tightened to the required torque using drillpipe tongs fitted with jaws to catch the drill collar. The drill-collar clamp is removed, and the assembly is picked up a short distance to pull the slips. Then it is lowered in the hole until the top tool joint is about 2½ ft above the rotary.

This procedure is repeated until all of the drill collars have been run in the hole. The drill-collar equipment is set aside, and the drillpipe running equipment is installed. The hydromatic or dynamatic clutch is engaged to help control lowering the increasing weights and conserve the drawworks brakes.

The stands of drillpipe can be picked up and connected much faster than the drill collars because they are lighter and smaller, and they are relatively easy to move and hold. Drill-collar clamps are not used.

Using the pullback line, the derrickman moves the top of a stand of drillpipe to the front of the monkey board. Below, the driller pulls the elevators up through the mast while the derrickman moves the top of the stand into position to catch it with the elevators. As the traveling block passes by the derrickman, the top of the stand of drillpipe is laid into the pipe guide, located on the side of the spring swivel below the traveling block. The pullback line is dropped, and the derrickman catches the handles of the moving elevators and latches them onto the stand of drillpipe below the tool joint. This is called catching the stand on the fly.

As the traveling block continues its upward movement, the stand is picked up until the bottom is above the tool joint in the rotary. The derrickman steadies the top of the stand. A rotary helper or pipe racker holds the bottom of the stand to prevent it from swinging and moves it toward the rotary. The tool-joint pin is stabbed, or set into the tool-joint box in the rotary. The stand is connected by using either a spinning chain or hydraulic tongs, and the tool joint is tightened to the correct torque.

After the connection has been tightened, the assembly is picked up a short distance to pull the slips. The stand is then lowered into the hole, and the slips are set when the tool joint on the top of the stand is about 2½ ft above the rotary. The elevators are unlatched and are pulled up to pick up another stand of drillpipe, which is then connected and lowered into the hole. The procedure is repeated until all of the drillpipe has been run into the hole.

The last stand is run in the hole, and slips are set. Singles that were laid down before the trip are picked up and run in the hole. The kelly is picked up from the rathole and is connected to the drilling assembly. In the meantime the derrickman has climbed down from the mast and has aligned the main pump for drilling by opening and closing the correct valves. This procedure is called placing the pump on line. It includes closing the fillup line valve and opening the main kelly standpipe valve. The pump suction valves are changed so the pump can take mud from the main suction mud pit. The pump is started slowly, and the flow line is watched to ensure that circulation is established. The pump rate is slowly increased to the normal pump rate.

The new bit must be drilled in or set in to start drilling. The bit is set in by starting to drill with a low bit weight and rotary speed and gradually increasing these to normal. This allows the bit to form a drilling pattern on bottom and to begin drilling without causing tooth breakage or other damage to the bit. More precautions must be taken in the drill-in procedure when using longer-toothed bits, diamond bits, and drilling in harder formations. In the normal drill-in procedure the bit should be at full operating conditions of bit weight and rotary speed by the time the bit has drilled from 1–3 ft.

Precautions During Tripping

The last 3–5 stands of drillpipe should be lowered into the hole at a slower-than-normal rate. Drill cuttings and caving material can settle to the bottom of the hole when mud circulation is stopped to trip the assembly. As the assembly is tripped out of and in to the hole, it can scrape and dislodge small pieces of formation from the walls of the borehole. This contributes to hole fillup, accumulations of formation debris in the bottom of the hole. If the bit is run into this material without circulating, the bit jets can become plugged and another trip will be required to unplug them. In the worst case the bit can become stuck, causing a fishing job.

If the last bit pulled from the hole is undergauge—worn to a reduced diameter—there is a good possibility that the last part of the hole drilled by the undergauge bit is undergauge. The lower hole section will be tapered from full gauge to some diameter less than full gauge. The new bit will be full gauge. If it is lowered into this tapered section by running into the hole too fast, the bit can be pinched, breaking the bit shanks or damaging the bit cones. This could reduce the drilling life of the bit and could leave the bit cones or shanks in the hole, requiring a fishing job. Or the bit could become stuck, also causing a fishing job.

The best method to eliminate problems associated with cuttings fillup and tapered hole is to ream the bottom 50–100 ft of hole after the trip. The circulating mud washes out any cuttings that may have accumulated. The rotating bit drills out the undergauge hole section to full gauge.

Tripping speeds are the rate at which the assembly is pulled out and run back in the hole. Excessively fast tripping speeds can cause a number of problems. The main problem while pulling the assembly out of the hole is *swabbing*. Swabbing occurs when a large assembly, relative to the inside diameter of the borehole, is pulled from the hole. There is insufficient clearance for the mud to bypass the assembly if it is pulled out fast, and the mud tends to flow out of the hole with the assembly. This is evidenced at the surface by mud returns in the flow line and possibly by increased apparent drilling assembly weight while pulling the assembly. If high-permeability formations are exposed in the open hole, they can be swabbed in. If a sufficient volume of mud in the hole is displaced by these lighter formation fluids and the formations have sufficient pressure, a blowout can occur.

A *balled-up* bit or other part of the downhole assembly can cause swabbing. The ball cuttings stuck to the assembly, can be removed by spudding the assembly and maintaining good mud properties. Swabbing action can also be controlled by regulating the rate at which the assembly is pulled from the hole.

When the assembly is tripped in the hole too fast, one problem is the increased pressure in the mud column below the assembly as a result of the plunger effect as the assembly is lowered in the hole. In the minimum case this can cause pressure surges, which tend to decrease wellbore stability and cause increased caving. In more severe cases pressure surges cause lost circulation, resulting in the attendant problems to restore circulation.

As the drilling assembly is run into the hole, it displaces an equal volume of mud that overflows out the flow line. This should always be monitored on tripping to ensure that the hole is full of mud. Any excessive flow should also be noted since this can indicate a flow of formation fluids into the wellbore and a pending blowout situation.

While tripping in the hole, the mud may flowback over the top of the drillpipe as the assembly is lowered in the hole. This is caused by running the drilling assembly in the hole too fast, running an oversized (large diameter) drilling assembly compared to the inside diameter of the wellbore, and using mud in poor condition. Flowback sometimes occurs when the assembly has been out of the hole for a long period of time and the mud in the hole has not been circulated and conditioned. Flowback creates hazardous working conditions for the drilling crews and can also lead to a blowout condition.

Bleed plugs can be used to help control flowback or to permit running the assembly in the hole without creating a hazard for the drilling crews. Bleed plugs are similar to drill-collar buttons, except they have threads for either drill collars or drillpipe. The top of the plug is sealed except for one small opening fitted with a valve. In operation after the stand of drillpipe is connected, the derrickman screws the bleed plug into the tool joint on the top of the stand of drillpipe and closes the bleed-plug valve. The stand is then lowered into the hole. After the slips are set and the elevators are removed, the bleed plug valve is opened to release any pressure before it is disconnected and removed from the tool joint.

Normally, two bleed plugs are used. While one plug is being disconnected from the tool joint in the rotary, the other plug is being tied to the elevators and carried to the derrickman to be used on the next stand of drillpipe. It takes more time to run the assembly in the hole when bleed plugs are used. After the drilling assembly has been run in the hole several thousand feet, flowback may be prevented by circulating the mud at that depth. If this does not stop the mud flowback, a weighted barite pill similar to a slug pumped into the drilling assembly sometimes prevents mud flowback.

When tripping the drilling assembly, the operator must always watch for excess drag, keyseats, bridges, and other downhole obstructions. Excess drag is indicated by increased weight pulling out of the hole and decreased weight running into the hole. Keyseats are indicated by excess drag, often at periodic intervals of about 30 ft as the drillpipe tool joints tend to hang up. Bridges are generally encountered on a trip in the hole. If these obstructions occur, it may be necessary to ream or drill them out because they can cause stuck pipe and a possible fishing job. The risk can be reduced by taking the preventive action noted.

SURFACE HOLE DRILLING PROBLEMS

The surface hole section normally is drilled faster and with fewer problems than the other hole sections. The surface hole is deep enough so all of the drill collars are in the assembly. This supplies sufficient bit weight to obtain a maximum penetration rate. The formations are usually softer and more drillable than the deeper horizons. Deviation problems may occur, but normally they are less severe than the problems in either the conductor hole or deeper hole.

The unconsolidated gravels and conglomerates that occur in the conductor hole are either absent in the deeper surface hole or are consolidated and can be drilled faster and easier. There is less tendency for bit balling to occur. When it does, it can be controlled without excessive problems.

Excess Cuttings in the Mud

Sometimes the fast drilling rates attainable in the surface hole can overload the mud system with drill cuttings. This can be easily controlled by reducing the drilling rate, but this control method is used last since the objective is to drill the hole as fast as possible in accordance with prudent operations.

Excess cuttings in the mud cause flowback while making drillpipe connections. Cuttings can overload the shale shaker and cause a loss of mud. They increase the apparent density of the mud and can contribute to lost circulation. They also increase the tendency for accumulating by balling on the bit and drilling assembly. In extreme cases excess cuttings in the mud can cause a stuck drilling assembly. When this problem occurs, it is controlled by increasing the circulation rate if possible.

If the problem is caused by a high cutting-slip velocity, the mud viscosity may be increased. The cutting slip velocity is the rate of fall of the cuttings in the mud. If there is a high cutting-slip velocity, the cuttings can be retained in the mud for a longer period at a given circulation rate. The slip velocity is reduced by increasing the viscosity. The overall time that the cuttings are retained in the mud system is further decreased by increasing the circulation rate or annular velocity, the calculated rising velocity of the mud in the annular space based on the volume of mud being circulated and the cross section of the annular space.

Another problem with formation drill cuttings that sometimes occurs, usually at deeper depths, is the accumulation of cuttings in a washed out or out-of-gauge section of hole. Mud tends to erode some formations, and this in combination with possible caving can create a large-diameter section of hole. As the mud is circulated upward in the annular space, the annular velocity is reduced in an out-of-gauge hole section, sometimes called a *possum belly*. Formation drill cuttings can accumulate in these as a result of the decreased annular velocity. The cuttings may be retained for a period and then fall into the hole, frequently when the mud pump is shut down. This causes an excess cuttings accumulation in the mud and in severe cases can cause stuck drillpipe. The cuttings accumulation in this case is not necessarily caused by fast drilling, although the resulting action may be similar.

Viscous mud sweeps remove the cuttings from an out-of-gauge hole section and carry them to the surface (Fig. 7–10). The viscous mud sweep is mud that has been treated to increase the viscosity. Frequently, fibrous lost-circulation material is added. The viscous mud is normally mixed in the trip tank or a similar

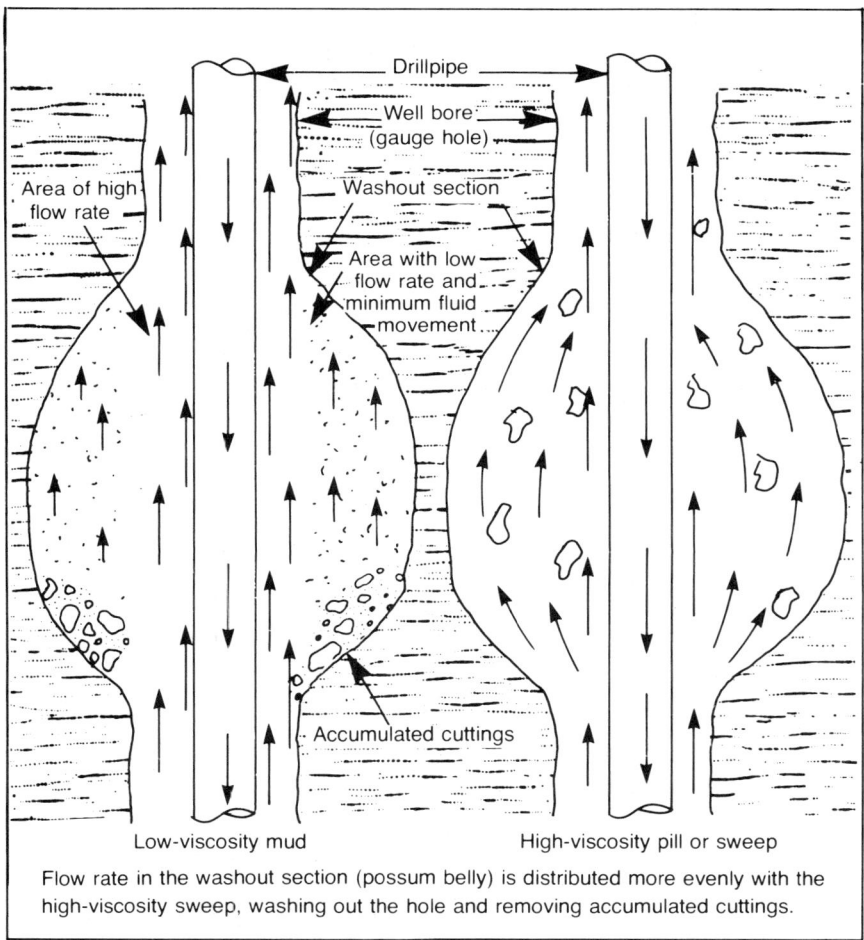

FIG. 7–10 Viscous mud sweeps

mixing tank. Mud from the mud system is pumped into the tank. Gel, fibrous lost-circulation material, etc., are added and mixed. The viscous mud is then pumped into the hole, followed by regular mud. As the viscous mud circulates through the mud system, it tends to pick up and carry any accumulated drill cuttings. The drill cuttings and fibrous lost-circulation material, if used, are separated from the mud at the shale shaker, and the viscous mud is then mixed with the mud in the system.

Shallow Gas Flows

One of the most dangerous surface-hole drilling problems is shallow gas flows. Fortunately, these are uncommon. However, when encountered, they can be extremely dangerous. They can normally be handled or controlled with adequate planning and using the correct equipment and operating procedures.

In some areas natural gas accumulations occur at shallow depths. They usually have a limited volume. The absolute pressure is normally relatively low but may be abnormal for the depth. With the lightweight muds used to drill the surface-hole section, the gas may have sufficient pressure to push the mud out of the hole. After the mud flows out of the hole, the gas can continue to flow, carrying formation debris, rocks, and gravel. The debris and flowing gas stream cause the surface equipment—the drilling lines, tong lines, and stands of pipe in the mast—to blow about and move rapidly. Metal striking metal can cause a spark and ignite the flowing gas stream. This is a blowout.

Although all blowouts do not catch on fire, many do. The forces involved and the short amount of time in some cases to clean the hole are almost unbelievable. In some cases the drilling assembly has been blown out of the hole. More commonly, though, the mud is blown out of the hole and the gas catches fire.

Blowouts from shallow depths are doubly dangerous. When the blowout occurs from a deeper depth, there is normally enough warning while the mud is flowing out of the hole to close the blowout preventers and prevent the blowout. When the blowout occurs from a shallow depth, there is only a limited amount of warning, and frequently, blowout preventers have not been installed because they would not be effective. For example, the surface-hole section is frequently drilled below shallow conductors set at 50–100 ft. If preventers are placed on the conductor pipe and there is time to close the preventers, the blowout could possibly blow the conductor pipe out of the hole if it was not fitted with a bypass line.

The normal and recommended procedure is to use a diverter on the conductor casing above the flow line. The diverter can be a rotating head, bag-type preventer, or similar equipment. If a blowout occurs, the mud and gas will be diverted out the flow line. In case of a fire, there is a reasonable chance that there will be no personnel injuries and only minor equipment damages.

Mud-Cake Buildup on Porous Sands

Thick, porous sands are frequently encountered while drilling the surface-hole section. Usually it is not necessary to maintain a high-quality mud while drilling the surface-hole section. In fact, native mud is often used. Low-quality muds generally have high mud solids and tend to build a thick, spongy mud cake. When drilling porous sands or after the sands have been drilled, a thick mud cake can build up on the surface of the sand. In some cases the wall cake buildup can become excessively thick and can restrict the inside diameter of the hole. This is seldom enough constriction to affect the drillpipe tool joints, but frequently it affects the passage of the drill collars or bit. On a trip in or out of the hole, the drilling assembly can become stuck in these restricted sections.

The problem can be handled by various methods. One method is to improve the quality of the mud so a thin, tough wall cake is formed. This mud is sometimes expensive, but it is certainly less costly than the fishing job when the assembly becomes stuck. Usually the problem can be handled by tripping the drilling assembly out of and into the hole at a slower rate and watching for the restriction as indicated on the weight indicator.

If the condition occurs on a trip out of the hole, the assembly can be worked past the zones by careful operations. If mud-cake buildup is encountered on a trip in the hole, it can be removed by reaming. Sometimes the problem can be detected and eliminated during drillpipe connections. Before making the connection, the assembly is pulled up to the extreme limit of the mud hose and is rerun to the bottom. If no excess drag is encountered, it can be assumed that the hole section below the top of the drill collars is not severely restricted. The accumulation usually tends to decrease with time so there is less risk of a hole restriction above the top of the drill collars.

Running and Cementing Surface Casing

After the hole is drilled, it must be cased and cemented prior to drilling deeper. Running and cementing casing is an important and sometimes critical part of the drilling operation. All precautions must be taken to ensure the casing is run and cemented correctly. The following detailed procedure generally applies to all casing strings. In some cases such as running shallow casing strings, it may not be necessary to perform all of the steps as listed in the procedure. However, the operator should be aware of the various steps and the reasons for them. He can then judge their applicability to the specific casing job being considered.

The first cementing job performed with the casing in the hole is called the primary cement job. This is the only time the entire casing and annular space is open so fluids, including cement slurry, can be pumped with a minimum of restriction and displaced correctly into the annular space. The various factors that contribute to a good primary cement job include the following:

1. Drilling a straight, vertical, clean, in-gauge hole
2. Using drilling mud in good condition
3. Rotating or reciprocating the casing while pumping and displacing cement

4. Having a well-planned cementing program including the correct volume and type of cement and additives
5. Using good mixing water and pretesting the cement and water
6. Using good operating practices and procedures while cementing casing.[1]

These objectives may not all be obtainable, but casing can still be cemented satisfactorily. For example, it is possible that the casing cannot be rotated and reciprocated while being cemented. However, if the mud and hole are in good condition, usually the cementing job is good. Out-of-gauge holes may be a problem, but a good cement job can usually be obtained by having good mud, allowing for the out-of-gauge hole section when calculating the cement volume, and working the casing (i.e., rotating and reciprocating while cementing).

The general cementing program is designed and included in the drilling program. After the hole has been drilled and logged, the actual depths and hole sizes from the caliper logs (preferably a 3-arm caliper) are used to calculate the final cement volumes. Bottom-hole temperatures from the logs may be used as a guide to adjust the final amount of cement retarders or accelerators. Cement additives are needed to allow sufficient time to mix and pump the cement and to ensure that the cement hardens within a reasonable period. If the depths are such that the hydrostatic head of the cement column exceeds the fracture pressure of the formation, expanders such as bentonite or vermiculite may be used to increase the volume and effectively reduce the cement density. Fine-mesh sand may be added to increase the strength of the cement.

The volume of cement slurry required is a simple calculation. For example, assume a string of 9⅝-in. casing is run in a 12¼-in. drilled hole and the bottom 2,000 ft of the casing is to be cemented. The actual diameter of the 2,000 ft of hole to be cemented as determined from the caliper log is 15½ in. The volume of cement required in this case would be 1,609 cu ft. A 20% excess would increase the cement requirements to 1,931 cu ft. Each sack of cement will yield a fixed volume of cement slurry. The volume per sack would depend on the amount of mixing water used and the cement additives used. The cementing service companies have standard tables that give the yield volumes of various cements with various additives.

Shallow conductor and surface casing holes are not normally logged. These hole sections are frequently out of gauge with large washed-out sections. The entire casing section is usually covered with cement to circulate cement. Common practice is to use about 100% excess cement or more, depending on experience in the area.

[1]Dwight K. Smith, *Cementing,* No. 4 (Dallas: American Institute of Mining, Metallurgical and Petroleum Engineers, 1974).

Cement pumpability tests should be run late in the drilling operation when the hole is approaching the casing point depth. Pumpability tests are also called thickening or hardening tests. The purpose of the test is to determine how long the cement slurry will remain pumpable after it is mixed. This test allows final planning on the actual cement operation and ensures that there is sufficient time to mix, pump, and displace the cement into its final position, based on the volumes to be pumped and the pumping rates achievable by the rig and cement pumps. The sample mixing water should be from the same source as the water that will be used to mix the cement. The cement sample should be taken from the same silo (cement storage) where the cement to be used on the approaching cement job is stored. Cement additives should be included in the correct portion, and the sample should be properly mixed and tested.

When the cement is ordered out for the job, the additives are blended into the dry cement that is loaded into bulk transport trucks. Pumpability tests should be run on cement samples taken from the transport trucks. These tests verify the earlier results and ensure that the correct cement and amount of additives have been selected and mixed. At least three dry cement samples should be collected from the bulk trucks and saved. At least three wet samples of the cement slurry should be collected from the mixing tank during the cementing operation: one near the start of the job, one in the middle, and one near the end of the cementing job. The wet samples give an idea of how the cement is setting up or hardening after making allowances for temperature. The dry samples are saved for additional testing in case future operations indicate that there may have been something wrong with the cement.

Sometimes, two different types of cement are used in the same cementing job. For example, it is common to use a lightweight, low-density cement followed by common cement. Each type of cement, including volumes and additives, is calculated separately. The two types of cement are run consecutively in one cementing operation or stage.

Centralizers and scratchers are frequently used on casing. Centralizers are designed to hold the casing away from the wall of the wellbore so cement can be displaced completely around the casing to improve the quality of the cement job. Centralizers are conventionally installed on the bottom two or three joints of casing and may be installed on every joint or on alternate joints through the prospective productive horizons. The number of centralizers needed can be calculated based on such factors as pipe weight and hole angle.

Scratchers scrape the wall cake off the borehole while reciprocating and rotating the casing. They provide a better bond between the cement and the formation. Some scratchers are designed to scratch the formation during reciprocation; others are used during rotation; and some are universal. The use and placement depth of centralizers and scratchers are included in the final casing-cementing program.

Receiving Casing at the Rig

The casing is usually received on the location several days to several weeks before it is to be run. The rig normally is drilling or sometimes is being rigged up in the case of conductor or shallow surface casing. The casing is unloaded on an empty pipe rack or on top of pipe that will not be used before the casing is run. It is unloaded with the collars pointed toward the rig. Although casing is normally loaded on transport trucks with the collars pointed toward the cab of the truck, on tight locations there may not be sufficient room to turn the trucks around to unload the casing correctly. In this case the shipper may be requested to load the casing with the pin or male end forward, toward the cab of the truck.

Casing is unloaded with a crane or is rolled over boards that bridge the gap from the truck to the pipe rack. It must be unloaded carefully to prevent damage. This is especially important with high-grade steels that are more subject to damage because of their brittle, hard nature. It is also important to protect the threads, even though the casing is normally fitted with thread protectors on both ends before it leaves the mill.

Casing is unloaded on the racks in layers. Each joint is visually inspected. Bent joints are easy to detect as they are rolled across the pipe rack from their telltale wobble. Bad joints have obvious defects such as bends, flattened areas, gouges, and damaged threads. They should be returned on the same truck, or they should be separated from the good casing and marked with red paint or soft lined.

Any piece of large, heavy-duty twine such as a strand from the catline is called *soft line*. The soft line is wrapped around the joint several times and is tied securely. It is interesting to note that this practice of denoting unservicable joints by soft lining was started early in the history of the drilling industry. It is still a convenient method of marking bad joints since there is almost always a piece of soft line conveniently available around the rig, whereas paint may not be as readily available.

The joints are counted and numbered with chalk as they are unloaded. If the casing occupies more than one layer on the pipe rack, it is measured after each layer is unloaded and the measurements are recorded in the pipe tally book. This procedure is commonly called tallying the casing or measuring the casing on the rack. The casing is measured in feet to the nearest one-hundredth of a foot. When measuring the casing, two people should always read the tape. The person recording the measurement reads the measurement aloud and records it in the tally book. The second person reads the measurement at the same time and makes corrections or requests a double check of the measurement if needed.

Casing is measured at the factory and is sold with *threads-on* measurements. The threads-on measurement is the measure of the entire length of the joint, including the male thread. For drilling purposes, casing is measured excluding the length of the male thread or the *thread-offs* length. The length of the male thread is known as *makeup* and is not counted in the actual casing measurement since it is screwed into and actually included as part of the collar or box end. The threads-on measurement is about 1½–2½% longer than the threads-off measurement, depending on the length of the joint and threads. The casing measurement should be checked to see if it agrees with the shipping measurement after allowing for the difference. It should also be checked with the casing program to ensure that sufficient casing is on hand.

About 3–5% excess casing based on joint count should be allowed for casing damage during running. Casing is normally received in joint lengths of 30–45 ft. All of the casing should be about the same length within a few feet. This makes it easier to connect the joints when the casing is run.

At a convenient time when personnel are available, the casing threads should be cleaned and greased. This treatment prevents delays when running the casing. The thread protector should be removed and the casing threads cleaned and greased with the correct lubricant. Then they are replaced but not tightened.

Inspecting Casing

Casing is inspected at the mill, although special inspections may be made at the storage yard. In some cases the casing is inspected and tested at the location before being run. The shallower casing strings usually have large safety factors and are not tested at the wellsite. However, they should be given a thorough visual inspection.

Deeper casing strings, especially in the larger sizes, are frequently inspected at the wellsite. These are subject to more severe conditions and sometimes run at relatively low safety factors. Wellsite inspections range from an end-area inspection, where the threaded ends are inspected, to a complete body inspection, including measuring the drift diameter and wall thickness and checking for structural defects such as mill scale or slag inclusions. Sometimes the casing is burst tested by pressurizing it with water to or near the rated burst pressure.

Conditioning the Mud and Hole

The mud and hole must be in good condition to run casing. Otherwise, casing may become stuck at a shallower depth above the total depth. Casing is susceptible to wall sticking, hanging up on a bridge, and sticking because of drill cuttings and cavings accumulation around the casing. Once it becomes stuck, there is very little chance of releasing it. Usually the only alternatives are (1) to cement the casing and drill ahead with the attendant high risk of not successfully drilling and completing the hole at the required depth, (2) to revise the drilling program to allow for running an additional string of casing, if this can be done, and (3) to abandon the hole and prospect or move the rig a short distance and redrill. The latter options are very expensive. They also emphasize the importance of conditioning the mud and hole so the casing can be run to total depth and allowing for an extra string of casing or alternative drilling program when the original drilling program is prepared.

Conditioning the mud and hole starts with observing the condition of the hole during trips before reaching the casing point, the depth at which casing is to be run. If there is pipe drag as indicated by fluctuating weights while pulling the pipe from the hole, the hole should be reamed on the next trip in the hole. If the drag occurs in the same hole section on consecutive bit trips, it may be advisable to run a reaming drill-collar assembly. Bridges encountered on trips in the hole should be thoroughly reamed. If the hole problems continue, it may be necessary to increase the mud weight slightly or take other action before running the casing.

The drilling mud should always be maintained in good condition. As the casing depth approaches, additional attention is given to the condition of the mud. It should be treated as needed to ensure that it has the required physical and chemical properties and is in good condition. It is also a good practice to run one or more viscous mud sweeps periodically to check for drill-cuttings accumulation in washed-out sections.

In the final step immediately prior to running casing, the hole is circulated with the bit on or near bottom (5–10 ft off bottom) for one or two circulations. After this, a short trip—sometimes called a *wiper trip*—is made by pulling about 20 stands of drillpipe out of the hole and rerunning it. If pipe drag or bridges are encountered while making the wiper trip, or short trip, the hole is circulated for a short period and the trip is repeated. After the last short trip, the hole is circulated to remove all cavings and cuttings. At this point the mud and hole should be in condition to pull the assembly from the hole to run casing.

Pulling out of the Hole to Run Casing

After the mud and hole have been conditioned, the drilling assembly is pulled out of the hole before running casing. The assembly may be pulled and racked in the mast in a manner similar to tripping, or it may be laid down depending on several conditions.

If the drillpipe or drill collars are to be used to drill the next section of hole, they are racked in the mast in the same manner as tripping. A few operators believe that rotating the drilling assembly while disconnecting the stands may cause excess cavings. In this case hydraulic tongs are used to disconnect the stand, or it is *chained out of the hole*. In this procedure after the tool joint is loosened with the pipe tongs, a spinning chain is wrapped counterclockwise around the bottom of the stand. The end of the chain is connected to a pulling line from the cathead, and the tool joint is disconnected by spinning the stand with the chain. Obviously, this is slower than disconnecting the tool joint by rotating the assembly with the rotary.

If the next section of hole is to be drilled with different size drillpipe and/or drill collars, the drillpipe or collars are laid down by the drilling crew or by the drilling crew and a casing crew. Casing crews are extra personnel who help pick up and lay down long strings of drillpipe as well as pick up and run casing.

Sometimes in critical, high-pressure holes the time factor is important so there is not enough time to lay down the drillpipe and collars and safely pick up and run the casing. In this case the drillpipe and drill collars are pulled from the hole and racked back in the mast. After the casing has been run, the drillpipe is laid down in the mousehole or is run into the casing and laid down, depending on the size of the drillpipe tool joints and the inside diameter of the casing. Laying the drillpipe down in the mousehole is a slow, time-consuming procedure that is avoided when possible. The drill collars are normally too large to fit in the mousehole. When this happens they are usually laid down, even if the time factor is critical.

Another factor to consider in deciding if the drillpipe should be laid down is the risk that the casing will not run to bottom. If this happens the casing must be pulled out of the hole and the drillpipe run to clean out the hole before rerunning the casing. If this occurs and the drillpipe has been laid down, a considerable amount of extra time is needed to pick up the drillpipe.

Laying Down the Drillpipe and Collars   Before laying down the drillpipe, a barite slug is pumped so the pipe can be pulled dry. The pump is aligned to keep the hole full of mud. After setting the kelly back, the elevators lift the assembly until the lower tool joint of the first drillpipe single is about 2½ ft above the rotary. Slips are set, and the single is disconnected in a manner similar to that when making a trip with the assembly. A thread protector is screwed over the pin end to protect the threads. The bottom of the single is pushed out of the V-door as the driller lowers the single. When the elevators are about 5 ft above the floor level, they are unlatched and the joint slides out the V-door, down the pipe ramp, and onto the catwalk. It is stopped by a bumper block fastened near the end of the catwalk. The single is then rolled off the catwalk onto the pipe rack. The bumper block is positioned so it stops the sliding joint in a position where the joint can be rolled onto the pipe racks. Otherwise, the joint may slide too far and have to be pulled back.

The elevators are unlatched from the single and are latched on the drillpipe in the rotary. Then the process is repeated until all of the drillpipe has been laid down. Sometimes the slips are left in the rotary or are allowed to drag.

A laydown trough may be used. This is a heavy metal trough mounted on legs and set at an angle. One end is near the rotary and the other end extends over the edge of the rig floor. After the single is disconnected and the thread protector is screwed on, the end of the single is set in the trough and slides down the trough and out the V-door. This eliminates the work of pushing the bottom of the single from the rotary over to the edge of the rig floor.

A laydown line performs the same function as the laydown trough. The line is a steel cable that has one end clamped to the mast and the other end clamped to the end of the catwalk. A pulley rides on the cable, and the bottom of the pulley is fitted with a hook. The bottom of the disconnected joint of drillpipe is set into the hook. As the joint is lowered, the end of the single is carried out the V-door. The pulley or trolley may be fitted with a spring-loaded counterweight so it returns to the floor after the joint of drillpipe is laid on the catwalk.

The drill collars are laid down in a similar manner as the drillpipe. Since they are heavier, they must be handled with the catline or other lifting lines. Lift nipples or drill-collar buttons are used. The drill-collar safety clamp is always used when handling the collars. The drill collars are laid down in the reverse order of the procedure used for picking them up.

In the laying down procedure when the end of the drill collar reaches the catwalk, a *pipe dolly* is used to lift the end of the collar so it can slide out the catwalk. The pipe dolly is similar to a conventional two-wheel furniture dolly, except the bottom is fitted with a hook that is inserted in the end of the drill collar. The end of the drill collar is lifted by pushing down on the handle of the pipe dolly. The dolly wheels act as a fulcrum and allow the collar to slide onto the catwalk.

Rigging Up to Run Casing The first step in rigging up to run casing is to clean the work area. If the drillpipe has been laid down, the laydown equipment should be dismantled and moved off the floor or out of the way. Casing slips are set on the rotary; elevator bails are changed if necessary; and the casing elevators are hung. The casing pick-up line is clamped to the traveling block. The stabbing board is picked up, set in the mast, and clamped in place. The belly and safety lines are tied in the mast. The hydraulic casing tongs are picked up and hung on the makeup drillpipe tong line. The jaws or links on the lead tongs are changed for use as backup tongs. A flexible fillup line is connected to the standpipe. Other changes are also made as needed to run the casing.

All of the equipment should be visually inspected to ensure that it is in good condition. For extra-heavy casing loads the running equipment can be checked by ultraviolet light or sonic methods several days in advance. An additional double line may also be strung between the crown and the traveling block for heavy casing loads. When the casing is being run through blowout preventers, one set of the preventer rams is changed to fit the casing. A safety valve swaged to fit the casing is placed on the rig floor ready for use in case of a kick.

Running Casing

It takes a crew of 8–10 people to run casing. The driller operates the drawworks and supervises the overall operation. The derrickman works on the stabbing board to stab or align the casing and to latch the casing elevators around it. Three or four floormen are needed for the floor operations, and two or more people work on the pipe rack.

All of the connections including both ends of the collar on the first three joints of casing are cleaned and dried. When the joints are connected, they are coated with epoxy cement to lock the joint to prevent the joint from backing off while drilling out below the cemented casing. Sometimes the joints are tack welded, but the use of epoxy cement is becoming more common.

A cementing shoe is connected to the bottom of the first joint to be run, and a cementing collar is connected to the top of the joint. This *shoe joint* is picked up with a lifting line, pulled into the V-door, and lifted a short distance so the elevators can be latched below the cement collar. The joint is picked up, and the shoe is set in the slips and tightened with the tongs. The joint is picked up to remove the slips. Then it is lowered in the hole until the top of the joint is about 2½ ft above the rotary. The elevators are unlatched and picked up, and the cementing collar is tightened with the tongs.

In the meantime another joint of casing has been pulled into the V-door with a lifting line. The pickup line is latched around the joint immediately below the collar, and the joint is picked up into the mast. As soon as the casing joint has been pulled into the mast, another casing joint is placed in the V-door with a lifting line and the collar thread protector is removed.

The joint on the lifting line is picked up slowly until the bottom end clears the top of the pipe ramp. A tailback line (snub line) is used to hold the bottom of the joint and move it slowly toward the rotary. The pin thread protector is removed, and the bottom end is moved toward the center of the rotary. The joint is lowered slowly while the rotary helpers guide the pin end of the descending joint into the casing collar of the joint in the rotary.

At the same time the derrickman holds the top of the joint steady and loosens, but does not remove, the pickup line. He aligns the top of the joint with the joint in the rotary so the two joints can be connected. This is called stabbing. A casing tong is latched onto the top of the joint sitting in the rotary and is used to hold backups so the bottom joint cannot turn. A hydraulic tong is latched onto the

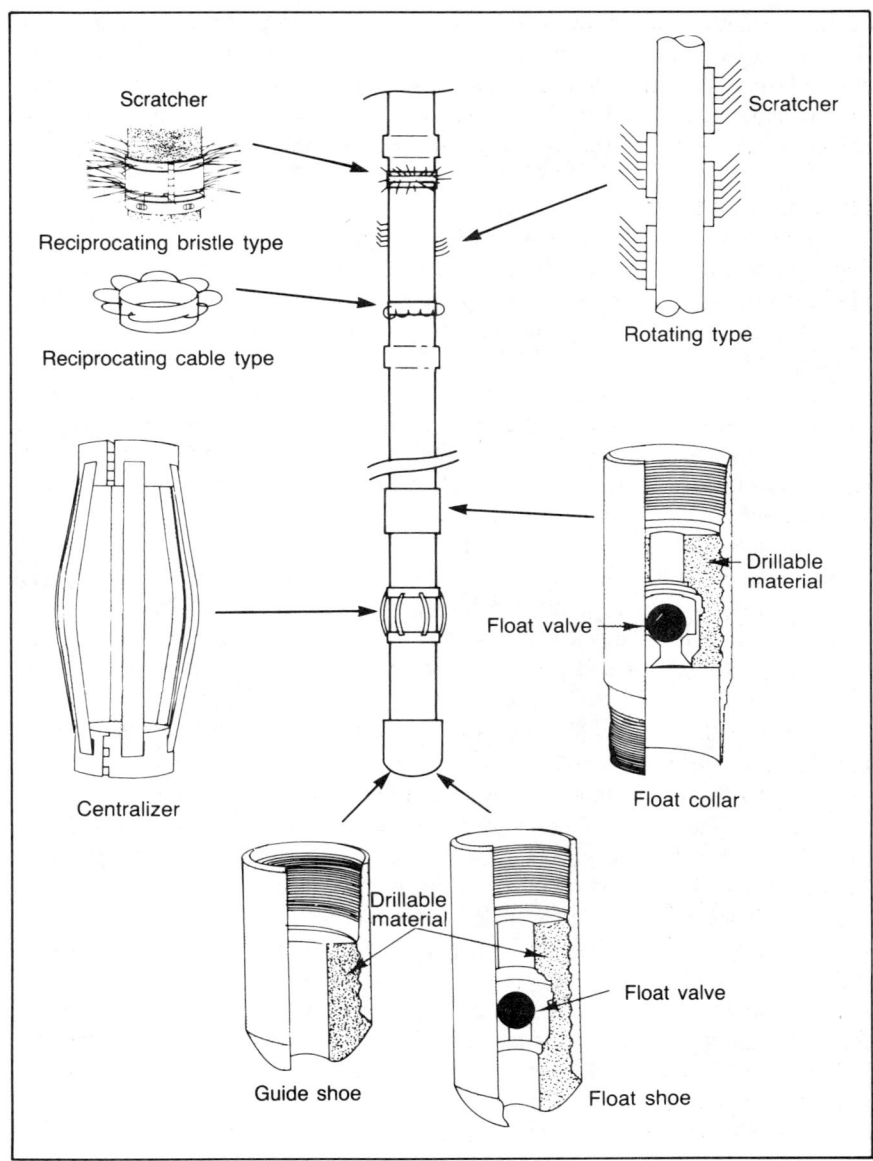

FIG. 7–11 Casing float equipment. Normally only one type of scratcher is used. The guide shoe is used on shallow casing, and the float shoe is used on deeper casing.

FIG. 7–12 (a) A joint is rolled onto the catwalk, picked up, carried into the V-door on a cable trolley, and raised with a pickup line connected to the traveling block. (b) The derrickman on the stabbing board aligns the joint and connects it to the joint sitting in the rotary. (c) The connected joint is lowered into the hole. (d) The joint in the casing elevators is lowered. The casing elevators are unlatched and the traveling block is started upward to run another joint of casing.

bottom of the upper joint to rotate it to make the connection. When the joints are partially screwed together and while completing the connection, the derrickman removes the pickup line from the top of the joint. The elevators are lowered and latched below the collar of the joint. The casing joint is tightened to the recommended torque rating with the hydraulic tongs.

A spinning line or endless rope can be used to rotate the casing joint to make the connection. The spinning line is a moderately long piece of rope with the two free ends spliced together. About four wraps are made around the casing clockwise looking down. The right or upper line is pulled by wrapping around the cathead. After the joint is spun up, casing tongs are used to tighten the joint with a jerk line. One end of the jerk-line rope is tied to the tong handle, and the other end passes over the drawworks cathead to pull the tong handle and tighten the joint. Hydraulic tongs have replaced this procedure in most cases.

After the connection is made using either of the previously described methods, the joint is picked up a short distance to allow the slips to be removed. Then it is lowered into the hole until the top of the casing joint is about 2½ ft above the rotary. The slips are set, the elevators are unlatched, and the pickup line is latched over the casing joint lying in the V-door. This joint is picked up and connected to the joint in the rotary in the same manner. This process is repeated until all of the casing has been run.[2]

Various precautions must be taken when running casing. It should not be run in the hole too fast since this can cause excess surge pressures and can break down weak formations and cause lost circulation. Then the casing cannot be cemented properly. This situation is similar to running drillpipe in the hole, except it is more severe because the casing diameter is usually only slightly less than the hole diameter. Standard formulas can be used to calculate the rate at which the casing can be lowered, called the *running speed,* without causing excess pressure.

When automatic fillup equipment is used, the mud level in the casing should be checked periodically to ensure that the equipment is working correctly. If the equipment is not used, the casing must be filled with mud periodically to prevent possible casing collapse. Large-diameter casing is more susceptible to collapse than casing with a small diameter because in most cases it has a lower collapse pressure.

Bridges are points or sections in the hole where cuttings and cavings accumulate and fill or partially block the hole. If bridges are encountered, the casing should not be pushed down or spudded too hard. This can cause the casing to

---

[2] It should be noted that each piece of casing has been referred to as a *joint.* This is common terminology similar to referring to joints of drillpipe as singles. The casing connector is also sometimes referred to as a casing joint.

become stuck so it cannot be pulled from the hole. When bridges are encountered, the recommended procedure is to use a circulating head to circulate the bridge out. If this operation is unsuccessful, pull the casing out of the hole, run the drillpipe to clean out the bridge, and rerun the casing.

When the casing is near total depth, install the circulating head, connect a circulating line, and begin to circulate the casing. Circulation should be started very slowly and should be gradually increased to a normal circulating rate. The casing should be rotated and reciprocated while circulating and then gradually lowered until it touches the bottom of the hole, referred to as *tagging bottom*. If bottom is tagged without circulating, there is a risk of sticking the casing in cuttings accumulate on bottom. The casing should be rotated if possible and reciprocated while pumping for at least one full circulation before beginning the cementing operation. If the casing begins to drag during this period, it is moved closer to the bottom of the hole. Then if it sticks, it will be at the correct setting depth.

Cementing Operations

The cement trucks usually arrives on location prior to running casing. The pump truck used for mixing and pumping the cement is positioned. Water and mud supply lines and a cement slurry discharge line are connected to the pump truck. The bulk trucks are then positioned so they can discharge dry cement into the mixing system (cement hopper) of the pump truck (Fig. 7–13). Depending on the size of the operations, a bulk storage trailer may also be used. This is a large-capacity trailer positioned to discharge cement into the cement hopper. It is moved on location and is filled with cement several days before the job. The pump truck is moved in later and is connected as described. At this point the cementing equipment is ready to conduct the cement job.

After the casing has been circulated and the cementing equipment has been positioned and connected, the next operation is to cement the casing. The casing is circulated through a cementing head that contains the wiper plugs (Fig. 7–14). It also may be circulated through a circulating head. In this case the circulating head is removed and the cementing head is connected to the top of the casing. The flexible, high-pressure cementing line from the pump truck is connected to the cementing head, and the lines are pressure tested. The circulating valve is then opened, and the casing is circulated for a short period to ensure that the pumping equipment is operating satisfactorily.

The cementing job is started by pumping a 10–20-bbl spearhead water spacer to separate the mud and cement slurry. This spacer may also contain special chemicals to help clean the wellbore and improve cement bonding. The bottom mud-wiper plug is dropped or released in the middle of the water spacer

FIG. 7–13 Pump truck (lower left) and two cement hopper trucks (upper right)

by opening the appropriate valve on the cementing head. This plug serves to separate the mud and cement. At the same time it wipes the casing walls to remove the layer of mud that adheres to the wall. If this bottom plug is not used, the mud film is removed by the next plug and can accumulate ahead of the plug, causing improper cement placement around the bottom of the casing. The cement is finally positioned around the casing shoe to prevent the lower joints of casing from backing off during subsequent drilling operations.

After the water spacer has been pumped, cement is mixed and pumped. After all of the cement has been pumped, a 10–20-bbl water spacer is pumped if the cement is to be displaced with mud. If the cement is to be displaced with water, water is pumped directly behind the cement. The top plug, called the *wiper plug,* is released immediately behind the cement by opening the appropriate valve on the cementing head. This plug is used to separate the cement from the displacing fluid. The plug also seats in either the float collar or the float shoe, sealing the opening in the shoe and thus indicating that the cement has all been displaced from the casing. The casing pressure increases when the top plug seats, indicating the end of the cement job.

Cement is displaced with either mud or water using the rig pump or the cement pump truck, depending on the circumstances. The cement pump truck is

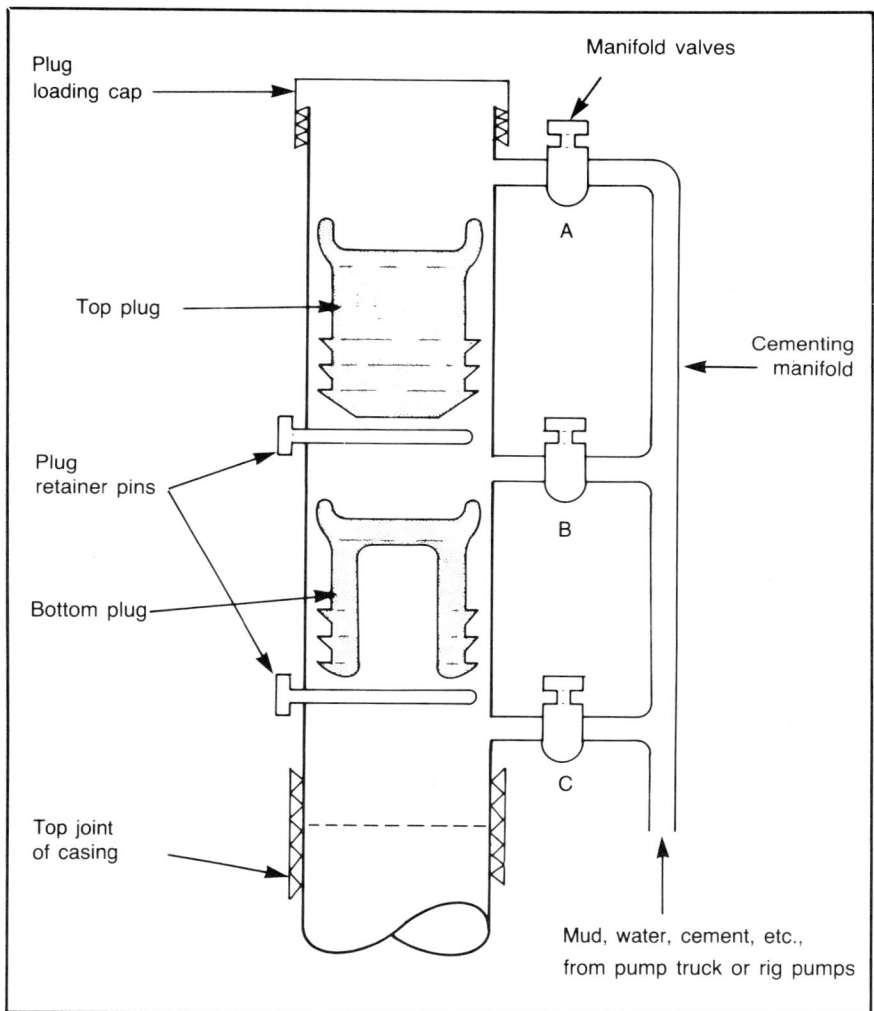

FIG. 7–14 Cementing head and manifold. The plugs are loaded in the cementing head, which is connected to the casing and mud circulating or cement lines. The casing is circulated through the manifold valve C. When the first plug is to be dropped, the retainer pin below the plug is pulled to release the first (bottom) plug. Valve B is opened, and valve C is closed. Fluid or cement entering through valve B strikes the top of the plug, pushing it out of the cementing head into the casing and downhole. The top (second) plug is released in the same manner.

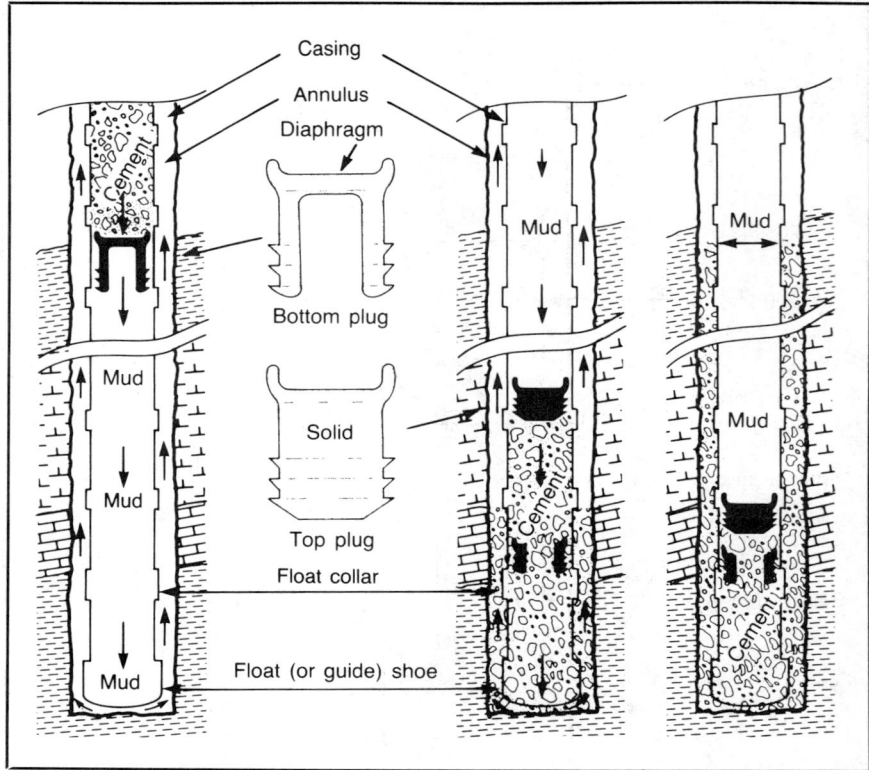

FIG. 7–15 (a) Cement is circulated down the casing; (b) cement is pushed up the annulus; and (c) cement reaches desired height in annulus, and pumping is stopped.

usually used on jobs where there is a smaller displacement and less pumping time is required. The pump truck cannot pump as fast as the rig pumps, but the displacement fluid can be measured more accurately in the tank on the pump truck. The faster pumping rig pumps are used for large-volume displacement jobs, and the volume displaced is measured by counting pump strokes. The rig mud pit level-measuring devices are used to check the displacement volume regardless of which pump is used.

When the bottom wiper plug approaches the float collar as determined by measuring the partial displacement volume, the pumps are either slowed down or are watched carefully so they can be shut down quickly. The plug is designed to rupture with a small pressure differential (about 50 psi) when the plug seats on the float collar. These plugs generally rupture without difficulty as indicated by a temporary, slight increase in the pump pressure. If the plug failed to rupture, the casing would become pressurized, or would pressure up, and could burst if the pump were not shut down immediately.

When the bottom plug ruptures, the cement is at the casing shoe. Continued pumping displaces the cement around the casing in the annular space until all of the cement is out of the casing. At this point the top plug behind the cement reaches the float collar and seals the cementing port through the float collar. This is known as bumping the plug and is indicated by a pressure increase. It is very important to keep an accurate measurement of the volumes pumped at this time. As the plug approaches the float collar, the pumps should be slowed down and then stopped immediately as the plug bumps the collar. If this is not done properly and carefully, there is a risk of overpressuring and bursting the casing.

The casing is normally reciprocated and may be rotated while pumping and displacing the cement. As the cement is displaced around the casing, casing drag increases. The casing is worked closer to the bottom of the hole at this time; if the drag becomes excessive, the casing is stopped within about 1 ft of bottom while continuing to displace the cement and bump the plug.

On shallow surface casing it is common to use excess cement and to obtain cement returns around the casing at the surface circulating cement. When the cement returns to the surface, it is diverted to the waste pit so it does not enter the mud pits and contaminate the mud. Cement returns are easily detected, especially when water has been used to spearhead in front of the cement.

After the plug is bumped, the pumps are shut down and the pressure is released to check for flowback. The float valves (check valves) in the float collar or float shoe should effectively seal the bottom of the casing. If it is not sealed, the higher hydrostatic pressure of the cement column outside the casing can cause fluid to flow from the casing, commonly called flowback. If the valves leak, the casing is closed in to allow the cement to harden, or set up, before operations are resumed. In normal operations the casing is sealed satisfactorily. As a precautionary measure, the casing is frequently shut in until the pumping time has expired before resuming operations.

As noted, the surface casing is normally circulated with cement. Sometimes cement is not circulated because of lost circulation, insufficient cement, or other reasons. In this case the upper surface casing section must be cemented. About 100–150 ft of joints of small pipe (1½ in. in diameter) is connected and pushed down the annular space between the casing and the hole or upper conductor casing. Cement is mixed and pumped through this pipe to fill the upper annular space with cement.

LANDING CASING AND NIPPLING UP

If the casing must remain suspended in the elevators until the cement hardens, this period is known as *cement time* or *waiting on cement*. The rig floor is washed and cleaned up, and all of the equipment used to run and cement the casing is dismantled and moved out. Preparatory work to landing the casing and nippling up can be done including washing out the inside of the mud riser or top of the conductor casing with water, disconnecting the flow line, and cleaning the blowout-preventer flanges prior to nippling up. Other work that may be needed includes changing pump liners, cleaning the mud pits, cleaning drillpipe tool joints—any general rigging up work that has not been done.

Landing Casing

If the casing is landed in the open hole, i.e., not inside another string of casing, it is suspended until the cement has gained sufficient strength to support the casing. Then tension on the casing is released, and the casing is set down. The cementing head is removed. If a mud riser has been used, the flow line is disconnected and the mud riser is cut off with a cutting torch and removed. The casing is then cut off with a cutting torch at the correct distance below the base of the rotary to allow room for the blowout preventers and other equipment. A bradenhead is then welded on the casing.

In some cases a landing joint is used when running casing. This is a short joint of casing that is the last joint connected when the casing is run. The joint is not connected too tightly so it can be disconnected later. When the casing is landed after cementing, it is spaced so the casing threads on the top of the first joint below the landing joint are at the correct distance below the bottom of the rotary to install the bradenhead. After the cement has hardened, tension on the casing is released and the landing joint is unscrewed, usually including the collar. This leaves a casing thread at the correct distance from the base of the rotary. A screw-on bradenhead is connected to the top of the casing and is securely tightened (Fig. 7–16).

Deep conductor or surface casing supports most of weight of the other casing strings. In deeper holes with heavy casing loads, a weight-support plate may be installed on the surface or deep conductor casing. This helps the surface casing to support the weight of the heavy casing strings to be run later. The weight-support plate is a large, thick, steel plate set on cement and fitted to the casing below the bradenhead.

Casing is landed by a different method when it is run through another string of casing fitted with either a bradenhead or a casing hanger spool. These can be on conductor, surface, or intermediate casing. The method used to land casing in

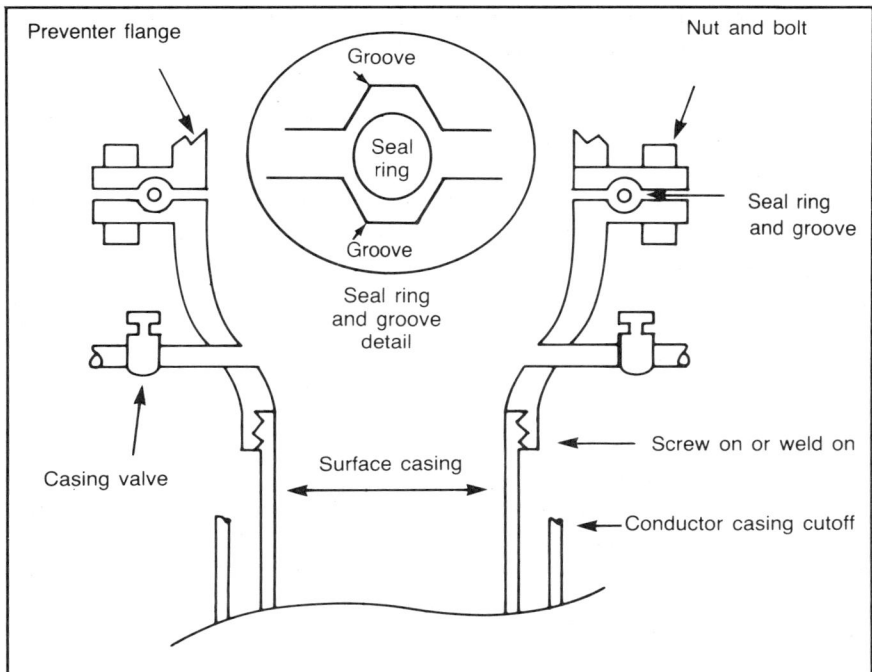

FIG. 7–16 Bradenhead

these cases is to set special casing slips that seat in the bradenhead or casing landing spool and that support the weight of the casing. The casing that has just been run must be suspended until these slips are set.

Before the slips are set, the casing must be lowered a short distance so part of the casing weight is supported by the cement and the remaining weight is supported by the slips. Casing landed in this manner is normally not cemented to the surface. The full weight of the casing is not suspended from the slips because this places unnecessary tension in the casing. In deep holes with heavy casing strings, such tension could overload the deep conductor or surface casing. Since the free, uncemented section of the casing must be suspended in tension, the slips are set to support a setdown weight.

The weight of the casing above the top of the cement is calculated with allowance for buoyancy. This provides for a free point at the top of the cement. Using a safety factor of about 10%, 90% of the calculated casing weight is the set-down weight to be supported by the slips. This effectively moves the free point into the cemented section. Sometimes instead of using 10%, the casing weight is calculated to give a free point a distance below the top of the cement equal to about 10–20% of the cemented section, and this amount is used as the set-down weight.

The slips can be set by several methods. The method most commonly used is to support the casing with casing running slips. The casing elevators are released, and a pickup line is passed through the hook and is fastened to preventers on top of the bradenhead or flange. The preventers are picked up with the traveling block and are suspended from the rotary beams with heavy chains. The pickup line is removed, the casing is picked up with the elevators, and the casing running slips are removed. The casing is lowered slowly until the correct set-down weight is shown on the weight indicator. The special slip seat and the facing casing are cleaned. Casing hanger slips are inserted and usually are sealed with a seal assembly. The traveling block picks up the casing weight so the casing running slips can be pulled. Then the traveling block is lowered slowly to transfer the casing weight from the elevators to the slip-and-seal assembly.

After the slips have been set, the casing is cut off a short distance above the slips with a cutting torch. The cut joint is pulled from the hole and is laid aside. The exposed upper stub of the casing is belled outward slightly so full-gauge tools can enter the casing easily without hanging up on the edge. The blowout preventers are then picked up with the pickup line, the chain is removed, and nippling-up operations are continued.

Nippling Up

Nippling up is the process of placing the wellhead control equipment (blowout preventers) on top of the bradenhead or casing hanger spool. It also includes installing all of the accessory equipment, as well as operating and pressure testing the entire system to ensure that it functions correctly.

As a review, the basic blowout preventer system includes two ram-type preventers—one to fit the drillpipe and one blank or blind ram—a bag-type preventer, kill line, choke manifold, and controls and an accumulator. A larger system has the basic equipment with additional ram-type preventers—usually two rams and possibly a shear ram—a larger complete choke manifold, and additional controls, including a remote control at some distance from the wellsite. The component parts are connected by large nuts and bolts or studs. Metal rings provide sealing for a flange-type connection.

In the nippling-up procedure the lowest preventer is set on the bradenhead, and the two flanges are connected and tightened. The next preventer is set on top of the first preventer and is connected in a similar manner. Additional preventers are connected in the same way. A rotating head, connected in the same manner, is frequently used on top of the preventers. A bell nipple may be used on top of the preventer stack and is fitted with a side outlet for the flow line leading to the shale shaker. Newer preventers have side outlets for the kill line and choke manifold line.

In the earlier preventers a cross or spacer spool with side outlets was used. One side outlet is fitted with a check valve and is connected to the mud system to serve as a kill line. This line is used to pump into the well during a blowout situation. Another side outlet is fitted with a line to the choke manifold. The choke manifold usually contains three outlets, one full opening, one fitted with a manually adjustable choke, and one fitted with hydraulically controlled choke. Then the accumulator and control panels with connecting lines are installed.

The preventers and choke manifold are first tested by actuating the various controls to ensure that they have been connected correctly and that each component works properly. The accumulator is checked to ensure that it supplies pressured fluid within a specified time, and the reserve nitrogen bottles are correctly pressured.

Pressure Testing the Preventers

In the pressure testing procedure a test plug is connected to a joint of drillpipe. The plug is designed to seat in the bottom of the preventers to seal off the lower casing. The casing must be isolated because the preventer test pressure is usually higher than the burst pressure of the casing. Various test sequences are used depending on the number of blowout preventers in the stack and operator preference.

Assuming two ram-type preventers and a bag-type preventer are used, one test sequence is as follows. A test plug is connected to the bottom of a joint of drillpipe. It is run in the hole and is seated in the base of the preventers. This plug seals the preventers from the lower hole section. A high-pressure, low-volume pump is connected to the fillup line or a test connection on the preventers. The lowest pipe rams are closed around the drillpipe and the annular area between the test plug, and the first set of pipe rams is pressurized to the required test pressure. This pressure is held for about 10 min, and any pressure reduction indicates leakage. The preventers are also visually inspected during this time. Any leaks are repaired, and the preventer section is retested until it tests satisfactorily.

The pressure is then released, the lower pipe rams are opened, and the next upper set of pipe rams is closed. The annular space from the second set of rams to the test plug is pressurized and tested. Repairs are made as necessary. The procedure is repeated until all pipe rams and the bag-type preventer have been tested. At this point the joint of drillpipe is backed out of the test plug and is pulled from the hole, leaving the test plug seated in the bottom of the preventers. The blind rams are closed, and the space between the blind rams and the test plug is pressurized and tested in the same manner. This also tests against the first choke manifold valve.

The pressure is released, the first choke manifold valve (the valve nearest the preventer) is opened, and the next valve is closed; then the test procedure is repeated. This procedure is continued until all of the choke manifold valves have been tested. At the end of the test, the single is run back into the hole and is screwed into the test plug. The test plug is then pulled from the hole, completing the testing procedure.

If any leaks are detected during the test sequence, the leak must be repaired, usually by tightening the flange. Then the section is retested. After the preventers have been tested, a bit is run into the hole to clean out and pressure test the casing. Then it drills out and tests the open hole before drilling ahead. If the drill collars and drillpipe are left in the mast, they are run in the hole. Usually, at least the drill collars are laid down to be replaced with smaller collars. The small collars and the necessary drillpipe are picked up to run the bit in the hole.

The pipe is run into the hole at a slower than normal speed. If there is any obstruction, such as cement left in the casing or a tight joint of casing, the slower running assembly can be stopped before it becomes stuck in the obstruction. If an obstruction is encountered, it is normally drilled out. One exception is partially collapsed casing. In this case the casing is usually opened by rolling or swaging it out, a special operation discussed in Chapter 10.

The mud in the casing may be somewhat gelled as a result of a long shutdown period without circulating. If this occurs, the hole is circulated at various intervals while running the assembly in the hole by connecting the kelly to the drilling assembly in the regular manner.

At this point the casing is pressure tested, plugs and cement are drilled, and drilling operations are resumed to drill the intermediate hole section.

CHAPTER 8
INTERMEDIATE HOLE SECTION

The intermediate hole section is that portion of hole drilled immediately below the surface casing. Shallow wells may not have an intermediate hole section, so the section below the surface casing will be the production hole section. Moderately deep wells usually have one intermediate hole section, and deeper wells may have two. Drilling operations in the second or deeper intermediate hole sections often are similar to drilling in the production hole section.

Operations in the conductor and surface hole sections are characterized by short, alternating periods of rigging up, drilling, running casing, cementing, and waiting on cement. In contrast, operations in the intermediate hole section are more stabilized and routine. Long periods are spent drilling, followed by trips to change bits with trip time increasing with depth. Formations become increasingly harder, so more time is devoted to bit selection and improving the drilling rate. A better quality of mud is needed, requiring more time to test, treat, and mix the mud and to work with solids removal equipment and degassers. Different types of formation problems are encountered. Deviation and crooked hole may require preventative or corrective drilling procedures. Higher pore pressures may necessitate more attention to mud properties, blowout preventers, and pressure control. All of these factors characterize drilling the intermediate hole section.

DAILY OPERATIONS AND CREW DUTIES

The crews settle into the daily or tourly routine operations while drilling the intermediate and deeper hole section. The term "routine" is a misnomer since nonroutine or out-of-the-ordinary operations and activities occur. However, there are many general duties that must be performed routinely to ensure a smooth, continued drilling operation.

A rig superintendent or tool pusher, commonly called *pusher*, normally is in charge of all rig activities and drilling operations on a 24-hr basis. The pusher is responsible for carrying out all of the instructions in the drilling program to drill and complete the well. Many problems arise in drilling operations, and the pusher must prevent, recognize, and correct these so the well can be completed satisfactorily. The tool pusher must be a highly motivated, experienced, and competent individual.

The tool pusher accomplishes the objectives by instructing and supervising the drilling crews, primarily the *driller,* who is in charge of the crew. The pusher also prepares and maintains records or supervises preparation of these records covering all rig activities.

A daily drilling report or *morning report* is prepared and submitted each morning to the drilling contractor, operator, or owner and in some cases to other companies or parties who have an interest in the well. Frequently, the operator or owner transmits the report to the other parties by telephone or radio. A more detailed report is sent by mail. This daily drilling report is a comprehensive, abbreviated record of all drilling activities for the past 24 hr, conventionally from 6:00 a.m. of the preceding day to 6:00 a.m. of the day of the report.

Different companies use different formats for the daily drilling report. The average report contains the following information:

- Total depth at report time
- Activity at report time such as drilling, tripping, or repairing equipment
- Footage drilled in the last 24 hr
- Summarized time breakdown such as 8 hr drilling, 3 hr working stuck pipe, 4 hr tripping, and 3 hr logging
- Summary mud properties such as weight, viscosity, gels, oil content, solids content, pH, and other chemical properties
- Daily and cumulative drilling costs—some companies also report daily and cumulative mud costs
- Abbreviated remarks and comments

The pusher orders supplies and equipment for maintaining and operating the rig. He also makes arrangements for third-party services such as welders, electricians, mechanics, pipe inspection services, mud materials and supplies, trucks, earth moving equipment, and drilling tools and equipment obtained from rental companies. The pusher handles personnel problems and either hires new crewmembers or coordinates this with the driller and the personnel department in the contractor's office.

The pusher generally consults with the owner/operator on all operating problems that affect the drilling operation. The owner/operator frequently assigns a drilling foreman, commonly called a *company man,* on the deeper, more critical, expensive wells. The drilling foreman is the company representative at the wellsite. The tool pusher or rig superintendent is still responsible for all drilling activities and operations and consults directly with the drilling foreman on those matters pertaining to drilling the well. The tool pusher and drilling foreman are familiar with the drilling contract and are guided by this in fulfilling the objectives of the drilling program.

Communications are very important in the drilling industry. Most rigs are equipped with mobile telephones, and the pusher and drilling foreman usually have mobile phones in their cars. Many rigs have trailer houses at the wellsite for the tool pusher and drilling foreman, who spend a major part of their time at the rig. On critical operations two tool pushers and two drilling foremen may be assigned to the rig to work alternate 12-hr periods. These personnel are relieved periodically.

The driller is in charge of the drilling crew of five or more people including a derrickman, a motor man, two floor or rotary helpers, and sometimes a third floor worker or cleanup man. The driller is in charge of all rig activities. He operates the rig controls, including the actual drilling and tripping operations, and answers directly to the tool pusher.

The derrickman works in the derrick while tripping. At other times he takes care of the repair and maintenance of the mud pumps and the treatment and conditioning of the drilling mud. (Fig. 8–1). The motor man is responsible for operating the motors, including routine maintenance and minor repairs, and he performs other work as needed. The floor workers or rotary helpers work on the rig floor during tripping. They perform routine repairs and maintenance for the various items of equipment and help the derrickman and motor man.

For many years rigs were operated by three crews working 8-hr shifts, commonly called tours. The day crew worked from 8:00 a.m. to 4:00 p.m., the evening crew from 4:00 p.m. to 12 midnight, and the morning or graveyard crew from 12 midnight to 8:00 a.m. The crews worked these shifts seven days a week until the well was finished.

Normally, the rig would be shut down for a period between wells so the crew could have time off. Many rigs are still operated in this manner. However, most contractors use an additional crew. The crew works five days and is off two. A contractor needs about four full crews to operate a rig on this schedule.

Excluding the costs of hiring and indoctrinating new personnel, there is a definite advantage to having the same crews working on the same rig. The crews are familiar with each other and are familiar with the equipment, and this improves overall operating efficiency. Rigs are frequently moved long distances, and the moves often occur at short time intervals of 1–3 months. This creates a problem with personnel who either must move their homes frequently or must drive long distances to work. High personnel turnover is a result.

FIG. 8–1 The derrickman periodically checks the mud weight and viscosity and mixes additives in the chemical barrel and mud hopper

To alleviate the problem, many rigs are operated by two crews working 12 hr/day for 5–10 days. The crews may live in a camp at the wellsite or in nearby towns, and they are paid subsistence. The crews are then off for 5–10 days. This work schedule is gaining more widespread acceptance.

When a crew is relieved, commonly called *changing tour,* the driller, derrickman, and motor man talk to their counterparts on the relieving crew about the operations. This helps provide continuity. The driller discusses overall activities, especially the condition of the bit, mud, and hole, suspected mechanical problems, supplies needed, special orders, and problems with connections and tight hole. The tool pusher often leaves special instructions to be reviewed. The derrickmen discuss the condition of the mud and the mud treatment being used or recommended by the mud company representative, usually called the *mud man.* The motor men discuss the condition of the various motors, routine maintenance, pending repairs, and fuel supply. These brief discussions between the crewmembers give the relief crew a good understanding of the current operations so they can be continued without disruption.

One main report filled out by the drillers is the driller's report or tour report. This contains an abbreviated record of all of the rig activities for the shift or tour. The data in the report is similar to the pusher's morning report, and the pusher actually makes his report from the tour report. In addition, the tour report carries the name of the different crewmembers, the hours worked, a list of all the pipe and other downhole equipment including the length of the downhole assembly, information on the condition of the mud, ton-mile record on the drilling line, notes on general maintenance and repair such as changing oil in the motors, and any other information that is pertinent to operating the rig or drilling the well. This is the most important report covering the overall drilling operation. It is signed by the drillers, the tool pusher, and the company man. Copies are distributed to the owner/operator, contractor, pusher or company man, and one copy is kept on the rig.

When the crews are changed at the end of a shift, the first job for the new crew is to complete or continue the job being done by the crew they relieved. For example, if a trip is being made to change the bit, the new crew continues the trip until the new bit is run to bottom and drilling operations are resumed. If some special downhole operation is being performed, it is completed.

Many duties must be performed for general maintenance. The equipment must be greased and oiled. Various chain oilers must be filled or oil levels checked. Bearings and other wear points are greased. Worn tong dies and slip segments are replaced. Wire-line clamps on the drilling line, tong lines, and other lines must be checked to ensure they are tight. Pump-rod oilers must be filled, worn pump valves must be repaired, and gland packing and head gaskets must be replaced as they become worn. Worn shale shaker screens also are replaced. Many different types of seals and packing must be maintained. Swivel joints on flexible connections are repaired. Water, mud, fuel, and air lines must be repaired or replaced as they develop leaks. Worn line valves are repaired or replaced. Accumulated mud is cleaned out of the rathole and mousehole, and drillpipe and drill-collar tool joints are cleaned and greased.

The tool house containing spare parts, supplies, and tools must be cleaned and everything must be replaced or rearranged, especially after rig moves in order to keep a running inventory of what is available and what is needed. The rig floor is washed after every connection, and the substructure, preventers, and other equipment below the rig floor are washed periodically to keep them clean and in working order and to prevent excessive corrosive damage. Drillpipe, casing, and other tubulars received on the wellsite must be measured and the tool joints or connections cleaned, greased, and inspected. Damaged tubulars are loaded and sent to the shop for repairs or to be junked. All metal surfaces must be cleaned and painted periodically.

These are part of the duties that must be done on a routine basis to maintain the rig and to keep it in a good operating condition. All of it is done on a spare-time basis when the drilling crew is not performing regular work such as drilling or tripping the assembly to change bits.

DRILLING THE SECTION

The first step in drilling the section is to determine the type of drill-out mud, the drill-out assembly, if the open hole is to be tested, and the test pressure. The foregoing applies when the section is drilled with liquid mud. Testing the open hole normally is omitted when the section is drilled with air or gas. In either case a conventional drilling assembly normally is used. This limber assembly consists of drillpipe, drill collars, and a drill bit.

In the air (or gas) drilling procedure about 1,000 ft of the assembly, usually including the drill collars and a few stands of pipe, is run into the hole. The kelly is connected, compressors are started, and all of the water above the bit is blown out of the hole. An additional 800–1,000 ft of drillpipe—depending on the maximum compressor pressure—is run into the hole, the kelly is connected, and the water above the bit is blown out of the hole. This procedure is repeated until the bit reaches the cement or top of the float collar. The hole is then blown (circulated with air or gas) dry. The cement and wiper plugs are drilled while circulating with air or gas, and the open hole is drilled in a normal manner.

In some cases the cement may be drilled out with fluid, the formation tested, and the water blown out of the hole to dry the hole before air or gas drilling. In these cases it is usually necessary to pull part of the pipe out of the hole before the water can be blown out because the pressure capacity of the air compressors is limited. This is the reason for blowing the water out of the hole in a stepwise manner.

Testing Casing and Drilling Out Cement

Cement contaminates most regular water-base drilling muds causing high viscosities and gel strengths and other undesirable characteristics. Therefore, the section of cement inside the casing between the float collar and the float shoe must be drilled with (1) water or old mud that is discarded, (2) mud that is either pretreated or treated later to reduce the adverse effects of cement contamination, or (3) a lime or oil mud that is not affected by cement contamination. Special muds are usually mixed before or while drilling the section.

The mud pits are usually cleaned to remove drill cuttings and sand after drilling the surface section. The better part of the mud, usually the upper part of the settling pit and most of the mud in the suction pit, may be saved to drill the intermediate hole section. Sometimes the pits are emptied, cleaned, and filled with water, which is used to start drilling the intermediate hole section. As drilling progresses, the clear water accumulates finer drill cuttings, making a native mud that is gradually treated and upgraded into a better quality mud as the hole is drilled deeper. This procedure often works well in areas where thick sections of bentonitic-type (montmorillonite) mud-making shales are drilled.

A bit is selected and run into the hole. Mills are sometimes used to drill long sections of cement in deeper, smaller casing strings, but roller bits normally are used in shallower, larger casing. The bit normally is selected primarily to drill the formations immediately below the surface casing. Only a small part of the bit life is used to drill cement; the remainder is used in drilling the formation. Usually the best bit for drilling cement is a medium-length, milled-tooth roller bit. However, since only a short time is spent drilling cement, the bit is selected based on the formations to be drilled.

There are exceptions to this rule. Diamond bits are never used to drill cement because the cement causes excess wear, breakage, and premature failure. If the formations must be drilled with a diamond bit, a milled-tooth bit is used to drill the cement and then a diamond bit is run.

When the formations are hard and abrasive, insert button-type bits are used. An earlier practice was to drill the cement with a milled-tooth bit, then make a trip to change bits to run an insert bit to drill ahead. The idea was that the insert bit would not drill the cement and rubber wiper plugs efficiently. Recent experience has indicated that the insert bit can drill the cement and rubber wiper plugs. It is not highly efficient, but it drills fast enough. Therefore, running a special bit to drill cement is not justified unless there is a long column of cement.

In normal practice the bit is run into the hole to a depth about 100 ft above the top of the cement. The 100-ft section is reamed and drilled until the bit is on top of cement or the first wiper plug. In some cases when a long hole section is filled with poor-quality mud, the bit is stopped at an intermediate point to circulate the mud. This is known as breaking circulation and is done to condition or replace part of the poor-quality mud in the hole with good mud. If the bit is run to bottom in this case without stopping to circulate, an excessively high pressure may be needed to break circulation because of the high gel strengths of the poor-quality mud in the hole.

After circulating for a short period, the pipe rams are closed and the casing is pressure tested. The test pressure is usually about 80% of the rated burst pressure of the casing or the maximum pressure expected during the subsequent drilling operation plus 20%, whichever is smaller. The driller must allow for the hydrostatic pressure of the mud in the casing when pressure testing. If the casing does not hold pressure, the leak is located and squeezed in a special operation.

The cement and wiper plugs are drilled with a minimum bit weight and rotary speed. The main objective is to drill out the wiper plugs and cement by running the drilling assembly as smoothly as possible. Excess torque and variation in torque and rotary speeds are avoided. This action creates excessive right-hand torque on the bottom of the casing. If the casing and joints are not well cemented, this right-hand torque can cause one or more joints to back off, become unscrewed and loose in the hole.

One or two *junk subs,* sometimes called *boot baskets,* may be run in the drilling assembly immediately above the bit when drilling cement, to remove large cement particles. The junk sub has a recessed area located above a basket container. As the mud flows up the annular space past the junk sub, the flow rate

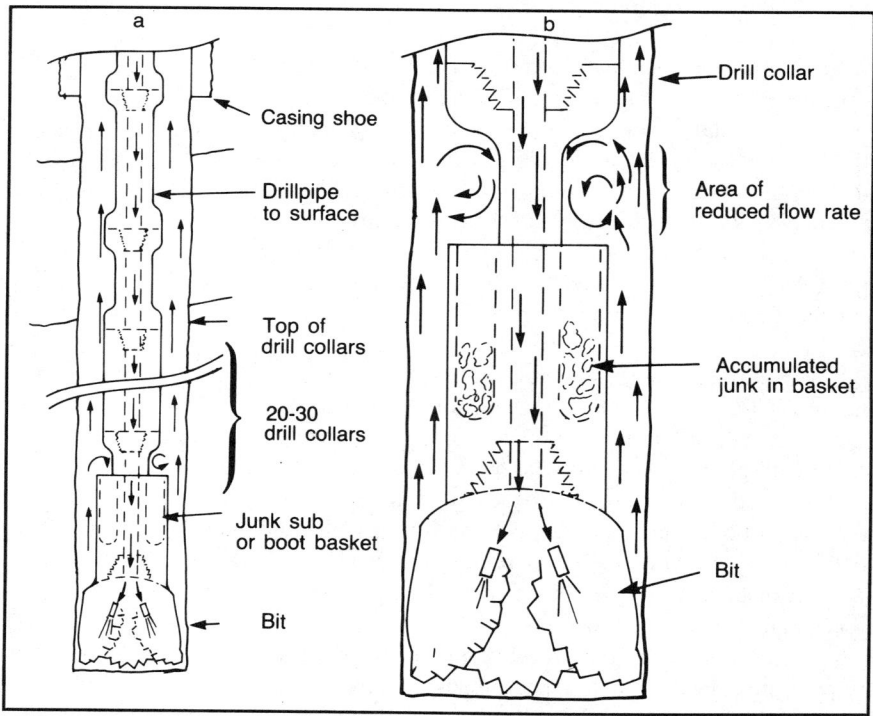

FIG. 8–2 Drill-out assembly (a) drilling assembly and (b) bit and junk sub

is reduced and a turbulence effect is created in the recessed area. These combined actions allow large particles to settle out into the basket or container of the junk sub where they remain until the junk sub is pulled from the hole. Catching the large particles in this manner prevents them from falling back to the bottom of the hole where they must be redrilled into smaller particles before they can be circulated out of the hole. Bit life is extended since it is not necessary to redrill the particles (Fig. 8–2).

Testing the Open Hole

One reason for running intermediate casing is to separate the hole into sections that can be drilled with a reasonable range of mud weights. The range of mud weights depends on the increase in pore pressure, depths, and other factors. If the intermediate hole breaks down while increasing the mud weight and drilling deeper, the most likely place for the formation to break down and lose mud will be in the upper section of the open hole at the base of the surface casing. The term *break down* indicates that the formation parts, fractures, or otherwise opens and accepts (or drains) mud from the borehole. The pressure at which this occurs is the breakdown or fracture pressure. Therefore, it is important to determine if the formation will hold the projected mud weight before drilling ahead.

For example, the lower part of the hole may have an oil, gas, or water zone that requires a higher mud weight to hold the fluids in the formation and prevent them from flowing into the borehole. If the shallow open-hole section cannot hold the higher-weight mud, other precautions must be taken. If these precautions are not taken, there is always the risk of a blowout when the high-pressure formation is drilled. This illustrates the need for and the importance of testing the open hole below the surface casing and below other casing strings where the hole will be drilled deeper.

The same bit is used to drill 10–30 ft of new hole below the surface casing. The hole is circulated, and the bit is picked up until it is inside the casing. Pipe rams are closed, and pressure is applied slowly with the mud pumps to a predetermined pressure. The pressure is equivalent to filling the casing with a heavier-weight mud.

One precaution should be noted in using water for pressure testing. Water is a penetrating fluid. It tends to enter the formation and cause it to break down at a lower pressure than when using mud, a nonpenetrating fluid. Since the hole normally is exposed to mud when it is subject to breakdown, mud should be used for testing.

If the formation breaks down while being tested, it may be squeeze cemented in a manner similar to squeezing off lost-circulation zones.

Drilling Ahead

The intermediate hole section is drilled with the same general drilling procedures used in the upper hole sections with some modifications. The upper part of the intermediate section normally drills relatively fast, and the rate decreases with depth. Hook loads continue to increase as more drillpipe is added to the drilling assembly. This load tends to increase the difficulty of handling the assembly. However, the hoisting equipment is designed for this weight, and the driller takes precautions accordingly. The bit, drill collars, and often the drillpipe are smaller than that used in the upper hole sections. The smaller equipment has less strength both in tension and torsion. The driller allows for this in handling the assembly so it is not damaged, possibly causing a fishing job. Less drilling bit weight is used. Safe limits must be recognized and provided for to prevent losing bit cones prematurely, causing a fishing job, and possibly damaging other parts of the drilling assembly.

There usually is less clearance between the drilling assembly and the walls of the hole in the deeper hole than in the shallower hole. This increases the risk of sticking the drilling assembly because of accumulated drill cuttings or formation caving and sloughing. Sticking the drilling assembly can be prevented in most cases if the driller recognizes the problem and takes immediate corrective action. As the hole is drilled deeper, lag time increases correspondingly, thus requiring more time to clean the hole. This is allowed for by increasing the circulating time when good hole cleaning is required.

The intermediate hole section is the first long section of open hole and may penetrate many different types of formations, thus increasing the risk of hole problems. The longer drilling assembly has more connections, each of which is subject to failure. The tool joints must be inspected periodically, greased (doped), and made up or tightened with the correct recommended torque.[1] In summary, deeper drilling requires continual vigilance and monitoring by the operating personnel to prevent problems and to handle them correctly when they occur.

Most modern rigs are equipped with an automatic driller, a hydraulic or air-actuated tool that maintains a predetermined weight on the bit while drilling (Fig. 8–3). After a connection is made, the automatic driller is connected to the drawworks brake handle. The rotary is adjusted to the desired speed, and the automatic driller is set to maintain a predetermined amount of bit weight on the formation. As the hole is drilled, the automatic driller increases or decreases the

---

[1]The recommended torque for making up various tool joints and connections is included in the API *Recommended Practices and Procedures Bulletins*.

FIG. 8-3 Control panel of an automatic driller

weight on the brake arm, which in turn allows the drilling line to unspool faster or slower to maintain the predetermined bit weight. Although the driller must carefully monitor the tool, it relieves him of the tedious minute-by-minute adjustments needed to maintain the proper bit weight. It also allows the driller more time to monitor other instruments and activities, even though he cannot leave the drilling console.

A good practice is to post a copy of the projected rate-time curve from the drilling program on the rig bulletin board. The actual drilling time can then be plotted on this curve and compared to the projected drilling time. Another approach is to compare an actual rate-time curve from a nearby well to a plot of the drilling progress of the well being drilled. These actions keep the drilling crews informed of overall operations and timing, help install a sense of competition, and contribute to the overall efficiency of the operation.

Factors Affecting the Drilling Rate

All efforts are concentrated on drilling the hole as quickly and safely as possible, i.e., at the maximum efficient penetration rate. Many factors affect the drilling rate: selecting the proper bit, operating it correctly, and using the correct mud with minimum weight and good physical and chemical properties. The bit normally is operated within the manufacturer's specifications, but these often vary widely. Therefore, the operator has considerable discretion as to the bit weight and rotary speed used.

Most drillers try different combinations of bit weight and rotary speed. For example, the operator may drill with 40,000 lb of bit weight and 50 rpm for 10–30 min. Then he may decrease the bit weight to 35,000 lb and increase the rotary speed to 60 rpm and see how this combination drills for a period of time. The bit weight and rotary speed may be changed again, and the results are often noted on the drilling time recorder. The driller continues to vary the bit weight and rotary speed within safe operating limits until he finds the combination that runs the smoothest and gives the fastest drilling rate. The procedure is repeated periodically as different formations are drilled to ensure that the hole is being drilled in the fastest manner.

When different formations are penetrated, the procedure is again repeated since different formations may drill faster with a different combination of bit weight and rotary speed. Drilloff tests may also be run to help select the best drilling rate. These tests are not standardized and usually involve running a higher bit weight, locking the brake down temporarily to stop downward movement of the upper part of the drilling assembly, varying the rotary speed, and noting the rate at which the bit weight is reduced as the bit drills.

The drilling rate can be strongly affected by the mud hydraulics and the type, weight, and condition of drilling mud. Mud hydraulics should be checked periodically to ensure that the correct size of pump liners or plungers is being used, the pump is operating at the recommended volume and pressure, and the proper jet sizes are being used if jet bits are being run.

The mud must have good physical and chemical properties to obtain a maximum penetration rate. This emphasizes the importance of checking the mud frequently and treating the mud as necessary. Normally the mud weight and viscosity are checked hourly, and the fluid loss and volume of mud solids are checked once on each 8-hr tour. Mud checks may be made more frequently when drilling with heavier muds or when formation problems or other conditions require more frequent testing. Standard mud-treating equipment is illustrated in Fig. 8–4.

In most cases the penetration rate increases with a reduced mud weight. Therefore a minimum mud weight should be used to obtain a maximum pene-

FIG. 8–4 Mud-treating equipment

tration rate and to reduce mud treating costs. Heavier muds cost more to prepare and maintain. When possible, the mud weight should be such that the hydrostatic pressure of the mud column is near the formation pore pressure or in some cases at a lower pressure, such as when drilling with air or gas.

Some formations such as the Wasatch and Mesa Verde in the Rocky Mountains have relatively high pore pressures and very low permeability. In many cases these formations can be drilled at a substantially higher penetration rate by using a very low-weight mud (below the equivalent pore pressure) and drilling the well in a highly underbalanced condition with gas flowing out of the hole with the mud. A rotating head is used to prevent the gas and mud from flowing up through the rotary and creating a fire hazard. A gas separator connected to the flow line separates the gas from the mud. The mud is returned to the mud pit, and the gas flows through a flare line and is burned at a waste pit located a safe distance from the rig. Drilling in this manner means relying heavily on the blowout preventers, but substantially higher penetration rates can be obtained.

This procedure is known as *underbalanced drilling*. The term underbalanced indicates that the hydrostatic head of the mud column is less than the pore pressure. *Overbalanced drilling* is the reverse; the hydrostatic head from the mud column is higher than the formation pore pressure. *Balanced drilling* indicates that the hydrostatic head of the mud column is approximately equal to the reservoir pressure. The penetration rate normally is increased as the mud weight decreases so near-balanced or underbalanced drilling should always be considered. However, this cannot be done safely if high-pressure formations with good permeability are expected.

Penetration rates also appear to increase with increased fluid loss and decreased gels and volume of mud solids in the mud. High water loss is usually associated with high *spurt loss*. The spurt loss is the initial mud loss into a freshly drilled formation while the mud solids are screening out on the face of the formation to form a wall cake. Then this cake reduces the mud filtrate rate of loss into the formation.

In general, oil-base muds are a less-efficient fluid for drilling purposes than water-base muds. Penetration rates with water-base muds are often two to four times higher. The reasons for this are not clearly defined and may be attributed to the low fluid-loss characteristic of oil muds, the improved lubricity, the reduced abrasiveness, and the fact that oil muds are often weighted. Oil muds normally cost considerably more than water-base muds, and in some cases the crewmembers are paid an additional bonus when working with oil muds. Recent developments in oil muds may eliminate some of the objections and generally increase their effectiveness.

Drilling costs, expressed in dollars per foot, can be used to evaluate bit performance and other operations. The formula for calculating drilling costs is as follows:

$$\text{Drilling costs/ft} = \frac{(\text{trip time, hr} + \text{bit life, hr})(\text{operating costs, \$/hr}) + \text{bit costs, \$}}{\text{footage drilled, ft}}$$

This formula can be used to calculate drilling costs per foot and is especially helpful in comparing the performance of various types of bits.

For example, the operator may be considering using a medium-formation standard roller bit that costs less, drills faster, and has a shorter bit life than a hard-formation, premium roller bit that costs more, drills slower, and has a longer bit life. The cost of each bit is known, and the footage and bit life for each bit can be estimated. The other parameters in the equation are known. The drilling costs per foot can be calculated for each bit and then used as a guide to select the bit that will drill the hole for the least cost.

Sometimes intangible factors must be considered. For example, the bit with the shorter life causes more tripping, which is a higher-risk operation than drilling. Or one type of bit may leave more junk downhole than another type bit. All of these factors must be considered in addition to standard drilling costs.

TRIPPING AND RELATED ACTIVITIES

The general procedures used in shallow holes to pull the drilling assembly from the hole and run it back are the same for tripping the assembly in the deeper hole with some modifications. It is important to keep the drillpipe in the hole as much as possible. Obviously, it must be pulled to replace worn bits, to core, to drill-stem test, and to perform other necessary operations. However, the pipe should be pulled as fast as is safely possible and run back into the hole as soon as the driller can. Keeping the drillpipe in the hole is like an airplane pilot preferring to fly at a higher altitude: if there is a problem, he has time to react. If the drillpipe is in the hole and a blowout occurs, the operator has a much better chance of controlling and killing the blowout than if a blowout occurs when the pipe is out of the hole.

As the hole is drilled deeper, more time is required to make a *round trip*—to pull the assembly out of the hole and run it back in. There is more pipe in the hole. Also, the pipe must be pulled from the hole at a slow enough speed to prevent swabbing the hole, and it must be run in the hole at a slow enough speed to prevent pressure surges.

If the pipe is pulled from the hole too fast, the large bit and drill collars may tend to act like a swab or piston and literally swab or lift the mud out of the hole, creating a possible blowout condition. Therefore, it is very important to pull the pipe at a rate slow enough to prevent swabbing the hole.

When the drilling assembly is run into the hole too fast, the plunger effect from the large bit and drill collars can create excess pressure in the hole—commonly called surge pressure. If this pressure is high enough, formations can break down. Then mud is lost from the borehole to the formations. This is known as induced lost circulation. When mud is lost to the formation and is not replaced at the surface, the mud level in the hole drops. This reduces the hydrostatic head, similar to the swabbing action. In turn, it reduces the hydrostatic pressure so high-pressure fluids can flow into the wellbore, causing a blowout.

Swab and surge pressures can be calculated with standard formulas using the dimensions and lengths of the downhole assembly; the inside diameters and lengths of the wellbore, including the cased hole; and the physical characteristics of the mud. For a given set of tripping speeds out of and into the hole, swab and surge pressures are less if the mud is in good condition and has lower viscosity and gel strength than high surge and swab pressures with mud that has high viscosity and gel strength. This is an important reason for having the mud in good condition before making a trip.

The hole must be full of mud while pulling the pipe out of the hole. As each stand of pipe is pulled, the mud level in the hole drops an amount equivalent to the displacement volume of the pipe. If the fluid level continues to drop, the hydrostatic head at the bottom of the hole will decrease. If the hydrostatic pressure reduces excessively, fluids can flow from the formation into the wellbore, creating a blowout situation. To prevent this, the hole must be filled periodically as the stands of pipe are pulled out of the hole. A measured amount of mud is pumped into the hole equal to the displacement of the pipe that is pulled. The amount of mud can be measured with a trip tank or by pump strokes. This is a very important consideration when pulling pipe.

Most rigs pull three joints at a time, called a thribble or stand. Every third tool joint is disconnected during the trip. This third joint is known as the *tool-joint break*. On succeeding trips the tool-joint break should be changed to another joint so all of the drillpipe tool joints are disconnected, visually inspected, cleaned, doped, and reconnected. The procedure, known as changing the break, helps detect loose or leaking tool joints. It also ensures that all of the tool joints are serviced periodically.

Drillpipe tool joints wear the casing walls as the assembly is rotated during drilling. This wear can range from minor to severe and can even wear holes in the casing, causing leaks. Tool joints are fitted with hard-facing material to reduce wear. However, this material causes excessive wear on the inner casing walls. Therefore, drillpipe rubbers, sometimes called pipe rubbers or casing protectors, are installed about 3 ft above the tool joint on alternate tool joints to protect the casing. In severe cases of crooked or deviated hole, rubbers may be run on every second joint, called double-rubbered pipe.

Rubbered drillpipe is added at the surface as the hole is drilled deeper. It is not run in the open hole because it creates excess wear on the rubbers, causing them to loosen and slip. To prevent this, part of the rubbered drillpipe is laid down as singles before a trip. On the trip back in the hole and before running rubbered pipe, unrubbered drillpipe is connected in the drilling assembly to replace the rubbered drillpipe that has been laid down. The rubbered drillpipe that has been laid down is then picked up during the drilling operation after the trip. Stands of rubbered drillpipe are also interchanged during the trip so the rubbered drillpipe is not run in the open hole.

The drillpipe joints immediately above the drill collars can be subjected to extra fatiguing action during the drilling operation, especially if heavier bit weights are used with a minimum number of drill collars. It is a good practice to change two or three stands immediately above the drill collars to position higher in the hole on trips so the excess wear can be distributed throughout the drilling assembly.

When a drilling bit becomes worn, it may become undergauge, drilling a slightly smaller diameter hole. This situation occurs near the end of a bit run and is sometimes called a *tapered* or *swaged hole*. A new bit can be damaged if it is forced into this tapered, undergauge hole and can cause premature bit failure. In severe cases the bit may be wedged in the tapered hole, causing sheared bit cones, cracked or broken bit shanks, or a stuck drilling assembly resulting in a fishing job. To correct a tapered hole 60–90 ft of hole should be reamed when running a new bit to eliminate the problem.

The drilling assembly should be measured periodically on trips to correct or verify depth measurements. All of the drillpipe should be counted on each trip to ensure that the total number of joints in the assembly is the same as the number on the drilling report. Counting the joints by counting stands when the drillpipe is standing in the mast is an easy way to verify the joint count. The drillpipe should always be measured on the last trip before logging or running casing.

When the drillpipe is suspended in the hole, it stretches 2–5 ft/10,000 ft of suspended pipe, depending on the weight and grade of the pipe and the mud weight. Therefore, the bit usually is drilling a few feet below the depth noted on the report. This small depth difference normally is unimportant, except in special, critical operations. The most accurate drillpipe measurement is made while the drillpipe is hanging on the hook. This procedure includes measuring each stand of drillpipe after it has been connected to the drilling assembly and picked up from the slips so the pipe is under tension. The extra time required to make these more precise measurements is normally not justified except in special cases.

A pipe wiper normally is installed on the drillpipe when the assembly is started out of the hole for a trip. This rubber wiper is positioned below the rotary by removing one of the split bushings to allow the pipe wiper to be pushed below the rotary and replace the bushing. The pipe wiper fits tightly around the drillpipe and expands to allow the drillpipe tool joints to be pulled.

The pipe wiper cleans the mud outside of the drillpipe, allowing most of the mud to run back into the bell nipple and into the circulating system. Cleaning the drilling mud from the pipe in this manner makes the pipe easier to handle and helps keep the working area clean so the crew can work safely.

The pipe wiper also serves another purpose. If any small metal objects such as hand tools and tong jaws are accidentally dropped into the hole while pulling the drillpipe, they will hit the top of the pipe wiper and fall away from the hole, not into the hole where they could cause the assembly to stick.

When the top of the drill collars is pulled up near the base of the rotary, the pipe wiper is removed. Most pipe wipers cannot be stretched over the larger drill collars and bit without being damaged. A good practice in this case is to replace the good pipe wiper with an old pipe wiper. Then the wiper can still prevent objects from falling into the hole while pulling the drill collars and bit.

Another good procedure is to pour about 1 bbl of water inside the drillpipe before it is pulled. This water runs down through the drillpipe and helps clean the mud from inside the pipe. This also helps prevent mud from dripping over the working area and in the tool joints when the stands of pipe are set back.

The blind rams should be closed as soon as the bit has been pulled through the rotary. This is another precaution to prevent dropping something into the hole. Normally, a good pipe wiper is not used when running the assembly back in the hole. However, again it is a good practice to use an old pipe wiper while running the assembly to help prevent a fishing job for lost tools.

Testing the Blowout Preventers and Choke Manifold

When the drillpipe is out of the hole, all of the preventer rams, including both the pipe and blind rams, should be worked—closed and opened—to ensure that they function properly. Note that the blind rams should never be closed with the drillpipe in the hole since they may either cut or bend the drillpipe, causing it to drop. The blowout preventers and the choke manifold are pressure tested periodically to ensure that they function properly and hold pressure satisfactorily in case of a blowout.

The preventers are normally tested after nippling up and at about four-week intervals during normal drilling operations. If there is any reason to suspect the preventers of being faulty or some change in conditions, they should be tested more often. Some regulatory agencies have rules and regulations regarding tests for blowout preventers. Service companies have equipment and personnel to test the blowout preventers and choke manifold, although some operators have the equipment and do their own testing.

Tool-Joint Inspection

The drillpipe tool joints are visually inspected while tripping the drilling assembly. Changing the break provides a time for all of the drillpipe tool joints to be visually inspected. If a drillpipe tool joint is damaged, the joint is replaced. Drillpipe and the tool joints may also be inspected during the rig move. Or the drillpipe may be laid down for a full inspection if drilling operations have been conducted for a long time. And sometimes the drillpipe is inspected before running heavy loads such as setting a liner.

Drill-collar tool joints are checked periodically at intervals from 100–500 drilling hr. The time interval between inspections depends on the size of the tool joints and the type of drilling service. For example, during smooth drilling the tool joints can be run longer between inspections. On the other hand, they must be inspected at shorter intervals when operated under rough drilling conditions, such as milling on junk with high, fluctuating torque.

One indicator of the correct amount of time between inspections is the number of tool joints that are found to be defective during the inspection. If few tool joints are defective, the interval between inspections can probably be extended. If a higher number of tool joints (6–8) are found to be defective during the inspection, the interval should be shortened.

Common tool-joint failures are swelled shoulders, stretched threads, and cracked pins or boxes. Swelled shoulders can be readily detected with either a straightedge or a caliper. Stretched threads can be detected with a standard thread gauge. Cracked pins and boxes are much more difficult to detect, especially small cracks.

Drill-collar tool joints may be inspected between rig moves, but they are also inspected during the drilling operation. They are normally inspected after the drillpipe has been pulled from the hole for a regular trip such as replacing a worn bit using one of two methods. The oldest and somewhat standard method is the *magnetic-particle inspection,* sometimes called the *black-light* method. The newer, *sonic* method is gaining wider acceptance.

In the inspection procedure, the drill collar is picked up through the rotary, a drill-collar clamp is installed, and the drill-collar tool joint is disconnected in the normal manner. The two parts of the tool joint are then thoroughly cleaned and dried and are checked for swelled shoulders and stretched threads.

In the magnetic-particle inspection method, the natural magnetism in the tool joint is measured. If this is not sufficient, magnetism is induced in the drill collar with several coils of heavy electric wire carrying a direct current. The thread areas of the tool joint are coated with a mixture of special oil and fine iron filings. The iron filings tend to orient themselves over a crack as a result of the induced magnetism. The concentration of iron filings can be detected with an ultraviolet light or black light. Defective tool joints are then replaced.

If the sonic method is used to test for cracks, a light lubricant is spread on the face of the tool-joint shoulder to ensure a good contact with sonic transmitter/receiver. The transmitter/receiver is moved slowly over the face of the tool-joint shoulder. Any cracks or pits in the thread area are indicated on a video display or digital readout. If the tool joint is defective, the drill collar is laid down so it can be sent to the shop for repairs.

Slip or Cut the Drilling Line

One of the most closely watched items of equipment is the drilling line. The drilling line is used to suspend and move all of the rotating equipment. The drilling line is subject to wear from very high hook loads. If the drilling line fails, all of the equipment suspended from the line is dropped. This can cause severe equipment damage, a fishing job, and personnel injuries. Therefore, the drilling line is carefully inspected and arrangements are made to change it before the failure point is reached.

Drilling line wear is measured in ton-miles. On each tour the rig spends a certain number of hours drilling, tripping, reaming, and running casing. Standard tables and calculations convert each of these work periods into ton-miles based on the time spent doing the work, the type of work, the hook load, the hole depth, the mud weight and all other factors that affect drilling-line loading. The ton-miles for each operation conducted during the tour are summed and recorded on the drilling report. This is added to the ton-miles at the end of the tour. The drilling line is either slipped or cut at periodic ton-mile intervals of wear.

Different practices are used by different operators. Some operators cut 50–75 ft of drilling line off every time a cut is scheduled, based on ton-miles. Some operators slip the drilling line 50–75 ft once and cut 100–150 ft the next time. Other operators slip the line twice and at the third interval cut off about 200–250 ft.

Operators depend strongly on ton-mile records, but they also make frequent visual inspections of the drilling line. Excess wear on the outer wires or broken strands indicate the need to slip or cut the line. Most of the wear occurs on the fast-line end of the drilling line—the line connected to the drawworks drum—and decreases toward the dead-line end—the line clamped to the dead-line anchor fastened to the substructure.

The drilling line normally is slipped or cut during an interruption of the tripping sequence. The most common practice is to pull the assembly out of the hole, change the bit or other tools, run the bit to the bottom of the casing, set the pipe in the slips, close the blowout preventers, and install a drillpipe blowout preventer on the drillpipe. The traveling block is set on the rig floor by pulling the elevators out toward the V-door with one of the lifting lines running through a snatch block or small pulley.

If the drilling line is to be slipped, the clamps on the dead line are loosened. The amount of drilling line to be slipped is reeled onto the drawworks drum; the same amount of line from the reserve drilling line reel or spool is allowed to slip through the loosened dead-line clamps. The dead-line clamps are then securely tightened, the traveling block is picked up by reeling the drilling line onto the drawworks drum, and operations are resumed. The amount of line that has been slipped stays on the drawworks drum, even when the traveling block is in the lowermost position. The slipped section of line on the drawworks drum is the most worn part of the line, but it is less subject to breakage since it stays on the drum and is under a negligible tension.

A similar procedure is used for cutting the drilling line with some modifications. The block is laid down on the floor and the dead-line clamps are loosened. Then the amount of line to be cut is reeled onto the drawworks drum while the same amount of line from the reserve spool slips through the dead-line clamps. After the required amount of line to be cut is reeled onto the drawworks drum, the dead-line clamps are securely tightened. The lower end of the fast line is snubbed or tied so it cannot unreel and run over the crown. The drilling line is then cut near the drawworks drum.

The cut-off piece of line on the drawworks drum is unreeled by pulling the free end away from the drawworks. When all of the cut-off line is unreeled, the drum clamps are loosened and the cut-off piece is pulled away from the rig floor to a junk pile near the edge of the location. The end of the fast line hanging down from the mast, which has been snubbed off, is securely fastened to the drawworks drum with the drum clamps. The fast line is then reeled onto the drum to pick up the traveling block. The first part of the line is reeled on the drum slowly and carefully, using hand hammers if necessary to help position the line firmly in the drum grooves until part of the traveling block has been picked up. The line must be holding sufficient weight to ensure that it remains in the grooves.

Slipping the line is faster than cutting the line. The bit is run in the hole to the bottom of the casing, the blowout preventers are closed, and a valve is placed on the drillpipe to secure the well completely before starting to work on the drilling line. The pipe is run into the hole. Then if the well begins to kick during the shutdown period to slip or cut the line, the kick will be easier to control with the pipe already some distance in the hole. The pipe is not run deeper into the open hole below the casing because of the possibility of the pipe sticking. The hole should also be checked periodically to ensure that it is full of mud during the shutdown period.

Blowout Drills

A blowout is one of the worst situations that can occur in a drilling operation, and the results are often disastrous. Blowouts occurred more frequently in the earlier history of the industry and in some cases caused tremendous losses in equipment and production, not to mention injury and death to personnel. Equipment and procedures were developed to prevent, detect, and control blowouts. The industry has concentrated on these items, and as a result very few blowouts occur.

However, the trend toward deeper drilling, where higher pore pressures are encountered, further increases the risk of a blowout. Deeper wells encounter a higher percentage of natural gas than oil. Blowouts caused by natural gas can occur faster and may be more difficult to control than those caused by oil. Therefore, all precautions are taken to prevent and control the ever-present risk of blowouts.

There is a higher incidence of blowouts while drilling the production hole section. However, blowout drills are started while drilling the intermediate hole section to train the crew.

The blowout drill is a practiced procedure to shut in and secure the well prior to killing the blowout. Most of the blowout-detecting devices use a loud horn or shrill, piercing whistle as an alarm signal when a pending blowout condition is detected. A blowout is more likely to occur while drilling because more time is spent drilling. Therefore, the average blowout drill is conducted while drilling, usually with the kelly about two-thirds down in the hole.

When the blowout alarm sounds, the driller immediately begins picking up the drilling assembly to pull the kelly out of the rotary. At the same time he shuts off the mud pump. Each crewmember goes immediately to a preassigned position. One experienced crewmember is assigned to the blowout-preventer control panel located on the rig floor, he also helps the driller as needed. Another crewmember is positioned at the choke manifold on the mud-pit side of the rig. Yet another crewmember is positioned at the driller's side of the rig at ground level and operates a remote blowout-preventer control if one is positioned in this area. These latter two crewmembers are also assigned the secondary duty of closing the manual blowout preventer controls if required. Another crewmember, often the derrickman, is positioned at the accumulator so he can also change the nearby pump manifold valves as needed to place one or more pumps on the hole.

The positions and duties of the individual crewmembers vary, depending on the rig equipment and the specific duties required. For example, after the pump is shut down, the crewman positioned on the rig floor may close the standpipe valve and open the fillup or kill line. The preventers are only closed on the driller's instructions.

It should be noted that the kelly is pulled out of the hole so the stronger pipe rams can be closed around the drillpipe. In a fast-acting blowout the bag-type preventer may be closed around the kelly to prevent mud from being blown onto the rig floor. In this case the driller continues to pull the assembly, with the kelly dragging through the closed bag-type preventer until it is out of the hole. At this time the pipe rams are closed. Closing the bag-type preventer in this case may be one of the alternative duties of the crewman on the floor.

The blowout drill is timed from the point when the alarm sounds until the preventers have been closed and the well has been secured. This blowout drill time is recorded on the drilling report. The average blowout drill time is 60–90 sec for a well drilling down to about 10,000 ft and an additional 3–5 sec for each 1,000 ft of depth below 10,000 ft. Additional time is required at greater depths because the drilling assembly is heavier and more time is needed to pull the kelly out of the hole.

Less commonly, blowout drills are conducted while tripping the drilling assembly. The drill is started by setting off an alarm in the same manner as a regular drill. When the alarm sounds, the drillpipe is either in the rotary with an open tool joint immediately above the rotary or in the elevators. This may require continuing to pull and disconnect a stand or lowering the pipe back in the hole if a stand is only partially out of the hole. In either case a safety valve, called a *drillpipe preventer,* is connected into the top of the drillpipe, and the drillpipe rams are closed. The valve is used to close the inside of the drillpipe and has an upper connection that fits the kelly. The safety valve should always be available on the rig floor ready to be picked up with a lifting line and connected. The safety valve and the pipe rams are closed to complete the drill.

Note that the drillpipe safety valve is open when it is connected to the drillpipe. It is important for the valve to remain open until it is connected. If there is a flow of mud out of the drillpipe, a closed valve cannot be connected without great difficulty as a result of the immediate accumulation of pressure. If the valve is open, fluid can flow from the drillpipe through the open valve while the valve is being connected. The valve can then be shut off to control the flow. A similar valve is kept available on the rig floor when running casing to help control a blowout if it occurs at that time.

DRILLING PROBLEMS

The intermediate hole section is the first long section of hole drilled, and it usually penetrates many formations. Mechanical problems such as equipment failures can occur. The risk of encountering hole problems understandably increases with the number of formations drilled and the length of the open hole. The time spent drilling the open hole is also critical; a few hours or days after being drilled a formation can begin to slough and cave—problems associated with fluid-sensitive shales. The risk of all of these problems occurring decreases with good maintenance and prudent operations, but such situations have not been eliminated.

Problem Formations

The major part of the geologic section is composed of shales. Many shales can be drilled without difficulty, but there are a number of problem shales. A fluid-sensitive shale has physical or chemical characteristics that are changed when the shale is contacted by mud filtrate. The most common alteration is called *swelling* shale because it tends to swell when certain types of mud filtrate penetrate the shale lattice—the molecular arrangement in the shale. As the shale swells, it tends to slough and cave into the wellbore, causing a tight hole or possibly a stuck pipe.

Sometimes the shales can be drilled safely by carefully controlling the physical and chemical properties of the drilling fluid. One method is to maintain a low fluid-loss mud so less filtrate is available to contact the shales. Another method is to select a mud with an inert filtrate that will not affect the shales, such as oil or invert emulsion mud. Inhibited mud systems such as phosphate and salt-saturated muds also may be effective.

Polymer muds effectively coat the formations so they cannot be contacted by the mud filtrate and, therefore, cannot cause a problem. In most cases shales can be drilled with common muds. However, when there is a severe shale problem, it may be necessary to convert to a more expensive special mud.

Fractured formations, or fractured shales, may be highly fractured so, after drilling, the formations are unstable and may fall into the borehole. These can often be drilled using good mud and a specialized drilling procedure shown in Fig. 8–5. The walls of the borehole often stabilize after a period of time. In some cases increasing the mud weight increases the hydrostatic pressure and helps prevent caving. In severe cases it may be necessary to cement the formation to stop it from caving. The cementing procedure used is similar to that used to help cure severe lost circulation.

Geopressured shales and sometimes other formations may cause a problem when drilling with a lighter-weight mud. The formations tend to slough, cave, and spall into the wellbore, creating a risk of sticking the drilling assembly. The two methods of drilling through these formations are available. Allow the shales to slough until they stabilize while carefully handling the drilling assembly to prevent it from sticking. If the shales do not stabilize within a reasonable period, the alternative is to increase the mud weight.

Lost-circulation zones are a common problem formation. The method of handling the problem often depends on the rate of mud loss, usually measured in barrels per hour. If the lost-circulation rate is low (5–15 bbl/hr), the circulations rate can be reduced temporarily, the mud weight can possibly be reduced, and good mud properties can be maintained. Often, minor lost circulation reduces with time as a mud filter cake is deposited on the formation face.

If there is a higher rate of loss, lost-circulation material can be added to the mud system. Then lost-circulation pills can be added as necessary, and drilling ahead should proceed if possible until the hole heals. If this does not correct the problem, lost-circulation material may be tried again in higher concentrations. If this is not successful, the next step is to plug off the lost-circulation zone with cement as described.

Intermediate Hole Section

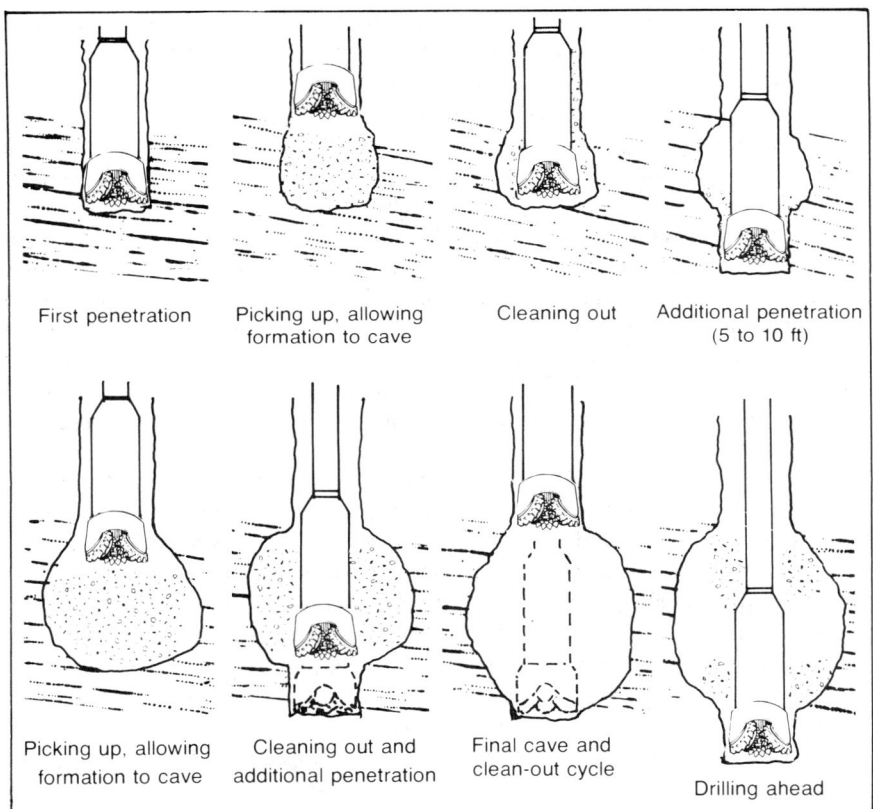

FIG. 8–5 Drilling fractured, caving, and sloughing formations

In severe cases of lost circulation, it may be necessary to run casing. Casing is run only as a last resort because of the expense of this operation. Also, if the casing is run, it may reduce hole size in deeper sections and, thus, may limit the maximum depth to which the well can be drilled.

Contaminating formations are another problem that may be encountered. One of the most common of these formations is anhydrite or gypsum. The anhydrite dissolves in the mud and severely contaminates most common water-based mud. The normal procedure is to treat the mud with chemicals. When long sections of anhydrite must be drilled, special inert mud systems may be used because the cost of treating common mud can be expensive.

Salt formations can be encountered that range from thin stringers to thick massive sections. Salt dissolves in freshwater mud systems and contaminates the mud. The dissolved salt leaves an enlarged borehole. The normal procedure is to drill salt sections with salt-saturated mud.

Many problem shales and other formations are encountered. These are penetrated by carefully using good drilling procedures, using the correct type of mud system, and maintaining good physical and chemical mud characteristics.

Caving material, usually shale, also can accumulate on top of the bit, drill collars, or other obstructions such as tool joints. The driller must be aware of a caving situation to prevent sticking the assembly. Sometimes caving is difficult to detect because the drilling mud that is circulating upward may tend to suspend the caving material so a large volume can accumulate. However, this situation may not actually cause a problem until the pump is shut down. Then the material settles around an obstruction on the drilling assembly and can cause the assembly to stick.

If the driller shuts off the pump and pulls up into these cavings, the drilling assembly may stick. Restarting the pump may not lift the caving material; however, this operation should be tried. The normal procedure when drilling in caving formations or if cavings are suspected is to pick up the complete drilling assembly periodically at least 30–45 ft. It is picked up carefully first with the pump running and then again with the pump shut off or reduced to a low circulation rate. This procedure warns the operator of a pending caving problem. The drilling assembly and pumping pressure can be varied to help remove the caved material so drilling operations can be resumed.

Good drilling practices and procedures are important when drilling problem formations. The driller must also know the strength characteristics of the drillstring and how much can be pulled if it becomes stuck while drilling problem formations. For example, assume a well has 9⅝-in. surface casing set at 2,500 ft and is drilling an 8¾-in. intermediate hole section at 7,500 ft. In this operation 4½-in. grade E, 16.60-lb/ft drillpipe is being used with a minimum yield or tensile strength of 331,000 lb.

The minimum tensile strength is the point at which additional loading causes the pipe to deform or to start stretching prior to failure by parting. The strength above the minimum tensile cannot be accurately predicted. Therefore, the minimum tensile strength is the commonly accepted, maximum usable strength of the pipe.[2] For example, a 900-ft limber drill-collar assembly weighs 70,000 lb in air, and 6,600 ft of drillpipe has an air weight of 109,000 lb for a total drilling assembly weight in air of 179,000 lb. The drilling mud weighs 10

---

[2]*Halliburton Cementing Tables,* Halliburton Services, Duncan, Oklahoma, Section No. 200 "Dimensions and Strengths," pp. 10–11.

lb/gal with a buoyancy factor of 0.85. Therefore, the actual weight of the drilling assembly is 152,000 lb in mud.

Strength characteristics are for new drillpipe. A safety factor of at least 20% reduces the maximum pull to 288,000 lb. If the drillpipe is used, it is further derated by a larger safety factor. Standard API tables can be used to determine how much the pipe is derated based on usage and inspections. For the purposes here it is assumed that the drillpipe is derated 20% so the maximum allowable pull is 264,000 lb.

Therefore, when the drillpipe is hanging in the hole filled with 10 lb/gal mud, the pipe weighs 152,000 lb, and the maximum allowable pull on the top joint is 264,000 lb. The difference between the maximum safe tensile loading and the actual pipe weight is 112,000 lb or the maximum safe overpull. In other words if the bit or any other part of the drilling assembly becomes stuck, it could be pulled with a total of 264,000 lb or 112,000 lb of force exerted on the formation at the point where the drilling assembly is stuck, assuming it is stuck near bottom.

The concept of overpull is very important in overall drilling operations. The weights and strengths given are for the drilling assembly below the rotary. The hook load is the actual weight plus the weight of the traveling block and swivel, kelly, elevator bails, and other equipment suspended from the traveling block. The weight of this surface equipment is constant for the equipment being used and is added to both the actual weight and the maximum safe tensile loading. In this case the actual weight shown on the weight indicator may be 167,000 lb, indicating there is 15,000 lb of weight above the top joint of drillpipe. Therefore, the maximum safe tensile loading will be 264,000 lb plus 15,000 lb or 279,000 lb. This weight is determined by the top joint of drillpipe.

When tapered drillstrings are used, it is also necessary to calculate the maximum safe tensile loading on the top joint of the smaller drillpipe. Normally when tapered drilling assemblies are used, the relative length of the two sizes of drillpipe are adjusted so a maximum safe tensile loading of the upper joint of the larger drillpipe is reached before the upper joint in the smaller drillpipe is loaded.

In the example 4½-in. drillpipe is used in an 8¾-in. hole below 9⅝-in. surface casing set at 2,500 ft. The same general principles apply with different drillpipe, casing, and hole sizes. In summary efficient operating techniques and an effective drilling mud must be used in problem formations. In severe cases it may be necessary to change the mud system to a different type of mud, to treat the formations such as cementing, and as a last resort to run casing.

Deviation, Dogleg, and Keyseats

Some formations cause the borehole to deviate from the vertical as it is drilled. This tendency to deviate is strong in some formations and less effective in others.

The actual reason why or how the formation causes the hole to deviate is not known. In general, the wellbore tends to deviate updip into the dip of the formation until the minimum angle between the borehole and the plane of the formation approaches 45°. At that point the hole begins to deviate downdip. Alternating hard and soft (layered) formations have a stronger effect on hole deviation.

Several theories have been proposed to explain how the formations affect wellbore deviation. It is important to recognize that the formations may cause a hole to deviate, sometimes with high deviation, even though the reason is obscure or unknown.

A dogleg occurs when the direction of the hole changes. This is caused by the natural tendency for hole deviation, the type of drill-collar assembly, and the methods or techniques used to drill the hole. For example, a limber drilling assembly often causes the hole to build angle, i.e., to increase the angle of deviation, sometimes known as an angle-building assembly. High bit weights and low rotary speeds may further increase the angle-building tendency of the limber drill-collar assembly. Low bit weights and a high rotary speed tend to drill a straight, vertical hole or maintain or reduce angle in a deviated hole.

Deviation and dogleg can cause problems while drilling the well and during the later producing life. The extent and severity of these problems increase with higher deviation and dogleg angles, hole depth, and usually with increased total drilling time. Deviation and dogleg cause drag when pulling the drilling assembly. It is not uncommon for deep holes to have 20,000–30,000 lb of drag in a crooked hole. High drag effectively reduces the amount of overpull available to the driller in case it is needed to overcome other hole problems such as caving formations.

High-angle deviation and dogleg can cause other problems during drilling, such as increasing the difficulty of cleaning the hole properly and decreasing the stability of the formations, which causes sloughing and caving. Running logging tools in crooked and deviated holes is more difficult than when run during normal operations. It is sometimes difficult to run casing into high-angle holes, especially crooked holes. Additional centralizers may be needed, and it may be difficult to obtain a good cement job on the casing. Casing set in deviated and crooked holes is subject to more wear and possible casing leaks, so extra drillpipe protector rubbers are needed when drilling below the casing.

If the dogleg is controlled so that it does not create a problem while drilling, it normally does not affect production operations. Production casing set in deviated and crooked holes is subject to more wear than casing set in straight holes. Pumping oil wells may require special conditions such as longer-stroke pumping units, reduced pumping rates (strokes per minute), and rod guides. In severely deviated holes a different type of artificial lift such as submersible electric pumps or hydraulic pumps may be needed.

A deviated or dogleg section is especially dangerous during drilling because it can cause a *keyseat*. A keyseat is a worn, cutout section in the side of the borehole. It can occur at a point where the hole changes direction (dogleg), and it is created by the rotating and reciprocating motion of the drilling assembly. Keyseats generally have a smaller diameter than the regular hole diameter. The drillpipe and tool joints can usually be pulled up through the keyseat section, although the tool joints may tend to hang up and cause the assembly to become stuck. The top of the drill collars, reamers, and stabilizers or a bit that has a larger diameter than the drillpipe and tool joints may become lodged or stuck in the keyseat. This can lead to a costly fishing job and possibly a lost or junked hole that must be sidetracked or redrilled.

Multiple keyseats may occur in long, open-hole sections. In this case the severity of the lower keyseat contributes to the growth of the upper keyseat, and this in turn increases the risk of sticking the assembly in the upper keyseat. Long keyseats may be formed at relatively low dogleg angles, in some cases as low as $2°/100$ ft. These keyseats tend to occur in holes with higher deviation and, fortunately, are not common.

As noted, keyseats cause severe problems. The elusive nature of the keyseat is another problem in itself. The keyseat may not be detected until the hole has been drilled hundreds or even thousands of feet below the keyseat point. It is not uncommon to stick the drilling assembly in a keyseat at a relatively shallow depth during a trip while drilling at a much deeper depth. If this causes a fishing job and a sidetracked hole, a long, deep, and usually costly section of original hole is lost.

Keyseats usually warn the operator that they are forming by the action of the weight indicator during trips and connections. However, these warning signals may be very subtle, and the operator must be vigilant to detect them. These signals include increased weight and drag and periodic weight fluctuations often at about 31-ft intervals as the assembly is picked up.

Theoretically, keyseats can be prevented by drilling a straight, vertical hole. However, this is almost economically impossible because of the time and expense involved. The best practical procedure is to limit deviation and dogleg so the risk of forming a severe keyseat is minimized. In a 10,000-ft hole this deviation is about $3°/100$ ft with a maximum deviation of $6°/100$ ft. A maximum deviation is set relative to keyseat formation because continued deviation at $3°/100$ ft would create a long section of hole favorable for the formation of a long keyseat.

Dogleg angle must be further limited in the upper section of deep holes because a higher horizontal force is caused by the longer, heavier drilling assembly used to drill the deeper hole. For a 15,000-ft hole, the upper 7,000–10,000 ft should be limited to a dogleg angle of about 2½°/100 ft and a maximum deviation of about 5°/100 ft. For a 20,000-ft hole the minimum and maximum should be 2°/100 and 4°/100 ft, respectively. These limits of angle are approximate and are subject to change based on experience and knowledge in the area.

Deviation and dogleg normally are limited or controlled by various drilling procedures described in the following section. However, if the formations in the area tend to cause high deviation and if experience indicates that this will not adversely affect drilling, completion, or production operations, higher limits should be allowed. Controlling deviation increases drilling costs, and the increase can be appreciable where the formations cause high deviation.

In some areas formations exhibit a strong deviation tendency, and the direction that the hole will deviate may be known. Moving the surface location may be justified in order to allow the hole to deviate normally, providing it enters the prospective producing formation in the correct location. Drilling the naturally deviated hole in this case may be more economical than the additional cost required to drill a straight, vertical hole.

High deviation should be avoided when possible. However, holes have been deviated to angles in excess of 60°, and experimental holes have actually been deviated until they drilled horizontally to drain gas out of coal beds.

Controlling Deviation and Dogleg An example of a drilling operation can be used to illustrate how deviation and dogleg are controlled while drilling. Consider a conventional hole with 9⅝-in. surface casing at 2,500 ft. An 8¾-in. hole is being drilled at 3,500 ft with a limber drilling assembly including 7-in. drill collars and 4½-in. drillpipe. This is an angle-building assembly. However, it is a common drilling assembly and often does not cause the hole to build angle, especially in soft formations. Further assume that the hole is being drilled at 30 ft/hr with 40,000 lb of bit weight and a rotary speed of 70 rpm. The average hole deviation is about 2°/100 ft.

As drilling progresses, the hole gradually gains angle until the deviation is 4°/100 ft at 3,800 ft. Sometimes the hole angle will build up to some reasonable value—for example, 8–10°—and then remain constant. The operator may allow the hole to build angle. However, in the normal case where the hole builds 2° of angle in the last 300 ft (from 3,500–3,800 ft), the standard procedure would be to start controlling the hole deviation.

The first deviation control procedure is to decrease the bit weight and increase the rotary speed. Theoretically, this should give about the same penetration rate but allow the bit to cut more on the low side of the hole, i.e., downward toward the vertical, thus maintaining or reducing the deviation. If this is successful, the operator will probably continue with the same procedure. The bit weight may be further reduced and the rotary speed increased up to a safe limit if the earlier changes are not effective or do not drop the hole angle satisfactorily. Generally, the penetration rate is affected more by bit weight than rotary speed so continued reductions in bit weight reduce the overall penetration rate.

If the inclination is reduced with this procedure but the penetration rate is limited by the lack of bit weight, the operator may elect to run a *packed-hole assembly*. On this assembly stabilizers and reamers are run in the drill-collar assembly near the bit. These large-diameter tools cause the bit to drill in the same direction as the upper hole, thus reducing deviation. There are various types of packed-hole assemblies using different combinations of reamers, stabilizers, and short drill collars. The operator normally tries to drill ahead until the bit is worn out or the drilling rate is reduced to the point that the trip is justified. The limber drill-collar assembly is pulled from the hole, the packed-hole assembly is made up on the bottom of the drill collars, the worn bit is replaced, and the assembly is run back into the hole.

This assembly normally must be run in the hole carefully if it is the first packed-hole assembly because the full-gauge diameter of the stabilizers and reamers could cause the assembly to hang up and become stuck. If sticking or hanging up is indicated by a reduction of weight on the weight indicator, the kelly is put on and the section is reamed. Normally the assembly is run to about 100 ft off bottom, and the last 100 ft of the hole are reamed.

Normal bit weight and rotary speed (40,000 lb and 70 rpm) are used to drill ahead. The packed-hole assembly increases the overall stiffness of the drilling assembly and is often referred to as a stiff assembly. In theory, the assembly tends to drill a straight hole in the same direction as the earlier hole because it is stiff and rigid. Normally it can drill a relatively straight hole. However, the formations may tend to cause the hole to deviate. This tendency also occurs with a stiff assembly, but the angle of buildup will be much less than that with a limber assembly. Sometimes the rate of change of deviation can be reduced by as much as 75%. For example, if the hole tends to build 2°/300 ft, the packed-hole assembly could reduce this to 2°/1,200 ft. This reduction helps prevent doglegs.

Several types of stiff assemblies are available. The main differences in these assemblies are the length of the lower drill-collar section that is packed or stiffened, the outside diameter, and the number and position of the stabilizers or

If deviation continues to increase, the next procedure is to drill with a pendulum assembly (Fig. 8–6). The pendulum drilling assembly uses one or

Fig. 8–6

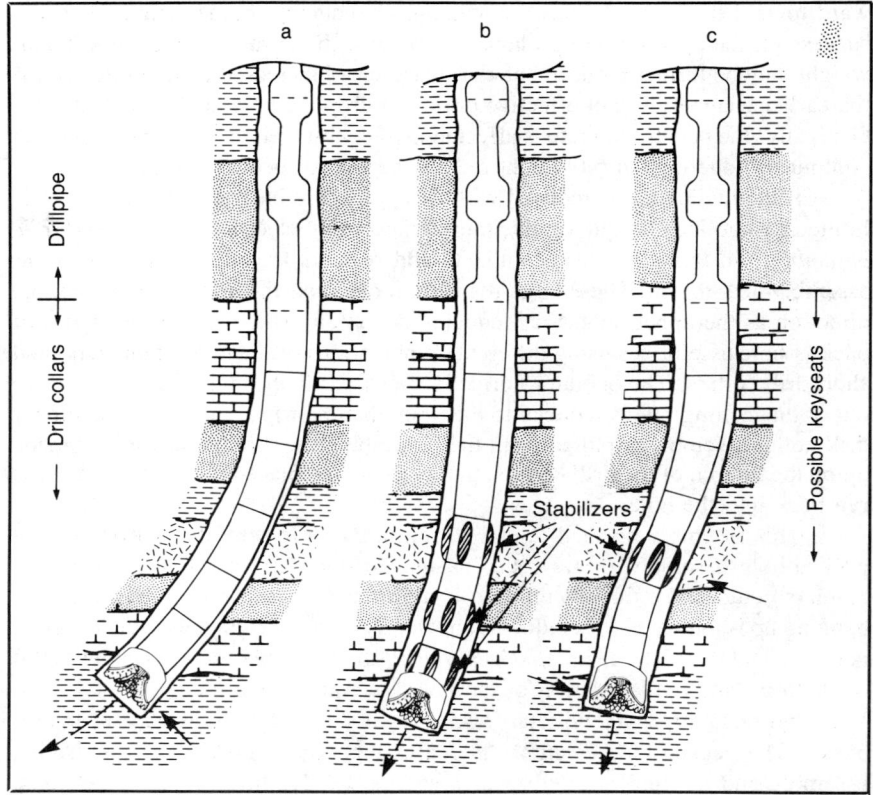

FIG. 8–6 Crooked-hole (a), packed-hole (b), and pendulum (c) assemblies

more stabilizers and reamers (as stabilizers) located 60–90 ft above the bit as a fulcrum. The weight of the drill-collar assembly below the fulcrum acts in the vertical direction to cause the bit to drill on the low side of the hole and reduce the hole angle. It also can maintain the deviation while drilling in formations that tend to increase the hole angle. The distance between the bit and the fulcrum point is precisely determined based on hole size, drill-collar size, bit weight, and other parameters.

The pendulum is made up by adding equipment to the lower drill-collar assembly. It is run into the hole, and drilling is resumed at a specified bit weight and rotary rpm. A packed pendulum assembly is used where more wall support is needed for the fulcrum, such as soft formations and out-of-gauge hole. It also reams out crooked holes and reduces the risk of doglegs.

Drilling the hole within certain limits of deviation and dogleg may require measuring the hole angle many times. The regular procedure of measuring the hole angle after the bit run may not provide enough measurements. Running the hole-angle measuring device on a wire line is time consuming, and there is a risk of sticking the assembly. Tools have been developed that can be run on the lower drill-collar assembly and that provide a surface reading of the hole angle in a relatively short period of time with a minimum risk of sticking. Such an operation is called *measurement while drilling* (MWD). The tools used are well justified when severe deviation and dogleg problems are encountered while drilling.

To illustrate how hole deviation occurs and keyseats are formed, detected, and cleaned out, consider the drilling well previously described. The hole angle increases to the point that action must be taken to control it. It would not be unusual to drill as much as 500 ft while testing the various procedures before finding one that controls the angle in a satisfactory manner. There would then be a section of hole from about 3,500–4,500 ft that could have several points where the dogleg angle ranged from 2°/100 ft to over 6°/100 ft. Normally the angle is not much higher than this because the operator would intentionally deviate the hole to straighten it and reduce the dogleg angle.

These points with a high dogleg angle become potential areas for the formation of keyseats as the hole is drilled deeper. Most keyseating occurs at the top of the drill collars, but the bit, reamers, and stabilizers on the drill-collar assembly and any point where the drill-collar diameters change can also become stuck in a keyseat. It is a common practice to run larger drill collars on the bottom of the assembly. The point of change from the smaller drill collars to the larger drill collars can become stuck in a keyseat.

The keyseat zone (from 3,500–4,500 ft) would probably not cause a problem until the bit was about 900–1,00 ft below the bottom of the zone. The drill-collar assembly normally is about 900–1,000 ft long, and the hole usually must be drilled until the top of the drill collars is below the keyseat zone to allow time for the keyseat to develop. This occurs from the rubbing-scraping action of the drilling assembly while drilling and tripping.

On trips when the drilling assembly is pulled out of the hole, obstructions on the drilling assembly such as drillpipe tool joints, the top of the drill collars, stabilizers, reamers, and the bit tend to drag as they are pulled through points where a keyseat is developed. The temporary increase in drag is 2,000–5,000 lb for a minor keyseat and 10,000–30,000 lb for a moderate to severe keyseat.

In very severe cases the drag increases until the drilling assembly is stuck. It is more difficult to detect a possible keyseat while running the assembly into the hole. However, in some cases short weight reductions occur as the restricting keyseat tends to hold up the downward moving assembly. These indicators are often relatively small, but they can be detected by a vigilant operator. It is important to detect the keyseat at an early stage of growth so it can be removed before the assembly becomes stuck.

A keyseat wiper should be installed on top of the drill collars any time a keyseat is suspected. The tool can be used to help remove or wipe out keyseats. It may also be used to help release the drilling assembly if it becomes stuck in a keyseat. The blades on the keyseat wiper scrape the smaller-diameter keyseat and enlarge it. This allows the drill-collar assembly to be pulled through the keyseat without sticking.

In operation, the assembly is lowered and picked up several times so the keyseat wiper blades can wipe the keyseat. The tool works fairly well to remove keyseats in soft formations but is less efficient in harder formations. It also helps release the assembly stuck in a keyseat at the top of the collars in both soft and hard formations.

The bit may be used to ream the keyseat section. The remaining procedure is similar to that described for reaming the hole after a connection except that a longer section must be reamed. Severe keyseat sections usually cannot be reamed efficiently with the bit. Sometimes the foregoing procedure keeps the keyseat wiped out to the point that the well can be drilled satisfactorily. If the keyseat continues to cause a problem, it can be reamed with a keyseat wiping assembly.

If this procedure is not effective and the keyseat section continues to cause a problem, it may be reamed with a string reamer. The string reamer is a conventional type of reamer with tapered ends designed to be run in the drillpipe string. It is connected in the drillpipe several hundred to a thousand feet above the top of the drill collars, depending on the location of the keyseat section relative to the bottom of the hole. The assembly is run in the hole until the string reamer is positioned near the top of the keyseat section, which is then reamed.

This is a higher-risk procedure than normal reaming. The heavy pipe and collars suspended below the reamer are rotated freely and have considerable momentum. If the reamer hangs up, the lower assembly can back off because of its momentum and drop down the hole, causing a fishing job. Both ends of the string reamer are subject to higher stresses, increasing the risk of a connection failure and a resulting fishing job.

The string reamer may be positioned in the assembly so it is near the top of the keyseat section when the bit is on bottom. The reamer wipes out the keyseat as the hole is drilled. On subsequent trips to change bits, the reamer can be repositioned as necessary to make another reaming pass through the keyseat section. This procedure of reaming while drilling is also a high-risk operation.

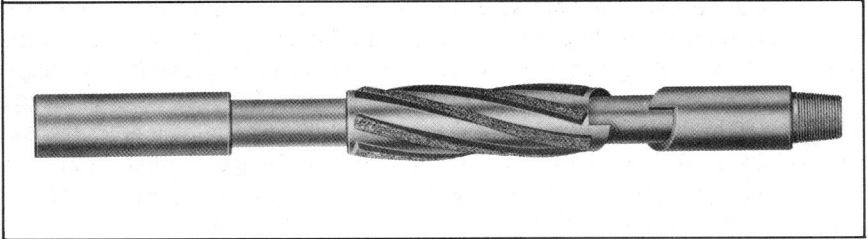

FIG. 8–7 Keyseat wiper (courtesy Henderson Tool Company)

Reaming before making connections while drilling is a straightforward, simple procedure known as reaming on bottom. Reaming off bottom indicates reaming operations conducted in the upper, open-hole sections. The main reasons for reaming off bottom are to clean out bridges of accumulated cuttings and caving material to open up an undergauge hole section, and to straighten a crooked hole section to wipe out keyseats or prevent them from forming. Reaming off bottom can cause higher stresses in the drilling assembly, which in turn increases the risk of equipment failure.

There is also a risk of sidetracking the hole when reaming off bottom. A hole is sidetracked when a new hole is drilled out into the formation through the wall of the old hole and at some distance above the bottom of the old hole. When sidetracking occurs, the drilling tools frequently do not enter the old hole, which is then lost.

There have been cases where a sidetracked hole has been drilled to an appreciable depth and later the drilling tools reentered the old hole. Often, the assembly cannot be forced back into the sidetracked hole. Therefore, it is very important to conduct the reaming operations off bottom to prevent sidetracking.

Various procedures help prevent sidetracking, such as using a duller bit or positioning the reamers in the drilling assembly so the bit will be in a straight section of hole while reaming. The best method in this case is to ream the hole so the drilling assembly is lowered at a rate of 3–10 times faster than when the hole was drilled. This drilling speed reduces the risk of sidetracking.

The severity of a keyseat problem may be further increased by *wall sticking*. Wall sticking occurs when the drilling assembly lies against the wall of the borehole, usually opposite sands or permeable sections. Under certain conditions a surface area on the assembly and a corresponding area on the wall of the borehole can become sealed as a result of the accumulation of wall cake around the periphery of the area. The hydrostatic pressure caused by the mud column is normally higher than the pore pressure of the formation. This differential pressure—the hydrostatic mud pressure less the formation pore pressure—is applied to the sealed area and effectively sticks the drilling assembly to the wall of the borehole. This sticking action is similar to a flat stopper over a drain in a bathtub, except it is applied in the horizontal direction. The weight of the water in the tub holds the stopper in place much like the weight of the mud column holds (sticks) the drillpipe against the wall of the borehole, but the force of the mud column is many thousands times greater. This wall-sticking condition can occur very rapidly under favorable conditions.

When the keyseat occurs in a permeable zone, all conditions generally are favorable for wall sticking the assembly. This action complicates and increases the difficulty of releasing the keyseated assembly because the tremendous forces exerted hold the assembly against the side of the wellbore or in the keyseated section. The assembly can become wall stuck in a straight, vertical hole. However, the risk of wall sticking increases with increasing hole deviation, and there is a higher incidence of wall sticking in deviated holes.

STUCK ASSEMBLIES AND FISHING

A blowout is unquestionably the worst thing that can happen in a drilling and completion operation. The second items on this infamous list and standing far above any other problems are stuck assemblies and fishing operations. A stuck assembly, or stuck pipe, occurs when the assembly in the hole becomes stuck so it cannot be pulled. The assembly can become stuck as a result of keyseating, wall sticking, sloughing and caving around the assembly, and many other reasons. *Fishing* is broadly defined as any operation or procedure conducted to release, remove, or recover the stuck assembly, part of the stuck assembly, or other material that adversely affects the drilling and completion operation.

Stuck pipe and fishing are severe problems. They increase costs that far exceed the cost of any other drilling and completion problem, including blowouts. Blowouts are classed higher in severity because of the risk to human life. Stuck pipe and fishing result in extra operating expenses to release the stuck assembly or to fish it out of the hole. Total rig operating expenses while conducting these operations can be 50–75% higher than normal drilling operations. In addition, it may be necessary to sidetrack the hole if the fish cannot be recovered. Also, there is a higher risk of losing the hole so that it must be completely redrilled. Both alternatives are costly.

Stuck pipe and fishing are caused by many factors. Most of these are controllable, but some are either uncontrollable or unknown. The factors that cause stuck pipe and fishing can be generally grouped into two classifications: (1)

inadequate drilling program design and (2) faulty operations. In some cases the cause cannot be determined because of a number of contributing factors.

The drilling program is prepared by competent personnel after studying all of the data available. But sometimes a drilling program cannot anticipate all of the problems because information is unavailable. Nevertheless, the program usually can provide for an additional string of casing if it is needed. This is an expensive remedy but a much better alternative than losing the hole. An inadequate number and size of casing strings, drilling too small of a hole at deeper depths, not limiting deviation and dogleg, and selecting the wrong type of mud and properties are some of the causes of stuck pipe and fishing. These situations are attributed to the drilling program design. Therefore, the drilling program must be carefully prepared after investigating all sources of information. It should also delineate areas where the program is weak and generally should provide a reliable, detailed guide to the operating group.

Faulty operations can also cause stuck pipe and fishing. The operations group bears the brunt of the problem, but this same group also successfully drills and completes many wells. The complexity of the problem of preventing stuck pipe and fishing is easily demonstrated by considering a 10,000-ft drilling assembly that contains over 300 tool joint connections or over 600 individual boxes and pins. Each one must be constructed by precise machining. Each must be selected by type and size to withstand the high torque and tensile forces involved in drilling operations. Each must be cleaned, greased, screwed together, tightened to precise torque limits, inspected periodically, and operated within designed limits. All of these precautions must be observed on an almost constant basis as the drilling assembly is used. Failure to observe these precautions can lead to a tool-joint failure and subsequent fishing job.

As another example the operator is faced with pulling the drilling bit when it becomes worn. If a bit is pulled too soon, called *pulling a green bit,* the maximum efficient drilling life has not been obtained. If a roller bit is run too long, it may leave cones in the hole and cause a fishing job. A leak in the downhole assembly may cause a slight reduction in pumping pressure. If this reduction is not detected immediately, it can cause a washed out tool joint and can lead to a fishing job. The operator is frequently faced with drilling ahead under minor caving conditions and risking stuck pipe or spending more time cleaning out with associated higher costs. A minimum mud weight must be selected to optimize the penetration rate. However, if the mud weight is too low, there is the risk of a blowout. Inspection procedures at the pipe yard may be inadequate, so the operator uses pipe that fails prematurely. Formation dips may change abruptly causing crooked hole that is not detected immediately. While drilling deeper, a keyseat develops in the crooked hole section causing stuck pipe.

These are only a few of the many factors that can cause stuck pipe or a fishing job. They emphasize the importance of conducting operations in a prudent manner with experienced, competent personnel. The best method of eliminating problems resulting from stuck pipe and fishing is to prevent them.

Stuck Assemblies

A downhole assembly is considered to be stuck when a restriction in the hole prevents the entire assembly from being pulled from the hole. Assemblies become stuck as a result of such problems as wedging the bit in a tapered hole, caving formations or bit cuttings settling around the assembly, sticking the assembly in a keyseat, or malfunctioning downhole tools.

As soon as it has been determined that the assembly is stuck, operations are begun to work the pipe. Working the pipe includes reciprocating, rotating, twisting, and circulating actions performed individually or in combination to release the stuck pipe. It is important to begin working immediately after the assembly is stuck and to continue working until it is released or until the driller determines that continued working will be ineffective and other actions must be taken.

The assembly is worked to release it so operations can be resumed. Working the pipe frequently accomplishes this end by wearing and eroding the material that has settled and stacked around a tool joint, securely sticking the assembly. As the pipe is pulled upward, the tool joint exerts an upward force on the pieces of formation. When the pipe weight is reduced or slacked off, the tool joint exerts a downward force on the caved formation particles. As this upward-downward stressing force is repeated, it can break down the particles into smaller pieces and dislodge them to release the pipe. Rotation and torquing may accomplish the same purpose. If the hole can be circulated, the circulating mud helps to erode and break up the pieces of formation and carry them out of the hole. These pieces of formation are often considered to be relatively soft, friable, and easy to break into smaller pieces. However, this is not always the case. Many of the deeper formations are extremely tough and hard, sometimes harder than the concrete used to construct sidewalks.

To illustrate the working procedure, consider a well being drilled at 10,000 ft with a $4\frac{1}{2}$-in. drillpipe and 7-in. drill-collar assembly. Pipe often becomes stuck while making a connection, so assume that the pipe sticks after it is picked up while adding another single to the assembly. After adding the single, when the pipe is picked up off of the slips, it is found to be stuck. The first step is to start the pump slowly to see if the hole can be circulated. If so, there is a very good chance of releasing the pipe.

The pipe is first worked by pulling 20,000–30,000 lb over the pipe weight (overpull) and lowering the pipe an equal amount below the pipe weight. The overpull and extra setdown weight are applied by the pipe to the formation at the point where the pipe is stuck. The formation is stressed upward with 20,000–30,000 lb. Then the stress is reversed, and the formation is stressed downward.

If the pipe is not released in a short period, the next step is to remove the extra single and connect the kelly. This step is because, if the pipe does begin to move in the upward direction, it would be very difficult to remove a single joint that is high in the mast. Depending on conditions, the pipe can be worked at this force level of 20,000–30,000 lb for several hours. Torque can also be tried by either rotating the pipe with the kelly in the rotary or using the pipe tongs. If the pipe can be rotated even a small, restricted amount, this will help wear out and break up the pieces of formation that are sticking the pipe. Circulation also can be very helpful.

Any action that helps to release the pipe or that causes it to start moving slightly is repeated. For example, if the pipe moves upward a distance of 1 in., less weight is set down. In some cases the same amount of weight may be set down on the pipe to help establish pipe movement in both the upward and downward directions. If the pipe can be moved a small amount, continued working will often release it.

Since the bottom of the pipe is stuck, the drillpipe must stretch in order to allow pipe movement at the surface. This movement is accompanied by an increase in the pull or tension on the pipe. The drillpipe literally acts as a very heavy spring and elongates as additional pull is exerted. The pipe at the surface is normally marked at a fixed weight, usually the original weight of the free-hanging assembly. After working the pipe for a period of time, the pipe weight is adjusted until the weight indicator shows the original hanging weight. The movement of the reference mark on the drillpipe relative to the original location of the mark shows whether the pipe has been moved, in what direction, and how much.

If the pipe is worked for several hours without any detectable movement at the stuck point, the working force levels are increased. For example, the pipe may be pulled up to an overpull of 50,000 lb and set down a corresponding amount. This action increases the stresses on the formation. The pipe is worked at this level for a period, and the stress levels are increased again to the maximum safe working strength of the drillpipe.

The maximum safe working strength is based on the weight and grade of the drillpipe, tubing, or casing. The pipe may be derated for usage as described earlier. The amount of weight that can be set down is not as well defined and is usually limited to about 50,000 lb for 4½-in. OD drillpipe and is slightly higher for bigger drillpipe. Excess set-down weights can cause a failure as a result of buckling—when the pipe literally bends and breaks, usually deep in the hole because of excess weight.

Sometimes there is a tendency to exert a maximum pull on the drillpipe shortly after it becomes stuck. In most cases this is not recommended since it can cause the drilling assembly to be securely wedged so that it cannot be worked down and released. This procedure can be adjusted to work the pipe (1) to establish movement and circulation if circulation is blocked, (2) to move the pipe upward, or (3) to move the pipe downward. The method of working the pipe depends on how the assembly is stuck. If the bit is wedged in a tapered hole, it should be worked upward. If this is not successful, it is worked hard in both directions. If the assembly is stuck with the bit on bottom, the only alternative is to work the assembly upward. If the bit is run into a ledge or cavings, it should be worked upward. If there are cavings on top of the bit or drill collars, the assembly should be worked downward first.

Drilling Jars and Bumper Subs   Drilling jars and bumper subs are frequently run on drilling assemblies and usually are run on fishing assemblies. These tools are designed to deliver a sharp blow to a stuck assembly or other equipment suspended below the tools (Fig. 8-8). Drilling jars and bumper subs are run in the drill-collar assembly about 3–4 drill collars below the top. If the assembly is stuck below these tools, they are used to help release it.

The drilling jar is designed to deliver a sharp upward blow. The tool has a mandrel that allows the two ends of the tool to separate by the distance of the mandrel travel or stroke length. In operation the drilling assembly above the tool is lowered to close the tool. The pipe is then picked up. When a predetermined amount of force is exerted on the tool, a tripping mechanism releases and allows the tool to open so the top section moves upward rapidly through a distance equal to the stroke length of the tool. The moving mass is stopped abruptly at the end of the stroke length, thus imparting a sharp, upward jarring blow to the equipment connected to the bottom of the jars. The tool is often efficient in releasing stuck assemblies that would normally be released by moving in the upward direction, such as a bit in a tapered hole or a bit and drill collars stuck on a ledge or in accumulated cuttings.

The bumper sub is run below the drilling jars; it has a free-floating mandrel with a knocker or stop at the end and is sealed so mud can be circulated through the tool. The mandrel travel or stroke length ranges from about 18 in. to over 36 in., depending upon the size and type of tool. In operation the upper drill-collar assembly is picked up until the tool is completely expanded. Then the assembly is lowered rapidly to impart a downward blow as the tool closes. This blow is

Intermediate Hole Section

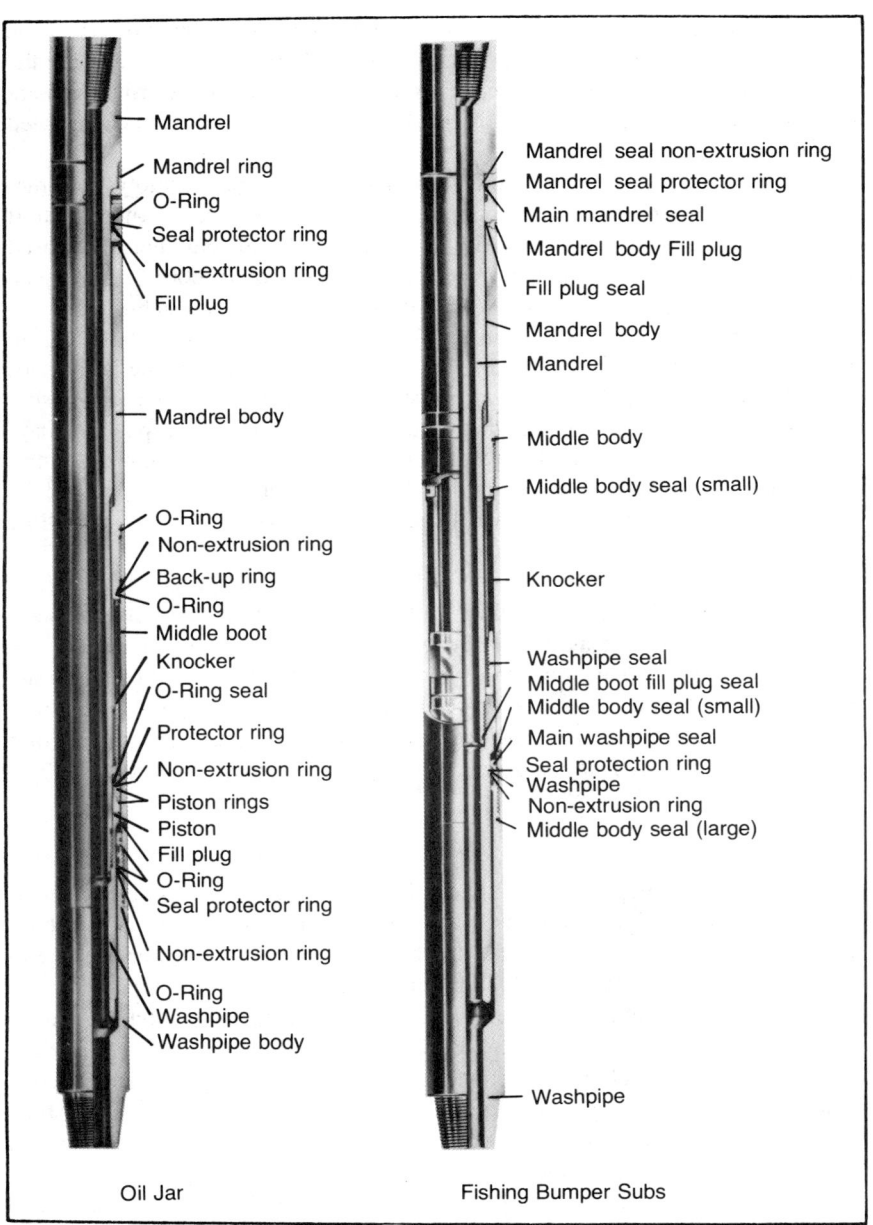

FIG. 8–8 Jars and bumper subs

transmitted to the stuck assembly or other tools connected to the bottom of the bumper sub. The action is more efficient than the slower action of lowering the pipe. It can be very efficient in releasing stuck pipe where downward movement is needed, such as an assembly stuck in a keyseat or pulled up into accumulated cavings.

The high magnitude of the forces that stick a downhole assembly are sometimes difficult to envision. For example, there have been cases where a short section of the assembly (less than 15 ft) has been stuck and could not be released after being jarred for 6–8 hr with repeated jar strokes of 70,000 lb. There is a better chance of releasing the stuck pipe if circulation can be established.

Sometimes the jet nozzles on jet bits become plugged with shale and bit cuttings, thus preventing circulation. Under favorable conditions the jet nozzles can be blown out of the bit using a string-shot charge (Fig. 8–9). The string shot uses a primer cord (an explosive) that is run through the drillpipe, positioned on top of the bit, and detonated. The force of the charge blows the jet nozzles out of the bit but normally does not harm the drill collars or bit. If the plugged jet nozzles restricted circulation, blowing the jet nozzles out of the bit may allow circulation to be established.

If the annular area around the drill collars or drillpipe is packed off—usually with caving formation and bit cuttings—then circulation may be established by perforating the assembly above the bridge with a perforating tool run through the drillpipe. This procedure generally is not as effective in releasing the stuck assembly because the assembly is normally stuck at the point where it is bridged. However, if it is also stuck at shallower depths, establishing circulation may help to unstick it. This can permit recovering a longer section of the fish by a backoff procedure.

Wall Sticking One special procedure utilizing circulation helps release wall-stuck assemblies. Wall sticking is caused by a higher pressure in the wellbore acting against a sealed area of the downhole assembly that has a lower pressure in the formation, as described. Releasing fluids, such as diesel oil mixed with special additives or oil mud, can be circulated down the hole and positioned opposite the section of pipe that is wall stuck. Excess releasing fluid is used, and the fluid is displaced past the wall-stuck section at a rate of about 1 bbl/hr. This fluid helps dissolve the mud-cake seal and allows the pressure to equalize around the stuck member. The procedure often releases wall-stuck pipe, and it is sometimes known as soaking the pipe loose.

The soaking-releasing process may take 6–12 hr. It is a good practice to work the assembly to release it during the soaking period. The releasing fluid has a higher lubricity than the normal clay-base mud, which also helps.

Intermediate Hole Section

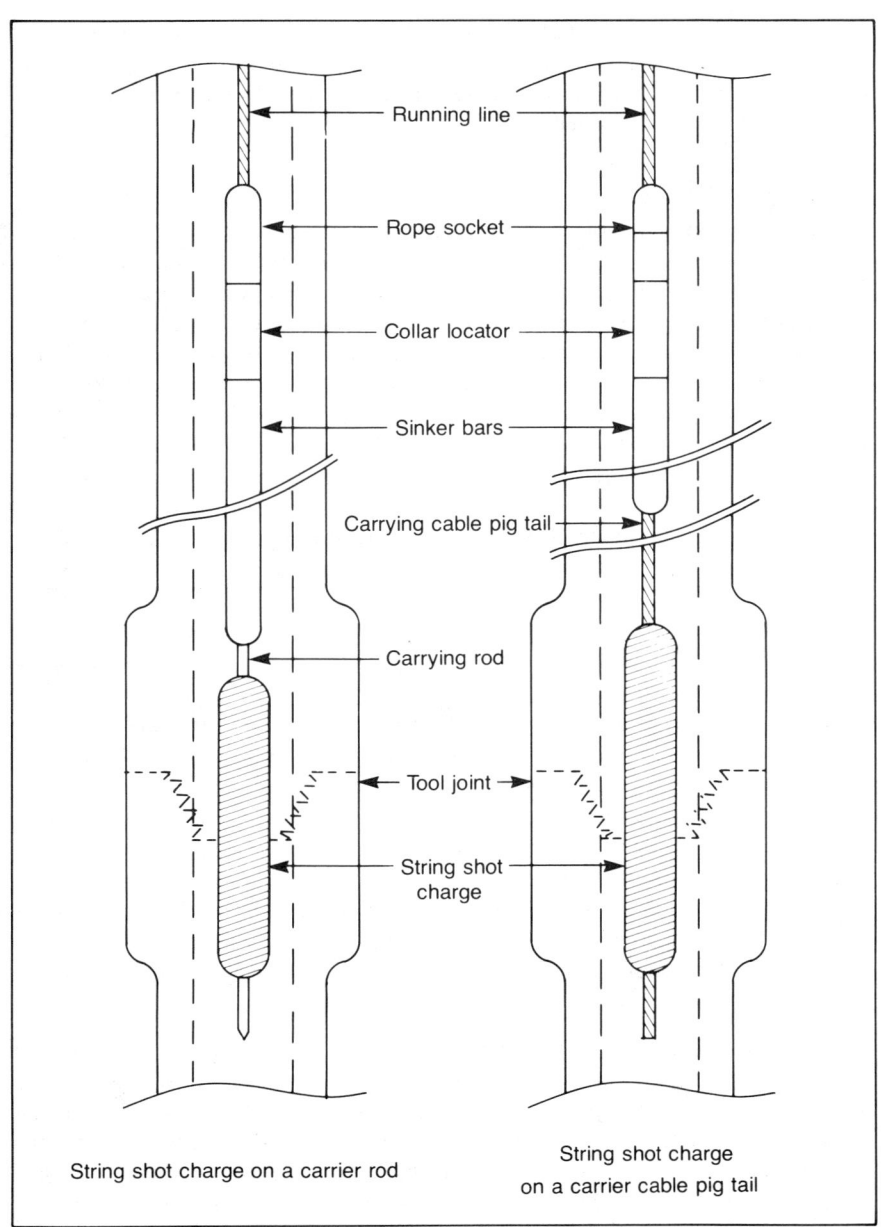

FIG. 8–9 String-shot backoff

Dissolving agents can also be used effectively in special circumstances. If the pipe is stuck by salt, fresh water may be circulated to dissolve the salt and release the pipe. Hydrochloric acid has been circulated to release pipe stuck in carbonates such as limestone.

Fishing

A *fish* is any foreign object in the borehole that restricts or prevents normal drilling and completion operations. The fish is usually metallic and in most cases is part of the downhole assembly. However, it may also be part of the rig equipment such as slip segments and tong heads that are accidentally dropped or lost in the hole. The size and configuration of a fish can range from a small item such as a preventer bolt to a very large object such as a complete downhole assembly. The fishing operation starts when the fish is lost in the hole or when a stuck assembly cannot be released.

Fishing is a specialized operation. Most operators use a fishing specialist called a *fisherman* or *fishing hand*. This person works either as an individual consultant or for a company that supplies fishing tools usually on a lease or rental basis. Some of these companies supply a few special tools and other companies have a wider assortment of tools. Other companies supply accessory services such as free pointing and backing off, which require special equipment.

Fishing operations are frequently hampered by other downhole problems such as caving formations, high-pressure zones, or lost-circulation sections. When possible, the best procedure is to cure the downhole problem before conducting the fishing operation since fishing incurs a risk of losing the hole and can lead to additional fishing. It is not uncommon to stick the fishing assembly, requiring fishing for a second or even third fish above the original fish. Working stuck pipe as described is an important part of fishing because, when the fish is caught, it is frequently stuck and must be worked loose.

Fishing operations are conducted in both cased and open holes. The same general procedures are used in both cases with some special applications in one case or the other, depending on the type of fish and the fishing tools and procedures used.

Fishing Tools Fishing tools are special tools run on drillpipe, tubing, or wire lines and are designed to help remove or recover the fish. Normally, each tool is designed to perform a separate function (Fig. 8–10). For example, an overshot catches a fish; a safety joint connects or disconnects the fishing assembly from the overshot connected to the fish; jars and bumper sub help work the fish; drill collars provide weight and increase the efficiency of the working action of the jars and bumper subs; and drillpipe or tubing is used to run and retrieve the tools and fish. This entire combination of tools is called a *fishing assembly*.

Some fishing tools are designed to be run on drillpipe or tubing. These tools are operated by either a reciprocating motion, a rotating action, or a combination of the two. Other fishing tools are run on wire lines and are operated by a reciprocating motion. These tools cannot be rotated with the wire line.

Some fishing tools are designed to catch the fish. Fish are caught by one of four methods, depending on the size and configuration of the fish:

1. *Screw-in*—If the top of the fish has a box or pin looking up, a screw-in sub or regular matching box or pin connection is run on the bottom of the fishing assembly and is connected to the fish by screwing in, similar to making a surface connection.
2. *Outside catch*—Die collars and overshots are run over the fish to make an outside catch. Die collars have internal, machine-type threads that are screwed down over the top and outside of the fish (which can be irregularly shaped) to make the catch. The overshot has internal slips that ride on a tapered inner body or use a spiral or basket grapple to make the catch.
3. *Inside catch*—Tapered taps and pipe spears are used to catch the fish with an inside catch. Tapered taps have external, machine-type threads that are screwed into a hole inside the fish (which can be irregularly shaped) to make the catch. A pipe spear has external slips that are inserted in a hole inside the fish where the slips engage the fish to make the connection.
4. *Swallowing*—Junk baskets, wash pipe with an internal pipe spear, and wash pipe with a full-opening overshot are used to swallow or completely cover smaller fish.

Overshots are one of the more important and widely used fishing tools. Various types of overshots are available for catching different types of fish. A short-catch overshot has the slips or grapple near the bottom of the overshot to catch a very short fish. Long-catch overshots have a long bowl section so the overshot can go over the fish a considerable distance and catch the fish 2–6 ft from the top of the fish. Single-, double-, and triple-bowl overshots have each bowl dressed with a different-sized slip or grapple so the combined overshot tool can catch more than one size of fish.

Large, permanent magnets can be run on pipe or a wire line to catch small pieces of junk such as bit cones and hand tools. A wire-line fish can be caught with a wire-line spear or a grab run on pipe or a wire line.

Mills are used for a wide variety of downhole cutting jobs. They have a heavy metal body shaped for the different milling jobs that the tools are designed for. The mill has one or more cutting or milling faces coated with a hard, durable cutting material such as particles of tungsten carbide embedded in a matrix material. Different sizes of particles are selected based on the type of milling operation.

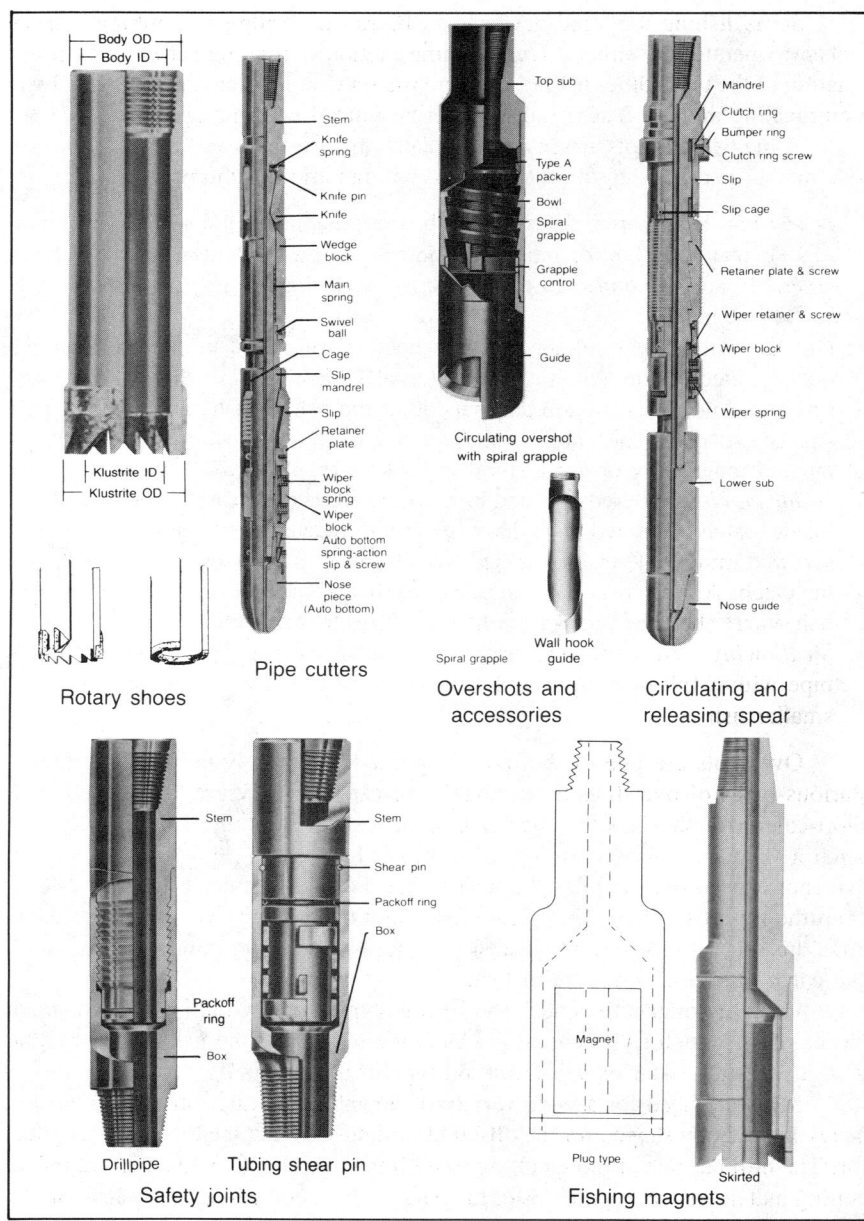

FIG. 8–10 Fishing tools

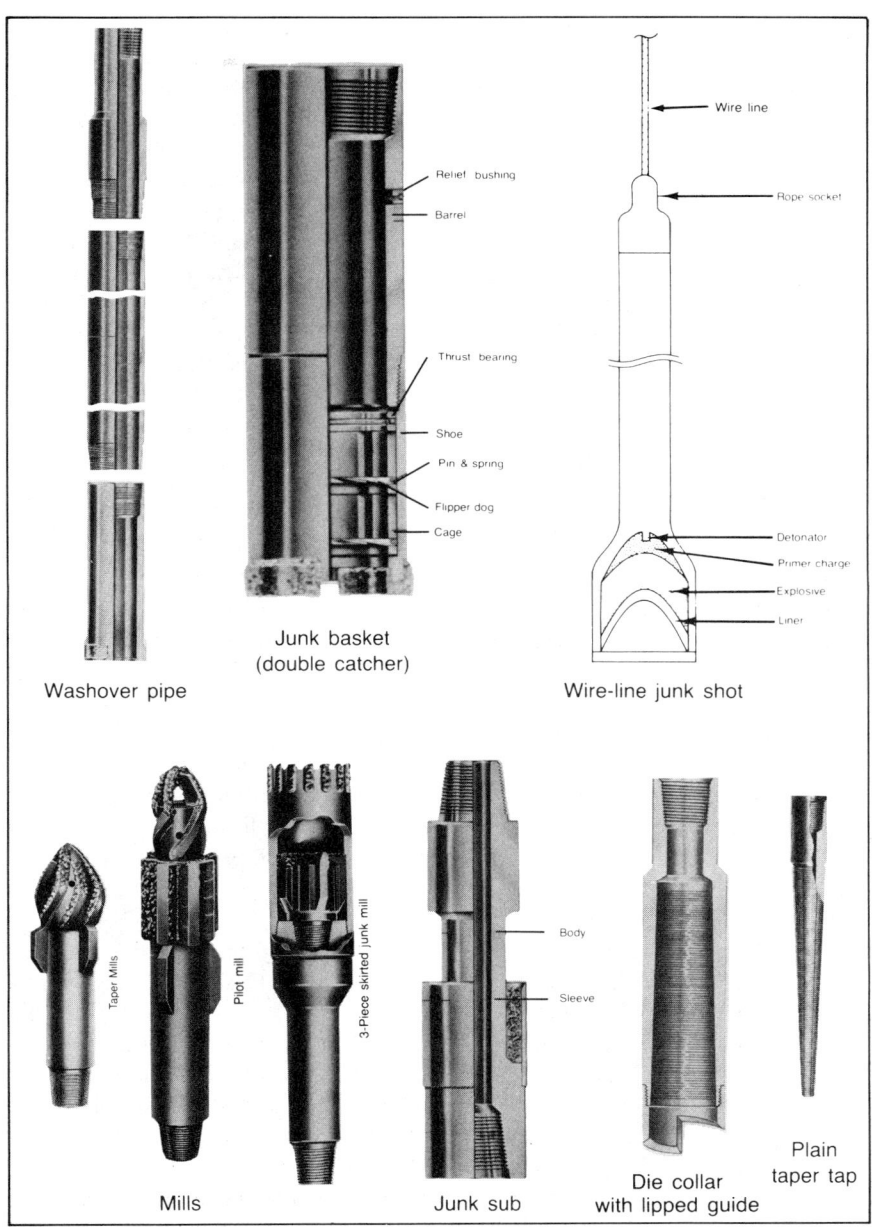

FIG. 8–10 *Continued*

Many fishing jobs and other operations require cutting or drilling metal, and regular drilling bits cannot cut metal efficiently. So mills are designed to cut metal and drill cement, especially in small holes where there is a higher risk of a bit failure. Different types of mills are designed for different cutting or milling jobs. Junk mills are used to drill a variety of shapes and sizes of fish (junk) left in the hole, ranging from bit cones to drill collars. A concave mill is a special type of junk mill that is highly efficient for drilling small pieces of junk such as bit cones. A pilot mill or flat-bottom mill can smooth or dress off a ragged fish top so it can be caught with fishing tools. Tapered mills can cut a window in casing for sidetracking, and the window may be enlarged with a reaming mill. Collapsed casing may be opened to full gauge (full inside diameter) with a reamer mill. Skirted or guided mills cut a fishing neck on a large fish so it can be caught with fishing tools. Cement mills drill out cement inside casing.

A variety of mills are available for different applications. They are highly efficient tools when used appropriately. In many cases mills are used almost as a last resort and often are the only way to complete the fishing operation successfully.

Fishing Operations The operations or methods selected to remove or recover a fish depend on the size, length, and overall configuration of the fish; the depth and size of the hole; the type, hardness, and stability of the formations; the formation problems; and other related factors. When a fishing job occurs, one of the first steps is to obtain all of the information about the fish and plan the overall fishing job making both intermediate and long-term plans. Use flexible planning and allow for alternative procedures. A key element of the plan is when to quit fishing and take other actions such as sidetracking or redrilling. This is especially important in severe fishing jobs and where the costs of fishing are very high.

The basic methods of removing or recovering a fish in the hole are one or a combination of the following:

1. Mill or drill up the fish into small pieces and *Wall-off* the pieces. Then catch them in a junk sub or circulate them out of the hole. This procedure generally applies to smaller fish such as bit cones, bits, packers, short pieces of wire line, hand tools, and other objects dropped in the hole. In some cases the smaller pieces of metal may be recovered with magnets or junk baskets.
2. Recover the fish by catching it and pulling all or part of the fish out of the hole. This procedure is used with long sections of downhole assemblies such as drill collars, drillpipe, tubing, and long wire lines.
3. If the fish cannot be recovered, start a deviated hole above the fish and drill around or sidetrack the fish to the original objective horizon.

4. If the fish cannot be recovered or sidetracked because of the shallow depth or for other reasons, the alternative is to abandon the hole, move the rig 10–100 ft, and redrill the well.
5. An obvious but seldom-used alternative is to plug the hole and abandon the entire drilling project.

Recovering lost bit cones is the most common fishing operation. The procedure is relatively simple and straightforward. Ordinarily, the work is supervised by the operations personnel without using a fishing specialist.

In order to explain the fishing operation to recover bit cones, consider a well with 7-in. drill collars and 4½-in. drillpipe drilling and 8¾-in. hole at 10,000 ft. The bit has been running smoothly, drilling steadily at 10 ft/hr with 35,000 lb of bit weight at 55 rpm. The bit then begins to run rough, exhibiting a high, fluctuating torque and varying rotary speed. The penetration rate drops immediately to 1–3 ft/hr. This indicates that the bit bearings have failed. If the bit life is approaching a normal bit run, the bit should be pulled immediately. Otherwise, the bearings will wear out completely and allow the bit cones to fall off. However, the same drilling symptoms can also be caused by drilling into a different formation. Chert nodules and dolomite stringers can also cause high fluctuating torque. Therefore, the driller may elect to drill ahead for a short period, especially if the bit has been in the hole for less than a normal run.

If the condition continues for 15–30 min, the bit should definitely be pulled. However, this also allows time for badly worn bit bearings to fail and to drop off cones. In this example the drilling assembly is pulled, but three bit cones are left in the hole. The bit cones must be removed or recovered (Fig. 8–11). Otherwise the next bit will have a reduced life and could lose its cones drilling on the cones left by the prior bit.

Various courses of action are available, and the method selected depends on a number of factors. In medium or soft formations, a junk basket may be run. The tool is connected to the bottom of the drill collars and is run into the hole. After the basket is on bottom, the pump is started and the basket is carefully drilled over the bit cones using reduced bit weight and rotary speed. Two to three ft are drilled with the junk basket to leave a plug of formation in the bottom of the basket and to help prevent the cones from falling out when the tool is pulled from the hole.

Junk-basket recovery can range from three cones to none. Assume one cone is recovered, the operator might elect to run a magnet. The tool is connected to the bottom of the drill collars and is run into the hole. After reaching bottom, the pipe is rotated for a few minutes and is pulled from the hole. One cone is good recovery for the magnet.

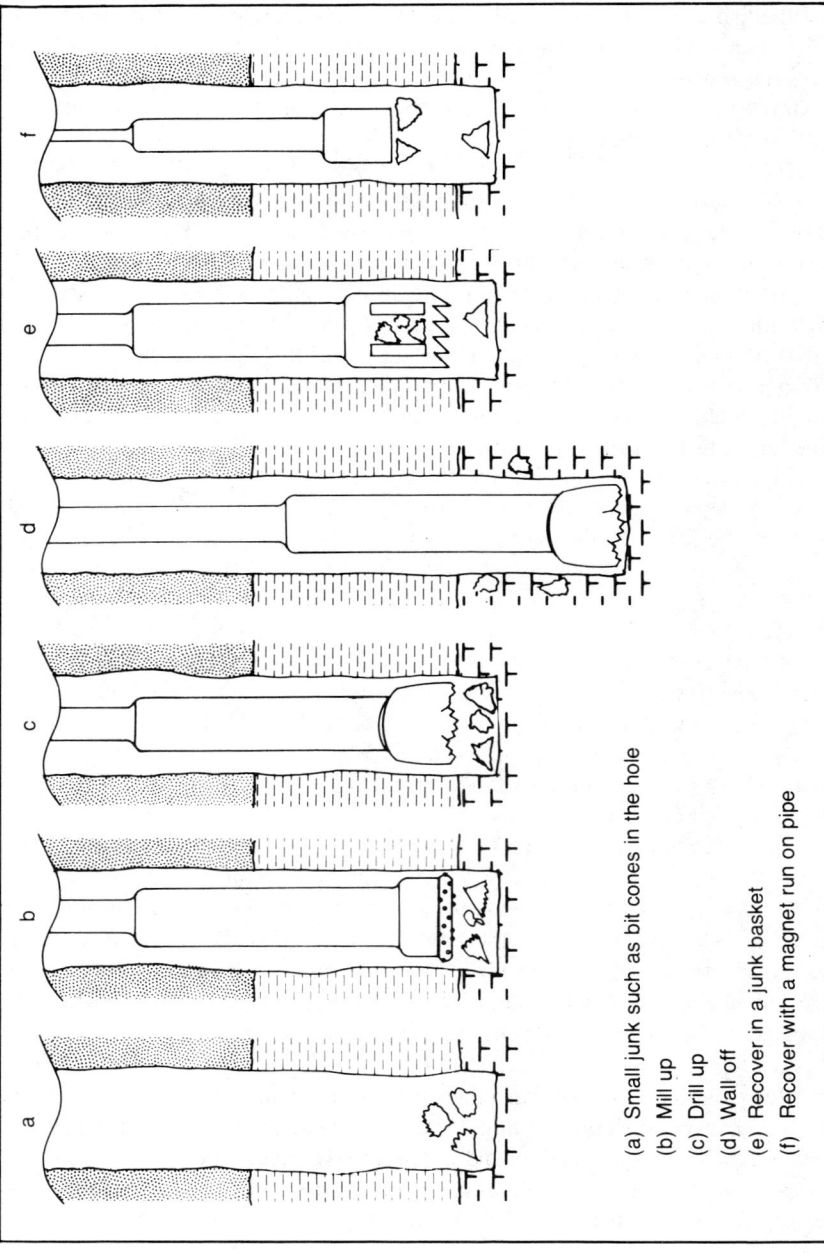

FIG. 8–11 Recovering or removing bit cones and small junk

(a) Small junk such as bit cones in the hole
(b) Mill up
(c) Drill up
(d) Wall off
(e) Recover in a junk basket
(f) Recover with a magnet run on pipe

There is still one cone left in the hole. In soft formations a short, mill-toothed bit can drill up the cone or can wall it off. The bit is connected to the bottom of the drill collars and is run into the hole. On reaching bottom, drilling operations are started with a very light bit weight, about 5,000–10,000 lb, and a slow rotary speed of about 20 rpm. When a bit is drilling on a cone, it will have a high, fluctuating torque and an uneven rotary speed. The bit action tends to become smoother as the cone is broken and drilled up and pieces are pushed or buried into the formation adjacent to the wellbore (walling off).

The bit is normally pulled if it continues to run rough for over about 1–2 hr. If it is run longer under these conditions, there is a risk of losing cones from the bit. Normally, the bit will start running smoothly in about 30 min, indicating the cone has been broken up and walled off. Bit weight and rotary speed are gradually increased to normal drilling conditions, and the bit is drilled until it becomes dull, even though it may not be the best bit for drilling this formation. As soon as the bit is dull, it is pulled from the hole. It will not make a normal bit run because of the excessive wear drilling on the cone. After the bit is pulled, the correct bit for the formations is run and normal drilling operations are resumed.

Some operators prefer to use mills in all cases for removing bit cones. Different operators prefer different types of mills, but generally a flat-bottom, ribbed mill is used in softer formations because it tends to drill up and wall off the junk. A concave-type mill tends to hold the junk under the mill and drill it up. These mills are preferred in harder formations.

In operation, one or two junk subs are connected to the bottom of the drill collars and the mill is connected to the bottom of the junk subs. The mill is run into the hole, and the cones are milled or drilled up with a slow rotary speed and low-to-medium bit (mill) weight. The mill will run rough with high, fluctuating torque and an uneven rotary speed. When the mill begins to run smooth, either the mill is worn out or the cones are completely milled up and walled off. The mill runs for about 6–10 hr before it wears out. One mill run normally removes 1–1½ bit cones. A second mill run is usually needed to remove all three cones. Normal drilling operations are then resumed, but only at the expense of losing 1–3 days' rig time.

It is a good practice to inspect all of the drill-collar connectors (tool joints) using either the iron-particle or sonic inspection technique after a heavy milling job. The high, fluctuating torque incurred in milling operations can excessively tighten the drill-collar connectors, causing cracked pins and boxes, swelled boxes, and stretched threads, so the connectors may fail in subsequent drilling operations and cause another fishing job.

Any tool joint is subject to failure and can cause a fishing job. In fact, the entire drill-collar assembly can twist off. This problem would be indicated by a loss of pipe weight, reduced drilling torque, decreased pump pressure, and decreased penetration rate. In a case like this, the drilling assembly less the fish is immediately pulled from the hole. The length of the pipe pulled determines the depth to the top of the fish. Therefore, it is very important to know the exact diameters and lengths of all of the equipment in the drilling assembly. This enables the operator to select the correct type and size of fishing tool to recover the fish.

Prior to pulling out of the hole, the operator would have a good concept of the fish left in the hole because the length of the fish can be approximated by the reduction in pipe weight. He would immediately call a fishing specialist so the proper fishing tools could be brought to the rig while the remaining part of the drilling assembly was pulled from the hole. This is important for two reasons. Rig downtime is minimized, thus minimizing cost. Also, and more importantly, if the fish can be recovered quickly, the job is usually much easier.

In this case the best tool to catch the fish is an overshot (Fig. 8–12). The fishing assembly is made up by connecting fishing jars on the bottom of 15 drill collars. A bumper sub is connected to the bottom of the fishing jars, and the overshot is connected to the bottom of the bumper sub. The fishing assembly is run in the hole until the overshot is immediately above the fish. After circulating for a period, the overshot is rotated slowly and lowered over the fish. After the overshot is over the fish, rotation is stopped and the fishing assembly is picked up slowly and carefully. An increase in weight over the weight of the fishing assembly indicates that the fish has been caught.

If the fish is not stuck, it is pulled from the hole. The fishing assembly is not allowed to rotate while being pulled from the hole since the rotating motion could cause the overshot to release its hold on the fish and drop it. Each stand is disconnected with a hydraulic or power drillpipe tong. If the rig is not equipped with these tools, a chain is used to unscrew each stand after it has been loosened with the drillpipe tongs. This is known as chaining out of the hole.

Usually the fish will be stuck. Then it is worked, including jarring upward in the same manner as working stuck pipe. The fish is worked free and is pulled from the hole. If the fish cannot be worked free, it must be washed over to release it before it can be pulled. The overshot is released from the fish by bumping down while holding right-hand torque, and the fishing assembly is pulled from the hole.

Washpipe is large-diameter, thin-walled pipe with flush-joint connectors. The overshot is removed, and several hundred feet of washpipe are connected to the bottom of the jars. A washover shoe is connected to the bottom of the washpipe and the washover assembly is run into the hole. The washover assembly is drilled down over the fish slowly and carefully while circulating. This washover procedure removes any formation cavings sticking the fish. After washing over

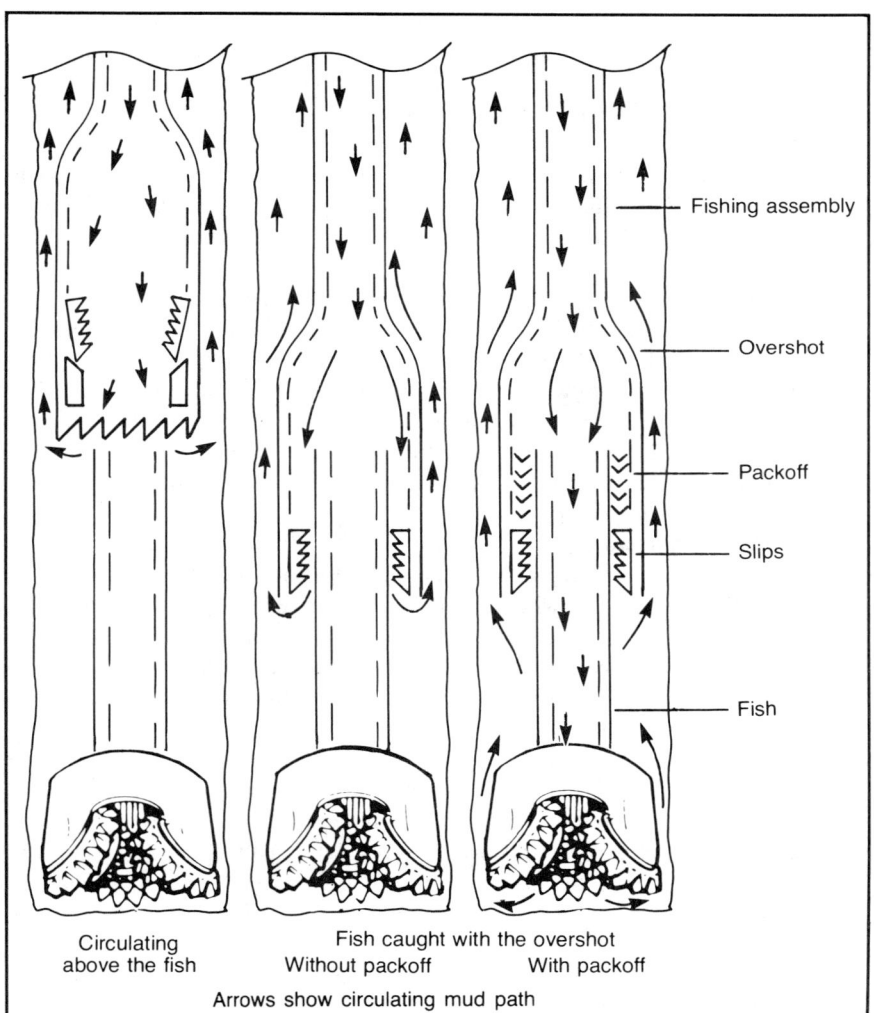

FIG. 8–12 Fishing with an overshot

the fish, the washpipe is pulled from the hole. The overshot is reconnected to the bottom of the bumper sub and run into the hole. The fish is caught, is worked free if necessary, and is pulled from the hole.

Washover operations must be conducted very carefully. The washover pipe is thin walled, structurally weak, and easy to twist off. Heavy torque is often encountered because high friction exists between the washpipe and the fish and walls of the hole. It is important to have good mud and a clean, well-circulated hole. If water-base mud is used, it is sometimes necessary to change to oil mud, which has a higher lubricity.

Large drill collars are often used, and in some cases there is not sufficient clearance between the drill collar and the wall of the hole to permit an overshot to be run over the collars. In this case a mill can be used to cut a fishing neck on the drill collars. The drill collars are then recovered by an overshot catch on the fishing neck.

Sometimes stuck assemblies cannot be worked loose. When this occurs, the assembly must be fished out of the hole. When drillpipe is stuck, the pipe above the stuck point can be stretched or elongated by pulling on the pipe at the surface. The amount of stretch is directly proportional to the amount of extra pull required to stretch the pipe a given distance. The stretch length can be measured at the surface, and the amount of extra pull can be read from the weight indicator. These values can then be used to calculate the amount of free pipe in the hole, which in turn gives the depth at which the assembly is stuck. Stretch data can also be entered into standard pipe-stretch tables to give the length of free pipe. This procedure is known as using pipe stretch data to locate the stuck point.

The 4½-in. drillpipe with 7-in. drill-collar assembly is used to illustrate how a stuck assembly is fished from the hole. Assume the 8¾-in. hole is drilled to 10,000 ft, and the pipe becomes stuck while making a connection. This situation is not uncommon and is usually caused by bit cuttings or caving formation falling in around the drilling assembly, by wall sticking, or by keyseating. Bit cuttings also frequently plug the annular space, preventing circulating and further increasing the difficulty of working the stuck assembly free.

The assembly would first be worked using gradually increasing pulls up to the maximum safe overpull, until it is determined that the assembly cannot be released by working. Stretch data indicate that the assembly is stuck in the drill collars. Stretch data are not precise, but they usually indicate the stuck point within a few hundred feet. Near the end of the working period, a service company specializing in free pointing and backing off would be ordered to the location to be available when needed.

If the stuck assembly cannot be worked free, the next step is to determine precisely the point where the pipe is stuck with a free-point tool. Then back the pipe off above the stuck point with a backoff shot, pull the free pipe out of the hole, and run a fishing assembly with a bumper sub and jars so the fish can be worked harder and more efficiently in order to release it.

The free-point tool is run on a wire line similar to a logging tool. Depending on the surface equipment, it is either run through an open tool joint with the pipe caught by the elevators, or through the kelly gooseneck connection. The free-point tool is designed to be run inside the pipe and uses magnetic and electrical methods to measure a very small amount of stretch or torque in a short section of the drillpipe or drill collars.

When the pipe is stretched or torqued at the surface, this stretch and torque are distributed throughout the drilling assembly from the surface to the stuck point. That part of the assembly below the stuck point does not stretch or torque because these motions cannot be transmitted below the stuck point. The free-point tool is run in the hole to about 1,000 ft above the stuck point as determined from pipe stretch data. The pipe is stretched and torqued from the surface, and the small increment of stretch and torque in the pipe is measured by the free-point tool. At this point the pipe is considered to be 100% free in stretch and torque.

The tool is moved further downhole, and the stretch and torque measured again. This procedure is continued until the free-point tool cannot detect any stretch or torque. At this point the hole is noted as 100% stuck. The operator uses this procedure to determine the interval from where the pipe is completely free until it is completely stuck. The length of this interval can range from a few feet to several hundred feet.

The next step is to back off the pipe immediately above the stuck point with a backoff shot. The backoff shot consists of a collar locator, weight bars, and a rod or cable (pigtail) to which an explosive charge, usually a primer cord, is attached. The tool is run in the hole with electrical conductors on the same cable used to run the free-point tool. The collar locator determines collars or tool joints in the downhole assembly by electrical induction methods. A depthometer is used to measure the wire line as it is run in the hole so the depth of the tool joints can be located precisely.

The tool joints are screwed together and are tightened with drillpipe tongs. The combination of torque and friction in the tool joint threads and shoulders hold the tool joints securely together. The tool joint is released with the backoff shot by first applying left-hand torque and then detonating the explosive charge. This charge jars the tool joint and allows the left-hand torque to release the joint. The string-shot explosive must be positioned precisely opposite the tool joint in order to release it.

The string-shot tool is run in the hole to the measured depth immediately above the point where the pipe has become stuck. The measured depths are not sufficiently precise to be used to position the explosive on the string-shot tool exactly opposite the tool joint. The collar locator is used for this purpose. The tool joint is precisely located with the collar locator. The distance from the collar locator to the center of the explosive charge is known. After the tool joint is located with the collar locator, the string-shot tool is then raised the exact distance from the collar locator to the center of the explosive charge. This positions the explosive charge precisely opposite the tool joint (Fig. 8–9).

The pipe weight at the surface is adjusted so the pipe free point or the point of neutral weight—not to be confused with the free-point tool—is at the point where the pipe is to be backed off. This neutral point is the point in the pipe that is neither in tension nor compression, commonly called the free point. The assembly is then rotated carefully in the left-hand direction a limited amount in order to increase the left-hand torque. The amount of rotation ranges from 3–10 left-hand turns, depending on the pipe weight and depth to the neutral point.

The backoff charge is detonated, and the jarring motion releases the friction hold on the tool joint allowing it to unscrew. The string-shot tool is pulled from the hole. The pipe is rotated in the left-hand direction a small amount if necessary to release it. Unrestricted pipe movement and the indicator weight can be used to determine if the pipe has been backed off. If it is not backed off, the entire backoff procedure is repeated.

The released section of assembly is pulled from the hole. Assuming the drillpipe has been backed off, additional drillcollars must be picked up for the fishing assembly. The fishing assembly is similar to that described earlier, except a drillpipe-pin connection is used on bottom instead of an overshot. This screw-in connection is the best method of catching the fish in this case.

The fishing assembly is run into the hole and is screwed into the fish. The fish is worked using the more efficient action of the jars and bumper sub. If the fish can be released in this manner—and it often is—then it is pulled out of the hole and a drilling assembly with a new bit is run to drill ahead.

In this case assume that the fish includes a full string of drill collars and ten joints of drillpipe. After screwing back into the fish and working it, the fish remains stuck. In many cases working the fish can unstick an additional section of the fish. The free-point tool is run in the hole to check this by finding the point where the pipe is stuck and comparing this point with the original stuck point. Assume the free-point indicator shows that the pipe is now stuck 200 ft below the top of the fish. A backoff shot is then used to recover six additional joints of drillpipe, about 186 ft.

The same fishing assembly is checked and run back in the hole. Then it is screwed into the remaining fish, and the pipe is worked. If the fish is not released after working it vigorously, the free-point indicator is run again to see if an additional section of fish is free. If a new section of the fish is free, it is backed off and recovered in the same manner as the prior section.

The combinations of working and backing off or working, washing over, working, and backing off are repeated until all of the fish is recovered. This is a time-consuming process with attendant high costs. It is not uncommon to spend at least two weeks and sometimes more than two months on a severe fishing job, especially where the fishing operations are hampered by various formation problems. Therefore, all precautions possible should be taken to prevent a fishing job.

LOGGING, RUNNING, AND CEMENTING CASING

The intermediate hole section may or may not be logged, depending on whether the well is a wildcat or a production well, whether drilling shows are encountered during drilling, and whether prospective zones occur in the interval. By its very name the intermediate hole section provides a means of drilling the production hole section. Therefore, it does not normally penetrate prospective productive formations and, consequently, does not need to be logged primarily for defining productive formations.

A complete suite of logs is normally run in the intermediate hole section on wildcat wells, especially in deeper intermediate hole sections. Shallow sections are seldom logged on production or development wells but may be logged on wildcat wells and on the deeper production or development wells. Lithology logs are often run on these wells in deeper horizons, especially if the wells are drilled on wide-spaced units, to correlate formations with adjacent wells, and to help locate the intermediate casing point. Since open-hole logs provide information about the formations that cannot be obtained with cased-hole logs; any open-hole section that penetrates prospective productive formations normally is logged.

The same procedures and operations as described earlier are used to design, run, and cement casing in the intermediate hole. The hole section can be cased with a liner. The casing run in this section generally is considerably longer and heavier than that run in the shallower hole sections. Therefore, some modifications may be required for casing and cementing operations, especially in deeper holes.

Additional precautions are taken to ensure that the mud is in good condition and that the hole has been circulated clean and is free from obstructions. Running casing is a time-consuming operation and may require relief casing crews.

Precautionary measures are taken to detect and control a kick if it occurs while running and cementing the casing, even though the hole has not penetrated high-pressure formations. These measures include placing a safety valve with a connection to fit the casing on the rig floor so it is available and can be installed immediately if the casing begins to flow—if fluid flows out of the casing. Rams sized to fit the casing are installed in the blowout preventers.

The drillpipe or drill collars may be laid down prior to running casing, depending on hole and casing sizes. An extra two lines may be strung up between the traveling block and the crown, depending on the total weight (hook load) of the casing.

A float shoe and float collar are almost invariably used as part of the float equipment. Usually two or more centralizers are used on the bottom joints to center the casing and ensure a better cement job near the casing shoe. The casing is run into the hole slowly and carefully. Running rates are calculated to prevent excessive surge pressures.

In some cases the casing is run to bottom and then circulated. In other cases, generally in deeper holes, circulation is started with the casing one or two joints off bottom. The last joints are added later and are circulated down, somewhat similar to a reaming operation while drilling but without rotation. The casing is reciprocated and circulated for at least one circulation to remove bit cuttings and formation cavings that may have been dislodged from the walls of the hole while running the casing. It is very important when cementing casing strings to have a clean hole filled with a quality mud with good flow properties.

The casing is cemented in a conventional manner using both bottom and top plugs and is reciprocated while cementing. Casing drag increases substantially as the heavier, thick cement flows upward in the annular space from the casing shoe. This increases the weight required to move the casing. Therefore reciprocation may be limited as a result of the following:

1. The casing tends to work uphole because increased weight is required to move the casing down
2. The pull required to move the casing upward may approach the strength limitations of the casing

When these conditions occur, casing reciprocation is limited and may be stopped before all of the cement is in place. Pumping rates are reduced as the top plug approaches the float collar in order to prevent overpressuring and possibly bursting the casing when the plug bumps.

An alternative to cementing long, large-diameter casing strings through drillpipe is illustrated in Fig. 8–13.

The normal procedure after cementing the casing is to release the pressure and to watch for flowback to ensure that the check valves in the float collar or float shoe are holding. Most operators begin preparations to nipple up at this time. Some prefer to wait until the cement has taken an initial set as a precautionary measure, in case the back pressure valves in the float collar and float shoe both fail.

Intermediate Hole Section

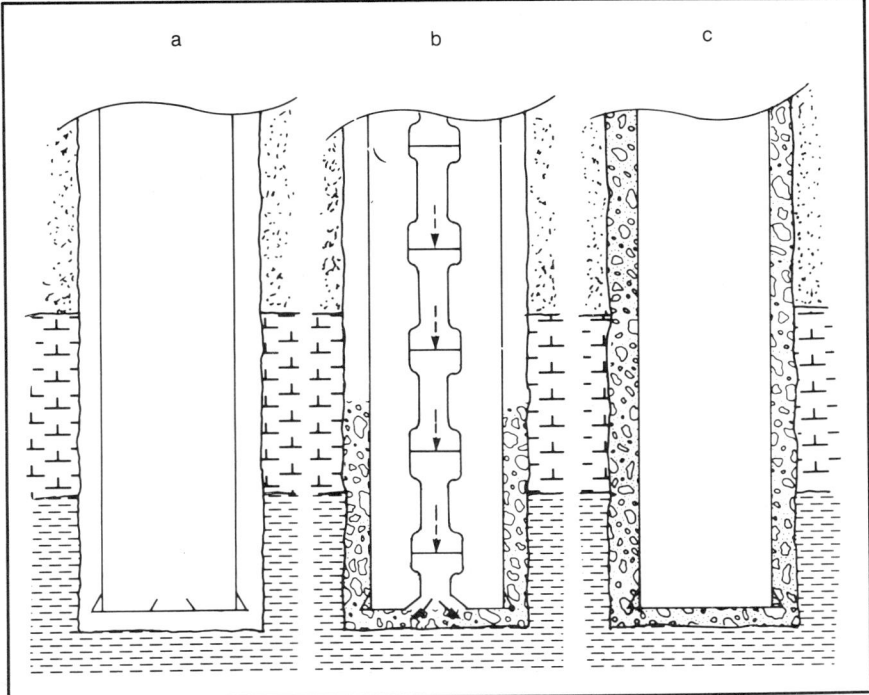

FIG. 8–13 Cementing large-diameter casing through drillpipe. The casing (a) is run and landed. Drillpipe is run inside the casing and is connected to a special fitting in the bottom of the casing. Cement is mixed and circulated (b). The drillpipe is released from the casing and is pulled from the hole, leaving the cemented casing (c).

After the cementing job is completed, the casing running and cementing equipment is disconnected, loaded, and moved off location. The rig floor and preventers are cleaned with high-pressure water hoses. The casing is landed in the bradenhead or a casing hanger flange with a precise amount of weight (Fig. 8–14).

Normally by the time the casing is landed, the cement has set up so it can support the cemented section of casing. The type and size of preventers to be used for drilling the production hole section are nippled up with the correctly sized pipe rams. The preventers are then tested, and the drilling assembly is made up or picked up to drill out through the intermediate casing and begin drilling the production hole section.

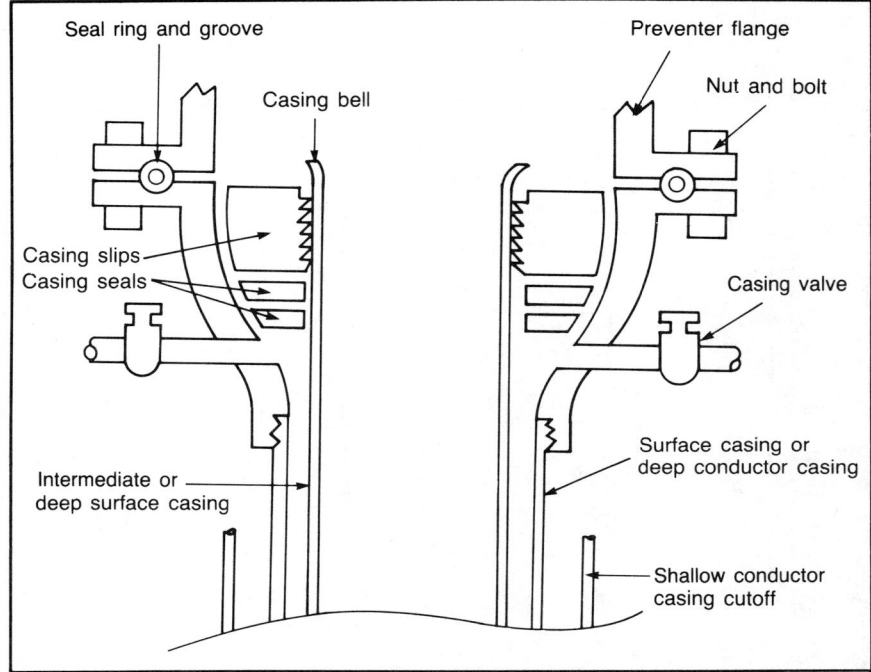

FIG. 8–14 Casing landed in the bradenhead

CHAPTER 9
PRODUCTION HOLE SECTION

The production hole section differs from the other sections in that it is the deepest and smallest-diameter hole section drilled. In many cases it is the longest openhole section. The formations are generally harder and more abrasive than those in the other hole sections. Productive zones with associated higher pore pressures are also encountered. These zones are frequently cored and drillstem tested. A full suite of logs is run followed by plugging or casing the hole. A good cement job is important to isolate the productive zones properly for perforation, stimulation, and production.

Special problems are encountered in the production section such as deep drilling with long, small-diameter drilling assemblies, hard, abrasive formations and associated bit selection, problem formations including high pressures, cores and drillstem tests, and fishing in small-diameter holes. The problems are generally more severe than those in the shallower hole sections.

A large expenditure has been made drilling the hole to this depth. If the hole is lost as a result of a fishing job or for other reasons, the equivalent monetary loss is higher than it would be under similar circumstances at a shallower depth.

DRILLING THE SECTION

The production hole section is the last and deepest hole section drilled. It extends from the bottom of the last casing, either surface casing in shallower wells or intermediate casing in deeper wells, to total depth (TD). In Chapter 8 the intermediate hole was drilled, cased, cemented, and nippled up. The next operation is to drill the production hole section.

The cement is drilled out of the lower intermediate casing in a normal manner using water, discarded mud, or pretreated mud. The casing is normally pressure tested with the rig pump after cleaning out cement to the top of the float collar and again after cleaning to the top of the float shoe. After drilling a short section of formation, the open hole is tested to the highest projected mud weight to be used in drilling the production hole section.

Diamond Drilling

Assume that the production hole, or at least the upper part, is drilled with diamond bits. The cement is drilled out with a short-toothed or insert-type hard-formation bit run below one or two junk subs. Diamond bits cannot be used to drill cement or the float shoe and collar. These are drilled with a roller bit that is most effective for the formation below the casing shoe. The junk subs catch large pieces of cement, the float collar, and the shoe as they are drilled. The bit is run until it becomes dulled to utilize its efficient life since bits are seldom rerun in deeper drilling operations. The extra drilling and circulating time also ensures that all of the cement and float equipment debris are circulated out of the hole.

The diamond bit must be run in a clean hole to prevent damaging the diamonds. It is normally run on a stiff, stabilized assembly. The deeper hole allows these tools to be buried below the intermediate casing shoe, thus reducing the risk of backing off the lower casing joints by rotating the stiff tools in the casing.

When the hard-formation bit dulls, it is pulled and the stabilized diamond bit is run to drill ahead. This bit is drilled in or seated by starting drilling with a minimum bit weight and rotary speed and gradually increasing these to normal operating conditions over a period of 10–20 min, depending on formation hardness. This bit break-in procedure is used with all bits. It provides time for the bit to create its own tooth or diamond pattern on the bottom of the hole so the bit weight is distributed evenly over the cutting face of the bit. This weight distribution minimizes the risk of damaging the bit teeth or diamonds.

The hole generally is drilled in a manner similar to other drilling operations. Bit selection and operation become more important because of the reduced drilling rate caused by harder formations. Used bits are examined carefully to determine the cause of failure and to help select new bits to be run. For example, if the bit bearings are completely worn and the cutting structure is only partially worn, this could indicate that a longer-toothed, softer-formation bit would drill more efficiently. In the reverse, if the cutting structure is completely worn and the bearings are only partially worn, a harder-formation bit may be needed. If the bit is severely worn in the gauge area, a bit with more gauge-cutting structure is needed. Drilloff and other drilling tests using different combinations of rotary speed and bit weight are run to obtain the most efficient penetration rate.

General Operations

Planning and attention to detail are key elements in an efficient drilling operation. The standby pump can be placed on line during or before a connection when the main pump begins to malfunction. The pump is then repaired while drilling operations continue with the standby pump. Many small repairs and equipment changes can be made during a connection without hindering the drilling operation. Small details such as using a pipe wiper to keep junk from falling in the hole prevent a fishing job or premature bit failure caused by drilling on junk. Metal cuttings in the mud can indicate excess metal wear in the system, such as drillpipe wearing on casing set in a crooked hole. Double-rubbered drillpipe may be needed through some sections of the cased hole (Fig. 9–1). This pipe can prevent wearing a hole in the casing.

Bit cuttings should be carried out of the hole at a constant rate with the mud. The size and shape of the cuttings help indicate pending formation problems. Large, angular, blocky, or tabular-shaped cuttings and a larger-than-normal flow of cuttings can indicate caving or sloughing formations. Well-rounded, small-sized cuttings can indicate excessive retention time in the hole as a result of an out-of-gauge hole or high slip velocity of the cuttings and improper hole cleaning.

The mud must be checked frequently and treated as needed to maintain good physical and chemical properties. A good solids-removal program with well-maintained and properly operated equipment is important. Although desilters and desanders often require considerable maintenance, proper operation is very important to a good mud system.

Mud characteristics may indicate pending formation problems. Bentonitic mud-making shales often cause increased viscosities. Drilling anhydrite and gypsum causes deteriorating mud properties. Increased chlorides can indicate salt formations or possible saltwater flows. The fact that the formation can flow some salt water into the mud system indicates an underbalanced condition.

The gas content in the mud is also an important indicator of underbalanced or overbalanced mud. This content refers to free gas that may or may not flow into the mud, depending on the mud weight and reservoir pressure and permeability. A higher gas content in the mud indicates underbalance.

This gas is not to be confused with drilled gas—gas contained in the formation that is drilled out. Drilled gas can cause several tenths of a pound per gallon reduction in the mud weight under favorable conditions and does not necessarily indicate an underbalanced mud system. It can be distinguished from gas flowing from the reservoir by stopping drilling for a short period while continuing to circulate. Drilling is then resumed, and the gas content of the mud

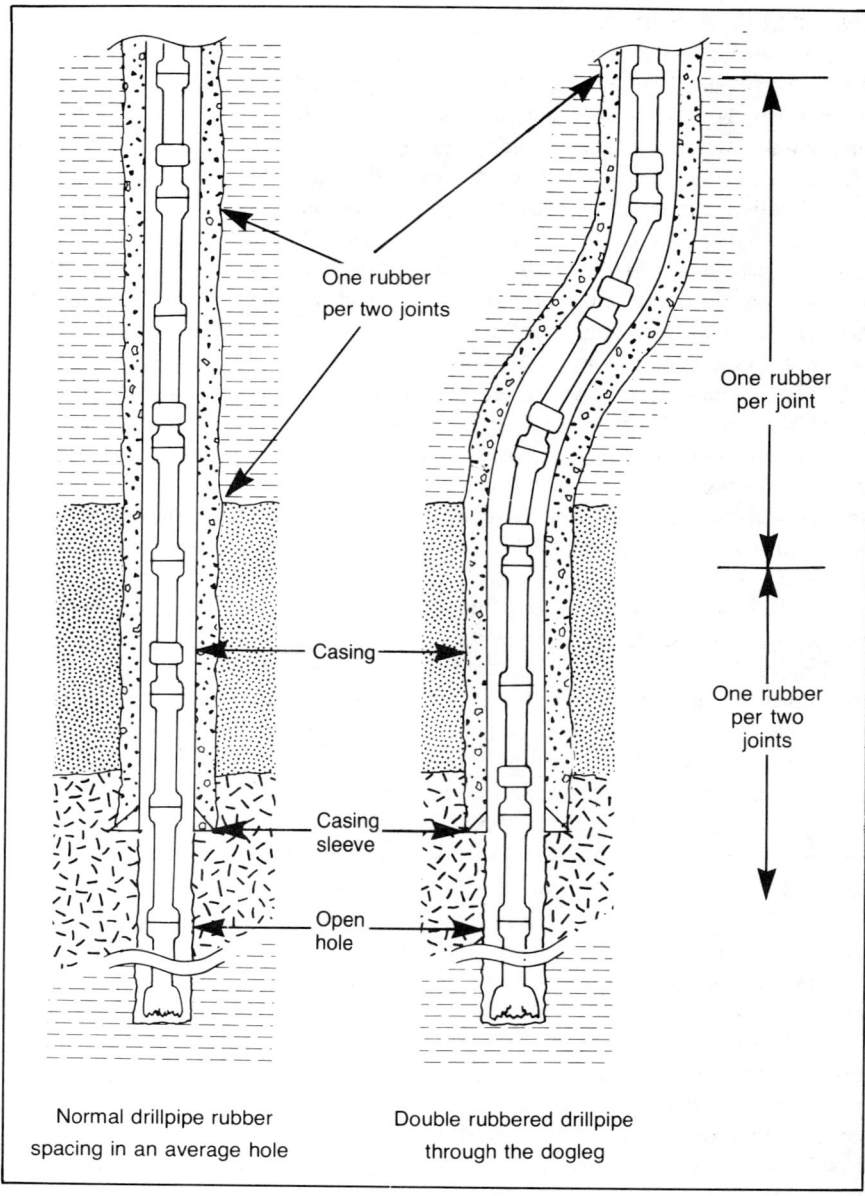

FIG. 9–1 Rubbered drillpipe

circulated without drilling is checked when it reaches the surface. If this mud contains gas, it is gas flowing in from the formation since the mud has been circulated without drilling. If the mud circulated without drilling does not contain gas but the mud circulated while drilling does contain gas, drilled gas is being recovered.

The mud program should be designed so a minimum safe mud weight is used to obtain a maximum drilling rate. Also, the mud hydraulics should be checked periodically to optimize the drilling rate.

Some operations are no longer necessary in modern drilling operations. For example, the hole is not automatically reamed before making a connection unless high pipe drag and rotary torque indicate the need. This will save a substantial amount of time over the drilling of a well.

Do not make periodic wiper trips when making long drilling runs with diamond bits except when special formation conditions occur. This also applies to wiper trips made before logging and running casing. The need for a wiper trip can often be determined by observing the drag while making connections when the lower section of hole is being drilled.

Drag as previously described is caused by reduced hole size (undergauge hole), dogleg, and high deviation so movement of the downhole assembly is restricted. Drag is indicated on the weight indicator by increased assembly weight when the assembly is moved downward. Therefore, if drag is more than normal, a wiper trip may be needed to remove developing obstructions.

It is not always necessary to make a complete circulation to clean the hole before making a bit trip, especially when drilling in hard formations. Small amounts of cuttings in the hole can be drilled out easily with the new bit in a much shorter time than circulating them out of the hole.

Recognize worn roller bits in time to pull them out before leaving cones in the hole and causing a fishing job. If the mud pressure drops slightly (10–15%) and the pressure drop cannot be traced to the fault of the mud pump or surface circulating system, pull the drilling assembly to look for a leak in the pipe or a washed-out tool joint. Either can cause premature bit failure as a result of inadequate circulation, possibly a stuck bit, or a fishing job if the tool joint washes out to the point where it loses strength and fails.

Other procedures also detect possible problems. Watch the weight indicator during the trip to detect areas where keyseats may be forming. Use a pipe wiper while pulling and running the pipe. Look for dried muddy tool joints that indicate the tool joint was leaking. If tool joints show wear or fluid erosion, replace them. Watch for bent drillpipe as the pipe is rotated while disconnecting a stand. Visually check the pipe protector rubbers to ensure that they are not excessively worn or have become loose and slipped farther away from the tool joint where they are less effective. Work all of the preventer rams while the assembly is out of the hole. On the trip back in the hole, stop about 60–100 ft off bottom and ream the remaining distance to the bottom. This operation is especially important if the last bit was heavily worn.

On deep holes run a base casing inspection log on the last string of casing after the first two or three bit runs. Repeat the log periodically at 4–8-week intervals to check the casing for drillpipe wear. Run double-rubbered drillpipe in areas where excessive wear occurs.

Be especially alert on the morning or graveyard tour and while tripping. About 75% of all problems occur during these two periods, which covers about 50% of the total drilling time. In freezing weather check the rig winterization equipment to ensure that it is functioning properly and that all trace lines are carrying heated fluid to prevent mud and water lines from freezing and plugging with ice.

In summary, concentrate efforts on drilling the hole at the most efficient rate. Schedule repairs and other activities so a maximum amount of time can be spent drilling.

Downhole Mud Motors

Normal drilling operations are conducted by rotating a bit and drilling assembly with the rotary. Although this is the best method of drilling devised, it is inefficient. Additional energy is required to rotate the drilling assembly. The rotating action causes wear on the assembly, especially at the tool joints. Casing wear occurs, and some formation problems are attributed to the rotating assembly. Also, it is more difficult to drill a deviated hole and control the direction while rotating the entire drilling assembly. So various drilling methods have been tested or proposed to increase penetration rates and overcome the other objections of conventional drilling techniques.

Penetration rate is related to rotary speed. Normal rotary speeds are limited to a maximum of about 150 rpm in shallow holes to 75 rpm or less in deeper holes. Excluding the bit, the factors that limit the maximum rate of rotation are the mass of the rotating assembly, excessive wear on all of the equipment, and other related factors such as synchronous rotary speeds that subject the drilling assembly to excessive loading, which may cause the assembly to fail.

Turbines or mud motors have been developed to overcome many of the disadvantages of conventional drilling, to improve penetration rates, and to deviate holes. The tool is connected to the bottom of the drill collars, and a bit is fastened to the bottom of the tool. The turbine uses the energy of the circulating mud via a pressure drop to rotate the bit at high speeds. Then drilling operations are conducted in a semiconventional manner, although drilling mud is circulated with a slightly higher pressure in order to operate the downhole rotating device.

Mud motors have sinusoidal shafts and can run 25–30 hr under good conditions. Turbines have blades, and runs of over 300 hr are common. The motor and turbine use a low bit weights of 10,000–20,000 lb and rotary speeds of 150–400 rpm. They are very efficient with a properly designed bit.

The tools are obtained on a rental basis and are expensive, but the higher cost may be justified. The driller needs to evaluate the drilling costs per foot with the tool or by conventional techniques to determine if the use of the tool is justified.

DRILLING PROBLEMS

A variety of problems are encountered in drilling operations, particularly mechanical failures. Most rigs are equipped with two pumps. When one fails, the second pump is placed in service while the first pump is repaired. In some cases the second pump may fail before repairs have been completed on the first pump. There is a high risk of sticking the downhole assembly if circulation is not maintained. Therefore, one of two procedures is used when both pumps fail.

The assembly can be pulled uphole until the bit is inside the last string of casing. The well is closed in or buttoned up by closing the drillpipe rams and connecting a safety valve or the kelly to the drilling assembly. There is a minimal risk of sticking the downhole assembly, and the well is safely shut in to prevent a possible blowout. The hole must be filled as the pipe is pulled, but this can usually be done even with a damaged pump. The pipe is run back in the hole after the pumps are repaired, and drilling operations are resumed. This is a safe procedure, but it often requires considerable time, especially when the pipe must be pulled up a long distance. Operators hesitate to use this procedure except when an extended shutdown period is expected.

An alternative procedure may be used when a short shutdown time is expected. The kelly and pipe are pulled up as high as possible, usually limited by the length of the mud hose. Normally there is a full kelly, one single, and part of another single above the rotary. The kelly and pipe are then lowered a distance of 1–2 ft at 5–10-min intervals while repairing one of the pumps. If the pipe begins to stick, it is indicated by excess drag as the pipe is lowered. In this case the pipe is pulled up into the casing, meaning all of the pipe including the bit is in the casing. After the pipe has been lowered to within about 15 ft of the bottom of the hole, it is picked back up. The periodic lowering procedure is repeated until the pump is repaired and drilling operations are resumed.

The same procedures are used when other failures prevent drilling. A few operators leave the kelly in the rotary and rotate the pipe slowly. However, moving the pipe is generally accepted as a safer procedure to prevent sticking.

Another surface mechanical failure that causes a severe problem and poses a risk of injury or death to personnel is dropping the traveling blocks. The drilling line can break, causing the block to drop. The block can also drop if the drilling line is not properly secured at the dead-line or fast-line end on the drawworks drum. The drawworks braking system, including the rods, equalizer bar, or brake bands, can fail by parting and can cause the blocks to drop. Worn crown and traveling-block sheaves can fail, allowing the drilling line to slip to the crown or traveling block shaft where it invariably fails as a result of the reduced curvature.

If the pipe is suspended from the traveling block when it drops, the entire assembly falls to the rig floor. The elevators normally break on impact and the entire assembly falls into the hole. This creates a severe fishing job. The dropped assembly is also bent and broken so it must be repaired or junked after being fished out of the hole.

The dropping block can also damage other equipment on the rig floor such as the kelly, rotary, or drawworks. The drilling line normally unspools, and the loose, whipping ends can severely damage the bracing in the mast or derrick. This line, which is normally about 1 in. in diameter, also damages equipment as it falls to the rig floor. Any falling equipment can kill or maim operating personnel. Dropping the block is a severe, sometimes disastrous problem. However, it can be prevented with proper line-slipping and cutoff practices and equipment maintenance.

In summary, surface equipment failures often create very severe problems. They generally are caused by inadequate maintenance, improper operations, or human error. All efforts should be made to eliminate the causes and prevent their occurrence.

Formation Problems

One of the most persistent and common formation problems is obtaining a satisfactory penetration rate. This can occur even with the best bit selection, good hydraulics, and a favorable mud system. For example, very hard, highly abrasive, tough, sandstone sections are drilled most efficiently with a hard-formation, insert type bit at rates of 1–2 ft/hr with a bit life of about 40 hr. The first thought is to use a longer-life diamond bit, but frequently the formations are too hard and abrasive for diamond drilling—especially with the occurrence of pyrite and chert nodules. The cost of drilling these formations can be high. Until technology develops a better answer, the operator's only alternative is to live with the problem.

Most of the formation problems described, such as fractured and sloughing formations, high-pressure sections, lost-circulation zones, massive salt sections, and saltwater flows, may occur while drilling the production hole section. They are generally handled as previously described with modifications for the deeper hole and reduced hole size. When formation problems occur, the standard procedure is to work the drilling assembly upward until the bit is above the zone causing the problem. This prevents sticking the assembly.

There is always a high risk of sticking when working with problem formations. In some cases the drilling assembly must be left deeper in the hole, even at the risk of sticking. This occurs most commonly when drilling high-pressure formations. The bit must be left deeper in the hole to circulate heavy mud downhole to control the high pressures. Pumping heavier mud down the annular space from the surface is used only as a last resort because of the possibility of losing mud in the weaker formations higher in the hole or of sticking. Efforts are concentrated on controlling lost circulation with special materials.

Compound problem formations can create hazardous operating conditions. Curing one problem is often difficult. Trying to cure two problems simultaneously can be extremely difficult because the procedures for solving one of the problems may adversely affect or increase the severity of the second problem.

One of the most difficult but relatively common combination formation problems is the occurrence of high-pressure formations and lost-circulation zones in the same wellbore. A minimum mud weight is needed for lost-circulation zones to prevent mud loss. This reduces the hydrostatic head, which in turn allows the high-pressure formation to flow into the wellbore, creating a kick or potential blowout condition. If the mud weight is increased to control the high-pressure formation, it may be lost into the lost-circulation zones. This renews the cycle, causing the fluid level to fall in the annulus and allowing high-pressure fluids to flow into the wellbore from the high-pressure zone. The general procedure for curing the problems is to use a large volume of lost-circulation material in a low-weight mud. The mud weight is gradually increased to control the high-pressure zone while plugging off the lost-circulation zone.

Another danger is sloughing resulting from hole instability. Abrupt changes in mud weight and fluctuating fluid-column heights can make the hole unstable as a result of the fluctuating pressure on the wall of the hole. This condition may lead to sloughing. With the combination of high-pressure zones and sloughing sections, it is still necessary to keep the drilling assembly deeper in the hole to cure the high-pressure zone. The risk of sticking the assembly may still occur because of sloughing and caving. However, increasing the mud weight to control the high-pressure zone normally does not increase sloughing or caving, and in some cases it may help reduce them.

Kicks

Technically, a kick occurs when formation fluids flow into the borehole. However, in practices there are cases where the formation fluids can flow into the wellbore without creating a serious condition such as gas seeps and minor water flows. These conditions indicate a near-balanced or possibly underbalanced condition and often occur in low-permeability formations. The flow of fluid into the wellbore usually is restricted by the low permeability and the low pressure differential into the wellbore. Therefore, a more practical definition of a kick is a condition where formation fluids flow into the wellbore at rates detectable at the surface and create an imminent blowout condition.

Detection The first step in controlling a kick is to detect it either very shortly after it occurs or before a large volume of formation fluid has flowed into the wellbore. Prior mud-weight and pore-pressure plots, shale density, drilling rate, and modified D-exponent—a special method of plotting drilling data—indicate if the hole is being drilled through formations that have an increasing pressure gradient. If this is the case, the operator should be alert for a pending kick.

Drilling breaks are immediate indicators of a possible kick. Formations with higher permeability usually have higher porosity and may drill at a faster rate than in adjacent formations. The penetration rate during a drilling break can be 1½–4 times faster than the penetration rate prior to entering the drilling break. If the formation has sufficient permeability and pressure, formation fluids may flow into the wellbore causing a kick.

It is standard practice on many drilling operations to drill 5–10 ft into the drilling break, then shut down drilling, and circulate bottoms up, i.e., circulate mud from the bottom of the hole to the surface. This practice limits the amount of formation exposed to the wellbore and possibly the severity of the kick. The hole is being circulated with the kelly above the rotary, which further reduces the amount of time required to shut the well in if a kick does occur.

The severity of the kick depends mainly on the following:

- Pressure differential between the formation and the hydrostatic pressure caused by the mud column
- Formation permeability
- Amount of formation exposed to the wellbore
- Rate of fluid flow into the wellbore before the well is shut in
- Type of fluid flow into the wellbore, i.e., oil, gas, or water

The severity of the kick is indicated by the shutin drillpipe pressure (SIDP) and the pit volume gain, measured in barrels. For example, a kick may be referred to as 400-psi SIDP and a 12-bbl pit gain.

Most rigs are equipped with sensors and instrumentation to measure the volume or fluid level in the mud tanks, the total mud volume in the 2 or 3 mud pits, and the volume of mud flowing through the flow line. When a kick occurs, formation fluids flow into the wellbore and increase the total fluid volume in the mud system. This additional fluid shows as an increase in the pit volume. It also causes an increased flow in the flow line. Mud pit volume increases and increased flow rates in the flow line are primary indicators of a kick. Sensors normally are set to sound an alarm when unusual increases occur.

Control The operational procedures to close in the well as described for blowout drills are started as soon as the monitors indicate a kick. The well should be completely shut in with all personnel at preassigned stations within 60–90 sec after the alarm is sounded.

The kick is controlled by first closing in the well. Then lighter kick fluids must be circulated out of the hole and replaced with a heavier mud. This overbalances the formation pressure and holds the formation fluids in the formation so drilling operations can be resumed, known as *circulating out the kick*. Operations must be conducted carefully to contain the pressures and limit the amount of pressure used in order to prevent breaking down other formations or causing lost circulation and other problems.

To illustrate the procedure for circulating out a kick, assume that a hole has 9⅝-in. casing set at 3,000 ft and an 8¾-in. hole is being drilled at 12,000 ft using a conventional 7-in. drill collar and 4½-in. drillpipe assembly. The mud weighs 11.5 lb/gal, and the slow pump rates recorded on the report are 1,600 psi at 40 strokes/min and 700 psi at 20 strokes/min. Slow pump rates are at about one-third to two-thirds of the normal pumping rate and are taken at least once or twice during each drilling tour. To obtain the information, the pump is first slowed down, in this case to 40 strokes/min. After the flow rate has stabilized (about 30 sec), the pressure and strokes per minute are recorded. The procedure is repeated at 20 strokes/min. These rates are recorded on the daily tour report.

In our example, drilling operations are fairly routine and blowout drills have been conducted periodically. The pore-pressure plots indicate an increasing pressure gradient. There also has been a gradual increase in the gas content of the mud, indicating a balanced mud-weight/reservoir-pressure condition. The operator elects to drill with a reduced mud weight to obtain a maximum penetration rate.

A kick may be indicated by a drilling break, but this does not necessarily occur. For this case assume the hole is being drilled normally and one of the sensors sets off the alarm. The driller immediately shuts off the rotary and starts to pick up the pipe. The pump is shut off while the pipe is moving upward. The upper drillpipe rams are closed as soon as the bottom kelly connection is above the preventer rams. This usually occurs when the kelly drive bushings are lifted out of the rotary 1–3 ft. In the meantime each crewmember goes to his appointed station, and the well is completely shut in 50 sec after the alarm first sounds. The entire procedure goes exactly like a well-performed drill, which it should.

It is significant to note that when the alarm sounds the driller immediately begins pulling the assembly up and lifting the kelly above the preventer section. He does not question the alarm but immediately starts the procedure to close in the well.

After the pipe has been picked up and the well is shut in, each crewmember looks at the equipment near his area or station to see if there are any leaks. The driller checks and finds that the drillpipe has 500 psi pressure and the annulus or casing pressure is 800 psi. These are recorded on the report as the shutin drillpipe and casing pressure.

Note that if check valves are used in the drilling assembly, there is a possibility that the shutin drillpipe pressure will not be the full well pressure. Then the driller would start the pump very slowly and watch the pressure gradually increase until it stabilized. This point would then be recorded as the shutin drillpipe pressure. The reduced pressure initially shown on the drillpipe is known as the trapped pressure and is not the correct shutin drillpipe pressure.

At this point the well is completely shut in and is safe, assuming there are no leaks. Excluding the possibility of sticking the drilling assembly because there is no assembly movement, the well is in a safe condition and could probably remain for several hours without creating an additional hazard. However, the best procedure is to circulate out the kick immediately. This allows drilling operations to be resumed and reduces the risk of sticking the drilling assembly.

In general for the procedure for calculating the kill mud weight and circulating the kick out of the hole, the driller first ensures that the well is safely shut in. Since the mud weight must be increased, the driller may decide to begin adding weighting material (barite) immediately to the mud in the suction pit. If heavyweight, premixed mud is available, it may also be used to help build the weight of the mud in the suction pit. As an added safety precaution, additional weight material may be ordered for delivery to the rig as soon as possible. The tool pusher and other supervisory personnel usually are on the rig floor at this time or can be called to review the situation.

Then the mud weight required to kill the well is calculated. First the mud weight equivalent of the 500 psi SIDP must be calculated:

Production Hole Section

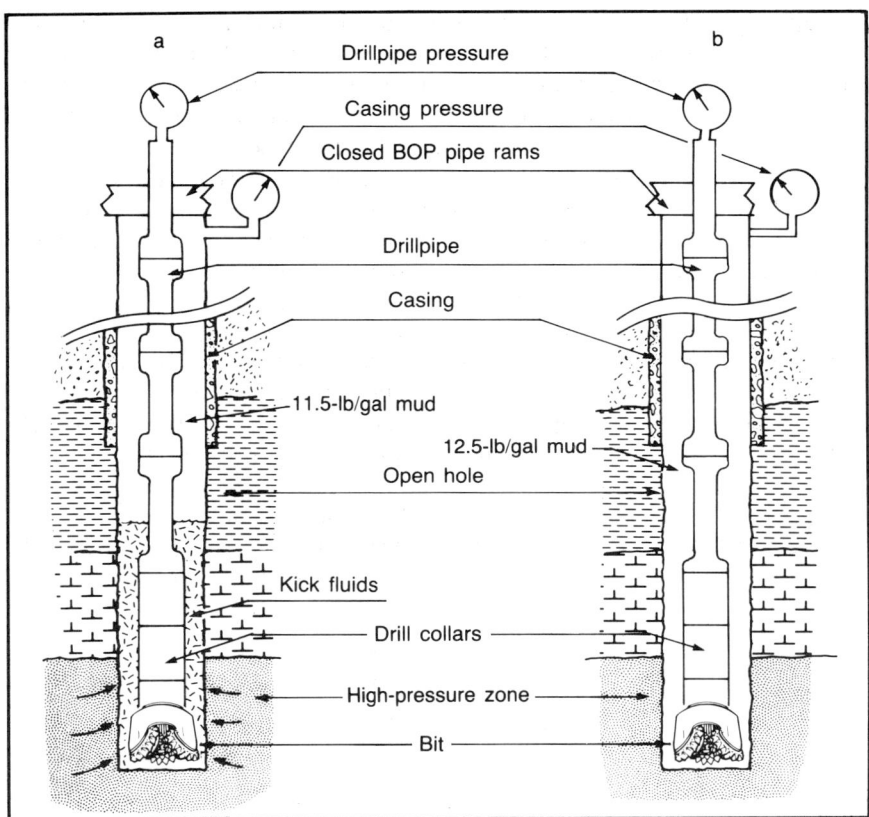

FIG. 9–2 Blowout control. The well was kicked and has been shut in (a). The drillpipe is filled with 11.5-lb/gal mud, and the annulus has a column of kick fluids and mud. The drillpipe and annulus (casing) are pressurized with the pressure contained by the blowout preventers. (b) The well is dead (no pressure) after circulating with 12.5-lb/gal mud.

$$\text{Mud weight equivalent} = \frac{(500 \text{ psi})}{12,000 \text{ ft } (0.052)} = 0.80 \text{ lb/gal}$$

The mud weight must be increased by 0.80 lb/gal to balance the extra pressure. Therefore, the extra mud weight required to balance the well would be:

$$11.5 + 0.8 = 12.3 \text{ lb/gal}$$

A mud weight of 12.3 lb/gal would balance the formation pressure. The operator normally elects to add a couple of points (one point = 0.1 lb/gal) to overbalance the formation pressure slightly. Therefore, the kill mud must weigh 12.5 lb/gal. This mud weight, which is used to circulate the kick out of the hole, can then be adjusted depending on the action of the well at that time.

The shutin drillpipe pressure is used to calculate the required weight for the kill mud. The mud in the drillpipe is clean mud—no gas cutting—and represents a solid fluid column from the surface to the bottom of the hole. The weight of this fluid column can be predicted, and any additional pressure such as the shutin drillpipe pressure represents additional formation pressure that must be overbalanced. The casing pressure is higher because that part of the column of fluid in the annulus comes from the formation and at this point has an unknown density. The length of the fluid column can be approximated by using the difference between the casing and drillpipe pressures and the volume of fluid gained in the mud pits. However, this measurement is less accurate than the shutin drillpipe pressure.

Note that the full name of shutin drillpipe pressure is used here. The term "drillpipe pressure" commonly designates the circulating drillpipe pressure. The densities of the kick fluids in the hole cannot be precisely calculated but range as follows:

$$\text{Gas} = 2\text{--}3 \text{ lb/gal}$$
$$\text{Oil} = 4\text{--}6 \text{ lb/gal}$$
$$\text{Water} = 8 \text{ lb/gal}$$

The mud in the suction pit is weighted to 12.5 lb/gal by adding barite material. Normally the suction pit does not contain sufficient mud volume to circulate out the kick. However, most of the mud in the hole can be circulated through the pits. Then the weight of the suction-pit mud is monitored, and weighting material is added as necessary to maintain 12.5-lb/gal mud being pumped into the hole.

Raising the mud weight 1 lb is a relatively large mud-weight increase. The operator may elect to use a mud weight of about 12.1–12.3 lb/gal and increase the mud weight in later circulations as necessary. Selecting the kill mud weight depends on the calculated weight required, the kick volume, and the experience in the area. It is assumed in this example that the kill mud is weighted to 12.5 lb/gal.

If the well is securely shut in, there is little danger of losing control. However, with time there is some risk of migration of the lower-density formation fluids uphole and an associated increase in casing pressure, depending on the density and volume of reservoir fluids in the annular space. This generally is not a problem but may require flowing a small amount of fluid from the casing side to reduce the pressure. Mud may be pumped into the drillpipe during this time if circumstances warrant its use. The kick must be circulated out of the hole under pressure to prevent additional formation fluid entry into the wellbore. If lower-density reservoir fluids are allowed to enter the wellbore and mix with the heavier mud, the net effect is a reduced mud weight, which will prolong and adversely affect circulating out the kick.

While the mud is being weighted, there is a high risk of sticking the drilling assembly. This risk increases with time. However, killing the kick has first priority, and all efforts are concentrated in that direction. In some cases the operator may elect to move the drillpipe to help prevent it from sticking, assuming that this can be done safely. With low pressures, the bag-type preventer can be closed and the drillpipe rams opened. The pipe can then be moved short distances periodically to prevent it from sticking. In these cases the kelly is normally left above the rotary. Then if the bag-type preventer begins leaking, the drillpipe rams can be closed immediately to contain the well. Another procedure for moving the pipe is to rotate or reciprocate it slowly at periodic intervals, allowing it to turn and slip in the closed drillpipe rams. This procedure is seldom used because it causes additional wear on the rams. Normally the pipe is rotated only inside the top drillpipe rams, and the BOP stack is usually equipped with two sets of drillpipe rams.

When the well is shut in, the driller closes the top drillpipe rams and the lower drillpipe rams are maintained in reserve. To illustrate how the lower drillpipe rams are used as a reserve, assume that the pipe has been worked slowly with the upper drillpipe rams closed. After some time the rams begin to leak. Then the lower drillpipe rams are closed. The upper rams are opened and changed and then are closed on the drillpipe. The lower pipe rams are reopened. This procedure is known as changing rams under pressure.

The kick is circulated out of the hole after the mud in the suction pit has been weighted to the correct density. Valves are opened and closed as necessary to connect the annulus to the hydraulically actuated adjustable choke in the choke manifold. The mud pump is started slowly, and the adjustable choke is opened slightly simultaneously. The mud pump speed is gradually increased until the pump is running at one of the slow pump rates selected earlier.

The pump speed is maintained at 40 strokes/min, one of the slow pump rates, and it is selected by strokes per minute, not pressure. At the same time the adjustable choke is opened and continually adjusted to restrict the annular flow rate and to maintain a drillpipe pressure of 1,600 psi, corresponding to the slow pump rate. The well is being circulated at the slow pump rate, but the pressure is controlled by the adjustable choke on the choke manifold.

The selection of the pump rate depends on operating conditions, such as the viscosity of the mud, the volume of formation fluids in the annulus, and the type of formation fluid—whether gas, oil, or water. One of the most important objectives in circulating out a kick is to maintain sufficient pressure in the wellbore so formation fluids are retained in the formation.

In the circulating process the reservoir fluids tend to expand as they move upward in the annular space. Water and oil expand very little and seldom cause a problem. However, gas is highly compressible and expands greatly with reducing pressure as it moves uphole. This increases the difficulty of controlling the flow with the adjustable choke.

The kick may be circulated out in one circulation, but in some cases it may require two circulations under pressure. After the hole has been circulated with heavier mud, the well is effectively dead and the preventers can be opened. At that time the circulation rate is increased to normal, and the hole is circulated to condition the hole and mud.

The drilling assembly is moved and worked as soon as the preventers are opened. If the drilling assembly becomes stuck while circulating out the kick, it is then worked to release it. If this is ineffective, the assembly must be fished out of the hole. Drilling operations are resumed after the kick has been circulated out, the mud is circulated and conditioned, and the hole is stabilized.

Blowouts

Blowouts seldom occur. But when they do occur, it is because they have not been prevented or because the equipment and operating procedures are not adequate to control the pressure and volume of the formation fluids. There can be many reasons for a blowout, but unfortunately most of these are probably caused by a lack of planning and preparation.

A large volume of formation fluid can flow into the wellbore before it is detected if the sensors or alarms malfunction. If this volume is large enough, it can cause a blowout before the well can be shut in. If the hoisting equipment fails, it probably will be necessary to close the well in with the bag-type preventer. These have a lower pressure rating than pipe rams. Therefore, the bag-type preventers have a higher risk of failure, leading to a blowout—especially under high-pressure conditions.

The weaker formations at the base of the surface casing can break down and begin to take fluid while circulating out the kick. This can cause an under-

FIG. 9–3 Blowout and fire destroy rig (courtesy *OGJ*)

ground blowout. Worn surface casing can burst while circulating out the kick, causing either an underground blowout or a blowout to the surface.

One of the first steps after closing in the well is to check for leaks. If a leak occurs, in most cases it can be isolated by closing valves. Then it can be repaired and operations can be resumed. However, if the leak occurs as the formation fluids reach the surface, a critical condition is created. A blowout may occur if the valve used to isolate the leak cannot be closed tightly. A blowout or a very difficult operating problem also can occur if one of the tool joints in the drilling assembly begins leaking.

A plugged bit during the circulating procedure can cause severe problems. A blowout is almost a certainty if the blowout preventers fail to function properly. The operation of the blowout preventers depends on a number of connections remaining tight and the proper functioning of the control panels, the accumulator, and the preventers. The foregoing may appear unreasonable, but blowout conditions have been caused by each item listed.

There are backup systems and alternative procedures. However, sometimes these are less-acceptable substitutes or there may not be sufficient time to initiate them. For example, specialists can be called to perforate the drillpipe in case the bit is plugged while circulating the kick out of the hole. However, 12 hr or more may be required for the perforating truck to arrive on location, and the long delay may lead to other problems.

Alternative procedures may require drastic action, but within reason almost any action is justified if it will prevent a severe blowout. For example, assume that a trip is being made to replace a worn bit and the well starts to blow out with a few stands of drillpipe and the drill collars in the hole. One stand in the derrick is still connected to the pipe in the hole. The normal procedure would be to immediately lower the stand back in the hole, connect a drillpipe blowout preventer to the tool joint on top of the drillpipe, and close the pipe rams. This would close the well in. Then preparations could be made to circulate the well dead.

Since the bit is at a shallow depth, the best procedure would be to run additional pipe in the hole. This might be accomplished by placing a check valve in the drillpipe, closing the bag-type preventer, and opening the drillpipe preventer. Then stands of drillpipe could be run in the hole through the bag-type preventer, called *stripping in the hole*. However, in most cases the pressures would be too high to permit this procedure.

If the pipe cannot be stripped in the hole, the next alternative would be to pump heavy mud through the kill line into the casing annulus and try to fill the hole with heavy mud. This is known as *bullheading* the mud in the hole or killing the well from the top. Generally, the procedure is at best only partially successful.

Assume the original blowout conditions exist with the stand lowered to the rig floor and the well flowing so strongly that the drillpipe preventer cannot be connected to the tool joint. In this case the well is effectively blowing wild and will probably catch fire within 2–10 min.

In this case any action that can shut in the well must be taken immediately. If the preventers are equipped with shear rams, they should be closed immediately. This cuts the drillpipe, allowing the lower section to drop and at the same time shutting in the well. The upper section of parted pipe is raised, and the blind rams are closed to ensure that the well is securely shut in. Heavy mud is then bullheaded (pumped) through the kill line in an attempt to kill the well from the surface.

If the preventers are not equipped with shear rams, the pipe is picked up a short distance to remove the slips, if possible, the elevators are knocked open by striking the latch with a sledgehammer or other heavy metal object. This allows the pipe to drop. Then the blind rams can be closed to shut in the well.

Dropping all or part of the assembly is a drastic action and certainly should not be taken except as a last resort. For example, if the pipe is not dropped and the well does not catch on fire, other methods may be used to install a valve on top of the pipe and shut in the well.

The foregoing lists only a few of the many ways that a well can blow out and some of the procedures that must be initiated immediately to shut it in. In many cases the formation is depleted or expends most of its initial energy. Then the flow subsides to the point where actions can be taken to shut in the well.

When the well blows wild, the high rate of flow carries pieces of formation that may create a bridge of formation material in the hole and effectively shut in the well. When this occurs, the well is later reentered and cleaned out, and junk is fished out of the hole as necessary to resume drilling operations. If the well cannot be cleaned out, it may be necessary to sidetrack the junked hole.

If a well continues to blow out, it will often catch on fire. When the well is on fire, operating personnel cannot work near the wellhead because of the intense heat. One of the first procedures in this case is to put out the fire. This is normally done by detonating a large charge of explosives placed near the wellbore. The explosives are positioned with long cables suspended between two cranes, a long boom attached to a large caterpillar tractor, or possibly are dragged into place with a cable. When the explosives are detonated, they literally surround the fire with noncombustible gases that suffocate the fire.

If there is molten or hot metal around the wellhead, the flowing gas and oil reignite. To prevent this, any metal around the wellhead is removed with hooks and cables pulled by tractors before putting out the fire. Large volumes of water are sprayed over the area to help cool down any hot spots that might reignite the flowing oil and gas. Ditches and large earthen pits are constructed so the oil and condensate can be drained away from the wellhead. After the fire is put out, various techniques and specialized equipment are used to cap the well and shut it in.

The procedure for capping the well while it is flowing is extremely difficult and in some cases is almost impossible. If the wellhead is still intact, the valves and rams are usually inoperable. A large master valve with a base or flange to fit the wellhead may be used. It is suspended over the flowing stream with cranes while connecting the flanges. Then the valve is closed to shut in the well. Depending on the circumstances, the blowout well may first be killed by a special method—drilling a *kill well* (Fig. 9–4). In this procedure another rig is moved onto a convenient location a safe distance from the blowout. A deviated hole is drilled into the formation producing the blowout fluids as near as possible to the intersection of this formation with the wellbore that is blowing out. Large vol-

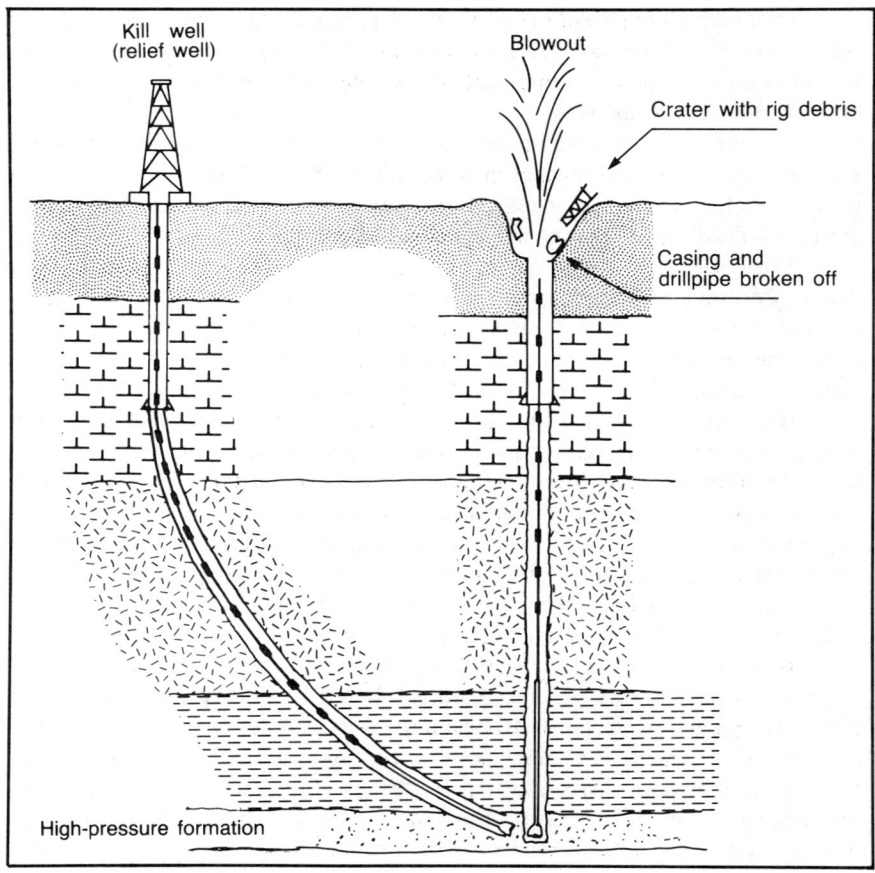

FIG. 9–4 Blowout and kill well

umes of weighted mud or cement are pumped into the formation at or near the wellbore of the blowout well. The natural pressure differentials in the formation tend to cause the weighted mud to flow into the formation around the wellbore of the blowout well; part of the mud also flows up the wellbore of the blowout well. This mud kills the blowout by creating a mud column with a hydrostatic head in the blowout wellbore. At the same time it plugs the formation so fluids cannot flow from the formation into the blowout wellbore.

The blowout well usually is cleaned out sufficiently to place cement plugs downhole to ensure that the wellbore is safely plugged. Then the well is abandoned. Depending on the condition of the casing, the upper part of the wellbore may be used to drill a deviated hole. The sidetracked well can either be plugged and abandoned, completed in the formation to replace the blowout well, or plugged back and sidetracked to make an offset producing well in the same formation.

DEVIATED HOLES

In the earlier history of the industry, all wells were drilled as straight holes. When instruments become available to measure deviation, it was found that most holes tend to deviate. Some holes, especially in flat formations, drill in a long, right-hand spiral. The tendency for the bit to drill a hole curved in the right-hand direction is known as *bit walk*. Bit walk is caused by right-hand rotation and the tendency for the side of the bit to climb or drag on the right side of the hole.

The first need for a deviated hole probably occurred when a fish was lost in the hole and could not be recovered. The only alternatives in this case were to abandon the hole, move the rig a short distance and redrill it, or plug the hole back and attempt to sidetrack the fish.

In cable-tool operations junk was sidetracked with limited success using a spud bit built with a long taper on one side. In earlier rotary operations various method of sidetracking were attempted with very limited success. The development of the *whipstock* provided the answer for sidetracking a fish and later for drilling a deviated hole (Fig. 9–5).

The whipstock is a wedge-type tool that directs the bit to the side of the hole, causing it to drill at an angle. Although the whipstock was the first positive method of deviating the hole, the tool had many disadvantages. At least one and often two or three trips were required to deviate the hole 2–3°. In some cases the whipstock would turn in the hole so it was difficult to drill in the desired direction. This was especially true when trying to set the whipstock upside down to reduce deviation. The tool was less efficient in harder formations. Sometimes the whipstock would deviate the hole too sharply, causing a dogleg. There was also a risk of leaving the pilot assembly in the hole, and it was not uncommon to leave the entire whipstock in the hole, causing a very difficult fishing job.

The increased demand for deviated holes led to the development of equipment and techniques for deviated drilling. Deviated holes have been drilled into reservoirs that are located on a horizontal plane almost 3 mi away from the surface location. Deviation angles of 50–60° are common. Vertical holes have been deviated until the well is drilled horizontally, a total deviation angle of 90°. (Note that deviation angles are measured from the vertical.) Multiple targets 100

428
DRILLING

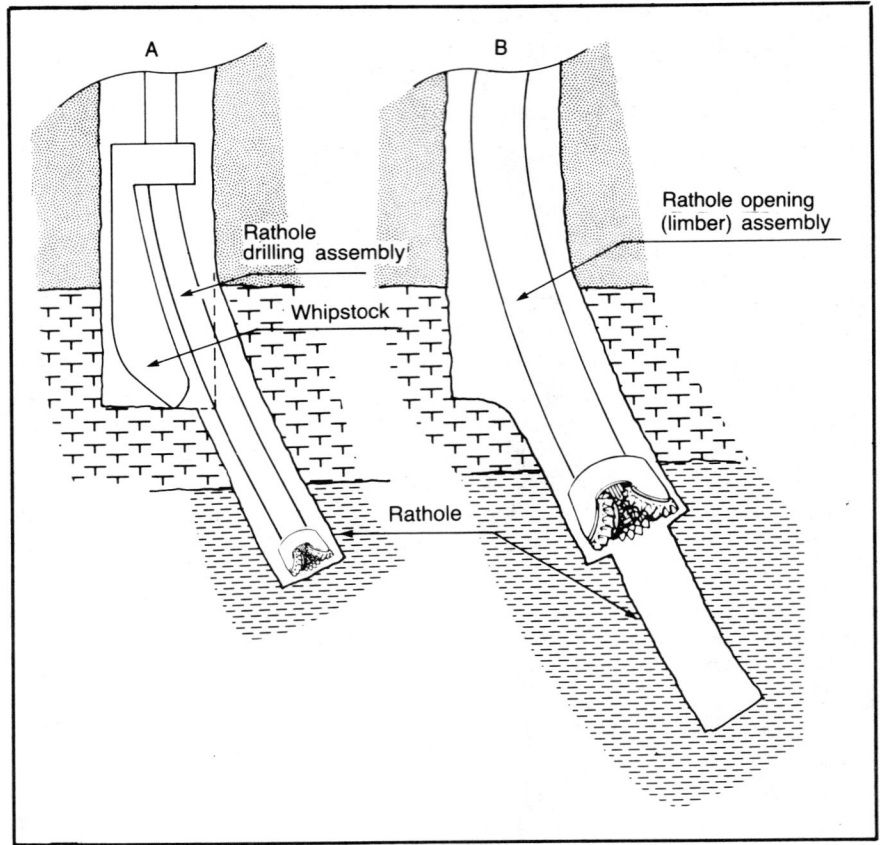

FIG. 9–5 Whipstocking. A rathole is drilled off the face of the whipstock for 10–20 ft in (a), and the assembly is pulled from the hole. A limber assembly is run to ream out the rathole (b) and drill ahead. Note: The rathole is sometimes opened with a pilot bit and reamer when high deviation is required or when formations make it difficult to deviate the hole.

ft in diameter at different depths and different relative surface locations have been penetrated by one deviated wellbore.

There is an increasing demand for deviated drilling. Probably the largest demand for deviated holes is in areas where the surface locations are limited or where the cost of building additional surface locations exceeds the cost of drilling deviated holes from one central location. The main example of this is offshore platforms. Other examples are man-made islands, swampy or mountainous terrain, tidewater areas, and insulated locations with deep permafrost such as the Alaskan North Slope.

Deviated holes are also used to develop reservoirs underlying areas where surface locations are effectively unobtainable. Examples of these are industrial and residential areas, public parks, recreation areas, rivers, and small inland lakes. Several major cities such as Long Beach, California, and Houston have oil and gas reservoirs underlying part of the city.

There are a number of miscellaneous uses of deviated holes:

- Sidetracking a fish that cannot be recovered
- Drilling into reservoirs under the edge of salt domes
- Deviating holes drilled across faults to a reservoir on the productive side of the fault
- Deviating holes updip into oil and gas zones
- Changing angles on wells drilled on small leases to stay within the lease lines
- Reducing the angle in areas where a pendulum assembly is not sufficiently effective to overcome the strong angle-building tendency of the formation
- Drilling kill wells

The angle of deviation or the direction of drilling (called the horizontal component) may be controlled in deviation drilling. The angle of deviation can frequently be controlled by selecting a drilling assembly to build, maintain, or drop angle. However, these assemblies normally do not change the angle as fast as is required. Therefore, an angle-deviating drilling assembly must be used to change the deviation rapidly or change the horizontal component of the holes direction.

The standard deviating assembly consists of a bent sub on top of a turbine or mud motor, which rotates the bit (Fig. 9–6). The bent sub connects to the bottom of the drill collars and changes the alignment of the drilling assembly (turbine or motor and bit) from the centerline of the drill collars. The angle between the centerline of the drilling assembly and the drill collars is the bent-sub angle. Bent subs are built with angles from $\frac{1}{2}$–$4°$ in $\frac{1}{2}°$ increments. Since the bent sub causes the angle to change continually as the hole is drilled, the change in deviation per 100 ft is considerably more than the angle of the bent sub. For example, a $\frac{1}{2}°$ bent sub causes an angle change of about 2–$3°/100$ ft, and a $2°$ bent sub causes a change of 6–$8°/100$ ft, both dependent on the type of formations and the manner in which the tool is run.

The deviation tool assembly has a receptacle to receive an instrument that measures the hole direction, the angle of deviation, and the tool face. The tool face is the direction the bent section of the sub is pointing. The direction is

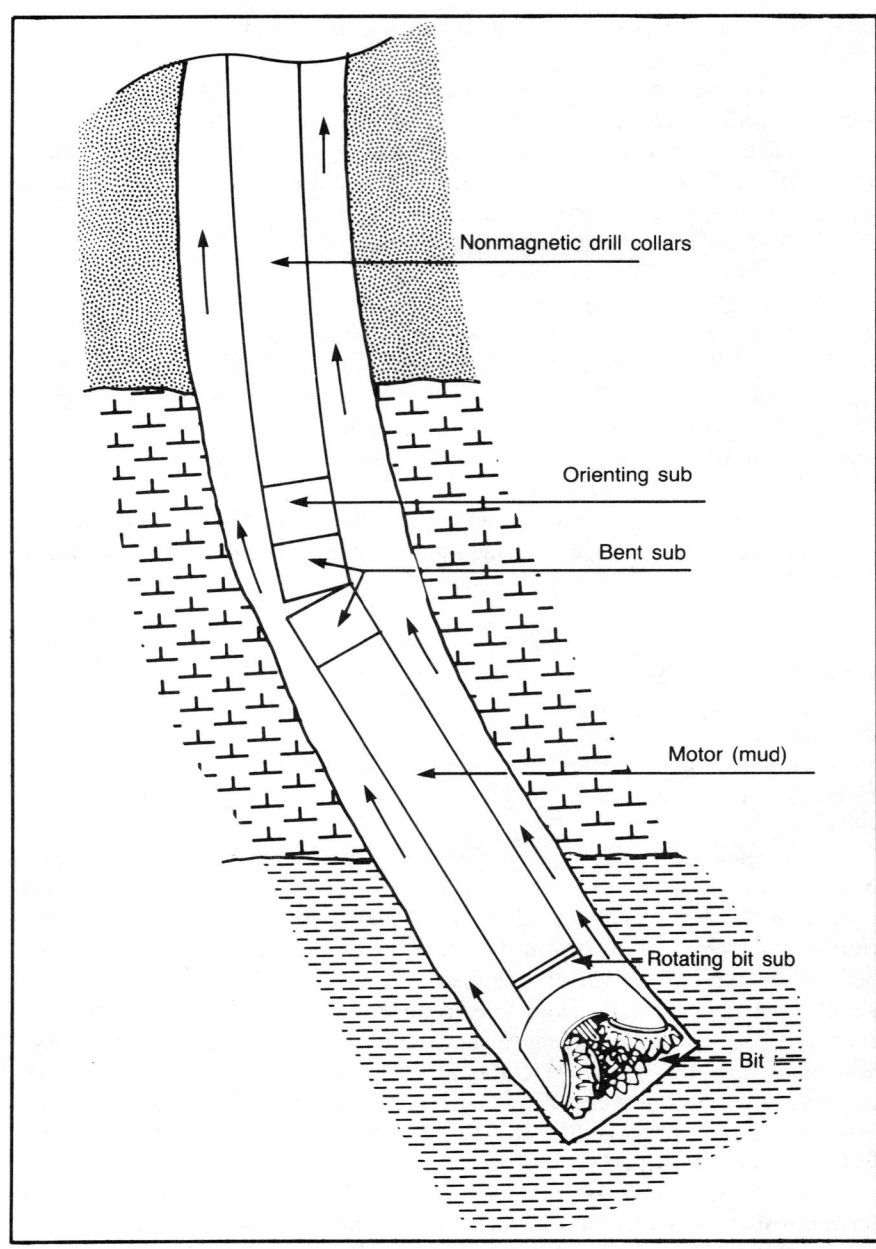

FIG. 9–6 Bent-sub, mud-motor deviating assembly

measured by a compass-type instrument that is actuated by the earth's magnetic poles. One or more nonmagnetic, monel drill collars are used above the bent sub since normal steel drill collars blank off the earth's magnetic field.

Although the drilling assembly must be at rest when angle and direction measurements are made, there are various methods of taking measurements. One is to run the measuring device on a cable with a single, insulated conduit. The angle of deviation, hole direction, and tool-face direction are displayed on readout equipment at the surface. A similar instrument can be run in the hole on a regular wire line without an electrical conduit. The readings are recorded in the instrument and are read after the instrument is pulled from the hole. In this case the instrument can take readings based on time or on a motion sensor that activates when the pipe is at rest for about 30 sec or longer. A magnetic sensor can be used that activates after it senses magnetic responses below the monel drill-collar section. Another type of instrument is activated by the pulses in the mud system to give a readout at the surface. The company supplying the tools and service prepares a diagram showing the course, direction, and deviation of the proposed deviated hole in accordance with the operator's recommendations.

To explain the procedure, assume that a hole is to be drilled vertically to 5,000 ft and deviated at about 3°/100 ft to a deviation angle of 30°, which is maintained to total depth on a course of 90° due east. The hole is first drilled to 5,000 ft with a conventional drilling assembly. At that point a deviated drilling assembly with a 1° bent sub is run. The tool face is oriented north 80° east with the measuring device, allowing 10° for bit walk. The hole can be drilled due east, but this requires more measurements and precise control. Normally the operator allows as much latitude as possible in order to expedite drilling.

With the tool face set at 80° (N80°E), the rotary is locked in place and the pump is started to begin rotating the bit. A bit weight of 10,000–15,000 lb is used. The hole is drilled slowly, and the drillpipe is not rotated. The course and direction of the hole is measured after drilling about 30 ft. Drilling is resumed if the hole is on course and the angle is building satisfactorily. If the course is not correct, the tool is reoriented. Drilling is continued with periodic measurements of the course and angle.

The operators can control the rate of angle buildup to a limited extent by the method of drilling. However, most angle buildup is controlled by the angle of the bent sub. If the hole angle is not building at the proper rate, the assembly is pulled from the hole and a different bent sub is run. A lower bent-sub angle is used to decrease the rate of angle buildup; a higher bent-sub angle is used to increase the rate of buildup.

Angle and course measurements are made and plotted on a graph showing the projected course of the hole. This is used to ensure that the drilled hole is following the course of the projected hole and as a guide to help determine future tool settings. When the hole angle has built up to about 20°, the deviation drilling assembly is pulled. Then a conventional hole-building drilling assembly with one or more monel drill collars and the seating device for the angle-measuring instrument are run. The hole is drilled in a conventional manner, rotating the drilling assembly, and periodic measurements of the course and direction are made. Normally this assembly continues building the hole angle but at a gradually reducing rate.

If the hole does not follow the correct course and angle, the assembly is pulled from the hole. The deviating assembly is rerun and oriented as needed to correct the course or angle of the hole. The hole-building assembly is pulled after the angle has built to about 30°, and a stiff assembly is run to drill ahead. Normally this assembly maintains a relatively straight hole.

If the hole angle must be increased or decreased, the stiff assembly is pulled and the required assembly is run. If more corrective action is needed, the deviation assembly with the required bent-sub angle is rerun to change the direction as needed. The hole is drilled to total depth in this manner.

Two depths are normally reported on deviated holes. The true vertical depth (TVD) is the true vertical distance from the bottom of the hole to the surface elevation or reference datum on the rig floor, usually the top of the kelly drive bushing. The *measured depth* (MD) is the actual measured distance along the axis of the borehole, the pipe measurement, or the actual footage drilled.

The operator can precisely control the course and direction of the drilled hole at all times. Extra control requires additional time. Normally only the required amount of control is used. For example, the direction usually is not important in sidetracking a fish, so only the deviation angle is controlled. Where precise control is needed, such as penetrating multiple targets at different depths for a well to be completed in multiple reservoirs or drilling a kill well, more time may be required to make course and angle measurements and to reorient the tool as needed. This method can deviate the hole in 1–3 days depending on the hardness of the formations. It may take a week or longer to perform the same operation with a whipstock, assuming none of the problems frequently associated with the whipstock occurs.

Deviated holes are subject to all of the problems associated with normal drilling. There may be additional problems associated with the deviated hole, such as caving, keyseats, and hole cleaning. It normally takes about 30–50% longer to drill an average deviated hole than to drill a straight, vertical hole to the same depth.

Deviated holes are logged and cased in the same manner as regular holes. When the holes have a high deviation, the logging tools may not fall freely. Then special pumpdown logging procedures are used. The logging tools are pushed to the bottom of the hole through pipe by pumping mud behind them. In special cases the logging tools may be run on small, wheeled carriers. In these cases usually the measuring instruments must also be pumped down during drilling. Some rigs may have a spare mud hose and pumpdown head to pump measuring and logging equipment down the hole.

The hole normally is not deviated near a casing point. Usually the bottom of the upper casing strings (surface and intermediate) is set in a straight section of hole even though the deeper hole may be deviated. Liners are also run in deviated holes. If the hole has a high deviation, a hydraulic liner setting tool may be used so the liner can be seated by pump pressure rather than rotation, which may be difficult in a deviated hole.

One modification of deviated holes is slant-hole drilling. In some cases the formations may be at shallow depths; the depth is not sufficient to use a reasonable buildup angle and still penetrate the formation at the desired point. In this case a special slant-hole rig is used. The rig is similar to a normal drilling rig, but it is usually smaller because of the shallow depth. The mast can be set at an angle less than vertical. The traveling block is guided or travels on a track carrier. A slant-hole rig can spud in at an angle of 30–45°, thus reducing the amount of angle buildup required to reach the same horizontal distance from the surface location.

Deviated drilling technology and equipment have been developed and improved continually, and now this technique is very reliable. Although it costs more to drill at an angle than to drill vertically, this operation has permitted an overall savings in special cases when vertical drilling is impractical or uneconomical. Deviated drilling also has led to the development of substantial oil and gas reserves.

OBTAINING RESERVOIR AND PRODUCTIVITY INFORMATION

Prospective productive horizons are penetrated while drilling the production hole section. Normally, the well logs that are run after the section is drilled are relied on to supply information about this section. In many cases important information can also be obtained while the formations are being drilled. For example, drilling information can indicate potentially productive fractures in a massive shale section. These sections may not be detected on the open-hole logs. After casing the hole, the zone can then be perforated based on information obtained while drilling the section. In one case a subsequent well potentialed over 5,000 bo/d and produced at a high rate for a long period. This emphasizes the importance of collecting and analyzing data while drilling.

Open-Hole Show Indicators

In the ideal case the exploration engineer would like to know everything about the formation immediately after it is drilled. At the present state of the art, this is impossible. There has been considerable experimentation on measurement-while-drilling (MWD) tools, but additional work is needed before most of them will be available on a commercial basis.

However, a number of different indicators that individually may not be significant can be combined with other data such as coring, drillstem testing, and well logs to provide meaningful information about the formation. One of the most prominent examples of this is a kick. The fact that the formation fluids can flow into the wellbore indicates that the formation has porosity, permeability, and mobile-fluids—all requirements for a reservoir.

Samples of kick fluids should be caught when possible as the kick is circulated out of the hole. If the samples contain oil or a large volume of gas, the reservoir may produce oil or gas. If only salt water is recovered, a sample should be caught and analyzed, including taking a resistivity measurement. The lack of good formation-water resistivity measurements causes problems in analyzing the well logs. A resistivity measurement from a kick sample can be very helpful and can mean the difference between evaluating a zone as hydrocarbon productive versus nonproductive as a result of high water saturation.

A mud logger or geologist analyzes cuttings on most wildcat wells and deeper development wells. This step may be omitted on shallower development wells; but unless the formations are well known, it is always a good practice to analyze the cuttings. Minimum mud weights should be used to obtain a maximum penetration rate. This also improves the quality of show indicators, especially for gas.

Cuttings are caught over the drilled interval of 10–20 ft in the less prospective hole section and 5–10 ft in more favorable sections. The samples are lag corrected by allowing for the time required to circulate them out of the hole. For example, assume that the samples are caught over 10-ft intervals, the hole is being drilled at 10 ft/hr, and the lag time is 1 hr. The samples caught over a 10-ft interval drilled in a 1-hr period would have originated from the 10-ft section of the hole drilled in the preceding hour.

Samples are washed, dried, and stored in cloth sacks or envelopes. They are properly labeled with the well name, date, footage interval, and other information as required. Clean drill cuttings are visually inspected under a low-power microscope. They are described in detail, including the type of formation (e.g., shale, sand, or limestone), color, and trace minerals.

Sands are of special interest since most reservoirs are found in sands. Sands include loose, unconsolidated sand grains and consolidated, hard, well-cemented sandstones. They are cemented naturally during or after deposition with either calcareous or silicious cementing material. The type of cementing material can be determined by placing a small droplet of dilute hydrochloric acid on the sand cutting. Calcareous cementing material bubbles or reacts with the acid. These sands may be acidized for mild stimulation during completion.

Sometimes void spaces can be observed indicating porosity. Sand-grain sizes range from very fine to large. Larger sizes such as pebbles, gravel, or boulders are seldom found in deeper wells. The size of the sand grains gives an indication of permeability, which increases with increasing grain size. Very fine grain size has a very low permeability and is seldom commercially productive except for low-viscosity gas in thick sections. The shape of the sand grains also gives an indication of porosity and permeability. Angular sand grains normally are densely packed and have low porosity and permeability. Subangular to rounded sand grains usually have better porosity and permeability. Well-rounded, large sand grains offer the most favorable porosity and permeability.

Some samples may have oil staining that can be detected visually or as fluorescent under an ultraviolet light. If oil mud is used as a drilling fluid, frequently the oil from the mud exhibits a different colored fluorescence from natural crude oil. Some cuttings bleed oil—oil evolves from the cuttings and can be detected as minute volumes are forced to the surface.

The gas content of the cuttings is determined by mixing the cuttings in a blender and analyzing the air in the container for liberated gas. A gas show is a favorable indicator of hydrocarbons. High methane content indicates a gas productive formation. A high content of pentanes plus indicates oil production.

Gas in the mud is measured in units, and the readings are not standardized. In some cases a trace of gas may read 10 units and in another case 20–30 units. Using the same relative measurement, saturated mud—where the gas bubbles out of the mud—may be recorded as 2,000 units on one instrument and 4,000 units on another. This sounds confusing, and it is to a limited extent. However, the relative difference in the gas measurements is the important factor.

Observing mud weight relative to the gas measurements also gives clues about the formation. Highly gas-cut mud can reduce the mud weight as much as 50% as a result of entrapped gas, but smaller reductions of ½–1 lb/gal are more common. The gas content of the mud is measured continuously and is plotted on a lag-corrected depth scale.

Mud is not circulated during connections. If gas flows into the wellbore in very minute quantities, it accumulates in the mud in a higher concentration when a connection is made than during normal circulation. This type of gas is called *connection gas*. A plot of connection gas versus depth is another good indicator of perspective formations.

Drilling breaks can indicate porosity. Therefore, a plot of the penetration rate indicates prospective zones.

Fractures sometimes provide a reservoir for oil and gas production. Generally they are not good reservoirs, but there are exceptions. Fractures are probably more effective as a reservoir when they are associated with sands, often low-permeability sands. The sands act as the reservoir, and reservoir fluids flow from the sands into the fractures to the wellbore. Larger fractures often cause a varying torque, and the rotary runs rough, i.e., rotary speeds vary rapidly. Cuttings show *slickensides* or *striations*—marks or scratches on the face of the cuttings caused by movement along the fault or fracture plain. *Calcite nodules,* a secondary deposition product, and *calcite-filled veinlets* may also indicate fractures where calcite has deposited. In these cases completion techniques often include stimulation by acidizing.

An accumulation of oil in the pits of water-based mud systems is a good show indicator. Quartz crystals may indicate porosity. *Oolites* and *reefing-type* material in carbonates indicate porosity.

Slow seeping-type lost circulation can indicate permeable sands or small fractures. Large losses of circulation frequently indicate large fractures. In some areas these are definite indicators of productive horizons and are often drillstem tested. In summary, it is important to note and record all types of show indications while drilling to help obtain the best completion.

Coring

Coring is the process of cutting, releasing, and recovering a sample of the formation, usually long and cylindrical called a *core*. Normally the core is large enough so it can be tested for porosity and horizontal and vertical permeability. Fluids in the core are recovered and analyzed. In many cases there may be a question whether the fluids in the core represent true formation fluids or are contaminated by mud filtrate. Nitrates do not occur normally in underground fluids and can be added to the drilling fluid. The presence or absence of nitrate in core fluids helps resolve the question of the source of the fluids. A visual examination of the core also is important.

Core permeabilities and, to a lesser extent, porosities measured at the surface often are higher than the actual or in situ permeability and porosity as a result of overburden loading, which compresses the formation (core) when it is in situ. Overburden is the composite weight of all formations from the core point to the surface of the ground. Water saturations may also affect results since the saturation at the surface may be less than that in the reservoir as a result of fluid escaping from the core. Pressurized core barrels help eliminate this problem. The pressurized core barrel seals the core after it is cut and maintains the core in a pressurized, natural state while it is recovered.

Cores are also used for capillarity and other tests needed for secondary recovery such as simulated flooding or carbon dioxide miscibility. A *core gamma,* essentially a special gamma log of the core, may be run so the precise position of the core in the hole can be determined by correlating the core gamma with the regular gamma log run during logging operations. This is often important since it may otherwise be very difficult to locate the core interval and depth precisely relative to the well logs.

Standard coring operations use a conventional core barrel (Fig. 9–7). The tool consists of an inner barrel to hold the core and a hollow core bit, commonly called a *core head,* to cut the core. A core catcher is fitted in the bottom of the core barrel to hold the core in the barrel after it is cut. Generally the diameter of the barrel is about 2–3 in. less than the diameter of the hole being drilled. Core barrels are available in various section lengths, from 10–20 ft. Two or more sections can be connected together to give a final core barrel length usually in the range of 30–60 ft.

Other types of core barrels are available, but they are used less frequently than the conventional core barrel. An oriented core barrel can cut and recover a core and mark one side of the core in a fixed direction. After the core is recovered, it can be oriented relative to the compass points. Oriented cores are used mainly to determine the direction of formation dip. Logging tools are now available to provide the same information in an easier, more convenient manner. Core barrels fitted with rubber sleeves are used to recover cores from unconsolidated formations. Wire-line retrievable core barrels are available but are seldom used because the recovered core has a relatively small diameter. Therefore, the core recovery is probably less than that with a conventional core barrel.

In operation, when a coring point is reached, the core barrel is made up on the bottom of the drill-collar assembly and is run into the hole. When the core head is near bottom, the hole is circulated to remove sloughing and other loose material that can plug the core barrel. The core is cut by operating the assembly in a manner similar to drilling but with much less bit weight and at a slower rotary speed. The coring operation is conducted as smoothly as possible to prevent breaking the core and possibly jamming the core barrel with rubble. If the core catcher is damaged, the core may fall out of the barrel on the trip out of the hole.

The core is broken off after the desired length of core has been cut. If it is not broken off, a long segment of the core may be pulled out of the core barrel,

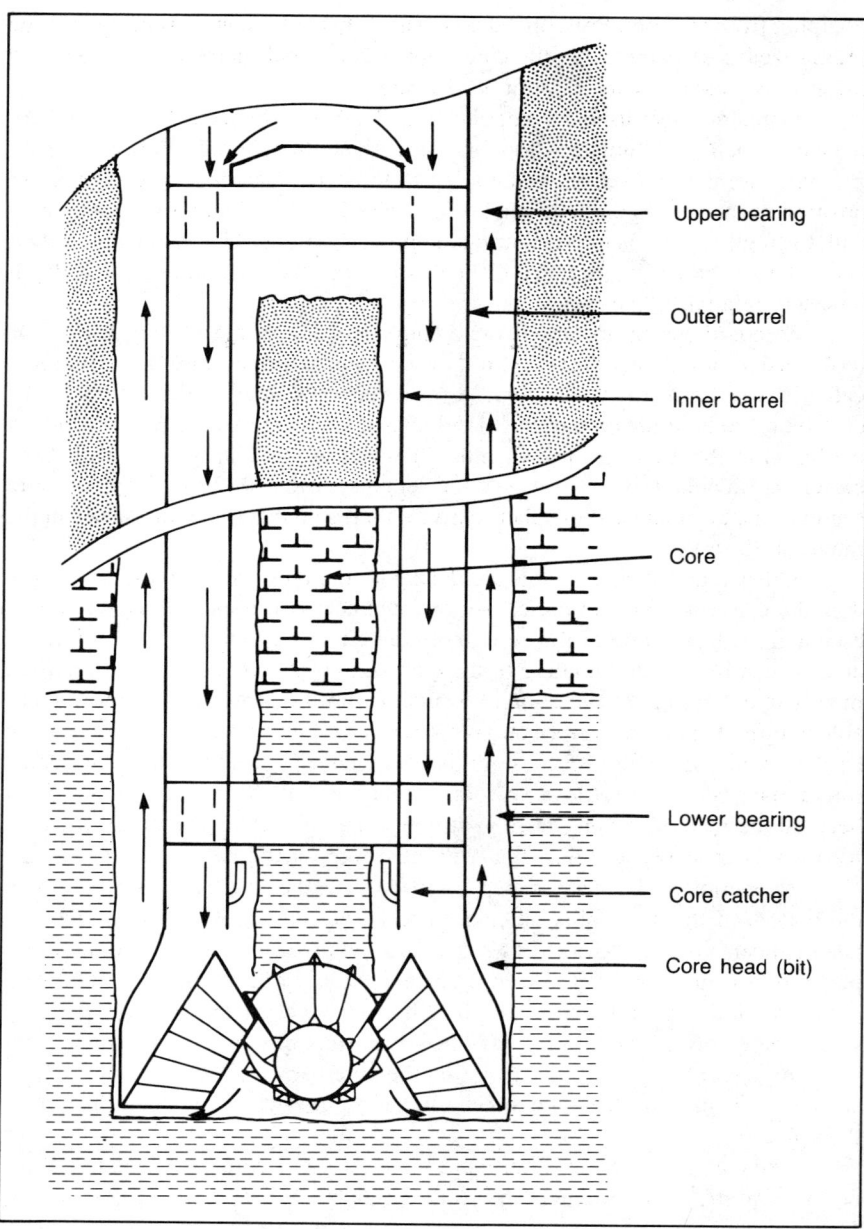

FIG. 9–7 Core barrel

breaking the core catcher when the core is pulled from the hole. The core is broken off using higher, alternating rotary speeds and reciprocating the pipe vigorously over a distance of about 2 ft. The core barrel assembly with the enclosed core is then pulled from the hole.

The core is removed and usually is placed in core trays where it is inspected and described. All or selected parts of the core are carefully wrapped, sealed, and sent to a laboratory for analysis.

Drillstem Testing (DST)

Drillstem testing, commonly called DST, is a procedure for testing formations in the open hole. The purpose of the test is to find out if the formation contains oil and gas and to indicate the well's productivity and pressure. Sometimes all of this information cannot be determined or some data give questionable results. However, in most cases the DST provides good usable information about the formation.

Another reason for running a drillstem test is to provide information that can justify running production casing and completing the well. An appreciable expenditure has been made in drilling the well to this point. However, this does not necessarily justify the cost of running production casing and completing the well unless there is evidence that the well can produce oil and gas in commercial quantities. A good drillstem test with favorable results is one of the main methods of determining this.

Two basic types of open-hole drillstem test tools are used (Fig. 9-8). The most common tool is used to test formations near the bottom of the hole. The test tool with inflatable packers is used for testing zones further uphole where it is impractical or unsafe to run a long string of tail pipe to set on bottom because it increases the risk of buckling or sticking the pipe. Both tools are open-hole drillstem test tools. However, in common oil-field usage the tool with compression packers is referred to as a drillstem test tool. The tool with inflatable packers is generally specified as an inflatable packer DST tool.

Both tools are equipped with chambers to catch samples of the reservoir fluids and recording gauges, commonly called pressure bombs, for recording downhole pressures during the test. The tools are equipped with circulating valves to reverse out the formation fluids and to circulate if necessary. They are normally run below jars and several drill collars so they can be jarred and released in case of sticking.

Test tools must have a good packer seat, the point in the wellbore where the packers contact and seal against the wall of the borehole. This seal is very important because it isolates the mud in the annular space above the packers from the section below the packers that is being tested. The packer seat must be in an in-gauge, competent section of the hole, preferably within 50 ft or less of the zone being tested. If the packers are seated in an out-of-gauge hole section, the hole

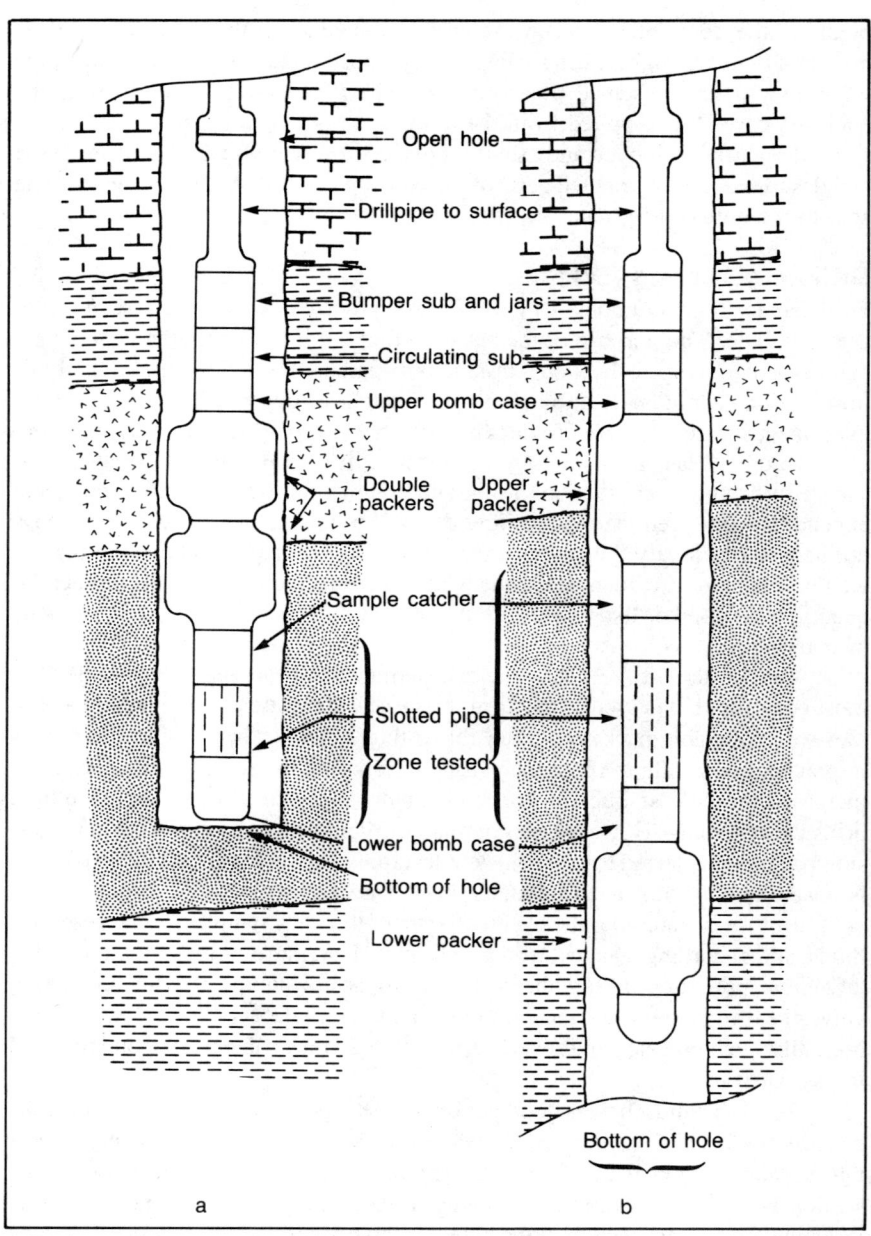

FIG. 9–8 Open-hole drillstem test. (a) DST tool and (b) inflatable packer or struddle test tool

diameter may be greater than the maximum expanded diameter of the packers. Then the lower hole section cannot be effectively sealed. A review of the description of the cuttings, the penetration rate log, and well logs if available helps select a good packer seat.

A valid question at this point is "How can the formation flow if the hole is full of weighted mud that exerts a pressure against the formation to prevent it from flowing?" The DST tool and test procedure solve this problem in a simple but ingenious manner. When a bit or other tool is run into the hole, the bottom of the assembly is open and mud flows into the inside of the drillpipe as the assembly is lowered. In contrast, the DST tool is run into the hole with the tool closed so mud or other fluid in the wellbore cannot flow into the drillpipe. Therefore, when the packer is seated—sealing off the annular mud column—and the tool is open, the full pressure of the formation is released or opened to the inside of the empty drillpipe, which provides a flow channel to the surface. This is known as the pressure differential into the test tool.

If the test tool were run completely empty, the pressure differential would create a tremendous, possibly destructive force when the tool opened.

Bottom-hole chokes are used to help contain and reduce this force. The average choke sizes range from $3/8-1$ in. in diameter. Smaller sizes are used for gas tests and higher pressured formations; larger sizes are used for oil tests and lower pressured formations.

The force of the released pressure differential is further controlled by use of a water cushion. When the DST tool is run into the hole, a predetermined amount of water is poured or pumped into the drillpipe. The water cushion is generally measured in feet of water. For example, 2,000 ft of water cushion would indicate that a 2,000-ft column of water was above the test tool. The remainder of the drillpipe above the water cushion is dry—filled with air.

The amount (length or height) of the water cushion depends on the projected bottom-hole pressure, the depth of the zone to be tested, the weight of the mud in the annular space, and the weight and grade of drillpipe. Most packers are designed to withstand a high pressure differential across the packer. If test conditions indicate a higher pressure, water is added to reduce the pressure differential. Dry drillpipe with no water cushion can collapse if it is run too deeply into weighted mud so that hydrostatic pressure of the mud column exceeds the collapse pressure of the drillpipe.

Nitrogen may be used in lieu of a water cushion or in combination with a water cushion for testing very high-pressure zones or for testing very low-pressure zones where the operator does not want to use a water cushion. The tool is run into the hole on drillpipe with allowance for drillpipe collapse and adding a water cushion if necessary. The tool is seated, filled, and pressurized to the required pressure with nitrogen. After opening the tool, the DST is conducted by flowing the nitrogen out of the drillpipe at a controlled rate to reduce the hydrostatic head and pressure inside the drillpipe, thus allowing the formation to flow.

In operation, the drillstem test is conducted by connecting the various pieces of the test tool together, called *making up the tool*. The tool with jars and several drill collars is connected to the bottom of the drillpipe and is run into the hole to a depth equal to the length of the water cushion. Once the drillpipe is filled with water to this point, the DST tool is run to the bottom of the hole.

The drillstem test manifold is a system of valves and flow lines connected to the top of the drillpipe to control test fluids. Conveniently, a double-wing manifold is used with adjustable chokes on each wing. A flexible metal hose with swivel-type joints is used to connect the test manifold to a test line. The test line connects to a testing system that may include a three-phase (water, oil, and gas) separator with meters to measure the volumes of fluid and gas produced during the test. Holding tanks are provided for the fluid, and the gas is normally burned in a flare. More or less metering and measuring equipment may be used, depending on the volumes and pressures expected during the test. A high-pressure pump truck may be connected to a kill line in case it is needed when testing very-high-pressure zones.

After the equipment has been installed and tested, the DST tool is lowered to seat the tail pipe on the bottom of the hole and to expand the packer or packers if combination packers are used. The tool is opened by a combination of a small amount of reciprocation or rotation of the drillpipe. This starts the drillstem test, which is normally conducted in four steps.

1. *Initial flow* (IF)—The opened tool is indicated at the surface by a blow, or flow of air. This may increase or decrease, depending on the formation pressure and amount of water cushion. An increasing flow rate indicates a good formation. The initial flow period is normally 5–15 min but may be as long as 60 min, depending on the reservoir pressure and permeability. The main purpose of the initial flow is to determine that the tool has opened correctly, to establish an initial flow into the drillpipe, and to ensure that the packers are holding. A static fluid level in the annulus shows that the packers are holding satisfactorily.
2. *Initial shutin* (ISI)—The tool is shut in for a period of 15 min to several hours to take an initial shutin pressure. This step is important because the pressure represents the original bottom-hole reservoir pressure. Surface shutin and flowing pressures are usually recorded, but final calculations are made using the pressures recorded by the bottom-hole recording pressure bombs.
3. *Final flow* (FF)—The tool is opened for a period of 30 min to 24 hr or longer to allow the formation to flow and to obtain flow rates, pressures, and vol-

umes. The flow rate is controlled by one adjustable choke on the flow manifold. The second choke is available in case the first choke plugs, cuts out, or otherwise malfunctions. It also can be replaced while flowing through the second choke. Surface flow rates, pressures, and volumes are measured and recorded, and fluid samples are taken for analysis.
4. *Final shutin* (FSI)—The drillpipe is again manipulated to close the tool. Then a pressure buildup is taken to obtain buildup and final shutin reservoir pressures. In a few cases the tool may malfunction so it cannot be closed or it will cut-out as a result of fluid erosion of the metal so the tool cannot be shut in. These conditions normally occur with very-high pressure, high flow-rate tests. If this occurs, the well is shut in at the surface and the kill line is connected to pump heavy mud down the drillpipe with the high pressure pump truck to kill the well.

After the normal four-step test procedure, any pressure in the drillpipe is released. Formations fluids in the drillpipe are reverse circulated out of the hole: the circulating valve is opened, the drillpipe rams are closed, and mud is slowly pumped into the annulus, allowing the fluids to flow out of the DST test manifold. The volume and type of fluid reversed out of the drillpipe are recorded and samples are taken. This is very important when the formation is not capable of flowing fluid to the surface during the final flow period.

After the drillpipe has been filled with mud, the test tool is unseated and pulled from the hole. The sample chamber pressure is recorded, and the gas, oil, and water in the sample chamber are recovered and measured. The sample chamber is designed to catch the last fluid flowing into the test tool at the end of the final flow period. The pressure bombs are removed from the test tool, and the charts are removed from the bombs. The chart pressures are read with a special tool at the wellsite to give the flowing and shutin pressures. They are then sent to the service company office for detailed analysis, including calculations for reservoir permeability, productivity, extrapolated shutin pressure, depth of investigation, and other important reservoir data.

A sample drillstem test chart is illustrated in Fig. 9–9. On the chart the test tool is being run in the hole at time interval A–B. As the tool is run deeper in the hole, the pressure increases as shown in the line from A to B. Each small step in the increasing pressure indicates lowering one stand of drillpipe. The test tool has reached the bottom of the hole at B, and the pressure reading from B to C represents the pressure resulting from the hydrostatic head of the mud column, commonly called the *initial hydrostatic* (IH).

Surface equipment is installed and tested during the time interval from B to C. The test starts at C when the tool is opened, releasing the formation pressure as the well starts to flow.

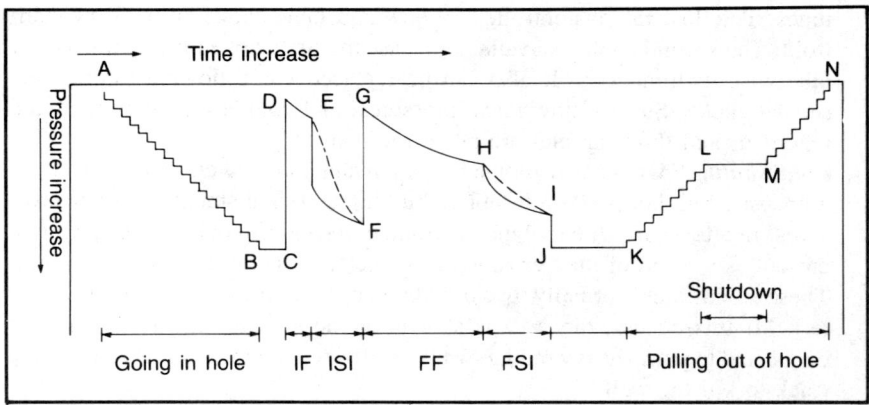

FIG. 9–9 Open-hole drillstem test chart

The first flowing pressure of the initial flow (IF) is D. The final flowing pressure of the IF is E. The initial shutin pressure (ISI) is taken during the time period from E to F. Since the well is shut in, the bottom-hole pressure increases from point E to point F. Point F is taken as the original bottom-hole pressure. The shape of the curve between E and F depends on the formation permeability. The solid curve roughly represents a high-permeability formation, and the dotted curve is more indicative of a low-permeability formation.

At point F the tool is opened for the final flow period (FF), which is run over the time period from G to H. Point G represents the initial flowing pressure of the final flow period, and H is the final flowing pressure of the final flow period. The tool is shut in at point H, and the final shutin (FSI) period is from H to I.

As with the initial shutin, the final shutin curve takes different shapes depending on formation permeability. The tool is unseated at I, and the pressure increases to the final hydrostatic pressure (FH) at J. As a check, the final hydrostatic and the initial hydrostatic pressures should be approximately equal. Preparations are made to pull the tool in the time period from J to K, and the tool is started out of the hole at K. The took is pulled from the hole in the period from K to N, and the pressure is reduced as the tool is pulled to the surface. The time period from L to M represents a shutdown for some reason such as equipment repairs. The total elapsed time to run the DST is measured from A to N.

Drillstem testing is a very important operation and can provide both positive and negative information. For example, if the test results are good on a formation, there is no question about committing the expenditures to run casing and complete the well. If a mechanically good drillstem test indicates that the zone is definitely not a commercial prospect, this information can save the cost of casing and completing the well.

When testing very low-permeability zones, the blow at the surface indicating fluid from the reservoir is entering the tool may be almost undetectable, which may mean a tool is plugged. It is a good practice to go through the testing procedure using limited time intervals so, if the tool is not plugged, test data will be available to show that the formation has very low permeability. This does not take long since the tool is already in the hole and may prevent running another drillstem test to verify that the formation is noncommercial.

In some areas there is a high risk of sticking the test tool because sloughing and caving formations are encountered. In these cases the test times must be limited to the minimum amount of time possible to obtain the formation. The approximate minimum times are as follows:

- Initial flowing pressure = 5 min
- Initial shutin pressure = 10 min
- Final flowing pressure = 15 min
- Final shutin pressure = 30 min
- Total test time = 1 hr

In other areas the tool can be left in the hole for several days to a week. These areas usually have harder formations, shallower depths, and lower temperatures and mud weights. The risk of sticking the DST tool increases with depth. Few tests are run below 15,000–18,000 ft.

The drilling-testing sequence of circulating, coring, and drillstem testing is a common method of evaluating prospective zones during a drilling operation. To illustrate the procedure, assume that a hole is being drilled and the bit is approaching a prospective producing horizon. A drilling break occurs where the penetration rate increases from 10 ft/hr to 25 ft/hr. After drilling 5 ft, the bit is picked up. The cuttings from bottom are circulated out while the gas content of the mud is recorded. The cuttings are recovered, washed, and analyzed. Assume that they show a medium-grain consolidated sandstone with oil staining and some free oil and gas in the cuttings. The gas in the mud also shows a substantial increase. The next step would be to pull the bit, run a core barrel, and cut and recover a core. If core results were favorable, the zone would then be drillstem tested. Some operators may elect to omit coring on a wildcat well. The theory here is that the most important objective of a wildcat well is to discover commercial production; reservoir information can be obtained on subsequent wells.

Logging

After drilling to total depth, the next step is to log the open-hole section. Basic logs as described in chapter 3 are normally run. Other types may also be run, depending on the area, the type of mud, the formations and other factors.

For example, potentially productive fracture zones may have been detected while drilling the well. A special log may then be run to help locate and define the fracture zones. In a wildcat well the exploration geologist may need additional information on the strike and dip of the deeper formations. A special logging tool is available to obtain this information. If the completion engineer is concerned about hole deviation and possible crooked hole sections, a log is available.

The interpretation of seismic data depends on the velocity of the sound (explosive energy) through the formation. In a wildcat area the geophysicist may have to use estimated data based on experience because of the lack of actual data. A velocity survey can be run in the wildcat well to obtain more accurate seismic velocities. The seismic data may then be reprocessed with the corrected velocity data to obtain more accurate seismic information.

When the well is ready for logging, the hole is circulated and conditioned, and the drilling assembly is pulled. If there have been formation problems such as caving or a tight hole, a wiper trip may be made. This includes pulling about 10–30 stands of drillpipe from the hole and running back to bottom to circulate and condition the hole and mud. Then the assembly is pulled to run logs.

A logging truck is ordered and normally arrives on location before or while the assembly is being pulled. The logging crew immediately begins rigging up. The truck is positioned about 100–150 ft in front of the rig on the V-door side.

A lubricator may be installed in the top of the blowout preventers if there are high-pressure zones in the well and some risk that the zones may begin flowing. The lubricator is a long piece of pipe that has an inside diameter large enough to allow the logging tools to pass. It is fixed in the top of the preventers by closing the bag-type preventer around the bottom of the lubricator. The top of the lubricator is fitted with a collet-type latch-in device that contains packoff rubbers to seal around the logging line in the event the well becomes pressurized.

In the rig-up procedure the logging line is run through the lubricator, and the lubricator is held up above the rig floor with the cat line. The end of the logging line is fitted with a rope socket that connects the logging line to the various logging tools. This socket also provides insulated electrical connections between the line and logging tools.

The logging line is checked to ensure that each electrical conduit is insulated—no electrical shorts—over the length of the cable from the logging head (rope socket) to the truck. The first logging tool is connected to the logging head. It is picked up slowly by the logging truck and is positioned in the hole so the

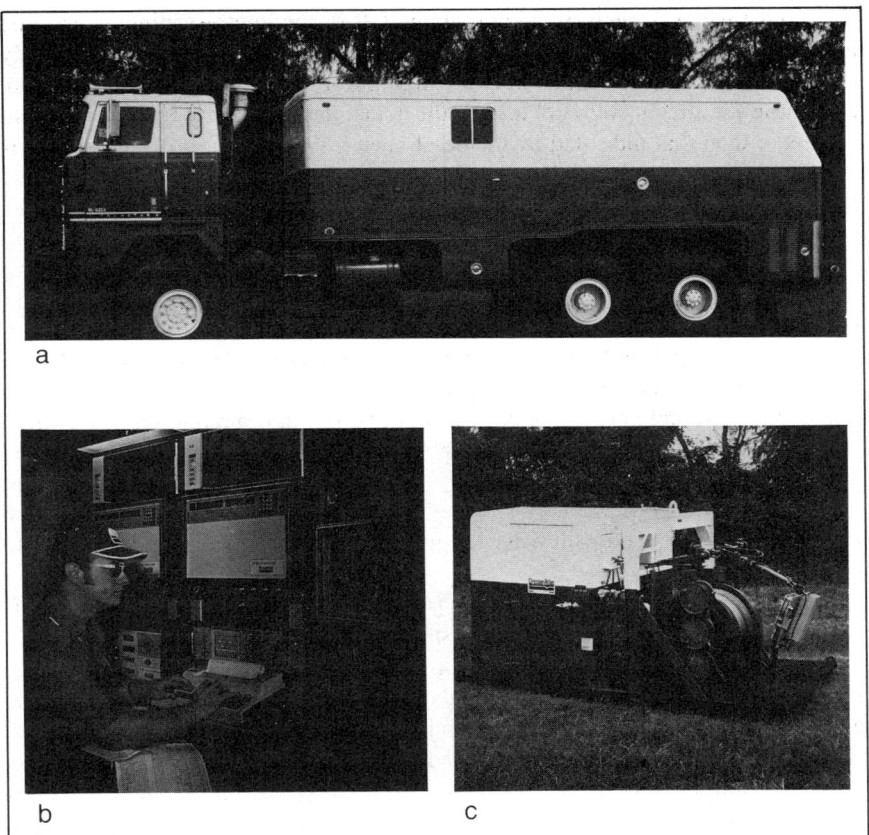

FIG. 9-10 (a) Truck-mounted logging-perforating unit, (b) engineer monitors and records logging-perforating operation, and (c) skid-mounted, portable, self-contained logging-perforating unit for offshore operations (courtesy Dresser Atlas)

depth reference on the tool is located about 1 ft above the rotary. The kelly bushing elevation datum point normally is the basic depth reference for all measurements in the well. The logging depth meters are reset to the zero position. If a lubricator is being used, it is lowered at this time and the bottom of the lubricator is secured.

The logging tool is then lowered into the hole slowly while the depth meters in the logging truck record the depth of the tool. One of the depth meters is corrected for line stretch by adding about 1 ft to the depth reading at periodic intervals as the tool is lowered. The stretch in the logging line is known precisely as a function of the amount of weight hanging from the logging line. If the depth correction for stretch were not added, the actual depth of the logging tool would be deeper than that indicated by the depth meter. One of the recording meters is not changed or corrected. The logging truck operator can determine the total depth correction by comparing the two meters.

The logging engineer normally reads the tool responses by means of electrical meters and a CRT screen. These readings are not recorded since they are not run at the proper speed. The depths also are not correct when the tool is run into the hole because of line slack.

A wire-line weight indicator shows the amount of weight suspended from the logging line. As the logging tool approaches bottom, the lowering speed is reduced correspondingly so the tool will not be run into the bottom of the hole, damaging the delicate electrical components. The descending tool is stopped when it touches bottom as indicated by reduced weight on the wire-line weight indicator. At this point the tool is picked up and lowered several times to ensure that it is on bottom and to allow the logging engineer to make final adjustments to the electrical equipment.

Then the logging tool is started out of the hole at a constant speed. Different logging tools are run at different speeds. As the tool is pulled from the hole, the electrical responses are recorded on a film with a depth scale and often on magnetic tape. The logging engineer also monitors the tool responses on meters and the CRT. Depth adjustments are made as necessary.

The taped data can be transmitted via telephone or radio to distant cities that have equipment to reproduce the log so it is available shortly after the log is run. Some of the more modern trucks also have computer equipment that calculates the logs to give a printout of reservoir characteristics such as porosity and oil and water saturations.

After the desired hole section has been logged, the tool is pulled the remaining distance at a faster but safe speed. This entire logging procedure with one tool is known as a log run or log trip.

Another tool is connected to the logging line to make another log run. Combination tools are frequently used so several different types of logs are obtained with each tool. Usually at least two or more log runs are made to record a full suite of logs.

After a log run has been completed, the film is developed and field prints are made so they can be reviewed and evaluated immediately at the wellsite. After all logging runs have been made and the film has been developed, the logging equipment is rigged down and stored on the logging truck to complete the logging job. It normally takes 8–12 hr to log a shallow hole to about 12,000 ft and 24–36 hr to log a deep hole (20,000 ft or deeper), depending on the number and type of logs run and the length of the hole section logged.

The foregoing description of a common logging job has omitted many problems that can occur. One of the most common is the failure of the logging tool to reach the bottom of the hole. This can be caused by ledges, caving, crooked hole, or high gel mud. If the logging tools do not reach bottom, they are pulled and the pipe is run to make a cleanout trip. This trip includes running the bit in the hole, reaming the section where the logging tool stopped, running the bit to bottom, circulating and conditioning the hole and mud, and pulling the assembly to run the logs again. In areas where severe formation problems are encountered, it is not uncommon to make one or more cleanout trips.

The logging tool can become stuck below wire-line keyseats when the wire line cuts into the side of the hole in crooked-hole sections or ledges. The hole can cave or slough on top of the logging tool, causing it to become stuck. The tool may also become stuck or the logging line can be kinked and broken if the tool hits an obstruction while being run or pulled at unsafe, high speeds.

If the tool becomes stuck, it is worked with the wire line to attempt to release it. This working procedure primarily consists of pulling on the tool to the safe working strength of the line or the maximum pull-out strength of the rope socket if it has been *crippled,* i.e., designed to break or pull the line out of the rope socket before the line breaks in another place, which could leave wire line in the hole. This working procedure is relatively inefficient. If the tool cannot be released, it must be fished out of the hole.

Several methods are available to fish the logging tool out of the hole. The line at the top of the tool can be cut with a wire-line cutter or pulled out of the crippled rope socket and released. Or a special overshot usually supplied by the logging company can catch the fishing neck on the logging tool.

Another method of fishing for the logging tool is *stripping over.* The special overshot is run on the fishing assembly. The logging line is cut and threaded through the inside of the fishing assembly as each stand is made up and run in the hole. This is a difficult, time-consuming procedure because the logging line must be held while threading the free end through each stand of drillpipe and the free end must be held while the stand is connected and run in the hole. After the overshot reaches the logging tool, the logging line is tightened. The rigid line helps guide the overshot over the logging tool. The logging tool is then recovered by pulling the fishing assembly out of the hole and reversing the stripping-in procedure. This type of fishing operation is used when there are indications that the top of the logging tool will be difficult to catch with a normal fishing run because the tool is laying over in a washed-out section or is caught under a keyseat.

Sometimes the logging line breaks and drops into the hole. Special fishing tools called wire-line spears and grabs are used to fish for wire lines. The spear tool is most commonly used. In the fishing operation the spear is connected to the bottom of a drillpipe fishing assembly and is run into the hole. The spear is pushed down into the tangled mass of wire line, which is caught on the barbs of the spear. The spear is then pulled from the hole, recovering part or all of the wire line sometimes with the logging tool still connected. If the wireline breaks below the spear, a section of wire line is recovered and the spear is rerun to fish out the remaining line. If the logging tool is stuck, the logging line frequently breaks at the rope socket and must be recovered by the spear. An overshot is then run to catch and recover the logging tool. If the logging tool cannot be recovered, it is pushed to the bottom of the hole or sidetracked.

The neutron logging tool has a long-life neutron source. Government regulations require that a diligent effort be made to recover this tool. If it cannot be recovered, it must be isolated by covering with cement. Then the hole must be sidetracked so the new borehole is located a specified safe distance from the lost neutron tool. The operator must pay the logging company for the value of any logging tools and logging line lost in the hole.

After examining the basic logs, the operator may also elect to run additional logs or surveys. Most logging companies have sidewall core equipment. The sidewall core tool has a number of small metal sleeves fitted into the body of the tool and held by steel wires. The tool is run in the hole in a conventional manner and is positioned at the correct depth. An explosive charge is detonated to drive the hollow metal sleeve into the wall of the borehole and fill it with a sample of the formation. The logging tool is then picked up slowly, and the core sleeve filled with a core of formation is pulled from the formation by the retaining wires. The coring tool can then be repositioned to take another core. The operation is repeated until the desired number of cores have been taken. The coring tool works well in softer formations, but it is less effective in harder formations.

Logging companies also have tools that take a miniature drillstem test. The tool is run into the hole and presses an orifice against the side of the wellbore. An internal valve in the tool is opened to allow formation fluids to flow through the orifice into a sample chamber in the logging tool. Pressures and volumes are measured. After the tool is pulled, the contents of the sample chamber are recovered for analysis.

PLUGGING AND ABANDONING A DRY HOLE

The open-hole show indicators often give the operator a good idea of whether a well's potential justifies completion. However, the final decision is never made until the well has been logged. Normally, the operator and representatives of the other owners are at the wellsite when the well is logged, or the well logs are delivered to them within a short time. The logs are evaluated, taking into consideration the open-hole show indicators.

After all of the data have been analyzed, the decision is made to complete or plug and abandon (P & A) the well. In the meantime the drillpipe is run back to circulate and condition the hole. If a decision is made to plug and abandon the hole, a cementing truck is ordered to the location. Regulatory agencies having jurisdiction are contacted and advised of the abandonment. Normally the abandonment requirements of the regulatory agencies are known, but they may be verified by final conversations. In some cases the regulatory agencies have a representative on location to witness the plugging and abandoning operations to ensure that the hole is plugged in accordance with agency regulations.

The hole is plugged by setting plugs of cement at various points in the wellbore for several reasons. If the hole is left open, salt water from deeper zones could flow up the wellbore to the surface and kill surface vegetation or ruin the land for agricultural purposes. Salt water could also flow into the underground potable (drinkable) aquifers—freshwater reservoirs—and contaminate them. The well did not have commercial hydrocarbons, but it could still have oil and gas reservoirs capable of producing small volumes. These can migrate to the surface and destroy the land or create a potential fire hazard. They can also migrate into the freshwater sands and contaminate them. Since oil and gas are under higher pressure, they could charge or pressure shallow reservoirs and in some cases have actually caused water wells pumped by windmills to blowout. Dangerous gases such as hydrogen sulfide also could create a hazardous condition.

The hole is filled with heavy mud, but this is not a permanent plug. Over a long period the solids (weight material) can settle out of the mud so it becomes equivalent to water. Cement is used to plug the wellbore permanently at various intervals. These plugs are also set in any casing left in the hole because over time the casing can develop leaks as a result of corrosion.

Prior to abandonment, the drillpipe is pulled from the hole and the drill collars are laid down. Sometimes part of the drill collars are laid down before running back into the hole after logging. For the final abandonment, open-ended drillpipe is run back in the hole. The normal abandonment procedure calls for setting cement plugs usually 50–100 ft long at different intervals in the wellbore.

This may include setting the plugs across certain known formations and above known geological intervals. Also, one plug is usually set half in and half out of the bottom string of casing.

Drillpipe is positioned with the end of the pipe at the point where the bottom of the cement plug will be located. The cement mixing truck is connected to the top of the drillpipe.

A calculated volume of cement (with about 15% excess) is mixed, pumped, and displaced until the cement is in position in the hole. The drillpipe is then pulled up to the depth for the next plug. Cement for the next plug is mixed, pumped, and positioned, and the procedure is repeated for subsequent plugs until all of the cement plugs have been placed in the borehole.

Sometimes casing is salvaged from the well prior to abandonment. This usually includes long strings of intermediate and surface casing. Casing that has been cemented cannot be recovered. This is one reason a minimum amount of cement is used for intermediate and some surface casing strings. Before attempting to recover casing, the operator should carefully evaluate the economics. It is expensive to recover casing with a drilling rig, and the recovered casing has a lower value because of its used condition. In many cases it is difficult to estimate the amount of casing that can be recovered. Theoretically, all of the casing that is not cemented can be recovered, but the uncemented casing may be stuck as a result of caving or gelled mud. Therefore, recovery is limited.

If the operator elects to leave the casing in the hole, cement plugs are set in the casing at points specified by the regulatory agencies. Usually the top 100 ft of hole is filled with cement. The blowout preventers and wellhead equipment are removed, the casing is cut off about 6 ft below ground level, and a steel plate is welded over the top of the casing. Then the hole is filled with dirt. The rig is dismantled or rigged down and moved off location. The location is restored by filling the pits, grading and leveling the location, planting natural grasses, etc.

An abandonment monument or dry-hole post is often left on the wellsite to identify the well and its location. A piece of pipe about 4–6 in. in diameter is placed into the upper part of the cemented hole and extends about 4 ft above ground level. A permanent plaque containing the company name, well location, depth, and other information as prescribed by the regulatory agencies is fixed to the piece of pipe.

Depleted or noncommercial production wells are generally plugged in a similar manner after pulling the tubing and all other recoverable equipment from the hole. In most cases casing is not recovered with the drilling rig because of the expense. If there is recoverable casing, the well may be plugged back to the bottom of the casing and the casing can be sold to a salvage company. These companies specialize in salvaging casing and plugging the well using small,

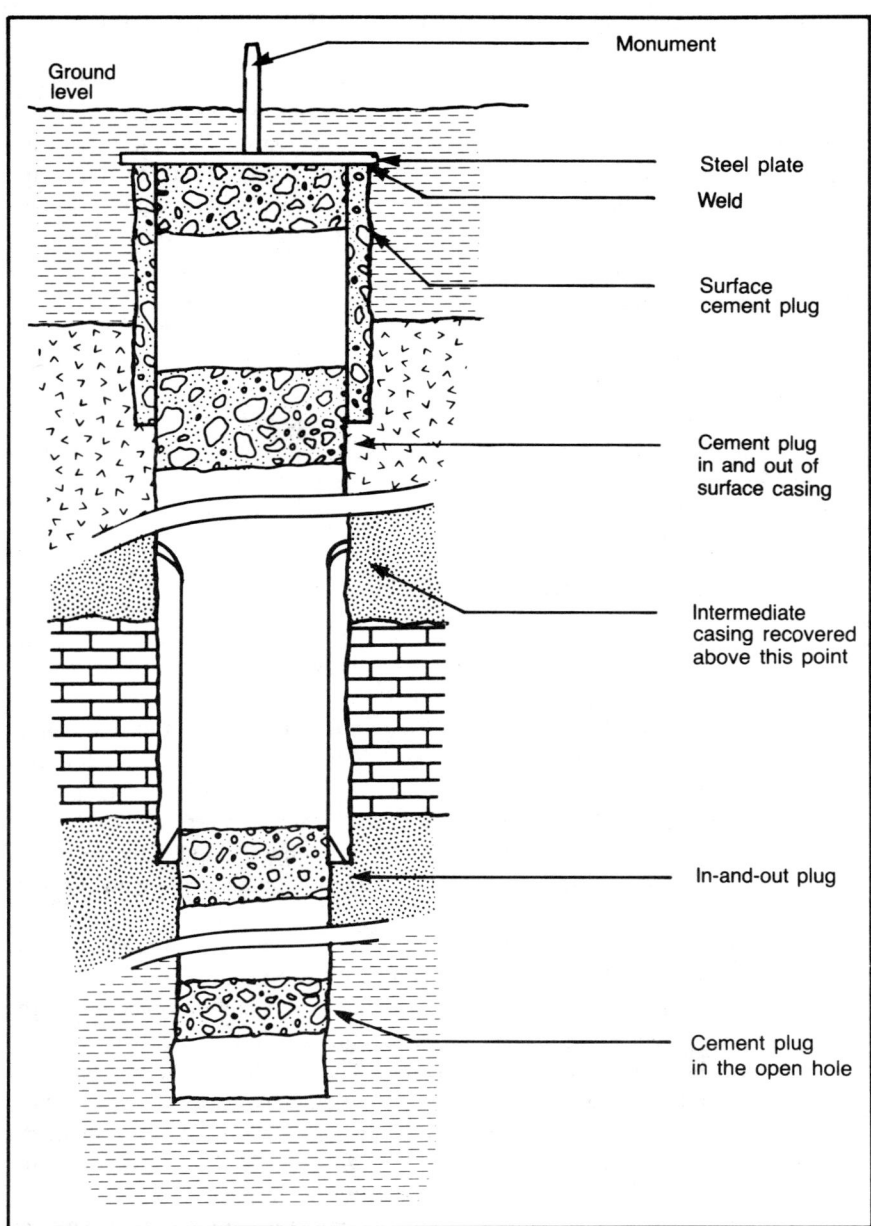

FIG. 9–11 Plugged and abandoned hole

specially equipped portable rigs. The rig is moved on the location and rigged up, which normally only takes a few hours.

The top of the casing to be recovered is caught with a casing spear or casing stub securely welded to the top of the casing. Special hydraulic jacks apply an upward force to the casing. Normally these jacks are strong enough to part the casing if needed. An explosive charge is made up on a wire line that is run in the hole and positioned above the top of the cemented casing. The charge is detonated to blow the casing apart. If the casing is free above this point, it moves upward as more force is applied to the jacks. Sometimes the casing is worked a small, limited amount by lowering and raising the jacks. If the casing is not released another explosive charge is run and detonated above the point where the pipe was first parted. This procedure is repeated until the casing is released. The casing is then pulled from the hole and laid down.

Small pipe such as $2\frac{3}{8}$-in. tubing is run in the hole to place the cement plugs that are mixed and pumped by the salvage company on its rig. The salvage company operator cuts the casing off or places a dry-hole monument over the hole and restores the location surface.

PRODUCTION CASING AND LINERS

The well is drilled to total depth and logged. If the evaluation of the logs and other shows indicate that the well has penetrated formations capable of producing hydrocarbons in commercial quantities, the next step is to case the hole. In most instances the well is cased by running a complete string of casing commonly called the *long string* or the *production string*. The production hole section may also be cased with a liner, possibly including a tieback liner. If the prospective productive horizons are located a considerable distance from the bottom of the hole, the hole may only be cased to a point several hundred feet below the zones. The general casing program for casing the production hole section is included in the drilling program.

Production casing is run and cemented in the same general manner as the intermediate and surface casing strings with modifications to allow for the extra length of the casing and the longer cement column. Precautions ensure that the casing is well cemented across prospective productive formations. Cement volumes are calculated in a conventional manner. They can be calculated accurately because caliper logs, which measure the hole diameter, are available. Cement additives must be carefully blended, and the cement must be pretested using rig mixing water. The cement thickening time is especially important since the cement must remain pumpable for a longer period of time to allow mixing, pumping, and displacing the larger volume of cement usually required. Dry and wet cement samples should be taken.

Centralizers are run on the casing and are positioned so they are spaced across the prospective horizons. This helps ensure that the casing is centralized in the hole so the cement can form a solid sheath completely around the casing. Scratchers are installed on the casing to remove the filter cake and help ensure a good bond between the cement and the wall of the hole. Both bottom and top plugs should be used to ensure that there is a minimum amount of mixing between the mud and cement interfaces.

Two-Stage Cementing

Sometimes special cementing procedures are required when a very long cement column is needed in a hole section with the prospective producing formations a long distance apart. Cement normally is considerably heavier than mud and exerts a comparably higher hydrostatic pressure. A very long (high) cement column can create excessive hydrostatic pressure and can break down weaker, less competent formations. Mud or cement can be lost, reducing the length of the production casing that is properly cemented. This could leave part of the upper production casing uncemented across prospective producing formations.

Therefore, it is important to have the production casing well cemented across all prospective producing formations. When the casing and formations are perforated in well-cemented casing, there is a direct channel from the inside of the casing to the formation. This allows the formation to be stimulated by pumping fluids into the formation and allows reservoir fluids to flow directly back into the casing. If the casing is not well cemented, the treating fluids or formation fluids can flow up or down the uncemented annular space. This is known as *channeling*.

Stimulating fluids may flow into the channels and not into the formation, thus losing the benefit of stimulation. Undesirable fluids such as salt water can flow from other formations through the channels and into the casing. This restricts the productivity of the producing formation. Fluids from the prospective producing formation can also flow through the channels to other formations. Therefore, it is very important to obtain a good cement job.

A procedure known as *two-stage cementing* is used when very long sections of production casing must be cemented and there is a question about the formations' ability to hold the higher hydrostatic pressure caused by a long cement column. The procedure includes running a stage tool, sometimes called a *cement diverting tool,* in the casing string so it is positioned where the bottom of the upper column of cement will be placed. Cement volumes are calculated for two individual cement jobs—one for the lower section of casing (first stage) and the other for the section of casing above the stage tool (second stage).

The casing is run in the hole and is circulated until the hole is clean. The first or lower stage of cement is mixed, pumped, and displaced in the conventional manner. After bumping the plug, pressure is released to ensure that at least one of the check valves in the float collar or float shoe is holding. A teardrop-shaped plug called a stage-tool opening plug is dropped in the casing and is allowed to fall until it is stopped at the stage tool. The casing is pressurized, and pressure forces the plug down a small distance to open ports in the stage tool. The upper section of casing is then circulated slowly while allowing the first or lower stage of cement to reach an initial set.

The upper section is then cemented with the second stage of cement in a conventional manner with a special top plug to close the stage tool behind the cement. This completes the cementing job on the production casing (Fig. 9–12).

If a conventional, single-stage cement job is used, the drilling rig frequently is rigged down and moved off. The well is then completed with a completion rig. Completion operations frequently require extensive periods of testing, squeezing, and waiting on cement. So it is more economical to perform this work with a completion rig, which costs less per day than the drilling rig. The

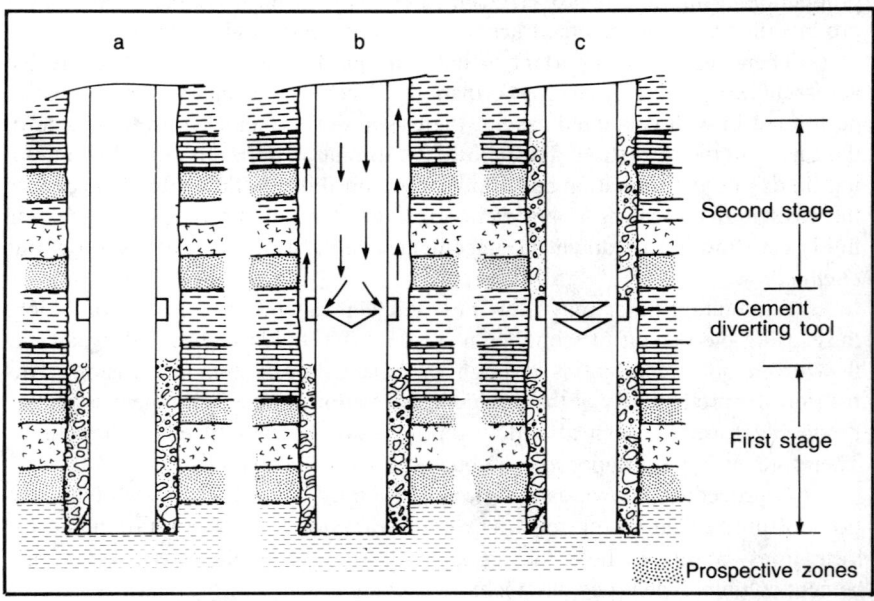

FIG. 9–12 Two-stage cementing. (a) The first stage of cement has been mixed, pumped, and displaced. (b) An opening plug is dropped to open the cement-diverting tool, and mud is circulated in while the first-stage cement hardens to an initial set. The second stage of cement is mixed, pumped and (c) followed by a closing plug.

first step in the move-out procedure is to land and cut off the casing. The blowout preventers are then removed, and the lower part of the Christmas tree is installed temporarily on top of the wellhead. The rig equipment is then dismantled, rigged down, and moved.

When the production casing has been cemented by the two-stage method, the stage tool and cement remaining in the casing must be drilled out before the well is ready to be completed. This may be done with either the drilling rig or a completion rig. For the purpose of continuity it is assumed that the drilling rig is used. The casing is landed and cut off in the conventional manner (Fig. 9–13).

A cleanout string including 6–12 drill collars and small drillpipe or tubing is picked up and run in the hole with a bit on the bottom of the drill collars. The cement above the stage tool is drilled out using a standard drilling procedure except with lighter bit weight and reduced rotary speed. After drilling out the cement and circulating the hole clean, the casing is pressure tested by closing the pipe rams and applying pressure through the fillup line. This test ensures that there are no leaks above the stage tool. The pipe rams are then opened, and the stage tool is drilled out including any small volume of cement inside the casing below the stage tool. The casing is again pressure tested. The stage tool may leak and must be squeezed using procedures described in chapter 10. If the stage tool is squeezed after the cement has set, it is drilled out and the tool is retested. The work string is then run to bottom as a cleanout trip to ensure that the casing does not contain obstructions. The casing is again tested, and the work string is pulled from the hole and laid down.

Normally the stage tool is drilled with a bit and casing scraper. Some operators prefer not to drill cement with a casing scraper in the cleanout string because of the risk of sticking the tools. In this case a special trip must be made with the casing scraper to ensure that the inside walls of the casing are clean and free of cement or other material. If the casing scraper is not used, there is a risk that any packer run into the hole will be damaged and may not function correctly. If production tubing is used in the cleanout procedure after the drill collars are laid down, the production tubing may be rerun and left in the hole suspended from a special tubing hanger assembly. The blowout preventers are dismantled (nippled down), the lower-half of the Christmas tree is installed on top of the wellhead, and the rig is dismantled, rigged down, and moved out. The well is then ready to be completed.

Liners

The production hole section can also be cased off with a *production liner,* commonly called a *liner completion.* The liner is similar to a long length of casing, but instead of extending from total depth to the surface, the liner extends from

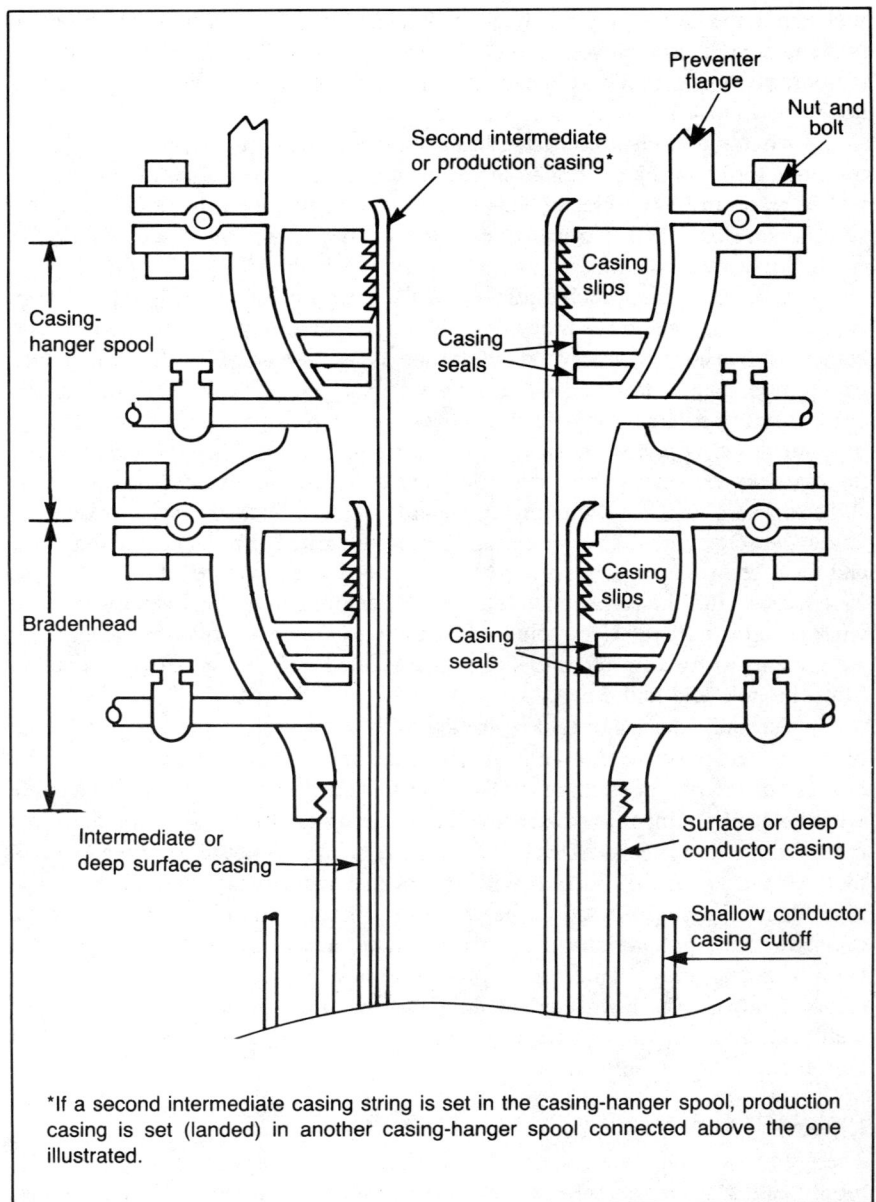

FIG. 9-13 Casing landed in a casing-hanger spool

total depth to inside the intermediate casing for a distance of 200–400-ft, the overlap section.

Liners are used for different reasons. One is to save running casing inside a section of the hole that may already be cased with intermediate casing. In deep, small completion holes a larger-diameter upper casing section may be needed to accommodate a larger tubing string required because of the depth. This can be accomplished by using a liner completion.

Liners are usually limited to shorter hole sections seldom exceeding 5,000–7,000 ft. Cement volumes for liners are calculated in a similar manner to that for production casing. The volumes are calculated as accurately as possible, and excess is limited to a minimum amount. Centralizers are used on the liner opposite prospective productive formations, and scratchers may be used depending on the operators preference and the manner in which the liner is handled during the circulating/cementing operation.

The hole is prepared for running a liner by circulating and conditioning in the same manner as running casing. The drillpipe is rabbeted on the last trip in the hole before running the liner to ensure that the the drillpipe is open. A rabbet is a short piece of pipe or rod that has an outside diameter slightly smaller than the inside diameter of the pipe to be rabbetted, in this case the drillpipe. The rabbet is dropped through each stand of drillpipe to ensure that the drillpipe is open and there are no tight places that would restrict the passage of plugs or other tools.

The liner is picked up, including a float shoe and float collar, and is run in the hole similar to picking up and running casing. After the correct length of liner is in the hole, a liner hanger connects the liner to the drillpipe by left-hand threads so the drillpipe can be released from the liner by normal right-hand rotation. The setting tool has a long tube called a *stinger* that runs through a seal assembly in the liner hanger. The bottom of the liner hanger contains both bottom and top wiper plugs with specially sized internal holes so they can be activated by a small-diameter, bomb-type plug run through the drillpipe. The liner hanger also has slips that can be activated to catch the inside of the last string of casing and support the weight of the liner.

The liner is connected to the drillpipe by the liner-hanger setting tool. It is run in the hole slowly carried on the bottom of the drillpipe. When the liner is in position at or near the bottom of the hole, a special cementing head and plug container are connected to the top of the drillpipe. A flexible steel hose is connected from the cementing head to the rig circulating system. Circulation is established through the liner by starting the pump slowly. The liner is circulated for a short time and then is hung off by rotating the drillpipe slightly to activate the slips on the liner hanger. These slips engage the inside of the casing and support the weight of the liner. The drillpipe is rotated to the right to disconnect it from the liner. A weight reduction indicates that the liner is hung off. The setting tool stinger is still in the liner-hanger seal, so circulation is maintained through the drillpipe and liner and around the liner shoe until the hole is cleaned of cavings and the mud is in good condition.

The flexible steel hose is disconnected from the rig circulating system and is connected to a cement-mixing/pump truck discharge. The bottom bomb is released from the cementing head and is pumped down the hole followed by cement. When the bomb reaches the liner hanger, it passes through the larger hole in the top cement wiping plug and seats in the smaller hole in the bottom cement wiping plug. Thus, the plug is forced out of the liner hanger down through the liner to serve as a standard bottom plug.

The top bomb is released after all of the cement has been mixed and pumped into the drillpipe. The cement is displaced down the hole with mud. When the top bomb reaches the liner hanger, it seats in the hole in the top of the top wiper plug, causing it to release and move downhole through the liner similar to a conventional top plug. The plug causes a pressure increase when it reaches the float collar, thus indicating that all of the cement is in place. The drillpipe is then picked up slowly to pull the stinger on the setting tool out of the seal assembly on the liner hanger. This completely releases the drillpipe from the liner, which is now hung and cemented.

Normally there is excess cement inside the pipe on top of the liner. Some operators reverse this excess cement out of the hole before pulling the drillpipe. Other operators pull the drillpipe out of the hole and clean the cement out later, since a cleanout trip must be made with either operation.

There have been a number of failures in running and cementing liners, and the overall procedure requires considerable planning and experience. The procedure outlined above is one of the more common methods of running and cementing liners, but some operators perform the job in a slightly different manner. For example, a few operators reciprocate the liner while circulating and cementing. Theoretically, this is the best method of obtaining a good cement job, but from a practical viewpoint there is a higher risk of a failure in the operation of the liner-hanger tool. There is also some risk that the drillpipe cannot be released from the liner after the cement is in place. If excess cement is used, it will be in the annular space around the drillpipe and can cause the drillpipe to be cemented in the hole. When this occurs, it can cause junked holes.

After the cement is in place, some operators pull the drillpipe up 500–1,000 feet to ensure that it is well out of the cement. Then the drillpipe rams are closed, and pressure is applied to the annular space. This pressure is held until the cement has taken an initial set.

The liner must be cleaned out and the top tested to ensure that it does not leak. A bit and casing scraper are run to clean out cement to the top of the liner. The liner top is tested with a tool that is similar in operation to the DST tool. This hook-wall test tool uses slips to catch the inner wall of the casing and frequently is fitted with hydraulic hold downs that lock the tool in place when pressure is applied.

The test tool is run in the hole on drillpipe and is seated a short distance above the top of the liner. The circulating valve is opened, and water is pumped to fill the drillpipe. The circulating valve is then closed, and the top of the drillpipe is opened to see if the well will flow. The water in the drillpipe exerts a minimum hydrostatic pressure against the liner top. Normally the formations have considerably higher pressure so, if there is a leak, fluid will flow from the formations into the drill pipe. This flow is reflected at the surface, and the procedure is known as an *inflow test*.

If the liner top leaks during the inflow test as indicated by a flow of water at the surface, the liner top is squeezed with cement using the procedure described in chapter 10. If the liner top does not leak, the circulating valve is opened and the water is reversed out of the drillpipe. The valve is closed, and pressure is applied to see if the liner top leaks under pressure. If it leaks, it is squeezed with cement. Assuming the liner top does not leak, pressure is released. Then the test tool is unseated and is pulled from the hole.

A smaller bit sized to fit inside the liner is run, and the liner is cleaned out to total depth. The entire hole is then pressure tested. After testing, the mud in the hole is normally displaced with water (treated water, salt water, etc.), and the pipe is pulled from the hole and laid down. The preventers are nippled down, the lower half of a Christmas tree is installed on top of the wellhead, and the rig is rigged down and moved. Then the well is ready for completion.

In some cases such as worn intermediate casing or high reservoir or treating pressures, a *tieback liner* is run to connect the liner top to the surface. The tieback is similar to another string of production casing, except it is connected to a polished bore receptacle—commonly called a PBR—in the top of the liner and is sealed by cementing. The tieback liner is landed in a conventional manner and is cut off. The wellhead is assembled similar to a standard production string completion.

The production hole section is drilled below surface or intermediate casing. Open-hole shows, cores, and drillstem tests give information on formation productivity. Logs are run, dry holes are plugged and abandoned, and wells with prospective formations are cased. Most of the work toward the primary objective of drilling—to complete a commercially productive well—has been accomplished, and most of the risks have been eliminated. Casing the well indicates it has high potential to become a commercial producer after completion.

CHAPTER 10
COMPLETIONS

A drilling prospect has been located by an exploration geologist. A drilling prospectus and drilling program have been prepared, and the land has been acquired by leasing and trades. A rig has been selected, moved in, and rigged up. The hole has been spudded; conductor, surface, intermediate, and completion hole sections have been drilled and cased. Drilling information, cores, drillstem tests, and well logs show that the wellbore penetrated formations expected to produce commercial quantities of hydrocarbons. The next step is to complete the well.

Completion includes testing the casing to ensure that it will withstand reservoir and treating pressures or providing other methods of containing these pressures. Individual formations are isolated by squeeze cementing if necessary. The well is perforated and tested, and the formation is stimulated as necessary. Then the well is placed on production and is tested to ensure that the formations are produced at the most economical rate. Artificial lift is used if necessary. Also, applicable secondary, tertiary, or enhanced recovery methods are evaluated, and a program is planned. All of these steps are included in a well completion.

A long period has been spent drilling and performing other operations to find hydrocarbons. There is a natural tendency to rush the completion stage, and it is easy to overlook the original objective of the program, which is to drill and complete a commercial well. When the hole is bottomed—reaches total depth—75–85% of the total drilling and completion funds have been spent. The relatively small amount of completion funds are the dollars that will greatly influence the amount of oil and gas revenues from the well. But the objective of drilling is not finished until the well is completed.

HISTORY

A review of the history of completions provides a basis for understanding modern completions. Early completions were relatively simple. All were made in the open hole, commonly called *barefoot* completions, i.e., no casing was set across the producing formation. The first wells were produced by bailing. Pumps were developed later. The first pumping wells used the rig and walking beam as a pumping unit. The rig was also used to clean out cavings periodically. Caving is one of the disadvantages of an open-hole completion. Therefore, barefoot completions are used only in hard, dense formations and then only to a limited extent.

When wells were drilled into high-pressure formations, the wells literally blew in, sometimes called *drilling the well in*. Dikes and drain ditches were used to collect the oil until the flow had subsided to a point where the well could be shut in or capped. Since oil is generally found at shallower depths, most of the earlier wells were completed as oil wells. Sometimes gas caps were encountered that prevented additional drilling. Then the well was allowed to flow gas until the flow rate subsided to the point where the well could be deepened into the oil column to recover the oil. This was a tremendous waste of reservoir energy, but it was the only completion technique available.

The history of stimulation—treating the well to improve productivity—is not well defined. One of the earliest, most successful stimulation techniques was to explode nitroglycerin in the wellbore opposite the producing formation. This created fractures in the formation. These fractures acted as flow channels so oil could flow into the wellbore at a faster rate, thus increasing production. Acidizing for stimulation reportedly was started when a chemical company used a depleted oil well for waste acid disposal. The well, completed in a carbonate (limestone), began to flow oil at a high rate.

These early completion techniques have been aided by improved technology and better equipment and have gradually evolved into the modern, efficient completion.

FACTORS AFFECTING COMPLETIONS

The basic principle of completions is simple—perforate the casing and the cement sheath to create a flow channel for the fluid. The pressurized fluids in the reservoir flow through the wellbore to the surface. In theory, the procedure is deceptively simple, so there may be a tendency to overlook, bypass, or shortcut operations that ensure the reservoir is produced at the maximum efficient rate (MER). The MER allows the maximum volume of hydrocarbons to be recovered economically from the reservoir before depletion. Depletion is the point in the well's producing life when hydrocarbons cannot be produced economically by conventional, primary production methods.

Basic reservoir engineering concepts apply to completions. Failure to recognize and observe these concepts can have adverse effects. The actual completion may be relatively simple, but the reasoning and evaluating process used to select the method can be detailed and complicated, requiring training and experience.

Many factors affect the completion process. These are a composite of the formation and wellbore conditions, the types of completion tools and equipment, and the various completion procedures and their limitations and applicability to the well being completed.

One of the first considerations is whether the well is a development or wildcat well. Stimulation and equipment requirements and production methods for development wells generally are established by other completed wells in the field that are already on production. This leads to a basic rule in designing and conducting completion operations: always review what has been done on other wells or in similar zones, and use these as a guide to improve your operations.

For example, data obtained during drilling may indicate that one of the prospective zones is highly questionable as a completion prospect. The same zone or a similar zone may have been tested, treated, or completed in one of the other wells. The results can be used as a guide on how to handle the zone in the well being completed. This may prevent the operator from overlooking a good zone that appears questionable on the data or may save the expense of trying to complete a zone that will be nonproductive. Both conditions do occur.

The wildcat well presents a different, often more difficult problem. The operator must depend more on experience since there are few if any nearby wells that can be used for correlative information. In general all of the prospective zones in a wildcat well are tested, starting with the lowest zone and working upward. Development wells are often tested by placing the well on production and conducting a production test with the produced fluids going to a sales outlet. The wildcat well normally does not have a pipeline sales outlet. In most cases it is necessary to test the well and determine its productivity to justify the expense of laying a pipeline to a sales outlet. Gas produced during testing must be flared. Oil produced while testing can normally be trucked to a sales outlet. Portable production test facilities are frequently used in either case.

The number of zones that must be tested can create a completion problem, but this normally is a good problem since it indicates that there is a much better chance of making a commercial well. The complexity of the completion increases with an increasing number of zones.

Downhole completion equipment can become very complicated if three or more zones must be produced separately. Often two or three zones are considerably better than the other zones.

A common practice is to complete and produce these zones before completing the lower-quality zones. Another procedure is to use a multizone completion technique to produce a number of zones through one string of tubing. More than 20 zones distributed over a gross interval of 3,000 ft have been perforated, stimulated, and completed in this manner. Additional surface equipment is required for multizone completions, especially where each produced flow stream from the well must be measured.

Other important factors are thickness, type of formation, and reservoir characteristics, such as porosity, permeability, pore pressure, fluid saturations and types of fluid in the reservoir. The type of formation to a great extent determines the type of stimulation. Carbonates (limestone) are commonly acidized, and sands and sandstones are hydraulically fractured with sand-laden fluids. Formation permeability and pore pressure to a lesser extent help determine the volume and type of stimulation. Low-permeability formations may need a large-volume, deep stimulation, whereas high permeability formations may only require a mild, shallow stimulation such as a wellbore cleanout.

The physical composition of the formation helps determine the types of completion fluids to use, including stimulation fluids. For example, sands that contain bentonitic shales can become sealed or plugged if they are exposed to freshwater fluids. Inhibited fluids such as water containing sodium or potassium salts must be used to prevent formation damage, which restricts production. Corrosion must be considered in all completions. Special casing and tubing may be needed because corrosion can damage regular tubulars, causing leaks and failures. Corrosion inhibitors also may be needed. Some types of corrosion, such as sulfate reducing bacteria, can plug the formation, which then requires special treatment.

When a well is completed, consideration should be given to pressure maintenance and secondary, tertiary, and enhanced recovery. In many cases if these processes are initiated early in the primary producing life of the well, the total amount of hydrocarbons recoverable from the reservoir will be increased considerably.

TYPES OF COMPLETIONS

A variety of completion techniques are available. The most commonly accepted methods are based on (1) the number of zones opened, (2) the type of production, and (3) the design purpose of the well. For example, a single gas well is a gas well completed in one zone. A dual gas well is completed in two separate gas sands with two individual flow streams to the surface. Oil wells are named in a similar manner, except the method of production—flowing or pumping—is often added. For example, a single flowing oil well flows oil from one zone. A dual flowing-pumping oil well flows oil from one zone, and another oil zone is pumped. The production method is not used in gas completions since all gas wells are flowing wells. A dual gas/flowing oil well produces gas from one zone and flows oil from another zone.

A triple completion is completed in three individual formations with each formation producing through individual tubing strings or the annular space. Wells have been completed from four or more individual zones, but single or dual completions are the most common. These are sometimes called multiple completions, or a multiple completion can be a commingled completion. A commingled completion has two or more zones individually completed, and the production is commingled in the wellbore. In most multiple completions only one zone is pumped, but completions can be designed to produce more than one zone by pumping.

A tubingless completion usually has small-diameter casing or tubing set as casing. A slim-hole completion is a well that is completed in a normal manner but with smaller casing. A permanent completion has all equipment run and installed before it is perforated and completed.

Some completions are named for the purpose for which the well will ultimately be used. Waste disposal wells are used to dispose of waste liquids; produced salt water is pumped into a formation through a saltwater disposal well. Other special wells include gas injection, gas storage, and water injection.

TOOLS AND EQUIPMENT

Most completion work is done with a completion rig, sometimes called a workover rig because the same general equipment is used for both procedures. Completion rigs are considerably smaller than drilling rigs since they do not support the heavy casing and drillpipe weights used in drilling operations. They generally are mounted on trucks or trailers for mobility. The mast is permanently mounted on the rig and is raised or lowered by the rig power unit. Most rigs use diesel engines for power requirements.

The amount of equipment supplied with the rig varies among the different service companies. Conventionally, the rigs are rented at an hourly rate with higher rates for bigger rigs or more accessories. The amount of equipment needed varies, depending on the job requirements. This ranges from a bare rig to pull and run tubing or rods to a fully equipped rig. Additional equipment is rented from service companies specializing in oil-field equipment rentals.

Completion rigs work on different schedules depending on the job requirements. Tours range from 10 hr/day with one crew to 24-hr operation with three crews.

Completion equipment as used in this text is equipment run and left in the hole to produce the well. This equipment may be changed for other equipment in subsequent workovers or recompletions. Artificial-lift equipment is actually part of the completion equipment, but in common practice it is considered as a separate type of equipment because of its specialized purpose.

The general approach to selecting completion equipment is to use the least amount and the simplest type of equipment that can satisfactorily produce the well. It is not uncommon for completion equipment to become stuck in the hole so that it must be fished out or milled up. Rubber elements on packers can become stuck, especially under high-temperature conditions. Formation debris, usually fine sand and silt, can be produced with the oil or gas, and part of this can settle on top of packers and around pipe, causing them to stick. High flow rates through the perforations can cut holes in pipe or can sever the tubing used to produce a lower zone, causing a fishing job if a special blast joint is not used. Sliding sleeves can become stuck open or closed or can be cut out by fluid flow so they cannot be closed. These and many other problems emphasize the importance of using a simplified, efficient downhole completion assembly.

Completion tools are the tools used to perform completion work. They are normally pulled out of the hole after the well has been completed. In many cases they are the same tools used later in the producing life of the well for a workover, a recompletion, or remedial work. Downhole completion and workover tools are run on either wire line, tubing, or work string. Special tools are run on wire lines and normally cannot be run on tubulars (tubing or drillpipe).

Completion tools and equipment are long and slender. They are seldom run less than one mile deep in the hole; often they are run several miles or deeper. They are relatively small, in most cases smaller than a common dinner plate. Therefore, completion tools must function properly, and completion equipment must be designed and installed to produce the well in the most efficient, safe manner.

Tubing is small-sized pipe normally used as part of the completion equipment. Common tubing sizes are 2⅜, 2⅞, and 3½ in., all measured as the outside diameter as with other oil-field tubulars. The tubing ends are fitted with threads, and tubing collars are used to connect the joints. Plain-end and external upset-end (EUE) joints also are used.

The end of plain-end tubing has the same outside diameter and is fitted with V-shaped threads, 10/in. The end of EUE tubing has a larger diameter and is fitted with round threads, 8/in., sometimes called 8-round tubing. Plain-end tubing strength is limited by the strength of the coupling and is usually used in shallow wells. EUE tubing joints are as strong as the tube (tubing body) and are used in deeper wells. This tubing offers a distinct advantage because normally the coupling is more subject to failure than the tubing body. Special or modified couplings with gasket-type seals or multiple-sealing shoulders are used in very deep wells with higher pressures. Tubing, like casing, is available in different sizes, weights, and grades (steel strength).

Tubing used as part of the completion equipment also may be used in the completion operations, providing that only a limited amount of tripping, drilling, and cleanout work is required. The body of the tubing joint has sufficient strength for most of this work. However, the threaded connections are not designed for repeated connecting and disconnecting (makeup and breakout) and for the fluctuating high torques that may occur in some operations. If the tubing is used for drilling and cleanout operations, it is commonly called a work string. However, heavier-duty, small drillpipe or special heavyweight tubing with tool-joint connectors—a stronger, more durable connection—is often used in completion and workover operations. After the operations are completed, the work string is laid down and production tubing is picked up and run in the hole as completion tubing, sometimes known as *tubing the well*.

Coil tubing is a special tubular used in completion operations. This small-diameter, thin-wall, endless tubing is stored on a large, truck-mounted reel in lengths of 8,000–15,000 ft. The truck also has special equipment that guides the tubing, runs it in and out of the hole, and unreels the tubing from storage. The shaft of the reel is fitted with a special connection so fluid can be pumped through the tubing.

Coil tubing is used in special situations for cleaning out, loading, and killing a high-pressure well and for similar operations. The tubing does not rotate. Any cleaning-out action is done with the jet of fluid flowing from the end of the tubing. Therefore, for cleanout purposes the tubing is limited to softer materials such as sand and dehydrated mud. The tubing usually is run in the hole under pressure through a seal assembly that is attached to the wellhead and that uses pressure seals fit around the tubing to contain the well pressure.

Small-diameter, conventional tubing, commonly called *macaroni tubing*, is also used as a work string in very small holes such as running inside $2\frac{3}{8}$- or $2\frac{7}{8}$-in. tubing. This small tubing can be fitted with a small bit and can be used to drill out harder sand or small plugs and pieces of junk. The drilling and cleanout operation must be conducted very carefully with limited bit weight and rotary speed because of the limited strength of the tubing.

This tubing also can be used to work under pressure or to run into a hole with a high pressure by using a *snubbing unit*. This process is known as *snubbing in the hole*. The snubbing unit uses small, double preventers mounted on top of the wellhead. A bit or other tool is made up on the bottom of the tubing, and a check or backpressure valve is fitted inside the tubing near the bottom. The tubing is run into the hole by alternately opening and closing the two sets of preventers separately.

Production tubing is similar to tubing used in the work string. Other miscellaneous equipment is used depending on the type of completion. Two common items are blast joints and sliding sleeves. Storm chokes are used extensively on offshore wells and to a limited extent on other wells. They are installed in the production tubing near the surface but are deep enough so there is no possible risk of the choke being damaged if the wellhead equipment is destroyed. This point normally is below the sea bottom in offshore wells.

The storm choke is contructed and adjusted to allow the well to flow at any normal, designed flow rate. If this rate is exceeded, the storm choke closes and shuts in the well. If an offshore structure is damaged during a storm or by any action that would destroy the wellhead, the well would blowout without the storm choke. With the storm choke installed as soon as the flow rate through the production tubing exceeds a predetermined rate—which would occur if the wellhead control were broken off—the choke closes and shuts in the well. The damaged wellhead and upper casing and tubing can be repaired with the well shut in by the storm choke. The well is placed back on production by opening the storm choke after the repairs have been completed.

Packers and Plugs One of the most often used items of equipment is the treating or test packer. This packer is similar to a retrievable production packer, and in some cases the two are interchangeable (Fig. 10–1). Test packers are run on a work string and are used to shut off the annular space above the packer. This

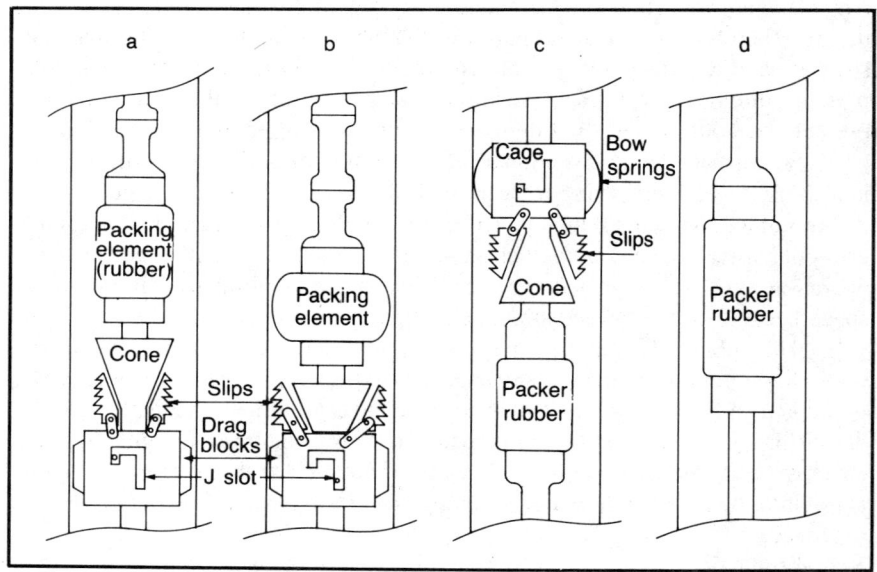

FIG. 10–1 Types of packers. (a) Hook-wall packer running, (b) hook-wall packer set, (c) tension packer running, and (d) compression packer running.

allows formation fluids to flow from the formation up through the tubing in a manner similar to the DST tool used in the open hole.

Three types of retrievable test packers are used. The *hook-wall packer,* sometimes called a *compression packer,* has slips below the packer element with teeth (serrated edges) on the slips pointing in the downward direction. The packer is seated by manipulating the tubing. This releases the J slot, or the packer latching-unlatching mechanism, and allows the slip teeth to engage the casing and stop the downward movement of the packer. The packer element then expands and seals against the inner surface of the casing as tubing weight is applied. Fig. 10-1 illustrates the operation of the J slot and the manner in which the packer is seated.

A *tension packer* has slips above the packing element with teeth pointed in the upward direction. Manipulation of the J slot allows the slips to engage the inner walls of the casing, and an upward pull on the tubing causes the packer element to expand and seal against the casing wall.

The conventional *compression packer,* not to be confused with the hook-wall packer, seats and expands the packer element as the tubing is lowered with the tailpipe setting on bottom (or an obstruction) in a manner similar to the open-hole DST tool.

Treating packers are used to isolate a perforated interval or other open section from the upper annulus so high-pressure fluid such as treating fluids or cement can be pumped down through the tubing and packer and into the perforated interval or open space. Pressures used in these situations may be considerably higher than the bursting pressure of the upper casing. The treating packer isolates the upper casing section so the pressures are confined to the tubing and the hole section below the packer.

This high pressure below the packer causes it to act like a plunger. If the upward force on the bottom of the packer exceeds the safe downward force exerted by the tubing, the packer can be forced upward. The tubing can be damaged and may fail as a result of compression and buckling if high treating pressures are used and the packer is not held down to keep it from moving. The tension packer can be used as a treating packer. The upward pointing slips prevent the packer from moving in the upward direction, and the packer can withstand high treating forces. This packer is known as an *upside-down packer* because the slip teeth are inverted from the normal downward direction. Problems have been encountered in unseating this packer to retract the slips so it can be pulled from the hole.

The most common treating packer is a heavy-duty, retrievable, hook-wall packer fitted with hydraulic hold downs. The hold downs prevent the packer from moving uphole when pressure is applied below the packer.

The cement retainer is another special type of packer used primarily for squeeze cementing (Fig. 10–2a). Squeeze cementing often requires pumping cement at high pressures. There is a risk of sticking a regular treating packer when squeezing under certain conditions. In most cases the cement retainer is a safer tool to use when squeezing. However, the retainer has some disadvantages.

It is a permanent tool in that it is not retrievable and must be drilled out if it is to be removed. The tool is made of drillable metal so it can be drilled out easily. It is a close-tolerance tool, so named because the outside diameter of the tool is only slightly less than the inside diameter of the casing. A gauge ring is run to ensure that the inside diameter of the casing is large enough to allow the retainer to pass and that there are no obstructions. Retainers have been accidentally set while running. Then they must be drilled out, and a new retainer must be run.

The retainer can be run and set on either the work string or a wire line (shielded conduit). A special seating assembly is run on the work string. It seats and seals in a special receptacle in the retainer. This seating process is commonly called *stinging in,* and the tool on the bottom of the work string is called a *stinger.*

After pumping cement through the retainer, the stinger is pulled out of the retainer and a check valve in the bottom of the retainer closes to prevent the cement from backflowing up the hole. Holding the cement is the big advantage of the retainer over the retrievable treating-test tool. If the cement backflows around a retrievable test tool and hardens, the test tool is stuck and must be drilled out. The test tool is made of hardened steel. Although it can be drilled with mills, it is much harder to drill out than the retainer.

Plugs are another important tool used in completion operations. Permanent plugs block off and isolate sections of the hole. For example, a lower zone may have been tested and found to produce salt water. A permanent plug could be run and set above the zone to isolate it from upper zones that may be productive.

Permanent plugs made of either drillable material or cast iron are used. The term permanent indicates that the plug cannot be removed except by drilling. The retrievable plug is run on and retrieved with the work string. Five to 10 ft of cement or sometimes sand are frequently dumped on top of permanent plugs to ensure that they do not leak when they are to be left in the hole for a long time. Both the drillable and cast iron bridge plug can be drilled out without difficulty (Fig. 10–2b).

Retrievable plugs are used for a temporary plugging action. A temporary plug may be set above a productive zone that has been completed and tested to separate this zone from an upper zone that must be completed and tested. After the completion and testing operations have been finished, the temporary, retriev-

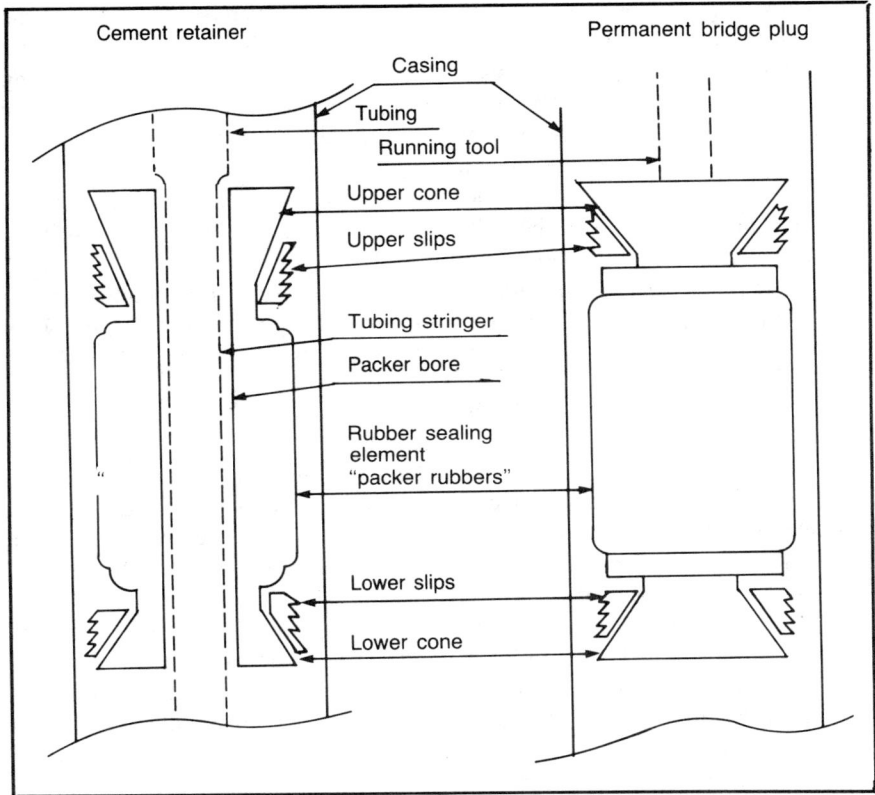

FIG. 10–2 (a) Cement retainer and (b) permanent bridge plug

able plug is pulled from the hole and completion equipment is run into the hole to isolate the two zones as a dual completion or to produce both zones together as a multizone single completion.

A variety of production packers are available. Many of these are very similar to test packers, and in many cases the tools are interchangeable. Packers are broadly divided into retrievable packers, similar to the test packers, and permanent packers.

The permanent packer is similar to the retainer used for cementing except it normally has a millout extension on the bottom of the permanent packer. The millout extension serves two purposes. Special seating nipples can be connected to the bottom of the extension or to a tubing nipple on the bottom of the extension.

Blanking plugs, bottom-hole chokes, and other devices can be run on a wire line and can be seated in the nipples.

For example, a plug can be set in the nipple to shut off a lower producing zone. The tubing can then be pulled for replacement or an upper zone can be completed. After the work is completed and the tubing has been rerun and seated in the permanent packer, the plug can be pulled with a wire line to replace the zone on production. Permanent packers are removed by drilling them out. A special tool is used to catch the millout extension and greatly facilitates drilling out and recovering the packer.

The permanent packer can be run on the work string, usually tubing, or on a wire line. In either case it is always a good practice to run a mud screen or junk pusher on the bottom of the packer to catch and remove or push any junk ahead of the packer that could cause the packer to become stuck while running it in the hole. The same procedure should also be considered when running other close-tolerance tools during the completion operation.

Retrievable packers are run on production tubing and are seated by manipulating the tubing. Some are seated with pressure as one of the final stages of the completion operation. The packer is recovered or retrieved later in the producing life of the well when the tubing is pulled for a recompletion, workover, or final plugging and abandonment.

Many other wire-line tools are used. Mud screens remove larger solid particles from the mud and prevent sticking a packer when it is run. Gauge tools ensure that the inside diameter of the casing or tubing is full gauge. Collar stops are set to land tools on. Swabs clean fluid from the hole. Gas valves are used in artificial-lift tools, open-and-close sliding sleeves, and various other tools.

Logging Tools A variety of cased-hole logging tools are available for completion operations. The tools are run in and out of the hole on single or multiconductor (insulated) steel cables called wire lines. Running and operation of the tools is generally similar to that described in open-hole logging operations. Pressurized wellbores are more common during completions than in open-hole operations. Therefore, a pressure lubricator is commonly used to confine wellbore pressures. The completion logging tool lubricator uses the same principle as the open-hole lubricator.

One of the most common completion logging tools is the cement log. After the casing has been cemented, it is important to know the location of the top of the cement, if there is a solid cement column from the top of the cement to the bottom of the hole, and the quality of the cement behind the casing.

Cement logs use acoustic principles transmitted to the surface by electrical means to locate the top of the cement, voids in the cement column, and the casing-cement and cement-formation bonds. Ideally, a formation that is to be completed should have good cement above, below, and through the formation interval. A formation that is well cemented in this manner is considered to be isolated. If the zone is not well isolated, injected treating fluids can move through channels or annular space that is not filled with cement into the underlying or overlying formations. Thus, the effectiveness of the treating fluid is lost, commonly called a lost treatment, and the formation is not stimulated properly. In these cases the zone is isolated by using squeeze cementing techniques.

Undesirable fluids such as salt water can flow from adjacent formations into the wellbore or into the producing formation if it is not properly isolated.

A collar locator log is run to locate casing collars. This is used to aid in positioning the perforating tools so the casing can be perforated precisely. The collar log tool uses electrical inductive methods to measure the mass of steel of the casing collars, which is larger than the amount of steel in the casing body. The tool is also run to position perforating tools. In many cases after the casing has been perforated, the collar log can be run to locate the perforations in the casing and to record these on a perforation log.

A gamma-ray log normally is run with the collar locator log so the formation lithology—the description of the formations—and the casing collars are printed on the same log display. This is used to help locate the intervals to be perforated relative to the casing collars so that they can be perforated precisely. The cement, collar locator, and gamma-ray logs can usually be made in one logging run and are considered as the base suite of logs used for completion.

A temperature log is the only logging tool that records data while being run into the hole; other logging tools record as they are pulled out of the hole. This log was originally used to locate the top of the cement. Heat is generated when cement cures and hardens. A normal well has a constant temperature increase (temperature gradient) from the top of the hole to the bottom. The cemented section has a similar temperature gradient, but it is hotter than the uncemented sections. The temperature log records the temperature as the device is moved through the wellbore. The top of the cement is characterized by an abrupt increase in temperature. Long cement voids may be detected, but usually these are relatively small. The tool does not have sufficient definition to distinguish them.

Temperature logs can also be used to detect points of fluid entry into the wellbore. Liquids such as oil or water may cause an increased temperature at the point where they flow into the wellbore. Gas may show an increased or decreased temperature, depending on the pressure and flow rate at the point of entry. Fluids flowing from the wellbore into the formation such as a lost-circulation zone, can sometimes be detected on the temperature log by the reduction in temperature since these fluids are normally cooler than the reservoir.

Different types of tracer logs are used for different purposes. For example radioactive material with a short half-life can be placed in the treating fluids during stimulation. Later, a gamma-ray log can be run as a tracer log to show where the radioactive material entered the formation. If there is a channel behind the casing, this may be detected by the tracer log.

Other similar tracer logs use a small amount of radioactive material ejected by the logging tool. A small amount of regular fluid is pumped into the hole, and the logging tool is then moved to determine the new location of the tracer fluid. The procedure of pumping and checking the position of the radioactive fluid is continued until the radioactive fluid seems to stop or disappear. This indicates that it has passed out of the wellbore into the formation and shows the exit point from the casing. In some cases the radioactive fluid can be moved outside the casing to indicate channeling. The use of radioactive material is a good, economical method to determine how many zones have been treated when a number of different zones, often widely spaced, are stimulated in one continuous treatment. A subsequent tracer survey shows which zones were stimulated and which did not receive treating fluid.

A radioactive material with a short half-life is used so the radioactivity will dissipate with time and not affect future logging operations. Tracer surveys can be an efficient tool in completion and subsequent workover or recompletion operations where they are applicable. They are probably not used as often as they should be, possibly because personnel are unfamiliar with the materials and equipment available.

Two logging tools that are frequently used together are the *spinner* and *densiometer surveys*. The spinner survey measures the volume of fluid moving in the wellbore past the spinner tool as the tool is moved through the wellbore. For example, if three zones are producing, the total production from the well is measured at the surface. The spinner tool can then be run into the hole to determine the percentage or volume of fluid flowing out of each producing interval.

The densiometer tool measures the density of the fluid. The three fluids normally produced are water, oil, and gas, each of which has a separate density. The production at the surface gives the total density of the mixture. The tool can be used to help detect the type and percentage of production from a zone.

One disadvantage of these tools is that they can only be used in flowing wells and generally only in the cased hole below the tubing. Sometimes the results are not definitive, but under favorable conditions the tools can be used to select zones that need additional stimulation and those that should be plugged such as a water productive zone. They also show those rare cases when cross flow occurs in the wellbore. Then fluid from a producing zone flows out of the wellbore into a lower-pressure formation.

A number of other specialized logging tools are available or are under development. There is a tendency to develop logging tools that can be run in the cased hole and give all of the information needed. Generally there is less danger of losing or sticking the tools in a cased hole than in an open hole. Also, there is a growing number of old, cased wells that are approaching depletion. These wells were not logged properly before casing was run, or the required logs were not available when the wells were drilled. Many may have potentially productive formations behind the casing, which cannot be detected by conventional cased-hole logging techniques. Since the major expenditure for drilling and casing the hole has already been made, there are obvious economic advantages of locating and completing these zones.

Perforating Tools Perforating tools perforate a hole through the wall of the casing and into the formation to create a channel or passage for fluid to flow from the reservoir into the wellbore. The first perforating tools, which are still in use, were bullet guns. These had a cylindrical steel carrier with the armor-piercing bullet (projectile) located around and along the longitudinal axis of the gun (Fig. 10–3). The bullet was expelled by a powder charge ignited by a blasting cap primer. Intervals up to about 20 ft can be perforated with one gun run.

Most modern perforating tools use shaped jet charges that are generally more adaptable to oil-field type perforating than the bullet guns (Fig. 10–4). The bullet gun literally fires the projectile through the casing wall and can damage the casing, usually by splitting where high-density perforations are used. The guns had to be sealed perfectly to prevent the powder from getting wet and causing misfires.

The jet charge uses a sealed, solid propellant with a high-energy content. It is less susceptible to misfires caused by the high pressures that can be encountered in oil-field perforating. The jet perforates the casing by literally burning a hole. There is correspondingly less shock loading on the casing. This reduces the risk of casing damage.

The jet perforator creates a slightly tapered hole with a maximum diameter at the casing wall. Any sand or perforating debris that enters the perforation tends to flow out of the perforation because the diameter is increasing. There also is less tendency for bridging and plugging of the perforations. The high-energy concentration in the shaped jet charge permits a deep perforation penetration.

An increasing percentage of perforating is done with small through-tubing perforating tools. Some of these are too small ($1^{11}/_{16}$-in. outside diameter) to use the bullet-type construction.

Jet charges generally are classified by their depth of penetration in a standard target formation (Fig. 10–5). The target uses a fixed thickness of steel,

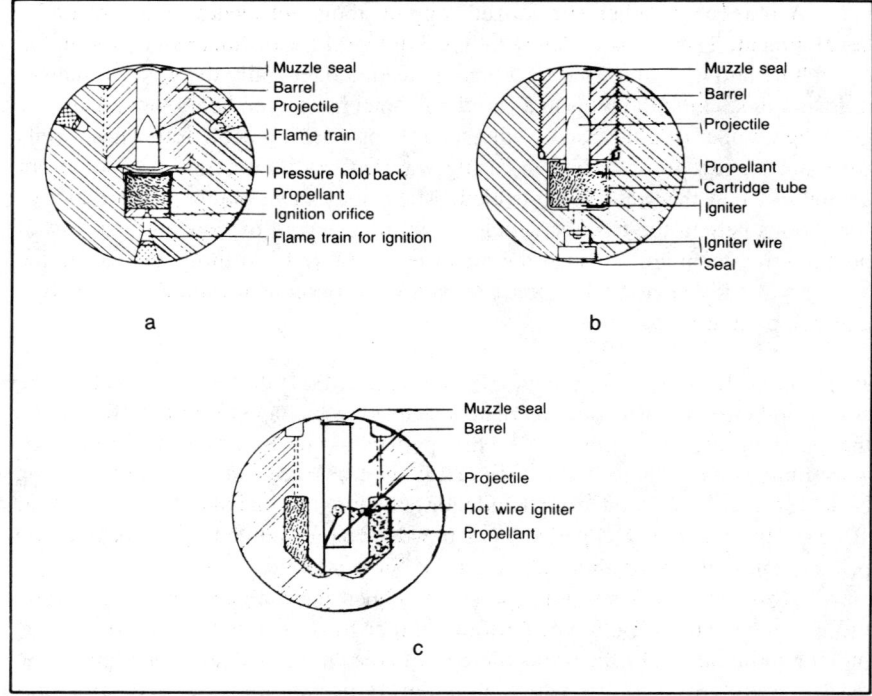

FIG. 10–3 Bullet guns (courtesy Dresser Atlas). (a) High-pressure gun, (b) intermediate-pressure gun, and (c) low-pressure, high-volume gun

cement, and sandstone. This is representative of the average downhole perforating conditions. Different charges fired into this standard target allow the operator to make a comparison of the different commercial charges available and to select the one that is most applicable for the well being completed. Table 10–1 shows the specifications of some of the jet charges available.

Jet perforating guns, commonly called jets or guns, are run centralized or decentralized. Centralized guns are normally sized slightly smaller than the casing and are run through the casing to perforate it. The size of the gun relative to the inside diameter of the casing provides the approximately correct standoff distance.[1] Centralized guns are often run in tubingless completions.

Decentralized guns are run through tubing to perforate the open casing below the tubing. These guns use a decentralizing device such as bow springs that

---

[1]Standoff distance is an important consideration. Standoff is the distance between the nose or front of the jet and the inner wall of the casing. This distance is relatively critical for obtaining a maximum depth of penetration.

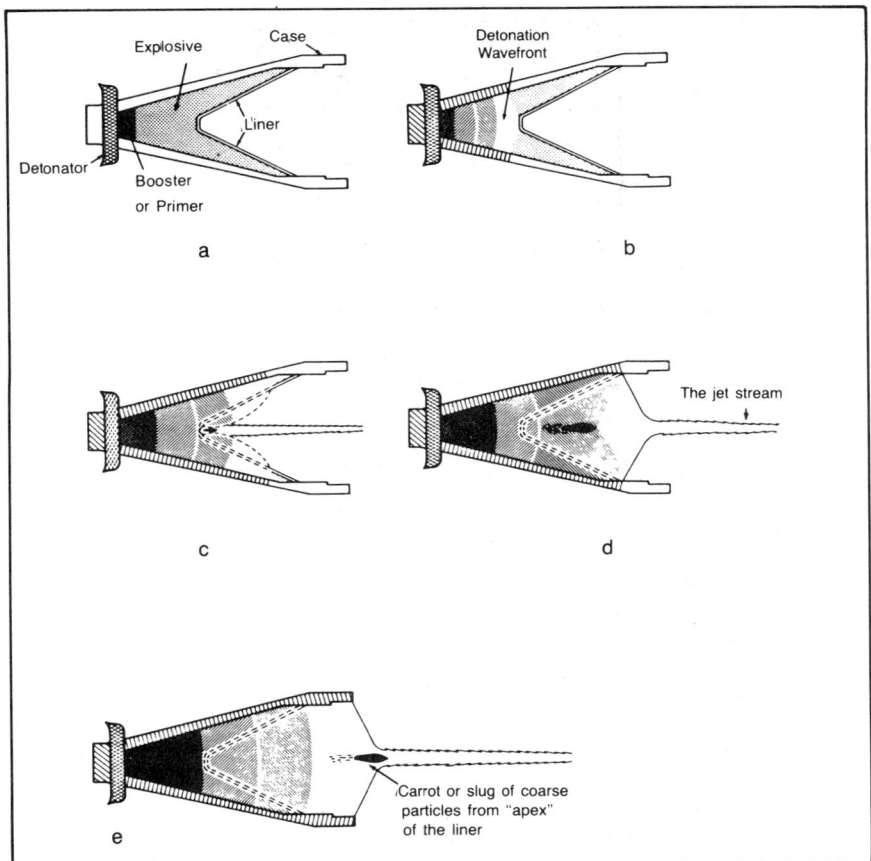

FIG. 10–4 Detonation stages of a typical shaped charge (courtesy Dresser Atlas). The major components of a shaped charge perforator are shown in (a). Detonation travels down the charge (b) and strikes the apex of the cone. The wavefront collapses the liner (c), and the liner's inner surface disintegrates to form part of the jet stream. An advancing wavefront (d) forms the jet stream. The outer surface of the liner forms a slug or carrot, which follows the jet stream. The fully developed jet stream (e) penetrates a target, and the carrot follows with no contribution.

can be compressed so the gun can run through the smaller diameter tubing. They expand in the casing to hold the gun next to the casing wall to give the correct standoff distance for perforating.

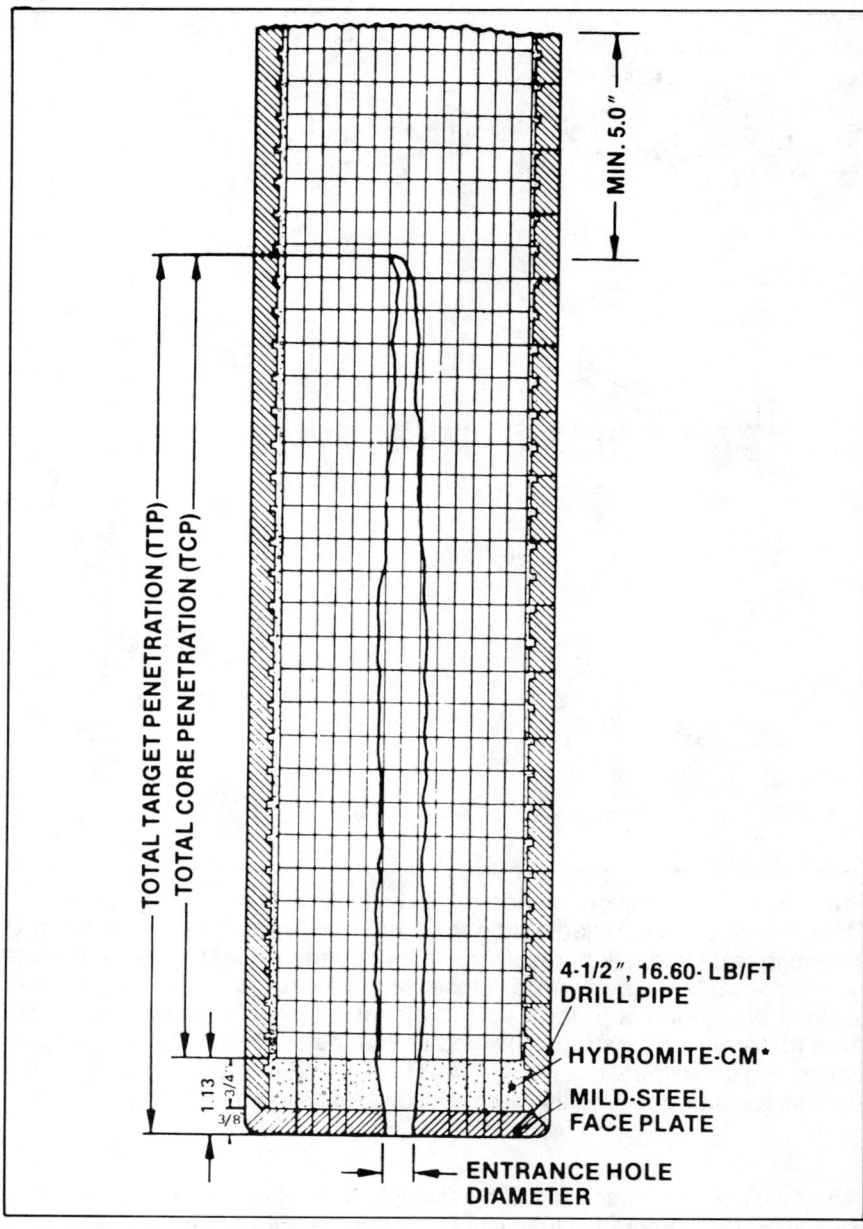

FIG. 10–5 Perforating test target (courtesy Dresser Atlas)

TABLE 10-1 Jet Perforating Test Data

Name	Gun Diameter OD, in.	Charge Weight Grams	Type Carrier	Min. Pipe Size ID	Concrete Target		Berea Target					Temperature, °F	Pressure, PSI
					Total Penetration	Entry Hole Diameter	Entry Hole Diameter	CFE	TTP	TCP	ECP (TCP × CFE)		
Slim Kone-Golden Jet	1½ / 1⅝	2.5 / 3	Steel	1.61 / 1.995	NA / 4.76	NA / .37	0.27 / 0.40	NA / NA	3.05 / 3.63	1.93 / 2.51	NA / NA	325*	20,000
Slim Kone-Hi Temperature	1 11/16	2.5	Steel	1.995	6.04	.23	0.26	0.76	4.81	3.68	2.80	425†	
Slim Kone-Golden Jet		2.5			5.34	.25	0.24	0.74	5.03	3.90	2.89	325*	20,000
Slim Kone-Jumbo Jet®		3.0			8.12	.25	0.29	0.81	5.48	4.35	3.52	325*	
Link Kone-Golden Jet	1 11/16	13	Aluminum	1.995	7.87	.33	0.41	0.88	6.93	5.80	5.10	300	10,000
Bar Kone		6	Steel bar		5.52	.43	0.45	0.68	5.29	4.16	2.81		12,000
Link Kone-Rigid Jet	2 1/16	19	Aluminum	2.441	8.83	.40	0.45	0.77	8.63	7.51	5.76	300	10,000
Link Kone-Golden Jet	2 1/8	22	Aluminum	2.441	9.59	.35	0.45	0.71	8.27	7.14	5.06	300	10,000
Bar Kone	2 1/8	11	Steel bar	2.441	7.08	.40	0.35	0.86	7.29	6.16	5.28	300	12,000
Slim Kone-Jumbo Jet	2 1/8	6	Hollow	2.441	7.96	.39	0.44	0.77	6.40	5.27	4.06	325*	20,000
Slim Kone-Jumbo Jet	2 5/8	12	Hollow	2.750	13.32	.36	0.36	0.78	10.36	9.21	7.15	325*	20,000
Kone Shot-Golden Jet	3⅛	11	Hollow	3.548	12.38	.30	0.32	0.86	8.74	7.61	6.52	325*	20,000
Kone Shot-Jumbo Jet		12			13.37	.32	0.40	0.83	10.46	9.33	7.77		
Big Hole Burr Free		11			6.25	.61	0.67	0.81	5.48	4.35	3.51		
Kone Shot-Jumbo Jet	3⅜	14.5	Hollow	3.826	16.26	.37	0.38	0.84	11.81	10.68	8.93	325*	20,000
Kone Shot-Golden Jet	3½	11	Hollow	4.026	12.87	.32	0.31	0.75	9.92	8.73	6.58	325*	20,000
Select Kone Golden Jet		11		4.026	12.87	.32	0.31	0.75	9.92	8.73	6.58		
Kone Shot-Jumbo Jet		12		4.026	13.47	.34	0.40	0.83	10.46	9.33	7.77		
Kone Shot-Golden Jet	3⅝	17	Hollow	4.026	15.40	.38	0.44	0.80	12.79	11.66	9.31	325*	20,000
Kone Shot-Jumbo Jet		22.5			18.40	.42	0.44	0.78	14.01	12.88	10.00		
Kone Shot-Golden Jet	4	17	Hollow	4.670	16.37	.40	0.44	0.80	12.79	11.66	9.31	325*	20,000
Kone Shot-Jumbo Jet II		22.5			23.68	.39	0.43	0.84	15.07	13.93	11.74		
Big Hole Burr Free		20			7.29	.65	0.76	0.79	6.38	5.25	4.14		
Kone Shot-Golden Jet	5	32	Hollow	4.670	16.10	.78	0.84	0.82	11.60	10.40	8.49	325*	15,000

Source: Dresser-Atlas, Division of Dresser Industries Inc.
*Charges available for temperatures to 525°F
†475° and 550°F available
CFE—Core flow efficiency
TTP—Total target penetration
TCP—Total core penetration
ECP—Effective core penetration

The depth of penetration of the bullet gun is more strongly affected by the type of material being penetrated than the jet gun is. For example, the same bullet gun penetrates considerably deeper in a soft formation than in a hard formation. The difference in penetration is less with a jet gun.

Shot density is the number of shots per foot measured along the axis of the wellbore. Conventional perforating guns can be loaded to shoot from 1–4 shots/ft, and some smaller guns may be loaded to 6 shots/ft. If more shots per foot are required, another gun run can be made to reperforate the same section.

Selective fire guns are also available. These guns can be loaded so several individual jet shots can be made. Sometimes the select fire gun is used for perforating two or more zones in the same gun run. For example, two sands 4 ft and 9 ft thick, respectively, may be perforated in one gun run with the select fire gun. The gun is loaded with a 4-ft and a 9-ft section. The gun is positioned, and one zone is perforated. Then the gun is repositioned to perforate the second zone.

Shot orientation refers to the horizontal direction of the perforations relative to each other and is commonly called *phasing*. The term "phasing" is not used consistently but generally refers to the horizontal angle made by the intersection of vertical planes between two adjacent rows of perforations. For example, 360° phasing is the same as placing all of the perforations in one vertical line; 180° phasing is two vertical lines of perforations 180° apart or perforating the opposite sides of the casing. The other common phasing angles are 120° and 90°, indicating three and four vertical rows of perforations, respectively.

Normally each lower shot is moved one phasing angle with respect to the upper shot. If 1 ft were perforated with 4 shots/ft on 90° phasing, the first shot would go out at one direction—north, for example. The next lower shot would go out 90° or to the east; the next lower shot would go out an additional 90° or to the south; and the fourth shot would go to the west. On 180° phasing the shots would alternate north, south, north, south. Centralized guns can shoot at all phasing angles subject to size limitations. Decentralized guns normally shoot in a single plane (360° phasing). A decentralized gun is reportedly being tested. The shots leave the casing in a single plane but alternate shots are phased 90° apart.

Oriented perforating is also available. In oriented perforating—not to be confused with shot orientation—the perforating tool is oriented so the perforations are pointed in the desired compass-point direction. This type of perforating is used where multiple completion strings are located in the same wellbore. The formation must be perforated in a specific direction to prevent perforating other tubing strings. A special tool is available to detect the direction of the other tubing strings, and the gun is oriented to point away from them.

Various types of perforating guns are available (Fig. 10–6). The most common is the steel carrier that can normally be loaded to the desired perforation density and shot orientation. This is basically a centralized tool, but the decen-

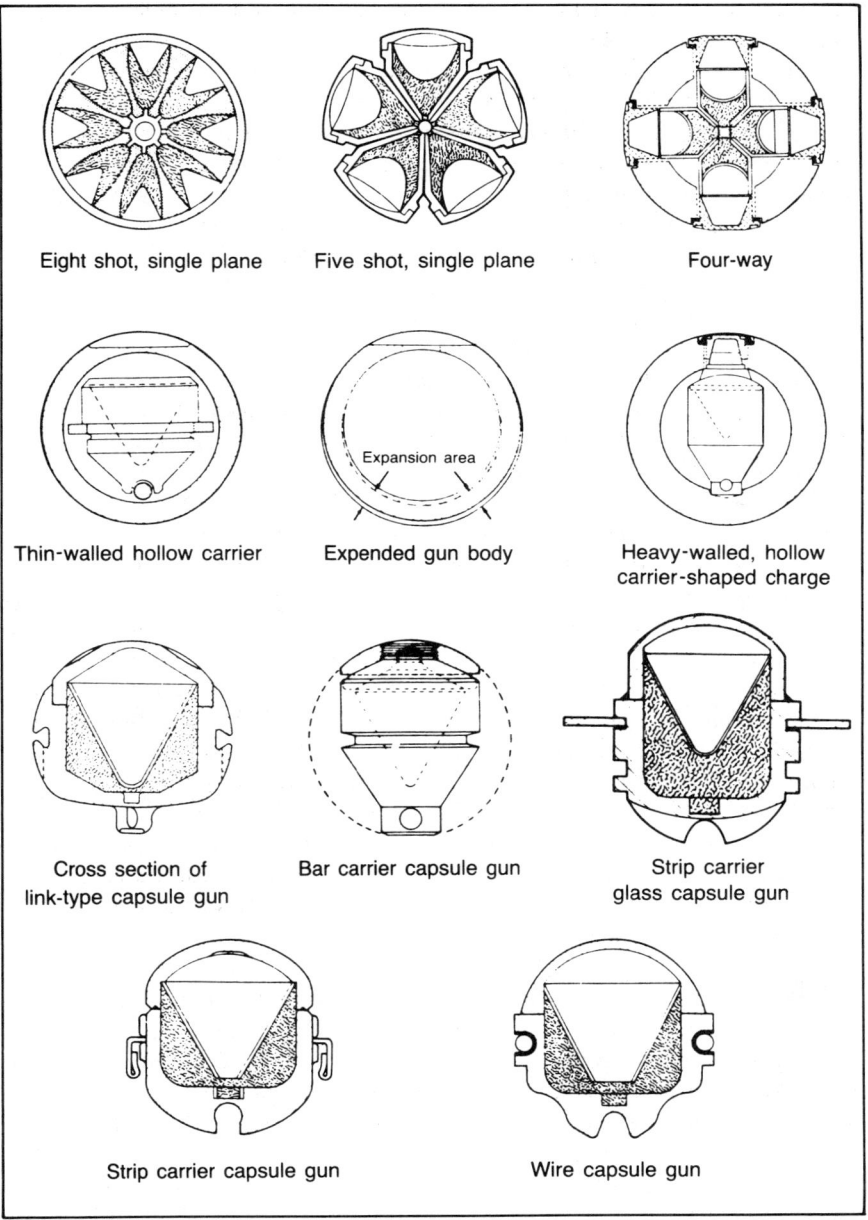

FIG. 10–6 Types of perforating guns (courtesy Dresser Atlas)

tralized tool is built in a similar manner. The steel carrier does not leave any junk in the hole except a minimum amount of jet debris. Other types of perforating tools include wire and strip carriers and swivel-type jets. The swivel-type jet is unique in that the jet shot is oriented vertically while the tool is run in the hole. The jet shot then swings or swivels to the horizontal position for perforating. This tool permits a larger (deeper-penetration) jet than with other jet charges for the same size tubing or casing. The latter three perforating tools—wire and strip carriers and swivel-type jets—normally leave only a small amount of junk in the hole after perforating. However, the movement and forces of perforating sometimes cause the carrier to break, leaving part of the carrier in the hole as junk.

There has been considerable discussion and some experiments on the best perforating density and shot orientation to use. Except for special cases, these measurements generally are not definitive. The type of carrier and size of jet charge depends on the specific application such as the condition of the well, the type of downhole completion equipment, and pipe sizes. As a general rule the operator selects the deepest penetrative charge that can be safely run. Generally, less shot density is used in massive, clean, consistent formations. Higher shot densities are used in layered formations, formations with lower vertical permeability than horizontal permeability, and fractured formations. Shot density usually varies from 1 shot/2 ft to 8 shots/ft. The effectiveness of a minimum amount of perforating is reflected by the fact that reperforating seldom increases production appreciably.

In operation the perforating gun is run below a collar locator on a single or multiple insulated conductor wire line (Fig. 10–7). The gun is positioned by using the collar locator and locating collars that have been prerecorded on a gamma log. The jet charges are detonated. The collar locator may be used to record the perforations, after which the gun is pulled from the hole. If a multiple perforating tool is used, the gun is repositioned after perforating the first set to perforate additional sections. If an orienting-type tool is used, the gun is positioned and oriented before perforating. In most cases a lubricator is used on the wellhead when perforating.

Completion Fluids Clay-base mud is seldom used in completion operations because of the high risk of plugging the formation. Salt water is the most common fluid used because salt gives additional weight needed to control formation pressures. Water containing 2–4% potassium chloride is also used. Oil and oil emulsion muds are used when heavier mud weights are required to control high pressures. These fluids do not damage the normal formation.

A completion fluid may be left in the annular space between the tubing and casing. Salt water containing corrosion inhibitors is commonly used in low to moderate pressure wells. Special, weighted oil muds are used for the completion

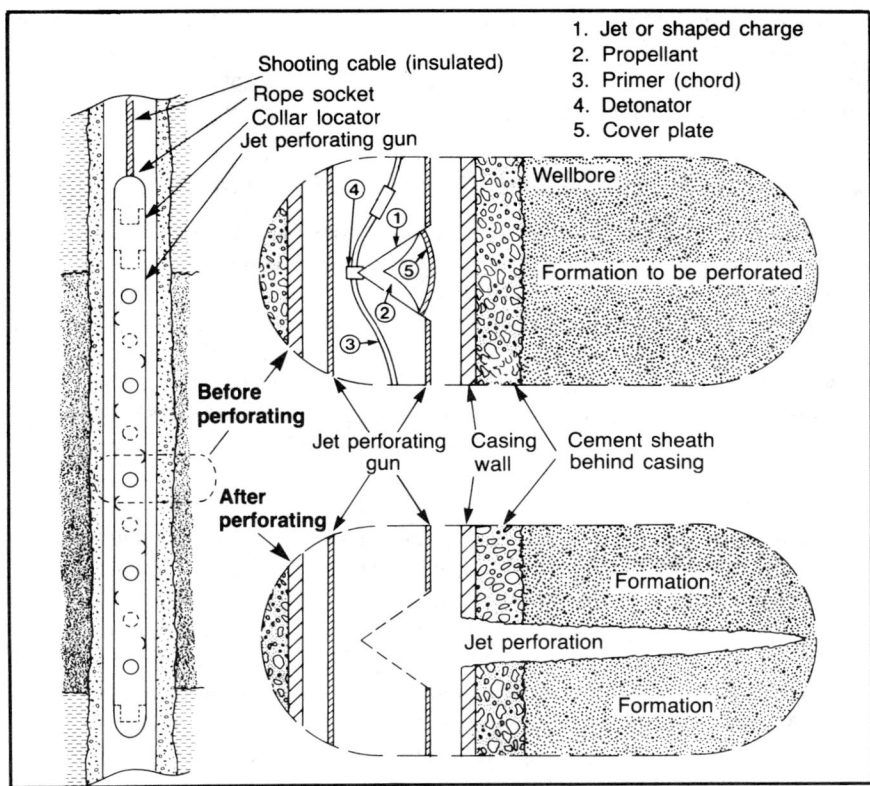

FIG. 10–7 Jet perforating. The jet gun is lowered into the hole with the shooting cable attached to the rope socket. The gun is precisely positioned opposite the formation to be perforated by locating casing collars with the collar locator. The gun is then fired, and all jets are detonated. A primer chord passes behind each jet charge at the ignitor and is connected to an electric detonator. The electric detonator is connected to the shooting truck by insulated electric wires passing through the shooting cable. In the firing sequence an electric current is passed through the wires and ignites the detonator (similar to an electric blasting cap). This causes the primer chord to detonate along its entire length and in turn sets off the ignitor in each jet charge, almost simultaneously. This fires the propellant in the jet charge, which burns at a very high temperature and fast rate. The hot gases are directed outward by the cone (shaped charge) and literally burn a hole (jet perforation) through the cover plate, the casing wall, the cement sheath behind the casing, and into the formation. The gun carrier is then pulled out of the hole.

fluid in high-pressure wells. When weighted fluid is used and the tubing develops a leak, the hydrostatic head of the fluid helps control the higher pressures. Special problems are encountered if the leak develops high in the hole at shallow depths, but the weighted completion fluid contributes to an extra margin of safety.

COMPLETION DESIGN AND PROCEDURES

Completion procedures are activities and operations performed during the completion process. Some of these, such as logging and perforating, are common to most completions and are logically included with a discussion of the tools and equipment. Others are specialized operations that often depend on formation conditions. Sand control is a less common completion procedure that normally is handled with a special completion design and treating techniques. Other procedures may be common to the general completion process but may be omitted on a specific completion depending on the requirements of the individual program.

Squeezing

Squeezing is the procedure of mixing, pumping, and displacing cement into the formation, usually through perforations in the casing. It is not to be confused with primary cementing where cement is displaced into the annular space between the casing and the wall of the borehole. Squeezing may involve pumping cement into the annular space, but this is normally termed a second primary cement job. The term squeezing usually implies displacing cement into the formation.

A quality cement slurry must be used in squeezing to ensure that the cement can be pumped and properly displaced. It is frequently pumped at high pressures and temperatures that may affect the pumping time. If the cement is not displaced properly, it must be drilled out and the squeeze job must be repeated. A severe fishing job and possibly a lost hole may result if the cement prematurely hardens while it is around tubing or other equipment. Therefore, when possible, the cement is displaced through tubing and open casing directly into the perforations.

Casing leaks generally are repaired by squeezing. After squeezing and cleaning out, the casing is pressure tested to ensure that the leak has been repaired or plugged with cement. This test also verifies that the casing is intact and capable of containing the reservoir pressure. Other methods of casing repair are often available, but squeezing is usually the most economical. In all cases the integrity of the casing must be maintained as the primary container of the reservoir pressure.

Leaking casing equipment such as cement-stage tools and liner tops can often be plugged by squeezing. Nonproductive perforations often must be plugged or isolated from the working casing section. If the perforations are in a lower section of the hole, they may be shut off by setting a plug above the perforations, often an easier procedure. However, they may be squeezed through a retainer, which is then left in the hole as a plug. Perforations lower in the hole are easier to shut off. Therefore, deeper zones are tested first. This procedure is known as *testing from the bottom up*.

The perforations or leak must be squeezed, cleaned out, and pressure tested if they occur higher in the working section of the casing. The general squeezing procedure is illustrated in Fig. 10–8. In this case a retrievable squeeze packer and retrievable bridge plug are used.

If the exact location of the leak is unknown, the retrievable packer and bridge plug combination can be used to locate the leak. The common procedure is to run the tools deep in the hole, seat and release the bridge plug, pull the packer up near the top of the hole, seat the packer, and pressure test the casing between the packer and plug. The leak will be indicated by declining pressure as fluid moves through the leak. The packer would then be unseated, moved downhole a

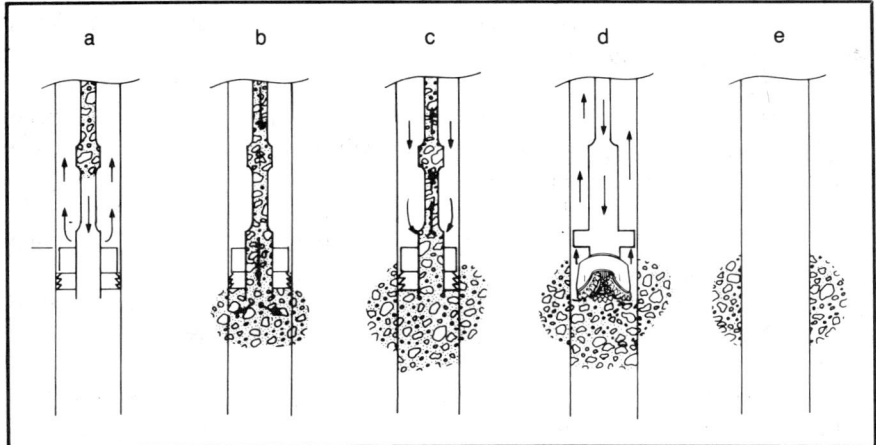

FIG. 10–8 Squeezing. (a) Circulating cement down the hole to the bottom of the tubing or drillpipe. Mud passes through the circulating valve and up the annulus. (b) Displacing cement into the formation, circulating valve closed, and continuing to pump down the tubing or drillpipe. (c) Reversing out excess cement, circulating with the valve open. Mud is pumped down the annulus and cement returns through the tubing. (d) Drilling out cement inside the casing with a bit and casing scraper. (e) Cleaning out casing after pressure testing to resume operations.

short distance, and reseated. The casing pressure is tested again. When the packer is seated below the leak, the casing should test satisfactorily, thus locating the leak. The packer is then unseated, picked up, and seated about 200 ft above the leak prior to the squeezing operation.

The first step in the squeezing operation is to pump fluid, usually water, through the leak to establish a breakdown pressure and pump-in rate and to ensure that the leak is open (Fig. 10–8a). The pressure and rate also help determine the amount of cement and the squeezing procedure needed. At higher pressures less cement is used, and it is displaced faster; at lower pressures more cement is used, but it is displaced more slowly.

In Fig. 10–8b the cement has been mixed and is being pumped down the tubing (or drillpipe) by circulating through the circulating valve. Water spacers separate the cement from the fluid in the wellbore. The water ahead of the cement is called *spearhead fluid,* and the water behind the cement is called *tail-in fluid.* All volumes are carefully measured. When the cement is near the circulating valve, the valve is closed and the fluids are pumped into the formation. Pressure increases at this time. In Fig. 10–8c part of the cement has been pumped through the perforation into the formation, and the remaining cement is in the casing and tubing. In Fig. 10–8d a small amount is left in the casing, and the remainder is displaced into the formation.

In the normal squeezing operation pressures gradually increase. The objective is to reach a predetermined squeeze pressure when all of the cement has been displaced into the formation except for a small amount in the casing. If the pumping pressures tend to increase too rapidly, the pumping rate is increased to displace the cement faster. If the maximum squeeze pressure is reached before all of the cement has been displaced, the cement remaining in the casing and tubing is reversed out of the hole. In the reversing procedure the packer is unseated, the preventer rams are closed, and the rig mud pumps are used to pump mud down the annulus and circulate the cement back up through the tubing and out of the hole.

The cement is reversed out in place of normal circulation to ensure that the cement remains inside the tubing. If the cement hardens, the tubing can be pulled from the hole to clean out the cement. If the remaining cement were pumped in the normal manner—down the tubing and up the casing—and the cement hardened as it was being pumped out of the hole, the pipe would become stuck. This would require an extensive milling and fishing job to clean out the hole. Therefore, the safer procedure is to reverse out the excess cement.

In many cases the pressures do not increase satisfactorily while the cement is being displaced into the formation. One procedure in this case is to reduce the pumping rate to allow more time for the cement to begin to harden. If this does not cause a satisfactory increase in squeeze pressures, the cement is staged.

Pumping operations are stopped for 30 sec to 5 min depending on conditions. A small volume of cement, usually about 1 bbl, is displaced, and pumping is again shut down temporarily. If the pressures increase too rapidly, the shutdown time is reduced accordingly. The experienced operator can normally obtain the required squeeze pressure by a combination of staging and increasing or decreasing the pumping rate.

In some cases a satisfactory squeeze pressure cannot be obtained even with staging. If this occurs, the remainder of the cement is pumped into the formation and is overdisplaced with part of the tail-in water. The packer is unseated and picked up, and operations are shut down for 2–4 hr. The packer is then reseated, and a second squeeze job is performed. In some cases multiple squeezes—usually two or more—may be required to obtain the desired squeeze pressure.

Some operators unseat the squeeze tool after a successful squeeze and wait 4–8 hr while the cement hardens before beginning to clean out the cement. Other operators unseat the packer, pick it up about 500 ft, reseat the packer, and apply pressure. This pressure is held for several hours before unseating the packer and pulling it from the hole. A bit is then run to clean out the remaining cement.

A casing scraper is run on the cleanout assembly. The bit drills the bulk of the cement, but pieces of cement adhere to the inner walls of the casing. The casing scraper scratches the walls of the casing to remove these cement particles. If the casing is not scraped, close-tolerance tools may hang up. Packers and other tools with rubber elements also can be damaged easily if they are run through a section of casing that has not been scraped. There is a minor risk of sticking the cleanout assembly when running the casing scraper and drilling long sections of cement. Some operators prefer to drill the cement without using a casing scraper and then make a separate scraper run.

After cleaning out, the squeezed perforations and casing are pressure tested. If they do not test satisfactorily, the squeeze procedure is repeated until a satisfactory pressure test has been obtained. After the section has been pressure tested satisfactorily, the retrievable bridge plug is recovered, and operations are resumed.

Squeezing is a moderate to high risk operation. After the cement has been started in the hole, all operations must continue until it has been displaced into the formation or reversed out. Any equipment failure during this interval that interrupts the sequence can allow the cement to harden, possibly causing a severe fishing job. There is also some risk in squeezing while part of the cement is in the tubing since any small leak can allow some cement to go outside the tubing on top of the packer. This invariably causes a difficult fishing job.

An alternative procedure is to seat the packer higher above the perforations so all of the cement is out of the tubing when the first part of the cement enters the formation. There is also a risk in safely reversing the excess cement out of the hole.

The squeeze procedure is applicable to medium-depth formations with moderate pressure. When higher pumping pressures are expected or the zone to be squeezed is deep in the hole, the best method is to squeeze through a cement retainer. The cement retainer is similar to a permanent packer with a check valve. It is seated in the casing, and the pipe is run and seated in the cement retainer.

The general squeezing procedure using a cement retainer is similar to that previously described; however, instead of unseating the packer, the tubing is picked up out of the cement retainer. A backpressure valve in the bottom of the retainer holds the cement in place. This reduces the overall risk considerably.

Another squeeze procedure is called a bradenhead squeeze. In this procedure the cement is pumped directly into the casing, usually below closed rams. All of the fluid in the casing is pumped into the formation as the cement is moved downhole. The cement is displaced into the formation by volumetric calculations, and the general squeezing procedure is followed. After squeezing the cement is cleaned out, and the casing is pressure tested.

Bradenhead squeezes are used at shallow depths and low pressures. Little risk is involved.

An alternative to the bradenhead squeeze is to run open-ended pipe to the perforations. Cement is displaced in the casing, and the pipe is picked up a safe distance above the cement. Pipe rams are closed, and the cement is squeezed into the formation. Squeeze pressures are restricted to less than the burst pressure of the casing.

Squeeze cementing techniques are also used for block squeezing for zone isolation. Sometimes the production casing is not properly cemented during the primary cementing operation. This can leave channels that cause treating and production problems. These problems can be alleviated or eliminated by block squeezing to isolate the productive zone. A short interval of 2–4 ft is perforated below and above the zone. Each set of perforated intervals is squeezed using the procedures described. After cleaning out and testing the squeezes, the producing formation is perforated and completed in a normal manner.

There also is a risk of plugging the producing formations when squeezing near such zones. This risk can be reduced by using smaller volumes of cement and lower pressures and by ensuring that the perforations are correctly placed.

Oil-water and gas-water contacts often occur in producing formations. Sometimes these are not recognized, or the water may encroach after a period of production. In either case if the zone is perforated in the section that is filled with water, water will be produced. The water can sometimes be plugged off by squeezing and reperforating higher in the zone. This procedure is known as *squeezing off water*.

Another method of using squeeze techniques to shut off water is the *gunk squeeze*. Fresh bentonite is mixed with diesel oil and is displaced into the water zone by the standard squeezing procedure. The diesel is used as a carrier for the bentonite. The bentonite is not affected by hydrocarbons, but when it contacts water, it swells and plugs the formation. Theoretically, the bentonite does not set-up except in the water zone. However, the gunk mixture is seldom used now and has been replaced with special plastics.

Bottom-Hole Pressure Test

Bottom-hole pressure testing equipment is used to take static bottom-hole pressures and to conduct buildup, drawdown, and multirate flow tests. Normally a well is only tested by one method, and in some cases different tests give the same results. These tests are usually run on flowing wells. They cannot be run on pumping wells except in special cases. Analyzing the test results is relatively complicated. The results obtained from the tests are summarized as follows:

- *Bottom-hole pressure*—The static bottom-hole pressure (BHP) or reservoir pressure is the average pressure throughout the reservoir in the area influenced by the well. This is used with other data to calculate reserves and other reservoir parameters. The flowing bottom-hole pressure is the pressure in the wellbore opposite the reservoir when the well is flowing. This pressure varies according to the flow rate, decreasing with increasing flow rate.
- *Drawdown*—The term drawdown is commonly used in two different ways. Drawdown pressure is the difference between the static reservoir pressure and the bottom-hole flowing pressure measured in psi. The term drawdown is sometimes used as the flow rate divided by the drawdown pressure and expressed in terms of bo/d/psi or Mcfd/psi.
- *Permeability*—The permeability obtained from these tests is known as the effective permeability and is the best permeability to use when calculating flow rates and other factors. The effective permeability is lower than the permeability obtained from measurements on cores. The effective permeability allows for compression caused by overburden, water saturation, and multiphase flow.
- *Flow efficiency*—Flow efficiency is a measure of the actual well productivity compared to the well productivity if the wellbore is neither stimulated nor damaged (partially plugged). Flow efficiencies are expressed in percentages or decimals. A well with a flow efficiency of 100% is neither stimulated nor damaged. If the well has a 50% flow efficiency, sometimes called completion efficiency, the wellbore is 50% plugged or otherwise damaged. If this damage is removed, the well returns to 100% completion efficiency. Good stimulations give completion efficiencies in the range of 200–300%. The measurement of completion efficiency is very important in determining if a well should be stimulated.

- *Skin damage*—Skin damage is related to flow efficiency and basically measures the same characteristic except it is generally expressed in psi.
- *Radius of investigation*—The radius of investigation is the radial distance from the wellbore analyzed by the tests. Longer tests give deeper depths of investigation.
- *Reservoir limits*—The distance from the wellbore to barriers such as pinchouts and faults can sometimes be calculated from the test data. Under favorable conditions the direction and distance to a barrier may be determined by testing two offsetting wells. These data can help the operator select a development drilling site relative to the wells that were tested.

Bottom-hole samples of produced fluids can be analyzed to provide information for calculating reserves. The bottom-hole pressure tool records pressure versus time. It is similar to the tool used in open-hole drillstem tests. In operation the tool is run through a lubricator, generally on a single-strand, small-diameter wire line, called a *slick line*. The tool is run in the hole to the reference datum and records the bottom-hole pressures while the well is flowed at one or more rates. The tool is shut in for a pressure buildup test, or it is opened to begin flowing for pressure drawdown. The amount of time required to test the well including the length of the flow and shutin periods depends on the type of test being run and various wellbore characteristics such as bottom-hole pressure and formation permeability.

Some tests can be run on certain types of pumping wells. A pressure recorder normally cannot be used because the pumping well has pump rods in the tubing. The bottom-hole pressures are obtained using a fluid sounder to measure the depth from the surface to the top of the fluid column in the annulus. This measurement is subtracted from the depth to the formation to give the height of the fluid column. The gas pressure in the annulus and the density of the fluid column can be used to calculate the bottom-hole pressure at datum. The measurements may not be precise but frequently provide usable information.

Pressure-Buildup Curves

Fig. 10–9 is a standard pressure-buildup curve; the recorded bottom-hole pressure (BHP) is plotted versus dimensionless time $(T + \Delta T)/(\Delta T)$ on semilog graph paper. Dimensionless time includes T, usually the stabilized flow period (minutes or hours) before the buildup test starts, and ΔT, the cumulative shutin time from the time the well is shut in to the time the BHP is recorded in consistent units. Therefore, T and ΔT must be in the same units, i.e., minutes (most common), hours, or days.

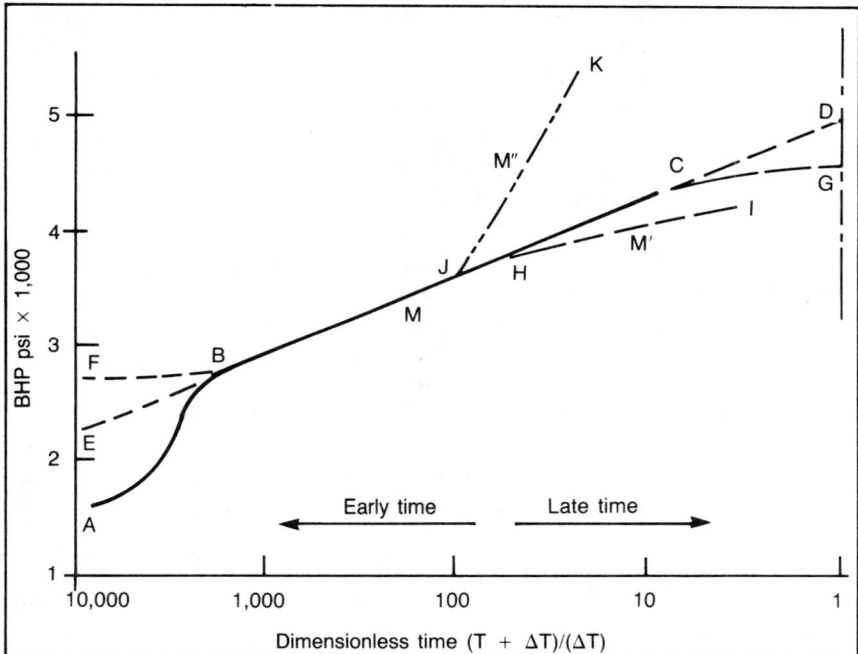

FIG. 10–9 Pressure-buildup curves

A common base curve is illustrated by the heavy line (ABC). A is the first point recorded, and C is the last. AB is the early time section and may represent wellbore effects or afterflow—flow into the wellbore after the well is shut in. The reservoir has not reached a true shutin condition. The straight-line section (BC) gives the slope of the curve M in psi/cycle used to calculate permeability and other factors. The curve is extrapolated by extending the straight-line section from C to D in late time to a dimensionless time of one. The curve will never reach the dimensionless time of one (infinite shutin time) because of the nature of the formula. For practical purposes the well is shut in until the straight-line portion is well established into late time. This is one major advantage of production buildup tests over drillstem tests, which may limit time because of the danger of sticking the test tool. Point D is the extrapolated shutin reservoir pressure, a very important value in all calculations including reservoir studies.

The straight-line portion of the curve is extrapolated backward into early time to indicate the bottom-hole flowing pressure without wellbore and afterflow effects. Theoretically, the true buildup curve would be a straight line (EBCD).

The two sections (AB and EB) can be compared, and calculations can be made to determine skin damage, flow efficiency, and Δp_{skin}. These factors are needed to determine if the well should be stimulated, the type of stimulation needed, additional productivity from stimulation, and other factors. For example, the section AB indicates wellbore damage. Assume the well is successfully stimulated and another buildup test is run. The early time part of the curve would probably be similar to FB and the remainder similar to the original curve (BCD).

If the buildup test were taken later in the life of the well or the reservoir were somewhat limited, the late-time section of the curve may show a new late-time curve (CG) with a lower extrapolated shutin reservoir pressure (G) compared to D. The difference between the two pressures and the production history help predict future production rates, pressure for artificial lift, and reserves.

Reservoirs are not necessarily consistent and may have areas away from the wellbore that have higher permeability (HI) or lower permeability (JK) indicated by different slopes (M', M''). The curve also supplies data to calculate the distance (not direction) from the wellbore—radius of investigation—to various points such as J or H where the permeability changes.

Stimulation

Stimulation is broadly defined as any activity or operation conducted on the wellbore or the formation to increase well productivity. First, a well is investigated or tested to determine stimulation requirements. This information is used to select the appropriate stimulation procedure, which then is performed. The orderly process of determining stimulation requirements and stimulating the well accordingly is necessary to ensure that the well will be produced at the maximum efficient rate and minimum cost.

Normally the well is drilled with mud. Fluids in the mud filtrate penetrate into the formation and can block some formations in the area adjacent to the wellbore. When the casing is cemented, filtrate from the cement can also damage the formations. Formation damage in these cases is usually blockage caused by swelling material in the formation.

The mud filtrate can also carry microscopic particles (mud solids) into the formation, further increasing formation blockage. Mud seepage into the formation or lost circulation can also damage the formation. This blockage or permeability restriction in the formation adjacent to the wellbore is commonly called *skin damage* or *wellbore damage*. The damaged section extends from a fraction of an inch to several feet into the reservoir.

Perforations often extend through the damaged section. This is partially verified in field practice when zones have been reperforated, in many cases without increasing the production rate. There is some evidence that the walls of the perforation can be partially plugged or damaged in the perforating process as a result of the force and pressure of the bullet or jet charge or the extreme heat from the jet charge glassing over the surface of the perforation. Skin damage normally is removed with a small, shallow stimulation at a moderate cost.

Assuming a minimum bottom-hole flowing pressure, the other main factors that affect well productivity are fluid viscosity and formation permeability. These cannot be economically changed with current technology. However, it has been found that in applicable cases, deep stimulations can increase well productivity.

Wellbore damage, effective formation permeability, and other data must be known in order to determine stimulation requirements. This information is obtained with well tests. The most common test is a stabilized flow rate followed by a pressure buildup. A properly run test may be expensive, and in some cases it is difficult to obtain good data. Also, it may be more economical to stimulate the well rather than conduct tests.

The basic method of stimulating is to create flow channels from the perforations into the formation. The flow channels are created by two basic methods, depending on the type of formation being stimulated. Carbonates such as limestone are treated with acid, and sandstones are stimulated by hydraulic fracturing with sand-laden fluid. In either case the depth of penetration is related to the amount of fluid used.

Penetration increases with increasing fluid volumes. The size of acid jobs ranges from 1,000–100,000 gal. Fracturing jobs are conventionally measured by the pounds of sand used. They range from a small job with about 20,000 lb of sand to massive hydraulic fracturing using over a million pounds of sand. Sand concentrations in the fracture fluid are from ½–4 lb/gal.

Methods have been established to calculate well productivity increases as a result of acidizing and fracturing. Hydrochloric acid is the basic treating agent in acidizing. It reacts with the solid limestone to form two liquids (water and calcium chloride) and a gas (carbon dioxide). These fluids flow out of the hole after the stimulation. The void space created by removing the limestone serves as a flow channel for the reservoir fluids. Acid concentration in the treating liquid normally is in the range of 10–20%. Acid may be injected at a high rate and pressure to fracture the formation for deeper penetration. Other acids, such as acetic acid, and additives may be used to improve overall treating efficiency.

Stimulation by hydraulic fracturing parts or splits the formation by fluid pressure. Fig. 10–10 shows an example of a formation hydraulically fractured with sand-laden fluid. Fracture orientation (direction) is determined by the earth's stresses. In the normal case shallow fractures are oriented horizontally. They begin to rotate at depths of 1,000–2,500 ft. Below these depths fractures generally are oriented vertically. The basic theory of fracturing is that the fracture is

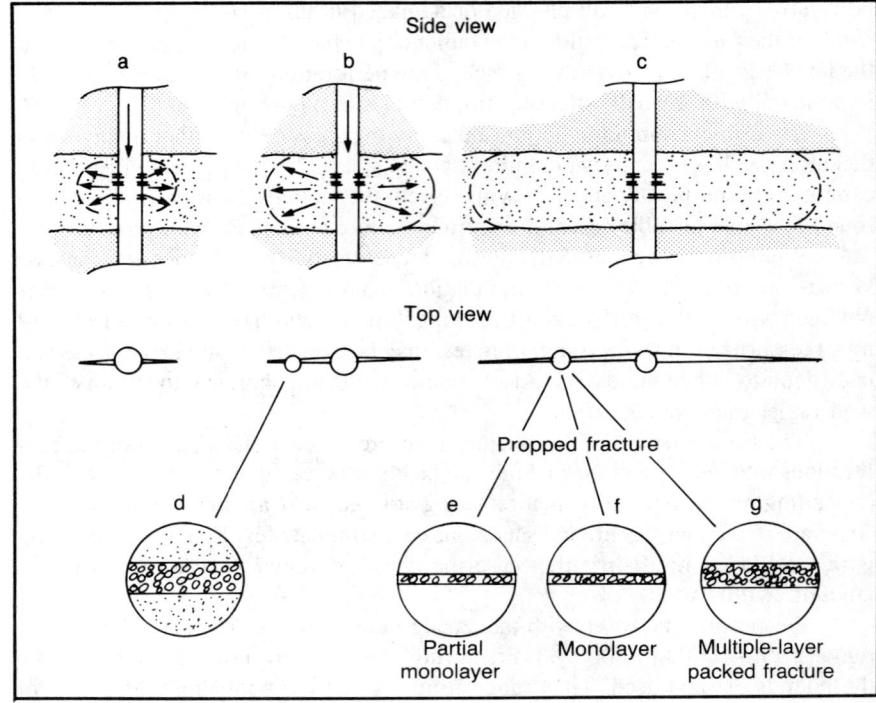

FIG. 10–10 Hydraulic fracturing. Sand (proppant) suspended in a gelled fluid is pumped into the formation to start a double-wing, vertical fracture (a). Gelled fluid is needed to suspend the sand so it can be carried down the fracture and not settle in the bottom near the wellbore. This settling would cause a screen out, which prevents further pumping. The fracture continues to grow (propagate) as additional sand-laden fluid is injected (b). Fracture growth in the vertical direction is limited to the shale barrier at the sand-shale interface (bed boundary), thus causing the fracture to stay in the sand. After the designed amount of sand-laden fluid has been pumped, the pump is stopped and the fracture closes (c). The sand grains are trapped between the closing walls of the fracture and prop it open. This creates flow channels for the formation fluid to flow into the wellbore. The fracture is wider during injection (d). Closure width depends on the type of packing (e, f, and g), which in turn is determined by the sand concentration—the pounds per gallon of sand in the injected fluid.

contained in the sand. The underlying and overlying shale beds limit vertical fracture growth so the fracture propagates radially away from the wellbore.

Special placement techniques are used in stimulating. It is important to inject stimulating fluid into all perforations, so they must all be open. This can be checked, and plugged perforations can be opened using a ballout procedure with perforation ball sealers. These are round balls usually made of a hard plastic and coated with rubber. Different sizes are available, depending on the size of perforations. The perf balls are injected into the treating fluid flowstream with a ball injector. The balls move downhole with the fluid and seat on the perforation as a temporary plug. This diverts the fluid to other perforations that are subsequently plugged with other perf balls. The ball action can frequently be observed at the surface as small, sharp pressure increases on the pressure recorder. Pressure is released after the treatment, and the perf balls either drop to the bottom of the hole or are carried to the surface with the returning fluid.

Limited entry is another method of distributing treating fluid among the perforations and different formations. When fluid is pumped through a hole at high rates (turbulent flow), the fluid volume passing through the hole is related to the pressure differential across the hole.

For example, ten holes are perforated in the casing, and fluid is injected at a sufficiently high rate to achieve a moderately high pressure differential across the perforations. All ten holes receive or pass approximately the same volume of fluid. Therefore, if an operator wants to ensure that a thick zone is stimulated entirely, hole sizes, pressures, and flow rates are calculated to determine the number of holes that can be used to give limited entry, i.e., equal fluid volume through each hole. This number of holes—for example, ten—are perforated at equal distances apart over the formation interval. The formation is then stimulated at the predetermined pressure and flow rate. The injected fluid is evenly distributed over the entire interval.

The same procedure can be used to stimulate two or more separate zones. In this case using the example of ten perforations, two holes may be perforated in each of two thin zones and six holes in a thicker zone. When the treating fluid is injected, one-fifth of the fluid goes into each of the two thin zones and three-fifths of the fluid goes into the thicker zone.

In some treatments it is desirable to divert the treating fluid in the open hole. For example, if a fracture is propagating from one set of perforations, the operator may want to stop the fracture and extend a fracture in another direction. This is done with open-hole diverting agents such as benzoic acid flakes, wax beads, or solid (granulated) salt. The materials are added in the fluid in controlled amounts. They cause a plugging action that tends to start another fracture in a different direction or location. The solid materials dissolve in the reservoir fluids and are flowed out of the well later. This type of treatment is not used extensively.

In some cases such as low-pressure wells, there is a problem recovering the treating fluid so the well can be cleaned up and placed on production. The fluids are often pumped or swabbed out of the hole. In some cases these methods cannot be used for mechanical reasons. Gases such as nitrogen or less commonly carbon dioxide can be used to help recover the treating fluid and restore the well to production in a fast, efficient manner. The gases are brought to the wellsite in a liquid state at low temperatures in insulated, pressurized tank trucks. They are injected into the well with the treating fluid in controlled volumes. Reduced pressure and formation temperature convert the liquefied gases to a gas. The combined stored energy in the gas and the reduced density help flow or expel the treating fluids from the formation and wellbore.

The stimulations described are generally applicable to shallow, medium, and deep stimulation treatments using different injection methods. Other treatments also are specifically designed as very shallow stimulations for wellbore cleanup.

One of the most common is an acid wash using hydrochloric acid with a small amount of hydrofluoric acid. A small volume is injected at low pressures and helps to remove mud solids. Other fluids such as surfactants, alcohols, and light hydrocarbons are also used as cleanup fluids. Special packers and fluids are sometimes used to flush the perforations.

One of the most important parts of the completion process is to determine the type of completion needed and to stimulate the well accordingly. The materials, equipment, and techniques for stimulating are well advanced, but new, more efficient methods are needed, especially for deep stimulation in low-permeability formations.

Fishing and Casing Repair

Fishing jobs can also occur while completing the well. Shallow casing strings may be damaged or fail while drilling deeper. Production casing and the shallow casing strings to a lesser extent can be damaged or can fail during completion. Generally these must be repaired as soon as they are detected before operations can be resumed.

Fishing and Sidetracking Many open-hole fishing tools are used to fish inside casing. Generally, the tools used in casing are smaller. The reduced areas often requires closer tolerance, which increases the risk of sticking. The smaller tools are more subject to failure and breakage, so working forces must be reduced accordingly. Also, fishing for a second or third fish is not uncommon. As a tradeoff all of the formation problems are eliminated in the cased hole except pressure control.

Fishing operations in the cased hole during completion use many different tools and techniques, some of which are similar to fishing in the open hole. Some special operations and tools include a packer plucker, which helps remove stuck packers; a reversing tool, which can be used to back off part of the fish; and mechanical and hydraulic pulling tools, which apply a very high pulling force (tension) to release stuck fish. A special, full-opening, overshot is used to recover coil tubing and other small-diameter, long fish.

If a fish cannot be recovered by conventional fishing operations or it cannot be removed by milling, the usual procedure is to sidetrack or redrill the hole. Plugging and abandoning the hole normally is not considered at this stage, except in very severe cases. The decision to redrill or sidetrack during completion depends on the size of the procuction casing, the depth to the top of the fish, the length of the section to be sidetracked, and the nature of the formations.

Two methods of sidetracking the hole out of the casing are available. One is to remove a section of casing 20–40 ft long by milling. The hole is then sidetracked through the casing window with a conventional sidetracking assembly. Another method of sidetracking is to run a whipstock-sidetracking assembly seated on a packer (Fig. 10–11). A special mill cuts a hole through the side of the casing and drills a short section of deviated hole, leaving the whipstock permanently seated.

After either method of sidetracking, the new hole is drilled to the required depth using directional drilling techniques. In most cases the sidetracked hole is cased with a liner. Sometimes another complete (smaller) string of production casing is run and cemented. The well is then completed in the normal manner.

Casing Failures and Repairs Four basic types of casing failures are leaks and collapsed, burst, and parted casing. When the failure occurs below the deepest prospective formations, the normal procedure is to isolate it with a plug set in the casing above the failure. Otherwise, it is repaired or isolated (Fig. 10–12). The point of failure is always considered as a point of weakness when casing has been repaired. This point generally can withstand normal pressures, but higher pressures must be used with caution.

Leaks occur with most casing failures. However, they are listed separately because they are the most common casing failure and can occur in combination with other casing failures. Leaks can be caused by drillpipe wear, malfunctioning slips on packers and liner hangers, excess wear during milling operations, corrosion, and other factors. They are often repaired by squeezing, and multiple squeezes may be required. Some casing leaks are repaired with a casing patch—a

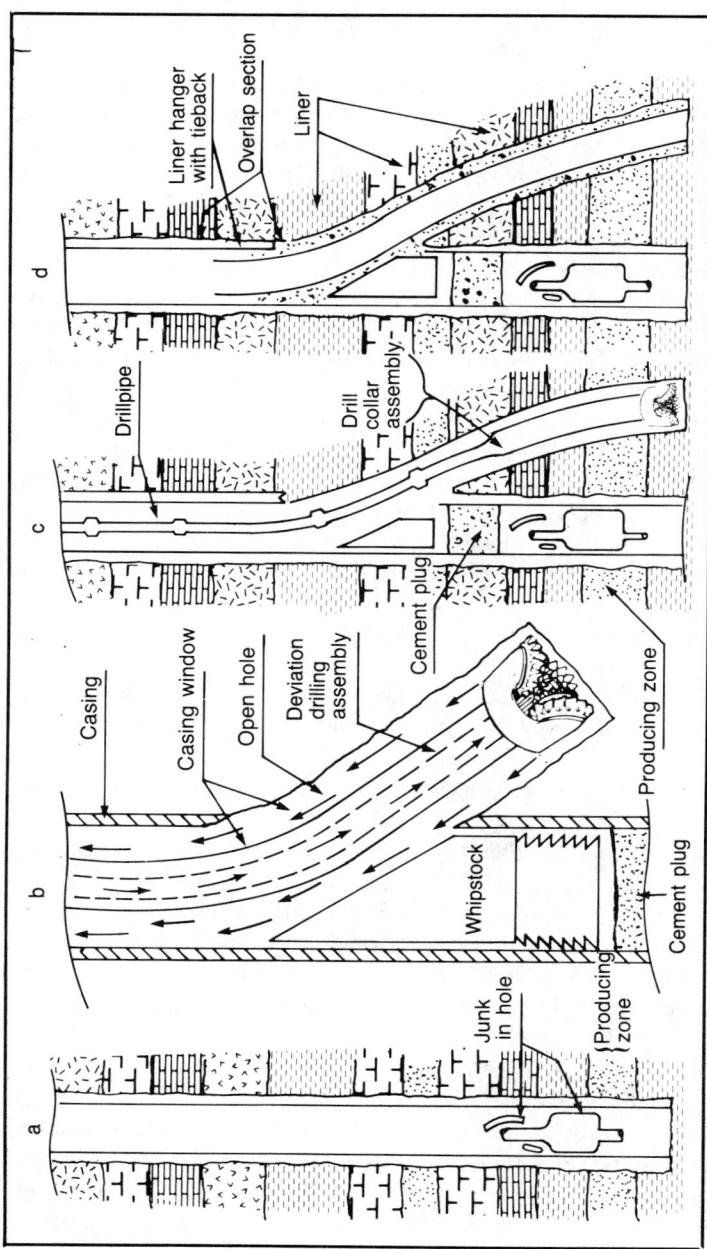

FIG. 10–11 Whipstocking around junk-liner completion. (a) Junk in hole plugs the wellbore so the well cannot be produced. (b) A whipstock is placed in the hole above the junk, and (c) a new deviated hole is drilled. (d) A liner is set in the deviated hole.

corrugated metal sleeve run into the well, positioned opposite the hole and expanded with the setting tool. The diameter of the repaired casing is reduced slightly, and operating pressures must be restricted.

In some cases leaks and other types of casing failures can be isolated. One method is to seat a packer below the leak and isolate the leak to the closed-in annular space. Another method is to straddle the section with two packers connected by tubing. A tieback or stub liner or another string of casing also may be used.

A leak or other type of casing failure in an uncemented section of casing may be repaired by unscrewing or cutting the casing below the point of failure, pulling the casing out of the hole, and replacing the damaged section. The casing is then rerun and reconnected to the casing cemented in the hole by screwing in if the casing is originally backed off, or a casing-bowl connection can be used. The casing bowl is a piece of casing with a slip and seal assembly. The rerun casing can be cemented in place by pumping cement through the casing bowl (seals omitted) or through perforations in the casing below the casing bowl. These connections make a strong casing repair that approaches the original strength of the casing.

Completely collapsed casing is repaired by milling out the collapsed section and repairing the resulting leak by one of the methods described previously. Partially collapsed casing is reopened to the original inside diameter using a casing roller for larger-diameter casing and a swedge for smaller-diameter casing or tubing. The repaired section is then pressure tested to ensure that it will hold pressure and that the casing did not crack or otherwise fail during the collapse and rollout operations. Leaks are then repaired by squeeze cementing.

Burst or split casing does not normally restrict the internal diameter of the casing. It is detected by the same method used to detect casing leaks. It is often repaired in a similar manner, but with more emphasis on pulling part of the casing out of the hole to replace the damaged section.

Parted casing is one of the most severe casing failures. The most common causes are excessive casing loads, overstressing while working stuck casing, shock loading caused by bumping cement plugs too hard, and defects that are not detected during inspection. The best method of repair is to pull the damaged section, repair the casing, and run a casing bowl to reconnect it. This type of repair is normally restricted to failures that occur inside another string of casing.

Otherwise, the casing is repaired in place using special alignment tools, squeezing, or packing off as previously described. An alternative is to run another string of smaller casing or a tieback liner or stub liner inside the parted casing if hole sizes and well conditions permit this type of repair.

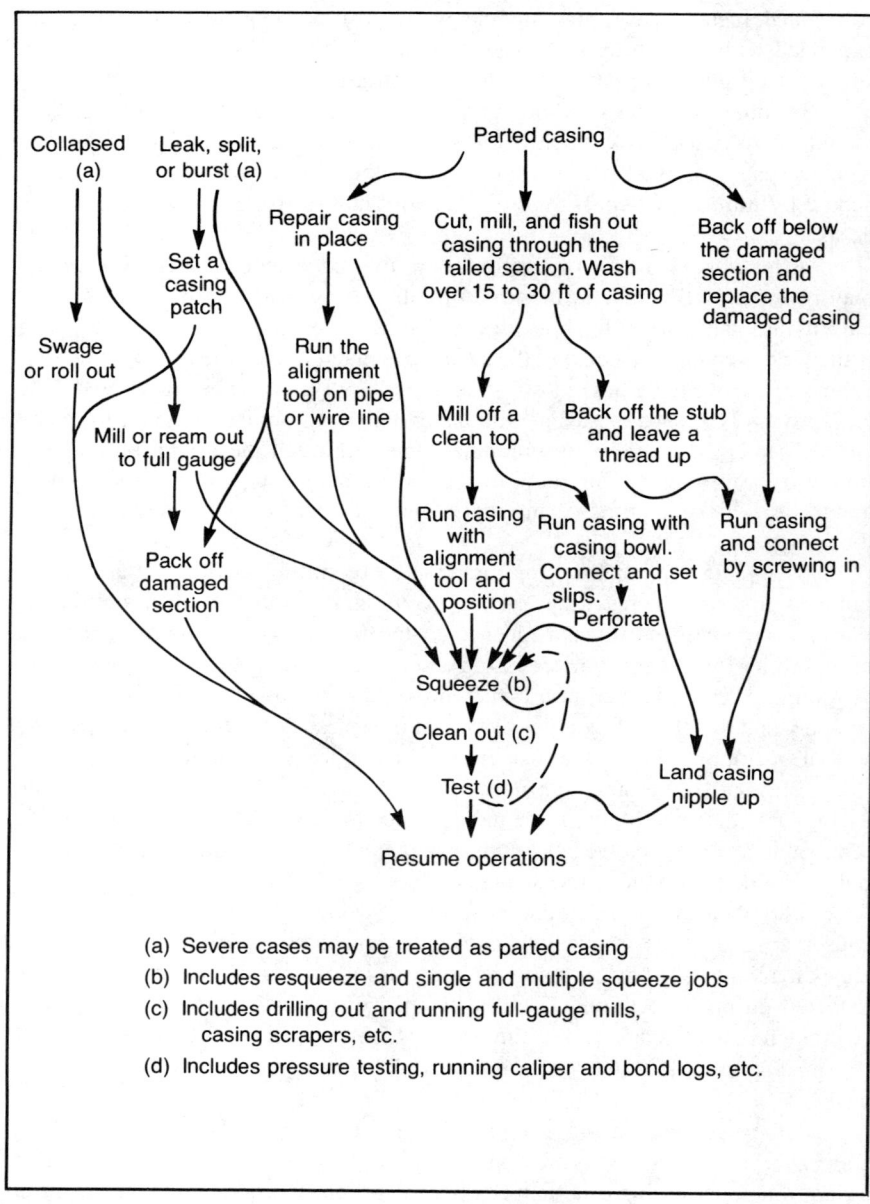

FIG. 10–12 Summarized casing repair flow chart

OPERATIONS

Completion operations include all activities necessary to complete the well, install the completion equipment, and place the well on production. Completion rigs normally are used in developing fields and other areas with general drilling and completion activities and when smaller, more economical rigs are available. Isolated wildcat wells and most marine wells are completed with the drilling rig, usually because of the additional time required and associated expenses of moving in the completion rig. The production equipment for permanent completions is often installed by the drilling rig since additional completion activities do not require a rig. Also, deep wells are often completed with the drilling rig because large, heavy-duty completion equipment is not available.

A completion program may be included in the drilling prospect. If so, it is finalized or a program is prepared after the hole is drilled and logged. It contains all of the information and instructions on completing the well, including zones to be perforated, type and number of perforations, testing procedures, stimulation program, and equipment specifications.

Most wells, especially those drilled into high-pressure formations, are produced through tubing run inside the production casing. All high-pressure fluids are confined inside tubing. It is easier, safer, and more economical to confine high pressures with smaller tubulars. Less metal is needed per linear foot of tubular. The production casing serves as a backup container in case the production tubing fails, e.g., if it develops a leak. Weighted completion fluids in the annular space provide an additional safety factor.

High-volume fluid flow can abrade and erode the confining tubular, thus reducing its strength characteristics. For the same flowing volume there is less erosion in the larger-diameter production casing. However, if the tubing is eroded to the point that failure is imminent, it can be replaced. If the casing is damaged, it is more difficult to repair or replace.

Workover and recompletion operations are also easier and safer to perform with tubing in the well. A *workover* is any type of remedial work on the well after it is completed. A *recompletion* is the operation to complete another zone later in the life of the well. These operations, especially workovers, can occur frequently, and they are a major reason that tubing is installed in completed wells.

Tubing can also help regulate the flow of fluids to the surface. Gas contained in the crude oil provides the major part of the energy necessary to produce the oil. This energy is utilized more efficiently in tubing with a smaller cross-sectional area than in the larger production casing. For the same flow rate of oil plus gas, the well continues to flow much longer through smaller tubing than through casing. Also, there is a greater tendency for the oil and gas to separate downhole in the production casing. Then gas can bypass the oil and flow to the surface leaving the dead (gasless) oil in the casing, which can ultimately kill the well due to the column of dead fluid. Using tubing helps prevent this bypassing. Tubing normally is large enough to allow the well to be produced at its maximum rate without restricting productivity. Therefore, except for special cases all wells are produced through tubing.

Preparing the Casing for Completion

The first step in completion operations is to prepare the casing. If the drilling rig is used, operations resume immediately. If a completion rig is to be used, the drilling rig is rigged down and moved off the location, generally to another well. In the rigging down process surface controls normally are installed temporarily on the wellhead. A completion rig, often called a workover rig, is then moved onto the well and rigged up including blowout preventers.

A cleanout trip is made to ensure that the casing is clean and open to the plugged-back total depth (PBTD). This depth normally is the top of the float collar, but it may be a bridge plug or cement top since it is always a good practice to leave 50–100 ft of open casing below the lowest prospective zone for junk.

The cleanout trip is made by running a bit and scraper on a work string. The work string, usually tubing, is picked up as the bit and scraper are run in the hole. If a considerable amount of cement must be drilled, the bit is run below 3–6 drill collars. If cement is drilled, the mud may be discarded or treated for cement contamination. If another type of working completion fluid is used, it may be displaced into the hole at this time.

A gamma-ray collar-locator cement log is run after the cleanout trip. This is the base log for correlating and perforating to determine the condition of the cement behind the casing. The log shows hole sections that require squeezing, such as channeled cement. Any cement squeezing required is done at this time. Then the hole is cleaned out, and the casing is pressure tested.

This completes the basic production casing preparation so that the well can be completed. A variety of completion techniques and procedures are available. The methods used depend on the specific requirements of the well being completed. These generally are less complicated for shallow, single-zone completions in developing fields. Complexity increases with depth, number of zones, and wildcat wells.

Single Completions

In a single completion all of the production from the formation flows to the surface through a single conduit, usually tubing. This category also includes commingled, multizone completions. Single-zone completions are the most common type of completion.

Open-ended Tubing The most common type of single-zone completion uses open-ended tubing that is hung or landed with the bottom of the tubing at or near the perforations (Fig. 10–13). An open-ended or packerless completion is uncomplicated and easy to install. It also is highly efficient for either oil or gas production and is easily adapted to various types of artificial lift.

In a gas well the tubing is commonly called a *siphon string* since it efficiently removes the varying amounts of salt water often associated with gas production. This type of completion is generally restricted to low-pressure wells because the entire section of production casing is exposed to formation pressures.

The open-ended tubing completion is often used in a developing field that has one, consistent, blanket-type formation with a wide areal extent. Completion procedures have been well established in such a field by prior well completions. The general procedure is to perforate, run completion tubing, and nipple up the

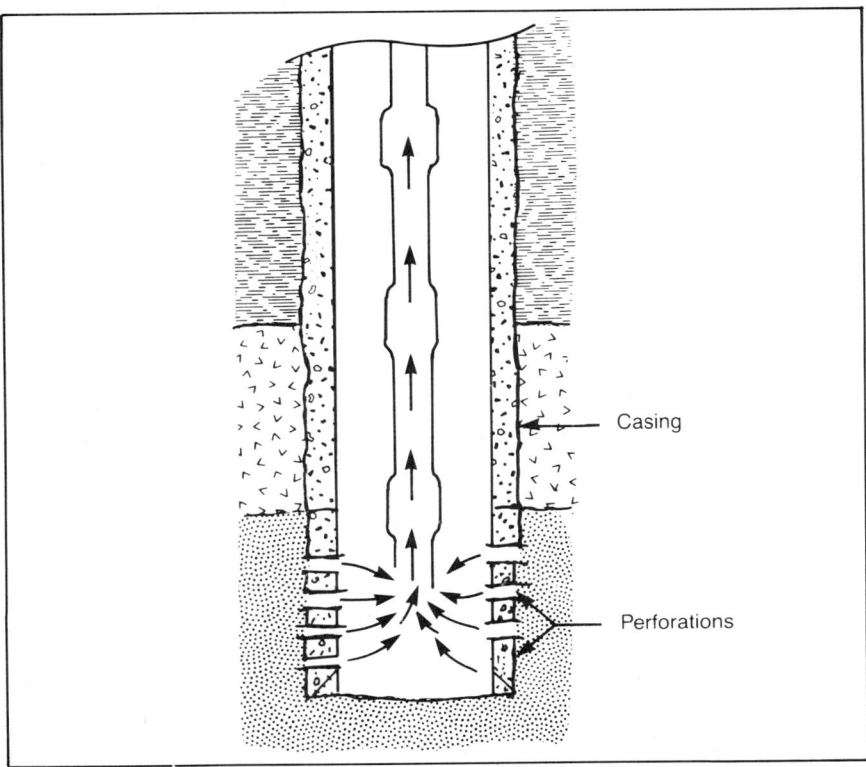

FIG. 10–13 Single completion using open-ended tubing

surface controls. The well is swabbed and cleaned up until it produces clean oil and gas. Then the well is placed on production and is tested. This is a natural completion that does not require stimulation.

A casing gun is used for perforating. A typical gun has a 12-ft section containing 4 jet shots/ft oriented to perforate at 90° phasing. The casing gun is selected in this case because it is the largest perforating tool available and gives the deepest perforation penetration into the formation. The 90° phasing is used to ensure penetration into most of the formation.

The operator normally selects the deepest penetrating gun to minimize the risk of reduced productivity resulting from the lack of penetration-perforation depth into the formation. In many cases the service company that runs the cased-hole logs perforates the well while the logging truck and equipment are still on location.

The perforating gun is made up with a collar locator and is run on a perforating line, in this case with the logging line. The gun is run through a lubricator. In this type of completion, the mud in the hole—usually salt water—exerts a sufficient hydrostatic head to contain the formation pressures. The lubricator is used as a safety factor.

The cased-hole log is correlated with the open-hole logs. Casing collars are located as reference points, and the perforating gun is positioned and fired to perforate the casing. A short strip of perforating log may be run with the collar locator to verify that the casing has been perforated correctly. The perforating gun is pulled from the hole while monitoring the casing fluid level to ensure that the well remains dead.

The next step is to run completion tubing. This frequently is the tubing used in the work string, or it may be a complete string of reconditioned or new tubing. If the work string is not used, it is run into the hole and then is pulled out and laid down. On small workover rigs tubing may not be stood back in the mast but is laid down after each trip. In this case the work string is moved to run the new string of tubing.

The next step is to connect the tubing to the wellhead. This procedure is called landing the tubing. Two common methods are used. In one case a tubing hanger seal is connected to the tubing, which is then lowered to seat the seal in a landing flange. As a safety factor, a blanking plug is seated in the tubing to shut off the tubing. The tubing landing joint is unscrewed from the hanger seal and is laid down. The blowout preventers are disconnected (nippled down) and removed. A Christmas tree—or wellhead equipped with shutoff valves, flow controls and chokes—is installed. The wellhead is pressure tested and connected to the production or test facilities.

In the second method the tubing is screwed directly into the bottom of the Christmas tree. To accomplish this, the blowout preventers are disconnected and picked up a short distance. The tubing is then caught with a tubing spider—a portable tubing bowl and slips—below the preventers. The preventers are picked up over the tubing and are set aside. A Christmas tree is connected to the top of the tubing. The tubing is picked up a short distance to remove the tubing spider and is lowered until the Christmas tree is sitting on the wellhead. A pressure seal, called a wraparound, may be installed in the tubing hanger. The Christmas tree is then connected and pressure tested.

The bottom of the tubing in the well may have a swedge to facilitate running wire-line tools back into the tubing if necessary. A mule shoe or saw-toothed collar may be run on the bottom of the tubing to clean out sand if necessary. A bar collar may be installed to catch swabbing tools and prevent them from falling out of the tubing in case a wire-line failure occurs during swabbing. A pump seating nipple may also be run near the bottom of the tubing to receive a rod pump at a future date.

The well is now ready to begin producing. If weighted mud is used, it may be circulated out of the hole by displacing it with fresh water pumped by the rig pumps. The lighter fluid reduces the hydrostatic head, and the reduction may be sufficient to allow the formation fluids to begin flowing into the well. The mud normally is displaced by reverse circulating—pumping clean water down the casing and allowing mud to flow out of the tubing. This procedure gives a cleaner interface between the water and mud, and it facilitates cleaning the mud out of the hole. The mud flows up through the tubing and into a waste pit or storage tank. This is the normal flowing direction of the producing well.

As formation fluids begin to enter the wellbore, they lighten the upward moving fluid column and decrease the amount of time required to start the well flowing. Pumping is stopped as soon as the well begins flowing naturally. The well flows at a gradually increasing rate as the fluid behind the tubing flows downward into the tubing and out of the hole. The fluid behind the tubing normally is displaced with gas from the production formation. Gas and oil singularly or in combination flow up through the tubing. As the well continues to clean up, the weight of the fluid column in the tubing continues to reduce and the well flow rate increases correspondingly.

The flow rate is controlled with a choke on one wing of the Christmas tree. Either positive or adjustable chokes are used. A positive choke has a fixed hole size through the choke. The diameter of the hole, commonly called the choke size, ranges from a few sixteenths to over 1 in. A larger choke must be placed in the choke to increase the flow rate. Conversely, a smaller choke reduces the flow rate.

Adjustable chokes are often used, especially when first bringing the well onto production. The adjustable choke has a cone and tapered seat constructed of very hard metal similar to that used in positive chokes. A screw is used to move the cone farther from or closer to the seat. This increases or decreases the opening through the choke and controls the flow rate.

The flow rate continues to increase as the well cleans up and is controlled with chokes. Chokes are used instead of regular valves because the high-velocity fluid can erode the metal in standard valves. The chokes are made of harder metal designed to resist fluid erosion. In some cases the well is allowed to flow at a maximum rate. In other cases the flow rate must be restricted to prevent formation damage. For example, some unconsolidated sands in the formation tend to move toward the wellbore, creating a plugging action in the perforations or causing sand accumulation in the wellbore at high flow rates. This is prevented by restricting the flow rate with a choke.

This overall procedure is known as bringing the well in. Pressure gauges installed on the tubing and casing are monitored during the cleanup process. Both pressures increase as water and mud are flowed out of the well. The casing pressure normally exceeds the tubing pressure since the tubing is partially filled with fluid. As the well cleans up, the difference between the two pressures decreases. There is little difference between the tubing and casing pressures in a clean gas well and a higher difference in an oil well as a result of the column of oil in the tubing. Monitoring these two pressures helps the operator determine the status of the cleanup process and when the well is clean.

The completion or workover rig is released, rigged down, and moved after the well is cleaned up. The well is then placed on production, and production is tested to establish an initial potential (IP). Many wells, especially those with low reservoir pressures, do not flow when the mud in the hole is displaced with water. Most of these low-pressure wells use water in the completion operations. Low-pressure wells are brought in or started to flow by swabbing to remove water from the wellbore.

Many smaller completion and workover rigs have double-drum drawworks. The main drum is used for the drilling line. The second or auxiliary drum contains the swab line, sometimes called the sand line. This is a multistrand steel wire line 7–15,000 ft long and normally $7/16$–$5/8$ in. in diameter. The line passes over a sand-line sheave in the crown. The free end of the sand line is connected to swab tools by means of a rope socket.

The swab tools consist of sinker bars, a check valve, and swab cups. The weight bars help pull the swab line into the hole through a column of fluid. Fluid passes through the check valve as the swab tools are lowered (Fig. 10–14). Swab cups are designed to collapse partially while running in the hole to bypass fluid. When the swab is pulled, the cups expand and provide a seal to the inner wall of tubing or casing.

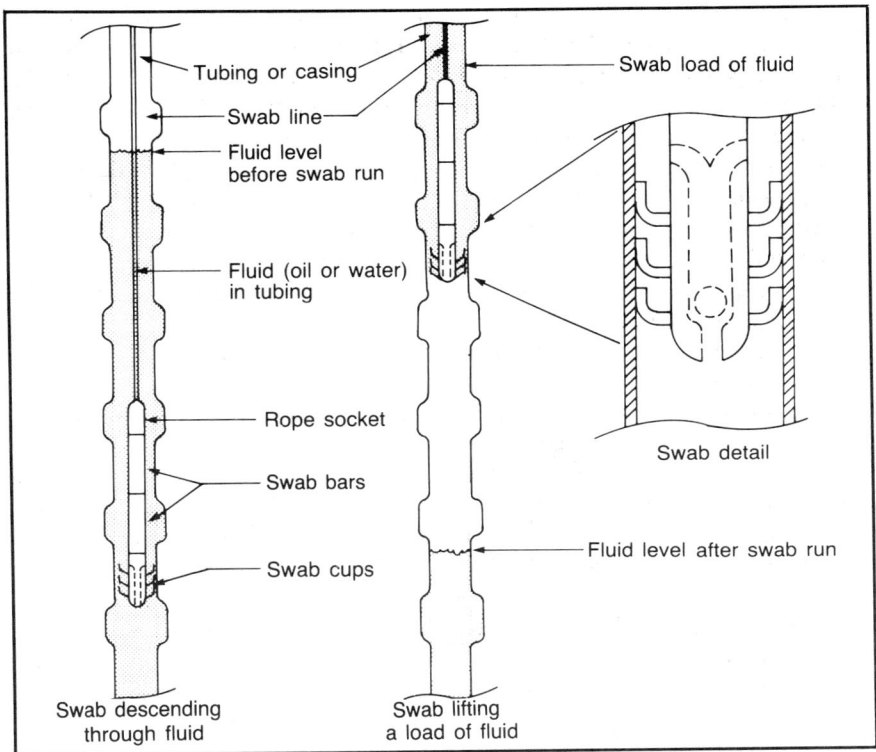

FIG. 10–14 Swabbing

In operation the swab is run through a pressure lubricator connected to the top of the Christmas tree. The tools are run through the tubing or casing and into the fluid column for a depth of 500–1,500 ft. As the swab tools are picked up, the check valve closes and the swab cups expand to seal against the inner casing or tubing wall. Fluid above the swab cups is lifted as the swab tool is pulled from the hole, thus removing or swabbing the fluid out of the well.

The top of the fluid in the wellbore and usually in the tubing is the fluid level. The fluid level drops as a fluid is removed by swabbing. Normally the swab must be run deeper on each successive swab run. The well begins to flow when the fluid level lowers to the point where the pressure exerted by the reducing height of the fluid column is about equal to or slightly less than the formation pressure. The principle here is the same as the U-tube effect.

In a normal swabbing operation the fluid level reduces until formation fluids start to flow into the wellbore. At that point the fluid level may remain static for a period, or it may start to move upward with continued swabbing. The fluid level moves upward at an increasing rate as formation fluids flow into the wellbore. At this point the well is almost ready to begin flowing, and the swabbing operation becomes more critical. If the swab is run too deeply in the fluid or the fluid is flowing upward at a high rate, the upward movement of the fluid can literally blow the swab out of the hole. The pressure lubricator helps control the flow of fluid. However, if the swab is relatively deep in the hole, the fast moving fluids can lift the swab, causing the swab line to become entangled above the swab. This can cause a fishing job.

Swabbing operations are slowed down when the fluid starts to move upward. The swab is run into the fluid for a shorter distance, and swab runs are made at longer time intervals. Experience and judgment are required to swab a well satisfactorily without causing a fishing job.

This procedure is known as swabbing the well in. The swab is left in the lubricator, which can be isolated from the wellhead by closing a valve when the well begins to flow. The well is flowed and cleaned up in the same manner as a well brought in by circulating. If a well flows naturally, it is also called a *natural completion*.

Most wells are stimulated to improve productivity. Such a completion is called a stimulated completion. Natural completions often have formation damage that must be removed by stimulation to obtain natural production. Productivity from a natural completion can often be doubled and sometimes tripled by stimulation, especially in low-permeability formations. Higher product prices often justify completing less prospective formations that must be stimulated to make a commercial completion.

To illustrate stimulation operations, consider the natural completion previously described. Assume that the standard stimulation for the area is a small frac job with about 20,000 lb of 20–40% sand carried into the formation at about 2½ lb/gal at an injection pressure of 3,500 psi and an injection rate of 10 bbl/min. Most new production casing can withstand these pressures safely. The well can be stimulated by closing preventer rams on tubing, but the best procedure would be to run the tubing and nipple up the tree as described.

The service company that supplies the materials and pumping equipment designs a fracturing program. Computers are used to help prepare the program. Fracture tanks are brought to the wellsite and are filled with water prior to fracturing. In this example about 400–500 bbl of water are needed. The special fracture fluid may be premixed in one of the tanks, or additives may be mixed with the fluid as it is injected during the fracture operation. On very large fracturing jobs sand also is brought to the wellsite prior to fracturing and is stored in a large sand-hopper tank.

High pressure pump tracks, sand blenders, and the sand are brought to the wellsite. All of the equipment is connected together with high-pressure, flexible steel lines for discharge lines and rubber suction lines from the tanks to the pump suctions. The equipment is pressure tested, and the well is circulated to ensure that the tubing and casing are full of fluid.

Fracture fluid is pumped into the formation to start a fracture, obtain the breakdown pressure and establish an injection rate and pressure. Sand is then added to the fracture fluid and is injected into the well through the tubing. The sand concentration is started low, about ½ lb/gal and is increased in half-pound increments to the design injection rate of about 2½ lb/gal. After all of the sand is mixed and pumped, the sand-laden fluid is displaced to the perforations with water or a treating fluid. The fluid may be overdisplaced by 5–10 bbl. The pumps are shut down, and the instantaneous shut in pressure (ISIP) is recorded. Injection pressures and volumes are also recorded during the fracturing treatment.

The well is shut in for 30 min to several hours depending on the type of fracture fluid used. The fracture fluid is thickened with chemicals. Therefore, the shutdown period must be long enough to allow the gel to break and to thin the fluid so the sand will not be carried back out of the fracture when the well is flowed back.

The well is flowed back at a controlled rate designed to recover as much fluid as fast as possible and at the same time retain the sand in the formation. Normally formation fluid begins to flow into the wellbore after the well has flowed back about half of the treating fluid. Formation fluids tend to lighten the fluid column, which in turn gradually increases the rate of flow back. The well is cleaned up and placed on production as previously described.

If the well does not flow back naturally, it is swabbed to recover the fracture fluid. Swabbing operations must be conducted in a careful manner since a small amount of sand may be carried out of the formation with the returning fluid. This sand can cause the swabbing tools to become stuck, thus causing a fishing job.

The process of recovering the treating fluid after stimulating the well is known as fluid cleanup. This procedure can be slow and time consuming in some cases such as low-pressure well and large stimulation treatments. This time delay is costly because the workover rig or a swabbing unit must be used. There is also the risk of a fishing job caused by such actions as sticking the swab with sand and equipment failures, with attendant higher costs.

Liquefied gases such as nitrogen or carbon dioxide can be injected with the treating fluid and help clean up the well much faster and safer. These gases also give other benefits such as acting as a fluid-loss agent and maintaining an acidic environment. Using these gases increases the cost of the stimulation, but the overall stimulation and cleanup may be economized.

Tubing-Packer Completion The tubing-packer combination is the second most common single-zone completion (Fig. 10–15). The tubing is set or landed on a retrievable or permanent packer. This arrangement has all of the advantages of the open-ended tubing completion, except it is slightly more complex with the packer. It is more commonly used to produce high-pressure formations.

Pressures are isolated from the casing, and the tubing-casing annulus can be filled with a weighted completion fluid as an additional safety measure. In case of a tubing failure, the packer holds the tubing and prevents it from falling to the bottom of the hole, causing additional damage and a severe fishing job. The completion is adaptative to various types of artificial lift or a permanent completion. Generally, it is easier to bring the well in—start it producing—with the tubing-packer completion than with open-ended tubing.

The general procedures for the tubing-packer completion are the same as that described for open-ended tubing, except a retrievable packer is run on the bottom of the tubing. A sliding sleeve in the open position is run above the packer on the completion tubing. After the Christmas tree is nippled up, the fluid in the well is displaced by circulating through this sliding sleeve. Special, weighted completion fluid can be left in the annular space as a safety factor. The sliding sleeve is closed with wire-line tools, and the operations generally proceed from this point in a manner similar to that for completing the low-pressure, single flowing well.

The single packer-type completion is also used to isolate the casing when the well is stimulated with pressures that exceed the safe burst strength of the casing. It is not uncommon to have production casing in a well that has a burst strength of 5,000–6,000 psi when the stimulation injection pressures are considerably higher. Producing well pressures may not exceed the safe burst of the casing so the casing must only be protected for a short period during the stimulation. The higher stimulation pressures normally are within the safe limit of most common tubing strings.

If the well is stimulated by injecting down open-ended tubing, the casing is exposed to these higher pressures and can fail in burst. The casing can be isolated with the tubing-packer combination so the higher injection pressures are contained inside the tubing and below the packer set near the top of the formation to be treated.

The normal hook-wall production packer usually cannot be used in this case. It is held down by the weight of the tubing. High treating pressures below

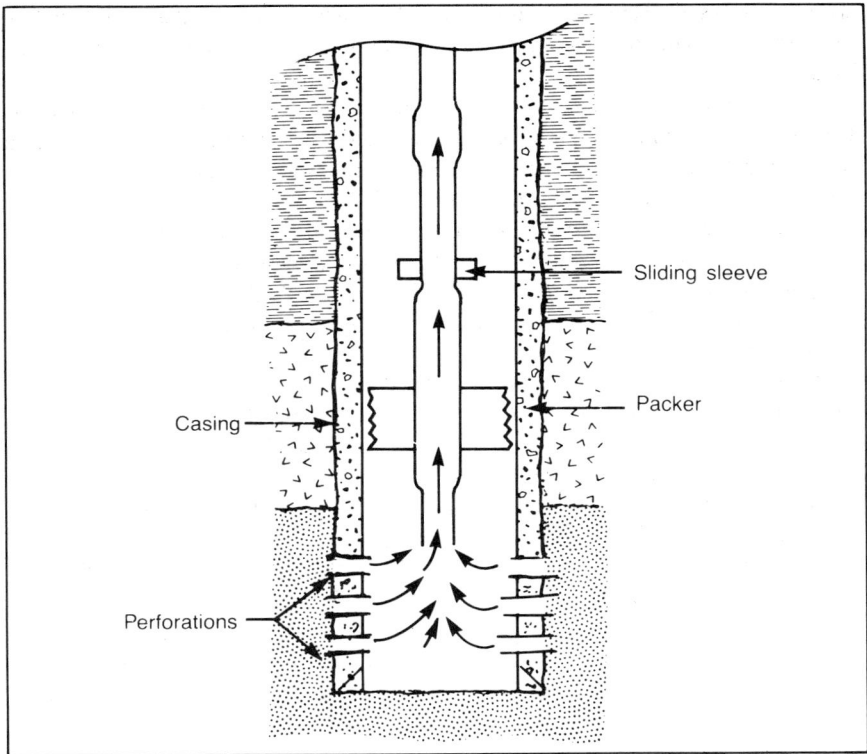

FIG. 10–15 Single completion using tubing on a packer

the packer create an upward plunger force on the bottom of the packer and can push it uphole and damage the tubing. This can be prevented by using a packer fitted with a hydraulic hold down or reverse-type slips. Most operators prefer the tool with hydraulic hold downs.

Another precaution must be observed because both ends of the tubing are fixed—one end at the packer and the other in the wellhead. Cold treating fluids can cause tubing contraction. In extreme cases the tubing can contract to the point that it fails in tension with severe consequences. The amount of tubing contraction is obtained from standard tables for various conditions. This must be checked to prevent a possible failure.

Several methods of correcting or alleviating the problem are available. Additional tubing weight can be set on the packer when it is seated. Heated fluids also can be used during the stimulation. Permanent packers with long sealing elements can allow for additional tubing movement during contraction in extreme cases.

Except for these precautions, the well is stimulated in the normal manner. This method of completing the well is often used when high-pressure stimulation is required because it is more economical than using higher-strength, heavier casing, which has a higher burst rating.

A treating pressure problem can occur with the wellhead. In the previous example the well could be fitted with a Christmas tree rated at a 5,000-psi working pressure and 10,000-psi test pressure. Then the well cannot be stimulated directly through the tree because the injection (working) pressure exceeds the working pressure rating of the tree. A special tool called a *tree saver* is available for these cases. The tool consists of a high-strength steel tube that is run through the Christmas tree and seals in the tubing. This isolates the Christmas tree from the high treating pressure. Then fluids are injected down through the tree saver. After the formation has been stimulated and flowed back, the tree saver is removed and the tree is exposed to normal pressures that are within its pressure rating.

The tubing-packer completion can also be used in some cases to stimulate the well at pressures that exceed the burst rating of the tubing. Assume the well previously described is fitted with 2⅞-in., N-80, 6.50-lb/ft, EUE tubing with a burst rating of about 10,500 psi.[2] Projected stimulation pressure is 12,500 psi. This pressure exceeds the tubing burst rating and would cause a tubing failure if applied.

Pipe burst ratings are based on the pressure differential across the pipe wall, in this case at about 10,500 psi. The sealed tubing-casing annulus is pressured to 3,000 psi. With 12,500 psi in the tubing, the pressure differential across the tubing wall is 9,500 psi, a 1,000-psi safety factor. The stimulation is conducted in the normal manner.

Permanent Completion The permanent completion is an efficient, safe method of completing a well when the procedure is applicable. This type of completion is used most frequently when the producing characteristics and stimulation requirements of the formation are known, in high-pressure formations and single completions, and when the formation is the lowest zone in the wellbore. In this completion all well equipment is installed before perforating the formation.

The equipment is frequently installed by the drilling rig. The casing is first prepared for the completion. A permanent packer is run on a wire line and is

---

[2]Cementing Tables, Halliburton Services (Duncan, March 1972), section no. 200, pp. 4–5.

seated, usually 30–60 ft above the formation. The packer is set slightly higher above the formation for several reasons. A decentralized gun is often used for perforating, and there must be room below the bottom of the packer so the gun can be positioned. The packer also is well above any possible casing split that may occur during perforating.

The next step is to run the tubing. The tubing is often pressure tested after it is run. Some operators also elect to pressure test the tubing while it is being run. A tubing tester is available in lengths of 30–90 ft. This tool has seals on each end to close off inside the tubing. It is raised and lowered by a line run over the crown and operated by a special small winch.

After 1–3 joints of tubing have been connected and run, the tool is run inside the tubing and pressure is applied between the seals on the end of the tubing tool. The pressure is held for about 15 sec. Any pressure decline indicates a tubing leak. If the tubing leaks, it is repaired and retested. This procedure is known as testing below the rotary.

An alternative is to test the tubing above the rotary. After the test tool has been run inside the tubing, the tubing is picked up until the section of tubing to be tested is above the rotary. Slips are set, and the tool is pressured up. In this case leaks can be detected by a decline in pressure and a visual inspection of the tubing. The tubing is then lowered, and the tool is pulled from the tubing. The procedure is repeated until all of the tubing has been run.

The tubing is spaced out. The seals are seated in the permanent packer and are latched in with the correct weight on the packer. Unlike the retrievable packer, the permanent packer is preseated and there is a fixed distance between the packer and the surface.

Tubing joints are about 31 ft long, and it is unlikely that the distance between the packer and tree will be exact multiple of 31 ft. Therefore, the tubing must be spaced out. That is, the tubing string length is adjusted by using different lengths of tubing nipples—special shorter lengths of tubing sometimes called pup joints—to give the precise length of tubing between the packer and the tree. The tree is then connected to the tubing, seated on the wellhead, and nippled up.

Sometimes completion fluid is displaced into the hole immediately before landing the tubing or by circulating through a sliding sleeve after the tree is installed. The sliding sleeve is closed, and the tubing and casing are pressure tested individually. If there are any leaks, the assembly must be pulled from the hole to repair them.

The rig is rigged down and moved out. The formation is perforated in the conventional manner with a decentralized gun. A light-duty portable mast supplied by the perforating company is used to support the shooting line. The well is flowed and cleaned up in the normal manner. It also can be swabbed with a portable swabbing unit if necessary. The formation is stimulated when needed, taking precautions for tubing contraction. The well is cleaned up and placed on production. Then the production is tested.

The permanent completion is very applicable to high-pressure formations because it can be performed safely and reduces the risk of formation damage. It is an important procedure in multiple completions.

Marginal Completions In a marginal completion the results of the completion are unknown. The value of the resulting production may not permit the operator to make a profit on investment. The cost of the completion is the basic expenditure considered in evaluating the economics to justify the completion. The original intent is to drill and complete an economic well. At that time the operator is faced with the total expenditure for drilling and completing the well. However, all of these costs have been expended and for all intents and purposes cannot be recovered. Therefore, the decision is whether projected income will economically justify the incremental costs of the completion.

To illustrate this, consider an isolated wildcat drilled to test a prominent formation. Testing the formation requires setting casing, and the formation is ultimately found to be nonproductive. The well also intercepts a thin sand that could be a low-volume producer. If this is an oil sand and the oil can be economically produced and transported, then the zone should be completed. However, if it is a gas sand, it probably should not be completed because the productivity would not justify the cost of a long transmission line to carry the gas to market. This decision would be further reinforced if the productivity of the sand would not economically justify drilling additional wells. However, if the well is located near a transmission line, then the zone probably should be completed if the net income will generate a reasonable rate of return on the cost of the completion.

Both deliverability (sustained production rate) and reserves must be evaluated in the most economical manner possible to determine if marginal formations should be completed. As an example a questionable zone is drillstem tested in the open hole and is produced at a rate of 1,000 Mcfd for a short period. The hole is cased, and the zone is completed and tested at a sustained rate of 50 Mcfd, which is uneconomical because of drilling and transmission-line costs. This type of behavior is characteristic of high-pressure, low-permeability reservoirs with limited recoverable reserves.

Other completion techniques are also available to improve marginal formations, such as underbalanced perforating. In this method a retrievable packer is run and set above the top of the formation before perforating. Fluid is swabbed out of the tubing. Normally the fluid level is lowered so the hydrostatic pressure caused by the fluid column is approximately one-half of the formation pressure. Some very marginal formations are perforated dry—the fluid level is lowered until it is near the top of the formation. This gives a positive pressure differential from the formation into the wellbore when the casing is perforated. Adequate wellhead controls are installed including a lubricator, and the formation is perforated with a through-tubing gun.

There is always a danger that the perforating gun will be blown uphole because a rushing flow of fluid occurs. This flow can cause the perforating line to become tangled, bunched up, and stuck, causing a fishing job. The risk is reduced by perforating the formation with the wellhead control closed so the in-rushing fluid is stopped when the void space in the tubing is compressed. The procedure must be performed with caution.

If the formation is productive, this procedure should allow sufficient production to obtain a preliminary estimate of the unstimulated formation productivity. Gas production is measured with testing facilities. If the formation produces oil, it is fluid productive. However, the pressure resulting from a full hydrostatic head of oil may restrict the flow rate.

The formation can be swab tested. Then the tubing is swabbed at a constant rate running the swab to the same depth. The depth to which the swab is run is determined by how often it is run and by the fluid entry into the wellbore. Swab testing normally is used to test oil-productive zones. However, some gas-productive zones have a high water saturation and in many cases produce salt water with the gas. The swab test gives a good indication of the amount of gas and water produced.

A gas productive zone is a very questionable completion prospect if it produces enough salt water to require swab testing to remove the water. However, if the well does have sufficient productivity with mechanical removal of the water, there are methods of completing the well and lifting the water mechanically to maximize gas production.

Insufficient formation penetration is always suspect when a prospective zone is perforated and does not flow or only flows a limited amount. Another question concerns the number of perforations that will sufficiently allow the formation to produce at its maximum rate. If the perforations extend into the undamaged reservoir the formation probably can be produced. A high-pressure drop across the perforations, called *perforation friction*, restricts production from the reservoir. However, field practice shows that reperforating seldom increases well productivity. Ten to fifteen average perforations are more than adequate to allow the normal formation to produce without restriction.

In the normal case average perforations extend past the damaged zone. However, there are enough cases where the perforations end in the damaged zone so there is a reasonable doubt when the well is perforated and does not flow or flows at a low rate. Reperforating normally is not a good solution if the well has been perforated adequately the first time. The new perforations usually do not extend past the old ones.

The best solution in this case is to ball out the perforations with a non-damaging fluid such as water containing 3% potassium chloride or a weak cleanout acid to ensure that the perforations are open. An average treatment is about 25–50 gal of fluid/perforation based on about 20% excess perforations. The fluids are injected into the well with perf balls spaced throughout the fluid. The balls seat over perforations taking fluid, allowing increased pressure to open other perforations. Eventually all perforations are either opened or proved to open.

Carbonates are less subject to formation damage than sands, but a similar treatment using 10–15% hydrochloric acid and about 100 gal/ft of formation may be used. Some operators prefer to treat sands with a small sand-fracturing treatment using about 1,000 lb of sand/ft of formation limited to about 25,000 lb of sand. All formation damage should be removed by these treatments; they should stimulate the formation so it will produce at about twice the rate of an undamaged formation.

The production rate obtained from a well that is properly treated and cleaned up should provide a good indication of the reservoir productivity. A well seldom produces at over twice the rate obtained from the shallow stimulations described, even with a deep, large-volume stimulation. A stabilized flow rate and pressure-buildup test should be taken if the well productivity and flow conditions permit. This provides information to determine if additional stimulation is justified. A deep stimulation is also probably not justified if the well does not flow at a sufficient rate to run the flow and pressure-buildup test or if an extensive period is required to clean up the well.

In conjunction with this, the use of flow and pressure-buildup tests to evaluate good completions should not be overlooked. For example, a submarginal zone with a natural production of 5 bo/d or 30 Mcfd may be given a deep, large-volume stimulation that would increase production threefold to 15 bo/d or 90 Mcfd. By comparison, a good well with normal productivity of 50 bo/d or 300 Mcfd stimulated at the same cost with the same results (threefold increase) would produce 150 bo/d or 900 Mcfd. The net productivity increase from the good well is 100 bo/d or ten times the 10 bo/d increase from the submarginal well, obtainable at about the same cost.

Multizone Commingled Completions A multizone, commingled completion produces from two or more distinct, isolated horizons that produce from the wellbore in one common flow stream. There is an increasing tendency to use this type of completion because improved technology and equipment allow more zones to be produced without using the more complex multicompletion technique and the mechanics of reservoir fluid flow are better understood. Also, the more productive zones help clean up the less productive zones for an overall improved efficiency, and in many cases this method is the easiest, most economical way to complete the well.

Multizone commingled completions are similar to the various single-zone completions, except multiple-zones have been opened. Open-ended tubing or tubing-packer completions are selected based on the reasons used in single-zone completions. Various completion techniques are used, the method selected depends on the well condition and the operator's preference.

One completion procedure is the packer-plug method, which is summarized as follows:

- Run a treating packer and retrievable bridge plug (RBP) on a work string or on the tubing.
- Seat the RBP below the lowest zone and release it.
- Pick up the packer above the RBP, seat the packer, and pressure test the RBP. This step is necessary to ensure that the plug is properly seated and will hold pressure.
- Unseat the packer, pull it up above the zone, and reseat it.
- Perforate the zone with a through-tubing gun. This operation can be done in the conventional manner, or it can be performed after circulating light fluid into the tubing through the packer circulating valve. Also, part of the fluid can be swabbed out of the tubing. Then, the perforating is done underbalanced.
- Flow and test the zone, swabbing it if necessary.
- Depending on the results of the flow test in the previous step, stimulate the well, flowing and testing as necessary.
- Open the tubing circulating valve and circulate the well dead after the test is completed. Close the tubing circulating valve and unseat the packer.
- Lower the packer. Then circulate and clean off any debris from the top of the RBP.
- Latch onto the RBP and release it.
- Pick up the packer and the RBP to the base of the next zone to be treated and tested.
- Repeat the procedure. Then move to the next zone and repeat again until all of the zones have been stimulated and tested.
- Pull the packer and RBP out of the hole after all of the zones have been

stimulated and tested. Run a completion assembly, nipple up the Christmas tree, and bring the well in using the same procedures as for a single-zone completion.

One alternative to this procedure is to run the RBP back to the bottom of the hold and release it, commonly called *shucking the plug*. Then the packer can be pulled above the perforation and reseated and the well can be completed.

One objection to this procedure is that it requires considerable manipulation of the packer and RBP. As a result it is seldom used for more than 2–3 zones. Sometimes the method is used, but 2–3 zones are covered in one step, stimulating the zones by limited entry techniques. A number of zones can be covered in 2–3 packer-plug settings using this method. Another objection to the procedure is that the stimulated zones may be damaged while killing the well to move the packer and plug. One method of partially overcoming this problem is to use a small, shallow ball-out stimulation of all zones after the completion equipment is installed.

One modification of the packer-plug completion procedure is to perforate all zones with a casing gun. Then test each zone and stimulate if necessary using the packer-plug method.

Another method of completing is to perforate all zones with a casing gun. Then stimulate all of the zones in one stimulation treatment using limited entry or ball-out methods. This latter method can be incorporated into a permanent completion by running the completion equipment first and perforating with a through-tubing gun followed by stimulation.

The ball-out stimulation procedure, modified from the standard ball-out, is to perforate each zone with the same number of feet and perforations per foot. The balls are then injected in groups behind larger stages of fluid designed to ball out either one or two zones simultaneously rather than individual perforations. This method has been used to stimulate as many as 20 zones successfully over a 2,000-ft gross interval with better than 90% zone stimulation efficiency.

Another completion procedure is available for use in specialized cases. A packer is run in the hole and seated above the top zone to be perforated. The zone is then perforated, tested, and stimulated as necessary. When this is completed, the hole is circulated through the packer circulating valve. The packer is unseated, lowered to a point above the next zone, and reseated. The perforation-stimulation procedure is repeated. The entire procedure is then repeated on each succeeding lower zone until all zones have been stimulated. The packer is picked up above the top zone and is reseated. The well is completed in the normal manner.

This is a fast, uncomplicated method of stimulating each zone individually. The main disadvantage is that the packer must be seated below zones that have recently been stimulated. If the zones are stimulated by sand fracturing or tend to flow formation debris, there is a high risk that this material will settle on top of the packer, causing it to become stuck and resulting in a fishing job.

Crossflow occurs when fluid from a high-pressure formation flows into a lower-pressure formation, called a thief zone. Surveys that detect fluid movement in the wellbore can locate the zone taking fluid. The zone must be isolated or plugged by squeezing with the packer-plug method followed by cleaning out. A through-tubing bridge plug can be used to plug off the thief zone if it is the bottom zone in the hole.

A survey that measures the density of the fluids flowing in the wellbore is used to locate zones producing water. These zones must be plugged or isolated if water production is excessive or restricts production from the other zones.

Open-Hole Completion An open-hole or barefoot completion produces from the open, uncased hole (Fig. 10–16a). This type is one of the earliest completions, but it is seldom used today except in hard, consolidated formations. It is especially applicable to fracture-type production when it is difficult to locate the fractures for perforation.

In the normal procedure the hole is drilled into the top of the formation, and production casing is run and cemented. The hole is then drilled into or through the formation, and tubing is run to wash the hole clean. The tubing is hung off, similar to the open-ended tubing completion, or it is pulled from the hole.

Slotted or Perforated Liner Completions A slotted or perforated liner completion is illustrated in Fig. 10–6b. This completion is distinct from the standard liner-type completion in that the liner is slotted or perforated before running and it is not cemented. The liner usually is slotted horizontally or vertically, or it can be perforated. A special but relatively common liner that is slotted vertically and wrapped with wire is sometimes called a wire-screen or wire-wrapped liner. The size of the slots and screen are determined by the size of the formation sand grains.

The basic purpose of this completion is to prevent sand production. Many formations contain very fine, unconsolidated sand that tends to flow into the wellbore with the formation fluid, even with very small perforations. The openings in the liner must be very small to strain out the sand and to allow clean formation fluid to be produced. Since the opening must be very small, there must be a large number of openings to obtain a maximum efficient flow rate. This is accomplished with the slotted or wire-wrapped liner.

In severe sanding conditions specially sized sand is displaced into the annular space between the liner and the walls of the formation before the liner is seated. This process is known as *gravel packing* and, when applicable, is an

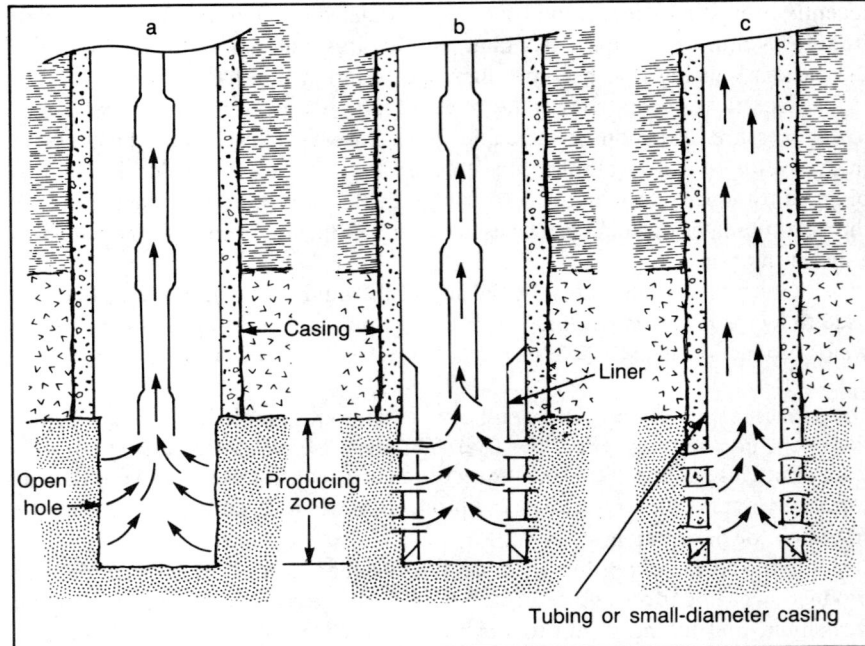

FIG. 10–16 (a) Open-hole or barefoot completion, (b) liner completion, and (c) tubingless or slim-hole completion

efficient method of preventing sand production. Special plastic materials are also pumped into the formation. When this material hardens or cures, it tends to break into a multitude of very fine, hairline cracks or fractures that serve to screen the sand out of the reservoir fluid during production.

A tubingless or slim-hole completion uses small diameter production casing or tubing as production casing (Fig. 10–16c). This type of completion does not use tubing inside the casing and is generally limited to shallow, low-volume wells.

A variety of single-zone completions are available to fit many production requirements. These methods are uncomplicated, easy to install, and allow the well to be produced efficiently.

Multiple Completions

In multiple completions two or more producing zones are completed and produced individually through separate tubing or annular flow channels in the same wellbore. In many cases they either save the cost of drilling another well or permit additional production from the well. Dual completions are relatively common, and triple and quadruple completions have been made.

The trend to dual and multiple completions is increasing. This is attributed to several factors. Drilling depths are increasing so more prospective producing horizons are encountered in each wellbore. The increased sale price for gas and oil helps economically justify completing zones that have a lower productivity. Multiple completions are easier to install because new developments have been made in equipment and in stimulation, completion, and production technology.

It is logical to wonder why all formations are not produced by the simpler, commingled, multizone, single-completion technique. This completion is always considered as an alternative, and a well is put on production as a single completion whenever possible. However, the single method cannot always be used for various reasons. Relatively high-pressure fluids from one zone can flow into a lower-pressure zone, thus reducing overall productivity. If the flowing bottom-hole pressure is low relative to the shutin pressure, the two zones often can be commingled if there are no other restrictions. A flowing oil zone invariably restricts or prevents a low-pressure, low-volume gas zone from producing. This phenomenon is known as *drowning out the gas zone*.

Some regulatory agency rules prevent commingling different formations. Surface separation may be a problem. In some cases there may be different ownerships at different depths, and it is necessary to produce the zones individually so production can be accurately measured and sale receipts and royalties can be allocated accordingly.

The completion design should always be flexible, especially when there are two or more prospective productive formations in the wellbore (Fig. 10–17). It is relatively simple to modify a single-zone completion design to accommodate a dual completion. Equipment and procedures are well established.

The main problems with multiple completions are the complexity of the downhole equipment, the difficulty of stimulation, the possible formation damage while installing the completion equipment, and the difficulty of installing the equipment and conducting artificial-lift operations. All of these must be considered when designing equipment and conducting multiple completion operations.

Both standard and special purpose packers (dual and triple) are used in multiple completions. Standard tubing is used when possible, but sometimes it is necessary to use smaller tubing or tubing with semiflush-joint connectors or tapered collars. Blast joints also are used. They have heavier metal bodies or are coated with abrasive resistant material. They are positioned in the tubing string opposite producing perforations and are designed to withstand the high, jet-type

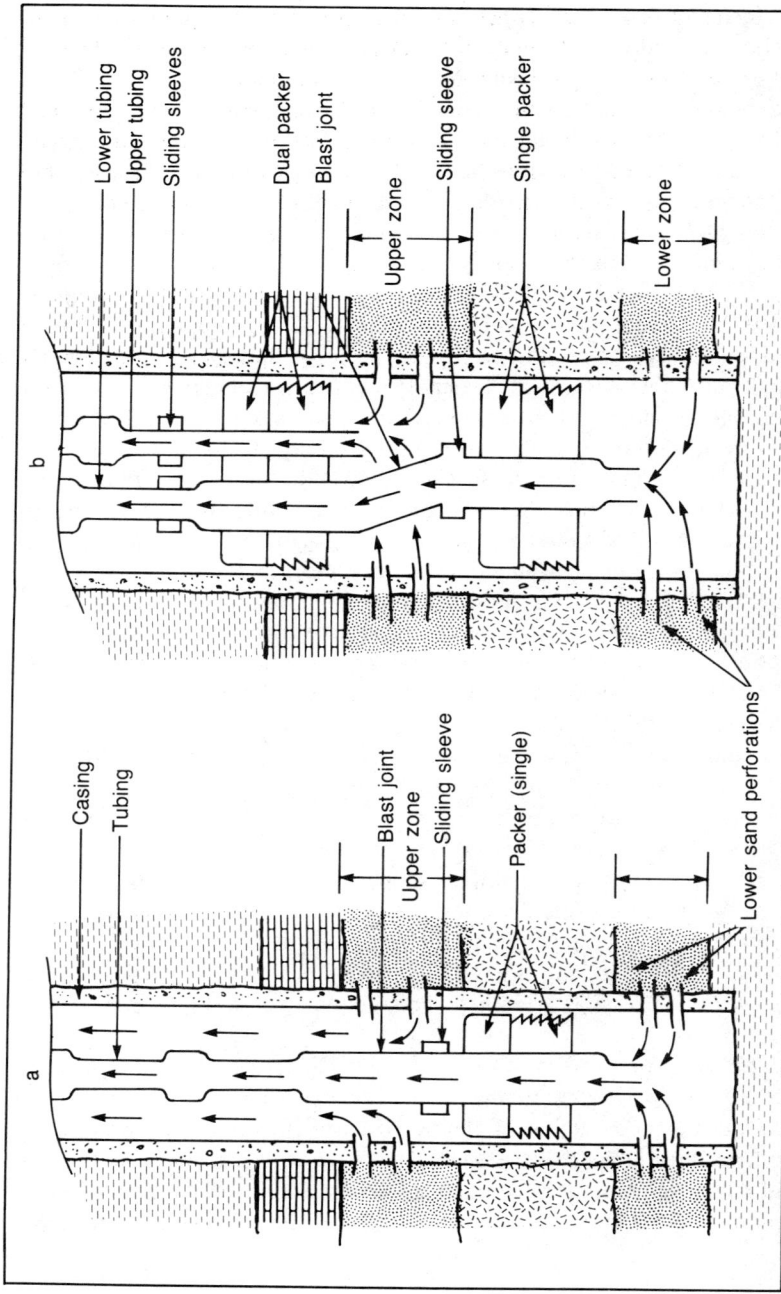

FIG. 10–17 Dual completions (a) tubing and annulus, (b) parallel tubing string, (c) tubing annulus with crossover packer, (d) dual tubingless completion (oriented perforations), and (e) concentric tubing string

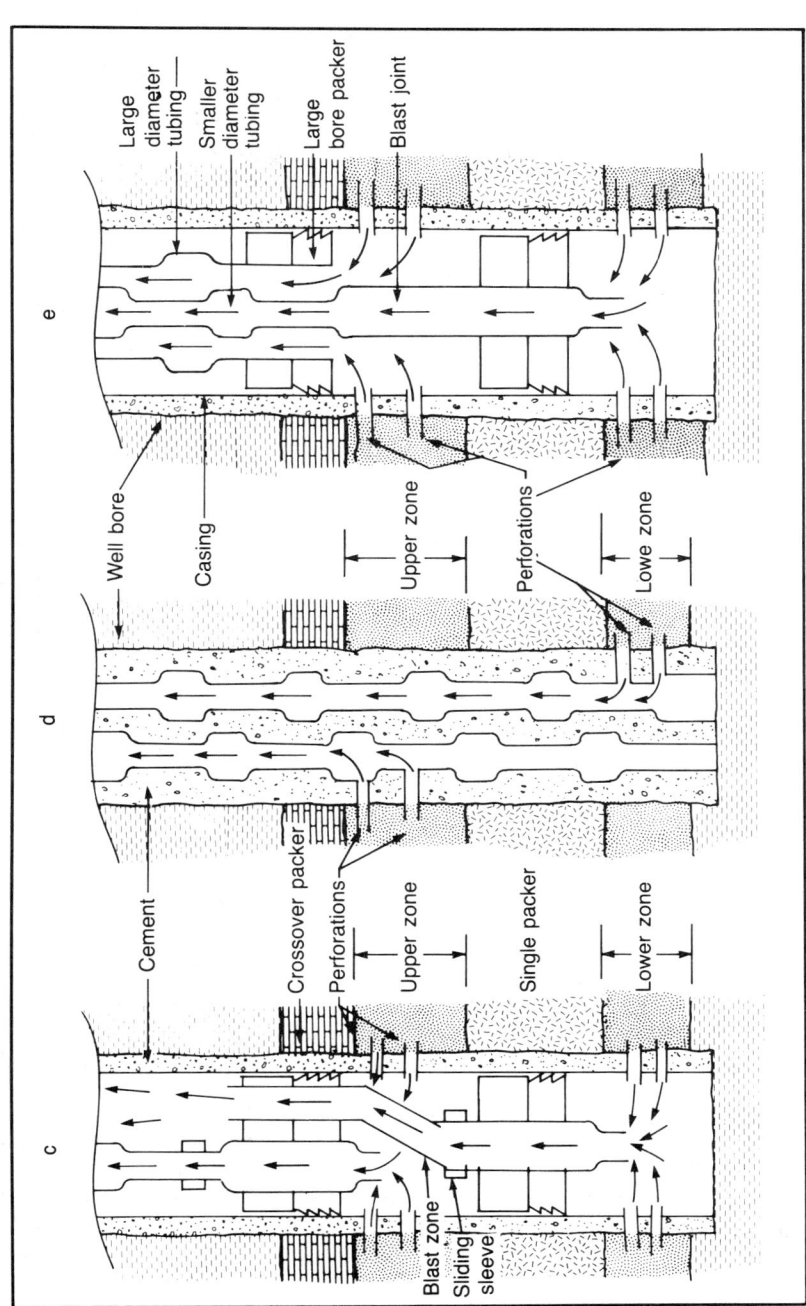

FIG. 10–17 *continued*

flow of formation fluids into the wellbore without becoming abraded or fluid cut. If this occurs, the zones cannot be isolated properly and there is a possibility of reducing the strength of the tubing until it parts, leaving a fish in the hole.

Sliding sleeves and plug receptacle nipples are used extensively in multiple completions and to a lesser extent in some single completions. The sliding sleeve (Fig. 10–18) is run in the tubing string. The tool can be opened to allow flow between the tubing and tubing-casing annulus, or it can be closed to isolate the tubing.

Several sliding sleeves can be run in the same string of tubing. Each sleeve uses a different combination of opening-closing grooves. The sleeve is opened or closed with a tool run on a wire line (slick line). This tool has lugs that precisely

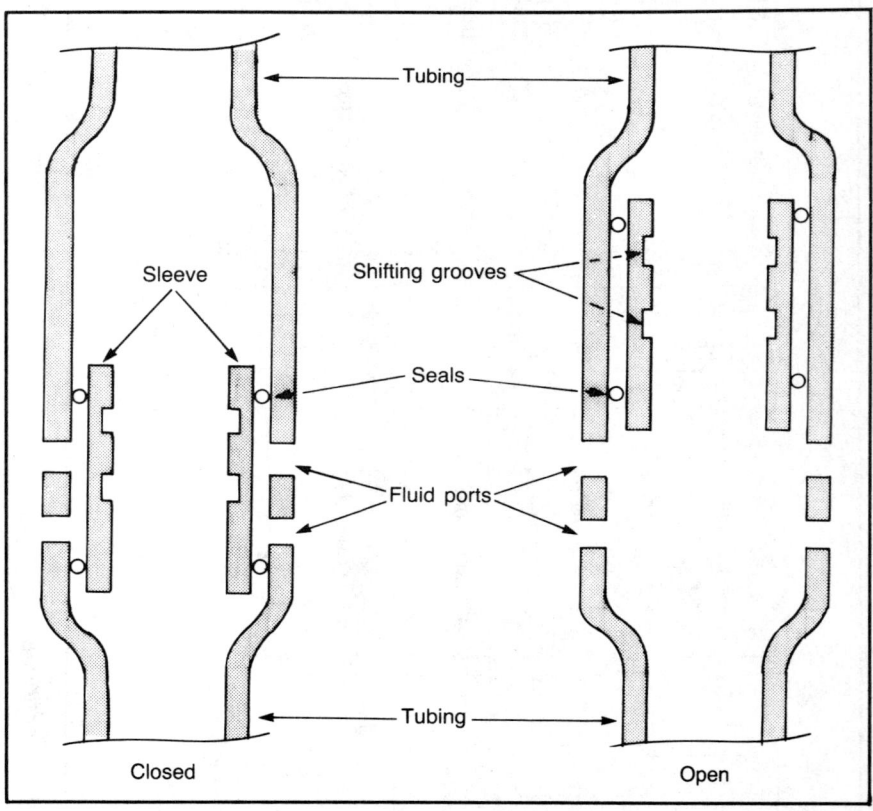

FIG. 10–18 Sliding sleeve

match the grooves of the sliding sleeve to be opened or closed. The lugs on the tool engage the recesses on the sliding sleeve, and the sleeve is then moved up or down to open or close the sleeve ports.

The running tool uses a shear-down or shear-up releasing mechanism. After the sliding sleeve is opened or closed, the lugs recess into the running tool so it can be pulled from the hole.

A plug receptacle nipple has a polished bore section and recessed grooves similar to those on the sliding sleeve. These are used primarily to seat blanking plugs that seal the tubing at that point to contain high-pressure formation fluids or to isolate zones for stimulation. These plugs are frequently equipped with an equalizer bar. The plug and wire-line tools can be blown uphole and stuck causing a fishing job if there is a high-pressure differential across the plug when it is pulled. The equalizer bar prevents this by opening a port through the plug that allows the pressures to equalize before the plug is released and pulled.

Downhole chokes, backpressure valves, storm chokes, and similar tools can also be seated in the plug-receptacle nipple. Several plug-receptacle nipples can be run in the same string of tubing. The tool to be seated in the nipple has expandable lugs that fit into the recesses to hold the tool securely in place. Seals on the tool act as a pressure seal in the polished bore section of the nipple to prevent leakage across the plug or other tool.

All of these tools are full opening. Full-opening tools have an inside diameter that is equal to or greater than the inside diameter of the tubing string or drillpipe on which the tools are run. This is an important feature because it allows full-gauge tools such as perforating guns to be run through the entire string of tubing.

Dual Completions The most common type of dual completion includes a packer run on tubing and seated between two producing formations. The lower zone is produced through the tubing, and the upper zone is produced through the tubing-casing annulus. A retrievable packer with a circulating valve can be used in low-pressure wells. Higher-pressured wells may have a sliding sleeve in the tubing string near the top of the packer, and a blast joint may be positioned opposite the upper producing perforations. The circulating valve or a sliding sleeve is used to circulate the well during the completion operations and to kill the well when it must be worked over or recompleted. The general operational procedure is summarized as follows:

- Prepare the well.
- Perforate, flow test, stimulate, flow, and test each formation individually using the methods described for multizone, single completions.

- Run a completion packer with a circulating valve on tubing and seat the packer between the two formations, usually near the bottom of the zone.
- Open the packer circulating valve and circulate the mud out of the hole displacing with low-weight completion fluid. Close the circulating valve and nipple up the Christmas tree.
- Clean up the well by flowing the lower zone through the tubing and the upper zone through the tubing-casing annulus. Place the individual, isolated flow streams on production and test after clean formation fluid (oil or gas) is produced.

This is one of the simplest dual completion operations. It is limited to moderately pressured formations when little if any additional (mild) stimulation is required and when there is a reasonably good chance that the zones will begin flowing without difficulty. In most cases a more common completion includes a plug receptacle nipple above the packer and a sliding sleeve about 6 ft above the nipple. This allows better control. Additional completion operations can be conducted if required. In the final step the well is circulated with low-weight completion fluid. The commingled zones are both flowed and cleaned up through the tubing. If the well cleans up and produced satisfactorily, the sliding sleeve is closed to isolate the formations and flow streams. Then the individual flow streams are placed on production and tested.

Sometimes it is easier and faster to start both zones flowing together as previously described, especially when one zone may be weaker and slower in cleaning up. This often allows the energy from one zone to be used to help clean up the other zone. The well can also be swabbed if it does not flow naturally. In the normal case the casing side is closed at the surface so all fluids are produced through the tubing. A longer swabbing period may be required when completing lower-pressure formations to recover a substantial part of the large volume in the tubing-casing annulus backside. This reduces the hydrostatic head so the well can begin to flow. Usually the closed-in casing side begins to develop pressure as the well approaches a flowing condition.

If the formations do not begin flowing satisfactorily, they can be brought in individually with the following procedure:

- Close the sliding sleeve with wire-line tools run through a pressure lubricator. Flow the tubing zone, swabbing if necessary. If the zone does not flow and clean up, stimulate it with a small, shallow cleanout treatment such as washing with acid to remove shallow formation damage. Then flow and clean up the zone. Swab the tubing if necessary.

- Run a blanking plug on a wire line through a pressure lubricator and seat the plug in the nipple above the packer to shut off and isolate the cleaned, productive lower zone. Open the sliding sleeve. Flow and clean up the upper (casing) zone through the tubing. Stimulate with a mild cleanout treatment if necessary and swab the zone. Do not use sand fracturing in this stimulation because it will stick the packer.
- After the casing zone produces clean formation fluid, close the sliding sleeve and remove the blanking plug to isolate the zone. Produce the upper casing zone through the casing and the lower tubing zone through the tubing. If the well must be worked over or recompleted in the future, open the sliding sleeve and circulate the well dead with weighted mud. Pull the completion equipment as required and proceed with the workover and recompletion operation.
- The order of zone completion can be reversed by setting the blanking plug first and completing the casing zone. Then close the sliding sleeve, pull the blanking plug, and complete the lower zone through the tubing. Note that if the casing zone must be swabbed in, this must be done through the tubing. The order in which the zones are completed depends on the formation pressures and productivity, the estimated ease of completion, and the type of downhole completion equipment.

An alternative is to use a single-packer dual completion to complete the lower zone first with permanent completion procedures as described previously. Then set a plug in the plug receptacle nipple, pull the tubing, and complete the upper zone with one of the applicable single-zone completion techniques. Run completion tubing with a seal assembly, a sliding sleeve, and a blast joint; seat the seal assembly in the permanent packer; and bring in the well with the same procedure as in the single retrievable-packer dual completion.

This completion is common and is especially applicable when the lower zone has a high pressure. It has the distinct advantage common to permanent completions in that the lower zone is cleaned up one time after stimulation and is isolated with the blanking plug. Therefore, it is not exposed to damage by the mud in the hole while completing the upper formation. The lower zone may be used to help clean up the upper zone, thus expediting the overall completion.

Note that both the tubing and casing zones can be sand fractured in this type of completion. Any sand above the packer can be circulated out just before latching the tubing into the packer in the final step.

Single-packer dual completions are relatively straightforward, and the operations normally do not present a high risk. However, many problems can and do occur. Some of these include equipment malfunctions while running; equipment leaks, especially in the packer, the sliding sleeve, or the blanking plug; failure of the sliding sleeve to open or close; stuck blanking plug; and additional stimulation requirements.

When these failures occur—and one or more of them are not uncommon—the equipment must be pulled out of the hole, fishing if necessary. The equipment is repaired and replaced, or remedial steps are taken as needed. The completion equipment is rerun into the hole, and the individual steps necessary to place each zone on production are repeated. When this occurs, there is a higher risk that a moderate stimulation will be needed so the formations can be produced at the maximum efficient rate.

Single packers are used for most dual completions. Two packers are not used unless for a specific purpose. One of the main reasons for using two packers is completing high-pressure zones when the upper casing must be protected from the higher pressures. Then each zone is produced through a separate tubing string. The upper zone is isolated from the tubing-casing annulus with a packer and tubing, the same as the lower zone. This procedure requires a single and a dual packer and two tubing strings—*parallel tubing-string completion*.

Note that two strings of tubing are connected to a dual packer. The term "dual packer" is often used to mean two packers, i.e., a dual (two) packer completion, when the upper packer is a dual packer and the lower packer is a single packer connected to one string of tubing.

The most common two-packer completion for high-pressure wells uses a permanent lower packer without the latch-in between the two formations and a retrievable, dual-completion packer above the formation. The general operation for a two-packer completion is as follows:

- Prepare the well.
- Run a permanent packer fitted with a mill-out extension and a plug receptacle nipple. Seat the packer above the lowest formation to be completed.
- Run tubing with a latch-in device and seat the tubing in the permanent packer. Perforate the lower formation with a through-tubing gun. Test and stimulate as necessary. Flow and test the well until clean formation fluid is recovered.
- Set a blanking plug in the plug-receptacle nipple to shut off the lower formation. Unlatch the tubing and pull it out of the hole. Perforate the upper formation, test, and stimulate as necessary. Clean up and test using one of the applicable single-zone completion methods. After the zone produces clean formation fluid and is tested satisfactorily, circulate and kill the well. Pull the equipment from the hole.

- Run the lower tubing or long string with the completion equipment. Position the blast joint in the tubing string so it will be opposite the upper perforations when the tubing is landed. Position the upper packer on the tubing string so it will be a short distance above the upper perfs when the tubing is landed.

A latch-in is not used in this case, although there are cases where it might be used depending on the type of upper packer and the distance between the two packers. The mud in the hole can be circulated out before landing the tubing, but generally the best procedure is to do this through the sliding sleeves. Run the seal assembly into the permanent packer and place the correct weight on the permanent packer. Measure and space out the tubing by adding tubing nipples as required so the lower tubing string has the correct length from the permanent packer to the Christmas tree. Run the seal assembly into the permanent packer and place the correct weight on the permanent packer. Seat the dual-completion packer and land the tubing in a split bushing hanger.

Pressure test the lower tubing string to the required test pressure. This is normally the lower pressure of either 80% of the rated burst pressure of the tubing or the maximum pressure to which the tubing is expected to be exposed in future operations plus 20%. The tubing may or may not have been tested while being run, but this final pressure test is always conducted. The sliding sleeves are run in the closed position in anticipation of testing the lower tubing string at this point in the operation. The tubing could be tested at a later time after running the remaining completion equipment. However, if it developed a leak, it would be necessary to pull the other equipment before pulling the lower tubing string for repairs. The lower tubing string is easier to repair at this time if it develops a leak.

The upper tubing string or the short string is run with tapered collars. The sliding sleeve is in the closed position. The tubing is spaced out, the seal assembly is inserted, and the tubing is latched into the dual-completion packer. Then the tubing is landed in the second half of the split hanger bushing. The lower tubing is hung in the first half.

Next, the Christmas tree is nippled up, and the tree connections are pressure tested. A blanking plug is run and seated in the plug-receptacle nipple. The upper tubing string is then pressure tested. The blanking plug could have been run in place while running the tubing, but this procedure may have required filling the tubing to prevent possible tubing collapse or filling the tubing before the pressure test. Therefore, the method described is normally used.

The next step is to test the pressure of the tubing-casing annulus, taking precautions to ensure that the test pressure does not exceed either the collapse pressure of the tubing or the burst pressure of the casing. Note that in the pressure-testing process the pressures are confined to sealed volumes. For example, the tubing-casing annulus between the two packers is not tested because it is not necessary and because the upper zone also could be damaged by plugging with mud.

The last step is to circulate weighted completion fluid into the dual tubing-casing annulus and displace the remaining mud in the hole with light fluid. Then the well can be brought in.

To bring in the well, the sliding sleeve in the upper tubing string is opened. Fluid is circulated down the short tubing string and up through the tubing-casing annulus to displace weighted, treated completion fluid into the tubing-casing annulus space. The completion fluid is displaced with unweighted fluid, usually water, to the sliding sleeve on the bottom of the upper tubing string. The sliding sleeve is closed, and the upper tubing string is pressure tested to ensure that the sleeve is closed and sealed properly. The blanking plug is pulled from the upper tubing. At this point the upper tubing should begin to flow. If flow begins, the process of flowing and cleaning up the upper zone should be continued.

The sliding sleeve is open in the lower tubing. Mud is circulated down the tubing, through the sliding sleeve, and up through the upper tubing string and is displaced with light fluid, usually water. If the upper zone did not come in earlier, it should begin flowing at this time. If it does not flow, the blanking plug in the lower tubing string can be pulled, and the pressurized lower formation can help clean out the water in the hole. The zones are allowed to flow and clean up. Then the sliding sleeve in the lower tubing string is closed to isolate the zones. The zones are swabbed if they do not flow naturally. The formations may also be stimulated with a small acid cleanout treatment if necessary.

The completion equipment can be pressure tested, and weighted fluids can be displaced by different methods. The main objective is to test the equipment adequately. The fluids should be displaced in the most convenient manner with minimum downhole work. Then the open, perforated formations can be exposed to a minimum pressure to ensure that the well pressures are safely controlled. The basic completion steps described are of course abbreviated. Various routine procedures have been omitted.

The equipment described is the basic downhole equipment used in dual completions with dual packers. Other types of packers may be used, but the general equipment design is similar to that illustrated. The general overall completion procedures also are similar to those described.

The completion normally is used for higher-pressure formations, but it is also used for another special application. Oil wells producing from a relatively low-pressure formation or with a low gas-oil ratio may tend to load up and restrict or prevent production when producing through a large cross-sectional area such as the tubing-casing annulus. The gas tends to separate from the oil, thus leaving a long oil column that exerts a back pressure on the formation and restricts its productivity. If these wells are flowed through a smaller cross-sectional area such as tubing, there is less tendency for the gas to separate from the oil. The buoyant force of the gas helps to produce the oil.

The same loading problem can occur with low-pressure gas wells that produce relatively large volumes of liquids such as condensate or salt water. Many of these wells continue to flow if they are produced through tubing, assuming the gas and liquid flowing volumes are sufficient. If two of these zones occur in the same wellbore, a dual completion using two packers may be used to produce the wells for a longer period.

Shallow formations normally have lower pressures than deep formations. In many cases these lower productivity formations cannot produce efficiently through the tubing-casing annulus, but they can produce efficiently through tubing. At the same time the deeper, higher-pressure formations can produce efficiently through the tubing-casing annulus. The single-packer dual completion cannot be used in this case because the upper zone does not flow through the tubing-casing annulus. Dual packers with dual tubing strings can be used, but a simpler, more economical method is available. This procedure includes using two packers with one single string of tubing.

The upper packer is called a *crossover packer*. The lower zone flows through the tubing connecting the packers and up the tubing-casing annulus. The upper zone flows through the tubing. A sliding sleeve and plug-receptacle nipple usually are run above the lower packer and a sliding sleeve is run above the upper packer. A blast joint may be used depending on flow rates and pressures. The general equipment installation and completion procedures are similar to those used for other dual completions with modifications to allow for the single tubing string.

This type of completion is less common. Normally the well is placed on artificial lift when the productivity decreases to the point where this type of completion is needed.

Dual tubingless or slim-hole completions can be made by running two strings of tubing in the hole and cementing them in place. The individual tubing strings must be perforated with an oriented perforating gun to prevent perforating into the adjacent tubing string. Otherwise, the completions are similar to the equivalent single-zone completion. By convention, the lowest zone produces through the lower tubing string and the upper zone through the upper string.

Dual completions with concentric tubing strings are used only for special circumstances. For example, if there is a casing failure, the outside tubing can be run and seated on an upper packer to provide an effective inner production casing string. The inside tubing is then run, and the overall completion is similar to the tubing-annulus dual completion.

Triple Completions The basic triple completion is an extension of the two-packer dual completion (Fig. 10–19). The two lower zones produce through parallel tubing strings, and an upper zone produces through the tubing-casing annulus. The general procedures for completing the formations, installing the equipment, circulating the well, and bringing it on production are similar to those for a dual completion with allowances for the extra casing zone.

Triple completions with three packers can be designed, and a few have been run. The middle tubing string is run through the top packer and seals in the top packer (triple packer) and the middle packer (dual packer). A third tubing string is run and seals in the top packer.

A quadruple completion is an extension of the triple completion with three packers. A casing zone above the top packer produces through the tubing-casing annulus. These completions are seldom used. The equipment is difficult to install and position correctly. Also, equipment failures cause a severe problem. All of the equipment must be pulled from the hole when one zone is reworked, and it is frequently necessary to fish the equipment out of the hole. Formation debris or a small amount of sand can stick the downhole equipment causing a fishing job. There also is a high risk of damaging the formations while fishing or reworking the well. Then all formations must be shut in while reworking one formation.

The two-packer triple completion can be installed with a moderate risk of malfunction or failure. The risk increases appreciably when additional packers and tubing strings are added. From a practical viewpoint if there are more than three zones—counting commingled multizones as one zone—the best completion procedure is to produce the three lower zones until they are depleted and then complete the upper zones or drill another well to produce the upper zones.

The single and multiple completions described probably cover over 95% of the completions made in the industry today. A few other types of completions have been designed for special applications. In most cases these are modifications of the completions described here. However, they are sometimes called by a different name. This may cause some difficulty since naming completions is not standardized. For example, multiple completions have been described using a single zone in each stage of the completion. However, each stage could consist of two or more commingled zones. Therefore, a single-packer dual completion could actually have multiple zones above and below the packer. Then the completion would be a multizone, commingled single-packer multiple completion.

Therefore, in some cases additional specific information must be known to classify the type of completion properly.

Injection well completions are a fairly common classification. In most cases these are single completions below a packer. Injection fluids are pumped down the tubing and into the formation below the packer. In a few cases tubingless completions are used. Then fluids are pumped down the casing and into the formation.

Wildcat wells, especially those with multiple prospective producing horizons, may present a problem in completion. The actual completion design may not be known until the individual formations have been tested and stimulated. However, the data obtained during drilling and the well logs delineate the formations that should be tested and give a reasonable idea of which are the most prospective. These data are used to design a general, flexible completion procedure. The individual zones are then tested, usually using the packer-plug method starting with the lowest zone. The final completion equipment installation can be designed as the test results are obtained on the individual formations. This normally does not present a problem.

Many oil wells are placed on artifical lift at completion. Most other oil wells are placed on artificial lift later during the producing life as the reservoir pressure declines. The artificial-lift equipment can be incorporated in the completion equipment, or the completion may be modified.

In summary, completions are designed so all prospective formations can be stimulated, tested, and produced in the most efficient manner. This includes providing for isolation between zones, control of pressures and flow rates, safety precautions, and provisions to reenter the well for remedial work. Completion designs for vertical and deviated holes and for marine and onshore wells generally are similar to those described.

SURFACE EQUIPMENT

The surface equipment includes all of the equipment located on the surface that is used to control the pressures and flow rates, to separate the different reservoir fluids (gas, gas liquids, oil, and water), to meter and measure gas and liquid volumes, and to provide storage and a means of transporting the oil and gas to market. The type, size, pressure rating, and amount of equipment depends on wellhead pressures and volumes, the types of fluid produced, the separation requirements, the distance to market, and other related factors.

Surface equipment can be divided into wellhead equipment and production facilities. Production facilities can be subdivided into oil and gas equipment.

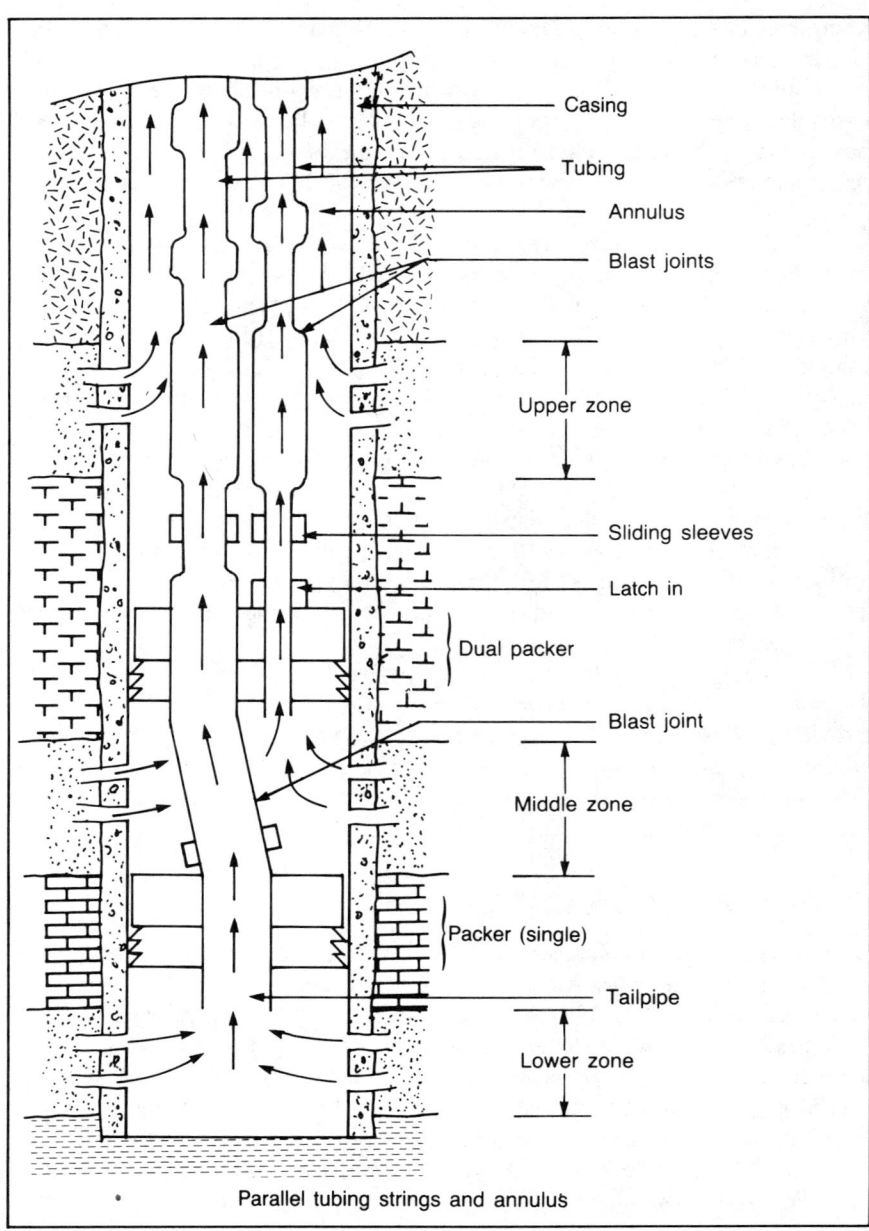

FIG. 10-19 Triple completions

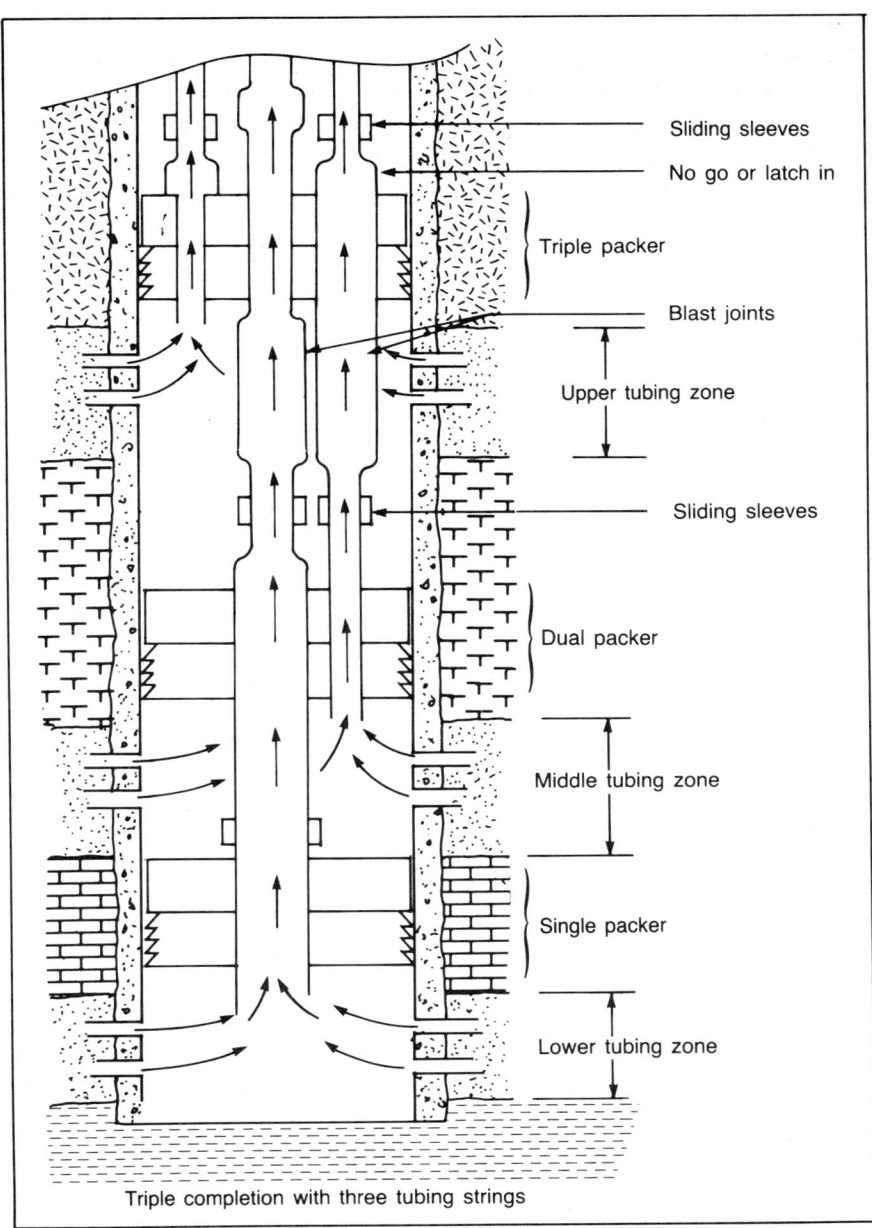

FIG. 10–19 *continued*

The Wellhead

The wellhead serves many functions. First and foremost, it must be strong enough to contain the maximum pressure expected in the tubing or tubing-casing annulus after allowing a substantial safety factor. The wellhead provides a means of landing the tubing, bringing in the well, killing the well, and reentering the well when necessary. The wellhead also serves to control and regulate the flow of fluids (liquids and gases) from the wellbore. High pressures are reduced by the wellhead to a safe working level compatible with the pressure rating of the surface facilities. The wellhead design allows full-gauge tools to be run under pressure.

Wellheads are made of large, heavy-duty castings with machined surfaces such as flange faces and ring grooves. They are available in various sizes with different pressure ratings in each size. The component parts of most wellheads are connected together with bolt-and-flange connections with metal sealing rings set in ring grooves. This provides a much stronger, durable connection in large sizes (diameters) than screwed connections, which are used on a few low-pressure wellheads. Special steels are used in some wellheads for severe operating conditions such as a corrosive hydrogen sulfide environment. The same basic wellhead is used for either flowing oil or gas wells, and wellheads have been standardized by the American Petroleum Institute.

The wellhead has six basic parts (Fig. 10–20). Named in the downstream direction of flow, these are the bradenhead or casing-hanger spools, the tubing-hanger spool, the master valves, the cross, the wing valves, and the choke.

Casing strings are landed and sealed in the bradenhead and one or more casing-hanger spools. Two outlets on these spools are used to pump fluid in or flow fluid out of the annular space. The tubing is screwed into a tubing hanger, commonly called a *doughnut,* which is landed in the tubing-hanger spool. The tubing hanger is held in place by the weight of the tubing and the locking studs. It seals the annular space between the tubing and the walls of the tubing spool, which in turn seals the annular space.

The master valve is the primary control for closing in the well. The cross diverts the well fluids into the Christmas tree wing(s) and provides access for running wire-line tools into the hole. The wing valve is a working valve to shut in the well. The choke can be a positive, adjustable, or choke-beam type. It is used to adjust the flow rate and pressure. High pressures are retained upstream from the choke, and low pressures are kept downstream. Either a flanged or threaded connection can be used at the choke outlet. Normally all other downstream connections below the choke are threaded. Pressure gauges are used to monitor the tubing pressure (upper gauge) and casing pressure (lower gauge).

The Christmas tree in Fig. 10–20 is correctly classified as a single-wing tree with single master and wing valves. The tubing is screwed into the master

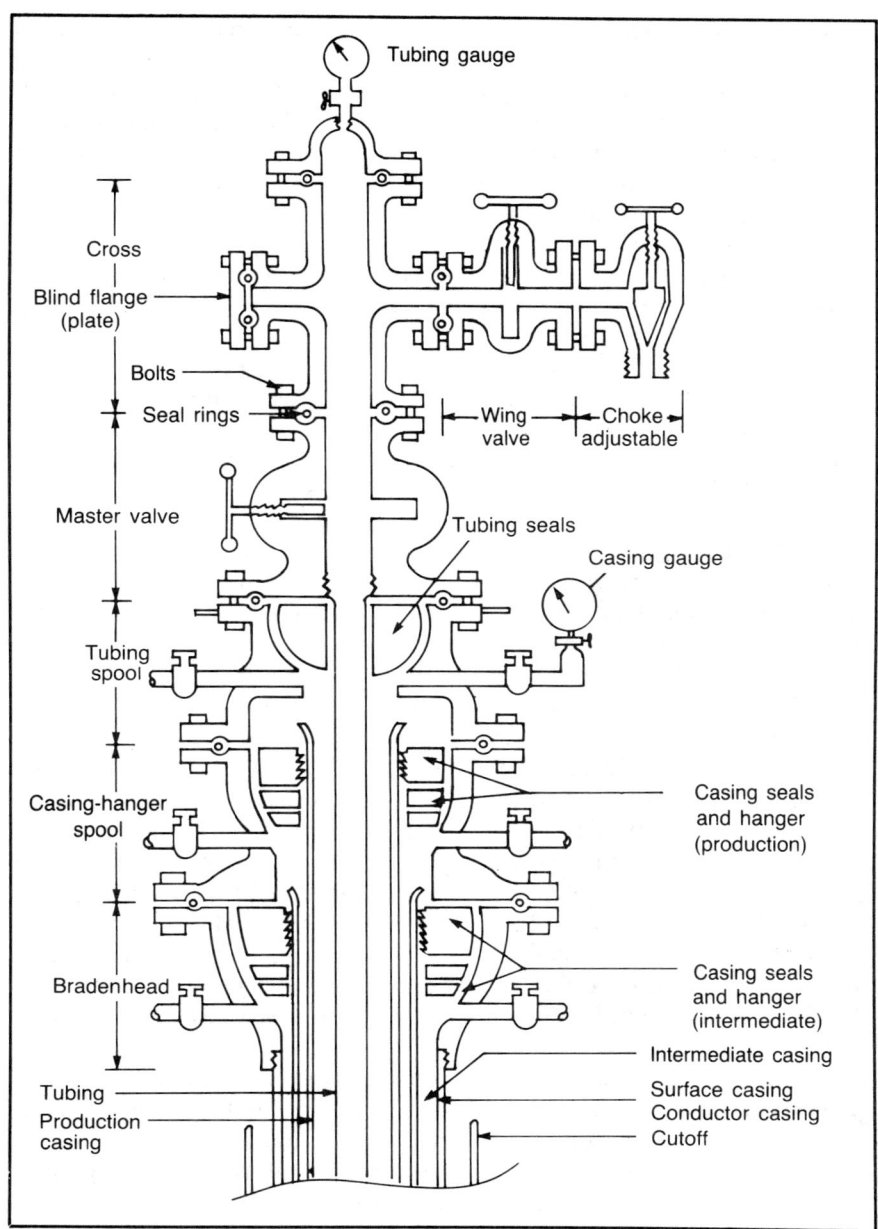

FIG. 10–20 Single-wing Christmas tree

valve, and the tubing seal allows the well to be reentered under pressurized conditions.

For example, the circulating valve on certain types of packers is opened by either picking up or rotating the tubing. When the well must be reentered, a joint of tubing can be screwed into the top of the cross with the master valve closed. The flange on top of the tubing hanger is disconnected. This allows the Christmas tree to be picked up while the tubing slides through the pressure seal, which retains the pressure in the annulus. The action of picking up the Christmas tree and tubing opens the packer circulating valve. Then the well can be circulated dead by forcing kill mud down the tubing and out the casing valve. After the well is circulated dead, the Christmas tree can be removed and preventers can be installed to rework the well as needed.

A lubricator is connected to the top of the cross with either a flange or screw-in connection when wire-line tools are run. The master valve is closed, the wire-line lubricator is installed, and the master valve is opened to allow wire-line tools such as perforating guns, blanking plugs, and swabbing tools to be run into the tubing.

The component parts of the Christmas tree can be rearranged or additional parts added to make different types of trees, depending on the tree requirements needed to produce the formation. For example, high-pressure wells often use double master valves and may use single or double wing valves on a single- or double-wing tree as illustrated in Fig. 10–21.

In operation, master valves are closed and opened *under pressure,* and wing valves are closed and opened *against pressure.* Closing against pressure means closing the valve on a flowing stream of fluid. For example, if a well is flowing and needs to be shut in, the wing valve would be closed first, followed by closing the master valve. The valve is more subject to fluid erosion when being closed against pressure than when closed under pressure. The reasoning here is to conserve the master valve as a primary safety shutoff. If the wing valve becomes fluid cut and starts to leak, it can easily be changed after closing in the master valve. However, if the master valve begins to leak and requires changing, the well must be killed or a plug must be set in the tubing to shut in the well. Therefore, when one valve must be subjected to extra wear, the valve that is the easiest to replace is used.

The Christmas trees can be modified slightly to serve as trees for a single-packer dual completion or dual-packer dual completion in which the upper packer is used as a crossover packer. The only modification necessary is to add a wing to one or both of the tubing-hanger spool outlets. This permits the tubing-casing annulus to be produced through a standard wing consisting of a wing valve and a choke.

FIG. 10–21 Triple-completion tree with double master valves and single wing valves

Christmas trees designed to produce through parallel tubing strings for dual and triple completions with two packers use the same basic components modified to provide for the two flow streams. The two tubing strings are individually run and hung in separate split-tubing bushings. The bushings are seated in the tubing spool and are locked in with locking studs to isolate and seal the tubing-casing annulus. A special dual-cavity companion-type flange is used to seal and separate the two tubing strings. The master valves and crosses are modified to allow for the close spacing, and standard wing valves and chokes are used. The tubing-hanger spool outlet also can be fitted with a standard wing to accommodate a triple completion with the upper formation producing through the tubing-casing annulus (Fig. 10–22).

Low-pressure wellheads are simple, economic heads normally limited to low-pressure, shallow wells. They are frequently used on pumping wells where the tubing is fitted with the pumping tee, and they are commonly called pumping heads, not Christmas trees.

FIG. 10–22 Dual completion tree with dual master valves

Testing and Production Facilities

Testing and production facilities include all of the surface equipment downstream from the wellhead (choke) to the point where the gas and oil are sold, such as a transmission line or other type of sales outlet. The sales outlet is also the point of custody transfer where the responsibility for the oil and gas (losses, equipment maintenance, etc.) is transferred from the operator or producer to the transmission company, or refinery. They continue to move, separate, refine, and otherwise process the oil and gas in its journey from the reservoir to the consumer or end user.

Testing and production facilities include all of the equipment necessary to separate the oil, gas, and water produced from the well, to dispose of the salt water, and to process the oil and gas as necessary. Contaminants and other undesirable materials, commonly called *basic sediment and water (BS&W)*, must be removed. A limited volume of oil storage is provided onsite. The clean oil is piped or trucked, and gas is piped to the sales outlet.

Many testing and production facilities are common to both oil and gas wells, and others are used specifically for one type of production. For example, both types of wells commonly use separators. Oil wells require oil storage, and gas wells may need a gas compressor.

Oil Testing and Production A large number of shallow, low-pressure, low-volume oil wells produce a very small amount of gas. Many are isolated or located on one-well leases with a minimum amount of production facilities. The oil is produced directly into the top of a tank that is used for both storage and oil-water separation (Fig. 10–23).

Most oil wells produce a small amount of salt water with the oil. The oil and water separate because they have different densities. The water-drain valve is located a few inches from the bottom of the tank, and the oil-sales valve is located about 12–18 in. above the bottom of the tank. Oil and water are produced into the tank. Water is drained off periodically and is run to disposal. Water disposal may be an injection well, an open pit where the water evaporates, or a storage tank where the water is carried off by tank truck.

After the tank is filled with oil, the water-drain line is checked to ensure that all of the water has been drained from the tank and the water-oil level is below the oil-sales valve. A transport truck arrives at the tank, connects a hose to the sales valve, and pumps the oil from the tank into the transport truck. The oil may be measured at the tank through the *thief hatch*, by a meter on the truck, or at the pipeline or refinery terminal when the truck unloads. This gives both completion testing and production volumes.

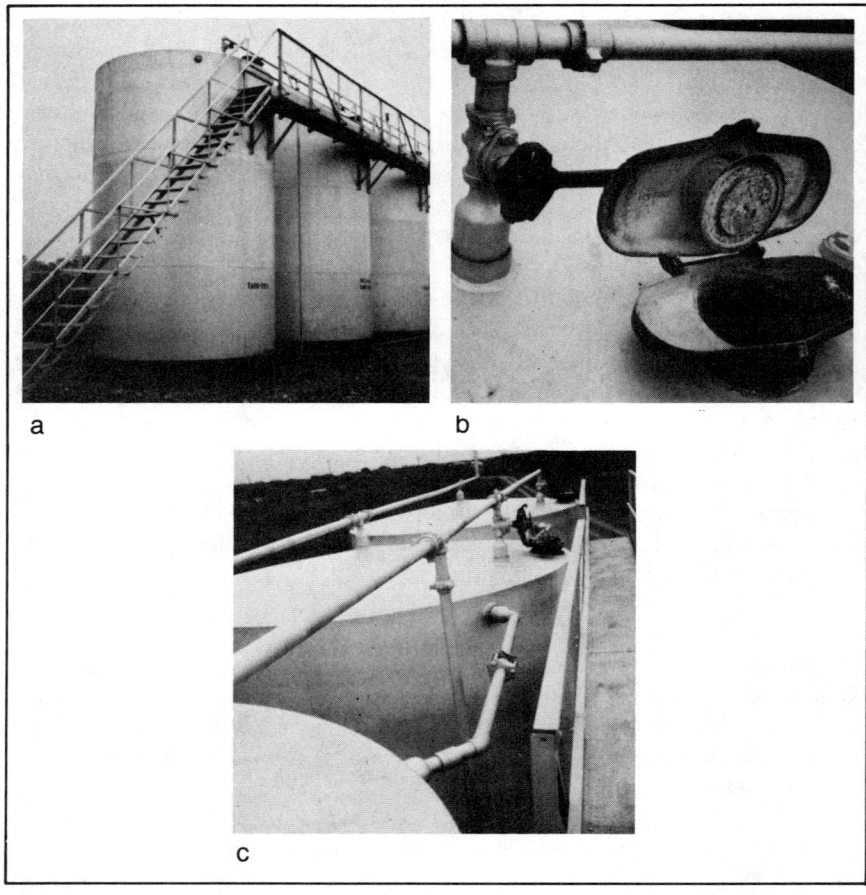

FIG. 10–23 (a) Production battery; (b) oil inlet line and thief hatch; and (c) oil inlet line, flare-gas line, and overflow line (on side of tank)

A stairway provides access to the top of the tank. The thief hatch is a small opening with a hinged cover. It can be opened so a measuring tape with a plumb bob on the bottom can be run into the tank to measure the amount of oil in the tank.

Most companies that purchase oil require that the oil have a BS&W content below a fixed level. The sample is obtained with an *oil thief,* hence the name thief hatch. The oil thief is a closed container that is lowered into the oil and opened to catch a sample of the oil in the lower part of the tank for analysis.

The analysis for BS&W includes placing a fixed volume of oil in a special test tube, which is then spun in a centrifuge called a *grind out*. This accelerates the separation of oil, water, and sediment. The volumes of water and sediment are read directly on the centrifuge tube in percentages.

In some cases a well may produce too much gas to flow directly into a tank. Then an oil-gas separator is installed because the tank cannot withstand over a few pounds of pressure without becoming distorted and possibly rupturing.

Oil, gas, and water flow from the well and enter an oil-gas separator. The oil and water fall to the bottom of the separator because they have high densities. The gas moves upward and out a gas line. The bottom of the separator is equipped with a dump valve actuated by a float inside the separator. As the water and oil accumulate in the bottom of the separator, the float rises on top of the liquid. A dump valve opens when the float reaches a predetermined height. Oil and water pass out of the separator through the dump valve. When the fluid level is lowered to a specific point, the dump valve shuts off.

Separators normally operate in the pressure range of 25–125 psi. Therefore, they are of heavy-duty construction in accordance with specifications for pressurized vessels. Separator pressure is maintained with an adjustable backpressure valve on the gas exit line. Various types of valves are used, but a weight-loaded valve is most common.

A safety release popoff valve vents the separator in case the weight-loaded backpressure valve malfunctions. The separator is also equipped with a pressure-relief plate that ruptures and safely vents the separator in case both of the valves fail to function correctly.

High-productivity, high-pressure wells with a high gas-oil ratio may require two stages of pressure reduction and oil-gas separation. In this case high-pressure and low-pressure separators are used. The operation of both separators is similar. The main difference is that the high-pressure separator has a heavy-duty construction to withstand the higher pressure. The gas, oil, and water from the well first enter the high-pressure separator so part of the gas can be separated. The gaseous fluid passes through a dump valve into the low-pressure separator so the remaining gas can be separated.

Large-volume wells with high water cuts are tested and produced through a gun-barrel tank. The oil and water from the low-pressure separator enters a gun barrel. The gun barrel is a tall, large tank used to separate the oil and water. Water falls to the bottom, and the oil rises. The height of the oil-water contact in the gun barrel is controlled by a water leg. Normally the oil-water contact in the gun barrel is maintained at a relatively low level to allow a larger volume of oil in the gun barrel and, correspondingly, more retention time to permit more efficient oil-water separation. Cleaner oil overflows into the tanks.

The flow of oil into the tanks is controlled by shutoff valves. Normally production is flowed into a tank until it is full. Then production is transferred to another tank. While the second tank is filling, the first tank is emptied. Large volumes are delivered to an oil pipeline with a pipeline pump installed near the tank battery.[3]

Large production facilities may transfer oil to sales by a lease automatic custody transfer (LACT) unit. This is an automated transfer system in which the oil is transferred as the tanks are filled. Volumes, gravities, BS&W, and other measurements are made as required.

In some production the chemistry of the oil and water is such that the mixture forms a very tight oil-water emulsion that is hard to break. Special chemicals that are available to help break the emulsion can be injected into the production flow stream with a small chemical injection pump. The pump may be located at the wellsite or at the production battery. In some cases the tanks are circulated with a small oil-transfer pump to help mix the chemical and to allow a longer retention time for the chemical to act to separate the oil and water. This is commonly known as *rolling the tank*.

Emulsions are frequently easier to separate if the oil-water mixture is heated. A special unit called a heater treater is used in place of the gun barrel. The oil and water emulsion from the separator enters the heater treater and is heated and retained for a period. A firebox and fire-tube combination supplies heat to the mixture. Oil separates and rises, and water falls to the bottom. The water-oil contact is maintained at a relatively low level similar to that in the gun barrel with a water leg. The oil overflows out of the heater treater into the tank battery.

Production well testing is an important part of the completion and production process. These tests are needed during the completion procedure to evaluate individual zones and during the producing life of the well to ensure that the well is producing at the maximum efficient rate. In large batteries production from the wells flows through gathering lines to the production manifold, the separators, and the heater treater and into the tank battery. The production manifold also contains a test line. Normally the valves leading to the test line are closed, and all of the production passes through the production line.

To test a well the valve to the production line is closed, and the valve to the test line is opened. This directs the production from the well into the test unit consisting of a test separator and heater treater. The production from the test unit is directed into one tank where it is measured.

Gas The type and amount of surface facilities needed to test and produce a gas well depend on the pressure, volume, and amount of water or condensate pro-

---

[3]By convention, oil is moved in an *oil pipeline,* and gas is moved in a *transmission line.*

duced. Additional equipment such as a desulfurization unit may also be needed to remove contaminants and corrosive agents (Fig. 10–24).

Gas and entrained liquids flow from the well into a separator. The gas and liquids are separated. If the liquid is primarily salt water, it is discarded to disposal. If the liquid is primarily condensate, it is passed to a storage tank. Water is drained off of the tank from the lower water-drain valve, and the condensate is transported to sales. If the well production does not contain sufficient fluid to justify the separation stage, it is bypassed.

If the gas contains sufficient liquids to require separation, the gas from the separator outlet probably is *wet* and requires dehydration. Gas directly from the well may be sufficiently wet to require dehydration even though it does not contain sufficient liquid volume to justify the separation step. The gas is passed through a dehydrator that removes the moisture. The most common dehydrator uses glycol to absorb the water out of the gas. The diluted glycol is regenerated by heating to remove the water. Then it is recycled through the gas. The dehydration step can be omitted if the moisture content of the produced gas is less than the maximum allowed in the gas sales contract.

The gas is passed through a meter run to measure the volume (Fig. 10–25). The gas flows through a precisely sized orifice. Pressure on both sides of the orifice and the temperature of the gas are recorded on a chart and used to calculate the gas volume. The gas is always measured at the well except when a number of wells are on a lease. Then the gas production from the various wells is combined and measured through a single meter run.

At this point the gas passes into the gas sales line or transmission line. If the gas pressure is lower than the transmission line pressure, it must be compressed to increase the pressure.

High-pressure gas wells are frequently fitted with a high-low safety shutoff valve. The high-low valve is normally installed on the wing of the Christmas tree between the wing valve and the choke. The valve is actuated by pressure and shuts in the well if the pressure becomes too high or too low.

Wells are normally checked periodically but are not attended on a 24-hr basis. A broken line or ruptured pressure vessel such as a separator can cause a dangerous condition and a possible fire hazard because of the escaping gas and oil. A large amount of valuable production may also be lost. Line breaks or ruptured vessels cause a reduction in pressure. This pressure drop is detected by the high-low valve, which automatically shuts in the well.

The higher well pressures are confined by the choke when the well is flowing in a stabilized condition. However, if a gathering line or other line in the production facilities becomes plugged or some automatic valve fails to open and release pressure, fluid continues to flow out of the well through the choke and pressurize the lines. However, in most cases the maximum wellhead pressure is

FIG. 10–24 Gas well and production facilities. (a) A line heater warms the gas and prevents "freeze ups," (b) a dehydrator removes moisture from the gas, and (c) a gas compressor raises the pressure of the product gas to meet the pipeline pressure. The compressor control panel has complete instrumentation and safety shutoffs for 24-hr unattended operation.

considerably higher than the bursting strength of the gathering lines and production facilities.

For example, the normal test pressure rating of a low-pressure separator is in the order of 150–200 psi. Most wells have much higher shutin pressures. In this case the separator has safety devices such as the popoff valve and the pressure-relief plate. However, if these are actuated or ruptured, fluids are still lost. The high-low pressure shutoff detects increasing pressures. When these exceed a predetermined limit, the valve closes and shuts in the well. The high-low shutoff is an important overall safety device on high-pressure wells.

Artificial Lift

The well production rate is directly related to the difference between the static reservoir pressure (pore pressure) and the flowing bottom-hole pressure, assuming that all other factors are constant. This is commonly referred to as the flowing bottom-hole pressure differential or the *pressure differential*. Artificial lift is broadly defined as a method of reducing the flowing BHP, which increases the pressure differential and the well productivity, correspondingly. This increase is accomplished by supplementing the natural reservoir energy to lift the fluid column in the wellbore artificially. The same basic principle has been used for years in water wells pumped by windmills. In fact, the first pumping oil wells were produced by water-well pumping methods.

Artificial lift is normally not applied to gas wells except in special cases. Gas wells that produce a large volume of water may be pumped by conventional oil-well pumping techniques to remove the water. This reduces the flowing bottom-hole pressure and allows the well to produce gas at a maximum rate.

Economics is the primary criteria for determining when a well should be placed on artificial lift. Artificial lift is economical when the present worth of the future net revenue less the cost of installing and operating the artificial lift exceeds the present worth of the flowing well. Practical considerations, such as difficulties in installing lift equipment, and possible formation problems, such as sanding up due to the increased flow rate, also must be considered. Many oil wells are placed on artificial lift during completion; others are produced for a period before lifting. Again, economics is the main criteria.

Four common artificial lifting techniques are tubing and rod pumps, hydraulic pumps, submersible (electric) pumps, and gas lift. Each method has advantages and disadvantages, and the lifting method selected depends on formation and wellbore conditions, depths, location of the well site, availability of electric power or engine fuel, and other relevant factors.

Tubing and Rod Pumps

Tubing pumps were the first type of pump used to pump oil wells, and they are

550
DRILLING

FIG. 10–25 Gas is measured through a meter run (a). The gas passes through an orifice (b), and pressures and temperature are recorded on a gas chart (c), which is used to calculate the volume of gas.

still in use today. They are simply constructed and highly reliable. However, they tend to lose efficiency with depth.

The tubing pump has a pump barrel run on the bottom of the tubing. A gas anchor is fitted to the bottom of the pump barrel and is used to separate gas and oil downhole. The tubing is landed in a low-pressure wellhead and is fitted with a pumping tee. A plunger is run on sucker rods and is positioned so it can move up and down in the pump barrel. The top of the sucker rods are connected to a polish rod that reciprocates through a packoff, which serves as a low-pressure seal. The top of the polish rods are connected to the pumping unit, which reciprocates the rod string.

The tubing pump is a single-action pump. On the upstroke fluid flows through the open standing valve into the pump chamber. The traveling valve located on the pump plunger is close, and the entire column of fluid in the tubing is lifted upward. On the downstroke the standing valve is closed. The downward moving plunger displaces the fluid in the lower part of the pump barrel, forcing the fluid to flow through the open standing valve and into the rod-tubing annulus.

The efficiency of the pump depends greatly on the lower pump chamber being completely filled as the plunger moves upward. Gas is highly compressive and decreases pump efficiency. Therefore, the gas anchor is used to separate the gas and oil and to allow gas-free oil to flow into the pump barrel. Another cause of inefficiency is tubing movement caused by the reciprocating action of the pump rods. This action increases with increasing pump depth. A tubing anchor that has slips similar to a packer can be run on the tubing. The anchor seats with slips engaging the casing and prevents tubing movement during pumping.

The rod pump is an improved version of the tubing pump. It is completely self-contained including the pump barrel and is run through the tubing on the rods. The rod pump is held in place with a top-lock or bottom lock hold down. The hold-down nipples are run in the tubing string, and a collet-type lock is run on the pump. Another definite advantage of the rod pump is that the entire pump can be pulled with the rods for replacement or repairs and then rerun without pulling the tubing.

A standard procedure and a good practice is to run a pump-seating nipple in the tubing string when a flowing well is completed. Then the well can be placed on pump with a rod pump at a future date without pulling the tubing. It is only necessary to change the wellhead equipment, install a pumping unit, and run the rod pump.

Hydraulic Pumps

The hydraulic pump has a power piston (engine) connected to a pump piston. The complete pump is fitted inside a case with an outside diameter less than the inside diameter of the tubing. The hydraulic pump is driven by power fluid pumped by a power pump, usually a triplex plunger pump, located with the other surface facilities.

The pump is installed by first running a pump cavity on the bottom of the tubing, which is run in the hole and landed. A special wellhead is installed and connected to the power fluid pump. The wellhead has a lubricator-type assembly that is used to insert and remove the hydraulic pump. The hydraulic pump is inserted and pumped down the hole to seat in the pump cavity. The power fluid then acts through a series of ports, check valves, and reversing mechanism to drive the power piston or engine back and forth at a fast rate. The power fluid is discharged through ports in the side of the pump cavity and returns to the surface through the tubing-casing annulus.

The reciprocating movement of the power piston is transmitted to the pump piston. The pump piston pumps oil up the tubing-casing annulus. The produced oil and power fluid are separated with the oil going to the production battery and the power fluid recycling downhole.

If the hydraulic pump fails to operate properly, it can be circulated out of the hole by reversing the flow of the power fluid. If the pump sticks in the cavity, it can be recovered with a wire line. If this is not successful, the tubing must be pulled to recover the pump. Various pump designs are available for special applications. Hydraulic pumps do not have the problems associated with reciprocating rod strings and can be used in crooked and deviated holes.

Submersible Electric Pumps Submersible electric pumps are also used to pump oil wells. A long, slender, specially designed electric motor is connected to a turbine-type pump. The entire assembly is cased in a housing connected to the bottom of the tubing. An insulated three-conductor cable that carries electricity to the pump is clamped to the side of the tubing as the tubing is run in the hole. The tubing is landed in a special wellhead that provides a means of sealing both the tubing and cable. The electric motor drives the pump, which in turn pumps oil to the surface. Different sizes and horsepower ratings are available for a variety of applications.

Gas Lift The artificial lifting methods previously described use mechanical action to help lift or pump the oil to the surface. Gas lift utilizes a different principle. The engery stored in compressed gas is used to help lift the oil to the surface.

Various well parameters are used to calculate gas-lift design, which includes the spacing for gas mandrels in the tubing string, the type of mandrels, and the gas pressure in the casing required to produce the well. Gas-lift mandrels with valves in place are installed in the tubing string at predetermined intervals as the tubing string is run into the hole. The tubing is spaced out and landed in a packer, and the Christmas tree is installed. A source of high-pressure gas is connected to the tubing-casing annulus. Gas flows into the tubing through the gas-lift valves after the fluid has been displaced out of the tubing-casing annulus.

The gas enters the tubing and begins to move upward, expanding as the pressure decreases. This provides a lifting effect, which reduces the bottom-hole flowing pressure and permits a larger volume of oil to enter the wellbore. The produced gas and oil are separated with the oil moving to the production battery. Gas is sold or is compressed and reinjected into the well.

Information obtained from drilling, well logs, and offset well data are used to design completion programs. Factors such as hole size, depth, formation type, pressure, and number of prospective producing zones must be considered in the design. Various completion tools, equipment, and procedures are available.

Single completions are the most common, and some wells are completed in two separate formations as duals. Open-hole, slotted-liner, permanent, and multizone commingled completions are incorporated in single and dual completions.

Completion operations include preparing the production casing, perforating, testing, stimulating, cleaning up, installing equipment, and testing production. Fishing and casing repair operations may be required. Wellhead and production facility designs depend on the type of completion, produced volumes and pressures, and liquid or gas production. Then artificial lift is installed in oil wells as soon as it is economically justified.

Most of the funds assigned to drilling and producing an oil and gas well are spent long before it is time to complete a well. However, this phase is the final, important step to producing the energy needed to move the world.

APPENDIX

SERVICE COMPANIES

As the industry matured and operations become more involved and complex, more specialization was required. Frequently, one company would not have enough operations to justify the cost of maintaining a specialist in one activity or operation that was only used to a limited extent. The problem was still more acute when a high capital investment in equipment was required.

This led to the formation of service companies that specialize in one operation or activity and supply this service to the entire industry. Many supporting services furnish materials, supplies, and technical services. They are generally familiar with overall operations and are very knowledgeable in their specific fields.

The largest single group of third-party services is consultants. In a manner of speaking, all third-party services are consultants, but the term is generally reserved for those individuals who utilize their technical expertise to advise the owner-operator and essentially do not include equipment and materials in the services they furnish.

Engineers, geologists, landmen, operations supervisors, and specialists in all other facets of the industry are consultants. These groups range from a small one-man operation to large, fully staffed companies. The larger companies can take a drilling prospect and design the drilling program, obtain all permits, contract a drilling rig, order all materials, and provide supervision during the drilling and completion operations. They can provide all or part of these services, depending on the requirements of the operator. The smaller consultant groups tend to provide specialized services in only one field such as drilling, fishing, completions, workovers, or mud treatment.

As one can see, many supporting personnel and services provide third-party services to the industry. Many supply more than one service, either under one company name or various subsidiary company names. The owner-operator must be able to select the best services, materials, and equipment suppliers for the particular drilling operation. This appendix may give the reader some indication of the kinds of services the operator may choose from. The particulars of these services are detailed in the preceding text.

Bit Suppliers

Bit suppliers sell drilling bits, usually through sales representatives or bit salesmen. In the past a bit salesman would leave bits at the rig, and the bit was not sold until it was used. The current practice is to sell the bit before it is delivered to the drilling site. If the bit is not used, then it is returned for credit.

Bit companies keep extensive records on bit performance and are usually a competent source for recommendations. They also can supply a recommended bit program for a proposed well. This includes recommended bit sizes, types of bits, operating conditions, bit life, footage, and bit costs per foot. It is common practice to obtain a recommended bit program from several bit companies before a well is drilled.

Blowout Preventer Tests

Blowout preventers must be test operated to ensure that they function correctly. Then they are pressure tested to ensure that they will hold the required pressure. Some companies have the equipment and personnel to perform these tests and make repairs as needed, usually with the help of the rig crews. At the end of the test, they submit a complete test report, which is often required by regulatory agencies.

Blowout Specialists

A blowout is uncommon but does occur. Special equipment and techniques are required to kill blowouts and bring them under control. Several companies have personnel who are experienced in this procedure.

Casing Crews

Some drilling operations such as running and cementing casing and laying down drillpipe require extra men and special equipment. Casing crew companies fill this demand. They supply the personnel and equipment, such as hydraulic tongs, pick-up or lay-down lines, special thread protectors, and casing stabbing board.

Cementing Services

Cementing services include cementing casing, squeezing, plugging, and other operations in which cement is pumped into the hole. Various companies supply cementing services, ranging from large companies operating on an international basis to small independents that operate in a localized area. Cementing companies usually supply a complete service including:

- Engineers to design and write cement programs as requested by the owner and to supervise on-site cementing operations
- Laboratories and technicians to test different cement slurries and additives to ensure that they will perform satisfactorily in the specific application for which they are designed
- A source of cement supply with equipment to dry-blend additives, bulk hopper trucks to transport the dry cement to the well site, cement storage at the wellsite if needed, mixing equipment to mix the cement on site, and pumping equipment (pump trucks) to pump and displace the cement into the well

Larger cement companies also provide similar facilities for marine operations.

Most cementing companies also supply cementing equipment to be run on the casing, commonly referred to as float equipment, such as guide shoes, float shoes, float collars, automatic fill-up equipment, cement baskets, centralizers, and scratchers. This equipment is also available from other suppliers.

Directional Control Companies
In drilling operations the drilled hole often tends to deviate, causing a crooked hole. Sometimes the hole must be deviated intentionally to sidetrack a fish, to drill multiple holes into one reservoir from a central surface point, and to drill a well to kill a blowout. These companies usually have both the technical personnel and the equipment to conduct wellbore deviation operations.

Fishing Tool Companies
Fishing tool companies specialize in providing technical supervision and the tools and equipment needed to recover pipe and other items left in the borehole. A few companies supply only the personnel; other companies supply only the tools. But the majority of companies supply both personnel and tools. These companies also frequently provide related equipment and services such as milling and sidetracking operations.

Formation Testers
Formation test companies provide the special equipment, tools, technical supervision, and analysis facilities to conduct formation testing in open or cased holes. The companies have technical personnel who recommend the testing procedure and type of equipment to be used. They also provide the testing equipment and technical supervision to operate the equipment and conduct the tests. After the test is completed, they provide a report with test results, including bottom-hole flowing and shutin pressures, pressure-buildup analyses, well productivity and flow rates, and analyses of the produced fluids.

Machine Shops
Machine shops provide many special services needed in the drilling operation. Most of the work is done in the machine shop because the heavy shop equipment is not portable. For example, repairing damaged tool joints, modifying or building tools for special conditions, and similar operations are usually done in the machine shop. Some machine shops also have portable equipment that can be used at the wellsite. For example, drillpipe can be hard banded to prevent excess wear at the drilling site.

Mud Suppliers
The main function of mud suppliers or mud companies is to sell drilling mud and/or mud additives. All of the companies sell mud additives. However, some do not supply technical services. Others supply technical services at an extra charge, and some supply technical services and include the cost in the sale price of their products. The companies range in size from large companies that operate internationally and have thousands of employees to small companies with a few employees who service a limited geographical area. These latter companies are sometimes referred to as independent mud companies. Most of the larger mud companies have a research lab and can troubleshoot to solve specific mud problems.

Most mud companies supply a recommended mud program for a well to be drilled. It is common practice to request a mud program from several companies. The program gives the recommended type of mud and mud properties to be used in the well at various depth intervals. More detailed mud programs give the recommended mud hydraulics—jet velocities, annular and cuttings slip velocities, and pressure losses in the circulating equipment. They may also give information on contaminating formations, geopressured zones including pore pressure plots, and other hole problems that may be encountered. These programs are usually well detailed and are very helpful in drilling the well, especially if the mud company is experienced in the area.

The mud companies are also the main suppliers of various items of equipment for mud treatment, solids removal, gas separation, mud-mixing units, and mud storage tanks. Other companies also supply this equipment but may not supply mud additives. Some mud companies sell liquid mud to drill the well and buy back the mud at the end of the well.

Mud Logging
Mud logging is the process of examining and analyzing the mud and cuttings circulated out of the hole during drilling. Mud-logging units are supplied by mud-vending companies, equipment supply companies, and independents spe-

cializing in mud-logging services. The units are usually operated by a geologist or geological technician. They can be operated from 8–24 hr/d and in some cases with two people on each shift, depending on the services required. The testing-sampling-monitoring services can include the following:

- Bit cuttings analysis and description
- Mud and bit cuttings gas
- Information on gas and oil shows
- Penetration rate curve
- Circulation rate, pump strokes, pressure, and lag time
- Flow line and mud pit levels and pit volume totalizer
- Weight on bit
- Rotary speed and torque
- Pore pressure plots
- Other specialized information as may be required

Mud-logging units are used on the deeper, more complex wells, especially wildcats.

Pipe and Equipment Rentals

Many types of pipe and equipment including special tools are used in oil-field operations. For example, if a producing well is to be reworked, normally the pipe and equipment are rented. Many common drilling tools such as drill collars, stabilizers, reamers, and shock subs are commonly rented. Pipe and equipment rental companies supply this equipment. Some companies have a wide range of rental equipment. The operator is normally responsible for the equipment from the time it leaves the rental company yard until it is returned. The operator must reimburse the rental company for any equipment damaged or lost in the hole.

Pipe Testing and Inspection

It is often necessary to inspect and test pipe to prevent failure. Some companies supply a complete range of pipe testing and inspection services, while others specialize in one type of service. For example, some companies specialize in inspecting drilling assembly tool joints using either the black light or sonic method. The equipment is portable, and these services are provided either at the company yard or at the wellsite. Most companies that provide complete pipe testing and inspection services—including the connector and body or tube—have portable equipment and can perform the testing and inspection services wherever needed. For example, it is relatively common to test casing at the wellsite before running it, to test tubing while it is being run, or to test drillpipe when the rig is being moved. Many of these companies also have pipe straightening equipment.

Pipe Vendors

Pipe vendors sell tubular goods used in drilling operations. These include drillpipe, conductor pipe, casing, liners, tubing, line pipe, and special-purpose tubulars. Pipe can be obtained in various sizes, lengths, weights, grades of steel, and different types of connections (connectors). The more common types of pipe are normally available from stock. Other less common items are special ordered, which usually causes some delay in delivery. The actual specifications of the pipe to be ordered and used depend on the specific conditions and requirements of the well. Some vendors, usually the larger ones, have a full line of tubular goods. Others specialize in one type such as drillpipe suppliers. Some vendors specialize in used pipe.

Most major pipe vendors provide a casing design service. When the operator plans to drill a well, the pipe vendor is supplied with all of the necessary data relating to the well to be drilled. This is used to calculate a recommended casing program. The program includes the lengths and specifications such as safety factors of the various sections of pipe needed for each individual casing. Cost data are also included.

Safety Equipment and Monitoring

When hazardous drilling conditions are encountered, such as drilling in formations containing hydrogen sulfide, special safety precautions must be taken. Safety equipment companies are available to supply all of the equipment and technical guidance including on-site supervision needed in these cases.

Stimulation Services

Newly completed wells are frequently stimulated to obtain maximum production, and older wells are often stimulated to increase well productivity. Well stimulation requires technical planning, materials and supplies to stimulate the well, wellsite storage, and large-volume, high-pressure pumping equipment. Stimulation companies recommend stimulation treatments and provide the technical supervision, materials, and equipment to perform the job. Most of these companies also conduct cementing operations since similar equipment is used.

Supply Stores

Supply stores are the "country stores" of the oil industry. They are located in areas of activity, and they stock items commonly used in that area. They are generally units of a large company that manufactures a variety of equipment and supplies. The supply store also can order and deliver equipment from other suppliers.

Training Courses and Seminars

Many private companies, universities and colleges with petroleum engineering curriculum, service companies, and some of the larger oil and gas companies have training courses and seminars covering all facets of the oil and gas industry. These courses range from one day to several weeks and cover almost every type of equipment and operational procedure in the industry.

Trucking Contractors

Oil-field equipment is usually large and heavy. Special, heavy-duty trucks and pipe trailers are needed to move the equipment. A number of oil-field trucking contractors are located in the major areas of activity. These contractors specialize in moving drilling rigs. They also have large cranes to lift equipment onto the rig floor when rigging up or off of the rig floor when tearing down.

Well Logging and Perforating

Various companies perform all types of logging, perforating, and specialized services. Other companies operate in certain geographical areas. These companies, which are usually smaller, may only offer services that are more commonly used in that area of operations. Some services have very limited demand, are very specialized, and require special equipment. In these cases only a few companies may supply the service, and some companies only provide one specialized service.

INDUSTRY TRADE JOURNALS, MAGAZINES, AND NEWSLETTERS

Business Research Publications Inc.
  87 Terminal Drive
  Plainview, New York 11803

Drilling Contractor
  3737 Westcenter, Suite B
  Houston, Texas 77042

Energy Management Report
  P.O. Box 1589
  Dallas, Texas 75221

Energy News
  P.O. Box 1589
  Dallas, Texas 75221

Energy Week
  P.O. Box 1589
  Dallas, Texas 75221

Federal Programs Advisory Service
  2120 L Street N.W., Suite 210
  Washington, D.C. 20037

Geothermal Resources Council Bulletin
  P.O. Box 96
  Davis, California 95617

Geotimes
  American Geological Institute
  5205 Leesburg Pike
  Falls Church, Virginia 22041

Appendix

Gulf Coast Oil Reporter
660 Bannock Street, Suite 160
Denver, Colorado 80201

Hotline Energy Reports
P.O. Box 2934
Casper, Wyoming 82602

Journal of Metals
The Metallurgical Society of AIME
420 Commonwealth Drive
Warrendale, Pennsylvania 15086

Journal of Petroleum Technology
1701 Brun Street, Suite 101
Houston, Texas 77019

MediaCom Industry Reports
Dept OT
4545-SN Industrial Street
Simi Valley, California 93063

Montana Oil and Gas Journal
3915 Palisades Park Drive
Billings, Montana 59104

New Mexico Oil and Gas Reporter
1420 Carlisle N.E., Suite 201
Albuquerque, New Mexico 87110

Northeast Oil Reporter
660 Bannock Street, Suite 160
Denver, Colorado 80201

Northwest Oil Report
4204 S.W. Condor
Portland, Oregon 97201

Ocean Oil Weekly Report
P.O. Box 1260
Tulsa, Oklahoma 74101

Offshore
P.O. Box 1941
Houston, Texas 77251

Oil & Gas Journal
1421 South Sheridan Road
Tulsa, Oklahoma 74101

Oil Patch
2620 Fountainview, Suite 115-B
Houston, Texas 77057

Oregon Oil and Gas Report
P.O. Box 393
Portland, Oregon 97207

Pacific Energy News
P.O. Box 40045
Portland, Oregon 97402

Petroleum Engineer International
800 Davis Building
Dallas, Texas 75221

Petroleum Equipment News
20 Community Place
Morristown, New Jersey 07960

Petroleum Information Corporation
P.O. Box 2612
Denver, Colorado 80201

Petroleum Information International
4150 Westheimer
Houston, Texas 77001

Pipeline and Gas Journal
P.O. Box 1589
Dallas, Texas 75221

Society of Petroleum Engineers Journal
5115 South Vandalia, Suite E
Tulsa, Oklahoma 74135

Southwest Oil and Gas News
P.O. Box 25847
Scottsdale, Arizona 85252

Washington Oil and Gas Report
P.O. Box 393
Portland, Oregon 97207

Well Servicing
6060 North Central Expressway, Suite 538
Dallas, Texas 75206

Western Oil Reporter
660 Bannock Street, Suite 160
Denver, Colorado 80201

World Oil
9301 Allen Parkway
Houston, Texas 77001

FEDERAL REGULATORY AGENCIES

Army Corps of Engineers
20 Massachusetts Avenue N.W.
Washington, D.C. 20314

Department of Energy (DOE)
Forrestal Building
1000 Independence Avenue S.W.
Washington, D.C. 20585

   Federal Energy Regulatory Commission (FERC)
   825 North Capitol Street N.E.
   Washington, D.C. 20454

   Bartlesville Energy Technology Center
   P.O. Box 1398
   Bartlesville, Oklahoma 74005

   Grand Forks Energy Technology Center
   P.O. Box 8213
   University Station
   Grand Forks, North Dakota 58202

   Laramie Energy Technology Center
   P.O. Box 3395
   University Station
   Laramie, Wyoming 82070

Note: These centers have a number of field offices and also maintain copies of the maps, reports, etc., on file in various public and university libraries.

Department of the Interior
Eighteenth and C Streets N.W.
Washington, D.C. 20240

Bureau of Land Management (BLM)
Interior Building
Washington, D.C. 20240

Bureau of Mines
2401 East Street N.W.
Washington, D.C. 20241

U.S. Geological Survey
National Center
Reston, Virginia 22092

Note: The Geological Survey distributes maps and reports from a number of different centers.

Department of Transportation (DOT)
2100 Second Street S.W.
Washington, D.C. 20590

U.S. Coast Guard
2100 Second Street S.W.
Washington, D.C. 20593

U.S. Department of Commerce
National Technical Information Service
5285 Port Royal Road
Springfield, Virginia 22161

U.S. Environmental Protection Agency
401 M Street S.W.
Washington, D.C. 20460

U.S. Synthetic Fuels Corporation
1200 New Hampshire Avenue N.W.
Washington, D.C. 20586

STATE REGULATORY AGENCIES

Alabama
State Oil and Gas Board of Alabama
   Ernest A. Mancini, Oil and Gas
      Supervisor
   P.O. Drawer O
   University, Alabama 35486

Alaska
Alaska Oil and Gas Conservation
   Commission
   C.V. Chatterton, Chairman
   3001 Porcupine Drive
   Anchorage, Alaska 99501
State of Alaska
   Department of Natural Resources
   Division of Geological and
      Geophysical Surveys
   P.O. Box 80007
   College, Alaska 99708

Arizona
Oil and Gas Conservation Commission
   A.K. Doss, Executive Director
   1645 Jefferson, Suite 420
   Phoenix, Arizona 85007
Arizona Bureau of Geology and
   Mineral Technology
   Larry D. Fellows
   845 North Park Avenue
   Tucson, Arizona 85719

Arkansas
Arkansas Oil and Gas Commission
   William E. Wright, Director
   314 East Oak Street
   El Dorado, Arkansas 71730
Arkansas Geological Commission
   Norman F. Williams, State Geologist
   3815 West Roosevelt Road
   Little Rock, Arkansas 72204

California
Department of Conservation
   M.G. Mefford, Oil and Gas Supervisor
   1416 Ninth Street, Room 1310
   Sacramento, California 95814
California Department of Conservation
   James F. Davis, State Geologist
   1416 Ninth Street, Room 1341
   Sacramento, California 95814

Colorado
Oil and Gas Conservation Commission
   Douglas V. Rogers, Director
   1313 Sherman Street, Room 721
   Denver, Colorado 80203
Colorado Geological Survey
   John W. Rold, Director and State
      Geologist
   1313 Sherman Street, Room 715
   Denver, Colorado 80203

Delaware
Delaware Geological Survey
   Robert R. Jordan, State Geologist
   University of Delaware
   101 Penny Hall
   Newark, Delaware 19711

Florida
Department of Natural Resources
   Charles H. Hendry Jr., Administrator
   903 West Tennessee Street
   Tallahassee, Florida 32304

Georgia
Department of Natural Resources
   William H. McLemore, State Geologist
   19 Martin Luther King Drive, S.W.
   Atlanta, Georgia 30334

Hawaii
Department of Land and Natural
  Resources
Robert T. Chuck, Manager–Chief
  Engineer
P.O. Box 373
Honolulu, Hawaii 96809

Idaho
Oil and Gas Conservation Commission
Gordon C. Trombley, Director
Statehouse
Boise, Idaho 83720

Oil and Gas Conservation Commission
W.R. Pitman, Petroleum Engineer
P.O. Box 670
Coeur D'Alene, Idaho 82814

Illinois
Department of Energy and Natural
  Resources
George R. Land, Director
704 Stratton Office Building
400 South Spring Street
Springfield, Illinois 62706

Department of Energy and Natural
  Resources
Richard H. Howard, Geologist
Natural Resources Building
615 East Peabody Drive
Champaign, Illinois 81820

Indiana
Department of Natural Resources
Homer R. Brown, Director
State Office Building
Indianapolis, Indiana 46204

Department of Natural Resources
John B. Patton, State Geologist
611 North Walnut Grove
Bloomington, Indiana 47405

Iowa
Iowa Geological Survey
  Donald L. Koch, Director and State
    Geologist
  123 North Capitol Street
  Iowa City, Iowa 52242

Kansas
Corporation Commission
R.C. Loux, Chairman
State Office Building
Topeka, Kansas 66612

Kentucky
Department of Mines and Minerals
Henry M. Morgan, Director
Box 690
Lexington, Kentucky 40586

Louisiana
Department of Natural Resources
Ray T. Sutton, Commissioner
Box 44275
Baton Rouge, Louisiana 70804

Maryland
Department of Natural Resources
Kenneth N. Weaver, Director
The Rotunda
711 West 40th Street, Suite 440
Baltimore, Maryland 21211

Michigan
Department of Natural Resources
  Howard A. Turner, Director and
    Supervisor of Wells
  Box 30028
  Lansing, Michigan 48909

Department of Natural Resources
  R. Thomas Segall, Acting State
    Geologist
  Box 30028
  Lansing, Michigan 48909

Minnesota
Minnesota Department of Natural
 Resources
 Division of Minerals
 Centennial Office Building
 658 Cedar Street
 St. Paul, Minnesota 55155
Minnesota Geological Survey
 1633 Eustis Street
 St. Paul, Minnesota 55108

Mississippi
State Oil and Gas Board
 Clyde R. Davis, Supervisor
 Box 1332
 Jackson, Mississippi 39205

Missouri
Missouri Department of Natural
 Resources
 Wallace B. Howe, Administrator
 P.O. Box 250
 Rolla, Missouri 65401

Montana
Department of Natural Resources and
 Conservation
 Charles G. Maio, Administrator and
  Petroleum Geologist
 2535 St. Johns Avenue
 Billings, Montana 59702
Montana Bureau of Mines and Geology
 Butte, Montana 59701

Nebraska
Oil and Gas Conservation Commission
 Paul H. Roberts, Director
 Sidney, Nebraska 69162

Nevada
Department of Conservation and
 Natural Resources
 Pamela G. Cosby, Administrator
 Nye Building, Capitol Complex
 Carson City, Nevada 89710

Nevada Bureau of Mines and Geology
 University of Nevada—Reno
 Reno, Nevada 89507

New Hampshire
Office of State Geologist
 Robert I. Davis
 117 James Hall, University of New
  Hampshire
 Durham, New Hampshire 03824

New Mexico
Energy and Minerals Department
 Joe D. Ramey, Director
 P.O. Box 2088
 Santa Fe, New Mexico 87501
Institute of Mining and Technology
 Frank E. Kottlowski, Director
 Socorro, New Mexico 87801

New York
Department of Environmental
 Conservation
 Robert F. Flacke, Commissioner
 50 Wolf Road, Room 404-A
 Albany, New York 12233
State Geological Survey
 State Geologist
 Room 3140
 Cultural Education Center, Empire
  State Plaza
 Albany, New York 12230

North Carolina
Department of Natural Resources and
 Community Development
 Stephen G. Conrad, Director
 512 North Salisbury Street
 Raleigh, North Carolina 27611

North Dakota
North Dakota Industrial Commission
 Wesley D. Norton, Chief Enforcement
  Officer
 900 East Boulevard
 Bismarck, North Dakota 58505

North Dakota Geological Survey
University Station
Grand Forks, North Dakota 58202

Ohio
Department of Natural Resources
Andrew G. Skalkos, Chief
Fountain Square
Columbus, Ohio 43224

Oklahoma
Oklahoma Corporation Commission
Hamp Baker, Chairman
2101 North Lincoln Boulevard, Suite 228
Oklahoma City, Oklahoma 73105

Oklahoma Geological Survey
Charles J. Mankin, Director
830 Van Vleet Oval, Room 163
Norman, Oklahoma 73019

Oregon
Department of Geology and Mineral Industries
Donald A. Hull, State Geologist
1005 State Office Building
Portland, Oregon 97201

Pennsylvania
Department of Environmental Resources
John A. Ifft, Chief
1205 Kossman Building
100 Forbes Avenue
Pittsburgh, Pennsylvania 15222

South Carolina
South Carolina Geological Survey
Eugene A. Laurent, Director
Harbison Forrest Road
Columbia, South Carolina 29210

South Dakota
Supervisor of Oil and Gas
Department of Water and Natural Resources
Division of Water Quality
36 East Chicago
Rapid City, South Dakota 57701

South Dakota Geological Survey
State Geologist, Merlin J. Tipton
Science Center, University of South Dakota
Vermillion, South Dakota 57069

Tennessee
Tennessee Department of Conservation
Robert E. Hershey, State Geologist
701 Broadway
Nashville, Tennessee 37203

Texas
Texas Railroad Commission
James E. Nugent, Chairman
Drawer 12957
Austin, Texas 78711

Oil and Gas Division
Bob R. Harris, Director
Drawer 1967
Austin, Texas 78711

Bureau of Economic Geology
The University of Texas at Austin
University Station, Box X
Austin, Texas 78712

Utah
Natural Resources and Energy
Cleon B. Feight, Director
1588 West North Temple
Salt Lake City, Utah 84116

Utah Geological and Mineral Survey
606 Black Hawk Way
Salt Lake City, Utah 84108

Vermont
Natural Gas and Oil Resources Board
Pavilion Office Building
Montpelier, Vermont 05620

Vermont Geological Survey
Charles A. Ratte, State Geologist
State Office Building Post Office
Montpelier, Vermont 05620

Virginia
Department of Labor and Industry
  Robert F. Beard Jr., Commissioner
  219 Wood Avenue
  Big Stone Gap, Virginia 24219
Virginia Division of Mineral Resources
  Robert C. Milici, Commissioner and
    State Geologist
  P.O. Box 3667
  Charlottesville, Virginia 22903

Washington
Department of Natural Resources
  Raymond Lasmanis, Oil and Gas
    Supervisor
  Olympia, Washington 98504

West Virginia
Oil and Gas Conservation Commission
  Thomas E. Huzzey, Commissioner
  1613 Washington Street East
  Charleston, West Virginia 25311

Wisconsin
Geological and Natural History Survey
  M.E. Ostrom, Director and State
    Geologist
  1815 University Avenue
  Madison, Wisconsin 53706

Wyoming
Wyoming Oil and Gas Conservation
  Commission
  Donald B. Basko, Oil and Gas
    Supervisor
  P.O. Box 2640
  Casper, Wyoming 82602
The Geological Survey of Wyoming
  Gary B. Glass, State Geologist
  University Station, Box 3008
  Laramie, Wyoming 82071

BIBLIOGRAPHY AND SUGGESTED READINGS

Adams, Neal, and Marsha Frederick. "How to Estimate Well Costs." *OGJ*, December 6, 1982, pp. 131–136; December 13, 1982, pp. 104–111; December 27, 1982, pp. 194–198.
A Dictionary of Petroleum Terms. second edition, Austin: Petroleum Extension Service, University of Texas, 1979.
Allen, James H. "Computer Optimizes Operations." *OGJ*, October 10, 1978, p. 128.
───── "Determining Parameters that Affect Rate of Penetration." *OGJ*, October 3, 1979, p. 94.
API Recommended Practice for Care and Use of Casing and Tubing. API RP 5Cl, 10th edition, March 1975.
API Specification for Casing, Tubing and Drill Pipe. API Spec 5A, 33rd edition, March 1976.
Baker, Ron. *A Primer of Oilwell Drilling.* second edition, Austin: Petroleum Extension Service, University of Texas, 1979.
Bowman, Glenn R. "Linear Leaks: How They are Detected and Isolated." *World Oil*, December 1982, pp. 53–58.
Bradley, W.B. "Drill Pipe Rubbers Reduce Rotating Wear at High Contact Loads." *OGJ*, January 20, 1975, p. 87.
───── "Factors Affecting the Control of Borehole Angle in Straight and Directional Wells." SPE Reprint No. 5070, 1974.
Bush, H.E. "Current Techniques for Combating Drill Pipe Corrosion." API Reprint No. 906-11-C for API conference, March 16–18, 1966.
Coffeen, J.A. *Seismic Exploration Fundamentals.* Tulsa: PennWell, 1982.
"Corrosion Control in the Well Bore." *Petroleum Engineer*, August 1976, pp. 50–62.
Desk & Derrick Standard Oil Abbreviator. Tulsa: PennWell, 1973.
Dewald, Omar E. "Severe Storms Still a Major Threat." *Offshore*, June 5, 1981, pp. 60–63.
Drilling Assembly Handbook. Drilco, 1977 edition.
"Drilling Equipment Plays Vital Role." Technology Conference, *Drilling Contractor*, March 1982, pp. 56–64.
Drilling Manual. ninth edition, International Association of Drilling Contractors, June 1974.
Eaton, Ben A. "Fracture Gradient Prediction and Its Application to Oil Field Operations." *JPT*, October 1969.
"Effect of Mud Column Pressure on Drilling Rates." *Trans.*, AIME, 1955.
Fertl, Walter H. "Knowing Basic Reservoir Parameters First Step in Log Analysis." *OGJ*, May 22, 1978, pp. 98–118.
Forgotson, J.M. "Indication of Proximity of High Pressure Fluid Reservoir, Louisiana and Texas Gulf Coast." *AAPG Bulletin*, Volume 53, January 1969, pp. 171–173.

Bibliography and Suggested Readings

Hammer, Sigmund. "Airborne Gravity is Here." *OGJ*, January 11, 1982, pp. 113–126.

Hooks, R.A., L.W. Cooper, and B.A. Payne. "Air Drilling." Drilling Technology Conference, International Association of Drilling Contractors, March 16–18, 1977.

Horvitz, Leo. "Near-Surface Evidence of Hydrocarbon Movement from Depth." American Association of Petroleum Geologists, 1980, pp. 241–270.

Howard, G.C., and C.R. Fast. *Hydraulic Fracturing*. Monograph Volume 2, Dallas: Society of Petroleum Engineers of AIME, 1976.

Kading, W.H. and O.J. Shirley. "Temperature Surveys: the Art of Interpretation." *JPT*, November 1966.

Knowles, Ruth Sheldon. *The Greatest Gambler*. second edition, Norman: University of Oklahoma Press, 1978.

Landman's Legal Handbook. Denver: Rocky Mountain Mineral Law Foundation.

Leonard, Jeff, and Robert D. Dudley. "Guide to Drilling, Workover, and Completion Fluids." *World Oil*, June 1981, pp. 77–124.

Lindsey, H.E. Jr. "Running and Cementing Deep Well Liners." *World Oil*, November–December 1974, January 1975.

Lubinski, A., W.A. Althouse, and J.L. Logan. "Helical Buckling of Tubular Goods." *JPT*, June 1962.

Material Classification Manual. Bulletin No. 6, Council of Petroleum Accountants Societies of North America, revised May 1971.

Maurer, W.C., et al. "Down Hole Drilling Motors." *Technical Review*, U.S. Energy Research and Development Administration, Contract No. EY 76 CV24037.M001.

McNair, Will L. "SCR Rigs Save on Fuel, Maintenance, and Rig-Up Costs." *World Oil*, June 1980, pp. 223–227.

———. *The Electric Drilling Rig Handbook*. Tulsa: PennWell, 1980.

Montgomery, M.S., and W.H. Marshall. "A Simple Introduction to Solids Control." *Drilling Contractor*, February 1982, pp. 21–32.

Moore, Preston L. *Drilling Practices Manual*. Tulsa: PennWell, 1974.

Murphy, Don R. "A Practical Engineering Approach to Running Bits." *OGJ*, February–March 1969.

Nesvail, Verne R. "Multi-leg Hole Successfully Drilled for Degasification." *Coal Age*, September 1980, pp. 144–150.

Nicholson, Robert W. "Charts Find Acceptable Dogleg-Severity Limits." *OGJ*, April 15, 1974.

Olsen, Odel A., et al. "Deepwater Operations Demand Safe." *Petroleum Engineer International*, May 1982, pp. 56–90.

"Proper Tubular Makeup Depends on Right Torque." *Western Oil Reporter*, May 1978, pp. S15–S20.

Rehm, W., and R. McClendon. "Measurement of Formation Pressure from Drilling Data." SPE Paper No. 3601, 1971.

Ruseska, I., J. Robbins, and J.W. Costerton. "Biocide Testing against Corrosion-Causing Oil-Field Bacteria Helps Control Plugging." *OGJ*, March 8, 1982, pp. 253–268.

Bibliography and Suggested Readings

Short, J.A. *Drilling and Casing Operations*. Tulsa: PennWell, 1982.
_____ *Fishing and Casing Repair*. Tulsa: PennWell, 1982.
Smith, Dwight K. *Cementing*. Monograph Volume 4, Dallas: Society of Petroleum Engineers of AIME, 1976.
Stratigraphic Oil and Gas Fields—Classifications, Exploration Methods, and Case Histories. edited by Robert E. King. APPG Memoir 16, SEG Special Publication No. 10, Tulsa: American Association of Petroleum Geologists and Society of Exploration Geophysicists, 1972.
Striegler, John. "ARCO Finishes Fourth Horizontal Drain Hole." *OGJ*, May 24, 1982, pp. 58–61.
Suman, George O. Jr., and Robert E. Snyder. "Primary Cementing: Why Many Conventional Jobs Fail." *World Oil*, December 1980, pp. 59–66.
Suman, G.O. Jr., and R.C. Ellis. *Cementing Handbook*. Houston: Gulf Publishing, 1977.
Timmerman, E.H. *Practical Reservoir Engineering*. Tulsa: PennWell, 1981, 1982.
Webb, Glenn W. "Mechanical?—DC?—SCR? Which Rig is Best for You?" *Drilling DCW*, May 1977, pp. 43–120.
Weintritt, D. "Corrosion Inhibition in Drilling Operations." *Oil Patch*, November 1980.
"Well Planning for Drilling in Hydrogen Sulfide Formations." SPE Paper No. 6655, Sour Gas and Crude Symposium, 1977.
Williams, Michael P., and Lawrence L. Hoberock. "Solids Control for the Man on the Rig." *Petroleum Engineer International*, October 1982, pp. 56–84; November 1982, pp. 102–110; December 1982, pp. 50–56.
Wilson, Jerry. "Factors to Consider for Selecting the Proper Bottom-Hole Drilling Assembly." presented at the 1979 Drilling Technology Conference, Denver: International Association of Drilling Contractors, March 6–8, 1979.
Woods, H.B. and A. Lubinski. "Practical Charts for Solving Problems in Hole Deviation." *API Drilling and Production Practices*, 1954, p. 56.

INDEX

INDEX

A

abandonment, 451, 474: monument, 452; regulations, 451; well, 60
abnormal pressure gradients, 88: *see also* pressure gradients
abstract, 124
accelerators, 148, 321, 344, 368
access road, 245–246
accumulator, 176, 223–224, 257, 321, 368
ac current, 200
acetic acid, 495
acidizing, 466, 495: cleanout, 518, 532; wash, 498
acre-feet, 106
active solids, 159
ad valorem tax, 122
aerial photographs, 34, 53
A-frame, 176, 178, 183, 258, 261
afterflow, 493
ages (geological), 21
air: compressor, 176, 299; drilling, 8, 159, 228, 300; head, 229; mist, 159; weight, 302
airplane, cargo, 289
alarm, signal, 368, 418
Alaskan North Slope, 429
alcohol, 498
all-weather road, 245
alternate drilling program, 330
alternating current, 200
alternators, 200
ambient temperature, 78
angle: beam, 179; buildup, 130, 431, 433; hole, 124, 374; iron, 179; of deviation, 132; of repose, 282
anhydrite, 17, 21, 58, 159, 371, 409
annual rental, 125
annular velocity, 208, 322
annulus, 176, 212–214, 322, 341, 536
anticline, 25–26, 45–46
API gravity, 40, 97
aquifers, 130, 135, 451
archeological study, 243

area of interest, 21, 54, 120
artificial lift, 375, 463–464, 494, 512, 523
assembly, 141, 149, 156, 159: angle-building, 374, 376; deviation, 409, 431–432, 500; drill-out, 352; drilling, 141, 149, 153, 156, 159, 160–162, 275, 306; keyseat wiping, 380, 386; limber, 352, 376–377; packed hole, 378; pendulum, 377–378, 429; pilot, 427; reaming, 330; small diameter, 407; stiff, 377, 408
assistant drilling superintendent, 170, 172
athey wagons, 253
atmospheric pressure, 86
atomic disintegration, 78
attapulgite, 160
auger, 246
authority for expenditure (AFE), 121, 163
automatic driller, 356–357

B

back in, 126
back off, 72, 296, 354, 380, 390, 400–402
back pressure, 527: valve, 527
back-up systems, 424
baffles, 215, 219
bag-type preventer, 344–345, 368, 421–422, 424: *see also* preventer
bailing, 201, 413
balanced drilling, 359
balled up (bit), 302, 320
ball out, 497, 518, 520
barge, 174, 233, 290, 292: cargo, 29; double, 234; rig, 234, 291
barite, 218, 282, 418: pill, 321; slug, 231; storage, 77, 210
barrel, oil-field, 107
barrier island, 233
base, rig, 176, 183
basement rocks, 34, 36
basic sediment and water (BS&W), 543
bay bottom, 32
bed boundary, 63, 496
bedding plane, 23

bell nipple, 172, 214–215, 229, 286, 344–345
benching, 48
bent drillpipe, 244, 328
bentonite, 56–57, 218, 265, 326, 352, 490
bent sub, 429–432
benzoic acid flakes, 497
big hole, 232
bit: action, 397; block, 270, 273; bounce, 277; break-in, 408; clearance, 146; collar, 268, 274–275; drill, 176, 187, 200, 205–206, 213–214; dull, 12; failure, 134, 263; gauge, 7; horsepower, 10; junk, 154; life, 11, 150–151, 154–156, 295, 360, 395; loading, 295; parts, 7, 150–153, 307, 319–320, 393, 408; record, 154; run, 395; selection, 17, 154, 347, 408; stabilization, 154; types, 7, 11, 146, 150–155; walk, 427, 431; weight, 2, 77, 149, 151–155, 273, 358, 408, 437, 469
black light, 365
blanking plug, 474, 506, 529
blast hole, 232
blasting cap, 477
blast joint, 468, 523–524, 536
bleed plugs, 321
block, traveling, 185
blooie line, 229
blowdown, 100, 102
blowout, 8, 14, 85, 138, 201, 219–220, 320, 345, 364, 367–368, 442, 451: alarm, 368; condition, 307, 316, 320, 362, 425; detection equipment, 163, 368; drills, 163, 228, 368, 417; fluids, 425; *see also* preventers
board platform, 245
board road, 245
boiler, 9–12, 230
bomb, 279, 459–460
bonus, 125
booster compressor, 229
boot basket, 354
bottom-water drive, 103
boxed beam, 179
brackish water, 65, 67
bradenhead, 176, 223, 342–346, 406, 539
brakes, drawworks, 189, 191, 356, 358, 414
breakdown, 355
breaking circulation, 353
breaking tour, 257
bridges, 89, 330, 336, 388, 477
bridge plug, 472, 487, 504
bright spots, 32, 36
brine wells, 1–2

British thermal unit (BTU), 111
bubble point, 101
budget, 106
buildup test, 491–492
bulk density, 69
bulldozers, 252–253
bull heading, 424
bull wheel, 7
bumper block, 331
bumper sub, 386–387, 390
bumping down, 398
bump the plug, 456
buoyancy factor, 300, 343, 372
Bureau of Land Management (BLM), 243
burial process, 20
burning pit, 176, 223
burst, 139, 144, 501: pressure, 138, 144, 329, 345, 353, 471; test, 147
bus bar, 13, 200
butterfly valve, 215

C

calcareous cementing material, 21, 37, 435
calcite (filled veinlets), nodules, 436
calcium chloride, 495
caliper logs, 61, 69, 71, 97, 149, 326, 454
canal, 234
capillarity, 94
capital expenditure, 123–124
capital investment, 122
carbon dioxide, 46, 101, 134, 143, 495, 498, 511
carbonated beverages, 101
carbonates, 466
carboxyl-methyl-cellulose (CMC), 160
carried working interest, 126
casing, 135, 137, 146: burst, 341; centralized, 147; clearance, 144–145; collapse, 136, 394; connection, 336; crews, 183, 331; design criteria, 139; drag, 341; extra string, 143–144; flush joint, 144, 271, 287; grades, 58, 136, 206; hanger spool, 176, 226, 342, 405, 539; inspection, 147; inspection log, 72; leak, 374; long string, 137; measurements, 328; patch, 499; point, 77, 127, 136, 330; point election, 126; scraper, 457, 461, 487, 504, 528; seals, 406; seamless, 136; setting depth, 17; slips, 332–333, 406; special design, 146; tables, 147; tensile strength, 136; threaded and coupled (T&C), 135; wear, 72, 412; weight, 136; window, 499–500
cathead, 10, 176, 262, 331–334, 446:

automatic, 190, 197, 278; break-out, 189, 227; makeup, 189, 270; spinning, 189, 196, 278
cat line, 176, 185
catwalk, 176, 183–184, 285, 298, 331–332
caverns, 57–58, 160
caving, 18, 138, 157, 265, 306, 319, 386
cellar, 176, 178, 223, 248, 266
cement, 148, 283, 456, 487: additives, 148, 326–327, 454; bond, 72, 146, 296, 337, 474–475, 505; bulk trucks, 337; circulate, 341; column, 326, 341; contamination, 352; diverting tool, 455; grind, 148; hopper, 337–338; low-weight, 148; plugs, 426, 432, 500; pumping time, 148, 327, 342, 486; pump trucks, 283, 285, 291, 337; retarders, 148, 326; second stage, 149, 453; sheath, 146; slurry, 148, 283, 325, 327, 337; stage tool, 455, 457, 486; thickening time, 148; top, 71, 504; voids, 475
cementing: first-stage, 149, 455; head, 285–286, 337, 339, 459; large-diameter casing, 405; manifold, 329; program, 148; puddle, 23; two-plug; 2, 337
centipoise, 84
centralizers, 147, 327, 340, 374, 455
centrifugal force, 216–217
centrifuge, 156, 176, 211, 267, 545
chain out of the hole, 331, 398
changing the break, 362
channel, 20, 103, 137, 455, 475, 490
channeling, 72, 455
chart, gas, 443, 550
chemical balance, 212
chemical tank, 176, 211, 213
chert nodules, 395, 414
chlorides, 156, 409
choke, 507: adjustable, 422, 442, 507, 539; beam, 538; bottom-hole, 441, 479, 507; control, 176, 222–223; hydraulic, 345; line, 176, 222–223; manifold, 163, 225, 230, 344, 368, 421; manual, 345; positive, 507; storm, 469–470, 527
Christmas tree, 457, 506–507, 512, 538–539, 541
chromatograph, 77
circulating: bottoms up, 416; out kicks, 417, 422; out shows, 124; to clean hole, 324, 408
clay particles, 159
clean sand line, 63–64
cleanup: man, 172, 349; problem, 161; trip, 449, 457

clockwise rotation, 206
closed cycle, 212
closure, 45, 48
coastal areas, 174
coast guard, 243
collapse, 336, 502: pressure, 144, 441; strength, 136
collar: locator (log), 72, 389, 400, 445, 484–485, 506; stops, 474; tapered, 523, 531; thread protector, 474
colloidal suspension, 159
commercial producer, 18, 43
commercial prospect, 445
company: contact, 127; man, 170, 341, 351; representative, 350; rig, 171
compensated logs: density, 70; neutron, 70; sonic, 69
competent rocks, 56, 91
completion: efficiency, 491; engineer, 85, 446; equipment, 468, 473; fluids, 466, 484, 503; program, 18, 502; report, 122; rig, 456, 467, 503–504; tools, 468
completion types: barefoot, 139, 463, 521–522; commingled, 467, 504, 523, 534; commingled multizone, 534; dual, 466, 473, 523–524, 542; liner, 457; marginal, 516; multiple, 466–467, 482, 504, 522; multiple commingled, 518; natural, 510; open-hole, 521–522; packerless, 505; parallel tubing, 530; permanent, 467, 512, 520; quadruple, 523, 534; single, 504–505; slim hole, 467, 522, 533; slotted liner, 521; stimulated, 510; triple, 523, 534, 536; tubingless, 467, 515, 522, 524, 533, 535
compound, 146, 194, 196, 198, 212
compressibility factor, 111, 113
compression, 296
compressor, 287, 302: booster, 479; control panel, 547; gas, 543, 548; suction, 112
computer, 19, 30–35, 60, 77, 122, 510: bit selection, 154; data banks, 119; equipment, 445; maps, 19; modeling, 108; plotters, 32; programs, 119, 139
computerized log, 76
concrete, 148, 246: bases, 184; reinforced slab, 246
condensate, 42, 110
conductor, 65
conglomerate, 21
coning, water, 104
connate water, 41
connectors, 135, 150, 153: left-hand 203–204; semiflush joint, 144

consolidated sands, 37
consultants, 160
contamination, 159: formations, 82; material, 160; water, 164, 241
continental drift, 48
Continental Shelf, 54
continuous operations, 125
continuous phase, 92, 94, 158
contour maps, 24, 34
contract depth, 172
contract gas sales, 547
contractors, 171
contracts, 163, 242
control panel, 227–228, 423
control stations, 163
conversion constants, 87
core, 55, 68, 83, 113, 153, 436, 445: bit, 437; bleeding, 445; conventional, 59, 437; drill, 232; gamma log, 437; head, 437; holes, 1; oriented, 437; penetration, 480; pressurized, 436; sidewall, 59, 450
coring, 18, 106, 134, 158, 407, 434
corrosion, 72, 82, 158–159, 184, 357, 466: agents, 143, 156; electrolytic, 143; fluids, 72
cost, 516: AFE, 121, 163; completed well, 163; completion, 463, 516; intangible, 163–164; of money, 122; operating, 463, 516; tangible, 163–164; transportation, 125
coupling, 136, 138, 192, 468
cranes, 235
crater, 220
crawler tractor, 252–253
Cretaceous, 22
crews, 161, 349, 352, 418
crippled rope socket, 449
crooked hole, 55, 279, 374, 378–379, 381–383, 409, 449
cross beams, 119
cross flow, 476, 521, 539
crossover subs, 264, 268, 277
crossplots, 69
cross sections, 27
crown, 1, 185, 257, 261
cumulative costs, 63
cumulative production graphs, 113
current income, 122
cut off, 367, 457
cutting practice, 191, 414
cuttings, 322, 409: fill up, 300; slip velocity, 322, 409
cycle skipping, 69, 71, 75
cyclones, 216: desander, 176, 210; desilter, 176, 210

D

daily: drilling costs, 163; drilling report, 348; tour report, 417
damaged zone, 517
darcies, 82
data, 492
daughter products, 34
day: crew, 349; rate, 171; work, 171
dc motor, 13
dc rig, 13, 200
dead line, 176, 185–188, 191, 366, 414
deep hole, 143
degaseated mud, 215, 218
degassers, 156, 176, 209, 218, 347
dehydration, 547–548
deliverability, 123
densiometer surveys, 476
density, 156
depleted, 59, 112, 142, 477
deposition process, 20
depth of invasion, 66
depth of investigation, 67, 69, 443, 492
depth of penetration, 478–479
depth rating, 162, 173–174
derrick, 175, 185, 253, 414
derrickman, 172, 246, 261, 317–318, 349–350
desander, 156, 216, 267
desert rigs, 253
desilter, 156, 216, 267, 409
desulfurization unit, 547
detectors, 69
detonator, 478, 485
development: drilling, 55; exploration geologist, 54–55; prospect, 117, 121; well, 121, 134, 238
deviated: conductor hole, 275; drilling, 132, 429–432, 425–428, 477; high angle, 132; hole, 13, 55, 91, 139, 292; surveys, 73, 137, 149, 277
deviation clause, 2
deviation maximum, 131
diamond, 153, 408: bit, 152–154, 414; commercial, 153; core head, 153; drilling, 414
diesel, 174, 191–192, 194–195, 388, 490
diesel rig, 12, 174, 192
differential erosion, 90
dimensionless time, 492–493
dip, 24–28, 47–48
direct current, 12, 199–200
discontinuity, 47
discounted future net revenue, 122
discovery well, 237

dissolved gas drive, 100
dissolving agents, 388
diverter, 286, 324
doghouse, 176, 179
dogleg, 131: absolute, 130; angle, 130, 133
dolomite, 38-39, 395
Donald Duck, 266
double: drum, 140; loading, 255; rubbered drillpipe, 362, 407, 412; substructure, 259, 262
downdip wells, 103
downhole assemblies: *see* assemblies
downhole explosion, 159, 288
downhole motors, 157
downtime, 242
drag, 276, 302: excess, 413; fluctuating, 302; folds, 47, 51; normal, 306
Drake well, 1, 2
drawworks, 174, 184-185, 190-199, 310, 414
drift diameter, 146, 329
drillable prospect, 54
drill collar, 171, 176, 206-207, 295, 439
drill-deeper prospect, 117
driller, 172, 348-349, 351, 356: console, 176, 183, 191; report, 350-351; side, 368
drilling: break, 76-77, 416, 434, 445; contractor, 249, 348, 364; department, 170; engineer, 126, 170; foreman, 170, 245; fund, 120; line, 6, 176, 179, 188-191; line loading, 366; permit, 128, 242; platform, 233-237; program, 119; rate, 76, 229, 275, 347; rig horsepower, 175; shows, 403, time curve, 125
drillpipe, 10, 171, 176, 206-207, 363: collapse, 441; connection, 263; measurement, 363; preventer, 366-367
drillstem test, 59, 141, 361, 407, 443-444
drive mechanisms, reservoir, 98
drive pipe, 282
driver fluid, 12, 103
dry drill, 202
dry hole, 25, 30, 97, 452
dual lateral log, 67
duplex pump, 208-209
dynamatic, 176, 187-190, 198, 318
Dynamometer Survey, 73

E

earth-moving contractors, 246
economics, 119, 122
edgewater drive, 103
effective weight, 300
efficient production rate, 85
electric rigs, 12, 262

electric logs, 60
elevators, 176, 187, 268, 306, 315, 332
embedment depth, 153
empirical relationship, 68
emulsion, 546
enhanced recovery, 59, 85, 105-108, 463
enlarged hole, 56
entrapped gas, 218
environmental impact statement, 130, 243
epoches, geological, 21
epoxy cement, 296, 333
equatorial bulge, 30
equipment rentals, 163, 467
equipment failures, 381
equivalent energy, 111
eras, geological, 21
evaluation, risk, 122-123
evaporates, 2, 19
expendables, 171
expert witness, 243
exploration, 463: engineer, 434; geologist, 15, 118, 446; well, 237
explosive, 30-32, 401, 425, 454, 479
explosive limits, 288
extrapolation, 108, 113

F

facies change, 22-23, 25, 51
farmin, 119
farmout, 126
faults, 35, 48-51, 57
field: extension well, 117; rules, 243; spacing, 243
fire, 12, 161, 198, 359, 425, 547
fireflood, 105
fishing, 57, 154-157, 235, 288, 320, 375, 390, 398, 407, 409
float equipment, 148, 229, 252, 285, 334, 404
floor man, 192
flow capacity, 83
flow channels, 208, 496
flow efficiency, 491, 494
flowing well, 476
flow line, 137, 176, 179, 215, 229
flow mechanism, 94
fluid cleanup, 475
fluid erosion, 151, 411, 443
fluid gradient, 87
fluid level log, 73
fluid loss, 156, 160, 360
fluid phases, 59, 92-93, 95, 111
fluid sensitive formations, 156-157, 160-162, 369, 422, 439
fluid sounder, 492

fluorescent, 435
foam, 157, 161, 230
foam drilling, 230, 288
footage contract, 170
footage rate, 171
formation: abnormal pressured, 157; crooked hole, 132, 275; damage, 466, 494; dip log, 71; shallow gas, 137; testing, 106; volume factor (FVF), 108, 115
fracture, 436, 496: gradient, 141; log, 73; orientation, 495
fracturing, 141, 157, 326, 495–496, 511, 518
free pipe, 72, 400
free point, 296–297, 343, 390, 400
free-point log, 72, 400
freeze up, 548
freshwater sand, 64–65, 67
frost line, 78
funnel second, 156

G

gamma gamma density log, 61, 71–73, 484
gamma rays, 36, 60, 69, 75, 504
gamma ray log, 475–476
gas, 55, 101–103, 159, 192, 215, 218: column, 46, 75, 98; compressed, 99, 543, 548, 552; condensate, 42, 110; contact, 417, 445; cycling, 102; deadly, 46; detectors, 36, 69; dissolved, 98, 101–103, 106, 108; drilled, 409; drilling, 229; equivalent, 17; flows, 137–138; free, 100–101, 218; gradient, 112; gravity, 106, 110; inert, 105; injection, 467; lift, 552; liquids, 110; noncombustible, 425; reservoir, 45–48, 54, 99, 113; sales contract, 547; separation, 222, 359; storage, 467; wells, 123, 143, 548
gas cap, 46, 59, 75, 97, 100, 102
gas in place, 113
gas-oil contact, 75, 100
gas-oil ratio, 106, 108
gas-water contact, 55, 68, 75, 490
gel, 148, 156–159, 265
generating unit (light plant), 176, 200
geochemical surveys, 18, 36, 118
geological poles, 33
geological sections, 21, 78
geological studies, 12, 59
geological time, 21–23, 34, 88
geologist, 434
geophone, 29, 30–32
geophysical, 446

geopressure, 17, 56, 59, 88, 90
geothermal drilling, 2
glycol, 547
gold, 20
gravel, 56
gravel packing, 521
gravimetric, 19, 32
gravity drainage, 59, 218
ground cushion, 231
ground water, 34
group shoots, 118
growth-oriented companies, 122
gun-barrel tank, 545
gunk squeeze, 490
gusher, 9, 59
gypsum, 55, 58, 134, 156, 371
gypsum mud, 160

H

half-life, 70, 476
hammer tool, 153
hard-facing material, 362
hard streak, 68, 74
headache post, 251
heater treater, 546
heat measurement, 111
heavy hydrocarbons, 44
heavy oil, 17, 43, 84
helicopter rig, 231, 289
helium, 46, 77
hexanes plus, 77
high-risk operation, 360
high temperature, 148, 152, 156
high-temperature mud, 82
high-viscosity pill, 323
highway permits, 130
history matching, 108
hole: course, 130; inclination logs, 71; instability, 415; original, 375, 381; restriction, 325
hook loads, 159, 184–185, 187–188, 190–191, 356, 365
hot area, 119
human error, 414
hydration, 265
hydraulics program, 242
hydrocarbon: deposits, 32, 46; light, 498
hydrochloric acid, 495, 498
hydrocyclones, 209, 216
hydrogen, 77
hydrogen sulfide, 34, 45–46, 143, 451, 538
hydromatic, 10, 190, 198, 318

hydrostatic head, 214, 307–308, 362
hydrostatic pressure, 141, 220

I

icebergs, 238
ice islands, 245
igneous rocks, 19, 36, 43
impermeable, 40–42, 44, 51
inclination survey, 71, 146
independents, 120
Indian agency, 243
Indian land, 130, 243
induction log, 67
inert solids, 157
infield development, 117, 121, 127, 162
inflection point, 75
inflow test, 461
infrared photography, 35
inhibitors, 158, 484
inside diameter, 146
in situ combustion, 105
inspections, 163, 329, 502: black light, 244; body, 147, 329; end area, 147, 329; magnetic particle, 365; service companies, 244, 348; sonic, 244, 365
insurance, 144
intangibles, 122
interest rate, 122
internal-combustion engine, 12, 192, 199
intrusives, 19, 33–34
invoicing, 170
ionic charge, 57
iron fillings, 215, 365
isolated locations, 289
isolate water, 138
isometric projections, 119
isopach maps, 24, 27, 55
isotopes, 35–36

J

jacks, 235
jack shaft, 10, 189, 196–197, 278
jars, 7, 314, 387, 390, 439
jet, 151, 247: bits, 151, 157; nozzles, 151, 241, 388; perforating gun, 478; perforating test data, 481; velocity, 208, 275
joint venture, 119–120: agreement, 54, 120; drilling unit 119
junk, 394, 409: basket, 391, 398; in-hole, 500; metallic, 155; pusher, 479; shot, 393; subs, 354, 394, 397, 408
junked hole, 157, 375, 394, 425, 460

K

kelly, 176, 187, 203–204, 263–264
kelly bushing elevation, 447
keyseat, 132–133, 276, 375, 378–380, 432: wiper, 207, 315, 380
kick, 307, 367, 403, 415, 434
killing a blowout, 227
kill wells, 71, 429

L

lag time, 76, 208, 215, 356, 434
landing strip, 231
landowner, 119, 124
land position, 120
land trades, 119
land-use plan, 243
large holes, 13
lateral log, 67
leaks, 423, 457, 466, 501–502
lease, 120, 123, 125–126
lease automatic custody transfer (LACT), 546
ledge, 276, 302, 386, 449
lessee, 124–125
lessor, 125–126
life-saving equipment, 143
lifting line, 184, 186, 268, 278, 366
light hydrocarbon, 43, 47, 112
limited entry, 497, 520
liner, 211, 309, 433, 499, 500, 554: hanger, 459; overlap, 459; setting tool, 459–460; stub, 502; tieback, 500; top, 486
liquefied gas, 498, 511
liquid petroleum gas (LPG), 112–113
lithology, 18, 22–23, 475: log, 403
locations, 241, 255, 257, 428–429
log, 18, 55, 60–61, 73–75, 433–434, 463
log analysis, 73
logging, 60, 66, 432, 446–448, 474: engineer, 448
logistics, 231
lost circulation, 140, 158, 212, 336, 370, 413: material, 148, 218, 322; partial, 282; zones, 17, 71, 138, 265, 285
lost treatment, 475
low-gravity oil, 44
low-gravity solids, 216–219
lubricity, 158, 161, 360, 388

M

magnetic, 19, 33–34
magnetite, 70
manifold, 212, 442–443
margin of error, 86

marine organism, 39
marine waters, 232
marker bed, 61
mast, 175–179, 181–185, 220–223
material balance, 108, 113, 115
material supervisor, 170
microcircuitry, 30
microseismicity, 35
migration theory, 40
mill, 391–394, 400, 472, 499
millidarcies, 82
milling, 364, 499, 502
mineral owner, 124–125
mist drilling, 288
mobile fluid phase, 92, 95
monitors, 417
monkey board, 176, 183, 230, 310
monolayer, 496
montmorillonite, 56, 157, 352
morning report, 350–351
motion compensator, 240
motorman, 172, 349–350
mousehole, 176, 246, 250, 262–263: connection, 309
mud, 155, 178, 211, 265–266: additives, 156, 158, 177, 211, 219; chemistry, 216; cleaner, 156, 267; column, 84, 132, 141, 208; equipment, 13, 162, 176, 211, 216; filter cake, 146, 160, 214, 218, 338, 465; filtrate, 62, 159–161, 360, 436; filtrate resistivity, 66; gas cut, 306, 435; gels, 57, 156, 353; guns, 176, 211, 219; hydraulics, 358, 411; mixing plant, 210–211; pits, 13, 190, 210, 230–232, 246; problems, 214; returns, 77; solids, 156, 214, 217–218, 358; storage, 156, 177, 211, 312; supplies, 168, 350; thinners, 160; treatment, 156, 216, 358; weight, 11, 156–157
mud logging, 18, 73, 76, 212
mud motor, 160
mud systems: clay-base, 484; dispersed, 160; high pH, 160; inhibited, 57, 161; invert, 57, 101; lime-base, 160–161; low fluid loss, 160; native, 234, 265, 352; nondispersed, 160; oil, 57, 158; oil emulsion, 161; polymer, 161, 370; salt, 160–161; spud, 265; water-base, 158
multirate flow tests, 491

N

natural gas, 367
net revenue interest, 125–126

neutral point, 402
neutron log, 60, 70, 450
nitrates, 436
nitrogen, 46, 345, 441, 498, 511
noise pollution, 198
nondrilling clause, 124
nonporous formations, 76
normal resistivity log, 67
nutrients, 39

O

offset, 55, 427: location, 55, 150–151; test, 55, 121, 242
offshore, 19, 32, 174, 237, 470
oil: column, 43, 46, 97, 100, 102; content, 156, 348; displacing mechanism, 103; equivalents, 111, 113; finder, 58; gravity, 106; minerals, 124; original in place, 102, 106–108, phase, 43, 92; recovery mechanism, 107; residual, 42; seeps, 18; shows, 77; shrinkage, 106; storage, 176; traps, 13, 47–48, 54–55, 97, 103, 176; wells, 143, 466
oil-gas contact, 70
oil-water contact, 25, 68, 103–104, 490, 545
oil wet, 92, 94
on structure, 55, 117
oolites, 39, 436
open hole, 139, 369, 407: log, 60, 66, 403, 433, 474; tools, 71, 433, 471
operating agreement, 119
operations department, 170
oriented perforating, 524
outcrops, 19, 28, 48, 103
out of gauge, 58, 71, 326, 379, 439
outside diameter, 135
overburden, 491
overpull, 206, 373, 384–385
override, 126
overriding royalty interest (ORRI), 125
overshot, 390–393, 398–399, 400, 449
oversize permit, 243
owner, 169–170, 242

P

packer, 394, 467, 470, 487, 513: combination, 492; compression, 439, 470–471; crossover, 525, 533; dual, 524, 530, 532, 536; hook-wall, 461, 470–471; inflatable, 439; large bore, 525; permanent, 473, 512, 529; production, 512; retainer, 487; retrievable, 470–474, 489; single, 524;

test, 470; treating, 471–472, 549; triple, 584; upside-down, 471
packer plucker, 499
packer-plug method, 521
packing factor, 37
parallel tubing, 524, 536
payout, 122, 126
penetration rate, 85, 155–156, 275, 377, 395, 412, 414, 441
percussion tool, 229
perforating, 388, 407, 475, 477, 517
perforating guns: bullet, 477–478; capsule, bar carrier, 483; casing, 506; centralized, 478, 482; decentralized, 478, 482, 515; high-pressure, 478; intermediate pressure, 478; jet, 478; multiple, 484; oriented, 553; select fire, 482; shaped charge, 477, 479; strip jet, 483–484; through tubing, 517, 520
perforation, 72, 107, 468, 486, 505: ball sealers, 497, 518; density, 477, 482; log, 72, 475, 506; orientation, 482, 484; phasing, 482, 506
permafrost, 429
permeability, 40, 466, 491, 493–494: effective, 491; horizontal, 42, 484; low, 466, 510; vertical, 42, 484
permits, special, 150
petroleum geologist, 18, 26, 28, 36, 48
pH, 348
phases, 92–94, 482, 506
phosphates, 161, 370
piles, 233, 282
pilot hole, 281
pinchout, 50, 51
pipe drag, 276, 302, 330, 411
pipe stretch tables, 400
pit volume totalizer (PVT), 228
planimeter, 45
plug, 470, 472–473
plugged and abandoned (P&A), 60, 394, 407, 451, 474
plunger force, 320, 512
point of weakness, 499
pore channels, 94–95, 99
pore pressure, 58, 77, 141, 160, 346, 359: gradient, 141; plot, 77, 113, 141, 220, 416–417; reversal, 68–69, 141
pore space, 37–38, 66, 70, 88–89, 92, 100
porosity: communicated, 37; effective, 37, 70; fracture, 39, 73; logs, 68; primary, 39; secondary, 39, 61, 70, 75; uncommunicated, 37, 70; vugular, 29, 38

potassium chloride, 484, 518
potential difference, 66
power transmission, 12, 175
Precambrian rocks, 36
present worth, 122
pressure, 85, 99, 511: absolute, 86; atmospheric, 86; bottom-hole, 73, 491, 493; breakdown, 488, 511; bubble-point, 101; buildup, 120, 138, 341, 491–495; differential, 94, 98, 112, 381, 441, 499; drawdown, 491–492; drop, 214; gradient, 87–88, 91, 104, 417; hydraulic, 85; hydrostatic, 87; overburden, 19, 89; subnormal, 88; surge, 57, 350, 361, 404; shutin drillpipe, 417, 420
preventers (BOP), 130, 137–138, 163, 178, 180: annular (bag), 176, 221, 223–226; blank (blind), 221, 225–226, 315, 344–345; controls, 176, 222–223; inside, 201, 204; pipe rams, 176, 221–226, 333, 418; shear rams, 344, 424
primary recovery, 108, 115
prime rate, 122
primer chord, 388, 485
production: battery, 233–234, 322, 544, 548, 552; efficiency, 102, 115; graphs, 108, 115; life, 106; logs, 72; primary, 105, 108; schedule, 123; test, 108
promoter, 120
prospects: generating, 118, 199–200, 257; high-risk, 117, 123; low-risk, 117–118, 123
protected waters, 233, 235
pumps, 154, 246, 257, 263: centrifugal 216; hydraulic, 375, 551; injection, 230; mixing, 176, 214–215, 219, 228; mud, 196, 218, 228, 257, 262; plunger, 209, 212; quadraplex, 209; standby, 228, 244, 409; submersible, 375, 552; triplex, 216
pumping oil wells, 374, 463, 492
pyrite, 414
P/Z curves, 113

Q

quartz, 20
quebroxine, 160

R

radar, 34
radial flow, 59, 96
radioactive fluid, 476
radius of investigation, 492, 494
rate-time curve, 113
rate of penetration (ROP), 13, 76, 128–129

rate of return (ROR), 122–123, 516
rathole, 176, 246, 250, 262
reamers, 315, 375, 377
reaming, 71, 212, 214, 277, 301
recomplete, 468, 474, 476, 503
recoverable reserves, 106–107, 112, 516
recovery efficiency, 102, 108, 115
recovery factor, 104
redrilling, 55, 57, 375, 382, 394
refinery terminal, 543
regional surveys, 33
regulatory agencies, 128, 243, 249, 364, 451–452
releasing fluid, 222
rental agreement, 163
repair facility, 172
repressurization, 105
reserves, 104, 113, 119, 123
reservoir, 44–45, 73, 75–76, 99: analysis, 108, 113; drives, 59, 99; energy, 59, 85, 115; engineers, 85, 98–99; fluids, 37, 106; modeling, 115; multiple, 432; pressure, 98–103, 112
resistivity, 60, 65, 434
return on investment, 121
revenue, 122
reversing out, 445, 487, 489
rework, 143
rig down, 13
right of way, 119
rig manager, 172
rig types: barge, 233, 290; big hole, 232; cable tool, 4–9; diesel electric, 12, 174, 192, 199; drillship, 15, 174, 238–239; floater, 232, 238–239, 292; helicopter, 289–290; jackup, 15, 235–237, 292; marine, 174, 178; mechanical, 12, 174, 192–199, 262; platform, 15, 232, 291–292; SCR, 13, 200; semisubmersible, 174, 237–238; spring pole, 3–4; steam, 9–10, 174; rotary, 9, 74; workover, 467, 504, 511
risk, 380
rollover, 48
rotary helper, 172
rotary table, 7, 154, 177, 179, 180
rotating head, 137, 215, 226, 344, 359
roughneck, 172
royalty interest, 125
rules and regulations, 364

S

safety, 130, 144, 188, 230, 548
safety lines (geronimo), 178
salt domes, 21, 32, 91, 429
salt sections, 17, 21, 65, 159, 475
salvage company, 452
sample, 215, 492
sand: control, 486; grains, 36, 435, 496; lenses, 52; production, 521; trend, 118
sandy formations, 39, 62
satellite imagery, 32, 34
saturation, 68, 92–94, 104–108, 434, 436
scour channel, 238
scratchers, 146, 327, 334, 459
sea floor, 233, 289
seagoing tub, 235
secondary enhanced recovery, 113, 115
secondary gas cap, 102
secondary recovery, 59, 102, 463
sediments, 19, 21, 22, 36, 48
seismic, 19, 26, 28–32, 174, 441
separator, 442, 545, 549
service companies, 467: see also appendix
setting up, 148, 286
shale, 20, 37, 49, 56–58, 134, 138
shale barrier, 496
shale shaker, 113, 156, 177–179, 211
shallow investigation, 67, 69
shaly formations, 21, 61–62, 67, 74
shock loading, 201, 502
short string, 531
short trip, 283, 330
show indicators, 434
sidetrack, 55–56, 133, 157, 277, 296: a fish, 427, 432; operation, 381, 394; well, 427
silica flour, 148
silicon controlled rectifier (SCR), 13
single-line loading, 185, 188, 190
siphon string, 505
skin damage, 492, 494
slant hole, 237, 433
sliding sleeve, 468–469, 512–514, 524–526
slim hole, 54
slip and cut method, 191
sloughing, 56, 138, 356
slow pump rate, 417
slug, 307–309, 312, 321
snubbing in the hole, 469
soap, 159, 228, 230, 288
sodium chloride, 62, 67
soil strength, 233
solids removal, 156, 216, 347, 409
solution channels, 39
solution gas, 102–103
solution-gas drive, 59

source beds, 36, 40, 44
specific gravity, 40, 89
spherically focused logs, 67
spill over, 46, 49
spud, 125, 156, 250, 265–266: assembly, 266, 274; bit, 427; conference, 249; mud, 265
spurt loss, 360
squeeze, 355, 472, 486–490, 499–502, 521
stabilizers, 207, 279, 315, 375, 377–379
staging area, 289
standard conditions, 84, 110
standard derricks, 162, 231
standard target (perforating), 477
stand off (perforating), 478–479
stand pipe, 179, 211–214, 285
state land commissioner, 242
state taxes, 122
steamflood, 105
steam rig, 9, 10
sticking, 71, 156, 158, 412, 437
stimulation, 119, 409, 463, 466, 494–495: fluids, 455, 466; pressure, 139; program, 503; shallow, 466
stock-tank barrel (STB), 106–107
strat holes, 54
stratigraphic traps, 45
string shot, 388–389
string up, 257
stripping in the hole, 424
structural closure, 45
structure maps, 24–26, 30
stuck pipe, 322, 363, 370, 382–386, 390
stuck point, 72, 385, 400
submittals, 120–121
subs, 179
subsea wellhead, 240
subsidence, 80
substructure, 113, 177, 179–181, 183
subsurface geology, 18
subsurface structure, 34
surface owner, 124
surface tension, 94–95
swabbing, 190, 220, 320, 474, 508–510
swivel, 177, 187, 201–202, 210
synchronous rotating speed, 412
syncline, 47

T

taking a picture, 279
tank battery, 546
tannins, 160
targets, 130, 134

tax: carry-forward, 124; credit, 124; federal income, 123–124; local, 124
tectonic forces, 35, 56
telluric surveys, 35
temperature: ambient, 78; flow line, 77; formation, 82; gradient, 475; logs, 71; reservoir, 106; wellbore, 82
tensile strength, 136, 144, 206, 296, 383
tertiary recovery, 59, 105, 463
testing, 134, 158, 306, 345
test well, 55
thief zone, 521
third party, 54, 170: services 292; tools 171
thread: makeup, 328, 356; protector, 270, 284, 333, 362; round, eight, 468; saver sub, 204, 274; V, 468
thribbles, 12, 296, 362
tidal flats, 233–234, 429
tieback liner, 137, 139, 454, 501
tight hole, 370
tiltmeter, 35
title check, opinion, 119, 124
tongs, 263, 306: backup, 7, 272, 277, 302, 312; break out, 312; casing, 333; chain, 315, 318; hydraulic, 303, 319, 333; lead (breakout), 177; makeup, 177, 315
ton miles, 351, 366
tool joints, 7, 11, 135, 153, 203, 268–269: box, 263; box shoulder, 206; flush joint, 523; hard banding, 244; semiflush joint, 523; upset, 138, 206–207
tool pusher, 347, 351, 418
torque, 192, 193, 199, 203–204, 270: converter, 193; fluctuating, 364; high, excess, 13, 354, 385
tour, 257, 349
tour report, 350–351
tracer log, 476
trailers, 229, 252–255, 337, 467
transmission system, 123, 547
trap, hydrocarbon, 44–45, 47–52, 54–55, 117
traveling block, 159, 177, 185–187, 190
trip tank, 307–310, 322, 362
trucks, 172, 176, 241–242, 249–250
true vertical depth (TVD), 432
tubing, 72, 390, 467–468, 473, 488: coil, 468–469; completion, 506, 529; external upset end (EUE), 468; lower, 532–533; macaroni, 469; parallel string, 524, 536; production, 457, 469, 474; pup joints, 515; upper, 533
tug, 235, 289
tungsten carbide, 393

U

turbines, 207, 412, 429
turbulent flow, 497
turnkey contract, 172
twist off, 276, 397

U

ultradeep holes, 13
unconsolidated sand, 37
unconventional recovery, 105
underbalanced drilling, 307, 359
underbalanced perforating, 516, 519
under gauge, 71, 276, 315, 319, 361
underground blowout, 423
unit operating agreement, 119
unprotected water, 5, 232
upgrading, 17
ultraviolet light, 76, 333, 365, 435
uranium, 34
U-tube, 97–98, 103, 308, 509

V

vacuum, 99, 218
valence bond, 159
V-door, 332–333, 366, 446
velocity survey, 446
vermiculite, 326
vertical closure, 45
video display, 365
viscosity, 495
viscous sweep, 103, 322–323, 330
vugular porosity, 29, 38, 57, 69

W

wall cake, 160, 327, 360, 381
wall off, 394, 396–397
wall stuck, 329, 381–382, 388
wash out, 322, 326, 330, 332
washover, 162, 391–392, 398, 402, 467–468, 474, 476, 503–504, 518
waste disposal well, 467
water: cushion, 441–442; cut, 545; disposal, 543; drive, 59, 103; encroachment, 103; flows, 71, 134, 137, 143, 265; injection, 467; production, 112; source well, 232, 249; table, 177, 183, 258
waterflood simulation, 437
water-gas contact, 70
water-in-oil emulsion mud, 161
wave action, 233, 235, 238
weather, 246
weight in air, 300
weight indicator, 177, 187, 235, 300, 302
weighting material, 216, 218, 418
welded connection, 184, 296, 333
welders, 172, 179
well: conductivity, 83; production, 495; production maps, 119; records, 127; spacing, 99, 100, 104, 106–107, 112
wellbore: cleanout, 466; damage, 494; effects, 493
wellhead, 48, 344
wet string, 309
wettability, 98
whipstock, 427, 499–500
wildcat, 12, 28, 54–55, 127, 134, 403
wiper plugs, 337–341, 352–354, 455, 459–460
wiper trip, 411
wire-line fish, 391
wire-line keyseat, 449
wire-line test, 450
working interest, 125–126
working pipe, 285, 326, 384–385, 388–390
work string, 143, 457, 469, 472, 504, 506

Y

Yellow Boy, 266

Z

zone isolation, 72, 490

SCIENCE, POLITICS, AND INTERNATIONAL CONFERENCES

SCIENCE, POLITICS, AND INTERNATIONAL CONFERENCES

A Functional Analysis of the Moscow Political Science Congress

Richard L. Merritt
Elizabeth C. Hanson

GSIS Monograph Series
in World Affairs

THE UNIVERSITY OF DENVER

Lynne Rienner Publishers • Boulder & London

Published in the United States of America in 1989 by
Lynne Rienner Publishers, Inc.
1800 30th Street, Boulder, Colorado 80301

and in the United Kingdom by
Lynne Rienner Publishers, Inc.
3 Henrietta Street, Covent Garden, London WC2E 8LU

©1989 by Lynne Rienner Publishers, Inc. All rights reserved

Library of Congress Cataloging-in-Publication Data

Merritt, Richard L.
    Science, politics, and international conferences : a functional
analysis of the Moscow Political Science Congress / by Richard L.
Merritt and Elizabeth C. Hanson.
    Bibliography: p.
    Includes index.
    ISBN 1-55587-134-8 (alk. paper)
    1. International Political Science Association. World Congress
(11th : 1979 : Moscow, R.S.F.S.R.)—Cost effectiveness.
2. Political science—Congresses—Cost effectiveness. I. Hanson,
Elizabeth C. II. International Political Science Association.
World Congress (11th : 1979 : Moscow, R.S.F.S.R.) III. Title.
JA35.5.I77   1979z   88-18322
320'.06'01—dc19   CIP

British Library Cataloguing in Publication Data
A Cataloguing in Publication record for this book
is available from the British Library.

Printed and bound in the United States of America

The paper used in this publication meets
the requirements of the American National
Standard for Permanence of Paper for
Printed Library Materials Z39.48-1984.

Dedicated to:

Liette Boucher
Karl W. Deutsch
Georgii Shakhnazarov
William Smirnov
John E. Trent

who made the Moscow World Congress work.

Contents

	List of Tables and Figures	ix
	Preface	xi
1	International Scientific Congresses: A Functional Approach	1
	The Scientific Consociation	3
	Growth of International Science	4
	Functions of International Scientific Congresses	9
	Perceptions and Realities: A Design for Research	19
2	International Political Science: From Paris to Moscow (and Back)	29
	Internationalization	30
	Politics of Site Selection	36
	Some Questions for Research	47
3	Who Went to Moscow and Why?	57
	Profile of Respondents	57
	Why Attend Professional Meetings?	63
	Value of International Scientific Congresses	64
	Deciding to Go to Moscow	72
4	An Occasion for Learning	77
	Expectations and Experiences	78
	Soviet Government and People	84

	International Scientific Networking	87
	Modifications in Scholarly Behavior	90
	Improving Scientific Communication at World Congresses	93
	Individual Learning	96
5	Politics at Play	101
	To Go to Moscow or Stay Home?	102
	Decisionmaking About Sites	106
	Politics on the Floor	112
	Impact of Politics	114
6	Functions of International Scientific Congresses	119
	Individual Political Scientists	119
	Research Organizations	124
	Scientific Associations	124
	Political Functions	127
	Some Future Directions	132
	Appendix A Breakdown of Registrants by Country	143
	Appendix B Questionnaires	145
	1. Questionnaire for Registrants	145
	2. Questionnaire for Nonregistrants	161
	Appendix C Scales of Professionalism and Internationalism	169
	Bibliography	175
	Index	181
	About the Book and the Authors	185

Tables and Figures

Tables

1.1	Growth and Geographical Distribution of International Meetings, 1950-1983	8
1.2	Size and Composition of Sample	21
1.3	Nationality of Responding Registrants	22
2.1	International Political Science Association, 1950–1985	33
2.2	Participation at IPSA World Congress, 1979	37
3.1	Variations in Professionalism and Internationalism	61
3.2	Intercorrelation of Scores on Scales	63
3.3	Perceived Value to Self of National, Regional, and Subdisciplinary Meetings	65
3.4	Chief Functions of International Scientific Congresses in Political Science	66
3.5	Registrants' Views on Functions of International Political Science Congresses	68
3.6	Importance of International Political Science Congresses	69
4.1	Expectations and Experiences: Mean Scores	80
4.2	Expectations, Experiences, and Professional Activity	83
4.3	Informal Contacts Established at IPSA World Congress, Moscow, 1979	89
4.4	Efforts to Keep Up with Colleagues Originally Met in Moscow	90

4.5	Impact of World Congresses on Participants' Scientific Work: A Comparison of Psychologists, Sociologists, and Political Scientists	92
4.6	Suggestions for Improving Future World Congresses: Views of Psychologists, Sociologists, and Political Scientists	95
5.1	Reasons Given for Attending or Not Attending the IPSA Moscow Meetings	105
5.2	Political Grounds for Evaluating IPSA's Decision on Moscow	108
5.3	Characteristics of Proponents and Opponents of Meetings in the USSR	109
5.4	Alternative Sites for IPSA World Congresses: Views of Registrants at the Moscow Meeting	111
5.5	Reasons Cited for Terming a Panel Session the Best or Worst	114
A.1	Breakdown of Registrants by Country	143
C.1	Professional Activity Scale	169
C.2	Professional Status Scale	170
C.3	International Research Competence Scale	171
C.4	International Activity Scale	172
C.5	IPSA Experience Scale	173

Figures

1.1	A Model of the Effects of an International Scientific Association	6

Preface

A peculiarity of human beings is that they carry through life a bundle of unexamined assumptions on the basis of which they act. Some of us do this more than others, of course, and there is always the possibility that, someday, one or another of us will look at what we are doing and ask why we are doing it and whether or not what we are doing will help us attain the goal toward which we are striving.

This is the case, we suggest, with respect to international scientific congresses (ISCs). We as scientists start with the assumption that effective science requires communication—an assumption that seems fair enough in its own right. Since many scientists are also teachers, and hence accustomed to verbal exchanges with students as well as colleagues, we quickly slip into the additional assumption that seminars and workshops are useful vehicles for scientific communication. If small meetings are good, then large meetings must be better since they bring together experts covering a wider range of topics. International conferences must be even better since they permit the expression of a still broader set of perspectives on the topics of interest to us. Ergo: International scientific congresses are vital to the advancement of science.

We assume other things about ISCs as well—that they are useful for transferring technologies (especially from developed to developing countries), that hosting them garners international respect for the sponsoring country, that attending them enhances the professional status of individual scientists and their institutions, and so forth. In a more general sense we sometimes assume that, by contributing to the international organization of science, ISCs facilitate the creation of world order and what Robert Angell (1969) called the "march toward peace."

But how do we know all this? Each individual who has attended an ISC doubtless returns home with some notion of how useful the particular meeting was in terms of learning something new, establishing contacts for possible future research, or simply visiting an interesting part of the world. We nevertheless have little generalized knowledge about the impact of ISCs on individual scientists, collective bodies (such as the international scientific associations that organize such meetings), a particular branch of science, or the establishment of world order. In short, we have never really examined our assumptions and tested them empirically.

Perhaps these unexamined assumptions are not overly problematic. If in doubt, we rhetorically ask ourselves no less than possible funding institutions, *why not* hold an international scientific congress? We *think* some good comes of them and, besides, they are not excessively expensive or difficult to organize.

There are several answers to such a rhetorical question, all seeking to balance the benefits of such congresses against their various kinds of costs. Without firm knowledge that ISCs are actually beneficial for advancing science, world order, or some general goal, we encounter severe difficulties in justifying their continuation in a world of scarce resources available for science.

For one thing, ISCs are indeed expensive in monetary terms. The budget of the 13th World Congress of the International Political Science Association (IPSA), held in July 1985 in Paris, was $150,000. Some 1,763 political scientists attended that meeting (not to speak of spouses, children, and others accompanying the participants). If we assume conservatively that the average participant spent $700 for transportation and $50 per day for six days for lodging and meals in Paris, paid $100 for IPSA membership and the congress registration fee, gave out $100 in incidental expenses, and devoted a week of salaried time to the meeting (at a rate of $20,800 per annum), then the real cost for that participant was approximately $1,600. Add to these sums approximately $150,000 for the direct costs of the program chairman and meetings of the international program committee, and we find that the total cost of the IPSA World Congress (deducting an amount for double-counting, but ignoring the association's general administrative expenses and the donated time of its officers) was in the neighborhood of $3 million—a major portion of it paid by governments, foundations, and universities in the form of individual travel grants.

Second, it is not clear that all scientists learn equally well from verbal interaction at large (or even small) conferences. If the question is the assimilation of new material, then the answer for many is clearly that an hour of reading refereed journals is more usefully spent than the same hour listening to someone making a formal presentation. If we broaden the question to more contextual factors, such as seeing how a productive scientist responds to challenges, or sharpening one's thinking skills by participating in the exchanges that follow the presentation of papers, or establishing contacts that may serve as the basis for future research, then we are in muddier waters. There is virtually no research on the actual effect of these activities, and none that we know of relating characteristics of conferences to individual styles of learning.

Third, can we comfortably believe that all, or some, or even a few ISCs contribute to the internationalization of science and that this, in turn, enhances world order? If we look at the political uses to which some international conferences have been put, then we might well respond negatively. Some writers have also feared what they term a "UNESCO-ization" of international scientific activity. By this they mean science run on the basis of "one nation, one vote," in which majority votes often based on political considerations can determine who is an acceptable invitee, what lines of inquiry are permissible, and which scientific results are to be credited.

It is important to obtain reliable data on such issues. If the benefits are low, especially in view of rising financial and perhaps political costs, then we may well want to consider new means of scientific communication to replace international scientific congresses (e.g., electronic mail). To continue unperturbed on an untenable path does not make good scientific (or fiscal) sense. Moreover, unless we have firm evidence that the benefits of ISCs significantly outweigh their costs in some general sense, then we as individual scientists who attend such congresses open ourselves to charges of academic profiteering—using the public purse for private gain.

This book reports on an effort to investigate systematically some of the uses and consequences of international scientific congresses. It seeks to improve the quality of scientific communication across national boundaries by examining the empirical basis of some of our assumptions. The procedure used in the study was to survey political scientists from nonsocialist countries who attended the 11th World Congress of the International Political Science Association (IPSA), held in August 1979 in Moscow, and also, by way of a control group, a random sample of political scientists from the United States and Canada who did not attend the Moscow meetings.

The assistance of several individuals and institutions enabled us to initiate and complete the project. Of particular importance was Judith Jones at the University of Illinois at Urbana-Champaign (UIUC), who coordinated all aspects of the project from the time we were considering the form of the questionnaires until the completed questionnaires were received and registered. Janie Carroll assisted during this period. Jutta Sebestik of the UIUC's Survey Research Laboratory helped us draft the questionnaires and resolve sampling problems. Barbara Hill, Department of Political Science at the University of Iowa, helped us develop the scales and process the data. Karl W. Deutsch of Harvard University and the Science Center Berlin, Jean A. Laponce of the University of British Columbia, Secretary General John E. Trent and former Executive Secretary Liette Boucher of the International Political Science

Association, Robert Alun Jones and Steven T. Seitz of UIUC, and Jerzy Wiatr of the University of Warsaw offered insightful comments that saved us from many an error. The UIUC's Research Board provided financial assistance at critical junctures. Anna J. Merritt of the Institute of Government and Public Affairs, UIUC, prepared the index. To all these individuals and institutions we are grateful.

Richard L. Merritt
Elizabeth C. Hanson

ONE
International Scientific Congresses: A Functional Approach

The world congress of an international scientific association is a peculiarly modern phenomenon. A Euro-American invention (Mead, 1968: 215), it brings together from all over the world scholars representing most aspects of a particular scientific discipline. It thus differs from a conference of experts on a single topic (see Capes, 1960) in the sense that the international scientific congress (ISC) presents a broad palette of sessions on a wide range of topics designed to appeal to both the specialist and the generalist, and to people with varying degrees of disciplinary sophistication.

Implicit in such a plan are several prerequisites. Some are physical and technical: conference facilities adequate for a large gathering, a staff trained in congress organization, inexpensive and convenient modes of transportation, and financial support from foundations, the participants' own universities, or elsewhere. Other prerequisites are structural and organizational. Among these are, most notably, a scientific discipline sufficiently developed that an international congress makes sense, an international scientific association representative of subdisciplinary concerns and the major countries where scientists identify themselves with the discipline, leadership capable of mounting such a congress, a communications network to facilitate the identification of topics of common interest, and the scientists who might have something important to say on them.

Perhaps the key prerequisite is a belief that an international scientific congress serves some function(s)—for the discipline, its members, or some other body.[1] But what are these functions? Or, more properly phrased, *who sees ISCs performing what functions, how, for whom, and with what effect?* It is this congeries of questions to which our study seeks at least partial answers.

The intuitively obvious answer to such a multifaceted question is that the world congress is an international communications medium to advance the state of science. That is, members of a scientific community (Polányi, 1951: 53-57; Shils, 1954) see the ISC as a means to learn more about the theories, findings, and methodologies in an area of scientific interest to them. Further reflection nevertheless reveals far greater complexity. Is the desire for scientific communication the sole motive underlying the individual scientist's decision to attend such a congress? If scientific communication is the goal, then is the ISC the most efficient or effective mechanism for achieving it? What interest do associational or governmental functionaries have in such congresses? Are there other functions so deeply embedded in the scientific enterprise that they all but escape the notice of casual or even experienced observers? As sociologists of science and scientific knowledge have shown us, easy answers to these questions do not lie at hand (Mulkay, 1979; Collins, 1983a, b; Jagtenberg, 1983).

And yet answers can be significant. They are germane to an emerging theory of science. As we shall see later, international scientific congresses play a role, implicitly or explicitly, in functional theories of international integration, structural theories of international relations (with their focus on center-periphery relations), and various approaches to the sociology and politics of science. If we would understand the nature of science and its relationship to society, then, information about the role played by ISCs can provide important clues.

On the more practical level of science policy, too, it makes sense to know what world congresses can and what they cannot accomplish. After all, an ISC is expensive in terms of time, money, and other resources, and hence constitutes a scientific investment. If its outcome in fact repays the investors by realizing their highest-priority goals, then they have sounder reasons to support future ISCs and perhaps even improve their financial and/or organizational foundation. If, however, world congresses do not accomplish what we intend them to, then we might think about restructuring them so that they will, or else consider alternative ways to meet these goals.

This chapter outlines a framework for analyzing the functionality of international scientific congresses. It looks first at the international scientific structure that has spawned ISCs in the first place. It then addresses the question of what individuals and other actors with stakes in a given scientific discipline consider the functions of ISCs to be. Finally, it describes a design for research aimed at answering some of the questions we have raised about the functionality of ISCs. The study explores such issues on the basis of a survey of political scientists from nonsocialist countries who attended (and some who did not attend) the 11th World Congress of the International Political Science Association, held in August 1979 in Moscow.

The Scientific Consociation

Our concern with science is, in fact, with some branch of science we may call a scientific discipline—chemistry, for instance, or physics or sociology.[2] Even so, the idea of a scientific discipline as such remains rather abstract. How can an observer identify, let us say, political science as a scientific discipline? What characterizes a scientific discipline best are, first, the distinctiveness of its subject matter and, second, the existence of a shared *myth* about its key dimensions. This myth, "the value-impregnated beliefs and notions that men hold, that they live by or live for" (MacIver, 1947: 4), lends integrative glue to a scientific discipline and sustains its activities.

"With the aid of authority," to follow MacIver's (1947: 42) line of reasoning, the "myth-conveyed scheme of values" determines the *order* of a scientific discipline.[3] What order myth backed by authority imposes on science has been the subject of intense dispute among theoreticians, and need not concern us here.[4] More to the point is the fact that various actors identify themselves with some version of the scientific order, make demands on it, and have certain expectations about its processes and outcomes, that is, they have *perspectives* on the scientific order (Lasswell and Kaplan, 1950: 25). Among these actors are not only producers of new knowledge but also those who organize facilities and raise funds to permit research to go on, see to the diffusion of knowledge throughout at least the discipline, apply it to practical problems, watch out for the interests of the scientists, and cover the financial and social costs of the particular scientific order. They are members of what we may call a disciplinary *scientific consociation*:[5]

1. Individual scientists who identify themselves with the scientific order of a given discipline, that is, generally accept its cognitive, normative, and behavioral standards, conduct research within its framework, and transmit these perspectives to their students

2. Research organizations—universities, government laboratories, research institutes in the private sector, or other—in which the scientists work, and which set priorities for their research

3. Disciplinary associations at various levels (local or regional, national, international), which seek to play an authoritative role in interpreting the myth of the scientific order, safeguarding standards, distributing rewards, and otherwise promoting the interests of the discipline itself as well as associational members[6]

4. National governments, which represent the societies that ultimately pay for science and expect it to be useful

To these some would add a fifth relevant actor:

5. Global society, to which writers have attributed functional needs

(ranging from hyperstability to a full-scale redistribution of resources) that science could, in their view, help meet

Each of these actors has a stake in the development of a scientific discipline; and, we would argue, a part of their perspectives on that discipline and its scientific order focuses on the functionality of international scientific congresses.

Regrettably, however, we do not know much about the place of ISCs in the diverse actors' perspectives. Some elements of this puzzle we may infer from empirically based research on other matters—the fact, for instance, that perspectives vary from one actor to the next. Variation occurs both within a single level of actor and across levels. Thus two scientists may have significantly different constellations of identifications, demands, and expectations; and the way in which those responsible for a research organization define research priorities may clash with the views of scientists who must conduct the research. It follows that behavior based on these perspectives, including developing such mechanisms as world congresses, may be mutually incompatible for those with an interest in the outcome. All this is a way of saying that what is functional for one actor may be dysfunctional for another, especially in a setting of competition for scarce resources.

For the rest, the research literature is curiously spotty. It has devoted considerable attention to some aspects of the structures that "represent" scientific disciplines—descriptive histories of national scientific associations, the operation and impact of scientific journals, and disciplinary reward structures—while leaving others virtually untouched. This is true with respect to disciplinary congresses at whatever level, and also with respect to international disciplinary associations as such. Brief attention to the latter may serve as an introduction to the problématique of their congresses.

Growth of International Science

The emergence of an organizational framework for international science has been an integral part of an even broader trend. The 20th century, particularly the era since 1945, has witnessed a dramatic increase in both the number of international nongovernmental organizations (INGOs)—organizations of private groups and individuals that share common interests across national boundaries—and the scope of their activities. Not surprisingly, this trend has stimulated considerable scholarly interest in the role that these nonstate actors play in an international system dominated by nation-states. Numbering less than 200 in 1909, they increased by mid-century to approximately 1,000 and by 1984 to 4,615 (Union of Intergovernmental Associations, 1984: i,1626).

International scientific and professional associations constitute nearly half the total number of INGOs. International *scientific* associations contribute to a growing body of knowledge and technology by aiding the communication of scientific findings, stimulating new research, improving methods of analysis, and facilitating international collaboration in research. International *professional* associations, serving the public with a body of applied systematic knowledge, seek to maintain high levels of competence and ethical standards. Our focus is the former, international scientific associations. Of particular interest to analysts have been the ways in which and extent to which these prolific transnational phenomena influence international policies and the political and social consequences of their activities in the aggregate.

From one point of view, international scientific associations are simply a means to pursue individual and group interests at the international level (Crane, 1981). They act as pressure groups exerting influence on intergovernmental organizations and governments to adopt certain policies. Establishing an international association with a permanent secretariat and procedures for periodic meetings helps national associations and individual members in various parts of the world to communicate, coordinate their activities, and cooperate for the purpose of furthering the goals of the discipline. Transnational activities can give status and visibility to the associations and individuals that pursue them. They can stimulate the growth and expansion of a discipline or profession by such means as encouraging the establishment of new national associations, curricula, research institutes, and training programs. Figure 1.1 presents William M. Evan's (1975) neat summary of an international scientific association's various activities and effects. More generally, the eminent French political scientist and first secretary general of the International Political Science Association, Jean Meynaud (1961: 12), concluded that, "in the tremendous and indispensable effort to convert the present-day world to the social sciences," international social scientific associations "played a very honourable part."

From another point of view, through their activities some of these associations have sought to influence international decisionmaking and the international system as a whole. Curtis Roosevelt (1970), for example, saw an important role for various types of nongovernmental organizations in the development of "constituencies" that can mobilize public opinion and support for world development goals. Several articles in the published proceedings of a conference on international scientific and professional associations (Evan, 1981a) emphasized their potential role in promoting the growth of transnational values and in advancing peace and justice. According to this view, these associations increase the level of integration among nation-states by multiplying international contacts and by creating a network of relationships that socialize members into the values of transnationalism. In his introduction the volume's editor (Evan, 1981b: 16, 23) argued that the

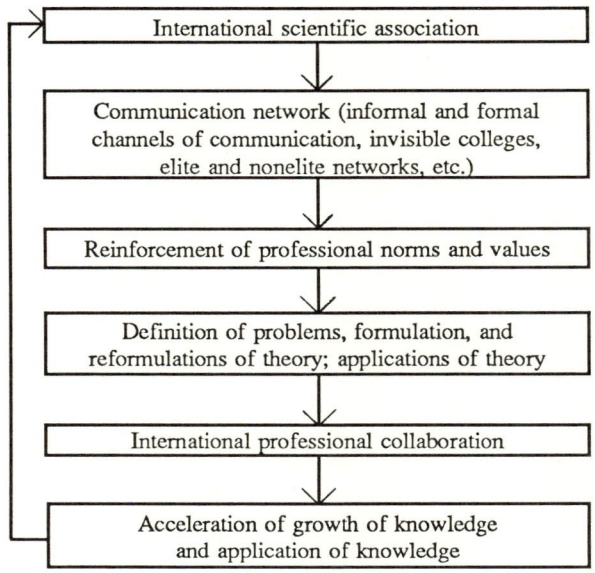

Source: Evan, 1975: 386.

Figure 1.1 A Model of the Effects of an International Scientific Association

activities of international scientific and professional associations reinforce the ethos of science and the commitment of scientists to the cultural values of universalism, communality, disinterestedness, and organized skepticism. If any segment of society can help the international system to move "from nationalist values to transnational humanist values," he concluded, it is the international community of scientists, engineers, and other professionals.

Louis Kriesberg (1981) suggested how the activities of international scientific and professional associations directly and indirectly contribute to the attainment of justice and the reduction of violence. One way is by providing a source of ideas for mutually acceptable solutions to intergovernmental problems and a setting in which new insights emerge from the discussions among private individuals from different countries. They can also provide less advantaged members with services and benefits aimed at helping them to solve their problems and advance their knowledge and skills. In the long term Kriesberg anticipated that these associations will aid the development of a common world culture by increasing awareness and tolerance of differences; and, as an extreme possibility, they may strengthen alternative transnational structures by providing quasi-governmental services.

A more ambivalent viewpoint emerged from a seminar, on "Social Science as a Transnational System," in which 15 social scientists from various regions of the world participated. The summary of the discussions (Alger and Lyons, 1974) emphasized the need for a global community of social scientists responsive to the growing number of global problems and the value of transnational activities for generating, disseminating, and applying social science knowledge. At the same time, some participants expressed concern that increased international cooperation in the social sciences under the present conditions will only strengthen existing patterns of dependency. International scientific associations, they pointed out, play an important role in establishing and maintaining contacts and communication among social scientists from the dominant centers; but in performing this role they tend to institutionalize the dependency relations of social scientists outside the dominant centers. Some empirical evidence of this tendency appeared in Kjell Skjelsbaek's (1971) analysis of INGOs. Citing the high density of INGO comemberships among the developed countries and the lower level of participation in these organizations by the less developed countries, he suggested that INGOs can lead to a higher degree of integration among the countries that are already dominant and thus help consolidate and enhance their position.

A main task of international scientific associations, however they are viewed, is to organize and convene international congresses and other meetings. These occasions as a rule involve more members than any other single activity of a given association, and they provide one of the most important mechanisms for conducting the association's work and implementing its goals. Hundreds of international scientific congresses occur every year, covering subjects from thermodynamics to the training of midwives. Table 1.1 shows that the number of international meetings for both nongovernmental and intergovernmental organizations increased between 1950 and 1983 by almost 6 percent annually. Although the international meetings of intergovernmental organizations (IGOs) are more frequent, those of INGOs involve more participants. For example, in 1978 only 250,000 people attended the 5,000 meetings organized by IGOs,[7] whereas 2 million participated in the 4,800 meetings sponsored by INGOs (Fighiera, 1984: 143).

Accompanying this growth in the number of international meetings has been their expansion to include an ever larger number and ever wider variety of countries. The geographical distribution of meetings shown in Table 1.1 indicates this development. The overwhelming majority of these meetings in the last two decades took place in Western Europe. As far as international scientific associations are concerned, a discrepancy has existed between the location of most scholars (North America) and the location of most

Table 1.1 Growth and Geographical Distribution of International Meetings, 1950-1983[a]
(Intergovernmental and Nongovernmental Organizations)

Region	1950 No.	1950 %	1955 No.	1955 %	1960 No.	1960 %	1965 No.	1965 %	1970 No.	1970 %	1975 No.	1975 %	1980 No.	1980 %	1983 No.	1983 %
Western Europe	578	80	843	75	1257	66	1115	63	850	63	1887	57	2915	61	2851	59
North America	70	10	85	8	160	8	193	11	133	10	441	14	633	13	691	14
Latin America	36	5	81	7	181	9	120	7	85	6	203	6	183	4	189	4
Asia/Pacific	23	3	66	5	143	8	166	9	152	11	376	11	590	12	668	14
Eastern Europe	8	1	22	2	87	5	111	6	104	8	276	8	339	7	296	6
Africa	9	1	28	2	71	4	63	3	29	2	137	4	152	3	169	3
Total	724	100	1125	100	1899	100	1768	100	1353	100	3320	100	4812	100	4864	100

[a]Meetings organized and/or sponsored by international associations listed in Union of International Associations (1983). Percentages may not add up to 100 because of rounding.

Sources: Union of International Associations Bulletin NGO, 4,2 (February 1952): 64-65; *Union of International Associations Monthly Review*, 8,4 (April 1966): 207-209; *International Associations*, 20,2 (February 1968): 92-95; *International Transnational Associations*, 24,1 (January 1972): 47-49; ibid., 29,1-2 (January-February 1977): 43-45; ibid., 34,1 (January-February 1982): 42-43; and ibid., 37,1 (January-February 1985): 58-59. Fighiera (1984: 145) has slightly different figures.

congresses (Western Europe). A simple explanation is the fact that international-minded people want to meet "abroad." Accordingly, the more North Americans there are in these associations, the less they want to meet in North America. It is also true that, by and large, North Americans have enjoyed greater access to travel funds. This lopsidedness is nevertheless changing. Other regions have shared in the general increase in numbers of international meetings. The proportion of meetings held in Western Europe has, in fact, declined substantially from 80 percent in 1950 to 59 percent in 1983. In part this reflects growth in the scientific establishments of other countries, in part a growing demand by these countries for scientific recognition. The result of these trends is that more non-European countries have had an opportunity to host an international meeting, and more of their nationals have been able to attend.

Functions of International Scientific Congresses

The periodic international congress has become a focal point of activity for most international scientific associations. It is typically a very large meeting which covers a wide range of topics and is attended by a geographically diverse group of participants. Although the subject matter varies from one discipline to another, certain practices and procedures have become common features. Advance planning, a not unproblematic logistical task, is carried out by the multinational executive and program committees together with the association's secretariat. These plans include decisions about the site and physical arrangements as well as the agenda, procedures, and participants.

The day for the congress arrives. Typically, an opening ceremony, sometimes including a welcoming address from a high political official of the host country, eases the participants into the theme and procedures of the congress and, afterwards, may provide an occasion for informal interaction. Participants must then choose for the rest of the week among a smorgasbord of smaller sessions, usually panels in which a particular topic is addressed from several points of view. In such panels, participants summarize orally the theoretical analyses and research results contained in their papers, which they have made available for distribution, and a brief discussion follows, sometimes led by designated commentators. Alternatively, roundtables eschew the formal presentation and critique of papers in favor of maximizing informal exchanges among the panelists. Additional discussion typically ensues in the corridors and more pleasant places. A grand reception and/or banquet and sometimes other events further facilitate informal interaction. The congress ends as it began, in a plenary session at which all relevant persons are thanked and the significance of the congress reaffirmed.

Just what international scientific congresses accomplish remains an open question. A considerable body of literature on intergovernmental conferences notwithstanding, little scholarly attention has been given to the international meetings of nongovernmental organizations in general or to those of international scientific associations in particular.[8] The gap in the literature is remarkable in view of the growth of this kind of transnational activity and the vast amount of human and financial resources that planning, conducting, and participating in these congresses absorb.

Given what participants and observers have written about the functionality of international scientific congresses, what is it we think we know? And what else do we want to find out? The following sections address these questions by focusing in turn on each of the five categories of actors identified earlier as having an interest in international scientific congresses: individual scientists, research organizations, disciplinary associations, national governments (representing their societies), and global society.

Individual Scientists

A relatively small portion of the membership of the relevant discipline actually attends a given international scientific congress. Why is this? One response is certainly that constraints of time, cost, convenience, and still other technical factors conspire to keep people away. If we assume, however, that these constraints are fairly constant across the membership of the discipline in any single country, we are left with a slightly modified version of our question. Why do some scientists in a country—or at a single university—choose to attend an ISC while their colleagues do not?

Personal characteristics. We might seek the answer to this question in terms of sociological differences that characterize scientists. Thus a comprehensive survey of the American professoriat by Everett Carll Ladd, Jr. and Seymour Martin Lipset (1978) confirmed widespread assumptions about the existence of an "academic jet set" elite and suggested certain hypotheses about its members. Of the faculty members they surveyed, 79 percent had never attended an international scientific meeting abroad; only 2 percent had traveled abroad for this purpose 10 or more times. These "high travelers" came overwhelmingly from major research universities, were heavily involved in research, and in the course of their careers most had published at least 20 professional articles.

Another answer lies in scientists' perceptions of what they might derive from attending such congresses. Even if an individual has been invited to present a paper at an ISC, expects to be free when it takes place, and has the requisite funding, the decision to accept the invitation entails an opportunity cost. The time, money, and energy that would be required could be spent in other ways. Assuming rationality, the individual must weigh the costs against the particular congress's functionality for meeting certain scientific, professional, and personal goals.

Scientific communication. We noted earlier the most commonly heard assumption about ISCs, that they are an important means of scientific communication. Thus the 13th report of the National Science Board to Congress on the status of science and technology in the United States asserts that international scientific congresses provide a faster method of communicating research results than publication, as well as immediate feedback and ideas from colleagues. The informal exchange of ideas, the report continues, can lead to "modifications of research, collaborative efforts, and elimination of duplicate work" (National Science Board, 1981: 41).

The extent to which international scientific congresses in fact advance scientific knowledge, however, has rarely been investigated scientifically. The most notable effort came from the Center for Research in Scientific Communication of Johns Hopkins University, which surveyed delegates to the International Congress of Psychology, held in 1966 in Moscow, and the World Congress of Sociology held in the same year in Evian, France. These studies sought to ascertain the information effects of an international congress by analyzing the scientific information exchange and interaction that resulted from these two meetings. The project's final report (Johns Hopkins University, 1968: 64) concluded among other things that:

> The research scientist today feels strongly that he should be free to disseminate his results to an international audience, to receive publications from and establish information exchange with colleagues in laboratories throughout the world, and to have personal interaction with experts in his field, regardless of the country in which an expert is located. The two international meetings studied to some extent facilitated each of these pursuits. . . . In the present day state of science, with scientific information accumulating in great quantity, the individual scientist relies on an international meeting every two or three years to bring himself up to date on new international developments in his field and to revitalize his international information-exchange network.

A closer examination of the project's data, as we shall see in Chapter 4, nonetheless reveals that these effects were far from universal among the participating psychologists and sociologists and far from uniform even among those reporting them.

Professional advancement. As long as opportunities to participate in an ISC are scarce, scientists will view them as part of their reward structure. For some it provides immediate gratification in the form of status vis-à-vis departmental and other colleagues who were not invited or could not raise the necessary funds. The act of presenting one's research before a panel of other scientists, distinguished at least by the fact that they, too, were invited to attend the congress, may also enhance one's professional reputation.

Doubtless more enduring is the recognition that one's research is sufficiently important to merit an international hearing. On the one hand, this recognition may make it more possible for the scientist to secure grants, publication outlets, professional offices, invitations to further congresses, and other preferments. On the other hand, it gains a potentially significant audience for the research itself. This may encourage scientists elsewhere to take the approach and findings more seriously, which in turn lends legitimacy to the particular research enterprise.

International scientific congresses also offer opportunities to the scientist interested in taking an active role in shaping the profession. They are an excellent occasion for learning about the association's power structure and who is who in it. Contacts with key individuals may in turn lead to appointments to the program or other committees, in which organizational decisions are made about the content of future conferences, structure of the association's publication activities, and the distribution of its rewards.

Personal enrichment. It is difficult to ignore the fact that the scientist who attends a meeting overseas gains something beyond scientific and professional growth. Other advantages include having opportunities for travel and recreation, enjoying the conviviality of shared experiences with old and new friends, sampling exotic foods, and enhancing one's status among friends and neighbors. Cynics have asserted that these benefits are indeed the primary factor moving scholars to seek to attend international meetings, and that any scientific reason volunteered by participants is more a justificatory ruse than an accurate assessment of their true motives. It follows, in this view, that the professional prestige gained has as its most important value attracting invitations to yet other such meetings. In his satire describing the "small world" and multifarious escapades of scholarly jet-setters attending international conferences, the English novelist David Lodge (1984: 238) wryly noted: "Afterwards, when they are back home, and friends and family ask if they enjoyed the conference, they say, oh yes, but not so much for the papers, which were pretty boring, as for the informal contacts one makes on those occasions."

We must not be too hasty in writing off such motives as frivolous. To be sure, given the scarcity of opportunities to attend ISCs, doubtless few foundations or universities will wish to support the participant who appears to be seeking pleasure alone. For the others, however, the ego gratification derived from presenting a paper, travel, and the like may have a strong scientific value as well. A Dutch "founding father" of the International

Political Science Association expressed this possibility well. The organization of international congresses, he wrote (Barents, 1959: 1090),

> by the ingenious systems of travel subsidies which accompany most of them, acts as an inexpensive travel agency for relatively poor people like university teachers. This is not a cynical comment. To some people regular visits to international congresses have an aspect of profiteering, but for many scholars, particularly of the younger generation, it would be quite impossible to travel all over the world without outside financial help. Travelling, if wisely done, being part of one's education, this function of international congresses may be something to be regarded with a watchful eye, but is in itself to be neither despised nor neglected.

The chance to stand up in front of an audience of peers to deliver a paper may similarly accomplish more than merely massaging the ego. As suggested earlier, and as the study of the world congresses of psychologists and sociologists showed (Johns Hopkins University, 1968: 64):

> Making a presentation placed an author in the "limelight" and attracted other workers in the field to him for discussion of areas of mutual interest. Information exchange as a result of having made a presentation at these meetings was frequent, had important effects on authors' work, and served as a stimulus for the establishment of continuing personal information exchange and, consequently, of the development or extension of informal, international communication networks.

And it may very well be that "the informal contacts one makes on those occasions" provide important new insights and the basis for innovative research.

Research Organizations

Universities, research institutes, and other agencies stand to benefit when their staff members participate in international scientific congresses, especially, of course, if these individuals are highly visible or successful in their activities. First, the participants may return from an ISC with enhanced knowledge, skills, and self-confidence that improve their productivity and hence augment their value to their home research organizations. Second, the participant, when presenting theories and findings, in a sense "represents" the research organization. The presentation tells other scientists what kind of work the organization deems significant as well as the quality of researcher it employs. A successful paper by one of its scientists may enhance the international credibility of the organization as a whole. Third, the organization whose scientists participate regularly in ISCs and whose projects gain an international audience may be able to retail such recognition in the form of greater attractiveness for potential recruits and improved access to research funds.

Disciplinary Associations

International scientific congresses may serve several functions for *national scientific associations* and their regional and subdisciplinary components.[9] For one thing, despite assertions that science has been internationalized, a residue of national competitiveness persists in many disciplines.[10] Certainly, scientific productivity and training facilities play the major role in assessing claims to precedence, but the number and quality of scientists national associations send to international meetings are also important. To the extent that the research frameworks of these scientists dominate the intellectual interaction, their national associations can be said to dominate the discipline.

Also, a national association seeking to gain greater prestige and enhance its role in the international association may try especially hard to send many of its members to the congress. Such behavior may be instrumental in obtaining yet another goal: a more important role in the international association itself. A greater presence at the congress may improve the likelihood that the national association's candidates will be selected to serve as officers of the association, its ideas given greater weight in organizing the succeeding congress, or its offer to host future meetings taken more seriously. At the same time, however, implicit in this process is occasionally the desire to control access to the international congress on the part of the country's scientists. While some national associations are content to let the international association's program committee determine whom to invite, others want to exercise a veto to ensure that the "right kind" of national scientists attend (for example, only those who are members of the national association or who toe a particular ideological or paradigmatic line). Especially important in this respect is the national association's ability to determine who receives travel grants or, in some countries, permission to travel abroad.

The national association that hosts an international congress may profit in several ways. The most obvious of these are, first, an enriched national discipline because of its improved opportunities for scientific communication, which in turn may strengthen the national association; and, second, the national association's greater visibility or prestige in the international disciplinary setting. It also certifies to its government the national association's relative importance in that setting. This in turn may have implications for future financial or other support for the association. A fourth outcome may be that the national association gains greater influence over the content of the congress. Although, to be sure, an international program committee usually determines what will be discussed, it is unlikely to ignore the favorite themes of the host association. Finally, hosting an international congress gives the national association opportunities to increase its sway over the members of the discipline in its own country. That is, it becomes quite visible as the national discipline's chief organizing force, in control of important resources such as access to the international communications network.

For the *international scientific association*, the congress it organizes also serves multiple purposes. Some of these meet general disciplinary needs: advancing the state of knowledge in the discipline, developing a sense of scientific community among its members, and focusing on the discipline international attention, albeit modest and fleeting. Then, too, a congress responds to some of the international association's structural needs. The fact that national associations and scientists from all over the world heed its call to convene is an important legitimation of the international association's existence and *raison d'être*. Its organization of the congress, through such organs as its secretariat and program committee, facilitates the association's control over the structure of the discipline as a whole, besides giving it greater standing in the international community of scientific disciplines. Moreover, we should not forget that association officers obtain their own rewards from a successful congress.

International scientific associations face an important strategic question that can affect the functionality of ISCs for them. Most scientific disciplines are moving toward a fragmentation of knowledge, in which specialists focus on ever smaller areas of research. An effect can be that the association becomes a holding company for a number of subdisciplinary interest groups. In such circumstances, how should the association organize its congresses? One answer points toward associational centralism. It embodies a "top-down" approach, in which an authoritative program committee identifies what it considers to be the discipline's core topics and integrates papers linked to these topics. Another answer stresses disciplinary pluralism—a "bottom-up" approach that envisions a broad palette of more specific research topics that may or may not fit well together. If it is true that the one approach runs the danger of promoting a sterile "official science," then it is also the case that the other could easily undermine the discipline's intellectual coherence. Steering a course between the two shoals, albeit far from simple, is necessary since either could make the association virtually irrelevant to its members, that is, deinstitutionalize it.

National Governments

Normally, a country benefits in numerous small ways from hosting an international scientific congress. Bringing to one of its cities several hundred or a few thousand foreigners has an impact on the country's tourist industry.

For a country short of foreign exchange, this can be a powerful incentive to encourage such congresses. The fact that many of these foreigners are first-rate scientists may improve the quality of the country's own scientific establishment. Not only will the participants exchange scientific information that may be of value to indigenous scientists and academic administrators, but their mere presence may inspire one's own scientists to greater achievements. A strengthened national scientific establishment in turn contributes to the country's international power capabilities.[11]

Such a conclave of distinguished scientists may yield other benefits as well: prestige for the country's scientific establishment and the government that supports it, an enhanced reputation as a place in which free scientific exchanges can take place, greater understanding on the participants' part of the country's political and other problems, and perhaps even more warmth toward the country itself and its inhabitants. All these in turn may serve the political interests of the government, indeed so much so that some governments may be tempted to manipulate the congress to produce the greatest possible political gains.

Normally, too, ISCs entail some costs for the host country. Its government usually must make available, or encourage its private foundations to provide, a substantial sum to facilitate the congress's organization. In the national budgets of most countries, the amount of money needed is usually trivial (however difficult it may be for congress organizers to extract it!). Another cost is the risk that the congress will be less than successful, and leave in the participants' mouths a bad taste toward the country as a whole. Some countries even fear that conferees may criticize the government (see Dickson, 1979) or else envision possible problems ensuing from too much interaction between foreign scientists and the local citizenry. They may thus refuse permission to a scientific body seeking to organize a congress on their soil or, if they grant it, impose restrictions on the press or isolate the congress itself in some idyllic spot far distant from potentially troublesome population centers.

The importance of international scientific congresses in the eyes of national governments makes them a legitimate object in interstate struggles. At one level governments can treat ISCs in an avowedly political fashion. Thus, the government of state X might declare that its national scientific associations should ensure that world congresses are not held in opposing state Y, or that it will refuse permission for its scientists to attend congresses in Y. Such open boycotts are known in sports as well as science. At another level, the scientific establishment of country X, with or without the backing of its government, may seek to influence country Y's behavior. It might declare that its members should not attend congresses in or otherwise cooperate with scientists in country Y because of some action by the latter's government (such as perceived violations of the "human rights" of scientists or others)—and then apply heavy pressure on individual scientists in X (and perhaps elsewhere) to comply with the injunction (Seltzer, 1978; Stone and Spilhaus, 1980).

Global Society

Insofar as holding or boycotting international scientific congresses has an impact on a nation-state's power capabilities, such actions pertain to the broader realm of international relations. They fit, first of all, into *systemic theories of international relations*, which posit alternative structures of the international system and the modes of interaction appropriate to each. A balance-of-power system, stressing flexible behavior among states that cooperate and confront as a balancing mechanism, is doubtless less likely to intermix politics and science than is a rigid bipolar system characterized by opposed ideologies or ways of life. Similarly, a state trying to ease relations with another may see scientific exchanges, including joint conferences and cooperation at international congresses, as a first step.

The latter possibility has led not a few writers to see an important role for science in their *functional theories of international political integration*. World congresses, this argument runs, help to internationalize science, and internationalized science can contribute to world order and peace. Referring specifically to international professional and scientific associations, Robert C. Angell (1981: 244) argued that "the very existence of a global association strengthens existing ties across national boundaries and inaugurates new ones." The consequence is not only improved scientific communication or expansion of professional norms. Rather, the elaboration of such webs of interaction invites a government to benefit from their existence and puts it on notice that political action to tear them apart entails at least some costs in terms of lost benefits. Angell thus concludes that such associations "have become working elements in the world institutional structure that is slowly fostering intersocietal and intercultural integration" (p. 254). More generally, "transnational participation," of which these associations are one form, "will gradually carry the world over the threshold of peace" (Angell, 1969: 28).

It is of course possible to make a contrary argument. A poorly conceived or implemented world congress may make national associations wonder about the utility of continuing to pay their annual assessments to the international scientific association, and sour individual scientists on the whole idea of such an association for their discipline. The attempt by a host country to use a congress for its own political purposes may create enough hostility to split the association. Or the congress organizers, by choosing papergivers according to criteria of geographic representativeness to the exclusion of considerations of quality, may trivialize the association for important

for those who organize and fund ISCs as well as to theorists of international science. Third, our evaluation of perceived functionalities for actors other than individual scientists must rely heavily on inferences drawn from the recorded responses and ancillary information.

Sample

To explore the impact of the Moscow meetings we sent separate questionnaires to two samples: registrants, or political scientists (and others) from nonsocialist countries who registered to attend these meetings; and nonregistrants, a control group randomly selected from the political science profession in the United States and Canada, but excluding any persons on the first list.

Registrants. In accordance with prior agreements among the congress's organizers, all participants from countries outside the ruble currency area were supposed to pre-register at IPSA's secretariat in Ottawa, Canada. Payment of a registration fee of $75 for professionals or $25 for students entitled the registrant to both an official invitation to the congress and the certification of registration necessary for obtaining a Soviet visa, as well as to materials provided at the congress itself. Participants from the ruble currency area were to pre-register or register with the Soviet organizing committee. We decided to limit our survey to scholars who pre-registered in Ottawa. Of the 1,027 registrants to whom we sent questionnaires, 420 (41%) completed and returned them (Table 1.2).

The list of registrants provided by IPSA's secretariat is not identical with the list of those from the nonsocialist world who actually attended the meetings in Moscow.[12] First, some who registered ultimately found it impossible to make the trip, for financial, time, health or other reasons. To the best of the organizers' knowledge, no one remained home because Soviet authorities denied them visas—although in some cases problems associated with obtaining visas led people to decide against going to Moscow. Of the 420 registrants responding to our questionnaire, 15 (3%) reported that they did not attend the congress and another did not indicate whether he attended or not.

Second, a few individuals, especially those such as Herbert Aptheker of the United States who were invited directly by the Soviet organizers, made their own arrangements with the latter's office in Moscow and hence do not appear on the registration lists prepared in Ottawa. Third, in one of those vagaries of international tourism, a group of perhaps a hundred Mexicans discovered that joining IPSA and paying the registration fee enabled them to take an unusually inexpensive guided tour of the Soviet Union; none, however, seems to have attended any IPSA function in Moscow or to have

Table 1.2 Size and Composition of Sample

Region	Number of Questionnaires Sent	Number of Questionnaires Returned	Percentage Returned
A. Registrants			
North America	305	217	71%
United States	(249)	(180)	(72)
Canada	(56)	(37)	(66)
Other West (incl. Israel)	432	157	36
Third World[a]	290	46	16
Total	1027	420	41%
B. Nonregistrants			
United States	428	214	50%
APSA members	(324)	(163)	(50)
Graduate dep'ts	(104)	(51)	(49)
Canada (CPSA members)	62	24	39
Total	490	238	49%

[a]Includes questionnaires sent to but not returned by approximately 100 tourists who registered for the IPSA meetings but did not attend; if the number of questionnaires sent is reduced by this figure (or if we take IPSA's official tally of 194 actual participants from the Third World), then the percentage-returned figure rises to 24 percent for the Third World and 45 percent for the entire sample.

filled out and returned our questionnaire. Table 1.3 shows the national origin of the 420 responding registrants (see Appendix A for details).

Nonregistrants. To ascertain the extent to which registrants at the IPSA congress were representative of the larger population of political scientists, we developed a second sample of professional political scientists in North America who did not register to attend the Moscow meetings. Anticipating a lower rate of response from these nonregistrants, we deliberately oversampled to ensure that the number of returned questionnaires from registrants and nonregistrants would be roughly equal. The sample was drawn from three sources. First, we selected from the *American Political Science Association Membership Directory, 1980* every 28th name with an address in the United States. Second, taking into account the fact that at least half of the country's professional political scientists are not members of APSA, and our observation that APSA's membership roster includes a large

Table 1.3 Nationality of Responding Registrants
(as indicated by place of work)

Country	No.	Country	No.
North America		Latin America	
United States	180	Mexico	8
Canada	37	Brazil	3
	217	Chile	1
			12
Other Western			
Germany (FRG)	26	Africa and Middle East	
Sweden	21	Turkey	7
United Kingdom	20	Ivory Coast	1
Netherlands	18	Nigeria	1
Norway	13		9
Denmark	10		
Finland	10	Asia and Pacific	
France	8	Japan	10
Israel	7	India	7
Spain	6	South Korea	4
Switzerland	6	Hong Kong	1
Australia	3	Malaysia	1
Greece	3	Philippines	1
Italy	3	Singapore	1
Belgium	2		25
Ireland	1		
	157	Total Third World	46
	Total respondents = 420		

number of graduate students, few of whom seemed likely to be knowledgeable about international conferences, we sent a questionnaire to every 45th faculty member listed in the APSA's *Guide to Graduate Study in Political Science, 1980*. Third, we sampled every fifth name from a membership list provided by the Canadian Political Science Association (CPSA). In each case we skipped over persons who had registered to attend the IPSA meetings in Moscow. Of the 490 nonregistrants to whom we sent questionnaires, 238 (49%) completed and returned them (Table 1.2).

Questionnaires

We designed the questionnaires after extensive consultation with North American colleagues who had attended the IPSA meetings in Moscow and with the expert advice of Jutta Sebestik of the Survey Research Laboratory, University of Illinois at Urbana-Champaign; we revised them after some pre-testing among professional colleagues. The 15-page questionnaire for registrants includes both open- and closed-ended questions which, we anticipated, would require approximately an hour to complete conscientiously. The eight-page questionnaire for nonregistrants contains (mostly closed-ended) questions taken from the questionnaire for registrants, and was designed to be completed in about 15 minutes. (See Appendix B for the questionnaires.)

The questionnaires include four kinds of questions. The first category sought to obtain biographical information about the respondents: professional background and status, attendance at national and international conferences, skill in foreign languages, and the like. Another set of questions aimed at eliciting in open-ended fashion opinions about the functions served by national professional meetings as well as international scientific conferences. A third set focused specifically on the issue of site selection—especially Moscow as the site for IPSA's world congress in 1979, but also Rio de Janeiro and Paris as future sites. Finally, several questions, deleted in the short questionnaire sent to nonregistrants, were intended to find out how those who participated in the Moscow meetings evaluated the congress as a whole and their own personal experiences.

The questionnaires were mailed to respondents during the latter part of August 1980 (with follow-up questionnaires sent in mid-January 1981 to those who had not yet responded). Members of the samples in the United States received postage-free envelopes in which to return the completed questionnaires, but those outside the United States were asked themselves to cover return-postage costs. There can be little doubt about the negative impact that the latter procedure had on the rate of response (and were we to conduct another such survey we would search for better ways to encourage responsiveness from political scientists outside the United States). In fact, however, our non-U.S. colleagues were very cooperative. Their rate of return was very respectable for surveys of this kind.

General Questions for Exploration

In the following chapters we shall concentrate on the evidence provided by individual respondents, with emphasis on the following concerns.

Individual characteristics and perceptions. Individuals attending the same meeting will often go away with widely differing explanations of what happened and why. This is due in part to the particular circumstances encountered by each individual, and in part to the qualities that participants bring to the meeting in the first place. Scientists with considerable professional and international activity and status, we might anticipate, will be more likely than others to stress the scientific aspect of an international scientific congress, to be skeptical about the effects of any possible politicization, and to understand better the argument for the international institutionalization of their scientific discipline.

Functions. Of particular interest are the uses participants make of an ISC, the functions they see it serving for other actors in the scientific consociation, and the rankings they give the various functions in terms of priorities. To take but one example from the array outlined earlier, we may presume scientific communication to be functional for the individual scientist's learning. What aspects of a congress facilitate or hinder this process? One way to approach the question is to compare the participants' prior expectations about what they will find at a congress with their actual experiences. Another is to examine the extent to which the experience of attending a congress changes the participants' scientific outlook or even their views of the host country and its people. A related approach is similar to Sherlock Holmes's "curious incident of the dog in the night-time," the dog that did not bark. What functions do analysts detect but congress participants do not mention? Are the analysts off-base in their conclusions, or are these in fact latent functions necessary for the effective operation of some actor?

Subsequent behavior. Do participants actually use the opportunities they have at an international scientific congress to expand their intellectual horizons? That is, do they establish contacts with colleagues from other countries, learn something that changes the orientation or otherwise improves the quality of their research, and/or subsequently keep in touch with colleagues they first met at the congress? Here we might hypothesize that scientists who are already well known in the profession and who frequently travel overseas to conduct research and attend conferences tend to go to an international congress for different reasons and to benefit in different ways than do their colleagues with less professional status and international experience.

Perceived functionality for others. Recognizing their limitations in this respect, we might expect that those who have attended an ISC have some notion of its utility for the disciplinary association that organized it, the country that hosted it, and, perhaps, global society as a whole. Of special interest, given the location of IPSA's world congress of 1979, is the question of political functionality: Does it make a difference—to the participants, to the association, to various national governments—where international scientific congresses take place? The answer to this question also has a bearing on the value such ISCs have for global society.

Organizational principles. Answers to such questions can not only contribute to the study of international scientific associations as transnational actors but can also provide information of practical value to organizers of future international scientific congresses. For example, what are the most effective organizational structures and procedures for facilitating scientific communication and exchange? Should we wish to maximize some other value, such as further international institutionalization of a scientific discipline, how might we go about doing it? What kinds of scientific communication problems occur in such congresses, and what can be done to eliminate them? Clearly, more comparative research on ISCs is needed. This study makes a start by providing empirical evidence aimed at clarifying the many functions of ISCs, identifying their advantages and disadvantages as a means of scientific communication, and suggesting areas in which they may be improved in the future.

This study probes the functionality of international scientific congresses for the various actors identified above. First, however, we must place the study in its context. Chapter 2 accordingly looks at the International Political Science Association and its world congresses, most particularly the Moscow world congress of 1979. Succeeding chapters address the issues outlined above. Chapter 3 profiles the respondents to the survey and what they perceive to be the functions of international political science congresses for themselves and others. The Moscow congress's scientific impact is the more specific subject of Chapter 4. It examines what participants expected the meetings to be like and how they actually found them, what they learned about the Soviet Union and its people, what they gained from scientific networking, and how the meetings affected their work. It also raises the question of how IPSA and similar associations might alter their congresses to enhance this function. Chapter 5 focuses on the functionality of ISCs for national governments, most particularly the political aspects of the Moscow world congress: how respondents dealt with the call to boycott the meetings, what they feel about holding meetings in the Soviet Union, and the extent to which they saw politics intruding into the congress's sessions. The concluding Chapter 6 asks what the analysis has taught us about ISCs and explores how we might improve them in the future.

Notes

1. We are using the term "function" in the sense of "the action for which a person or thing is specially fitted or used or for which a thing exists" (*Webster's New Collegiate Dictionary*). This usage includes Merton's (1957: 51) *manifest* functions, which "are those objective consequences contributing to the adjustment or adaptation of the system which are intended and recognized by participants in the system," and *latent* functions, or "those

which are neither intended nor recognized." Implicit in this terminology is our acceptance of the idea of functional analysis, which Merton (1957: 47) has defined as the "practice of interpreting data by establishing their consequences for larger structures in which they are implicated" (see also Cancian, 1968; and Levy, 1968).

2. By the term "science" we mean to suggest both an outcome (in the sense of "systematized knowledge," as defined in *Webster's New Collegiate Dictionary*) and a process (in the sense of Kuhn's [1970: 10] "normal science," that is, "research firmly based upon one or more past scientific achievements, achievements that some particular scientific community acknowledges for a time as supplying the foundation for its further practice").

3. A society in which a particular social myth about science as a whole and individual scientific disciplines dominates may well be able to institutionalize the order that myth specifies. "Full institutionalization" of the Western scientific order was achieved, in Bernard Barber's (1968: 96) words, "when universities, various governmental organizations, and many industrial firms recognized the great need for science, and established regular and permanent roles and careers for scientists."

4. Early explorations in the sociology of science (e.g., Merton, 1942) saw the "ethos" of science in terms of interlocked cognitions defining the research framework of the discipline, norms indicating what values scientists are "supposed" to hold as scientists, and behaviors appropriate for scientists acting both as individuals and as an aggregate. The argument that the entire scientific community shared this ethos gave it legitimacy and enabled the community to exert social controls on its behalf. "Strict implementation" of "sets of formal rules," these studies assumed, "guarantees an undistorted revelation of the real physical world" (Mulkay, 1979: 95). The revisionist position on the scientific order is pluralistic. It sees not one but an array of "scientific orders," any one of which may enjoy varying support in a given society (which does not imply equal validity). It thus shifts the focus of attention from (idealized) scientists trying to unlock the unambiguous secrets of nature to the interaction between society (with its social myths) and scientists seeking to interpret their findings. Just as "objects present themselves differently to scientists in different social settings," so, too, "social resources enter into the structure of scientific assertions and conclusions" (Mulkay, 1979: 5).

5. A "consociation" is "an ecological community with a single dominant" (*Webster's New Collegiate Dictionary*). It includes, as noted below, not only a scientific community or "practitioners of a scientific specialty" (Kuhn, 1970: 177) but also individuals acting on behalf of other structures for which the particular scientific discipline is significantly relevant.

6. We are treating the scientific discipline (e.g., chemistry) as an institution in the sociological sense, and disciplinary associations as the structures (or agents) that speak or would speak for them.

7. Since Fighiera extrapolated the data on IGOs from the number of meetings held by the United Nations, they are not directly comparable to those shown in Table 1.1.

8. The United Nations Educational, Scientific and Cultural Organization conducted from 1951 to 1953 a major interdisciplinary study of the management and operation of international conferences, which sought to find out more about the factors that encourage and inhibit effective international decisionmaking through the conference process (Sharp, 1950; Soddy, 1953; UNESCO, 1953). Only one of the systematic case studies in this project analyzed a conference of a nongovernmental organization. Another surge of enthusiasm for the subject came more than two decades later in response to the series of global conferences initiated by the UN General Assembly to consider various economic and social problems. These studies were concerned with the effectiveness of such massive gatherings for making decisions on international policies and the nature of that decisionmaking process (see, for example, Weiss and Jordan, 1976).

9. We shall not discuss either subnational and subdisciplinary associations or supranational regional associations. Our assumption is that their goals and behaviors resemble those of national associations, albeit on a different scale.

10. The function this competitiveness serves for the nation-state is discussed in the next section.

11. Thus, in Bernard Barber's words: "Scientific knowledge is power, that is, power to adjust more or less satisfactorily to the nonsocial environment and to the internal and external social environment. . . . Whatever their values in regard to science, [powerful modern industrial societies] feel an urgent need to use it to strengthen their national defense, promote industrial and agricultural growth, and improve the health of their populations. . . . Finally, the nonindustrial or underdeveloped societies of the present also push the acquisition of science for urgent instrumental needs, to cope with 'the revolution of rising expectations' in their populations" (Barber, 1968: 94; see also Storer, 1970: 89).

12. IPSA's official tally reports 1,027 registrants from the nonsocialist world, of whom 856 actually attended the Moscow meetings. Using the latter figure, our 369 respondents who both registered for and said they participated in these meetings represent a response rate of 43 percent.

TWO

International Political Science: From Paris to Moscow (and Back)

Political science as a scientific discipline has been international since its beginnings scarcely a century ago. Its subject matter, the scientific study of politics and political phenomena, is inherently comparative across time and across nations. The diffusion of the discipline's concerns, approaches, and methodologies also reveals its internationality. What we now call political science spread from Germany and France throughout the Western world, including North America, and then after 1945, in revitalized form, back across the Atlantic and throughout the rest of the world. Not until the late 1940s, however, was there an institutional framework for political science that was truly international rather than the overseas extension of a particular school or subdisciplinary interest. The creation in 1949 of the International Political Science Association (IPSA) aimed at realizing the discipline's international dimensions.

A chief tool envisioned by IPSA's founders as a means to internationalize the discipline was a periodic world congress of political scientists. It was not the only such tool they and their successors devised. Others include international roundtables of IPSA research committees and study groups, various publications (the quarterly *International Political Science Review*, the bimonthly *International Political Science Abstracts*, a newsletter entitled *Participation*, as well as edited symposia contained in "Advances in political science: an international series"), and participation in the various activities and fora of the International Social Science Council. Yet the triennial world congress remains the jewel in IPSA's crown as far as establishing and enhancing international scientific communication within the discipline is concerned.

The development of IPSA and its world congresses epitomizes key aspects of the problématique of international scientific communication. In one sense its growth and the problems IPSA encounters are like those of any other scientific discipline organized internationally. Yet, in another sense, IPSA is unique. Its members, true to the central intellectual concerns of the political science discipline, tend to be more alert to the political context and ramifications of their international meetings than others might be toward their own. Moreover, like its national counterparts, IPSA's meetings seem to attract a few people more interested in making than studying politics. IPSA and its world congresses thus bring together some unusual political as well as the usual scientific and ancillary aspects of international scientific congresses.

Internationalization

The origin of the International Political Science Association lies in efforts of the United Nations Educational, Scientific and Cultural Organization (UNESCO) to further international research collaboration in the social sciences. Its Second General Conference, held in November-December 1947 in Mexico City, called on UNESCO's director general "to promote a study of the subject-matter and problems treated by political scientists of various countries in recent research materials," with particular emphasis on methodological aspects (UNESCO, 1949c: 28). The reasons given were the recent emergence of political science as a discipline and "hence the need for political science to reach, in the shortest possible time, the level of adjacent disciplines," problems caused by national diversity in terminology and approaches, and, most important, the desire to further "the maintenance of peace through intellectual co-operation":

> Among the many reasons why human beings have slaughtered one another, bringing untold sufferings (the most frightful are too recent to need description), some have been, and some are, purely political reasons. Whether these reasons are primary or secondary, the present tension between nations, and within many nations, is tied closely to phenomena that political scientists should know and understand.

Implicit in the study was the idea that enhancing "the degree of clarity and accuracy" with which "citizens of various states perceive the significance of their political conduct" could ease international tensions (UNESCO, 1949c: 28).

After some preliminary explorations, a project director appointed by UNESCO convened a meeting in September 1948 in Paris. Political scientists from several countries were asked to provide guidance for the study, including developing a questionnaire to be sent to political science institutes and associations around the world. They went further, however. Concluding that the time was ripe to institutionalize opportunities for international cooperation, they called for an international conference in 1949, "with the aim of launching an International Political Science Association" that would strengthen "cultural ties in their particular field of learning" (UNESCO, 1949b: 66). The Preparatory Committee they set up then turned, with UNESCO's assistance, to the task of organizing such a conference to be held a year later in Paris.[1]

Some 23 political scientists from 17 countries met in September 1949 to shape the International Political Science Association. Their first step was to draft a constitution, which would establish firmly that IPSA "should have purely scientific objectives, not excluding, however, the promotion of a more intelligent understanding of the principles of political science by the general public" (UNESCO, 1949a: 82). Article 5 of the constitution enumerated several appropriate activities:

> encouragement and development of national political science associations in countries where none yet exist; steps to secure a fuller recognition of political science as [a] distinct academic discipline; the facilitation of personal contacts among political scientists of various countries; the holding of international "round-table" discussions; the provision of a documentary and research service for members of the Association; and the widest possible dissemination of information concerning significant developments in political science teaching and research.

The association was to comprise collective members (national and regional associations), associate members (initially other international and national groups with objectives similar to IPSA's, but later including academic institutions), and individual members. As far as governance was concerned, the conference envisioned a tripartite structure:

1. The authoritative *council*, proportionally representative of the collective members (now with the addition of a number of co-opted "representatives" of research committees, otherwise underrepresented groups, and the like), meets only during the triennial world congresses to elect an IPSA president, select from among its own membership a new executive committee, and make formal decisions on items proposed by the outgoing executive committee or possibly arising from the council floor;

2. The eighteen-member *executive committee* (including the president and immediate past president) meets occasionally, usually at least once a year, between world congresses to oversee preparations for the ensuing congress, nominate a president-elect, select sites for future world congresses, and otherwise conduct the association's ongoing business; and

3. A permanent *secretariat* under the direction of the secretary general carries out the instructions of the council and executive committee, prepares meetings and agendas, communicates with members, and so forth.

Finally, the conference elected a provisional executive committee that would govern until the following September, when IPSA would hold its first world congress.[2] The constitution entered into force in 1949 after the accession of four members: Canada, France, India, and the United States.

The association has grown both numerically and geographically. From the 4 original collective members IPSA expanded to 8 by the time of its first world congress (1950) and to 18 by the time of the second (1952).[3] Today, some 39 member associations (including two regional associations, for Africa and Pacific Asia) come from all parts of the world. Table 2.1 traces the increase in the number of individual members from a modest 52 in 1952 to 1,510 at the outset of 1986. Equally impressive have been a broadened representativeness of the individual as well as national members and increased involvement of non-Western scholars in the activities of the organization. Although India was among the four founding countries and Poland joined in 1950, the association during its first years was very much under Western influence. Not until the 1960s did Third World countries and socialist countries other than Poland and Yugoslavia begin to play a significant role in the organization. By 1986 members came from 55 different countries. The composition of the council also changed markedly: The proportion of members from developed countries dropped from 79 percent in 1967 to half in 1976, to somewhat more (56%) in 1985; and the percentage of council members from developing and socialist countries rose from 15 to 24 percent and from 6 to 20 percent, respectively.

IPSA World Congresses

Attendance at the triennial congresses has also reflected the growth and geographical expansion of IPSA's membership. Some 80 scholars, representing 23 countries participated in the first congress, held in September 1950 in Zürich; 14 years later almost 500 political scientists from 43 countries attended the Geneva congress; and over 1,000 from 60 countries went to the Montréal congress in 1973, the first to be held outside Western Europe. The decisions to hold the 11th World Congress in Moscow in 1979 and the following one in Rio de Janeiro in 1982 increased the number of participants from Eastern Europe and Latin America and symbolized the enlarged geographical scope of the association. The total number of registrants at both these meetings neared 1,500. The world congress in Paris in 1985 attracted 1,763 participants.

As important as the quantitative expansion of IPSA world congresses was their substantive growth. The Zürich congress of 1950 began with Quincy Wright's presidential address, in which he said (1951: 275):

> The conditions which have brought our Association into existence are the corruption of politics by inhuman tyranny and total war which

Table 2.1 International Political Science Association, 1950-1985
Membership, Attendance at World Congresses, and Countries Represented

Date	Number of Members	Congress Location	Number of Participants	Number of Countries Represented
1950		Zürich	80	23
1952	52	Hague	220	31
1955	232	Stockholm	275	36
1958	425	Rome	320	31
1961	442	Paris	425	46
1964	420	Geneva	494	43
1967	520	Brussels	745	56
1970	510	Munich	894	46
1973	450	Montréal	1044	56
1976	532	Edinburgh	1081	56
1979	687	Moscow	1466	53
1982	1211	Rio de Janeiro	1477	49
1985	1510	Paris	1763	66
1988		Washington		

Sources: Philippart, 1970: 17, 42-57; Scohy, 1977: 21, 41-58; IPSA, *Participation*, 1,1 (January 1977): 25, 4,1 (January 1980): 13, and 7,1 (Spring 1983): 21; and the secretary general's "Three year report, 1982-1985" (August 1985) as well as "Secretary-general's report, July-December 1985" (February 1986).

have brought and may again bring disastrous consequences to all sections of the world. The purpose which inspires our Association is to eliminate these corruptions by the universal application of scientific method in dealing with political problems.

The congress, convened jointly with the International Sociological Association's first world congress, then turned to three main themes:[4] minimum conditions for an effective and permanent union of nation-states; influence of electoral systems on political life; and the citizen's part in a planned society (UNESCO, 1950a). Each theme focused on three written papers, followed by general discussion.[5] Finally, a joint roundtable with sociologists dealt with the influence of a country's ethical structure on its foreign policy, conceived in the framework of UNESCO's project on tensions that affect international understanding.

Ensuing IPSA world congresses saw the association moving slowly toward a format that would be at once flexible and stabilizing. The Hague congress of 1952 offered 57 papers in 11 roundtables organized around 3 themes—local government as a basis of and training in democracy, the role of ideologies in political change, and the political role of women—and 2 roundtables on a fourth, the UNESCO survey of teaching in political science in 12 countries (for extensive rapporteurs' reports, see UNESCO, 1953). The Stockholm congress of 1955 developed a new pattern of plenary sessions on 5 main themes followed by related sessions for specialists (with 25 papers) and, at the end of the congress, a plenary session at which rapporteurs for each theme presented their summaries (Meynaud and Reynolds, 1956). The themes were metropolitan governance, political parties, political implications of economic development programs, large and small states in international organization, and political conditions of democracy. Subsequent congresses in Rome (1958) had 6 themes and 77 papers, in Paris (1961) 5 themes and 59 papers, in Geneva (1964) 6 themes and 94 papers, and in Brussels (1967) 9 themes and 96 papers (UNESCO, 1959; Philippart, 1970: 42-62).

By this time two trends were identifiable. One was a proliferation of topics. It suggested, at least symbolically, a fragmentation of the discipline into its diverse subspecialties. To counteract such an impression, future congresses sought to emphasize a very small number of overarching themes (two in Montréal, 1973; one in Edinburgh, 1976; three in Moscow, 1979; three in Rio de Janeiro, 1982; and four in Paris, 1985) while, through the allocation of sessions, recognizing differences in specific subject matters and approaches. The Edinburgh meeting was especially interesting in this respect. The central theme, "Time, space, and politics," provided a framework within which the executive committee organized 22 sections (two-thirds of them comprising two distinct sessions each). Later program organizers identified a number of section topics appropriate to each theme and published in *Participation* brief rationales for the selection of their themes and organization of section topics.[6]

A second trend saw the emergence of subsidiary groups within IPSA that sought time outside the regular program to conduct their own sessions. Specialists' meetings appeared for the first time at the Geneva congress (1964). Some of these turned into organizational meetings for more enduring networks that would subsequently distribute newsletters, develop periodic roundtables, and ask for reserved slots on congress programs. At the Munich congress in 1970 the executive committee officially recognized two such groups: the Research Committee on Conceptual and Terminological Analysis (COCTA) and the Research Committee on Political Sociology (a joint committee with the International Sociological Association). By 1976, when their number had grown to 14, the executive committee began the process of institutionalizing the research committees (Trent, 1978). Four years later it adopted guidelines for recognizing new research committees, ensuring that they were active on a continuing basis (that is, not appearing solely every three years to claim space on the program), terminating those that were inactive, and establishing the category of "study groups" as incipient research committees. The path of growth nevertheless continues. In 1986 IPSA had 24 research committees and an equal number of study groups.

In part to control this tendency toward diffused foci at its world congresses, IPSA established a program committee in 1976. Previous practice had the executive committee, under the direction of the IPSA president, organizing the congress program and supervising its realization. Effectively, however, the main organizational burden rested on the shoulders of the president, who also had a wide variety of other administrative and representational tasks to perform. The new president elected in 1976 at the Edinburgh congress, Karl W. Deutsch (United States), agreed to serve on the condition that he could share with an independent program committee the burden of the Moscow congress. Accordingly, and with the executive committee's approval, he appointed the first such committee and a program chair. Six years later the executive committee changed IPSA's constitution to incorporate this innovation, making the program chair an appointive officer of the association and ex officio member of the executive committee.

Moscow World Congress, 1979

It was in this fluid associational setting that the IPSA program committee undertook to organize the 11th World Congress, to be held in August 1979 in Moscow. On the one hand, the task was well defined. The committee had for its guidance the precedents of 10 previous congresses. Moreover, the association's president, preceding president, secretary general (all members of the program committee), and program chair had each been directly responsible for organizing either an IPSA world congress or an annual meeting of the American or Canadian political science association. This experience made it relatively simple to proceed in a straightforward fashion.

On the other hand, the Moscow congress posed some new challenges. Establishing the freedom of action of the newly created program committee was one of them. Not all members of the current or previous executive committees saw the need for such a committee; and a few even felt that its creation had taken organizational privileges out of their own hands. This matter was resolved by creating overlapping membership (with 8 members of the executive committee plus the IPSA secretary general on the 20-person program committee), bringing the executive committee as much as possible into the program committee's communications network, and ensuring that the executive committee had opportunities to review the program committee's decisions on important questions. The procedures developed in 1976–1979 worked sufficiently well that the executive committee institutionalized them for succeeding world congresses.

A second challenge was to set up a program that was intellectually first-rate, one that would attract the world's leading political scientists to Moscow in 1979. The program committee developed the congress around three main themes. One was the *politics of peace* (Alker, 1978; Shakhnazarov, 1978; Merle, 1978). Its nine sections (each comprising two sessions) dealt with such topics as conceptions of peace, relaxation of international tensions, arms races and arms control, and the domestic politics of peace and war. A second theme, on the *politics of development and system change* (Bose, 1978), included sections on socioeconomic structures and political systems in comparative perspective, planning and its implementation, the politics of unbalanced growth, and the politics of nonalignment as a factor of development. The third theme, *cumulative growth in political knowledge since 1949* (Semenov, 1978; Ludz, 1979), looked at what political scientists had learned since the year of IPSA's founding. Individual sections explored comparative macro- and micro-analysis, systems theory, normative political theory, information systems, and related topics.

The Moscow meetings attracted more registrants and active participants than any previous IPSA world congress. As many as 1,466 political scientists attended (Table 2.2). Under the 3 main themes, section convenors organized 56 sessions. The final program scheduled for these sessions 56 chairpersons, 212 papergivers presenting a total of 177 papers (abstracted in Merritt and Smirnov, 1979-1981), and 58 discussants.[7] The congress program also included 45 sessions organized by 17 research committees and 7 study groups, and 39 special meetings, which, together, accounted for 269 scheduled papers written by 326 scholars. In all, the program listed 846 active participants (some of them, of course, acting in a double capacity as papergiver in one session and discussant in another).

A third challenge was organizing the first IPSA congress to take place in a non-Western country: the Soviet Union. With the program chair residing in the United States and the IPSA secretariat in Canada, this meant at best attenuated lines of communication between them and the Soviet organizing committee. As it turned out, however, it was not communication but international politics that intervened to make this task especially challenging.

Politics of Site Selection

Selecting the site for an international scientific congress is not a trivial matter. A selection committee must first raise a host of questions about a potential site's facilities, accessibility, and demonstrated skill in organizing congresses. It might inquire about the host country's willingness to cover some proportion of the costs for administration, travel, and the like. Increasingly, it seems, the committee must also ask essentially political questions. It must take account of the host country's policy on admitting

INTERNATIONAL POLITICAL SCIENCE   37

Table 2.2 Participation at IPSA World Congress, 1979

Country	Participants No.	%	Program Comm. Sessions Con	Pap	Dis	Res/St Groups Sessions Con	Pap	Dis	Special Meetings Con	Pap	Dis	All Sessions Convener No.	%	Papergiver No.	%	Discuss. No.	%
Eastern Europe																	
Soviet Union	260	18	4	16	2	2	7	1	6	11	6	12	8	34	6	9	5
Bulgaria	50	3	-	-	1	-	-	-	-	-	-	-	-	-	-	1	1
Czechoslovakia	50	3	1	4	1	-	-	-	-	-	1	1	1	4	1	2	1
Germany (GDR)	50	3	1	5	3	-	-	1	-	2	-	1	1	7	1	4	2
Hungary	50	3	1	2	2	-	-	-	-	2	1	1	1	4	1	3	2
Poland	50	3	2	2	1	5	5	2	-	3	2	7	5	10	2	5	3
Romania	50	3	2	7	1	-	3	2	1	1	1	3	2	11	2	4	2
Yugoslavia	50	3	3	4	1	2	4	1	2	3	2	5	4	11	2	2	1
	610	42	14	40	11	7	19	6	9	22	13	30	21	81	15	30	18
North America																	
United States	229	16	10	59	15	15	64	15	13	48	11	38	27	171	32	41	25
Canada	51	3	3	19	3	4	8	3	4	10	1	11	8	37	7	7	4
	280	19	13	78	18	19	72	18	17	58	12	49	35	208	39	48	29
Western Europe																	
Germany (FRG)	64	4	3	12	2	2	7	5	3	11	1	8	6	30	6	8	5
France	42	3	3	2	2	-	5	1	2	5	4	5	4	12	2	7	4
Sweden	37	3	1	1	4	-	7	3	-	5	1	1	1	13	2	8	5
Spain	34	2	-	4	-	-	-	1	-	3	1	3	2	7	1	2	1
United Kingdom	34	2	3	4	-	4	7	5	3	3	1	10	7	14	3	10	6
Netherlands	32	2	1	4	-	-	3	2	3	3	5	-	-	9	2	2	1
Israel	29	2	1	4	-	1	3	1	-	2	-	4	3	9	2	2	1
Turkey	22	2	1	5	1	1	2	-	2	2	1	4	3	8	1	1	1
Norway	21	1	-	3	-	1	3	1	-	1	1	2	1	6	1	2	1
Finland	20	1	2	7	2	1	2	1	-	-	-	2	1	15	3	6	4
Belgium	15	1	-	-	-	1	3	-	-	6	3	1	1	3	1	1	1

Table 2.2—continued

Country	Partici-pants No.	Partici-pants %	Program Comm. Sessions Con	Pap	Dis	Res/St Groups Sessions Con	Pap	Dis	Special Meetings Con	Pap	Dis	All Sessions Convener No.	Convener %	Papergiver No.	Papergiver %	Discuss. No.	Discuss. %
Western Europe—continued																	
Denmark	12	1	1	4	–	–	4	–	–	–	1	1	1	8	1	1	1
Italy	12	1	1	1	–	–	7	–	–	1	1	1	1	9	2	1	1
Switzerland	11	1	2	2	–	1	1	1	–	2	1	3	2	5	1	1	1
Greece	4	*	–	–	–	–	1	–	–	1	–	–	–	2	*	–	–
Ireland	3	*	–	–	–	–	–	–	–	–	–	–	–	–	–	–	–
Luxembourg	2	*	–	–	–	1	2	1	–	–	–	1	1	2	*	2	1
Austria	1	*	–	–	–	–	–	–	–	1	1	–	–	–	*	–	–
	395 / 27		1/19	/53	/11	1/13	/58	/22	/13	/42	/20	2/45	/33	/153	/28	/53	/32
Asia and Pacific																	
India	32	2	4	10	3	3	4	4	–	9	4	7	5	23	4	11	7
Japan	30	2	–	6	2	–	5	1	1	2	–	1	1	13	2	3	2
Korea (South)	21	1	–	9	5	–	4	–	–	–	–	–	–	13	2	5	3
Australia	6	*	–	–	–	–	2	–	–	3	1	–	–	5	1	1	1
New Zealand	3	*	–	–	–	–	1	–	–	1	–	–	–	2	*	–	–
Thailand	2	*	–	–	–	–	–	–	–	–	–	–	–	–	–	–	–
Vietnam	2	*	–	–	–	–	–	–	–	–	–	–	–	–	–	–	–
Hong Kong	1	*	–	–	–	–	1	–	–	–	–	–	–	1	*	1	1
Indonesia	1	*	–	–	–	–	–	–	–	–	–	–	–	–	–	–	–
Malaysia	1	*	–	1	1	–	1	–	–	–	–	–	–	2	*	–	–
Pakistan	1	*	–	–	–	–	–	–	–	–	–	–	–	–	–	1	1
Philippines	1	*	–	–	–	–	2	1	–	–	–	–	–	2	*	1	1
Singapore	1	*	–	–	–	–	–	–	–	–	–	–	–	–	–	–	–
	102 / 7		1/4	/26	/11	1/3	/20	/7	1/1	/15	/6	/8	/6	/61	/11	1/24	/14

INTERNATIONAL POLITICAL SCIENCE   39

Africa and Middle East

Nigeria	4	*	-	-	2	1	-	2	3	-	1	-	3	2	6	1	1	1
Algeria	2	*	-	2	-	-	-	-	-	1	-	-	-	-	-	-	-	-
Cameroun	1	*	1	-	-	-	-	1	-	-	-	-	1	1	-	1	-	-
Ivory Coast	1	*	-	-	-	-	-	-	-	-	-	-	-	-	1	*	-	-
Jordan	1	*	-	-	-	-	-	-	-	-	-	-	-	-	-	-	-	-
Sierra Leone	1	*	-	-	-	-	-	-	-	-	-	-	-	-	-	-	-	-
	10	1	1	2	1	1	4	2	1	1	1	-	4	3	7	1	1	1

Latin America

Mexico	47	3	-	3	2	-	-	-	-	-	-	-	-	-	3	1	2	1
Brazil	11	1	3	3	-	-	4	-	-	1	-	2	3	2	8	1	3	2
Venezuela	5	*	-	-	-	1	-	-	1	1	-	-	-	-	1	*	-	-
Argentina	3	*	-	1	1	-	-	-	1	-	-	-	1	1	1	*	1	1
Cuba	2	*	-	-	-	-	-	-	-	-	-	-	-	-	-	-	-	-
Chile	1	*	-	-	1	-	2	-	-	-	-	-	-	-	2	*	1	-
	69	5	3	7	4	1	6	1	1	2	-	2	4	3	15	3	7	4

Other

| | 1466 | 100 | 2/56 | 6/212 | 2/58 | 2/45 | 5/184 | 1/55 | 1/41 | 2/142 | 1/53 | 2/142 | 1/100 | 13/538 | 2/100 | 3/166 | 2/100 |

Summary

Eastern Europe	610	42	14	40	11	7	19	6	9	22	13	30	21	81	15	30	18
North America	280	19	13	78	18	19	72	18	17	58	12	49	35	208	39	48	29
Other Western	382	26	18	48	10	12	59	22	13	45	20	43	30	152	28	52	31
Third World	194	13	11	46	19	7	34	9	2	17	8	20	14	97	18	36	22
	1466	100	56	212	58	45	184	55	41	142	53	142	100	538	100	166	100

Source: Participation, 4, 1 (January 1980): 13; and *IPSA World Congress Program,* 1979.

*Less than 0.5 percent; columns may not add to 100 because of rounding error.

nationals of certain countries and the possible political uses to which it may put a congress. Like the sports world, the scientific world has seen countries shamelessly use such events to tout the virtues of their own political systems or national character. IPSA officers are keenly aware that this kind of behavior can adversely affect the success of an international scientific congress.

Such political considerations play a role not only for members of selection committees but also for individual scientists who must decide whether or not to attend a particular congress. At one pole are those who are prepared to go any place where they can freely discuss their research and exchange information. They represent the traditional view that science is (or ought to be) above politics and that the value of scientific communication transcends ideological differences no less than national boundaries. The discipline of political science will also advance, according to this line of thinking, as perspectives broaden. It thus makes sense to hold meetings precisely in those places where scholars pursue lines of inquiry and use methodologies different from our own. Should they choose not to learn from us, we can at least learn from them and hence improve the breadth and quality of our own scientific work.

At the other extreme are scientists for whom political conditions are very important when choosing a site for a congress or deciding about their own attendance. One such point of view insists that scientific communication and exchange, which to flourish require an open setting, are ipso facto inhibited in a repressive political system. Accordingly, a decision to hold a scientific meeting in such a system does not serve the interests of science—but may legitimize and/or reward the repressive régime by providing tourist revenues and conferring status, as well as providing propaganda opportunities for a government that violates basic human rights. It may also put a stamp of international acceptance on a régime that pursues an aggressive foreign policy. It follows, in this view, that individual scientists should refuse to attend congresses held in these countries.

An alternative, and equally political, point of view argues that holding scientific meetings in such a country may contribute to its liberalization, to opening it up for free discourse. This stance rejects boycotts, which only increase the country's isolation in the international arena and encourage its repressive behavior. Enhancing communication through scientific meetings in such a country may, by contrast, help to reduce tensions and thereby contribute to a more stable international environment.

Not surprisingly, given its essential intellectual concerns, the International Political Science Association has seen international politics crop up in decisions about sites for its triennial world congresses. This was not a serious problem before the late 1970s. Some members questioned the appropriateness of particular sites—Munich in 1970 because of its historical associations, Montréal in 1973 because of a potential danger to Israeli scholars, and Edinburgh in 1976 because of the possibility that Scottish nationalists might disrupt the planned sessions on devolution—but their voices were few in number, the problems they foresaw manageable, and the tone of argumentation not very convincing. Decisions reached in 1976 to hold IPSA's next world congresses in Moscow and Rio de Janeiro raised a few more eyebrows.

What is political about selecting a site for an ISC? The answer to this query can shed light on the much broader question of the political ramifications of international scientific cooperation. As Francis Bacon and many after him have observed, knowledge is power. This truism has become more accurate as societies move toward ever more complex industrial bases that demand effective scientific research and development. Governments have had to decide for themselves when freedom of information endangers their competitive position vis-à-vis other states in the global arena or even their national security. A logical extension of government secrecy acts is the effort to prevent scientific interchanges that can unduly advantage one's competitors. Thus, during the hottest days of the cold war, governments did not grant entry to scholars from the "other side" and sharply limited the free flow of scientific publications. Nor were they prepared to grant passports to those among their own citizens who wished to study or attend conferences in hostile countries.

Steps toward international relaxation in the 1970s opened up new possibilities for international scientific cooperation, including joint U.S.-Soviet ventures in outer space. But the hostility of an earlier decade was not forgotten, and few were willing to interpret the new setting of peaceful coexistence and competition as one in which the superpowers had ceased jockeying for a position of scientific supremacy. The relaxation also turned what had been a practical constraint into opportunities laden with moral overtones. In the 1950s, as some scientists—not all of them in the West—pointed out, it had not been possible to attend a truly scientific international congress in the Soviet Union. Now that it was possible, was it right to go to a country which, in their opinion, controlled information, persecuted dissidents, and maintained its imperium over Eastern Europe? Other scientists raised similar questions about visiting the United States, with its legacy of McCarthyism, Vietnam, and the CIA, or countries in the Third World with records of military dictatorship and popular oppression.

Decisions by IPSA to hold its world congresses of 1979 in Moscow and 1982 in Rio de Janeiro forced political scientists to come to terms with such questions. They faced a condition, not a theory. Whether intended or not, the decisions certified that, as far as IPSA was concerned, Soviet and Brazilian political science had come of age. The decisions made clear the fact that IPSA would not impose political conditions on any one country, such as the

Soviet Union or Brazil, that it was unwilling to impose on others. What was less clear was the extent to which national associations and individual political scientists—especially those outside the socialist world—would accept the implications of these decisions. Did they, do they, view political principles as ascendant over the ideal of unbounded scientific cooperation?

In short, although the potential for conflict has always existed in IPSA decisions about where its world congresses should be held, and although rumblings of dissatisfaction had been heard before, it was not really until IPSA decided to go to Moscow for its world congress of 1979 that the political implications of such decisions became manifest. Virtually no dispute at all attended the initial decision. Subsequently, however, it erupted in a way that challenged some basic principles underlying the association's viability as a truly international scientific body.

Choosing Moscow

The International Political Science Association does not have a highly elaborated procedure for selecting sites for its triennial world congresses. National associations interested in hosting a congress submit invitations to IPSA's executive committee, which in turn decides which invitation to accept.

Before accepting an invitation, however, the IPSA executive committee ascertains that the host country can meet certain "normal" conditions. These include free access to the congress for all bona fide political scientists (at least those from countries that are collective members of the association), free discussion at the congress, agreement to the principle that the association and its committees make all programmatic decisions, and some minimal financial guarantees (such as providing travel funds for participants from the Third World and simultaneous translation for at least the plenary sessions). As noted earlier, the executive committee must also satisfy itself that the host country has appropriate facilities for the congress: an ample number of meeting rooms, available hotels at various price ranges (including, if possible, inexpensive dormitory space for graduate students and others), technical capabilities such as a staff that will be made available for organizing the congress, and, if necessary, formal support from the host country's government and/or national academy of science.

In practice, the process of selecting a site can entail complex negotiations. First of all, only rarely are national associations competing to host an IPSA world congress. The problem is more one of persuading national associations to submit invitations. The IPSA world congress in 1970 almost collapsed when the British association withdrew its invitation at virtually the last minute; only a hastily arranged agreement by the West

German association to host the meeting at Munich saved it. Typically, the IPSA executive committee must lobby its associational members to persuade them that the advantages of hosting a world congress—advantages such as the common weal of international political science, an opportunity for the host country's young scholars to interact with a vast array of experts, and prestige for the host association—are worth its costs.

Second, working out acceptable conditions can raise a number of delicate problems. Every national association must function within constraints set by its government; and, while the association may try to make necessary arrangements with its government, in the last analysis it does not control either what the government sets as policy or how petty officials such as customs agents interpret this policy. To take actual examples, neither the American Political Science Association nor the Canadian Political Science Association can absolutely *guarantee* that officials of its government will not bar entry to a communist or homosexual wishing to attend an international scientific meeting.[8] Were the IPSA executive committee to demand an ironclad guarantee of this sort, it would be demanding something it cannot obtain. What the executive committee can demand is an assurance on the part of the host association that it will do everything possible to secure the admission of bona fide political scientists—again, at least those from IPSA's collective members—and a recognition by both parties that, should the host association's government violate this principle, IPSA will cancel the meeting, even on short notice.

The complications entailed in locating an appropriate site have meant in practice that the task devolves upon the IPSA president and secretary general. Initiatives are discussed with the executive committee; and ultimately it is the latter body that makes the final decision to hold a world congress in a particular city. The mode of decisionmaking on the question has in the past been collegial—especially since the task has been more frequently to find a national association willing to host a world congress than to select one from among many proffered sites.

The Decision

At its meeting in summer 1974 the IPSA executive committee considered the question of where to hold its world congress of 1979. Only the representative of the Rumanian Political Science Association had submitted a tentative offer. The executive committee encouraged the Rumanian representative to pursue the question with his national association and government, and return to the executive committee meeting in summer 1975 with a formal invitation. The amount of time available proved to be too short for this to be done. (A Rumanian invitation eventually arrived, but only after the IPSA president had sent to executive committee members a request for advice on a Soviet invitation which had been submitted in the meantime.)

Without the Rumanian invitation in hand, the executive committee was seriously concerned when it met in summer 1975 about the site for its world congress a mere 4 years hence. A Brazilian member of the executive committee, Candido Mendes, indicated that he could secure an invitation from his national association to hold the world congress in Rio de Janeiro. When, however, Vladimir Toumanov proposed Moscow as a site, Mendes deferred with the observation that, if it proved to be impossible to hold the meeting in Moscow, he would again offer Rio as the site for 1979. Since previous informal offers from Soviet representatives had not turned into formal invitations, evidently few members of the executive committee expected that one would be forthcoming for 1979.

As it turned out, in December 1975 the IPSA president, Jean A. Laponce of Canada, received a formal Soviet invitation. With no others in sight, and with the Brazilian offer deferred, he circulated the invitation to executive committee members with the request that they consider it, discuss it with their national associations, and be prepared to vote on it at their next meeting in August 1976, on the day before the opening of the Edinburgh world congress. The executive committee voted unanimously at that meeting to accept both the Soviet invitation to hold a world congress in Moscow in August 1979 and the Brazilian invitation for 1982.

Questions were raised retrospectively about the way in which that decision was made. The answers are several. On the one hand, in summer 1975 the IPSA executive committee did not have a serious candidate to host the meeting in 1979. The Rumanian invitation had not yet materialized, and the suggestions of Rio de Janeiro and Moscow were really straws in the wind. Members of the executive committee were very conscious of the near collapse of their world congress in 1970; and the association's president and past presidents—Laponce, Carl J. Friedrich of the United States, and Stein Rokkan of Norway—had not met with success in their efforts to drum up a site. The receipt of the Soviet invitation was viewed with relief rather than apprehension. This was especially so since Soviet representatives agreed orally to provide travel grants for Third World participants equivalent to the amount given by the Canadian government for the Montréal world congress in 1973, and agreed in writing both to guarantee free access to the Moscow meetings for all members of national associations then affiliated with IPSA and to ensure that the meetings themselves would, as in the past, enjoy complete freedom of discussion.

On the other hand, IPSA executive committee members could not see any particular reason why the world congress should *not* be held in Moscow. If there were moral or other objections to the status of full equality enjoyed by the Soviet Political Science Association as a member association of

IPSA, these objections should have been expressed in 1955 when membership was granted and not two decades afterwards, when the Soviet association sought to exercise one of the rights of membership. Moreover, although executive committee members had long sought to hold a world congress in Eastern Europe, all previous efforts to do so had come to naught. The notion of geographic equity in site selection suggested that Moscow might well be an appropriate site. Besides, as some European members of the executive committee pointed out, postponing the meeting in Rio de Janeiro was a good idea on the grounds of cost alone, since the city is far distant from the universities and research organizations housing most of the association's members. Still others viewed an IPSA world congress in Moscow as a logical step in the light of the Helsinki accords and even as one that might further the process of global détente.

Negotiations and Confrontations

After the IPSA council, representing the member associations, had met and raised no objection to the executive committee's decision to hold the world congress in Moscow in 1979, the newly elected IPSA president, Karl W. Deutsch of the United States, set in motion the organizational work discussed earlier. The ensuing months also saw a number of practical issues arise as the principles agreed to in 1976 were made operational. Perhaps the most important of these was ensuring that all bona fide political scientists would secure visas from Soviet authorities early enough to make travel plans in a timely fashion. Recent difficulties, especially on the part of Israeli delegations seeking to attend scientific congresses in the Soviet Union, led to lengthy talks and a series of mutual assurances on the part of the IPSA executive committee and the Soviet organizing committee. Deutsch, speaking also on behalf of the executive committee, made it clear that denial of visas in violation of these mutual understandings would lead him to cancel the world congress scheduled for 1979 or else transfer it to another site. This turned out not to be necessary—despite a near breakdown in communications that delayed the Israeli visas until virtually the last minute.[9]

Far more serious was the dispute that erupted because of the Soviet treatment of dissident scholars. Soviet authorities arrested Yuri Fyodorovich Orlov and Alexander Ginsburg in February 1977 and Anatoly Sharansky a month later. Orlov was sentenced to a prison term in May 1978, Ginsburg and Sharansky in July 1978. The view that these arrests and trials were clear-cut violations of the Helsinki accords and basic principles of human rights led a number of political scientists in the West to question seriously the appropriateness of holding an international scientific congress in Moscow, and to ask other political scientists to search their consciences before deciding if they would attend such a meeting.[10] (It is quite likely, we might note in passing, that the substantive questions implicit in this line of argument would have arisen even in the absence of any such Soviet actions. What many saw as outright political persecution of scientists nonetheless lent virulence to the issue.)

The issue, plain and simple, focused on the political preconditions IPSA should impose on potential hosts of world congresses. Doubts were expressed about the sincerity of Soviet guarantees of free access and free discussion (and, indeed, even about whether there was a credible political science in the Soviet Union!). Some political scientists in the United States and elsewhere argued that the demands of solidarity with persecuted scientists, which would have IPSA cancel or at least have the national associations boycott the Moscow world congress, outweighed any potential benefits that could be derived from participation. Calling off or boycotting the meetings, in this view, would symbolically demonstrate to the Soviet Union that scientists in the rest of the world would neither tolerate Soviet persecution nor cooperate with the Soviet government when it engaged in such behavior, withhold from the Soviet Union the rewards attendant upon hosting an IPSA world congress, and possibly even force Soviet authorities to end their pressure on dissidents.

Others argued that withdrawing the world congress from Moscow was neither a desirable move nor an effective sanction on Soviet behavior. Indeed, they said, such a step could easily destroy the gains IPSA had made in its three decades of existence and perhaps even the association itself. At a minimum, a boycott would split IPSA along East-West lines—and it was by no means certain that the Third World would follow the West in recreating an *international* political science association. Would punishing or at least withholding rewards from the Soviet Union in fact affect Soviet behavior? Probably not, this line of argument responded, given past evidence about what losses in terms of world public opinion the Soviet government was prepared to accept if it saw basic political values at stake. The imposition of political preconditions would more generally set a precedent that might make it difficult in the future to find any acceptable site for an IPSA world congress. If the principle were accepted with respect to the Soviet Union, for instance, would not charges of "institutionalized racism" in the United States or suppression by the United Kingdom of "legitimate" Irish interests in Ulster rule out those countries as potential sites for future IPSA world congresses?

The issue reached a head in summer 1978, after the sentencing of Ginsburg and Sharansky. At its meeting in the last week of August in Rio de Janeiro, the IPSA executive committee drafted a lengthy resolution which, after taking note of "the trials and incarceration of political dissenters in the Soviet Union on issues which many consider to be tied to basic questions of human rights," went on to explain why it was nevertheless important to continue planning for the meeting in Moscow, and to cite the continued importance of "freedom of access, communication, speech, and debate" as conditions for actually holding the Moscow world congress.[11] During the following week the council of the American Political Science Association and its general membership meeting rejected resolutions calling for a boycott and accepted in their place one creating a special APSA advisory committee. Its tasks were to ensure that the conditions set by IPSA were being met in Moscow and to advise the APSA membership should the Soviet Union violate those principles.[12] (Other national associations did not discuss seriously the possibility of boycotting the Moscow world congress.)

The IPSA and APSA resolutions cleared the air in one sense: They established firmly that, barring some major and unforeseen event, the world congress of 1979 would take place as scheduled in Moscow. They did not, however, resolve the central issue at dispute. The sentiment persisted in some quarters that it was morally wrong to attend the congress. Doubtless some political scientists decided on this account not to apply for a place on the program; and a trio of leading political scientists in the United States even circulated a letter asking their colleagues to boycott the meetings. (As far as the IPSA program chair could discern, however, only one person who was listed in the preliminary program actually asked that, pursuant to the call for a boycott, his name be withdrawn.) Nor did the resolutions satisfactorily answer the question of whether or not decisions on sites for IPSA world congresses should or even could be used as sanctions. Least of all did the modus vivendi achieved in summer 1978 mean that participants in the Moscow world congress would later be of one mind about its conduct and consequences (see Chapter 5).

Some Questions for Research

Chapter 1 listed several questions of interest to theory and practice regarding international scientific congresses in general. The fact that the ISC studied here dealt with political scientists enables us to pose some of these questions a bit more precisely in their context. On the one hand, since in most respects IPSA world congresses conform to the basic format of other ISCs and perform the same tasks more or less imperfectly, the Moscow meetings can be seen as a case study of the more general phenomenon of international scientific congresses. On the other hand, however, IPSA's world congress in Moscow was sufficiently different that it invites special attention. Its underlying political tensions serve to highlight that particular aspect of other congresses as well.

Individual Scientists

Characteristics. The findings of Ladd and Lipset (1978) cited earlier regarding the American professoriat led us to expect that the participants at the IPSA congress in Moscow would be highly unrepresentative of the profession as a whole—to have traveled more, published more, and attended more international meetings than the average political scientist. To test this hypothesis we compared the biographical characteristics and attitudes of participants in the Moscow congress with those of a comparable sample chosen at random among North American political scientists who did not go to Moscow in 1979. We were particularly interested in the relationship between the participants' level of professional activity and international experience and their views on the purposes of ISCs and the impact of the IPSA congress in 1979.

An hypothesis our data did not allow us to test, but for which we shall assume prima facie validity, is that the participants, being political scientists, were more alert than others might be to political ramifications of the congress. Since we did not send questionnaires to scholars from other disciplines, and no directly comparable study exists, we cannot be completely sure that the assumption is accurate. It nonetheless stands to reason. Political scientists may be no more attuned than others to the overt politics of an international congress—whether or not, for example, the host country discriminates against scholars from some other country, and what practical consequences participants should draw from a finding that discrimination does exist. Yet the more subtle political behavior of a host country may be more apparent to trained political scientists than to others.

Perceived functions. We addressed this question by asking our respondents both why they did or did not participate in national and international congresses, especially the IPSA meetings in Moscow, and what value such congresses have for individual participants. "At first sight," a respected Dutch political scientist (Barents, 1959: 1092) noted of the Rome world congress,

> it seems paradoxical that hundreds of political scientists should have travelled to Rome to hear about pressure groups in Kamchatka or about the strange ways of public administration in Ruritania; yet I was time and again struck by the sudden flashes of insight and the unexpected doses of human interest that some of these interventions provided.

Such arguments and partial data led us to expect that a preponderance of respondents would both cite as the major function of ISCs some aspect of the process of scientific communication and indicate a scientific reason for attending the Moscow congress. We also expected that those who went to

Moscow would express a more scientific orientation than those who did not. We did not expect respondents to place great weight on the personal, nonscientific, or careerist advantages of attending the IPSA meetings in Moscow. Recognizing that questionnaires distributed by mail were not an appropriate instrument for ascertaining the participants' "true" motives, we did not press them too hard on this question. We simply asked them about the value of national and international congresses for the individual participant.[13]

Behavioral change. If it is true, as many people say, that the key function of an international scientific congress is scientific communication, then the experience of attending the congress should change in some way the participants' scientific outlook or modus operandi. Did participants actually use the opportunities they had in Moscow to expand their intellectual horizons? That is, did they establish contacts with colleagues from other countries, learn something that changed the orientation or otherwise contributed to the quality of their research, and/or subsequently keep in touch with colleagues first met at the congress? We hypothesized that those who were already well known in the profession and who frequently traveled to conduct research and attend conferences tend to go to an international congress for different reasons and to benefit in different ways than do their colleagues with less professional status and international experience.

Research Organizations

We did not expect our respondents to pay much attention to the value of the IPSA world congress for the universities and other organizations that employ them. The Western tradition of scholarship stresses the accomplishments of the scholar as an individual rather than as a representative of some organization.[14] Individual participants might note institutional affiliations but will not dwell on them to any great extent. (Had we circulated questionnaires among political scientists from socialist countries that stress a more collectivist orientation, we might have discovered significant differences on this score.)

Disciplinary Associations

Among the several scientific associations that stand to gain or lose from various aspects of an international scientific congress, the most prominent are the national association hosting the congress and the international association that organized it. The former may profit in terms of greater visibility or prestige. Archie Brown (1984: 322-323), outlining the struggles of Soviet political scientists to gain a measure of permanent status for their fledgling Soviet Political Science Association, wrote:

> A considerable stimulus to the development of political science in the USSR was the holding of the eleventh world congress of the International Political Science Association in Moscow in 1979. The fact that this was widely reported in the Soviet press (including a message of welcome to the participants from Leonid Brezhnev on the front page of *Pravda*), radio and television helped further to legitimise both the academic study of politics and the idea that there could be constructive and good-tempered discussion with Western scholars (who tended to be described, at least for the duration of the congress, as "non-Marxist" rather than "bourgeois" political scientists) even on such a sensitive subject.

(Leading political scientists in Brazil and, more generally, Latin America hoped that the IPSA world congress held in 1982 in Rio de Janeiro would have a similar effect.) An interesting question is whether or not the respondents from nonsocialist countries sensed the concerns and needs of the Soviet association. We expected that they would not. Without the concrete knowledge that comes from study of or direct contact with the Soviet association, most commenting on this point might simply be transferring their understanding of what their own national associations might gain or lose.

Moreover, since relatively few respondents were part of IPSA's decisionmaking structure, it seemed unlikely that many would probe deeply into the functionality for the association of its world congress. We nevertheless anticipated that respondents with substantial experience at IPSA world congresses would be more likely than others to recognize the value of the Moscow meetings for that body.

National Governments

Earlier we noted that the country hosting an ISC normally obtains certain advantages and incurs certain costs. The IPSA world congress in Moscow, of course, was not a "normal" international scientific congress in at least its political dimension. What distinguishes it was the opening it provided for national and international politics to intrude overtly into international political science. The possibility of a boycott, denial of visas to the political scientists of one or another country, or even some disruption at the meetings themselves lent a piquant political flavoring to the final year of preparations. Some officers of the association even feared that a pattern of political criteria might be set for future roundtables or world congresses—something that would effectively cripple IPSA. By the same token, bringing to Moscow from nonsocialist countries a thousand or more experts on politics presented the Soviet government with some unusual political risks and perhaps some opportunities.

Given the widespread attention before and after 1979 to the political nexus between congress and host country, we expected that respondents would virtually ignore other possible benefits accruing to the Soviet Union from its role as host country. Some of our questions about the functions of the Moscow world congress were nevertheless sufficiently open-ended to elicit a sense of the relative importance accorded economic, scientific, political, and other possible benefits. We also tackled the political issue directly. How important did the respondents regard political considerations to be in the selection of sites for ISCs? In particular, what were their views about the effects that hosting a political science congress by the USSR might have on Soviet domestic politics and on the international political environment? How important were these political considerations in determining the choices of individual political scientists to attend or not to attend the Moscow meeting? What did political scientists think about the utility of scientific sanctions, that is, limiting scientific communication and cooperation as a means to influence a target nation's foreign policy? Did they feel that the Soviet government had used the congress for propagandistic purposes and, if so, did these efforts inhibit scientific communication at the meeting? How much contact occurred between Eastern European scholars and their counterparts in the West? How did these contacts along with the experience of attending a meeting in the Soviet Union affect the participants' images of and attitudes toward the Soviet government and people?

This analysis of an ISC in which the political dimension was highly salient, namely, a political science congress in Moscow, should bring into sharper focus some of the political effects of all international scientific congresses, which are often obscured in the more homogeneous context of normal scientific discourse in Western countries. Important political consequences flow from the many decisions that are made concerning even the more technical aspects of any given ISC, such as forms of organization, operating procedures, and especially the choice of site. By virtue of their intellectual interest and expertise as political scientists, our respondents occupy a useful vantage point from which to consider the political factors that affect scientific communication at ISCs and the effects of this form of transnational activity on the international political environment.

Global Society

Our questionnaire did not specifically address the IPSA world congress's possible contribution to international science and world order. Indeed, we avoided questions we thought might elicit "idealistic" responses of the kneejerk variety. We nevertheless scrutinized the returned questionnaires, searching for views on two kinds of concerns. First, to what extent did the experience of political scientists who attended the IPSA congress in Moscow support the integrationist view of international scientific associations in general, that is, the view that they promote transnational values, peace, and justice? Second, and referring to the claims of dependency theorists, how much contact occurred between Third World scholars and political scientists from Western countries? Did the experience of the former at this particular meeting support the view that ISCs serve to reduce the knowledge and information gap between the developed and the developing countries?

Related to the first concern is a practical question: What impact did attendance have on the participants' images of and attitudes toward the Soviet Union, its people, and its scientific establishment? We would not expect a week's experience to produce far-reaching changes of a general nature; and even if they occurred, the social-psychological literature tells us, the passage of time would tend to diminish their effects. Marginal changes, such as making one's images more concrete through the addition of detail, were more likely. It was also likely that participants would obtain a much clearer notion of the interplay between politics and science in the Soviet Union. Those who had previously visited the Soviet Union or who occasionally, through their international activities, had dealt with Soviet scientists would, we anticipated, report encountering fewer surprises. Such findings are relevant to the broader question of whether or not ISCs contribute to the kinds of transnational interaction that promote peace.

Implications for Future IPSA World Congresses

Our analysis seeks not only to contribute to the study of international scientific associations as transnational actors but also to provide information of practical value to organizers of future international scientific congresses. What aspects of the congress itself facilitated or hindered scientific learning, for instance? Here we were interested not only in the participants' overall impressions of the congress, whether they found it satisfactory or not, but also in their views on its specific facets. A series of questions asked what they expected to find, for example, with respect to opportunities for informal interaction, and what they actually experienced. Further questions inquired what could be done to improve future IPSA congresses. Again we anticipated that the more professionally and internationally adept respondents would be those whose expectations and experiences were most closely matched.

Notes

1. The original members of the Preparatory Committee were Walter R. Sharp (United States), chair; John Goormaghtigh (Belgium), secretary; Raymond Aron (France); and William A. Robson (United Kingdom). They subsequently co-opted Angadipuram Appadorai (India) and Marcel Bridel (Switzerland). UNESCO's Third General Conference, held December 1948 in Beirut, authorized the director general to encourage the development of such international scientific associations.

2. Members of the provisional executive committee were Quincy Wright (United States), chair; Marcel Bridel (Switzerland) and Denis W. Brogan (United Kingdom), vice-chairs; and Jan Barents (Netherlands), Fethi Celikbas (Turkey), Maurice Duverger (France), Isaac Ganon (Uruguay), Elis W. Hastad (Sweden), H. Khosla (India), C. B. MacPherson (Canada), and Adam Schaff (Poland). The committee co-opted John Goormaghtigh (Belgium) as a member and François Goguel (France) as provisional executive secretary and treasurer. At the executive committee's meeting in February 1950, Jean Meynaud (France) was elected permanent executive secretary and treasurer (UNESCO, 1950b: 237); this title was changed in 1952 to secretary general and treasurer.

3. In almost all countries (excepting Canada, India, and the United States) it was necessary to organize a national association that could apply for IPSA membership. Those joining in 1950 were Israel, Poland, Sweden, and the United Kingdom; in 1951, Austria, Belgium, Greece, and Mexico; and in 1952, the Federal Republic of Germany, Finland, Italy, Yugoslavia, Japan, and Brazil (Philippart, 1970: 13).

4. Settling on a common terminology for the world congress has posed some minor problems. At the outset, program organizers identified themes, each with roundtables at which papers are presented and discussed. The terminology since at least the mid-1970s has focused on general *themes* (developed by consultants), broken down into a number of *sections* (organized by conveners), each of which typically has two *sessions* (presided over by chairs).

5. The papers and summaries of the discussions appear in UNESCO (1951).

6. For the Moscow congress, see Alker (1978), Bose (1978), Ludz (1979), Merle (1978), Semenov (1978), and Shakhnazarov (1978). Given the discipline's pluralism of subject matters and methodologies, of course, program organizers have not sought, nor could they effectively try, to police the structure of the individual sessions or the content of papers to ensure that they adhere to the stated themes.

7. Several of those listed on the program did not in fact appear in Moscow. Although the number of no-shows proved to be modestly disruptive and continues to be so, the problem may be intractable given current circumstances, namely, the need for funding and the weak mechanisms for social control enjoyed by international scientific associations.

8. In the late 1970s the Liberal government in Canada four times denied to the West German Marxist scholar, André Gunder Frank, visas to attend

conferences or give lectures (Scully, 1979). U.S. legislation dating from 1952 bans those "afflicted with psychopathic personality, or sexual deviation or a mental defect" from entering the country. Subsequent decisions by the Public Health Service to declassify homosexuality as a "mental disease or defect" (Treaster, 1979) and the Justice Department to turn back only those who openly declare their homosexuality (Pear, 1980) have modified the impact of the legislation, however, and courts are currently ascertaining its fundamental constitutionality.

In December 1985 the local organizing committee at Britain's University of Southampton, under pressure from the city council that would otherwise have withdrawn its financial support, voted to "disinvite" South Africans who planned to participate in the World Congress of Archeologists, scheduled for summer 1986. The International Union of Prehistoric and Protohistoric Scientists (IUPPS) accordingly cancelled its sponsorship and rescheduled the congress for summer 1987 in the West German city of Mainz. Southampton carried out its planned congress anyway, with as yet undetermined consequences for the IUPPS's future (Walker, 1986).

9. To secure the Israelis' visas in time for their scheduled departure for the Soviet Union, IPSA's secretary general, John E. Trent, had to fly first to Vienna to pick up the processed applications at the Soviet consulate and then to Israel to deliver them.

10. The question of civil rights violations in the Soviet Union became even more problematic for other scientific associations. In March 1979 some 2,400 American scientists pledged to suspend, as individuals, all professional cooperation with Soviet colleagues. The peak of organized political action came still later, with the mass campaign of the ad hoc organization, "Scientists for Sakharov, Orlov, and Sharansky" (SOS), which gathered signatures from 7,900 scientists and engineers from 44 countries on petitions pledging a six-month moratorium on scientific cooperation with Soviet scientists from May to November of 1980 (Seltzer, 1978; Stone and Spilhaus, 1980). The use of scientific cooperation as a lever to influence the domestic *and* foreign policies of the USSR gained momentum with the U.S. government's response to the Soviet invasion of Afghanistan. All official efforts at scientific cooperation under the intergovernmental agreements of 1972-1973 were suspended, and in an unprecedented action the National Academy of Sciences imposed a 6-month moratorium (subsequently extended for another 6 months) on all activities under its exchange program with the Soviet Academy of Sciences. As a result of these official and unofficial efforts to use scientific relations as a lever to influence Soviet behavior, symposia, visits, and even congresses were cancelled.

11. The executive committee also set up a standing "visa committee," charged with the task of investigating and reporting on complaints that a country hosting an IPSA meeting was violating the resolution's general principles. Guiding IPSA's (and, ultimately, APSA's) procedures was the memorandum offering "Advice to organizers of international scientific meetings," prepared by the International Council of Scientific Unions (ICSU, 1976). The most important correspondence and decisions on the proposed

boycott and conditions for holding the Moscow and future IPSA world congresses appear in the association's newsletter, *Participation*.

12. Several observers, such as the nationally syndicated columnist George F. Will (1978), found a delicious irony in the back-to-back decisions of APSA's general membership meeting to relocate the association's annual meeting of 1979 from Chicago (because the Illinois legislature had failed to ratify the Equal Rights Amendment) and not to boycott the IPSA world congress in Moscow.

13. To determine the extent to which respondents were willing to underline their own seriousness by attributing to others either frivolity or careerism, we asked them to distinguish the value of national meetings for themselves as against "other participants." Our expectations were that those who did not go to Moscow would be more apt to make such a distinction than those who did attend that congress, and that, among the latter group, respondents would attribute pretty much the same motives to others as to themselves.

14. The individual who is part of a closely knit research team may of course serve in a more representative capacity; our survey did not seek to ascertain the internal structure of research organizations.

THREE
Who Went to Moscow and Why?

Our survey of political scientists sought, first of all, to ascertain who participates in IPSA world congresses and why. Are they in effect a random sample of the international political science profession, or do they differ significantly from nonparticipants? To explore this point we questioned not only those who registered to attend the Moscow meetings in 1979 but also a control group of Canadian and U.S. political scientists who did not register. What general functions do respondents attribute to world congresses? To secure a baseline for evaluating such views, we initially asked respondents about the uses of national meetings. We then turned to the value of international meetings for individual scientists and other members of the political science consociation. Finally, we asked those who completed our questionnaire why they did or did not attend the IPSA world congress in Moscow.

Profile of Respondents

Registrants

Well over a thousand political scientists from 45 countries outside the socialist bloc registered to attend the IPSA world congress in Moscow. Not surprisingly, most of them are academics. Almost 9 in 10 (89%) indicated that their primary affiliation is with a university (with 75 percent of the entire sample holding a professorial title), while another 5 percent work for government agencies and 9 percent for private or governmental research organizations (multiple responses permitted). The median respondent, male (as are 87 percent of the registrants) and 41 years of age at the time of the Moscow meetings, embarked upon his or her professional career in 1967. By far the largest number (81%) finds research more interesting than teaching or administration.

The sample of registrants is also fairly accomplished in professional terms. Four in five (81%) hold a doctorate. Most (63%) with professorial duties enjoy the rank of full professor or equivalent; they constitute 49 percent of the entire sample of registrants. The registrants are active professionally: About half (49%) attend almost every annual meeting of their national political science association, and 69 percent say that they do so at least every other year. Well over half have published at least 3 books (62%) and 11 articles in scholarly journals (56%).

Finally, our registrants are oriented to international matters. As far as their main fields of research and teaching are concerned, 31 percent named international relations and another 27 percent comparative politics. Almost four in five (78%) speak at least one foreign language, half of them (40%) two or more. Four in five (80%) have attended at least one international scientific conference in addition to the IPSA meetings in Moscow; over half (54%) have attended five or more such conferences. Almost two-thirds (65%) have gone abroad at least once to conduct research and half (48%) at least twice.

Registrants vs. Nonregistrants

Is the degree of professionalism exhibited by those who signed up to go to Moscow in 1979 typical of the political science profession at large? To answer this question we may compare registrants from the United States and Canada ($n = 217$) with our control sample of political scientists from the same two countries who did not register to attend the Moscow meetings ($n = 238$). Before doing so, however, it should be noted that in most regards the North American registrants are quite similar in their professional characteristics to registrants from other regions. The two subsamples of registrants (those from North America and those from elsewhere) differ most markedly with respect to the latter's greater language skills and more frequent attendance at international scientific conferences.

Although the North American registrants and nonregistrants are roughly the same age, entered the profession at roughly the same point in their lives (at age 29 vs. 31), and are almost equally likely both to hold a doctorate (91 vs. 83 percent) and to work in an academic setting (91 vs. 81 percent), there the resemblance ends. The registrants are substantially more oriented to the political science profession than are those who did not attend the IPSA meetings in Moscow. In contrast to the three-quarters (74%) of the former who report attending at least every other annual meeting of the American or Canadian political science association (48 percent almost every year), only half (51%) of the nonregistrants do so (28 percent almost every year). Similarly, the former manifest greater interest in research than do the latter (77 vs. 55 percent). They have also been more productive: The median registrant reports having published 3 or 4 books and 11 to 20 scholarly articles, the median nonregistrant 1 or 2 books and 3 or 4 articles. Not surprisingly, in view of such behavior, the registrants enjoy higher academic status. Three in five of them (58%), as opposed to only a quarter of the nonregistrants (23%), are full professors.

Then, too, registrants have a greater range of international research skills and experience than do nonregistrants. They are more likely to be bi- or multilingual: Three in five registrants (62%) speak at least one foreign language (23 percent two or more), while only two in five nonregistrants (42%) do so (14 percent two or more). Three in five nonregistrants report never having attended an international scientific conference (59%) or taken a research trip abroad (60%). In contrast, about half that many registrants indicated that they had attended no more than one such conference (27%) and that they had no overseas research experience (31%).

These data suggest that our main sample of 420 political scientists who registered to attend the IPSA meetings in Moscow is far from typical of the profession as a whole. We are dealing with a special group of scholars—individuals who have been unusually productive, have advanced in their scholarly careers, and are active internationally as well as in their national associations. And yet there are differences even among them.

Professionalism and Internationalism

To highlight these differences more than the above profile can, we developed five simple scales to characterize the respondents: professional activity, professional status, international research competence, international activity, and experience in world congresses organized by IPSA. To some slight degree a couple of the scales overlap, but each by itself points to a significant dimension that may help to explain respondents' views on international scientific conferences in general and the IPSA congress in Moscow in particular. (Note: All scales, standardized to range from 0 to 10 points, are additive; for details, see Appendix C.)

Professional activity. Political scientists who are professionally active are likely to take a view quite different from that of the less active. Our scale of professional activity looked at the respondent's position on five questions: attendance at annual meetings of the respondent's national political science association; orientation to research; published books and monographs; scientific articles; and recent publications record. To score the maximum of 10 points on this scale the respondent would have to attend annual meetings almost every year, be strongly interested in research, and have published at least 5 books and 21 articles, 6 or more of them in the past 2 years.

Table 3.1 shows how the various categories of registrants and nonregistrants scored on the scale of professional activity and other scales. The main difference to be noted is between the two primary samples: The average scores for all registrants (5.9) and North American registrants (5.8) are well over two points higher than the score of our control sample of North American nonregistrants (3.6). Political scientists from Western countries other than the United States and Canada turned out to be the most active professionally, while Third World scholars scored slightly higher than North Americans. (Mean scores for nonregistrants are, for instructors in U.S. graduate departments, 4.2; CPSA members, 4.0; and APSA members, 3.4.)

Professional status. Among the many indicators of professional standing or status are academic accomplishments, title and position, and length of time in the profession. The person scoring 10 points on this scale would have a doctorate, be a full professor or director of a research institute or the equivalent, have been professionally active for more than 20 years, and be the author of 5 or more books and at least 21 scholarly articles.

Again we see an important difference between those who registered to attend the IPSA meetings in Moscow and those who did not. In this case, however, North Americans are highest on the totem pole of professional standing, and the category of other Westerners lowest. North American registrants enjoy a status two full points higher than their compatriots who did not sign up for the congress. (The mean score for nonregistering instructors in U.S. graduate departments is 5.2, that for CPSA and APSA members 4.1.)

International research competence. The ability to communicate and perform research in foreign languages and actual experience conducting research abroad are frequently touted as skills predisposing scholars to take part in international scientific activities and rendering them more able to make sophisticated judgments about the quality of such activities. The person judged in this survey to be most highly capable of conducting international research can communicate effectively in at least two foreign languages, actually uses a foreign language in his or her research, and has gone abroad at least three times for research purposes.

North Americans lag considerably behind political scientists from other parts of the world in their linguistic skills and overseas research experience. Respondents from the United States, with a mean score of 5.1, are in this respect substantially below Canadians, who have a mean score of 6.1. Even so, registrants from these two countries rank considerably higher than the

Table 3.1 Variations in Professionalism and Internationalism
Mean Scores Standardized on an Eleven-Point Scale[a]

	Registrants				Non-registrants
Scale	North American	Other Western	Third World	Total	
Professional activity	5.8	6.0	5.8	5.9	3.6
Professional status	6.3	5.7	5.8	6.0	4.3
International research competence	5.3	6.8	6.1	5.9	2.2
International activity	5.2	5.6	5.2	5.3	2.7
IPSA experience	1.9	2.2	0.7	1.9	0.3

[a]The original (additive) scales ranged from 6 points (0-5) to 11 points (0-10); see Appendix C for details.

comparable sample of nonregistrants (among whom CPSA members average 4.2 points, instructors in U.S. graduate departments of political science 2.5 points, and the general APSA membership 1.9 points). The most internationally competent researchers, according to these data, are from that complex of countries including Western Europe, Australia, and Israel (6.8 points).

International activity. The questionnaire includes several items designed to ascertain the extent to which respondents actively participate in international research. Do they focus their attention on foreign countries or international relations? Do they conduct research in foreign settings or meet together with foreign scholars? A political scientist scoring the maximum on this scale would have attended at least six international scientific conferences both at home and abroad, made three or more overseas research trips, specialized in either international relations or comparative politics, and given a global focus to his or her teaching and research.

Political scientists from Western countries outside North America once again emerge as the most international-minded. The gap between them and both North Americans and registrants from the Third World is nonetheless not great. The fact that Canadians score higher (mean score = 5.7) than U.S. respondents (mean score = 5.0) is doubtless attributable to the significance for political scientists throughout North America of both APSA meetings and research institutions and libraries in the United States. The difference between North American registrants and nonregistrants (a gap of 2.5 points) is more remarkable. (Canadian nonregistrants, with a mean score of 3.6 points, also manifest more international research activity than either instructors in U.S. graduate departments or APSA members, who score 2.7 and 2.5 points, respectively.)

IPSA experience. Still another way to characterize our respondents is according to the number of IPSA world congresses they attended before the meetings held in 1979 in Moscow. We asked about participation in previous triennial world congresses held in Geneva (1964), Brussels (1967), Munich (1970), Montréal (1973), and Edinburgh (1976). Respondents scored two points for each of these meetings they had attended.

Only 2 percent of the registrants faithfully attended all IPSA world congresses since 1964. West Europeans were more likely to have attended three or more of them than were North Americans—an observation that is not very surprising in view of the fact that four of these five meetings took place in Europe. Similarly, Canadians, who had the opportunity to attend the IPSA world congress of 1973 in Montréal, have a higher mean score than do their intellectual companions south of the border (2.5 as opposed to 1.7 points). Political scientists from the Third World are by and large newcomers to IPSA activities. As far as the difference between North American registrants and nonregistrants is concerned, it is apparent that a readiness to go to Moscow in 1979 went hand in hand with previous experience with IPSA world congresses. Among the nonregistrants, too, Canadians were more likely to have attended an IPSA world congress (mean score = 1.0) than were political scientists in the United States (mean scores of 0.3 for APSA members and 0.1 for instructors in graduate departments).

Intercorrelations among scores. Each of the five scales is positively correlated with the others (Table 3.2). This means that the more professionally active an individual is, the more likely that individual is to enjoy a high professional standing, be active internationally, have attended previous IPSA world congresses, and so forth. In only a couple of cases, however, are the correlation coefficients sufficiently high to warrant special attention. The first is between professional status and level of professional activity, where the two variables account for two-fifths of the variance in the data ($r = .63$; $r^2 = .40$). Thus, if we know how active a person is in the political science profession, we can, very roughly speaking, predict two out of five times how that individual scores on our scale of professional status. Second, levels of international competence and international activity are highly correlated ($r = .61$; $r^2 = .38$). It is also interesting to note that the higher an individual's professional status and the more active professionally that individual is, the more likely he or she is to have attended past IPSA world congresses.

What all these data indicate is that the political scientists who registered to attend the IPSA world congress in Moscow (those to whom we shall devote most of our attention) constitute a highly select group. Professionally, they are more active at home and abroad than their fellow scholars who planned to stay at home in 1979; and, internationally, they

Table 3.2 Intercorrelation of Scores on Scales[a]

	Professional Activity	Professional Status	International Competence	International Activity
Professional Status	.63			
International Competence	.17	.20		
International Activity	.29	.31	.61	
IPSA Experience	.35	.39	.23	.32

[a]Pearson's r; all coefficients are significant at the $p < .001$ level.

possess greater experience and skills for research. While the data show that high professional status is associated with attendance at IPSA congresses, they cannot tell us whether or not any country's leading political scientists, that is, those recognized as key disciplinary leaders, are the ones participating in these meetings. We may nevertheless reasonably conclude that our registrants represent the *leading stratum* of political scientists in North America and probably across the world.

Why Attend Professional Meetings?

Professional meetings are a standard component of contemporary scientific disciplines. By now, most national political science associations throughout the world hold annual or less frequent general meetings at which members read and discuss papers and transact the association's business. In large countries, such as the United States, there may be regional and even state associations, each organizing its own annual convention. Ever more subdisciplinary groups also hold regular meetings for specialists. Some of these conclaves take place over the course of several days and involve thousands of participants, while others are more modest in both respects.

The political scientists we questioned attend such meetings fairly regularly. Among those who registered for the IPSA meetings in Moscow, and whose national associations hold annual meetings, almost three-quarters indicate that they participate in the latter almost every year (51%) or approximately every other year (23%); only 7 percent report never doing so. Nonregistrants in North America were also asked about attending regional and subdisciplinary as well as national meetings (see Appendix B). About half (51%) attend APSA or CPSA meetings regularly, that is, approximately every other year or more frequently, and somewhat fewer regularly go to conventions of regional associations (45%) or subdisciplinary groups (42%).

(The numbers never attending amount to 13 percent for APSA/CPSA, 27 percent for regional, and 34 percent for subdisciplinary meetings.)

Well over half of the respondents see the main functions of these disciplinary meetings to be scientific (Table 3.3). The meetings are useful, they say, primarily for general communication and professional contacts, and also for learning something new. Personal goals such as seeing old friends and career development definitely take second place. Only the occasional individual mentions the value of subdisciplinary meetings for institution-building in the specialized fields. The distributions of responses are quite consistent across the various categories of respondents and conferences (with correlation coefficients, all highly significant statistically, ranging from .85 to .96).[1] Moreover, the respondents assert that these various meetings serve the same hierarchy of functions for most other participants as for themselves.[2]

Some distinctions among responses are nevertheless noticeable. North American political scientists who did not register to attend the Moscow world congress emphasize the scientific value of subdisciplinary meetings more than national and regional ones. They are also more apt to downgrade the importance of the meetings as far as they themselves are concerned (11 percent for each of the three kinds of meetings) than with respect to others (2-3%), and to attribute careerist motives to others (averaging 22 percent) rather than to themselves (averaging 12 percent). North American registrants are more oriented to the scientific functions of national meetings than are nonregistrants (68 to 58 percent, respectively)—but here it must be added that the registrants from outside North America are even more so (80%).[3]

These modest differences aside, a picture emerges of political scientists who are reasonably well tied into their national associations, as indicated by attendance at annual meetings, and who stress the scientific function of these meetings. Somewhat fewer of the North American registrants participate in regional or subdisciplinary meetings (even though they recognize the latter as being of more scientific value than either national or regional meetings).

Value of International Scientific Congresses

When it comes to international as opposed to national disciplinary meetings, political scientists see a more differentiated variety of functions being performed. Our questionnaires asked respondents to indicate what they considered to be the main functions of international conferences for (1) the individual participant, (2) the political science profession as a whole, and (3) the host country. The questionnaires went on to ask which of all these functions respondents think is the single chief value of international scientific congresses in the field of political science. This section reports the

Table 3.3 Perceived Value to Self of National, Regional, and Subdisciplinary Meetings
Views of North American Registrants and Nonregistrants

	Registrants		Nonregistrants					
	National		National		Regional		Subdisc.	
	N	%	N	%	N	%	N	%
Scientific								
General communication	76	21	55	20	35	15	51	25
Contacts	86	23	55	20	50	22	36	17
New information	80	22	38	14	38	17	36	17
Other	9	2	7	3	5	2	9	4
Total	251	68	155	58	128	56	132	64
Political								
Personal	–	–	–	–	–	–	–	–
Career development	39	11	33	12	30	13	22	11
Social; see friends	58	16	46	17	41	18	24	12
Travel	2	1	5	2	5	2	4	2
Total	99	27	84	31	76	33	50	24
Disciplinary	–	–	–	–	–	–	2	1
Other	–	–	–	–	–	–	–	–
Little or no value	19	5	30	11	25	11	23	11
No response	22	*	88	*	106	*	124	*
Total[a]	391	100	357	100	335	100	331	100

[a]Multiple responses recorded. Because of rounding, percentage columns (which omit "no response" category) may not add to 100 percent.

responses as they relate to *all three* categories of actors. We shall leave to Chapter 4, however, a more detailed consideration of ISCs' scientific impact on individual scientists, and to Chapter 5 the functions of ISCs for the host country.

Before exploring the perceived functionality of ISCs for individual scientists, the discipline, and the host country, we should note two areas of general interest. The first pertains to the distinction between registrants and nonregistrants (Table 3.4). The overall distribution of responses to the question about an ISC's chief function shows no statistically significant difference between those who had and those who had not registered to attend the IPSA world congress in Moscow.[4] Indeed, virtually the only point worth noting is the registrant's secondary concentration on its scientific value for the host country, while the nonregistrants split their secondary concentration

Table 3.4 Chief Functions of International Scientific Congresses in Political Science

	For the Individual		For the Profession		For the Host Country		Overall Chief Value	
	N	%	N	%	N	%	N	%
Scientific								
Registrants	589	86	360	63	123	22	243	69
Nonregistrants	247	81	150	65	20	10	125	76
Total	836	84	510	64	143	19	368	73
Political								
Registrants	1	0	12	2	413	74	13	4
Nonregistrants	4	1	3	1	149	73	2	1
Total	5	1	15	2	562	74	15	3
Personal								
Registrants	97	14	22	4	-	-	6	2
Nonregistrants	53	17	14	6	-	-	8	5
Total	150	15	36	4	-	-	14	3
Disciplinary								
Registrants	-	-	167	29	-	-	85	24
Nonregistrants	-	-	58	25	23	11	23	14
Total	-	-	225	28	23	3	108	21
Other								
Registrants	1	0	2	0	2	0	3	1
Nonregistrants	-	-	1	0	-	-	-	-
Total	1	0	3	0	2	0	3	1
Little or no value								
Registrants	-	-	6	1	18	3	3	1
Nonregistrants	1	0	6	3	11	5	6	4
Total	1	0	12	1	29	4	9	2
No response								
Registrants	13	*	26	*	57	*	67	*
Nonregistrants	58	*	73	*	98	*	74	*
Total	71	*	99	*	155	*	141	*
Total[a]								
Registrants	701	100	595	100	613	100	420	100
Nonregistrants	363	100	305	100	301	100	238	100
Total	1064	100	900	100	914	100	658	100

[a]Multiple responses recorded. Because of rounding, percentage columns (which omit "no response" category) may not add to 100 percent.

between the ISC's scientific and disciplinary value for the host country. This may nevertheless be a distinction without a difference: Response categories for the former category emphasized political science in the host country, while those for the latter stressed political science as an international discipline. Table 3.5 breaks down into finer subcategories the registrants' responses.

A second general question asked respondents how important they considered the functions they mentioned to be for the individual, profession, and host country (Table 3.6). Here, as in the case of the functions named, nonregistrants are less apt than registrants to respond, but, when they do, they also express a less positive orientation toward the overall value of international political science congresses.

Value of ISCs for the Individual

Our respondents make quite clear their view that the individual will gain most from these congresses' scientific aspects. In fact, 86 percent of those who went or had planned to go to the IPSA meetings in Moscow and 81 percent of the nonregistrants cite such values (Table 3.4). These include

- professional contacts (29% of the two samples together)[5]
- general communication (21%)
- broadened international perspectives (16%)
- acquisition of information about the field's new literature, approaches, and research techniques (17%).

Registrants display a much more specific orientation than nonregistrants (Table 3.5). They place the greatest weight on the contacts one is likely to make on such occasions, while those who did not register for the Moscow meetings look more at an expected broadening of international perspectives. Personal gain weighs far less heavily in their minds than scientific values (14 percent for registrants, 17 percent for nonregistrants):

- career development (8%)
- travel and tourism (5%)
- social activities or seeing friends (2%).

Finally, neither set of respondents is much taken by the prospects for political advantages to be gained through these meetings; and no one mentions disciplinary values that could be served.

Are these functions important for the individual participant? Registrants definitely think so and nonregistrants are only a bit more doubtful (Table 3.6). Almost two-thirds of the former group answering the question indicate that the functions served are very important and about half that number

Table 3.5 Registrants' Views on Functions of International Political Science Congresses

	For the Individual N	For the Individual %	For the Profession N	For the Profession %	For the Host country N	For the Host country %	Overall Chief Value N	Overall Chief Value %
Scientific								
General communication	150	22	172	30	49	9	124	35
New information	86	13	23	4	-	-	10	3
Broaden perspectives: gen'l	29	4	14	2	-	-	9	3
Broaden perspectives: int'l	86	13	47	8	12	2	49	14
Contacts	223	32	67	12	50	9	48	14
Facilitate collab./research	15	2	37	7	12	2	3	1
Total	589	86	360	63	123	22	243	69
Political								
International	1	0	12	2	-	-	13	4
National: image; prestige	-	-	-	-	178	32	-	-
National: economic	-	-	-	-	49	9	-	-
National: scientific	-	-	-	-	72	13	-	-
National: other	-	-	-	-	31	6	-	-
National: varies by country	-	-	-	-	83	15	-	-
Total	1	0	12	2	413	74	13	4
Personal								
Career development	57	8	17	3	-	-	4	1
Social; travel	40	6	5	1	-	-	2	1
Total	97	14	22	4	-	-	6	2
Disciplinary								
Prestige; disc. development	-	-	60	11	-	-	12	3
Improve pol. sci. knowledge	-	-	49	9	-	-	17	5
Build int'l discipline	-	-	58	10	-	-	56	16
Total	-	-	167	29	-	-	85	24
Other	1	0	2	0	2	0	3	1
Little or no value	-	-	6	1	18	3	3	1
No response	13	*	26	*	57	*	67	*
Total[a]	701	100	595	100	613	100	420	100

[a]Multiple responses recorded. Because of rounding, percentage columns (which omit "no response" category) may not add to 100 percent.

Table 3.6 Importance of International Political Science Congresses

	Registrants		Non-registrants		All Respondents	
	N	%	N	%	N	%
For the individual participant						
Very important	251	65	67	36	318	55
Somewhat important	126	32	96	51	222	39
Not very important	10	3	17	9	27	5
Not at all important	1	0	8	4	9	2
Varies or no response	32	*	50	*	82	*
Total[a]	420	100	238	100	658	100
Average score[b]	2.62		2.18		2.47	
For the political science profession						
Very important	199	52	66	36	265	46
Somewhat important	159	41	82	44	241	42
Not very important	21	5	27	15	48	8
Not at all important	7	2	10	5	17	3
Varies or no response	34	*	53	*	87	*
Total[a]	420	100	238	100	658	100
Average score[b]	2.42		2.10		2.32	
For the host country						
Very important	79	22	33	19	112	21
Somewhat important	143	40	59	35	202	38
Not very important	106	30	62	36	168	32
Not at all important	31	9	17	10	48	9
Varies or no response	61	*	67	*	128	*
Total[a]	420	100	238	100	658	100
Average score[b]	1.75		1.63		1.71	

[a]"Importance varies" and "no response" categories are not included in calculations for either percentage distributions or average scores. Because of rounding, columns may not add to 100 percent.
[b]Average scores are computed according to the following weights: Very important = 3; somewhat important = 2; not very important = 1; and not at all important = 0.

considers them somewhat important. The median response for nonregistrants is "somewhat important." Even among those who did not go to Moscow, then, seven in eight think these kinds of meetings are at least somewhat important.

Value of ISCs for the Profession

Disciplinary values are viewed as more important for the profession than for the individual. Asked what function international congresses play for the political science profession, almost two-thirds (64%) of the responses given by our combined samples mention scientific values such as those discussed above, but another 28 percent point to disciplinary values (and barely 1 percent indicates that international congresses have little or no value for political science as a discipline). Three disciplinary themes are dominant in these responses.

One focuses on developing the prestige and identity of the political science profession (10 percent of the combined sample). The mere fact that such international congresses are held gives the profession a cachet of prestige and, over the long run, a more deeply held sense of the profession's legitimacy as a scientific discipline. Similarly, bringing together scholars from different countries can enhance their awareness of the degree to which they share similar interests and approaches, or will at least be able to deal with alternative ones within a broad, common framework. Thus, congresses serve a latent integrative function[6] by enhancing political scientists' sense of community and professional identity.

A second theme emphasizes the role of international congresses in internationalizing political science (10%). Many respondents suggest that the "cross-fertilization of ideas" across national boundaries—a phrase that frequently appears on completed questionnaires—helps to identify transnational research problems and stimulates new lines of research. International congresses provide an opportunity to test the applicability across political systems of theories, findings, and conclusions, and facilitate the dissemination of new methodologies and bodies of data.[7] Other respondents point to more logistical and practical functions, such as facilitating the development of transnational study groups and international collaboration.

Third, a few responses (8%) stress the part played by international congresses in keeping the science in political science. "Tunnel vision" and ideological and cultural "blinders" are some of the metaphors used to convey what some respondents see to be unfortunate tendencies toward parochialism or ethnocentricity in political science, when assumptions and conclusions are not subjected to the challenges and criticisms of scholarly exchange in an international context. Others refer to the value of international congresses for improving the study of political science—its methods and substantive identity as a discipline.

Respondents are somewhat less confident that the functions international political science congresses serve for the profession are as important as those they serve for individual participants (Table 3.6). Again, nonregistrants express greater doubts than those who actually paid their money to register for the Moscow meetings.

Value of ISCs for the Host Country

International political science congresses are seen as being even less important for the host country. The median response of both registrants and nonregistrants to our question on this point is that the congresses are "somewhat important." Moreover, over two-fifths of the entire sample describe the functions served for the host country as not very or not at all important—and those closest to such congresses, that is, those who registered for the Moscow meetings, are not much more likely to term those functions important than are the nonregistrants.

Asked what functions the congresses serve for the host country, more than three in five (61 percent of the entire sample) respond in terms that have little to do with science. The most popular response (37%) stresses national prestige, public relations, or some variation on this theme. A point that merits attention here, although it will be discussed in greater detail in Chapter 5, is that only a very small number (6%) feel that the IPSA world congress gave the Soviet Union greater legitimacy in the eyes of the world, opportunities for propaganda, or some other political advantage. Far more important in this category of response is the perceived likelihood that foreign scholars would increase their knowledge and understanding of the country. About 1 in 11 (9%) cites economic advantages, most notably receipts from tourism. Another one in seven (14%) indicates that such domestic gains vary with the country hosting the congress, and roughly a third of these respondents note that the Soviet Union would gain advantages that other countries would not.

Most of the remainder—only 4 percent see little or no value accruing to the host country—point to expectable scientific gains. The largest number of these (19%) refer to the intellectual stimulation that international congresses provide for local political scientists, as well as opportunities to communicate with foreign scholars and to participate more actively in the profession. Some suggest that this improved communication and participation can encourage and improve indigenous research in political science. A few others (3%), all of them respondents who had not registered to attend the Moscow meetings, argue concomitantly that the organizational experience obtained by political scientists in the host country will assist the profession as a whole. A substantial number of respondents (13%) also notes that the increased visibility and prestige, both domestic and international, that an international congress provides political scientists in the host country enhances the status of their professional association.[8]

Chief Value of International Scientific Congresses

Not surprisingly, given the respondents' views on the value of international congresses for the individual scientist and profession, when they turn to identifying the single chief value (Table 3.4) they concentrate on the congresses' scientific contributions. One in three (34%) mentions general scientific communication, almost that many reducing parochialism either among individual scientists (18%) or in the discipline as a whole (13%), and a quarter mentions methods other aspects of the scientific process such as building contacts (13%) and enhancing knowledge or general perspectives (10%). Those registered to attend the Moscow meetings are almost twice as likely as nonregistrants to emphasize disciplinary values, while the latter focus more than the former on the scientific goal of broadening international perspectives.

Looking solely at registrants, we find remarkably few differences among the categories of respondents. Political scientists from North America, other Western countries, and the Third World are fairly well agreed on the chief purposes of international congresses in their field. North Americans are slightly more apt to identify disciplinary purposes and other Westerners scientific purposes, but the distinctions are not significant statistically.[9] The same conclusion emerges when we divide the registrants according to whether they are high or low in terms of professional activity, professional status, international activity, international competence, or IPSA experience. The relative importance of the various types of purposes is a very robust finding.[10]

In sum, political scientists believe that international congresses serve a variety of purposes: primarily scientific for the individual, scientific and organizational for the political science discipline, and primarily political for the host country. Differences between North American registrants for the Moscow meetings and nonregistrants are discernible but are not statistically significant. Nor does it matter much where the registrants live and work, or how active they are in the profession. The respondents consider the various purposes they identify to be important—more so, as we shall see later, for the individual and discipline than for the host country. It is the political scientist and the political science profession as a whole that benefit from international congresses. To the extent that a country stands to gain from hosting a congress, the gain is marginal rather than significant.

Deciding to Go to Moscow

Ultimately, however political scientists view the functionality of a particular ISC for themselves and others, they must decide whether or not to attend. This section examines the reasons given by those who registered and attended the IPSA world congress of 1979, those who registered but did not attend it, and those North Americans who did not sign up to go to Moscow.

Participants

Those who went to Moscow in 1979 say they did so primarily for either scientific or personal reasons. The most prominent explanations offered by those who attended for scientific reasons were contacts (14%), broadened international perspectives (11%), and general communication (10%). The predominant personal reason, career development, which accounts for 28 percent of the responses, centers on practical considerations such as the fact that the respondent was invited to present a paper or otherwise participate in the program. If we consider such responses as being related to the scientific enterprise, then we might conclude that close to three in four (73%) were seeking to develop further their skills as political scientists. About a quarter of that number (19%), however, simply cites the desire to visit Moscow or travel in the Soviet Union. (North Americans are a bit more likely than others to give this last response, but they are also more likely to say that they went to broaden their perspectives and learn about the Soviet system.) As we shall see in Chapter 5, political reasons did not play a significant role in participants' decisions on this matter.

A further question about which of the reasons given is the primary one elicited roughly the same distribution of responses: 41 percent scientific, 33 percent career development (again, essentially, the chance to participate in the program), 17 percent travel and tourism, and only a handful political.

Non-Attending Registrants

Fifty of those responding to our questionnaire registered but ultimately did not go to Moscow. Why? The main reasons they cite are personal. Over half (55%) refer to what might be called reasons of convenience—family obligations, illness, or the like—and another quarter (24%) to the dearth of financial resources. Only five of these responses (8%) mention problems of scientific communication and another six (10%) political restrictions in the Soviet Union. The subsequent question about the primary reason produced an almost identical set of responses (with personal reasons climbing to a total of 85 percent).

Nonregistrants

The sample of political scientists in Canada and the United States who did not register to attend the Moscow meetings also stresses personal considerations behind their individual decisions. The most prominent one is the lack of funds (41%), but lack of time or interest also played an important role (33%). Another set of respondents was either unaware of the meetings or had not been invited to attend them (6 percent each). Five and eight percent, respectively, refer to anticipated difficulties of scientific communication and political considerations. Respondents from the United States are somewhat more likely to mention political considerations (9%) than are Canadians (5%), who in turn are slightly more concerned about scientific communication, but both groups clearly identify the lack of funding, time, or interest as the main thing keeping them away from Moscow.

The data on reasons for attending or not attending the Moscow meetings are fairly unambiguous. Those who went did so mainly because they expected to learn something or develop their professional capabilities. A much smaller proportion saw the meetings primarily as an opportunity to visit a part of the world otherwise less easily accessible to them. It was not politics but mostly practical considerations that kept the rest at home.

Notes

1. The coefficients are: for national vs. regional meetings, $r = .96$, $p < .001$; national vs. subdisciplinary meetings, $r = .93$, $p < .001$; regional vs. subdisciplinary meetings, $r = .85$, $p < .01$; and for nonregistrants vs. North American registrants, $r = .90$, $p < .01$.

2. The coefficients for perceived value to self vs. others are: among registrants, for national meetings, $r = .82$, $p < .01$; regional meetings, $r = .83$, $p < .01$; subdisciplinary meetings, $r = .90$, $p < .001$; and, among all registrants, for national meetings, $r = .96$, $p < .001$.

3. Even so, the distribution of responses of North American vs. other registrants is remarkably similar, with $r = .87$, $p < .001$.

4. The correlation coefficients for the distribution of responses by registrants vs. nonregistrants are: function for the individual, $r = .87$, $p < .001$; the profession, $r = .77$, $p < .01$; host country, $r = .87$, $p < .001$; and chief value, $r = .88$, $p < .001$.

5. Although the desire to establish and maintain contacts could be construed as a career-related benefit, most of those who used the word "contacts" placed it in the broader context of scholarly exchange.

6. "Latent functions," in Merton's (1957: 51) terminology, refer to observable objective "consequences which are neither intended nor recognized." We shall return to this point in Chapter 6.

7. See Nuttin's (1974: 299) comment: "The main purpose . . . of

symposia in general international congresses should be a constructive 'confrontation' of approaches, and a comparison of theoretical points of view, in an effort to learn from each other and to escape from the narrow circle of problems and methods in which research in some countries is often confined for several years."

8. This finding corresponds with Archie Brown's (1984: 322-323) argument, cited earlier, that holding the IPSA world congress in Moscow provided "a considerable stimulus to the development of political science in the USSR."

9. The correlation coefficients by region are: for North American vs. other Western respondents, $r = .87$; North American vs. Third World respondents, $r = .85$; and other Western vs. Third World respondents, $r = .92$—all significant at the level of $p < .001$.

10. Dividing each distribution into equal halves, the correlation coefficients for the top vs. bottom halves are: for professional activity (0-6.4 vs. 6.5-10), $r = .96$; professional status (0-6.9 vs. 7-10), $r = .98$; international activity (0-6.1 vs. 6.2-10), $r = .99$; international competence (0-6.9 vs. 7-10), $r = .96$; and IPSA experience (0-1.4 vs. 1.5-10), $r = .97$. All these coefficients are significant at the level of $p < .001$.

FOUR
An Occasion for Learning

Although international scientific congresses are perceived as serving a variety of purposes, scientific communication remains the primary reason given to justify the expenditure of time and resources required to organize and participate in such meetings. This chapter investigates the validity of such claims. It addresses four kinds of questions. Two concern short-term learning. First, how do participants in IPSA's world congress in Moscow assess it as a context for scientific exchange? How did their experiences match up with their expectations before they went to the Soviet Union? Second, turning to a general area where learning seemed possible, we sought to determine what impact attending the conference had on images of and attitudes toward the Soviet Union and its people. Do IPSA members feel they gained a better understanding of life and scholarship in the Soviet Union?

The other pair of questions concerns the congress's effect on the participants' long-term environment for learning. Third, how did the experience of having attended the conference change their network of colleagues and interaction with them? Finally, in what ways did participation in the congress change the respondents' scholarship or other professional activities? Since some of our questions on these points stem from a survey conducted by the Johns Hopkins Center for Research in Scientific Communication of participants in the international congresses of psychology and sociology, both held in 1966, comparative data about the variable impact of international scientific congresses across time and disciplines enriches our findings.

It is worth reminding ourselves as we examine the data that, since the survey was conducted a good year after the end of the IPSA meetings in Moscow, we are dealing with recalled experiences. It would have been preferable to have sent out before-and-after questionnaires to tap changes in perceptions that might be attributed directly to the impact of the world congress, for we know that subsequent events affect the ways in which we recall the thoughts we had before those events occurred. Recalled experiences may nevertheless provide us with insights into the long-range effect of attending international conferences.

Expectations and Experiences

"What we anticipate seldom occurs," Benjamin Disraeli wrote in his novel, *Henrietta Temple*. And yet social psychologists have frequently pointed out that our expectations about a social setting can set up a chain of self-fulfilling prophecies, and that, once we commit ourselves to an image of what will occur, we are quite apt to "see" that outcome irrespective of the objective situation. For almost three-quarters of our respondents, the visit to the IPSA world congress in Moscow was their first in the Soviet Union. Given the facts that most students of political science have a keen awareness of the international political arena and that much has been written about Soviet life and behavior, it would be surprising indeed if our respondents did not undertake their visit with some sharply defined expectations about what they would encounter. What were some of these expectations? Was the reality they encountered anything like what the respondents expected?

To find out more about this dimension of conference outcomes, we gave respondents a series of 11 statements, and asked them 2 questions about each. As an example, one assertion was: "I would have a great deal of contact with Soviet scholars at the congress." The first question queried the extent to which the statement corresponded with the respondents' *expectations* about what they would encounter in the Soviet Union. Possible responses were "not at all what I expected," "only a little like I expected," "approximately what I expected," or "exactly what I expected." The second question inquired how closely the assertion in fact described what the respondent *experienced* in the Soviet Union. Here the response categories were "not at all what actually happened," "only a little like what happened," "approximately what happened," and "exactly what happened." In each case we assigned scores across the range of categories from 1 (= "not at all what I expected/what happened") to 4 (= "exactly what I expected/what happened"). The sample's mean expectation about contact with Soviet scholars (2.37) was between the categories "somewhat" and "approximately what I expected." The mean experience reported (2.48) was slightly above the anticipated condition.

For the sake of convenience we may group the eleven statements into four areas as follows:

AN OCCASION FOR LEARNING   79

Contact with East Europeans

    b. I would have little social contact with people in the USSR (other than Congress participants). (Note: In the analysis the scores for this question have been reversed to make the question more similar to the following two.)
    c. I would have a great deal of contact with Soviet scholars at the Congress.
    d. I would have a great deal of contact with other East European scholars at the Congress.

Professional Learning

    f. A substantial number of Congress participants would have research interests similar to my own.
    g. I would learn something professionally useful from papers and comments presented in the formal sessions by Soviet and East European participants.
    j. There would be opportunities for informal interaction with Congress participants from other countries that would contribute to my ongoing and future research.

Organization to Permit Interaction

    e. The organization of the Congress itself would facilitate interaction among participants.
    h. Participants in formal sessions would be able to raise any topics and present any points of view they wished (even on highly contentious "cold war" issues).
    i. The procedures followed by panel chairpersons would encourage uninhibited discussion and the presentation of a wide range of viewpoints.

General

    a. The accommodations (including hotel and provision for meals) would be adequate to my needs.
    k. Overall, my visit to the IPSA World Congress in Moscow would be a satisfying personal experience.

Table 4.1 arranges the four groupings in order of ascending expectations.

Table 4.1 Expectations and Experiences: Mean Scores

Statement	Expectation	Experience	Difference	Individual Corr.Coeff.
East European contacts				
b. Soviet people	1.97	2.09	+.12	.51***
c. Soviet scholars	2.37	2.48	+.11	.14**
d. East European scholars	2.45	2.59	+.14	.33***
Average	2.26	2.39	+.13	
Professional learning				
f. Partic. interests similar	2.55	2.81	+.26	.51***
g. Soviet/E.Eur. contribution	2.43	2.55	+.12	.31***
j. Informal interaction	3.04	2.87	−.18	.47***
Average	2.67	2.74	+.07	
Organization				
e. Facilitate interaction	2.72	2.55	−.17	.42***
h. Free discussion	2.74	2.70	−.04	.42***
i. Chairs facilitate disc.	2.77	2.66	−.11	.34***
Average	2.74	2.64	−.11	
General				
a. Accommodations adequate	2.99	3.03	+.04	.50***
k. Satisfying experience	3.30	3.32	+.02	.54***
Average	3.15	3.18	+.03	—
Overall averages	2.67	2.70	+.03	.41

Significance levels: ** = $p < .01$; *** = $p < .001$.

Thus respondents report having been far more confident that their general experiences at the congress would be positive (mean score of 3.30) than that they would have adequate opportunities to interact significantly with Soviet and East European scholars and citizens (mean score of 1.97).

What respondents experienced was not far removed from what they had anticipated. (A Pearsonian product-moment correlation for the mean scores of expectations and experiences shows much overall consistency: $r = .93$, $p < .001$.) Respondents were most pleasantly surprised (difference, or $d = +.26$) by encountering scholars with interests similar to their own, and most disappointed by limited opportunities for informal interaction with foreign scholars that might contribute to their research ($d = -.18$). They evidently had expected the organization of the congress to facilitate interaction to a greater degree than they actually found to be the case ($d = -.17$). They had expected that a wider range of viewpoints would be presented in the formal sessions and were unhappy that panel chairs did not generate more discussion ($d = -.11$). Respondents had expected that discussions, especially on contentious cold war issues, would be more free-flowing than they were ($d = -.04$).

Previous visits to the Soviet Union made a difference, albeit with apparently contradictory effects, in the degree to which what one expects is what one finds. On the one hand, in an overall sense those respondents who had been in the Soviet Union before the IPSA world congress had similar expectations to those who had not (mean scores across the 11 items = 2.68 and 2.67, respectively). What they experienced, however, was much more positive (with mean scores of 2.83 and 2.65, respectively). Previous visitors note particularly that they learned much more from Soviet and East European participants than they had anticipated ($d = +.45$) and that they found more colleagues with similar interests ($d = +.30$). First-time visitors to the Soviet Union had expected the organization of the congress to facilitate interaction more than it did ($d = -.25$) and found their opportunities for contacts with foreign scholars more limited ($d = -.21$). If, on the other hand, we turn to individual correlation coefficients for each of the eleven items, then we discover that the match between expectations and experiences is greater for previous visitors ($r = .52$) than for first-time visitors ($r = .38$). What these data suggest is that previous visits give individual assessments greater congruence, but leave the aggregate of those who made them a bit more pessimistic about the opportunities provided by the IPSA meetings than their subsequent experience warranted. The congress provided more opportunities for interaction than they had dared hope.

Looking at the home country of the participants does not add much to our understanding of the congress's impact. Overall, North Americans and participants from other Western countries report sets of expectations and experiences that are close to the average. The former indicate that they were somewhat better able to make contacts with Soviet citizens than they had expected ($d = +.17$)—possibly because of a concerted effort by some North Americans to meet with Soviet dissidents and Jews. Other Westerners found it easier than they had anticipated to meet with Soviet and East European scholars ($d = +.18$ and $+.22$, respectively). By contrast, political scientists from the Third World found things overall slightly worse than they had thought they would be. They are particularly disappointed that the congress's organization did not better facilitate interaction ($d = -.30$) and that they did not have more opportunities to meet foreign scholars ($d = -.25$). Correlations by region of aggregated mean scores across the 11 items indicate only slight differences, and virtually none between North Americans and other Westerners in terms of their expectations ($r = .98$) and experiences ($r = .97$).

Classifying respondents according to their professional and international skills and activity provides some additional clues. In no case, however, as Table 4.2 indicates, does a statistically significant degree of correlation explain much variance in individual responses. Perhaps the most interesting classification is the participants' record of previous attendance at IPSA world congresses. By and large, those who had attended several such congresses went to Moscow with fairly low expectations about what they would encounter, and are most positive about the professional contacts they actually made there and the organizational aspects of the congress itself. The more internationally competent participants are, the more likely they were to expect and experience opportunities for contacts with foreign scholars and, more generally, other participants with research interests similar to their own. Enjoying high professional status seems to be the best predictor of whether or not respondents will establish contacts with Soviet and East European scholars. Those with high professional status are also more likely to conclude that the organization of the congress facilitated informal interaction. In meeting Soviet scholars, possessing international research skills or a record of international scholarly activity seemed to help; while past experience with IPSA world congresses and high levels of professional activity were conducive to meeting East European scholars. Those with an international orientation, whether in the form of skills or activities (including attending IPSA world congresses), are least likely to have found the Moscow meetings a satisfying experience.

The data suggest *in fine* different but compatible findings according to whether we are looking at individual participants or at the aggregate of respondents to the survey. At the individual level, those who went to Moscow found pretty much what they expected to find. This is especially the case for those with prior experience in the Soviet Union, and with respect to such matters as contacts with the Soviet people, accommodations, and the Moscow meetings as a whole. But even these respondents report some surprises, most notably the fact that they learned more professionally from the Soviet and East European participants than they had anticipated. Further, levels of professional and international involvement are clearly associated with the individual respondents' expectations and experiences. Those higher on various scales tended to have lower expectations and more positive experiences than did the remainder.

At the aggregate level, expectations and experiences were both fairly high. Averaging the mean responses shown in Table 4.1 for the categories most directly relevant to scientific impact (that is, omitting contacts with the Soviet people, accommodations, and general satisfaction), we find that, overall, the experience of the congress (2.65) slightly exceeded expectations (2.63). Respondents were most dubious about their prospects for meeting East Europeans and, while the level of contacts they experienced was still not very high, it was substantially higher than what they had anticipated. They

Table 4.2 Expectations, Experiences, and Professional Activity
Correlation Coefficients ($p < .10$)

Participants who are very active internationally reported:	
Low experience of overall satisfaction	−.22***
Low expectation of overall satisfaction	−.14**
Low experience of adequate accommodations	−.11*
Low expectation of adequate accommodations	−.08*
Low expectation of prof. learning from Soviet/E.Eur. scholars	−.07*
High expectation of contacts with foreign scholars	+.07*
High experience of contacts with Soviet scholars	+.08*
Participants who are highly competent internationally reported:	
Low expectation of prof. learning from Soviet/E.Eur. scholars	−.10*
Low experience of overall satisfaction	−.08*
High expectation of participants with similar research interests	+.09*
High experience of contacts with Soviet scholars	+.09*
High experience of participants with similar research interests	+.09*
High experience of contacts with foreign scholars	+.14**
High expectation of contacts with foreign scholars	+.20***
Participants who have attended several IPSA world congresses reported:	
Low expectation of contact with Soviet people	−.16**
Low experience of contact with Soviet people	−.13**
Low expectation of prof. learning from Soviet/E.Eur. scholars	−.12**
Low expectation of overall satisfaction	−.12*
Low experience of overall satisfaction	−.12*
Low expectation of adequate accommodations	−.10*
Low experience of adequate accommodations	−.09*
High experience panel chairs encourage free discussion	+.07*
High experience of participants with similar research interests	+.08*
High experience of contacts with East European scholars	+.09*
High expectation Congress org. would facilitate interaction	+.10*
High experience of contacts with foreign scholars	+.11*
Participants who have been very active professionally reported:	
Low expectation of prof. learning from Soviet/E.Eur. scholars	−.09*
Low expectation of free discussion in sessions	−.08*
High expectation of contacts with East European scholars	+.08*
High experience of contacts with East European scholars	+.09*
Participants with substantial professional status reported:	
Low expectation of free discussion in sessions	−.08*
High expectation of participants with similar research interests	+.08*
High experience Congress org. facilitated interaction	+.08*
High expectation of contacts with East European scholars	+.11*
High experience of contacts with East European scholars	+.11*
High experience of contacts with Soviet scholars	+.13**

Significance levels: * = $p < .10$; ** = $p < .01$; *** = $p < .001$.

are disappointed that the organization of the congress did not facilitate interaction to a greater extent, and that there were inadequate opportunities for informal interaction with participants from other countries and insufficient formal interaction in the panel sessions. We shall return later in this chapter and in Chapter 6 to some organizational problems and suggestions for improvement.

Such problems notwithstanding, most respondents (91%) are either very or somewhat satisfied with the IPSA world congress as a whole. On a four-point scale (from "not at all satisfied" = 1 to "very satisfied" = 4), the mean score for the entire sample of those who went to Moscow is 3.38. North Americans are particularly pleased (3.46), while other Westerners and Third World participants are less so (3.30 and 3.28, respectively). The higher individuals score on our various scales of professionalism and internationalism, the more likely it is that they rate their experience in Moscow as very satisfactory: international activity ($r = .19, p < .001$); IPSA experience ($r = .15, p < .01$); professional activity ($r = .13, p < .01$); professional status ($r = .10, p < .10$); and international competence ($r = .09, p < .10$). All these figures correspond well to the findings reported in Table 4.1, both that respondents had high expectations regarding their visit to Moscow and that these expectations were by and large realized.

Soviet Government and People

A purpose of traveling abroad is to learn something about foreign lands and people. Political scientists attending a meeting in Moscow, we might expect, would be particularly keen to observe the Soviet political system at work. But how were they supposed to do this? Except for those with relatives or acquaintances in the Soviet Union whom they could visit, or those who had appropriate linguistic skills, possibilities for meaningful contacts were limited. Linguistic and perhaps other barriers made it difficult for political scientists from nonsocialist countries even to meet with their Soviet colleagues over breakfast or in the hotel lounge. It is no wonder that many participants lament the infrequency of contacts with Soviet citizens and scholars.

Even so, opportunities for a firsthand glimpse at life in the Soviet Union existed. At a casual level, interaction with hotel staff and tourist-shop personnel, along with observations made on the streets or in the subway, may have produced some useful insights. A number of participants made a point of visiting coreligionists or known Soviet dissidents. Some also took advantage of visits organized by the Soviet Political Science Association to government officials, daycare centers, and the like, or went on guided tours to other parts of the Soviet Union. Then, too, all were touched by the organizational aspects of the congress and probably most attended several panel sessions at Moscow State University. The impressions gained of Soviet life and society were doubtless fragmentary, but perhaps not more so than those of other countries obtained in the course of IPSA world congresses elsewhere.

Images

Most participants, we have hypothesized, went to the IPSA world congress in Moscow with a fairly concrete image of the Soviet Union, that is, some picture in their minds of what they believed to be true. Images correspond only more or less to reality. But, since we often form attitudes and take action on the basis of our images, the content of those images is an important variable in studying the political behavior of individuals and groups.

We asked our respondents to engage in an intellectual game: to try (in 1980) to recall what image they had held of the Soviet Union before attending the world congress in August 1979, and then to specify how the experience during the course of their visit affected that image. More particularly, did the experience reinforce the respondent's pre-existing image, leave it unchanged, or significantly change it? With what impressions did congress participants return to their home countries after a brief sojourn in the Soviet Union?

Exposure to life in the Soviet Union had a widely varying effect on the participants' images. Responses to the overall question of changed images fall roughly equally in all three categories: 30 percent think that being in the Soviet Union merely reinforced the images they already enjoyed, while the remainder are split evenly between those finding no change and those finding at least some change (35 percent each). On a three-point scale (ranging from reinforcement = -1 over no change = 0 to change = +1), the average score is +.05—a very slight change. Of the various categories into which we classified our respondents, the most interesting in this respect comprise participants from the Third World, whose images changed fairly substantially (average score = +.31), and those who had participated in many previous IPSA world congresses, who tend to find their earlier images reinforced (average score = -.26).

Those whose views were reinforced only infrequently report the reasons for this. Most say that they were already fairly familiar with the Soviet Union, either from previous visits or extensive reading, or else that the experience of visiting Moscow and perhaps participating in a tour enhanced the complexity of their images without changing their overall tenor. Of respondents in the latter category, most cite information critical of the Soviet

Union. Only three respondents reporting reinforced images cite positive developments. One expressed the belief that the Soviet Union is "more open to contact and discussion" than previously experienced, another reported the discovery that one's Soviet colleagues are "more sophisticated, more open-minded" than previously thought, and the third observed that "whatever their discontents, people seem to feel their leaders know how to control the economy and that life is growing gradually better."

The responses of those whose images had changed are much more variegated. Some were "impressed by the achievements of the country," or became "more aware of internal pluralism" or other developments. Others were, in the words of one of them, "surprised at how much freedom we were allowed compared to my previous visits." A majority, however, found the domestic circumstances in the Soviet Union even worse than they had imagined.

Attitudes

Subsequent questions asked whether and how attending the IPSA world congress changed the respondent's attitude toward the Soviet government and the people of the Soviet Union. About two in seven (29%) report attitudinal changes with respect to the government, and three in seven (43%) with respect to the people. The overall changes are nonetheless in opposite directions. Using a five-point scale (from an attitude change very much in a negative direction = -2 over neutral = 0 to an attitude change very much in a positive direction = +2), we find the average shifts to be negative for the government (-.28) and positive for the people (+.40). North Americans became much more positive about the Soviet people (+.49) than did participants from other Western countries and the Third World (+.29 each). Similarly, political scientists who attended many previous IPSA world congresses distinguish themselves from others who are highly professional in their orientation: Those with extensive IPSA experience are the only ones with a below-average negative shift in attitudes toward the government (-.26) and a substantially higher-than-average positive shift in attitudes toward the Soviet people (+.70).

Reasons offered by respondents to explain their change in attitude toward the Soviet government parallel those summarized above regarding changed images. Persons with more negative attitudes criticize what they saw as controls on the movement of people (including participants in the IPSA world congress), the prevalence of petty corruption, rigid adherence by Soviet scholars to the government's ideological line, excessive bureaucracy, and, more generally, a governmental system that, as one respondent says, "simply does not work." Others express more positive shifts in attitude. "I liked the emphasis placed on education, art, sports, health, housing projects (still below present needs), etc.," writes one political scientist. Another merely notes that the government enjoys "wide acceptance among people; multiracial society; sense of communism as an *indigenous* ideology widely accepted." But, for the most part, the written comments support the coded ratings reported earlier: A firsthand view of the Soviet government in action did not elicit positive views among political scientists attending the IPSA world congress.

Quite a different picture emerges with respect to the Soviet people. Some respondents criticize an apparent surliness and rigidity, but most express greater understanding for Soviet citizens' response to their lot in life. A few find the "Soviet people more freeminded and accessible" than expected; "they talk of liberalization and expect democratic communism." Other characterizations abound: obedient, warm, docile, hospitable, apathetic, decent, male-chauvinistic, generous, suspicious of foreigners, culturally rich, perversely pleased with their suffering, and withal friendly. Some who had toured the Caucausus or Soviet Asia also comment on the "national pride and distinctiveness" of the various Soviet peoples.

In short, taking into account the possibility that our respondents did not make the clear distinctions between "images" and "attitudes" that we had intended, and considering that what some saw as reinforcement of pre-existing images others interpreted as significant change, a fairly coherent picture of changes in perspectives becomes visible. Those without previous experience in the Soviet Union gained a substantial amount of information to fill out and temper their views. The nature of the situation doubtless precluded substantial change. After all, most congress participants were in the Soviet Union for only 1 or perhaps 2 weeks, and were exposed more to problems of daily life (such as unexpected controls and unresponsive bureaucracies) than to the larger picture of life and politics in the Soviet Union. Moreover, their contact with common Soviet citizens was for the most part superficial. What resulted was marginal rather than spectacular change. Congress participants as a whole became slightly more convinced of the shortcomings of the Soviet system of governance, and more sympathetic toward the common citizens (albeit not the *nomenclatura*) who must cope on a daily basis and throughout their lives with that system.

International Scientific Networking

A basic justification for scientific congresses is that they further scholarly activities by bringing together disciplinary leaders who can learn from one another. Indeed, it is the opportunity to develop and maintain contacts that respondents cite as the single most important function of national scientific congresses for individual participants (Table 3.5). It follows that international scientific congresses should facilitate cross-national learning. We would expect those attending such conferences not only to develop contacts especially with colleagues from other countries and to learn something from interactions with others, but also subsequently to keep in touch with new colleagues met at the congress. Networks of international contacts help the individual scholar keep up-to-date on new lines of research in other parts of the world. They also provide sources of valuable information, possibilities for future meetings, and perhaps even scholarly collaboration.

To what extent did the Moscow world congress facilitate this process? To answer this question we shall look first at the kinds of informal contacts that respondents report having had with various categories of other participants in the congress. Table 4.3 shows the aggregate scores for each category. It reveals that participants were most likely to meet informally with colleagues from their own country, and least likely to enjoy informal contacts with Soviet and East European scholars. The degree of respondents' international competence is modestly associated with Soviet contacts; and those with IPSA experience are the most apt to have informal contacts with East European scholars.

The regional origin of respondents shows some interesting diversity in this respect. Of the three regional groupings that received questionnaires, participants from the Third World are the most isolated in terms of both the contacts they have with others and the likelihood that North American and other Western scholars have contacts with them. A certain asymmetry in North-South communication is apparent. Whereas the aggregate Third World score for contacts with Western Europeans and North Americans stands at 3.33 (slightly under the score of contacts with scholars from their own country), the reverse scores are 2.65 for North Americans and 2.28 for Western Europeans. The relatively smaller number of registrants from the Third World (perhaps 194—but see Table 1.2—as opposed to 305 North Americans and 432 other Westerners) may account in part for this finding, for there were fewer Third World participants with whom the latter could meet. The lower part of Table 4.3 nonetheless suggests that other factors were at work as well. Highly professional and internationally active political scientists seem to gravitate more toward East Europeans and especially Westerners than toward Third World participants.

Do participants try to keep up contacts with colleagues they met for the first time in Moscow? As Table 4.4 indicates, considerable effort was expended. Correspondence with Western Europeans is most common: Half, that is, 184 of the 369 respondents who went to Moscow report such an activity. Almost as many (182, or 49 percent) corresponded with North Americans they met at the congress. In general, political scientists made the greatest effort to maintain contacts with colleagues in Western Europe (1.51

Table 4.3 Informal Contacts Established at IPSA World Congress, Moscow, 1979

Category of Respondent	Region of Those with Whom Contacts Established				
	Own Country	Soviet	Other E. Eur.	W. Eur. No. Am.	Third World
A. *Scale Scores*					
North Americans	3.64	2.44	2.45	3.29	2.65
Other Westerners	3.59	2.35	2.51	3.48	2.28
Third World	3.36	2.31	1.97	3.33	2.28
Total	3.59	2.39	2.43	3.37	2.52
B. *Correlation Coefficients*					
International Activity	−.13**	+.05	+.12*	+.23***	+.09*
International Competence	−.12**	+.08*	+.15**	+.19***	+.00
IPSA Experience	−.04	+.03	+.19***	+.19***	+.06
Professional Activity	+.04	−.00	+.12*	+.11*	+.01
Professional Status	−.04	+.05	+.15**	+.06	+.08*

Note: Aggregate scores based on the scale: No contact at all = 1; very little contact = 2; occasional contact = 3; great deal of contact = 4. Significance levels: * = $p < .10$; ** = $p < .01$; *** = $p < .001$.

contacts per participant), followed by those in North America (1.43), the Soviet Union and Eastern Europe (1.04), and, in last place, the Third World (0.69). Only 82 respondents (22%) report that they made no successful efforts to keep up with colleagues from Western Europe they originally met in Moscow, and only 54 (15%) failed to pursue any contacts from North America. The number of congress participants who made no effort to keep up with Eastern European contacts is about the same as that for the Third World (123 and 121, respectively, or 33 percent).

Figures for the number of contacts per participant, shown in Table 4.4, provide some indication of the extent to which the various groupings of participants maintain the contacts they made at the congress. Other Westerners are somewhat more active in this regard than are North Americans or participants from the Third World. Those who score high on the various scales of professionalism and internationalism have with but one exception (namely, the efforts of scholars with high professional status to maintain contacts with colleagues in the Third World) better than average records.

These results indicate substantial activity aimed at establishing and maintaining contacts: At least half of those attending the Moscow world congress pursued one or more contacts made there. Although most activity occurred among North Americans and other Westerners, a quarter of the political scientists report that they corresponded with at least one colleague

Table 4.4 Efforts to Keep Up with Colleagues Originally Met in Moscow

	Location			
	USSR & E. Eur.	Western Europe	Third World	North America
A. Modes of maintaining contacts, all participants				
I have not tried to keep up contacts	123	78	121	49
My attempts have been unsuccessful	23	4	9	5
Total: no contacts	146	82	130	54
Through correspondence	139	184	92	182
Scientific meetings, conferences	43	88	33	79
Visits or exchange of visits	46	61	24	58
Exchange of reprints, papers	111	151	74	145
Joint or parallel research	20	40	15	40
Joint publication	19	32	13	23
Other	6	3	2	2
Total: contacts	384	559	253	529
Contacts per participant (n = 369)	1.04	1.51	0.69	1.43
B. Contacts per participant for various participant groupings				
North American participants (n = 193)	1.07	1.34	0.70	1.34
Other Western participants (n = 140)	1.16	1.78	0.61	1.55
Third World participants (n = 36)	0.42	1.39	0.86	1.47
High international activity (n = 74)	1.27	1.91	0.88	1.74
High international competence (n = 153)	1.23	1.97	0.79	1.81
High IPSA experience (n = 23)	1.35	2.61	0.70	2.04
High professional activity (n = 88)	1.36	2.03	0.86	1.70
High professional status (n = 143)	1.32	1.59	0.67	1.48
Total (n = 369)	1.04	1.51	0.69	1.43

from the Third World whom they met for the first time in Moscow, and almost two-fifths claim they wrote to newly established contacts in the USSR and Eastern Europe. The fact that 139 (or 38 percent of the sample) managed to correspond with Soviet and East European scholars they met at the congress, that 111 exchanged papers, 46 exchanged visits, and 20 initiated joint research projects with them is notable in view of the various obstacles to such ties.

Modifications in Scholarly Behavior

How and to what extent did the experience of the IPSA world congress affect the work of those scholars who attended? A little less than a quarter (23%) of the respondents report that, considering the congress as a whole and all their interactions with congress participants, they received information that led to a major modification in their professional activities. Of these respondents, 75 percent indicate that the information came from informal contacts, 40 percent mention papers they read and almost that many papers they heard presented (35%), and a third (33%) cite floor discussion in panel sessions. Over three-quarters (77%) state that this new information modified their research, and somewhat fewer than a third see an impact on their teaching and plans for publication (32 percent each). Only a handful mentions applied work (6%) or administration (4%).

Three points stand out when we look at how various groupings of respondents consider their work modified by the experience of participating in the Moscow world congress. First, scholars from the Third World are the most likely to see modifications (38%), especially through reading papers or hearing them presented orally; each one reporting a modification refers to his or her research and almost half (46%) to teaching. Second, North Americans are more likely to report a major modification than are other Western participants (23 and 16 percent, respectively). Indeed, over half (45 of 81) of all those who experienced such an impact are from the United States and Canada. More than two-thirds of these North Americans (31 of 45) say that what they learned at the congress affects their research and half that many (16) refer to their teaching. Third, the grouping least likely to report any such modification comprises those who attended the greatest number of previous IPSA world congresses (8 percent, or 2 of the 24 participants who attended at least 4 of the 5 triennial congresses between 1964 and 1976).

Asked what the nature of these modifications is, those reporting them emphasize intellectual contributions. Thus 43 percent, or 37 of the 85 responses (up to 3 of which were recorded for each respondent), say that the congress gave them new insights, perspectives, or interests. Another fifth report either that the congress resulted in a change of emphasis, focus, or approach (11%) or that it helped them to clarify or better to understand certain issues (9%). About the same number (19%) points to new materials, techniques, or sources, while the remainder (17%) mentions such practical considerations as collaboration, publication, and travel opportunities.

A comparison of political scientists with psychologists and sociologists who in 1966 attended their own world congresses (Johns Hopkins University, 1968)—in Moscow and Evian, France, respectively—shows some interesting similarities and differences with respect to modifications in their work (Table 4.5).[1] The number of political scientists reporting major modifications in *any* area of professional activity (23 percent of all who attended) corresponds roughly to that of sociologists (21 percent of whom report changes in their subject matter area, 10 percent in some other area of activity) and psychologists (18 and 13 percent, respectively). Taking the number of

Table 4.5 Impact of World Congresses on Participants' Scientific Work: A Comparison of Psychologists, Sociologists, and Political Scientists

	Int'l. Psych. Ass'n 1966	Int'l. Sociol. Ass'n 1966	Int'l. Pol.Sci. Ass'n 1979
Activities Modified			
Research	22%	24%	17%
Teaching	5	3	7
Publication	1	<1	7
Applied work	2	2	1
Administration	–	<1	1
Theoretical work	2	–	–
Other and unspecified	2	2	–
Total	34	32	33
Sources of Modification			
Oral presentation	14	10	8
Written copy of paper	3	–	9
Session discussion	6	8	7
Informal contacts	8	11	16
Other and unspecified	6	≤1	1
Total	37	30	41
Nature of Modification			
New scientific knowledge:			
New insights	4	3	10
Theory construction	2	–	–
New emphasis	6	6	2
Subtotal	(11)	(9)	(12)
Enhancement of ongoing activities			
Clarification	4	6	2
Initiate new type of related work	9	7	–
New materials, techniques	11	5	4
Intensify present work	8	12	–
Subtotal	(33)	(31)	(6)
Other			
Publication, collaboration	1	<1	2
Conference visits	–	–	2
Other and unspecified	3	1	≤1
Subtotal	(4)	(1)	(5)
Total	48	41	23

Note: The IPA and ISA asked slightly different questions of authors vs. other attendants and reported the findings separately (see Johns Hopkins University, 1968: 11, 12, 21, 38, 39, 47); the IPSA questionnaire asked both authors and others the IPA/ISA's questions for attendants. The data reported here on activities modified and nature of the modification combine the two sets of responses for IPA and ISA respondents.

overlapping categories into account, these figures are all in the same range, that is, roughly a quarter. The following is a brief summary of some of the other findings.

1. Activities modified: Each of the three sets of respondents emphasizes major changes in their research as a result of attending world congresses. Political scientists are less likely than the others to stress research, and more likely to point to impacts on their teaching and publication.

2. Sources of modification: While some of the differences may be artifacts of alternative modes of scientific discourse—for example, the average length of papers, which tend to be longer in political science meetings, and hence the relative emphasis on reading papers vs. hearing them presented—the importance of informal contacts for political scientists stands out clearly.

3. Nature of modification: By grouping the responses into three categories—new scientific knowledge (new insights, new emphasis, theory construction), enhancement of ongoing activities (clarification, initiation of new type of related work, new materials or techniques, intensification of present work), and other (publication and collaboration, conference visits, and other)—we find a very sharp difference. Whereas both psychologists and sociologists are apt to stress the congress's contribution to thought and work in process (69 and 76 percent, respectively, of their responses), political scientists place greater weight on new insights and emphases (52 percent of their responses).[2]

Despite major similarities, then, political scientists see the impact of their world congress somewhat more in terms of their teaching and publication, stress the importance of informal contacts, and learned something they describe in more than incremental terms.

Improving Scientific Communication at World Congresses

To note that our respondents found the IPSA world congress valuable for scientific communication is not to say that they see no room for improvement. An indication of this came in response to our question about the best and worst panels respondents attended (see Chapter 5 for details). Almost a quarter of the responses (24%) citing a "best" panel emphasize the quality of discussion, interaction among scholars with diverse points of view, or some similar characteristic. Another 15 percent single out the benefits of exchanges between East Europeans and Western scholars in particular. Over half the reasons given in identifying the "worst" panel have political overtones. Concern with the inhibiting effect of politics and ideology on the scholarly exchange of views, however, points directly at the need for a firmer hand to keep the sessions on the proper intellectual track.

We then asked our respondents if they had encountered problems of scientific communication that could be corrected in future world congresses. Of the 369 respondents who went to Moscow, fewer than half (46%) report running into such problems, and they made some 204 suggestions for improvement. By contrast, psychologists and sociologists evaluating their own world congresses of 1966 provided many more suggestions of this sort (respectively, 2.5 and 2.7 times as many suggestions per respondent).

Table 4.6, which compares the reactions of the three sets of respondents, reveals rather different sets of concerns. (In fact, the Pearsonian correlation coefficients between IPSA participants on the one hand and IPA and ISA participants on the other are, respectively, .21 and .40; that between IPA and ISA participants is .77.) Some of these differences are fundamental. Political scientists, for instance, are happier with the general organization and format of their meeting than are psychologists and sociologists. They are far more distressed than the others, however, by the intrusion of politics into their proceedings. Other differences doubtless stem from technical issues. Thus the large number of suggestions regarding the distribution of papers at the political science and sociological meetings may indicate a particular problem rather than a general one.

Yet in another way the three sets of reactions, together with the suggestions made by these scholars, present a congruent picture. They constitute an inventory of structural, organizational, and substantive characteristics of the ideal world congress.

First, a good program must be well planned in advance. The most disgruntled in this respect are the sociologists, over half (51%) of whom make suggestions for improvement. Most emphasize the general importance of advance planning and distribution of the program, while 7 percent mention a need for more opportunities for informal discussion. Somewhat under half of the psychologists (46%) are concerned enough about organization and format to suggest alternative procedures. Almost half of these in turn (22%) want more opportunities for informal discussion or a larger number of small meetings for persons working in specific areas. This need evidently becomes more critical as the size of a meeting increases. Nine percent of the sociologists address the problem of size directly by suggesting that attendance be restricted. The remainder of the psychologists' and sociologists' responses in this general category are distributed over many practical suggestions, such as avoiding the simultaneous scheduling of sessions having related subject matter. By contrast, a much smaller proportion of political scientists (9%) singles out communication problems at the IPSA world congress. Other matters seem to concern them more.

Second, attention should be paid to the conduct of the panels, generally

Table 4.6 Suggestions for Improving Future World Congresses: Views of Psychologists, Sociologists, and Political Scientists

	International Psychological Association (1966)		International Sociological Association (1966)		International Political Science Association (1979)	
	% of Respondents (N=467)	% of Total Responses	% of Respondents (N=476)	% of Total Responses	% of Respondents (N=369)	% of Total Responses
Improve general organization and format	46%	35%	51%	34%	9%	16%
(More opportunities for informal discussion)	(22)	(16)	(7)	(5)	(4)	(7)
Improve paper distribution	8	6	32	22	14	25
Improve conduct of panels	37	28	30	20	11	20
(More panel time for discussion)	(10)	(8)	(7)	(5)	(3)	(5)
Reduce language difficulties	30	23	21	14	11	20
Reduce politics	2	1	1	1	8	15
Improve other facilities	10	8	13	9	2	4
Total	133%	100%	148%	100%	55%	100%

Note: Multiple responses permitted. Percentages may not add up because of rounding. IPA and ISA data recalculated from Johns Hopkins University (1968: 25-27, 50-52).

seen as the heart and soul of a scholarly meeting. The suggestion most frequently made urges a higher ratio of discussion to formal presentation in the panels. Other suggestions—stronger leadership, stricter time limits on presentations, and fewer papers per panel—aim at the same goal. Five percent of the psychologists and 1 percent of the sociologists, but none of the political scientists, call for quality controls to improve the papers; and a few in each group want improved presentations, especially urging that papers not be read (for a psychologist's suggestions, see Nuttin, 1974).

A third important concern is an improved procedure for distributing papers. Sociologists (32%) and political scientists (14%) stressing this point want a better supply of papers available at the congress (although a few urge that abstracts of all papers be distributed either before or at the congress so that participants might better gauge which panels would be most interesting).[3] Psychologists (8%), for whom this problem had apparently not been so serious, emphasize the need to distribute the papers before the congress.

Fourth, language barriers should be reduced. Most suggestions refer to the need for better capabilities for simultaneous translation, and, especially for political scientists, greater availability of headsets.[4] (Ten political scientists but no psychologists or sociologists argue that the language problem could be easily solved by using English only.) Six percent of the sociologists think it would be useful to extend simultaneous translation to the smaller workshops. An equal number of sociologists and 5 percent of the psychologists suggest interpreters for informal person-to-person discussion.

Finally, a world congress should stick to science, not politics. Some 8 percent of the IPSA participants express a concern with clarifying the role of the host country to avoid interference and ensure discussion. A much smaller number of psychologists (2%), who had also met in Moscow, recommend "separating ideological discourse from science" or "eliminating politicking." About half as many sociologists voice similar notions.

Individual Learning

The findings outlined above suggest that the impact of international scientific congresses on their participants' intellectual development and scholarly work is more cumulative and subtle than immediate and dramatic. Roughly a third of our respondents claim to have changed their image of the Soviet Union as a result of their experience in Moscow, but the changes reported are for the most part marginal rather than basic. All in all, the participants' attitudes became more negative vis-à-vis the Soviet government and more sympathetic toward the Soviet people.

Images, Attitudes, and International Understanding

A vast array of social-psychological studies on the effects of foreign travel, overseas study, and the like urges caution in interpreting such findings (see Kelman, 1965). For one thing, we should note that some 65 percent report no change in their images and roughly the same proportions record no changes in their attitudes toward the Soviet government (71%) or people (57%). This fact suggests relatively little movement in terms of images and attitudes. Still, a third or so—some 350 political scientists if we extrapolate our findings to all participants from nonsocialist countries—is a sizable number. This is especially so if we consider that, memory being what it is, many respondents filling out the questionnaire in fall 1980 may well have forgotten what they learned in summer 1979.

We must be particularly wary about overinterpreting the finding of improved attitudes toward the Soviet people. It need not mean that the respondents "like" these people more. Taken in conjunction with a more negative view of the government, it may only indicate that they have a greater understanding of the circumstances in which Soviet citizens live and hence look on them with greater admiration. International exchanges of various sorts ultimately aim more at understanding than affection. We have no reason to assume that a world political science congress would be any different.

By the same token, although visiting the Soviet Union in the course of attending the world congress undoubtedly broadened the intellectual map of some political scientists, it is not certain that this process had anything to do with the world congress as such. Visiting the Soviet Union in some other context might have had just as great an effect—or perhaps greater, if we take into account the amount of time that congress participants spent in scholarly sessions or with colleagues from their own country. In this sense, holding the world congress in Moscow was more an occasion for learning than its cause.

The argument that interaction among scientists at international congresses promotes international understanding and contributes toward a basis for peace is thus problematic, at least at the interpersonal level. Firsthand experience with the Soviet governmental apparatus did not lead many political scientists to view that government in a more positive light. What effect an enhanced understanding of the Soviet people's situation will have in the long run is a question our data cannot answer. Since in the future most congress participants will probably have little contact with the Soviet population, we might speculate that enhanced understanding alone will not yield much by way of concrete consequences for their attitudes toward the Soviet Union. But, then, a functional integrationist would expect ISCs to promote peace without necessarily changing attitudes toward governments. This fact raises the far more complicated question of the effect of interpersonal interaction at the scientific level.

Scientific Networks and International Understanding

The IPSA world congress provided different configurations of benefits for different groupings of participants. Third World scholars report a substantially greater-than-average degree of modification in their work activities as a result of attending the congress. But they appear to have gained least from the informal interaction of participants in Moscow. They are even more disappointed than other participants that the organization of the congress did not better facilitate this interaction. Their difficulties in becoming involved in networks of informal communication are further illustrated by the fact that other participants report fewer efforts to keep up with colleagues from the Third World than with those from Eastern Europe, Western Europe, and North America.

For veterans of previous IPSA world congresses, the reverse seems to have occurred. They report abundant opportunities for informal interaction and were the most successful grouping in establishing and maintaining new contacts. The Moscow world congress nevertheless seems to have had the least impact on the scholarly activity of these political scientists.

Generally, participants higher up the scales of professionalism and internationalism enjoy more informal interaction with colleagues from other countries than do those lower on these scales. Respondents with higher professional status are the only grouping that thinks the organization of the congress facilitated this interaction, and they report the most contact with Soviet and East European colleagues. Participants with good language skills and extensive overseas experience also report positive experiences with respect to interaction with foreign scholars during and after the congress.

Such findings suggest three points about the importance of world congresses for international scientific networking. First, by bringing new people into them and sensitizing others to their role, congresses serve to revitalize networks that have been established through overseas work and travel. Second, as the suggestions for improving world congresses amply indicate, it makes sense to modify their organization to enhance possibilities for discussion within panels and informal interaction without. Third, emerging within international political science are specialists in networking. These individuals, frequent participants in world congresses, seem to have acquired the knack for developing and maintaining contacts (although, possibly, at some cost to advances in their own research).

New Knowledge

Most participants feel that they benefitted at least to some degree from attending IPSA's world congress in Moscow. The overwhelming majority reports being at least somewhat satisfied with the experience of the congress, and almost half claim they are "very satisfied." This does not mean, of course, that they are bereft of suggestions for improving such meetings. In fact they named quite a few. Most suggestions underscore the participants' commitment to IPSA world congresses as a significant occasion for scientific communication, and many are cued directly to the development of international scientific networks.

For one of every four participants in the Moscow world congress, however, the occasion was more than a networking affair. They point to some specific way in which attending the congress significantly affected an aspect of their scholarly work. (The validity of the finding is strengthened by the fact that a similar figure obtains in other disciplines as well; in fact, we might reasonably assume that in each case the actual percentage is considerably higher, since people frequently forget the precise source of a new idea or research orientation.) About a sixth (17%) refer specifically to research as the aspect of work most significantly affected.

Although this may at first examination seem to be a modest result, it is of some consequence. Even one truly new idea or orientation can transform the face of a scientific discipline. The fact that informal contacts pursued during the congress were more important in changing these scholars' work than were the papers presented in the panels or even the formal discussions is instructive for the organization of future world congresses.

Notes

1. Some differences between the study conducted by the Center for Research in Scientific Communication of Johns Hopkins University (1968) and ours should be noted. First, of course, the IPSA world congress took place 13 years after the other two. From 1966 to 1979 the context of international science changed substantially, as East and West moved toward détente.

Second, they differed in size and composition. The 18th International Congress of Psychology, held in Moscow, was attended by 4,215 participants from 44 countries. Of these participants, more than a third (36%) came from the Soviet Union and a fifth (20%) from the United States. The 6th World Congress of Sociology, held in Evian, France, was with approximately 2,000 participants closer in size to the IPSA world congress. It was much less dominated than the psychology meetings by scholars from the Soviet Union (4%) and the United States (8%). In contrast, 18 percent of the IPSA world congress participants were from the USSR and 16 percent from the United States.

Third, the large number of psychologists meant using a sampling procedure, whereas the surveys of sociologists and political scientists used a universal sample (albeit, for the latter, only those from nonsocialist states); the surveys of psychologists and sociologists sent separate questionnaires to papergivers and other participants.

Finally, since the purposes of the studies differed, they pursued different lines of inquiry. In some cases, however, the questions were identical: on the impact of the congress on work activity, scientific communication problems encountered, and suggestions for preventing such problems in the future.

2. The unavailability of raw data from the IPA and ISA world congresses makes it difficult to specify this finding in greater detail. If, however, we view the individual response categories in Table 4.5 as percentages of the total set of modifications listed (which, of course, are also percentages), then the point made in the text appears even more clearly. For the psychologists 61 percent and, for the sociologists, 79 percent of the responses point to enhancement, whereas only 27 percent of those made by political scientists do so. By the

same token, whereas 55 percent of the political scientists' responses refer to new scientific knowledge, only 23 and 21 percent, respectively, of the responses made by psychologists and sociologists are in this category. This finding may say more about the relative degrees of acceptance of research paradigms in the three professions than anything else.

3. In fact, a book of abstracts was prepared for the IPSA meetings in Moscow (Merritt and Smirnov, 1979-1981); technical problems, however, prevented the Soviet organizing committee from producing copies for general circulation in time for the congress itself, so they were published afterwards. Such books of abstracts were distributed at the outset of the congresses in Rio de Janeiro (1982) and Paris (1985).

4. English and French are the working languages of IPSA sessions. Normally, simultaneous translation is provided only for plenary sessions. In Moscow, however, all sessions set up by the IPSA program committee (as opposed to those organized by IPSA research committees, IPSA study groups, or individual members) permitted Russian and hence required simultaneous translation. Competition for scarce headsets nevertheless became a source of irritation; on this and related points, see the exchange between Urban (1980) and Merritt (1980).

FIVE
Politics at Play

International scientific congresses, we postulated in Chapter 1, serve a multiplicity of functions for national societies. Not all these functions may be equally manifest. Moreover, the societies' governments may have widely varying perspectives on the importance of the functions they perceive. One may actively court ISCs in the hope of attracting prestige or tourist receipts, while another is blithely unaware that an international congress is taking place on its soil. One government may seek to prevent ISCs from going to certain countries, or bar its scientists from attending such meetings, while another pays scant attention to where its scientists are doing what.

From time to time the political function ISCs ostensibly serve becomes paramount in the thinking of government officials and scientists alike. This is not to say that other functions drop out of sight, and in Chapter 3 we dealt with our respondents' views on some of these other functions. The point is rather that, in certain circumstances, the political context of a world congress outweighs all others. The following are possible scenarios.

1. As during the hottest days of the cold war, implacable hostility between two blocs of states makes joint conferences impracticable and forces nonaligned states to make choices between competing congresses
2. An interest group—ethnic, political, terrorist, or whatever—below the level of the government finds the congress an opportunity to pursue its own goals
3. An issue of international political import seizes the imagination of participants, who turn the congress into an occasion to inveigh for or against the policies of given states
4. Specific quarrels, such as an association's decision to locate its congress in a particular country, restrictions set by the host country

on the association's freedom of action, refusal to grant entry visas to certain participants, or manipulation of the congress itself for the government's own purposes, arise to generate divisive strife.

Some of these actions stem from governments' intentional behavior, while others may not be under their control. Any of them, however, and still other occurrences, can cloud the scientific intent of a world congress, pushing it into the realm of international politics.

We saw in Chapter 2 that the Moscow world congress of the International Political Science Association became a political issue. In any circumstances, holding an international political science congress in Moscow is a political act. In the particular case, hostility to the Soviet Union in principle, the calls for a boycott because of the Soviet treatment of dissident scholars, struggles to secure visas for all bona fide political scientists, the view that the Soviet Union used the congress to score propagandistic points, and procedural aspects of the congress itself concatenated to focus attention on the potential or actual political functions of such congresses.

This chapter explores some central dimensions of that putative political functionality. It looks first at the extent to which political considerations played a role in our respondents' decisions to attend or not to attend the Moscow meetings. It then asks how they viewed IPSA's original decision to hold its world congress in Moscow and, more generally, whether it is ever a good idea to schedule scientific meetings in the Soviet Union. The issue of selecting sites is broadened by raising similar questions about IPSA's choices for its world congresses of 1982 and 1985. Finally, it examines participants' views about the intrusion of international politics in the actual proceedings of the congress.

To Go to Moscow or Stay Home?

As it turned out, a record number of political scientists, 1,466 in all, attended the Moscow meetings, and two-thirds of these were from nonsocialist countries. Doubtless each participant had his or her unique reason for making the journey. So did those who stayed home. With Chapter 3 having dealt with the full set of reasons for attendance or nonattendance, the question of concern here is the extent to which political scientists saw their decision to go or not to go to Moscow as an essentially political act. Among those for whom political considerations were important, was the decision an act of conscience, that is, a personal matter, or did they think that their behavior would have an impact on national governments engaged in the global arena?

Unheeded Call to Boycott

More specifically, did political scientists in nonsocialist countries, especially in the United States, boycott the IPSA world congress in Moscow for political reasons? If so, then we might expect that those who did not attend the meetings would cite political factors as the basis for their decision. To ascertain this we asked all respondents, registrants and nonregistrants alike, first whether or not they attended the Moscow world congress and, second, why they had made the choice they did.

Political reasons do not loom large in the responses of those who stayed away from the Moscow meetings. In fact, of the 62 reasons cited by those who registered for but did not attend the world congress (with multiple responses recorded separately), only 6 (10%) refer to domestic political conditions in the Soviet Union (2), its undemocratic government (2), the economic benefits that the congress would give that government (1), and the red tape entailed in getting to Moscow (1). None mentions Soviet foreign policy. The remaining respondents point to personal as well as scientific and disciplinary considerations. Asked what the *primary* reason for their decision was, an even larger fraction (91%) cites personal or scientific reasons, while only 9 percent mention political considerations.

The distribution of responses is similar for North Americans who did not register to attend the Moscow world congress. In this case, more than 11 in 12 (92%) give personal or scientific reasons for their decision (including the response that they had not heard about the world congress or were simply not interested in attending it), while only 8 percent refer to domestic or international political grounds. Half of these (4%) cite what they see as the repressive or nondemocratic character of the Soviet regime and its violation of human rights. Two percent give the Soviet government's foreign policy as the reason for not attending the congress, and another 1 percent say that they did not want to contribute to a Soviet propaganda effort.

Differences among the various subgroups in the samples exist but are insufficiently interesting from a statistical point of view to warrant much attention.[1] To put it in more technical terms: Applying the X^2-test, we were unable to reject the null hypothesis of "no difference" between nonregistrants and nonparticipant registrants in North America ($X^2 = 3.27, p > .10$), between nonparticipant North American and other registrants ($X^2 = 3.00, p > .20$), or between nonparticipant registrants from the United States and elsewhere ($X^2 = .13, p > .90$). What these relationships mean in practical terms, based on the tau-beta measure of association developed by Goodman and Kruskal, is that, even in the strongest case, between North American and other registrants, knowing the respondents' home base improves by only 3.2 percent our ability to predict what reasons they would give for not attending the Moscow meetings (T_b = .032). Knowing if a nonparticipant North American is a registrant or nonregistrant improves his or her predictability by 0.6 percent; and knowing whether a registrant comes from the United States or elsewhere does so by less than 0.1 percent.

In short, it was not political but primarily personal and other considerations that kept political scientists away from Moscow in 1979. This is not to ignore the fact that some saw their decisions politically motivated, still less to question the intensity of the feeling of those who shunned the Moscow meetings. The data merely show that the number of those expressing political motivations was relatively small, and suggest that their arguments did not impress many of their colleagues. Whatever the heat and smoke surrounding the matter, we can hardly speak of a serious boycott of the Moscow world congress.

Anticipated Impact of Attendance

Well, then, did political scientists think that by going to Moscow in 1979 for the IPSA world congress they would have some kind of political impact? If we look at the reasons they gave for going, then the answer to this question is a resounding "No." The proportion of participant registrants stating that their presence would make a political difference is even smaller than that of nonparticipant registrants who averred that their absence would do so. Only 2 percent of the responses given by those who went to Moscow cite either the congress's putative effect on the domestic political situation in the Soviet Union or else its more general international political consequences; about 1 in 40 participant registrants (2.6%) giving a primary reason for attending the meetings cites such considerations.[2]

Potential for Disruption

We can learn a bit more about the political scientists who expressed diverse reasons for attending or not attending the Moscow meetings by looking at their position on the five scales of professionalism and internationalism discussed in chapter 3 (see also Appendix C). As a matter of convenience, and since the data presented earlier reveal no substantial differences between them, we shall consider as a single set of responses those made by both participant and nonparticipant registrants. Table 5.1 shows separate breakdowns for *all* reasons given by those filling out our questionnaires (with as many as four responses coded per questionnaire) and the *primary* reason identified by each respondent. To facilitate comparison we shall use for each scale standardized

Table 5.1 Reasons Given for Attending or Not Attending the IPSA Moscow Meetings
Standardized Mean Scores for Scales

Scale	All Reasons Given		Primary Reason Given	
	Political	Nonpolitical	Political	Nonpolitical
Professional Status	6.1	5.9	6.9	6.0
International Competence	6.6	5.9	6.9	5.9
Professional Activity	5.1	5.9	5.7	5.9
International Activity	5.0	5.3	5.1	5.3
IPSA Experience	2.4	1.8	2.9	1.9
Average	(5.0)	(5.0)	(5.5)	(5.0)
Number of responses	19	743	13	362

Note: The actual distribution of scores on each scale was transformed into a standardized, 10-point scale by dividing the mean score of the scale by the number of points in it, and multiplying the quotient by 10.

scores ranging from 0 to 10 (Table 3.1). Table 5.1 indicates that, by way of example, for the category "all reasons given," the mean level of professional status held by those giving a political response (6.1) is slightly higher than the status of those responding in nonpolitical terms (5.9).

Several points of interest appear in Table 5.1. The most important is that respondents emphasizing political factors in their decisions on whether or not to attend the Moscow world congress generally enjoy higher levels of professional status, international competence, and IPSA experience than do those stressing personal, scientific, or disciplinary reasons. By contrast, those who underscore their scientific or disciplinary interests are more active both professionally and internationally.

These data point to a dilemma faced by those who would organize congresses for IPSA or other international scientific associations. On the one hand, the more active members tend to focus their attention on personal, scientific, or disciplinary aspects of international meetings and play down any political implications. But, on the other hand, the politically oriented scholars, though few in number, are because of their status and experience likely to be taken seriously even if their advice is not always followed. So long as an association holds its world congresses in noncontroversial countries this tension is not likely to be problematic. The dispute that broke out in summer 1978 among some members of the American Political Science Association nonetheless reveals that the tension, once unleashed, can be virulent and even disruptive.

Putting together the entire set of responses, presented both here and in Chapter 3, to the question of why political scientists did or did not visit Moscow in 1979 for the IPSA world congress points to a consistent pattern. Those who thought about it were attracted to the idea for essentially professional reasons (although tourism also played a role). What kept some of them from going were such personal factors as a lack of time, unavailability of funding, or conflicting commitments. The desire to make a political point either one way or another about the Soviet Union was important for only a very small minority. Even so, the professional standing and experience of this minority contain a potential for disruption to which IPSA organizers must be alert.

Decisionmaking About Sites

The previous section discussed the views of our respondents on an actual situation: whether or not they went to IPSA's world congress in Moscow and the reasons for their decision. Suppose, however, that we phrase the question in a somewhat more abstract vein. Given the international situation in the mid-1970s, was the decision of the International Political Science Association to hold its triennial world congress of 1979 in Moscow a good one? More generally, is it a good idea ever to hold scientific meetings in the Soviet Union?

Special Nature of Moscow as a Site

Looking retrospectively at the decision of the IPSA executive committee to hold a world congress in Moscow, by far the majority of the political scientists we questioned (80%) thinks it a good one. Again very little difference exists among the various subsets of the samples. The use of X^2-tests reveals statistically insignificant distinctions in the views expressed by those registrants who went to Moscow as opposed to those who did not, by North American registrants as opposed to nonregistrants, by North American as opposed to other registrants. Other tests indicate that knowing which category the respondents are in does not improve by much a more random procedure for predicting how they would respond to the question (T_b = .006, .005, and .007, respectively).

Political reasons play a somewhat stronger role in views on IPSA's decision than they did in the respondents' personal decision to attend or not to attend the Moscow meetings. Certainly, scientific and disciplinary reasons prevail in both cases. But about 4 in 10 respondents, irrespective of whether they think IPSA's decision good or bad, give political explanations for their views. (Proponents and opponents distribute their reasons among four general categories—scientific, personal and other, disciplinary, and political—about equally: $X^2 = 1.10, p > .70; T_b = .0005$.)

Those citing political considerations are almost four times more likely to anticipate positive political consequences than negative ones (Table 5.2). Among the 371 reasons given by respondents who think the decision to go to Moscow correct, 142 (38%) refer to domestic and international politics. Of particular importance are the prospects for international understanding that the congress might provide (49 responses, or 13 percent of all positive reasons given), expected positive social effects such as contributing to opening up or liberalizing Soviet society (8%), improved East-West communication (6%), and the need for the rest of the world to avoid isolating itself from the Soviet Union (4%). Respondents thinking the decision to go to Moscow incorrect are almost equally likely to give political reasons (38 of 106 negative responses, or 36%). The idea that an IPSA world congress could in some way help to legitimize a repressive regime weighs heaviest in their minds (17%). Almost as many cite the possibility that the Soviet Union would use the congress for propagandistic purposes (15%).

General Idea of Scientific Meetings in the Soviet Union

Roughly the same picture emerges from our examination of responses to a second, more hypothetical question: In general, is it a good idea ever to hold scientific meetings in the Soviet Union? In this case 78 percent of the 328 registrants giving definite responses (another 82 persons gave contingent responses) answered in the affirmative. Differences among participant and nonparticipant registrants from various regions and nonregistrants exist but are slight, and point in no particular direction. And roughly the same kinds of reasons for holding or not holding such meetings are offered as are used for explaining views about the correctness of IPSA's decision to go to Moscow in the first place.

A glance at Table 5.3, however, reveals another dimension of attitudes toward holding scientific meetings in the Soviet Union. Respondents who think the IPSA decision of 1976 wrong score higher on each of our five scales than do those who consider it correct; and those generally opposed to holding scientific meetings in the Soviet Union score higher on four of the five scales. The discrepancies are not great. They nonetheless underscore the point made earlier about the likelihood that political scientists will take seriously the views of opponents of such meetings if for no other reason than the fact that the latter are both active and experienced, and they enjoy high status among their professional colleagues.

Table 5.2 Political Grounds for Evaluating IPSA's Decision on Moscow

	Decision was			
	Correct		Incorrect	
	N	%[a]	N	%[a]
Reasons referring to situation in the Soviet Union				
Broad social effects: opening up or liberalizing Soviet society	28	8		
Rejection of proposition that IPSA world congress "endorses" régime	4	1		
Positive effects vis-à-vis dissidents	2	1		
Probable Soviet use of congress for propaganda or other political purposes			16	15
Repressive, totalitarian character of Soviet régime			10	9
Congress legitimizes régime			8	8
Red tape; régime discourages participation			3	3
Total	34	9	37	35
Reasons referring to international political relations				
Foster détente, international understanding, reduction of tensions	49	13		
Importance of East-West or international communication (among countries)	23	6		
Rejection of isolation or boycott strategy as ineffective or damaging	15	4		
Acceptable under prevailing conditions	12	3		
Recognition of importance of the Soviet Union in the international system	6	2		
Congress benefits West more than East	2	1		
Positive effect on Soviet foreign policy	1	-		
Aggressive nature of Soviet foreign policy			1	1
Total	108	29	1	1
Total political references	142	38	38	36

[a]Based on total number of consistent responses (371 positive, 106 negative).

Note: For the broad categories of Soviet domestic situation vs. international political relations, $\chi^2 = 67.66$, $p < .001$; $T_b = .376$.

Table 5.3 Characteristics of Proponents and Opponents of Meetings in the USSR
Standardized Mean Scores for Scales

Scale	IPSA Decision in 1976 Correct		Generally Favor Meetings in USSR	
	Yes	No	Yes	No
Professional status	6.0	6.1	6.0	6.4
International competence	6.0	6.1	6.0	5.8
Professional activity	5.8	6.0	5.7	6.5
International activity	5.3	5.5	5.3	5.4
IPSA experience	1.0	1.3	1.6	2.6
Average	(4.8)	(5.0)	(4.9)	(5.3)
Number of responses	328	84	255	73

Note: The actual distribution of scores on each scale was transformed into a standardized, 10-point scale by dividing the mean score of the scale by the number of points in it, and multiplying the quotient by 10.

Future World Congress Sites

Bearing in mind the fact that 80 percent of our sample of registrants think that the IPSA decision to go to Moscow in 1979 was a good one, we might ask how this level of approval compares with views on other sites. At the time of the survey, the next triennial world congress, held in Rio de Janeiro in August 1982, was almost 2 years off; and members of IPSA's executive committee were already moving toward placing the world congress of 1985 in Paris. How did our respondents react to Rio de Janeiro and Paris as future sites?

Political considerations play only a small role in perspectives on the Rio meetings in 1982. More than seven in eight (88%) of those who expressed their views indicate that they were thinking of going to Rio. Only 5 percent of them refer to political conditions (of which the most frequently mentioned deal with human rights issues in Brazil). None of the 47 respondents who think that they would not attend the Rio meetings mentions political conditions. About five in six of those thinking they might attend and of those doubting they would attend the meeting in Rio indicate that their decision will rest on personal considerations, especially the availability of funding.

A question about the appropriateness of Rio de Janeiro as the site for the

IPSA world congress in 1982 elicited similar responses in some respects. Again, seven in eight (88%) of those answering express positive views. About a tenth of those respondents give politically relevant reasons, the most important of which is the view that holding the congress in Rio would recognize the role of Brazil (or Latin America) in the international political system. A handful (1%) think such a step might help open the country to liberalizing influences. Substantially more responses stress scientific (17%) or disciplinary (37%) advantages of Rio as a site. Many of these (30%) underscore the hope that holding the meeting in Rio would help to internationalize political science through exposure to more Latin American scholarship. A few simply reject the idea of political criteria for selecting sites, insisting that only scientific considerations are important or that, even if political problems abounded in Brazil, "nobody is perfect." (In the words of one respondent, "Maybe the penguins in Antarctica do not violate human rights but. . . .")

Among those who think Rio de Janeiro an inappropriate site for an IPSA world congress—46 (or 12 percent) of the 376 respondents who addressed this question—14 (31 percent of the negative responses) find political conditions in Brazil objectionable; and another 10 (22%) express doubts about whether the quality of discourse expectable at such a meeting would warrant the time and expense involved in going there. Some respondents feel that holding the congress should be made contingent on political conditions not becoming worse or on guarantees of free access to and free discussion at the meetings. The largest percentage (36%) of negative responses, however, points to the inconvenience and cost of flying to Rio de Janeiro.

Paris as a potential site for the IPSA world congress in 1985 attracts almost the same level of support (87%). A few (5%) of those favoring the site refer to such political factors as France's democratic government and its importance in international politics. Many more (46%) cite rather such positive characteristics as Paris's charm, central location, cultural amenities, and cosmopolitanism. Variations on the "I love Paris" theme are bolstered by the argument that such an attractive and acceptable site would surely encourage maximum participation. Another sixth (17%) points to the capabilities of the French to host an international conference. Only one respondent of an opposing view mentions any political factors that IPSA organizers should take into consideration before settling on Paris as the site for the world congress in 1985—that the hosts would use the congress to enhance their own prestige. Most of those opposed simply prefer some other site to Paris.

A comparison of grouped responses to questions about sites—whether or not IPSA's decision to go to Moscow was correct, and the appropriateness of Rio de Janeiro and Paris as sites for IPSA world congresses—illustrates some of the points suggested above (Table 5.4). First, the degree to which

Table 5.4 Alternative Sites for IPSA World Congresses: Views of Registrants at the Moscow Meetings

	Moscow		Rio de Janeiro		Paris	
	No	%	No	%	No	%
Appropriate	328	80	330	88	355	87
For political reasons	(143)	(38)	(39)	(10)	(19)	(5)
For other reasons	(229)	(62)	(338)	(90)	(348)	(95)
Inappropriate	84	20	46	12	53	13
For political reasons	(42)	(37)	(19)	(32)	(1)	(2)
For other reasons	(72)	(63)	(40)	(68)	(51)	(98)
Other/No response	8	–	44	–	12	–
Total	420	100	420	100	420	100

Note: The calculation of percentages omits "other," "don't know," and inconsistent responses; multiple responses were recorded.

respondents think in political terms about the three sites stands out. In no case do political considerations dominate their thinking. Second, where such considerations play a role, it varies by site. Moscow raised the hackles of a substantial minority of our respondents. This is true even among those who approve of IPSA's decision in 1976 to go to Moscow. Rio is not unproblematic, but opponents are more apt to use political arguments to justify their position than are proponents to justifying theirs. Paris is almost uncontroversial from this point of view. (Respondents who did not register to attend the Moscow meetings express views about Rio de Janeiro and Paris that are similar to those of registrants.) Third, the fact that a substantial number of registrants who disapprove of the Moscow site attended the congress anyway, together with the fact noted earlier that those registrants who did not attend withdrew for essentially nonpolitical reasons, indicates that political scientists generally place their political reservations about sites behind other considerations when deciding whether or not to attend an IPSA world congress.

More generally, the data support the idea that sites for IPSA's triennial world congresses should be rotated on a worldwide basis. This is not only a political matter deriving from the national associations' desire for equal treatment. Congresses held at different sites serve different purposes. One located in a nontraditional site may help to develop political science in that region, and give political scientists from the rest of the world an opportunity both to learn about a very different political system and to check their research findings against a different real-world context. The preference indicated by respondents for the site of the next IPSA world congress after the Paris meetings of 1985 was Tokyo, followed by New Delhi and then Beijing. Over half of the North American registrants and 38 percent of those from other Western countries mentioned locations in the Third World. The comments as a whole suggest that a system of rotation providing for geographical, cultural, and political diversity has a sound intellectual basis as well as a more political one.

Politics on the Floor

Before they went to Moscow in 1979, many political scientists from the nonsocialist world questioned the extent to which ideological or East-West disputes would creep into the congress itself. Doubtless few feared the likelihood of sharp exchanges on an intellectual plane—the confrontation of Marxism-Leninism with other philosophies of politics. The problems they foresaw lay more in stylistic elements that might stifle free discussion or the possibility that the Soviet government would manipulate the congress for its own purposes. To what extent did participants feel that these problems materialized?

There is little doubt in our respondents' minds that the Soviet government sought to use the IPSA world congress for propagandistic purposes. Three in ten (30%) agree strongly and another 5 in 10 (49%) agree somewhat with such a complaint. Only a fifth disagree somewhat (13%) or strongly (8%). Placing the responses on a four-point scale (disagree strongly = 1 to agree strongly = 4) yields an average score of 3.01, that is, very close to the response category of "some agreement." North American participants, with an average score of 3.10, are sharper in this judgment than others, but political scientists from other Western countries (2.97) and the Third World (2.83) are not far behind. Position on our various scales of professionalism and internationalism has a negligible impact on perceptions of Soviet propagandizing: The only statistically significant finding indicates that the more professionally active tend less to see any propaganda ($r = -.08, p < .10$).

There is nonetheless considerable divergency in interpreting this perception. Asked why they say the Soviet government sought to use the IPSA meetings for propagandistic purposes, over two-fifths (42%) of the 254 respondents who answered simply cite examples or define what they mean by the term. An equal percentage (42%) respond, in effect, "So what?" It was to be expected, they note, or others do it as well, or the Soviet propaganda was useless and even counterproductive. Fifteen percent evaluate the propaganda negatively. Only a third of these—5 percent of the entire sample—argues that it hindered effective communication at the congress.

But what about the intrusion of politics into individual sessions at the congress? To ascertain views on this and related matters, we asked respondents both to identify what they consider to have been the best panel and the worst panel they attended and to explain their judgment. Some 195 respondents cited a "best" panel and gave us 246 reasons for thinking so, while 123 respondents offered 152 reasons for considering a particular panel the worst. As Table 5.5 indicates (and this finding corroborates the argument of the previous chapter), the main reasons for singling out sessions for praise focus on the interest of the topic and the quality of the papers, with the organization and conduct of the session following behind. In third place are responses pointing to East-West interaction (15%) and, more rarely, the fact that politics did not intrude (3%). By contrast, in identifying why a particular panel was the worst, respondents are less apt to list scholarly matters such as interest and quality of papers or organizational issues than reasons that carry a political overtone. These last include feelings that the sessions were inhibited generally by ideological considerations (21%) or more specifically by either the behavior of Soviet participants (13%) or Soviet propaganda (12%). All in all, only about one in four of those citing reasons for a "worst" panel—a group constituting about 8 percent of all respondents who attended the Moscow meetings—found Soviet individual or governmental actions a barrier to scholarly communication.[3]

Another question approached the same issue from a slightly different direction. We asked respondents if they had encountered scientific communication problems that they believe should be corrected before the next IPSA world congress. Of the 369 respondents, less than half (46%) answered affirmatively. We have touched on some of these findings earlier. The relevant point here is that, asked to specify what had hindered effective scientific communication in Moscow, only one in eight (13%) of those making specific complaints mentions such "political" issues as interference by the host country, restrictions on free discussion in the panel sessions, or the desirability of changing the site. The number giving political responses amounts to less than six percent of all respondents who went to Moscow in 1979.

As far as their overall evaluation of the IPSA world congress is concerned, we saw that most respondents are reasonably satisfied. Placing the responses on a four-point scale (disagree strongly = 1 to agree strongly = 4) yields an average score of 3.38, that is, almost midway between "somewhat satisfied" and "very satisfied." Significantly, however, the more likely respondents are to note the propagandistic uses made of the congress by the Soviet Union, the more disgruntled they are about the congress as a whole ($r = +.28, p < .001$). Those who strongly agree with the charge of propaganda

Table 5.5 Reasons Cited for Terming a Panel Session the Best or Worst

	Best Session		Worst Session	
	N	%	N	%
Subject matter, paper quality	119	61	34	28
Organization/conduct of session	74	38	36	29
East-West considerations	37	19	61	50
East-West interaction	(29)	(15)		
Unpolitical	(5)	(3)		
Controversial	(3)	(2)		
Inhibited ideologically			(26)	(21)
Soviet speaking monopoly			(16)	(13)
Soviet propaganda			(15)	(12)
Too political			(4)	(3)
Other	16	8	21	17
Total	246	126[a]	152	124[a]
Number of respondents	195		123	

[a]Multiple responses permitted.

(N = 102) have an average evaluative score of 2.09. The remainder have quite positive scores: Respondents agreeing somewhat (N = 165), disagreeing somewhat (N = 46), or strongly disagreeing (N = 28) have scores of 3.44, 3.61, and 3.68, respectively.

Although some participants found political problems besetting the IPSA world congress in Moscow in 1979, they generally do not see them as overwhelming. Doubtless few are oblivious to the pervasive undertone of East-West politics that characterized the meetings. Most, however, find it at worst a nuisance rather than a barrier to scientific communication; and some report finding it exhilarating, a reason in itself for political scientists to meet in the Soviet Union. Only a few concentrate their criticism on Soviet scholars or their government for violating the canons of normal scientific discourse.

Impact of Politics

The data surveyed in this chapter thus suggest that the political impact of world congresses is marginal at best. This is not to say that political issues are nonexistent. First, the process of selecting a site for an IPSA world congress is at best a complicated one, and occasionally one fraught with conflict and tension. It entails negotiations ranging from persuading a national association to host the meeting all the way to laying down and enforcing the ground rules for its conduct. As the operations of the association expand in coming years, we can anticipate that the negotiating process will become ever more complex. Even so, provided that the association sticks to its basic principles—principles developed during almost four decades of activity no less than those outlined in the late 1970s—nothing inheres in this process to make it necessarily disruptive.

Second, although most members of the association prefer to see decisions about sites for world congresses as an emotionally neutral issue, one restricted to questions of how best to facilitate the international flow of scientific information, subgroups within a discipline can always challenge the political morality of a particular choice. The same is potentially true about decisions on invitations and visas, designation of speakers for plenary sessions, and a host of other matters. This disruptive potential grows as the political morality at issue captures the attention of an ever broader palette of scholars, especially if they include persons recognized as disciplinary leaders. Barring a truly momentous incident, however, few political scientists seem willing to forego the advantages derived from world congresses for the sake of political issues, even as important as those that arose in 1978-1979.

Third, governments of host countries may use congresses to toot their own horns or to achieve other domestic (or international) political purposes. But how important is this? Participants in the Moscow world congress were aware of the propagandistic purposes to which the Soviet government sought to put the meetings, and many found this behavior irritating. Relatively few, though, saw it hindering scientific communication. Many more considered it useless or counterproductive. If the primary Soviet goal was propagandizing political scientists from nonsocialist countries, then their relative imperviousness must have made the IPSA world congress in Moscow a failure from the Soviet viewpoint. Moreover, given the welter of other events competing for world attention, we must question whether the costs of politicizing a world congress are worth any political benefits in terms of a host country's foreign policy.

We would of course be hardpressed to claim that the Soviet government's efforts were fruitless in the broader framework within which it works. Televising a Soviet-organized session on Lenin, which happened to take place on the congress's last afternoon, and claiming that the session was the culmination of the congress's endeavors may well have shored up the government's overall information program for its own citizens and perhaps others. Soviet press reports emphasizing the open exchanges that took place may symbolically have supported the Soviet Union's stress on its contribution to détente. It nevertheless seems that, for the international political science community, such measures fell on deaf ears.

It would also be incorrect to suggest that, inevitably, political scientists' distaste for politicized world congresses will lead them to ignore genuinely contentious issues. In fact, in the midst of the tug-of-war between those who doggedly insisted that politics has no place in an international scientific congress and those who were equally insistent that, whether one liked it or not, politics permeated the Moscow congress's organization, proceedings, and implications, occasions for potential disaster severely challenged IPSA's decisionmakers:

1. Had the proposed boycott by U.S. political scientists taken place, it might well have split the association itself.

2. Had the Israeli visas failed to arrive in time—and they literally reached the participants only hours before they were scheduled to leave for Moscow—IPSA's executive committee would doubtless have honored its commitment even on very short notice to cancel the congress.

3. Had bona fide political scientists from South Africa or Taiwan[4] registered for the congress, Soviet authorities might well have balked at issuing visas; and, while this would not have been a violation of IPSA's resolution of August 1978 (since neither country was a collective member of the association), it could have reopened the question of Soviet good faith.

4. Had the IPSA leadership insisted that a well-known Soviet dissident be permitted to address a working session and the Soviets persisted in denying him access to the halls of Moscow State University, where the congress was held, the effect would have been chilling.

5. Had any of the participants initiated a politically motivated demonstration in an IPSA session or elsewhere, or had the Soviet police been less lax about permitting participants to visit Soviet dissidents in their homes, untoward confrontations might have resulted.[5]

All these events were at one time or another very real possibilities. That they did not occur was due to the desire of Soviet authorities, IPSA officers, and the participants themselves to hold a "normal" scientific congress in Moscow. The Olympics of 1980 and 1984 nonetheless indicate that circumstances can change. It is quite likely, for instance, that an IPSA world congress in Moscow scheduled for 1980, that is, after the crisis in Afghanistan, would have collapsed entirely or at least in large part. It is just as likely that political considerations will affect participation in future IPSA congresses.

As new national associations join IPSA, and as an international ethic calling for a redistribution of the world's resources (including scientific information) grows, issues of national pride and international equity become increasingly important. A main argument for holding the congresses of 1979 and 1982 in the Soviet Union and Brazil, respectively, was to recognize the contribution made to international political science by these two countries. We may expect political scientists in still other countries not yet tapped for world congresses to feel that their own countries should be considered soon.

The issue of equity raises some implications for IPSA's future. On the one hand, to the extent that the argument in favor of distributing world congresses equitably runs counter to IPSA's need to have an invitation with financial guarantees from a national association, the prospect is for a serious reconsideration of the nature and organization of world congresses. On the other hand, the association may be able to deal with this problem by ensuring that its meetings between the triennial world congresses are distributed more evenly among the member associations. While this is possible for meetings of the IPSA executive committee, it may not be so for its research committees and study groups, the membership of which is normally not evenly distributed in this way.

These lines of argument point to one central fact: The political functionality of world congresses, albeit marginal to date, lies only slightly below the surface in any international scientific setting. The selection by an international scientific association of a site for an international meeting is essentially a political process. And so are such matters as the behavior of groups with points to prove no less than government attitudes toward ISCs held on their own soil or elsewhere. These are political processes that deserve further analytic attention.

Notes

1. This calculation rests on a general division of responses into political, scientific, and personal categories; respondents did not cite disciplinary reasons for going or not going to Moscow.

2. To be clear, we must add that one might believe that one's presence could have a political impact but not mention this as a reason for going to Moscow. Similarly, respondents might give political reasons for not going (e.g., as a matter of personal political ethics) without believing that their presence or absence would have any political impact on the Soviet Union.

3. A different way of looking at Table 5.5 would note that, of the total of 398 comments on both the "best" and "worst" panels, only 31 (8%) specifically criticized Soviet behavior.

4. At least one South African political scientist expressed interest in participating in a session at Moscow, but evidently did not follow up on the idea; and, in the week before the congress was to convene, a representative of Taiwan's Chinese Political Science Association telegraphed a U.S. scholar to regret that it was unable to send a delegate to Moscow, "owing delay of secretariat in issuing registration certificate." For a similar problem faced in 1986 with respect to the World Congress of Archeologists, see Walker (1986).

5. The only untoward incident known to us that involved an IPSA participant occurred on the day after the congress closed. A Canadian lawyer

and professor of law, who reports having gone to Moscow because it afforded him an otherwise unavailable opportunity to meet with Soviet dissident scientists and to present their case to Soviet officials, apparently violated Soviet law by journeying without formal permission beyond the city limits of Moscow; 3 hours after his arrest the police placed him on a flight to Canada (Giniger, 1979). Other congress participants who stayed on in the Soviet Union reported that official tolerance toward visits to dissidents ceased, sometimes rather abruptly, when the congress ended.

SIX

Functions of International Scientific Congresses

For close to four decades the International Political Science Association has sponsored triennial world congresses in the belief that they help political scientists meet certain goals. The prospect for the future is that IPSA will continue to organize such congresses. Our survey of political scientists, besides telling us something about those who did and those who did not attend the IPSA world congress of 1979, explains why they went to Moscow and the functions they see these meetings serving. In the data are implications for how the association might better plan for future world congresses and, more generally, how it might help political scientists to communicate internationally. This chapter outlines what we see as the key functions of international scientific congresses before turning to some suggestions for enhancing their usefulness.

Individual Political Scientists

Participants as a Scientific Elite

The comparative data on North American political scientists show clearly that participants in IPSA world congresses form part of the discipline's leading stratum. Compared to those who did not sign up to attend IPSA's world congress in Moscow, registrants have stronger records of scholarly activity, are more adept internationally, and enjoy higher professional status.

This fact is not surprising given that, in any academic discipline, activity and status generally reinforce each other. In this sense, international scientific congresses are merely another arena in which scientists can be active. Those who participate typically have histories of thriving on activity, and we might well suspect that they consciously seek out appropriate arenas.

Participants thus constitute a self-selected group. Scientists with other goals evidently decide not to spend their resources in attending ISCs or, as in the case of a substantial number of the nonregistrants surveyed here, are simply not alert to the fact that such meetings take place.

An international scientific association's procedure for choosing congress participants fundamentally reinforces the principle of self-selection. However broadly congress organizers cast their nets, they lean toward individuals with records of achievement or newer members of the profession whose promise is certified by those already enjoying status. Two aspects of the process of developing ISC programs—far from unique to the International Political Science Association—are germane here:

1. A strong emphasis on multinational participation means the possibility that the person organizing a session is not completely familiar with all scientists throughout the world who are conducting significant research on the specified topic. A simple rule of thumb in such circumstances is to fill the session with highly visible scientists known to the organizer or else younger scientists whom more visible colleagues recommend.

2. An alternative procedure would require the scientist seeking to make a presentation to submit a full-fledged paper that the organizer (or review panel) could then examine for quality. The constraints of time and exigencies of the worldwide communications system being what they are, this procedure is less practicable than it might be for a national association. The result is the temptation to accept and review not finished papers but abstracts, which usually turn out to be statements of intent rather than concise summaries of research completed.

Actual practice usually combines the two procedures. It thus sustains the existing network of scientists (and their likely heirs) while keeping open to outsiders who submit interesting proposals the possibility of membership in the network.

Rewards

Why do accomplished scientists want to attend world congresses? The obvious answer is that they find participation rewarding.[1] Phrased somewhat differently, from the individual scientist's vantage point, a key function of ISCs is to provide rewards. Now, then, what are these rewards? Chapter 3 revealed that our respondents identify them primarily in terms of *direct scientific communication* (Table 3.5).[2] Of signal importance are developing and maintaining contacts with other productive scientists (cited by 32 percent of the registrants), general scientific communication (22%), keeping up with new work, developments, and approaches (13%), and broadening intellectual, disciplinary, and international perspectives (13%). Chapter 4 added that, of the political scientists reporting some major modification in their work as a result of attending the Moscow meetings,[3] about one in three traces it to either the written papers (40%), their presentation (35%), and/or discussion in scheduled sessions (33%)—together accounting for three in five of the responses given.

If, however, direct scientific communication is the sole reward offered those considering whether or not they should invest their time and other resources in attending an ISC, then we (and, presumably, the scientists themselves) must ask whether or not the congress produces the greatest benefit for the costs incurred. It probably does not. Expanded publishing and abstracting services, electronic mail, teleconferencing, and even the telephone would go far toward providing a more cost-effective functional equivalent. Scientists with specific questions on their minds would doubtless find attending an ICS an expensive way to secure answers.

But in fact, as our capabilities for these other modes of scientific communication grow, the triennial world congress of political science seems to be ever more important. It attracts an increasing number of both participants and requests to hold sessions. This suggests that the lack of creative thinking on the part of political scientists is not at fault for failing to find functional equivalents for world congresses, but rather that ISCs provide other, perhaps less easily identifiable rewards.

One such reward is *serendipitous learning*. Bernard Barber (1968: 98) points to the "interesting suggestion" stemming from general sociological ideas and Menzel's (1959) research that scientists "sometimes think in terms of too rational a conception of the communication process; that is, they may be expecting too much from the journals and the formal meetings." Excessively goal-directed behavior can actually be dysfunctional for scientific learning. Participants who fill every minute with sessions to attend and papers to read, while leaving little time free for informal interaction with colleagues, may not be making the best use of the learning opportunities the congress affords them. By the same token, congress organizers who overschedule participants may not be doing them a favor.

This is not to say that such approaches to ISCs are useless. Presenting and listening to papers are important links in the transmission belt of scientific communication. But overscheduling may restrict participants' ability to be receptive to information the importance of which is not immediately apparent. "Scientists cannot always know precisely what they want and merely push a computer button to get it," Barber (1968: 98) points out. "Often, through 'milling around' at meetings, through chance visits, through indirect channels, they get essential information which they can recognize as essential only when they get it." Informal, unplanned communication at ISCs broadens access to a significant international network of political scientists—a network of like-minded individuals with whom an active scientist can talk about current developments, try out new ideas, scout for collaborators in future research, and sniff in the wind for research areas likely to break open in the near future.

Now, scientists do not go to an international congress with the explicit intent to learn serendipitously. Formal opportunities for exchanges of information doubtless dominate their planning—as they certainly do the calculations of administrators who must approve and/or fund the participants' travel requests. Nevertheless, their past experience tells many scientists that a certain probability of serendipitous learning exists.[4] It is the possibility of its reoccurrence that lends vitality to the very idea of attending the congress, and participants may promise themselves to be responsive should they encounter it again.

Related to and in fact enhancing serendipitous learning are the travel, conviviality, and other rewards classified in Chapter 1 under the rubric "personal enrichment." Although the typical respondent filling out a questionnaire received in the mail doubtless plays down the importance of such considerations, possibly in the belief that they may seem less "noble" than scientific communication, our feeling is that, used intelligently, these opportunities constitute an important element of the ambience that invigorates scientific communication. Sharing a pot of coffee at a riverside café or quiet dinner at a good restaurant may well provide scientists with the best occasion for hashing out different conceptualizations or outlining ways to solve a particular puzzle. Lodge's (1984: 238) biting observation— "afterwards, when they are back home, and friends and family ask if they enjoyed the conference, they say, oh yes, but not so much for the papers, which were pretty boring, as for the informal contacts one makes on these occasions"—points to an essential truth. The informal contacts *are* important.

Herein lies a key latent function of international scientific congresses. An ISC structured to facilitate informal communication, besides making serendipitous learning more likely, enhances the contextual framework in which participants seek to understand the scientific information communicated through more formal channels. *Informal communication makes formal communication in ISCs sensible.* Stripped of opportunities for informal communication, congresses are surely less cost-efficient for individual participants than are other means for exchanging information. Informal interaction helps scientists to catch nuances, clear up ambiguities, evaluate the depth and breadth of other scientists, and pursue substantive points until there is, if not agreement, then at least mutual understanding of where differences lie. Although perhaps only poorly articulated by participants, this latent function greases the skids for effective scientific communication.

International scientific congresses perform a second latent function for individual participants. *Their selection procedures confirm individual scientists as accepted (or candidate) members of an international disciplinary elite.* This has a double impact. In the short run, as we have already noted, ISCs reward the active by making it *ceteris paribus* easier for them to obtain places on the program. With invitations in hand, they can approach funding sources, make travel plans, and begin to anticipate some of the other rewards accruing to ISC participants. All this may boost the scientists' egos and lend prestige, both at home and abroad, to their work. Such considerations apparently undergird the desire for scientific communication as the immediate reasons leading scientists attend international scientific congresses.

More interesting is a second, long-range impact: Although scientists may not be aware of it at the time, successful participation in an ISC gives them a claim on the future. For one thing, such participation makes it more likely that they will be asked to join in the next congress and in myriad other activities spinning off from the international association's scientific program. For another thing, successful participation in a congress gives scientists a firmer basis for taking an active role in the international association itself. How scientists react to this associational recognition that they are members of the international disciplinary elite may vary widely, of course. Some will see it merely as another accolade, as confirmation of a status they always knew they deserved. Others will see it as an instrument to accomplish other goals, such as a broadened research agenda or international scientific organizational activity.

Still another latent function of the international scientific congress vis-à-vis participants pertains to their research. *ISCs help set individual research agendas.* IPSA world congresses unabashedly cater to research-oriented political scientists. Its process of selecting papergivers and discussants, we suggested earlier, gives great weight to those who have achieved a name for themselves through their research. Not surprisingly, then, over four-fifths of our respondents who registered to attend the Moscow meetings identify themselves as primarily researchers rather than teachers or administrators.[5] And, of those reporting a significant impact on their work activity, almost three in four (73%) specify research or publications as the type of activity modified while only a fifth (21%) cite teaching.[6] To be sure, we might assume that the confrontation of participants at the international congress with an array of diverse perspectives would stimulate some rethinking about the discipline itself and how it should be presented to students. The point is more that the selection process and the climate at the meetings strongly encourage a preference for research.

What is more, by selecting participants who focus on certain kinds of issues or work with particular paradigms, the ISC "certifies" the centrality of what they are doing. The effect is to tell scientists what kind of research is most likely to earn them a place on the program of the next world congress. Those doing something different have several options. They can change directions, live with the low likelihood of being invited to attend world congresses, or try to participate anyway, in the knowledge that theirs will be voices in the wilderness. Another alternative is to seek to influence the international scientific association's process for making decisions about the content of and participation in future sessions. Given the fundamental inertia of such associations, however, the last row would be a tough one to hoe.

Research Organizations

Our research design did not permit us to delve into the functionality of ISCs for the university, governmental agency, laboratory, or research institute sending scientists as participants. Chapter 1 suggested that such a research organization stands to benefit from ISCs in various ways: making the scientists it employs happier by according them opportunities for communication and travel, acquiring some part of the information these scientists obtain at the meetings, gaining the organization recognition for its research program and personnel, and attracting potential recruits. All these may augment the research organization's effectiveness and reputation. Although open-ended items on our questionnaire provided opportunities for respondents to volunteer pertinent views, in fact none did so. Research specifically designed to address this issue might survey both scientists and administrators at a sample of research organizations, and trace through (for example, by analyzing citations) the concrete impact attending an ISC has on the organization's published work. Such research could contribute significantly to our understanding of the role of international scientific congresses in advancing knowledge.

Scientific Associations

Chapter 2 noted our expectation that political scientists filling out the questionnaires would not be especially alert to the functionality of international scientific congresses for disciplinary associations *qua* associations. Only a few, we felt, would have sufficient knowledge about the activities and needs of their national association or IPSA to provide anything more than guesses about the associational impact of the Moscow meetings. This assessment turned out to be accurate. Asked about the chief functions of ISCs for the political science profession (Table 3.5), more than twice as many registrants (63%) emphasized the exchange of information and other scientific values as those who cited the entire range of disciplinary values:

enhancing the prestige of the discipline, creating a sense of professional identity, or institutionalization (11%); broadening the discipline's international dimensions or reducing parochial perspectives (10%); and improving the quality of political science, assessing the state of the art, or identifying new trends (9%).

International Institutionalization

And yet, reading between the lines of such findings, and taking into account the variety of issues that arose in organizing and carrying out the Moscow meetings, we find at the international level a number of associational functions meriting attention. The most obvious of these derives from the fact that the congress actually took place. IPSA issued a call for a world congress to be held in the Soviet Union and answering it were political scientists from around the world, including a substantial number from a country where some disciplinary leaders urged their colleagues to boycott the meetings. IPSA's organizing committees absolutely refused to permit political considerations to impede their progress; and individual scientists attending the meetings refused to be put off the track by what some saw as the intrusion of cold-war politics into the proceedings. The association thus met head-on and weathered the disruptive storm posed by a politicized environment.

Another potentially disruptive storm raised by the Moscow meetings forced the association to re-examine its intellectual and organizational priorities. IPSA had traditionally focused on North American and Western European concerns. To be sure, from its earliest years the association included collective and individual members from socialist countries and the Third World, its governing bodies embraced their representatives, and from time to time IPSA-sponsored roundtables met in those parts of the world. However, viewed in the framework of IPSA's overall activities, such countries were peripheral to the association's main thrust. Holding a world congress in Moscow meant that IPSA had to face its worldwide responsibilities. It had to come to terms with the possibility that scientists with non-Western perspectives would dominate the proceedings, and it had to deal seriously with such matters as blocked currencies, travel restrictions, limited access to reproduction equipment, and the like.[7] That it did so fairly effectively again indicates that IPSA had come of age internationally.[8]

However we measure its success, then, *the minimal fact that the congress took place without major incident bolstered substantially the disciplinary standing of the International Political Science Association*. The association's ability to master significant problems, if not to resolve them permanently, merely enhanced its authority. Faced with challenges to the association's control over its own meetings, the bulk of political scientists from at least nonsocialist countries identified themselves with the association, its aims, and its policies. And, of course, the icing on the cake was the fact that participants rated the congress highly successful in other regards. By far the majority went home feeling positive about effective scientific communication, fruitful contacts, lively interchanges, and so forth. If any doubt existed before 1979 that IPSA was an effective international scientific association with a global scope, the Moscow meetings put that doubt to rest.

Reward Structure

We spoke earlier of the functionality of international scientific congresses in providing manifest and latent rewards for individual participants. Through their programmatic decisions those organizing ISCs accord recognition to certain lines of research, confer status on individual researchers, and enhance the latter's access to the international disciplinary elite. The way in which such decisions are made and the rationale used to justify them constitute the international scientific association's reward structure. From the vantage point of the individual, the most important outcome of an ISC's reward structure may be the degree to which it reinforces the discipline's more general reward structure, that is, makes it more likely that activities meriting praise today will also do so tomorrow, that those enjoying high status today will continue to enjoy it tomorrow. What is important from the vantage point of the international association is *its ability to implement a reward structure* that individual scientists and national associations accept as legitimate. This in turn *contributes to the legitimacy of the international association.*

This latent function has identifiable consequences. First, to the extent that it strengthens the often inchoate cycle of professional activity and rewards, it clarifies that cycle, that is, makes it clearer what criteria are to be used in judging contributions, no less than who belongs to the international disciplinary elite and who does not. Such clarification can be functional for the association by making it easier for program organizers to identify and invite the discipline's leading stratum to participate in future ISCs. The participation of these disciplinary leaders in turn lends the meetings an added cachet of respectability, which then makes it more possible for the association to attract other participants and perhaps even to raise funds for travel and other purposes. Carried to an extreme, however, this self-reinforcing cycle could lead to an international disciplinary class structure that could doom the association to the dustbin of history.

Second, an international associational reward structure lends a degree of stability to the international scientific association itself. Whether this is good or bad depends on the use to which the association's officers put the stabilizing influence. They could use it as a basis on which to expand the association's services to and impact on the discipline—for example, redoubling efforts to raise travel funds for Third World participants, or encouraging the development of political science in countries where it is now weak or absent. Alternatively, they could capitalize on the stabilizing influence by providing more perquisites (such as increased international travel) for themselves and/or new staff and buildings for the association.

Related to these consequences is a third one linked to the discipline's future. A stable international reward structure gives the association controlling it substantial influence over the process of recruiting and socializing younger scientists. It tells them (and their employers) that an important path toward accomplishing their goals—access to key international communication networks, high professional status, and so forth—runs through the association and its activities, including congresses. Such a message will also implant in the minds of these younger scholars the scientific and other norms (for example, a strong orientation toward research) of the international disciplinary elite that makes decisions about participation. The negative side of this control is the possibility that the latter will seek merely to reproduce scientists like themselves, thereby diminishing the chances for intellectual innovation or upward mobility on the part of those who do not fit today's molds.

Political Functions

National Societies

The functionality of international scientific congresses for national societies, as represented by their governments, does not loom large in our respondents' imagery. They are aware that the country hosting such a congress may benefit in the form of the attention it receives, enhancement of its scientific establishment, or tourist revenues. They nevertheless consider these gains less important than what individual scientists and the scientific discipline take away from ISCs. Least of all do they see such benefits as contributing much to the state's international power position.

The argument that ISCs are available instruments that can or should be used for larger political purposes also carries little weight. In principle, states (or their citizens on governing councils of international scientific associations) could dangle before a country they wish to influence the prospect of locating an ISC within its borders. Few of our respondents commenting on this possibility, however, seem to feel that a country of the magnitude of the Soviet Union or indeed any other state that has hosted an IPSA world congress would rise to such bait. At best, a substantial minority expresses the view that, by participating in an ISC in a closed society, scientists might be able to aid those in the country seeking its opening.

In short, the data surveyed in this book make it clear that political scientists—and, by implication, other scientists as well—believe that political considerations should not be permitted to intrude into their scientific congresses. The problem is that sometimes they can hardly escape such intrusions. Selecting a site for a world congress may tempt national associations or their governments to take essentially political stances for or against a potential host country. It may be just as difficult to ignore individual scientists who, for their own reasons or acting on behalf of groups with which they identify, wish to make a political fuss. Then, too, IPSA can do little to prevent host countries from trying to turn congresses to their own advantage. Judging from their comments, however, participants feel that such actions, however irritating they may be, have little immediate impact.

Circumstances will determine whether or not such incidents have a broader impact. In a time of "normal" international discourse, when states have no particular axes to grind and subnational groups are not trying to make a point, none of the above incidents would make large waves. An international scientific association could rely on the disinclination of governments to block its activities, and on the intervention of national disciplinary leaders to persuade recalcitrant bureaucrats and political-minded colleagues to adopt a more expansive attitude. In a time of trouble, by contrast, states and their citizens may forget or ignore the earlier argument that freedom of scientific communication serves their own long-range interests. Partisan uproar focusing on some ISC is thus more an effect than a cause of general political turmoil.

International Systemic Structure

The political use of international scientific congresses is in this sense a function of the structure of the international system. Scientists themselves—if we may judge from the claims made by those from nonsocialist countries who attended the IPSA world congress in Moscow—generally adhere to the oft-postulated norm of scientific communalism (Merton, 1942): an "institutional conception of science as part of the public domain . . . linked with the imperative for communication of findings." At least some of them are nevertheless prepared to swing into line behind their governments when the structure of the international system changes.

The international system obtaining in 1976-1979 was one characterized by the efforts of the superpowers and their allies to achieve East-West détente. Accordingly, the main international actors were emphasizing cooperative endeavors and exchanges of all sorts. International scientific congresses bringing together scientists from all parts of the world were natural in such a climate. Government policy thus reinforced the scientists' more general proclivity toward "full and open communication" (Merton, 1942). Only a few political scientists, most but not all of them in the United States, felt that Soviet violations of human rights (in this case, the rights of dissident scientists) were a sufficiently serious breach of faith that they warranted retaliation in the form of a boycott.

Our data do not reveal how the nonsocialist community of political scientists would have responded to a changed international system. How many of them would have changed sides on the question of full and open communication? Evidence from other scientific fields after the Soviet incursion into Afghanistan suggests that a substantial number in a handful of countries would have expressed strong reservations about continued scientific cooperation with the Soviet Union and even cancelled their planned participation in congresses being held in that country. The half-life of such protests in the early 1980s nevertheless proved to be short and by now exchange programs broken off a half-decade ago are being resumed.[9]

Functional Theories of Integration

In the theories of some writers (e.g. Angell, 1969, 1981), success in internationalizing science will expand the web of international interdependence in ways that contribute to peaceful relations among nation-states. In such theories the functions of international scientific congresses are to promote international understanding and to encourage international networks among scientists. The burden of the argument, of course, is not solely on international science. It is but one strand in the putative web of interdependence. International science is nevertheless a crucial strand because of its symbolic significance for world order and because success in this area may well spill over into other fields of human activity.

Our survey does not provide conclusive proof on the validity of these points. Some networking did emerge from the Moscow meetings. Substantial numbers of registrants justified the Moscow world congress in terms of reducing parochialism at both the individual and disciplinary levels. And, although we included no items specifically on this point, respondents occasionally volunteered the view that the IPSA congress had a positive impact on international relations. Consider, for example, the question of whether or not IPSA's original decision to hold the world congress in Moscow was correct. Three in ten (29%) of the responses by those who approved it gave reasons referring to promoting positive international relations. Our hunch is that more specific questions would have yielded considerable acceptance of the proposition that international science promotes world order.

But can we be sure about this? Certainly the evidence regarding changes in the participants' views of the Soviet governmental apparatus does not suggest that familiarity breeds admiration or even acceptance. What effect an enhanced understanding of the circumstances of the Soviet people will have in the long run is a question our data cannot answer. This kind of evidence is nevertheless beside the point. The more important question is the extent to which fruitful interaction among scientists can produce an international scientific culture that influences policymakers to mitigate conflicts and strengthen transnational political ties. The proof of the pudding lies in part in the extent to which political scientists attending ISCs pick up an "international" perspective on political matters, carry it over into their research and classroom instruction, and conduct future exchanges with foreign scholars they meet at these congresses. The other part of the proof requires some indication of how policymakers might deal with such an international scientific culture.

Structure and Change in the International System

Of particular interest to students of international social change is the issue of how international scientific congresses influence the distribution of such scientific values as skill, knowledge, and rewards. ISCs in any field have the potential for transferring such disciplinary values from the "have" to the "have-not" states. The questions of concern are whether transfers actually take place, the extent to which transfers leave the "have-not" states in a position of greater autonomy vs. dependency, and the degree of seriousness with which the "have" states take into account the concerns and special needs of the others.

As far as scientific growth in the Third World is concerned, our data and more general analysis indicate that international political scientific congresses may be performing quite unevenly the function of transferring disciplinary values. First, their reward structure as outlined earlier places strong emphasis on highly visible scientists, most of whom work in a traditionally Western mode of analysis. Save for token representatives, participants selected from Third World countries tend to be those who have studied in the West and/or use predominantly Western paradigms—with both categories quite heavily relying on and citing the Western scientific literature. The exception to this generalization is the unusually large number of scientists from Latin America who presented papers at the Rio de Janeiro meetings 1982. Latin American participation nevertheless dropped off at the Paris meetings of 1985. This fact reinforces the argument made earlier that, at least until the problem of high transportation costs can be resolved, a regular rotation of sites provides an important means for transferring disciplinary values.

Second, at the Moscow world congress, Third World participants were fairly isolated. They indicate in response to our questionnaire that they were reasonably active in seeking informal contacts at the congress itself albeit less so in trying to maintain them afterwards. Participants from North America and other Western countries, though, put their Third World colleagues at the bottom of the list when it came to subsequent networking. At the same time, the Third World scientists report a substantially greater-than-average degree of modification in their work activities as a result of attending the congress. In a way, these scientists ended up playing a role more as recipients of information about substance and techniques than as active contributors to the scientific endeavor.[10]

The complex of problems implicit in an international scientific congress's functionality for redistributing scientific values on a worldwide basis is one that for some time to come will arrest the attention of IPSA's leadership and the organizers of its future world congresses. Available evidence does not suggest that the Moscow world congress did much to redress the disciplinary imbalance between North and South; nor are there clear guidelines about how this task could be accomplished and, in fact, whether or not the ISC is an appropriate instrument to that end. The International Political Science Association's critics sometimes complain that it is not taking the issue seriously enough. Others, however, have argued that structural conditions in the Third World (such as governmental distrust of political science as a discipline and the lack of funds for research) have prevented more rapid movement and that, indeed, the excursions to Moscow in 1979 and to Rio de Janeiro in 1982 moved the association too far from the central disciplinary concerns of political science.

Transferring disciplinary values between East and West is quite another matter. With a few exceptions, the political scientists from socialist countries who have been most successful in the international disciplinary framework have been those able to work in a Marxist tradition but sufficiently flexible in their thinking to communicate effectively with Western political scientists working in a non-Marxist tradition. The unusually large number of Soviet and East European scientists at the Moscow meetings (610 registrants) provided the first large-scale test of the proposition that East-West communication in political science is possible. The results were moderately positive, as we saw in Chapter 4. Some irritation about what they saw as their socialist colleagues' formalistic pronouncements in scheduled sessions notwithstanding, our nonsocialist respondents generally found the opportunities for informal contacts useful, and many sought to extend the contacts beyond the Moscow meetings. As the desert fox told Antoine de Saint Exupéry's little prince, building the mutual trust required for effective communication takes a while. Holding more meetings in Eastern Europe may facilitate such communication even if it does not produce a perfect meeting of minds on disciplinary values.

Some Future Directions

The data surveyed in this book point to several general conclusions regarding international scientific congresses in general and IPSA world congresses in particular. First, an ISC clearly serves several valuable functions for various members of a scientific consociation. But, and this is the second point, its ability to perform these functions is quite uneven. The typical ISC is better, for instance, at giving individuals opportunities for scientific communication than at trying to effect some political outcome.

Third, many of these functions are quite straightforward, there for all to see. Sessions at which scientists present their ideas and findings permit participants from around the world to learn about research under way in countries and intellectual settings other than their own, obtain feedback on their own work, and exchange information. Similarly, the international scientific association that organizes an ISC performs an obvious service for the profession. An IPSA world congress is an intellectual watering place for current and future members of the discipline's leading stratum, and a framework for developing international networks of scientists seeking to advance knowledge in some particular subfield.

Fourth, other functions are less apparent to casual observers and even participants. What is more, the ability of an international scientific congress to perform key latent functions goes far toward explaining its success with respect to more manifest functions. The free use of unorganized and sometimes recondite informal channels gives formally communicated information a context that makes it more meaningful to working scientists. In some ways an ISC's formal sessions are only a pretext that enables scientists to get together informally to discuss what affects them most directly. By the same token, IPSA's success in mounting a world congress strengthens its hold over the international scientific consociation we call political science. The disciplinary integration an ISC enhances redounds to the benefit of the international scientific association; and this in turn gives the association a firmer basis to develop further congresses and other activities that serve the discipline. Latent functions such as these drive the entire scientific communication process.

Possibilities nonetheless exist, we feel, for improving the ability of IPSA world congresses to carry out their multifarious functions. The following pages outline some possibilities aimed particularly at improving scientific communication and contributing to the growth of an international science of politics. A few suggestions require only minor modifications in standard operating procedures. Achieving other goals, however, may require a fundamental rethinking of the nature and organization of international scientific congresses.

Quality of IPSA World Congresses

Comments offered by our respondents indicate that IPSA world congresses have room for improvement. Some ideas along this line are essentially technical—the need for a professional on the IPSA staff to assist in organizing the world congresses; ways to structure the program and program committee, distribute papers, exhibit books, and schedule panels; expanded facilities for simultaneous translation—and need not concern us here. More interesting are those that go to the heart of an ISC as a medium for scientific communication.

One is the obvious point that national and international political considerations must not be permitted to intrude into the process of setting up and running a world congress. This concern is of course not peculiar to political science. At the more general level the International Council of Scientific Unions has spent considerable time seeking to realize the "well-established agreement that scientific meetings shall not be disturbed by political statements or by activities of a political nature."[11] IPSA, too, as we have seen, has developed procedures to minimize the possibility of political intrusions. These procedures proved to be reasonably effective during the Moscow world congress of 1979 and afterwards. But each world congress is a new test, and that to be held in August 1988 in Washington, D.C. is no exception.

Second, the scientific credibility of the IPSA world congress requires greater attention to quality controls. The tendency of organizers to emphasize balanced representation of papergivers by region rather than by the quality of their papers portends problems for the future. Though it is nowhere spelled out as such in IPSA documentation, the "ideal" panel set up by the program committee contains one papergiver each from North America, other Western countries, the socialist countries, and the Third World. Frequently this means turning down an interesting paper because that region's "quota" has already been reached, or accepting a weak paper because the panel lacks a "representative" of a particular region. The general practice of accepting for the program titles of papers rather than finished papers exacerbates this situation. In short, quality controls are rudimentary and applied at best unevenly.[12]

While such policies may serve an institutional goal, namely, broadening IPSA's membership and participation in its world congresses, they also entail a cost. If active scientists attend too many panels put together on some basis other than the quality of the research to be presented, then they may well lose interest in both the congresses and the association sponsoring them. It is precisely this loss of topflight scientists that *no* international scientific association can afford. Alternatively, such scientists may begin to ignore the prestigious, representative sessions to focus instead on the activities of smaller working groups either contained within the association or meeting concurrently with it. It is this alternative that suggests an important consideration for future IPSA world congresses, networking, to which we shall return later.

A third task is far more complex: ensuring the internationality of world congresses. Internationality means more than simply assembling participants from all over the world. It implies significant interaction among them. If nonscientific boundaries had no meaning, then we would expect networks emerging within some part of the discipline to be completely random in terms of the geographic distribution of their members.[13] Our study suggests that this is far from true today. Only in part can we attribute the salience of national boundaries to IPSA's federal structure, an association of national associations, with which each of its own members identify. More importantly, significant scientific discontinuities reinforce these national boundaries. If it is true, as the data indicate, that Western scientists slight networking opportunities with colleagues from the Third World, then the explanation surely lies less in national and other prejudices than in these scientists' tendency to avoid interaction with those whom they consider less substantively or methodologically proficient than themselves.

How can such nonscientific barriers to internationality be broken down? Easy answers are not at hand. Still, one point seems certain: More careful selection of participants rather than lowered quality controls is imperative. A Third World country might benefit most in the long run by sending its best scientists, even though they are not in tune with generally prevailing paradigms and approaches and even though they have language difficulties. Such a step would pose new tasks. For one thing, the international association would have to deal seriously with a polyglot world rather than assuming that all participants are thoroughly conversant in English and/or French. An increased demand for simultaneous translation services would in turn raise the financial cost of holding a world congress. For another thing, scholars from Western countries must be prepared to enter into dialogues in which they listen as well as talk. This is a matter of recruitment: searching out scientists who can contribute to and learn from a genuinely two-way dialogue. It also requires the ISC to give over program space to such focused—and, perhaps, risky—exchanges along with more traditional sessions at which papers are formally presented and discussed. This kind of restructuring is in its own way costly for an international scientific association.

Another way to strengthen IPSA's internationality is to broaden the geographic spread of potential sites for world congresses (and other meetings). The Moscow meeting greatly increased participation by East Europeans and the Rio meeting that followed attracted many Latin Americans. It stands to reason that, even if these upsurges in activity are not sustained as the congress site moves elsewhere, some portion of the newly won participants will maintain their ties with the association. Then why not simply institute a firm principle of rotating sites? Whatever support it finds among IPSA's leaders and our respondents, the principle raises sharply the question of financing.

Finances and World Congress Sites

The sensitivity of decisions about sites for IPSA world congresses argues for more attention to the way in which these decisions are made. A fundamental revision in procedures is clearly more easily proposed than accomplished. As it is now, IPSA's emphasis on voluntarism forces it to rely on a national association willing to undertake the organizational work and raise the necessary funds. These funds turn out to be vital. Thus the French government provided a grant of $150,000 to help finance the Paris meetings in 1985. And yet, if we take into account the real costs of an IPSA world congress (including the travel and per diem of participants), then we realize that this sum probably represents no more than 5 percent of the total outlay in 1985.[14] The effect of relying on this procedure is that the tail wags the dog.

Recognizing the financial straits in which doubtless all international scientific associations find themselves, and IPSA is certainly no exception, it may seem gratuitous to recommend searching for the means to divorce decisions about sites from financial considerations. A modicum of financial independence (or some other solution, such as greater self-financing of world congresses[15]) can nevertheless enable IPSA to seek an appropriate site rather than vice versa. One useful model is a congress site that does not offer external participants too many nonscholarly distractions or indigenous participants the opportunity to carry on their normal teaching or administrative responsibilities while the congress is in session. This may mean a single conference facility that can house 2,000 to 3,000 people, one far enough away from major cities that participants can keep their minds on the business at hand and yet close enough to attractive touristic or cultural sites that organized or informal visits can be arranged on a free afternoon and evening. Such an environment of limited mobility would strongly increase the opportunities for informal discussions and networking.

Informal Networking

What comes up time and again in our analysis is the importance of informal modes of communication. They seem to produce the greatest amount of learning, a sense of empathy with scholars working in other countries under other conditions, and possibly even a feeling that an international political science community exists, however circumscribed by political conditions and communication barriers it may be. A major task of an international scientific congress, or, indeed, any scientific congress, may well be to create a framework in which informal networks (or "invisible colleges") can emerge and thrive. The preparation of papers for presentation at formal congress sessions may serve other purposes (such as committing scholars to putting their thoughts down on paper). In this context, they nevertheless serve more than anything else as a device for helping scientists to identify others with similar interests, suggesting topics for informal discussions, and, more generally, legitimating the opportunities for informal networks of scholars to operate effectively.

IPSA's mushrooming number of research committees and study groups testifies to the importance of networking for political scientists. To meet the association's conditions, a research committee must include on its governing committee persons from at least a half dozen countries, provide some medium for exchange of information, and hold occasional conferences outside the congress setting. A research committee in good standing now has the right to organize two formal sessions at each IPSA world congress. (A study group becomes a research committee after it has functioned for several years and has organized an international congress for its members and others; upon request, it is accorded the privilege of setting up one formal session at the world congress.) Research committees and study groups raise their own funds, sometimes with the IPSA secretariat's assistance, and determine their own procedures. They thus have greater freedom of action than the parent body in such matters as selecting sites for meetings, and inviting participants on the basis of their scientific merit rather than merely for reasons of representativeness.

The research committees and study groups are rapidly becoming, at least in terms of scientific output, the most productive activity sponsored by IPSA. On the one hand, IPSA can enhance its overall effectiveness by encouraging the development and operation of these bodies. On the other hand, their centrifugal effect for IPSA as a whole cannot be ignored. Their increasing number means growing demands on the association for financial assistance, space in the program of the world congress, and a formal voice in IPSA's decisionmaking. The challenge facing the association is to accommodate the divergent interests of the research committees and study groups but at the same time maintaining some sense of what is central in the discipline of political science.

Another challenge for international scientific associations such as IPSA is to develop new and, one hopes, less cumbersome procedures for facilitating networking among scholars with related interests. Two ideas come to mind here. First, smaller workshops of individual networks may be one answer. Second, the rapid development of inexpensive facilities for computer networking makes it reasonable to explore Kochen's (1985: 299) idea of an "inquiring community," proposed to advance the use of social know-how in invention and innovation:

> Teams of social scientists, policy analysts, and policymakers would be structured both to take advantage of newer modes of communication and collective memory (computer conferencing), and to reduce the fear of making mistakes from which may come learning. Members could inject into the network a policy concern or proposed solution which others interested or with expertise in the issue could pick up. They could then pursue it in their own thinking and research, all the while remaining in constant contact with each other via teleconferencing, exchanging data, testing out ideas on relevant populations, and, more generally, seeking solutions to the policy problem.

To some measure groups of social scientists have already created mechanisms for such teleconferencing. It should be noted, however, that, even if access to the network is available to anyone interested in the subject matter, this kind of procedure for exchanging scientific information at least initially advantages scientists from developed countries where microprocessors are readily available. It may thus exacerbate the problem of "scientific dependency" that a number of writers have identified.

Workshops and teleconferencing, as well as IPSA's research committees and study groups, are most likely to work well within the framework provided by the association's more general world congresses. The ISC provides participants with a broad exposure to the discipline and hence a very different type of experience in communication and learning. It enables scientists to interact directly with specialists in a variety of subfields and from different scientific traditions. There is no substitute for face-to-face communication. In short, without the opportunities for cross-fertilization and recruitment of new members afforded by periodic world congresses, focused workshops and teleconferencing run the risk of self-encapsulation and decreasing relevance to the political science discipline as a whole. Developed appropriately, however, they hold out the promise of flexibility and significant scientific progress.

Diffusion of Research Paradigms and Results

Related to such measures is the need for IPSA to take a more active role in spreading research resources among its members. One problem of scientific communication is particularly acute. Although networking can ensure that experts on a particular topic share the latest paradigms and research results, and although annual meetings of national associations can keep generalists on top of other developments in their own countries, the 3-year gap between IPSA world congresses makes it difficult for political scientists to remain up-to-date on international research outside their own field of specialization. This is especially the case for scholars in countries without the resources to permit extensive travel or the purchase of books and journals from overseas. Moreover, insofar as some countries dominate the channels for scientific communication, the ideas of others are slow to gain attention. The long-term result may be the kind of "scientific dependency" that produces lopsided research and considerable resentment. The more immediate result is to entrench barriers to effective cross-national communication at IPSA world congresses.

To ease such problems, IPSA might develop occasional traveling seminars. By this we mean a team of scholars performing work at the forefront of some field within political science, and who are able to accept the invitation of countries or research institutes to conduct week-long seminars in their area of expertise. The goal would be two-way communication, that is, an exchange of information among peers rather than the creation of a master-pupil relationship. The visiting team would discuss its own research paradigms and results, but it would also listen carefully to those of its hosts.

Such traveling seminars could have a number of positive results.[16] First, if they are properly structured and carried out in good faith, learning will take place on both sides. Second, the resulting interaction may generate transcendent ideas to replace those deriving from more particularistic perspectives. The ideal is a political science of universal rather than merely parochial dimensions. Third, the seminars could spawn new networks of political scientists and expand existing ones. In this sense they would lay the groundwork for more effective communication at subsequent world congresses—two-way communication of the sort that will be required if the congresses are to be truly international.

A related project would aim at developing international data resources for research in political science. In recent years a number of teams have developed banks of reproducible data on political phenomena—public opinion, electoral behavior, civil strife, international wars and militarized disputes, trade and mail flows across national boundaries, and much more. In some areas, notably in North America and Western Europe, political scientists have set up inter-institutional consortia to clean, store, and disseminate such data. Enormous gaps in their coverage, both geographically and in terms of the types of data sets they include, nevertheless characterize these archives. By helping to expand the enterprise of collecting and archiving data, IPSA could greatly facilitate the exchange of scientific information and, again, better enable participants at world congresses to build on a common base of knowledge.

The experience of the meetings in Moscow, held in circumstances that maximized the potential for nonscientific disruption, demonstrates the importance of IPSA world congresses. Scientific communication triumphed in 1979. Some new ideas were generated, some networks developed. And, for the most part, the participants went home with a strong sense that the idea of an IPSA world congress is worthwhile. Room for improvement nevertheless exists; and it is possible for the International Political Science Association to undertake new activities that can greatly enhance the discipline's ability to communicate political science internationally.

Notes

1. We assume the scientists' rationality in the sense that they seek consciously to maximize their utilities. We thus ignore an entire range of questions dealing with unconscious behavior (e.g., the underlying motives that drive activity and hence the function activity serves for the personality).

2. Although respondents describe as their chief motive for attending ISCs the desire for scientific communication, their comments make it clear that personal circumstances—availability of funding, family commitments, timing, and convenience—are crucial in determining whether or not they in fact attend the meetings.

3. Almost a quarter (23%) of our respondents cite such a major modification; and surveys of psychologists and sociologists who participated in their world congresses of 1966 suggest that one-third may be the upper limit for such an explicitly recalled impact (Table 4.5). Although this is a modest result to show for the considerable time, energy, and money that many scholars put into the organization of ISCs, it may be more significant than it appears at first glance. Our question inquired about information that led to *major* changes only. It seems reasonable to assume that many more political scientists received information that was helpful or interesting but that did not fundamentally reorient their work. Then, too, scientists may well have received important new ideas in Moscow but were simply unable a year later to pinpoint their precise source. Something said in a corridor may not have struck home until many weeks or months later, and in a context that did not recall the original conversation.

4. Of the political scientists who report having significantly modified their work activities because of something they learned at the Moscow world congress (23 percent of the total), almost three-quarters (74%) cite as a source informal contacts; these constitute about two-fifths of the total responses given to this question (Table 4.5).

5. Thus 27 percent indicate that their interests are very heavily in research and another 53 percent that, while they are interested in both teaching and research, they lean toward the latter; by contrast, only one in seven

reports being mainly interested in teaching (2%) or leaning toward teaching (12%).

6. A study by the American Council on Education on the professional benefits to be derived from attending international scientific meetings supports this finding (Atelsek and Gomberg, 1981: 13). In a survey of department heads in the natural sciences at 760 colleges and universities in the United States, only 1 percent indicated that improvement in the quality of faculty teaching was the chief benefit of participation in these meetings.

7. Each of these matters led to protracted negotiations that yielded more or less adequate resolutions. In the case of participation, for instance, the Soviet organizing committee agreed to register no more than 500 scientists from socialist countries.

8. IPSA's Western vs. international focus continues to be a matter of concern in some quarters. We shall touch on these issues again.

9. The degree of governmental concern in such matters may well vary according to scientific field. Breaking off relations or maintaining secrecy in a technical area such as solid-state physics differs from similar steps in political science. From the vantage point of those who would limit the free flow of scientific information, barriers in the former aim at preventing the outflow of secret information useful for military and commercial purposes while barriers in the latter seek to keep out ideas that may contaminate domestic political discourse.

10. This consideration may explain in part the growing trend to hold international meetings in Third World countries (see Table 1.1).

11. This "Resolution on the Nonpolitical Tradition of ICSU," adopted in October 1966, appears in the Council's "Advice to Organizers of International Scientific Meetings" (ICSU, 1976).

12. Expressed in functional terms, the organ of a "living system" (cell, organism, organization) performing a particular function faces atrophy when it no longer can perform the function or when the function itself is replaced by another one (Miller, 1978: 83, 711). Accordingly, if we consider the formal session at a world congress to be important in its own right, and not just a fig leaf hiding an ISC's "true" (latent) function aimed at informal communication networks, then organizers must ensure that the formal session in fact performs the scientific communication function assigned to it. Attention to quality control is thus critical to maintaining the ISC as we know it today.

13. A complete model of interpersonal interaction would also take into account such variables as distance, the number of active scientists in each country, and diffusion rates of new knowledge.

14. As noted in the preface, we calculate the minimum real cost to be an average of $1,600 for each of the 1,763 participants, plus perhaps $150,000 for the costs of the program committee and related expenses; this amount, together with the $150,000 allocated to the French organizing committee (but eliminating double-counting), totals nearly $3,000,000.

15. If in 1985 IPSA had raised the registration fee by $100, and assuming that the elasticities of demand are such that this step reduced registration from 1,763 to 1,500, then the amount raised would have been $150,000—the sum

provided by the French government. The point is not to suggest that IPSA erred in accepting the French invitation. It is rather that IPSA cannot always expect such bounty and must be able to select congress sites to meet its own needs irrespective of the host government's ability to finance some of the institutional costs.

16. The cost need not be too great in financial terms if various national associations are willing to participate; more difficult will be the task of persuading some scholars in research-rich as well as research-poor countries that they should want to become involved in a true interchange of ideas with those who do not share their research paradigms, methodologies, and particular concerns.

APPENDIX A
Breakdown of Registrants by Country

	Official Tally	Questionnaires Sent	Questionnaires Returned	% Quest. Returned
North America				
United States	229	249	180	72
Canada	51	56	37	66
Total North American	280	305	217	71
Other Western (incl. Israel, Australia, and New Zealand)				
Germany (FRG)	64	67	26	39
France	42	41	8	20
Sweden	37	52	21	40
Spain	34	41	6	15
United Kingdom	34	42	20	48
Netherlands	32	31	18	58
Israel	29	31	7	23
Norway	21	23	13	57
Finland	20	27	10	37
Belgium	15	16	2	13
Denmark	12	14	10	71
Italy	12	16	3	19
Switzerland	11	13	6	46
Australia	6	8	3	38
Greece	4	5	3	60
Ireland	3	1	1	100
New Zealand	3	1	0	0
Luxembourg	2	2	0	0
Austria	1	1	0	0
Total Other Western	382	432	157	36
Third World: Latin America				
Mexico	47	123	8	7
Brazil	11	11	3	27
Venezuela	5	8	0	0
Argentina	3	6	0	0
Chile	1	1	1	100
	(67)	(149)	(12)	(8)

	Official Tally	Questionnaires Sent	Questionnaires Returned	% Quest. Returned
Third World: Africa and Middle East				
Turkey	22	19	7	37
Nigeria	4	10	1	10
Algeria	2	0	-	-
Cameroun	1	2	0	0
Ivory Coast	1	1	1	100
Jordan	1	1	0	0
Sierra Leone	1	1	0	0
Zambia	0	1	0	0
	(32)	(35)	(9)	(26)
Third World: Asia and Pacific				
India	32	33	7	21
Japan	30	39	10	26
Korea (South)	21	23	4	17
Thailand	2	2	0	0
China	0	1	0	0
Hong Kong	1	1	1	100
Indonesia	1	1	0	0
Malaysia	1	1	1	100
Pakistan	1	1	0	0
Philippines	1	2	1	50
Singapore	1	1	1	100
Taiwan	0	1	0	100
	(91)	(106)	(25)	(24)
Total Third World	190	290	46	16
Total	852	1027	420	41

APPENDIX B
Questionnaires

Quest. #

IntSci
7/80

1. Questionnaire for Registrants

University of Illinois

Department of Political Science

All information that would permit identification of the individual will be held in strict confidence, will be used only by persons engaged in and for the purposes of the survey, and will not be disclosed or released to others for any purpose.

(Please circle one answer code number unless otherwise instructed.)

1. How frequently do you attend the annual (or other regular) meetings of your national political science association (e.g. Indian Political Science Association, Deutsche Vereinigung für Politische Wissenschaft, American Political Science Association)?

Almost every year 1	6
Approximately every other year 2	
Less often than that 3	
Never 4	
No national political science association . 5	
National association does not hold meetings 6	

2. What value does attending the annual meetings of your national political science association have

   a. For you personally? _____ 7-8

   _____

   b. For most attendants (in your view)? _____ 9-10

   _____

3a. Are you able to communicate effectively in one or more foreign languages (that is, a language other than that of the country in which you are now working) with scholars from other countries on matters of mutual scientific interest?

Yes 1	11
No *(Skip to Q.4a)* 2	

   b. In what foreign language(s)? _____ 12

   13-19

4a. In what languages do you read journals, books, or other research reports in obtaining scientific information relevant to your research and/or teaching?

   (1) Native language(s) *(Specify which)* _____ 20

   21-27

   (2) Language(s) currently used in teaching, if different from native language(s) *(Specify which)*

   _____ 28

   29-35

   (3) Other languages read *(Specify which)* _____ 36

   _____ 37-43

APPENDICES 147

4b. During 1979, did you read any scientific works in these foreign languages?

                              Yes 1   44

                              No *(Skip to Q.5a)* 2

  c. For each scientific work read in a foreign language, please indicate (a) the language and (b) the type of work (such as journal article, book, manuscript):

      (a) Language   (b) Type of Written Work

      _____   _____   45–52

      _____   _____

      _____   _____

      _____   _____

5a. Have you ever attended any international scientific meetings or international conferences <u>within</u> the country in which you are now working?

                              Yes 1   53

                              No *(Skip to Q.6a)* 2

  b. How many such international conferences have you ever attended? *(Please give approximate number)*

                              _____   54

6a. Have you ever been outside the country in which you are now working on a trip for scholarly purposes which lasted more than three months?

                              Yes 1   55

                              No *(Skip to Q.7a)* 2

  b. How many times have you made such a trip? *(Please give approximate number)*

                              _____   56

7a. How many international scientific meetings or international scholarly conferences <u>outside</u> the country in which you are now working have you attended? *(Please give approximate number--if none indicate 0)*

                              _____   57

  b. What group(s) sponsored these meetings or conferences? _____   58–67

_____

_____

_____

8. It is sometimes said that international scientific congresses in the field of political science serve a variety of functions--for the individual participant, the political science profession as a whole, and the host country.
    a. What do you consider to be the chief functions of such international conferences for the <u>individual participant</u>?

    _____ 68-69

    _____

    b. What do you consider to be the chief functions of such international conferences for the <u>political science profession as a whole</u>?

    _____ 70-71

    _____

    c. What do you consider to be the chief functions of such international conferences for the <u>host country</u>?

    _____ 72-73

    _____

9. From your own perspective, how important are these functions . . .

	Very important	Somewhat important	Not very important	Not at all important	
a. For the individual participant? . . .	1	2	3	4	74
b. For political science profession? . .	1	2	3	4	75
c. For the host country?	1	2	3	4	76

10. Taking everything into consideration, what do you think is the single chief value of international scientific congresses in the field of political science?

    _____ 77-78

    _____

11. Did you attend the World Congress of the International Political Science Association held in

	Yes	No	
a. Geneva, Switzerland in 1964?	1	2	79
b. Brussels, Belgium in 1967?	1	2	80 1-2 | 22 3-5 | DUP
c. Munich, West Germany in 1970?	1	2	6
d. Montreal, Canada in 1973?	1	2	7
e. Edinburgh, Scotland in 1976?	1	2	8

APPENDICES 149

12a. Did you attend the 1979 World Congress of the International Political Science Association, held in Moscow, Soviet Union?

                                                      Yes 1   9

                                                     No 2

   b. Why did you decide (to attend/not to attend) the 1979 IPSA World Congress in Moscow? *(Please list all relevant reasons)*

      _____ 10-19

   c. If you gave more than one reason, which of these was the single most decisive reason for your decision?

      _____ 20-21

13a. In 1976 the Council and Executive Committee of the International Political Science Association decided to accept the Soviet Union's invitation to hold the 1979 IPSA World Congress in Moscow. Given the international circumstances of the mid-1970s, do you think that this decision was a good one?

                                                       Yes 1   22

                                                     No 2

   b. Please explain your response in a brief sentence or two.

      _____ 23-24

14a. More generally, is it a good idea to hold scientific meetings in the Soviet Union?

                                                       Yes 1   25

                                                     No 2

   b. Please explain your response in a brief sentence or two.

      _____ 26-27

(IF YOU DID NOT ATTEND THE 1979 IPSA WORLD CONGRESS IN MOSCOW, PLEASE SKIP TO Q.29a, PAGE 12)

15. Had you ever been to the Soviet Union before attending the IPSA World Congress in Moscow in August 1979?

                                                       Yes 1   28

                                                     No 2

16. We would like to know about some of your expectations with respect to the IPSA World Congress in Moscow and the extent to which your experience matched these expectations. For each of the following statements you are provided with a set of possible responses indicating how closely the statement describes your expectations, and another set indicating how closely the statement in fact describes what you experienced in the Soviet Union. Space is also provided for you to elaborate upon your feelings, should you wish to do so.

	Expectation				Experience			
	Not at all what I expected	Only a little like I expected	Approximately what I expected	Exactly what I expected	Not at all what actually happened	Only a little like what happened	Approximately what happened	Exactly what happened

a. The accommodations (including hotel and provision for meals) would be adequate to my needs 1 2 3 4 (29) 1 2 3 4 (30)

Comment: _____

b. I would have little social contact with people in the USSR (other than Congress participants) . 1 2 3 4 (31) 1 2 3 4 (32)

Comment: _____

c. I would have a great deal of contact with Soviet scholars at the Congress 1 2 3 4 (33) 1 2 3 4 (34)

Comment: _____

d. I would have a great deal of contact with other East European scholars at the Congress 1 2 3 4 (35) 1 2 3 4 (36)

Comment: _____

APPENDICES   151

	Expectation					Experience			
	Not at all what I expected	Only a little like I expected	Approximately what I expected	Exactly what I expected		Not at all what actually happened	Only a little like what happened	Approximately what happened	Exactly what happened
e. The organization of the Congress itself would facilitate interaction among participants . . .	1	2	3	4 37		1	2	3	4 38
Comment:									
f. A substantial number of Congress participants would have research interests similar to my own . . .	1	2	3	4 39		1	2	3	4 40
Comment:									
g. I would learn something professionally useful from papers and comments presented in the formal sessions by Soviet and East European participants . . .	1	2	3	4 41		1	2	3	4 42
Comment:									
h. Participants in formal sessions would be able to raise any topics and present any points of view they wished (even on highly contentious "cold war" issues) . . .	1	2	3	4 43		1	2	3	4 44
Comment:									

	Expectation					Experience			
	Not at all what I expected	Only a little like I expected	Approximately what I expected	Exactly what I expected		Not at all what actually happened	Only a little like what happened	Approximately what happened	Exactly what happened
i. The procedures followed by panel chairpersons would encourage uninhibited discussion and the presentation of a wide range of viewpoints . . . Comment:	1	2	3	4 45		1	2	3	4 46
j. There would be opportunities for informal interaction with Congress participants from other countries that would contribute to my ongoing and future research . . . . . . Comment:	1	2	3	4 47		1	2	3	4 48
k. Overall, my visit to the IPSA World Congress in Moscow would be a satisfying personal experience . . . . Comment:	1	2	3	4 49		1	2	3	4 50

17. About how much informal contact would you say you had with the following kinds of people during your visit to the 1979 IPSA World Congress?

	No contact at all	Very little contact	Occasional contact	Great deal of contact	
a. Colleagues from your own country	1	2	3	4	51
b. Soviet scholars	1	2	3	4	52
c. Other East European scholars	1	2	3	4	53

Colleagues from countries other than your own in:

d. West Europe and North America	1	2	3	4	54
e. Third World	1	2	3	4	55

18a. Considering the Congress as a whole and all interactions with Congress participants, did you receive information that led to a major modification in any of your work activities?

    Yes 1   56

    No *(Skip to Q.19)* 2

b. How did you receive this information?

    Oral presentation of paper . 1   57

    Reading copy of paper 2

    Floor discussion 3

    Informal contacts 4

    Other *(Specify)* _____ 5

c. What type of activity was modified by this interaction?

    Research 1   58

    Teaching 2

    Manuscript/publication plans 3

    Applied work 4

    Administration 5

    Other *(Specify)* _____ 6

d. Please describe the nature of this modification.

_____ 59-60

_____

_____

25b. Why? _____ 40-41

_____

26a. Some critics of the IPSA World Congress in Moscow complained that the Soviet government used the meeting for its own propagandistic purposes. Do you . . .

                                Strongly agree? 1   42

                                Agree somewhat? 2

                                Disagree somewhat? 3

                                Strongly disagree? 4

   b. Why are you of this opinion? _____ 43-44

_____

27a. Most of us, before we went to the IPSA World Congress in Moscow, had an *image* of the Soviet Union, that is, some "picture in our mind" of what we believed to be true (irrespective of whether or not we liked the image we saw). How did your experience during the course of your visit affect your *image* of the Soviet Union?

                        Reinforced existing image *(Skip to Q.28a)* 1   45

                        Did not change image *(Skip to Q.28a)*. 2

                        Significantly changed image 3

   b. In what way did your experience change your image of the Soviet Union?

_____ 46-47

_____

28a. Did attending the IPSA World Congress change your *attitude* toward (a) the Soviet government or (b) the people of the Soviet Union?

	Government	People
Yes 1	48	1 49
No 2		2

   b. How did it change your attitude?

	Government	People
Very much in positive direction 1	50	1 51
Somewhat in positive direction 2		2
Somewhat in negative direction 3		3
Very much in negative direction 4		4
Did not change my attitude 5		5

APPENDICES 157

28c. Please elaborate on how your attitude changed toward the . . .

Soviet government _____ 52-53

_____

People of the Soviet Union _____ 54-55

_____

29a. As you may have heard, the next World Congress of the International Political Science Association will be held in August 1982 in Rio de Janeiro, Brazil. Are you thinking of attending this meeting?

                              Yes 1   56

                              No *(Skip to Q.30a)* 2

b. On what circumstances does your decision to attend the 1982 IPSA World Congress depend?

_____ 57-58

30a. Do you think that Rio de Janeiro is an appropriate site for an IPSA World Congress?

                              Yes 1   59

                              No 2

b. Please explain your response in a brief sentence or two.

_____ 60-61

31. What three themes would you most like to see included in the official program for the 1982 IPSA World Congress in Rio de Janeiro?

a. _____ 62-63

b. _____ 64-65

c. _____ 66-67

32a. It has been proposed that the 1985 IPSA World Congress be held in Paris, France. Do you think that Paris is an appropriate site for this meeting?

                                Yes 1

                                No *(Skip to Q.33)* 2

   b. Please explain your response in a brief sentence or two.

_____

_____

33. In what city would you like to see future IPSA World Congresses held? *(Please list three cities rank-ordered in terms of your own preference.)*

   a. _____

   b. _____

   c. _____

34a. With what type of institution are your currently affiliated?

                Academic 1

                Government 2

                Industry 3

                Private research organization 4

                Governmental research organization 5

                Other *(Specify)* _____ 6

   b. In what country is this located? _____

   c. What is your title? _____

   d. In what year did you begin your professional career as a political scientist?

                                                                     _____

35. Within the discipline of political science, what is the <u>primary</u> focus of your teaching and/or research? *(Circle one only)*

        Political institutions and processes of your own country . . 1   10

        Public law (including jurisprudence) 2

        Public policy and administration 3

        Political behavior 4

        Normative or empirical theory, philosophy 5

        Comparative politics 6

        International relations 7

        Methodology 8

        Other *(Specify)* _____ 9

36a. Do your teaching/research interests have a geographic focus?

        Yes 1   11

        No *(Skip to Q.37a)*. 2

  b. Is this geographic focus regional or global?

        Regional 1   12

        Global *(Skip to Q.37a)*. . . . 2

  c. What is the specific region? _____   13-14

37a. What is the most advanced degree you have earned? _____   15-16

  b. At what institution was this degree awarded? _____   17-18

38. In what year were you born? _____   19-20

39. What is your sex?

        Female 1   21

        Male 2

40. Which one of the following best describes your interests?

    My interests are very heavily in research 1   22

    My interests are very heavily in teaching 2

    My interests are in both, but lean more toward research 3

    My interests are in both, but lean more toward teaching 4

    My interests are in neither research nor teaching but rather in
    (Please specify) _____ 5

41. How many books or monographs have you published or edited, alone or in collaboration?

None	1
1-2	2
3-4	3
5-10	4
More than 10 . . .	5

23

42. How many articles have you published in academic or professional journals?

None	1
1-2	2
3-4	3
5-10	4
11-20	5
21-30	6
31-50	7
More than 50 . . .	8

24

43. How many of your professional writings have been published or accepted for publication in the last two years?

One	1
Two	2
Three	3
Four	4
Five	5
Six-ten	6
More than ten . . .	7
None	8

25

THANK YOU VERY MUCH.

APPENDICES   161

PolSci
7/80

Quest. # _____

2. Questionnaire for Nonregistrants

University of Illinois
Department of Political Science

All information that would permit identification of the individual will be held in strict confidence, will be used only by persons engaged in and for the purposes of the survey, and will not be disclosed or released to others for any purpose.

(Please circle one answer code number unless otherwise instructed.)

1. How frequently do you attend the annual meetings of . . .

	Almost every year	Approximately every other year	Less often than that	Never
a. Your national political science association (e.g. American or Canadian Political Science Association)?	1	2	3	4
b. Your regional political science association (e.g. Midwest Political Science Association)?	1	2	3	4
c. A subdisciplinary association (e.g. International Studies Association)?	1	2	3	4

2. What value does attending each of the following have for (a) you personally and (b) most attendants?

	a. Personal value	b. Value for attendants
(1) Your national political science association?		
(2) Your regional political science association?		
(3) A subdisciplinary association?		

3a. Are you able to communicate effectively in one or more foreign languages (that is, a language other than that of the country in which you are now working) with scholars from other countries on matters of mutual scientific interest?

　　　　　　　　　　　　　　　　　Yes . . . . 1　　21

　　　　　　　　　　　　　　　　　No *(Skip to Q.4a)* 2

　b. In what foreign language(s)? _____　22-29

___

4a. Have you ever attended any international scientific meetings or international conferences <u>within</u> the country in which you are now working?

　　　　　　　　　　　　　　　　　Yes . . . . 1　　30

　　　　　　　　　　　　　　　　　No *(Skip to Q.5a)* 2

　b. How many such international conferences have you ever attended?
　　 (Please give approximate number)
　　　　　　　　　　　　　　　　　　　　　　　　　　　　　　　　31

5a. Have you ever been outside the country in which you are now working on a trip for scholarly purposes which lasted more than three months?

　　　　　　　　　　　　　　　　　Yes . . . . 1　　32

　　　　　　　　　　　　　　　　　No *(Skip to Q.6a)* 2

　b. How many times have you made such a trip?
　　 (Please give approximate number)
　　　　　　　　　　　　　　　　　　　　　　　　　　　　　　　　33

6a. Have you ever attended any international scientific meetings or international conferences <u>outside</u> the country in which you are now working?

　　　　　　　　　　　　　　　　　Yes . . . . 1　　34

　　　　　　　　　　　　　　　　　No *(Skip to Q.7a)* 2

　b. How many such international conferences have you ever attended?
　　 (Please give approximate number)
　　　　　　　　　　　　　　　　　　　　　　　　　　　　　　　　35

8. It is sometimes said that international scientific congresses in the field of political science serve a variety of functions--for the individual participant, the political science profession as a whole, and the host country.

　a. What do you consider to be the chief functions of such international conferences for the <u>individual participant</u>?

_____　36-37

_____

_____

8b. What do you consider to be the chief functions of such international conferences for the political science profession as a whole?

_____ 38-39

_____

c. What do you consider to be the chief functions of such international conferences for the host country?

_____ 40-41

_____

9. From your own perspective, how important are these functions . . .

	Very important	Somewhat important	Not very important	Not at all important	
a. For the individual participant? 1		2	3	4	42
b. For political science profession? . . . 1		2	3	4	43
c. For the host country? 1		2	3	4	44

10. Taking everything into consideration, what do you think is the single chief value of international scientific congresses in the field of political science?

_____ 45-46

_____

11. Did you attend the World Congresses of the International Political Science Association held in . . .

		Yes	No	
a.	Geneva, Switzerland in 1964? 1		2	47
b.	Brussels, Belgium in 1967? 1		2	48
c.	Munich, West Germany in 1970? 1		2	49
d.	Montreal, Canada in 1973? 1		2	50
e.	Edinburgh, Scotland in 1976? 1		2	51

12a. Did you attend the 1979 World Congress of the International Political Science Association, held in Moscow, Soviet Union?

Yes 1   52

No 2

12b. Why did you decide (to attend/not to attend) the 1979 IPSA World Congress in Moscow? *(Please list all relevant reasons)*

_____ 53-62

_____

_____

c. If you gave more than one reason, which of these was the <u>single</u> most decisive reason for your decision?

_____ 63-64

13a. In 1976 the Council and Executive Committee of the International Political Science Association decided to accept the Soviet Union's invitation to hold the 1979 IPSA World Congress in Moscow. Given the international circumstances of the mid-1970s, do you think that this decision was a good one?

                Yes 1   65

                No 2

b. Please explain your response in a brief sentence or two.

_____ 66-67

_____

14a. More generally, is it a good idea to hold scientific meetings in the Soviet Union?

                Yes 1   68

                No 2

b. Please explain your response in a brief sentence or two.

_____ 69-70

_____

15a. As you may have heard, the next World Gongress of the International Political Science Association will be held in August 1982 in Rio de Janeiro, Brazil. Are you thinking of attending this meeting?

              Yes 1   71

              No *(Skip to Q.16a)* 2

APPENDICES   165

15b. On what circumstances does your decision to attend the 1982 IPSA World Congress depend?

_____ 72-73
_____
_____

16a. Do you think that Rio de Janeiro is an appropriate site for an IPSA World Congress?

          Yes 1   74

          No 2

b. Please explain your response in a brief sentence or two.

_____ 75-76
_____
_____

17. What three themes would you most like to see included in the official program for the 1982 IPSA World Congress in Rio de Janeiro?

 a. _____   77-78

 b. _____   79-80
                         1-2 | 1 2
                         3-5 | DUP

 c. _____   6-7

18a. It has been proposed that the 1985 IPSA World Congress be held in Paris, France. Do you think that Paris is an appropriate site for this meeting?

          Yes 1   8

          No 2

b. Please explain your response in a brief sentence or two.

_____ 9-10
_____
_____

19. In what city would you like to see future IPSA World Congresses held? *(Please list three cities rank-ordered in terms of your own preference.)*

    a. _____ 11-12

    b. _____ 13-14

    c. _____ 15-16

20a. With what type of institution are you currently affiliated?

    Academic 1   17

    Government 2

    Industry 3

    Private research organization 4

    Governmental research organization 5

    Other *(Specify)* _____ 6

  b. In what country is this located? _____ 18-19

  c. What is your title? _____ 20-21

  d. In what year did you begin your professional career as a political scientist?

                                                                                                  _____ 22-23

21. Within the discipline of political science, what is the <u>primary</u> focus of your teaching and/or research? *(Circle one only)*

    Political institutions and processes of your own country . . 1   24

    Public law (including jurisprudence) 2

    Public policy and administration 3

    Political behavior 4

    Normative or empirical theory, philosophy 5

    Comparative politics 6

    International relations 7

    Methodology 8

    Other *(Specify)* _____ 9

22a. Do your teaching/research interests have a geographic focus?

                               Yes 1   25

                               No *(Skip to Q.23a)*. 2

22b. Is this geographic focus regional or global?

                                        Regional 1

                                  Global *(Skip to Q.23a)* . . . 2

    c. What is the specific region? _____

23a. What is the most advanced degree you have earned? _____

    b. At what institution was this degree awarded? _____

24. In what year were you born? _____

25. What is your sex?

                                              Female 1

                                              Male 2

26. Which one of the following best describes your interests?

    My interests are very heavily in research 1

    My interests are very heavily in teaching 2

    My interests are in both, but lean more toward research 3

    My interests are in both, but lean more toward teaching 4

    My interests are in neither research nor teaching but rather in
    (Please specify) _____ 5

27. How many books or monographs have you published or edited, alone or in collaboration?

                                          None 1

                                          1-2 2

                                          3-4 3

                                          5-10 4

                                          More than 10 . . . 5

28. How many articles have you published in academic or professional journals?

   None 1   38
   1-2 2
   3-4 3
   5-10 4
   11-20 5
   21-30 6
   31-50 7
   More than 50 . . . 8

29. How many of your professional writings have been published or accepted for publication in the last two years?

   One 1   39
   Two 2
   Three 3
   Four 4
   Five 5
   Six-ten 6
   More than ten . . . 7
   None 8

THANK YOU VERY MUCH.

APPENDIX C
Scales of Professionalism and Internationalism

C.1. Professional Activity Scale

	Registrants				
Score	North American	Other Western	Third World	Total	Non-registrants
0	1%	1%	0%	1%	10%
1	1	1	7	1	11
2	7	3	4	5	15
3	7	4	4	6	12
4	11	15	9	12	18
5	16	22	24	19	12
6	18	16	11	16	12
7	20	13	11	17	6
8	11	13	17	12	3
9	8	8	9	8	2
10	2	5	4	3	0
Total*	100%	100%	100%	100%	100%
Mean	5.8	6.0	5.8	5.9	3.6
Std. mean**	5.8	6.0	5.8	5.9	3.6

*Columns may not add up to 100 percent because of rounding.
**Mean scores standardized to a scale ranging from 0 to 10.

Scoring:
Attendance at national political science association meetings (2 = almost every year; 1 = every other year; 0 = less often or never)
Research vs. teaching interest (2 = heavily research; 1 = leaning toward research; 0 = other)
Published books and monographs (2 = 5 or more; 1 = between 1 and 4; 0 = none)
Published articles (2 = 21 or more; 1 = between 3 and 20; 0 = 2 or fewer)
Publications in last 2 years (2 = 6 or more; 1 = between 2 and 5; 0 = 1 or none)

C.2. Professional Status Scale

	Registrants				Non-registrants
Score	North American	Other Western	Third World	Total	
0	2%	3%	4%	2%	6%
1	2	6	2	3	12
2	6	8	9	7	16
3	12	17	13	14	19
4	18	14	22	17	18
5	19	17	13	18	12
6	17	14	9	15	8
7	15	17	15	15	8
8	10	5	11	8	1
Total*	100%	100%	100%	100%	100%
Mean	5.0	4.6	4.7	4.8	3.5
Std. mean**	6.3	5.7	5.8	6.0	4.3

*Columns may not add up to 100 percent because of rounding.
**Mean scores standardized to a scale ranging from 0 to 10.

Scoring:
Level of highest degree (1 = Ph.D. or equivalent; 0 = other)
Title (1 = professor, president, dean, director, senior researcher, or equivalent; 0 = assistant or associate professor, graduate assistant, researcher, or equivalent)
Seniority, as indicated by year entered professional career (2 = 1922–1959; 1 = 1960–1972; 0 = 1973 or later)
Published books and monographs (2 = 5 or more; 1 = between 1 and 4; 0 = none)
Published articles (2 = 21 or more; 1 = between 3 and 20; 0 = 2 or fewer)

C.3. International Research Competence Scale

	Registrants				Non-registrants
Score	North American	Other Western	Third World	Total	
0	10%	3%	0%	6%	45%
1	15	4	9	10	22
2	24	13	22	20	16
3	17	31	37	25	11
4	21	28	20	23	5
5	13	21	13	16	0
Total*	100%	100%	100%	100%	100%
Mean	2.6	3.4	3.1	3.0	1.1
Std. mean**	5.3	6.8	6.1	5.9	2.2

*Columns may not add up to 100 percent because of rounding.
**Mean scores standardized to a scale ranging from 0 to 10.

Scoring:
Number of foreign languages usable for scientific communication (2 = two or more; 1 = one; 0 = none)
Actual use of foreign languages for research (1 = yes; 0 = no)
Number of research trips overseas (2 = three or more; 1 = one or two; 3 = none)

C.4. International Activity Scale

	Registrants				Non-registrants
Score	North American	Other Western	Third World	Total	
0	0%	1%	0%	0%	18%
1	4	3	4	4	24
2	9	4	17	8	18
3	20	14	7	16	14
4	18	17	26	19	8
5	15	21	17	17	8
6	18	19	9	17	4
7	8	12	2	9	5
8	7	8	9	7	1
9	2	1	9	3	0
Total*	100%	100%	100%	100%	100%
Mean	4.6	5.0	4.7	4.8	2.4
Std. mean**	5.2	5.6	5.2	5.3	2.7

*Columns may not add up to 100 percent because of rounding.
**Mean scores standardized to a scale ranging from 0 to 10.

Scoring:
Number of times attended ISCs in own country (2 = six or more; 1 = between one and five; 0 = none)
Number of research trips overseas (2 = three or more; 1 = one or two; 3 = none)
Number of times attended ISCs overseas (2 = six or more; 1 = between one and five; 0 = none)
Field of specialization (1 = comparative or international politics; 0 = other)
Geographic focus of teaching and research (2 = global; 1 = regional; 0 = national or other)

C.5. IPSA Experience Scale

	Registrants				Non-
Score	North American	Other Western	Third World	Total	registrants
0	54%	47%	74%	53%	90%
1	18	24	15	20	6
2	17	12	11	14	3
3	6	10	0	7	1
4	4	5	0	4	0
5	2	3	0	2	0
Total*	100%	100%	100%	100%	100%
Mean	0.9	1.1	0.4	1.0	0.2
Std. mean**	1.9	2.2	0.7	1.9	0.3

*Columns may not add up to 100 percent because of rounding.
**Mean scores standardized to a scale ranging from 0 to 10.

Scoring:
Attended Geneva world congress, 1964 (1 = yes; 0 = no)
Attended Brussels world congress, 1967 (1 = yes; 0 = no)
Attended Munich world congress, 1970 (1 = yes; 0 = no)
Attended Montréal world congress, 1973 (1 = yes; 0 = no)
Attended Edinburgh world congress, 1976 (1 = yes; 0 = no)

Bibliography

ALGER, Chadwick F. and Gene M. LYONS (1974) "Social science as a transnational system: report of a seminar." *International Studies Notes*, 1,3 (Fall): 1-13.
ALKER, Hayward R., Jr. (1978) "The politics of peace." *Participation*, 2,3: 23-31.
ANGELL, Robert Cooley (1981) "Do ISPAs promote global integration?" pp. 237-254 in William M. Evan (ed.), *Knowledge and Power in a Global Society*. Beverly Hills, Calif.: Sage Publications.
———. (1969) *Peace on the March: Transnational Participation*. New York: Van Nostrand Reinhold.
ATELSEK, Frank J. and Irene L. GOMBERG (1981) *An Analysis of Travel by Academic Scientists and Engineers to International Scientific Meetings in 1979-80*. Washington, D.C.: American Council on Education.
BARBER, Bernard (1968) "The sociology of science," pp. 92-100 in David L. Sills (ed.), *International Encyclopedia of the Social Sciences*, vol. 14. New York: Crowell Collier and Macmillan.
BARENTS, J. (1959) "Vanity fair? or, international congresses reconsidered." *American Political Science Review*, 53,4 (December): 1090-1094.
BECKER, Howard S. and Irving Louis HOROWITZ (1972) "Radical politics and sociological research." *American Journal of Sociology*, 78,1 (July): 48-66.
BOSE, Nirmal (1978) "The politics of development and system change." *Participation*, 2,3: 43-55.
BROADHEAD, Robert C. and Ray C. RIST (1976) "Gatekeepers and the social control of social research." *Social Problems*, 23,3 (February): 325-326.
BROWN, Archie (1984) "Political science in the Soviet Union: a new stage of development?" *Soviet Studies*, 36,3 (July): 317-344.
CANCIAN, Francesca M. (1968) "Varieties of functional analysis," pp. 29-43 in David L. Sills (ed.), *International Encyclopedia of the Social Sciences*, vol. 6. New York: Crowell Collier and Macmillan.
CAPES, Mary [ed.] (1960) *Communication or Conflict: Conferences: Their*

Nature, Dynamics, and Planning. New York: Association Press.
COLLINS, H. M. (1983a) "An empirical relativist programme in the sociology of scientific knowledge," pp. 85-113 in Karin D. Knorr-Cetina and Michael Mulkay (eds.), *Science Observed: Perspectives on the Social Study of Science.* London and Beverly Hills, Calif.: Sage Publications.
——— . (1983b) "The sociology of scientific knowledge: studies of contemporary science." *Annual Review of Sociology,* 9: 265-285.
CRANE, Diana (1981) "Alternative models of ISPAs," pp. 29-47 in William M. Evan (ed.), *Knowledge and Power in a Global Society.* Beverly Hills, Calif.: Sage Publications.
——— . (1972) *Invisible Colleges: Diffusion of Knowledge in Scientific Communities.* Chicago and London: The University of Chicago Press.
——— . (1971) "Transnational networks in basic science." *International Organization,* 25,3 (Summer): 585-601.
——— . (1970) "The nature of scientific communication and influence." *International Social Science Journal,* 22,1: 28-41.
DICKSON, David (1979) "Brazil bans conference on Amazon biology." *Nature,* 277,2 (February): 340.
EVAN, William M. [ed.] (1981a) *Knowledge and Power in a Global Society.* Beverly Hills, Calif.: Sage Publications.
——— . (1981b) "Some dilemmas of knowledge and power: an introduction," pp. 11-25 in William M. Evan (ed.), *Knowledge and Power in a Global Society.* Beverly Hills, Calif.: Sage Publications.
——— . (1975) "The International Sociological Association and the internationalization of sociology." *International Social Science Journal,* 27,2: 385-393.
FIGHIERA, Gian Carlo (1984) "The geographical distribution of meetings throughout the world." *International Transnational Associations,* 36,3 (July-August): 142-159.
GINIGER, Henry (1979) "Canadian ousted by Soviet hints 'gulag justice' is not monolithic." *The New York Times* (3 September): A4.
HAGSTROM, Warren O. (1965) *The Scientific Community.* New York: Basic Books.
HANSON, Elizabeth C. and Richard L. MERRITT (1983) "International conferences as a mode of technology transfer." *World Policy,* no. 1: 1-8.
INTERNATIONAL COUNCIL OF SCIENTIFIC UNIONS (1976) *The Free Circulation of Scientists: Advice to Organizers of International Scientific Meetings.* Paris: ICSU Secretariat.
JAGTENBERG, Tom (1983) *The Social Construction of Science: A Comparative Study of Goal Direction, Research Evolution and Legitimation.* Dordrecht, Holland: D. Reidel.
JOHNS HOPKINS UNIVERSITY, Center for Research in Scientific Communication (1968) *Reports of the American Psychological Association's Project on Scientific Information Exchange in Psychology* (2 vols). Washington, D.C.: American Psychological Association.
KAPLAN, Norman and Norman W. STORER (1968) "Scientific communication," pp. 112-117 in David L. Sills (ed.), *International*

Encyclopedia of the Social Sciences, vol. 14. New York: Crowell Collier and Macmillan.
KELMAN, Herbert C. [ed.] (1965) *International Behavior: A Social-Psychological Approach.* New York: Holt, Rinehart and Winston.
KNORR-CETINA, Karin D. and Michael MULKAY (1983) "Introduction: emerging principles in social studies of science," pp. 1-17 in Karin D. Knorr-Cetina and Michael Mulkay (eds.), *Science Observed: Perspectives on the Social Study of Science.* London and Beverly Hills, Calif.: Sage Publications.
KOCHEN, Manfred (1985) "Social know-how and its role in invention and innovation," pp. 269-289 in Richard L. Merritt and Anna J. Merritt (eds.), *Innovation in the Public Sector.* Beverly Hills, Calif.: Sage Publications.
KRIESBERG, Louis (1981) "Varieties of ISPAs: their forms and functions," pp. 49-68 in William M. Evan (ed.), *Knowledge and Power in a Global Society.* Beverly Hills, Calif.: Sage Publications.
KUHN, Thomas (1970) *The Structure of Scientific Revolutions* (rev. enl. ed.). Chicago: University of Chicago Press.
LADD, Everett Carll, Jr. and Seymour Martin LIPSET (1978) "The Ladd-Lipset survey: faculty members who travel abroad." *The Chronicle of Higher Education,* 16,9 (24 April): 8.
LAKOFF, Sanford A. (1977) "Scientists, technologists and political power," pp. 355-391 in Ina Spiegel-Rösing and Derek de Solla Price (eds.), *Science, Technology and Society: A Cross-Disciplinary Perspective.* London and Beverly Hills, Calif.: Sage Publications.
LASSWELL, Harold D. and Abraham KAPLAN (1950) *Power and Society: A Framework for Political Inquiry.* New Haven: Yale University Press.
LEVY, Marion J., Jr. (1968) "Structural-functional analysis," pp. 21-29 in David L. Sills (ed.), *International Encyclopedia of the Social Sciences,* vol. 6. New York: Crowell Collier and Macmillan.
LODGE, David (1984) *Small World: An Academic Romance.* New York: Macmillan.
LUDZ, Peter C. (1979) "Cumulative growth in political knowledge since 1950." *Participation,* 3,1: 46-51.
MACIVER, Robert M. (1947) *The Web of Government.* New York: Macmillan.
MARCSON, Simon (1972) "Research settings," pp. 161-191 in Saad Z. Nagi and Ronald G. Corwin (eds.), *The Social Contexts of Research.* New York: John Wiley and Sons, Wiley-Interscience.
MEAD, Margaret (1968) "Conferences," pp. 215-220 in David L. Sills (ed.), *International Encyclopedia of the Social Sciences,* vol. 3. New York: Crowell Collier and Macmillan.
MENZEL, Herbert (1959) "Planned and unplanned scientific communication," pp. 189-212 of *Proceedings of the International Conference on Scientific Information*; reprinted as pp. 417-441 in Bernard Barber and Walter Hirsch (eds.), *The Sociology of Science.* New York: The Free Press, 1962.
MERLE, Marcel (1978) "The politics of peace." *Participation,* 2,3: 39-42.
MERRITT, Richard L. (1980) "Communication." *PS,* 13:3 (Summer): 390-392.
——— and William SMIRNOV [eds.] (1979-1981) *International Political*

Science Enters the 1980s: Abstracts of papers presented at the XIth World Congress of the International Political Science Association, Moscow, U.S.S.R., August 12-18, 1979 (2 vols.). Ottawa, Ont.: International Political Science Association.

MERTON, Robert K. (1957) *Social Theory and Social Structure* (rev. enl. ed.). New York: The Free Press.

——. (1942) "A note on science and technology in a democratic order." *Journal of Legal and Political Sociology*, 1,1-2 (October): 115-126; reprinted as "Science and democratic social structure," pp. 550-561 in Merton (1957).

MEYNAUD, Jean (1961) "International co-operation in the field of the social sciences: a tentative balance-sheet," pp. 7-14 in UNESCO (ed.), *International Organizations in the Social Sciences*. Paris: UNESCO, Reports and Papers in the Social Sciences, no. 13; reprinted as pp. 103-115 in Philippart (1970).

—— and P. A. REYNOLDS (1956) "Third congress of the International Political Science Association: Stockholm, 21-27 August 1955." *International Social Science Bulletin*, 8,1: 191-197.

MILLER, James G. (1978) *Living Systems*. New York: McGraw-Hill.

MULKAY, Michael J. (1979) *Science and the Sociology of Knowledge*. London: George Allen and Unwin.

——. (1977) "Sociology of the scientific research community," pp. 93-148 in Ina Spiegel-Rösing and Derek de Solla Price (eds.), *Science, Technology and Society: A Cross-Disciplinary Perspective*. London and Beverly Hills, Calif.: Sage Publications.

NATIONAL SCIENCE BOARD (1981) *Science Indicators 1980*. Washington, D.C.: National Science Foundation.

NUTTIN, Joseph R. (1974) "Scientific communication and information exchange in an international congress setting." *International Associations*, 26:5 (May), 296-299, 273.

PEAR, Robert (1980) "U.S. bars exclusions of homosexual aliens in most circumstances." *The New York Times* (10 September): 20.

PHILIPPART, André [ed.] (1970) *Synthesis Report on the I.P.S.A.: 20 Years Activities, 1949-1969*. Paris: International Political Science Association.

POLANYI, Michael (1951) *The Logic of Liberty*. London: Routledge and Kegan Paul.

PRICE, Derek J. de Solla (1967) "Nations can public or perish." *Science & Technology*, no. 70 (October): 84-90.

——. (1961) *Science Since Babylon*. New Haven and London: Yale University Press.

ROOSEVELT, Curtis (1970) "The politics of development: a role for interest and pressure groups." *International Associations*, 22,5 (May): 283-289.

SCOHY, Michele. (1977) "Supplement of 1970-1976 to the synthesis report on the activities of the International Political Science Association." Ottawa: International Political Science Association.

SCULLY, Malcolm G. (1979) "Canada lifts ban on Marxist scholar." *The Chronicle of Higher Education*, 18:21 (30 July): 5.

SELTZER, Richard J. (1978) "Science, world politics, and human rights." *Chemical and Engineering News*, 56,8 (20 February): 34-47.
SEMENOV, Vadim S. (1978) "Cumulative growth in political knowledge since 1949." *Participation*, 2,3: 56-60.
SHAKHNAZAROV, Georgii (1978) "Policy of peace and our time." *Participation*, 2,3: 32-38.
SHARP, Walter R. (1950) "The scientific study of international conferences." *International Social Science Bulletin*, 2,1 (Spring): 104-116.
SHILS, Edward (1954) "Scientific community: thoughts after Hamburg." *Bulletin of the Atomic Scientists*, 10:5 (May): 151-154.
SKJELSBAEK, Kjell (1971) "The growth of international nongovernmental organization in the twentieth century." *International Organization*, 25,3 (Summer): 420-442.
SODDY, K. (1953) "International conferences and international nongovernmental organizations." *International Social Science Bulletin*, 5,2: 391-396.
STONE, Jeremy and A. Frederick SPILHAUS, Jr. (1980) "Scientists and international politics: a debate on linking scientific relations to Soviet actions." *Chemical and Engineering News*, 58,16 (21 April): 37-46.
STORER, Norman W. (1970) "The internationality of science and the nationality of scientists." *International Social Science Journal*, 22,1: 80-93.
TREASTER, Joseph B. (1979) "Homosexuals still fight U.S. immigration limits." *The New York Times* (12 August): 20.
TRENT, John E. (1978) "International research committees and political science." *Participation*, 2,2: 40-62.
UNESCO (1959) "The fourth world political science congress, Rome, 16-20 September 1958." *International Social Science Journal*, 11,2: 288-304.
―――. (1953) "The technique of international conferences." *International Social Science Bulletin*, 5,2: 233-339.
―――. (1951) "The world congress of political science." *International Social Science Bulletin*, 3,1 (Summer): 273-420.
―――. (1950a) "The first world congress of political science: Zürich, 4-9 September 1950." *International Social Science Bulletin*, 2,4 (Winter): 545-547.
―――. (1950b) "International Political Science Association." *International Social Science Bulletin*, 2,2 (Summer): 237-238.
―――. (1949a) "International Political Science Association: summary report of the constituent conference held at UNESCO House, 12-16 September 1949." *International Social Science Bulletin*, 1,3-4: 81-85.
―――. (1949b) "International political science conference 1949." *International Social Science Bulletin*, 1,1: 66-67.
―――. (1949c) "The UNESCO project: methods in political science." *International Social Science Bulletin*, 1,1: 28-32.
UNION OF INTERGOVERNMENTAL ASSOCIATIONS [ed.] (1984) *Yearbook of International Organizations 1984/85* (21st ed). Munich: K. G. Saur Verlag.
URBAN, Michael E. (1980) "Communication." *PS*, 13:2 (Spring): 261-262.

WALKER, David (1986) "World archeology group in uproar over barring South African scholars." *The Chronicle of Higher Education*, 31:23 (19 February): 1+.

WEISS, Thomas G. and Robert S. JORDAN (1976) *The World Food Conference and Global Problem Solving*. New York: Praeger.

WILL, George F. (1978) "Such resolute political scientists." *The Washington Post* (14 September): 23.

WRIGHT, Quincy (1951) "The significance of the International Political Science Association." *International Social Science Bulletin*, 3,2 (Summer): 275-280.

Index

Afghanistan, as political issue, 116
Alger, Chadwick F., 7, 18
Alker, Hayward R., Jr., 36, 53
American Chemical Society (1977 in Cairo), 18
American Political Science Association: membership, 21–22; decision not to boycott Moscow World Congress, 47, 54–55, 105, 116 (*see also* Boycotts)
Angell, Robert Cooley, ix, 17, 129
Appadorai, Angadipuram, 53
Aron, Raymond, 53
Atelsek, Frank J., 140

Bacon, Francis, 41
Barber, Bernard, 26, 27, 121
Barents, Jan, 12–13, 48, 53
Becker, Howard S., 18
Bose, Nirmal, 36, 53
Boycotts, of ISCs, 16, 47, 54–55, 103–106, 116
Brezhnev, Leonid, message of welcome, 50
Bridel, Marcel, 53
Broadhead, Robert C., 18
Brogan, Denis W., 53
Brown, Archie, 50, 75

Cancian, Francesca M., 26
Capes, Mary, 1

Celikbas, Fethi, 53
Collins, H. M., 2
Communication: as goal of ISC, ix, 2, 68, 72, 77, 93–96, 120–122, 136–137; networking, 89–90, 91, 120, 136–137
Cost of ISCs (*see* ISCs *and* Site selection)
Crane, Diana, 5, 19

Data resources, need to develop, 138
Dependency theory, 18, 52
Deutsch, Karl W., 35, 45
Dickson, David, 16
Diffusion of research paradigms and results, 137–139
Disraeli, Benjamin, 78
Duverger, Maurice, 53

Evan, William M., 5, 6
Exupéry, Antoine de Saint, 131

Fighiera, Gian Carlo, 7, 8, 26
Frank, André Gunder, 53
Friedrich, Carl J., 44
Functional analysis, 1–4, 9–10, 24, 25–26

Ganon, Isaac, 53
Giniger, Henry, 118
Ginsburg, Alexander, 45–46 (*see also*

Soviet dissidents)
Global society, and ISCs, 17–18, 51–52
Goguel, François, 53
Gomberg, Irene L., 140
Goormaghtigh, John, 53

Hanson, Elizabeth C., 18
Hastad, Elis W., 53
Horowitz, Irving Louis, 18
Host country: functions of ISCs, 15–17, 50–51, 71, 127; cost of ISCs, x, 36, 51
Human rights, 16–17, 45–47, 116, 118

Inquiring community, 137
Integration, functional theories of, and ISCs, 2, 17, 52, 70, 97, 129–130
International nongovernmental organizations (INGOs), 4–9
International Political Science Association (IPSA): creation in 1949, 29–32; publications (review, abstracts, book series, newsletter), 29; and UNESCO, 30–31; world congresses, 32–34, 40–44, 48, 133–135 (*see also* Moscow World Congress, Paris, *and* Rio de Janeiro); research committees and study groups, 34–35, 136, 137; quality of world congresses, 125–126, 133–135
International professional associations, 4–6
International relations, structural theories of and ISCs, 2, 17
International scientific associations: functions of, 5–9; and ISCs, 7–9, 15
International scientific congresses (ISCs): and international communication, ix, x, 2, 77; costs, x, 20, 42, 135 (*see also* Site selection); functions of, 1–2, 9–19, 24, 48–49, 64–72, 132–139; for individuals, 10–13, 24, 48–49, 66, 67–69, 119–124 (*see also* Respondents); as medium for scientific communication, 11, 17–18, 29–30, 68, 72, 77, 93–96, 120–122, 136–137; as part of academic reward structure, 11, 120–124, 126–127, 131; for research organizations, 13–14, 49, 124; for scientific associations, 14–15, 49–50, 66, 68, 70–71, 124–127; for host country, 15–17, 50–51, 66, 68, 71, 127–128; for international system, 51–52, 128–132; participant characteristics, 10–11, 23–24, 48; politics and, 16, 101–102, 114–117, 127–132, 134; future directions, 132 (*see also* Moscow World Congress, Psychology, *and* Sociology)
Israeli visas, in 1979, 45, 116

Jagtenberg, Tom, 2
Johns Hopkins Center for Research in Scientific Communication, 11, 13, 77, 91–96, 99, 139
Jordan, Robert S., 27
Justice, 52

Kaplan, Abraham, 3
Kelman, Herbert C., 96
Khosla, H., 53
Knowledge, as power, 41 (*see also* Communication)
Kochen, Manfred, 137
Kriesberg, Louis, 6
Kuhn, Thomas, 26

Ladd, Everett Carll, Jr., 10, 48
Laponce, Jean A., 44
Lasswell, Harold D., 3
Levy, Marion J., Jr., 26
Lipset, Seymour Martin, 10, 48
Lodge, David, 12, 122
Ludz, Peter C., 36, 53
Lyons, Gene M., 7, 18

MacIver, Robert M., 3
MacPherson, C. B., 53
Mead, Margaret, 1
Mendes, Candido, 44
Menzel, Herbert, 18, 121
Merle, Marcel, 36, 53
Merritt, Richard L., 18, 36, 100
Merton, Robert K., 25, 26, 74, 128, 129
Meynaud, Jean, 5, 34, 53
Miller, James G., 140
Moscow World Congress (1979): xi, 2, 19, 32, 35–36; survey of participants, 19–25 (*see also* Respondents); choice of Moscow as site for meeting, 25, 36, 41, 42–47; attendees, 36–39, 48–49, 143–144; issue of dissidents, 45–47, 116, 118; possibility of boycott, 47, 50, 103–104; research organizations, 49, 124; disciplinary associations, 49–50, 66, 68, 70–71, 124–127; global society, 51–52, 128–132; Third World participants, 52, 60, 61, 88–90, 91, 131; participants' expectations and experiences, 78–84; impact on scholarly behavior, 90–96; problems emerging at conference, 115–117
Mulkay, Michael J., 2, 26

National disciplinary associations, and ISCs, 14–15
National governments, and ISCs, 15–17 (*see also* Host country)
National Science Board, report to U.S. Congress on status of science and technology, 11
Networking (*see* Communication)
Nuttin, Joseph R., 74–75, 95

Olympics, as political issue, 116
Orlov, Yuri Fyodorovich, 45 (*see also* Soviet dissidents)

Paris, as site of future IPSA world congress (1985), x, 32, 34, 109–111, 130, 135, 140–141
Peace, 17, 52
Pear, Robert, 54
Personal enrichment, 12–13
Philippart, André, 33, 34, 53
Polányi, Michael, 2
Politics: and Moscow World Congress, 45–47, 50–51, 102–109; impact on ISCs, 16, 101–102, 114–117, 127–132; of site selection, 36, 40–47
Price, Derek J. de Solla, 19
Professional growth, 11–12
Psychology, International Congress of, Moscow (1966), 11, 13, 91–96, 99, 139

Research organizations, and ISCs, 13–14, 49, 124
Respondents: registrants, 20–21, 57–59, 67–69, 73–74, 143–144; nonregistrants, 21–22, 58–59, 67–69, 73–74; professional status, 59–61, 62–63, 123, 169–170; international activity, 59–62, 62–63, 171–173; Third World participants, 60, 61, 88–90, 91; Western participants, 60, 61; reasons for attending professional meetings, 63–65; reasons for attending ISCs, 64–72; decision to attend, 72–73; political considerations for attending, 102–109; reasons for attending Moscow World Congress, 103, 119–124; expectations and experiences, 78–84; Soviet life, 84–87; attitude changes, 86, 96–97; post-congress networking, 87–90, 97–98; post-congress scholarly work, 90–93, 98–99; concerns raised, 93–96; views on Moscow as site, 106–109; Soviet propaganda, 112–114; as scientific elite, 119–120
Reynolds, P. A., 34

Rio de Janeiro, as site of future IPSA world congress (1982), 32, 34, 41, 44, 50, 109–111, 130, 131, 135
Rist, Ray C., 18
Robson, William A., 53
Rokkan, Stein, 44
Roosevelt, Curtis, 5
Rumanian Political Science Association, 43–44

Schaff, Adam, 53
Scientific consociation, 3–4, 9, 18, 26, 47, 132
Scientific discipline, definition, 3
Scientific knowledge, 27
Scohy, Michele, 33
Scully, Malcolm G., 54
Sebestik, Jutta, 23
Seltzer, Richard J., 17, 54
Semenov, Vadim S., 36, 53
Shakhnazarov, Georgii, 36, 53
Sharansky, Anatoly, 45, 46 (*see also* Soviet dissidents)
Sharp, Walter R., 27, 53
Shils, Edward, 2
Skjelsbaek, Kjell, 7
Site selection: financial considerations, x, 42, 135, 140–141; political considerations, 36, 40–47, 106–112, 134–135
Smirnov, William, 36, 100
Sociology, world congress of: Evian (1966), 11, 13, 91–96, 99, 139; Zürich (1950), 33–34
Soddy, K., 27
Soviet Union: Soviet Political Science Association, 44–45, 49–50, 75; dissidents, as political issue, 45–47, 54, 116, 118; participants' views of government and people, 84–87, 96–97; propaganda at Moscow World Congress, 103–104, 107–108, 112–114
Spilhaus, Frederick, Jr., 17, 54
Stone, Jeremy, 17, 54
Storer, Norman W., 27
Survey, 19–25, 145–168 (*see also* Respondents)

Technology transfer, ix, 18
Third World: needs of and ISCs, 18; and IPSA, 32; scholars, 52, 131 (*see also* Respondents); impact of ISCs, 72, 97–98, 125, 127, 130–131; networking with others, 88–90, 91
Toumanov, Vladimir, 44
Transnationalism, 5–6
Treaster, Joseph B., 54
Trent, John E., 33, 34, 54

UNESCO, and IPSA, 27, 30–31
Urban, Michael E., 100

Values: transnational, 6, 52; ISCs and exchange of, 130–132

Walker, David, 54, 117
Weiss, Thomas G., 27
Western Europe, international meetings in, 7–9
Will, George F., 55
Wright, Quincy, 32, 53

About the Book and the Authors

While there are many widely held assumptions about the impact of international scientific congresses (ISCs) on individual scientists, collective bodies, a particular branch of science, or even the establishment of world order, these assumptions have not previously been fully examined or tested empirically. Merritt and Hanson present here the results of their systematic investigation of the uses and consequences of ISCs. Deriving their data from a survey of political scientists from the nonsocialist countries who attended the 11th World Congress of the International Political Science Association in Moscow, as well as a control group of North American political scientists who did not attend the Moscow meetings, they explore the function of ISCs at both manifest and latent levels. Their results provide a solid basis for evaluating the costs versus the benefits of future scientific congresses.

Richard L. Merritt is professor of political science and communications at the University of Illinois, Urbana-Champaign, and served as program director of the 11th World Congress of the International Political Science Association. Elizabeth C. Hanson is associate professor of political science at the University of Connecticut.